COMPREHENSIVE SERIES IN PHOTOCHEMISTRY AND PHOTOBIOLOGY

Series Editors

Donat P. Häder
Professor of Botany

and

Giulio Jori
Professor of Chemistry

European Society for Photobiology

COMPREHENSIVE SERIES IN PHOTOCHEMISTRY AND PHOTOBIOLOGY

Series Editors: Donat P. Häder and Giulio Jori

Titles in this Series

Volume 1 UV Effects in Aquatic Organisms and Ecosystems
Edited by E.W. Helbling and H. Zagarese

Volume 2 Photodynamic Therapy
Edited by T. Patrice

Volume 3 Photoreceptors and Light Signalling
Edited by A. Batschauer

Volume 4 Lasers and Current Optical Techniques in Biology
Edited by G. Palumbo and R. Pratesi

COMPREHENSIVE SERIES IN PHOTOCHEMISTRY AND PHOTOBIOLOGY – VOLUME 4

Lasers and Current Optical Techniques in Biology

Editors

Giuseppe Palumbo
Dipartimento di Biologia e Patologia Cellulare e Molecolare "L.Califano"
Università di Napoli "Federico II"
Napoli - Italia

and

Riccardo Pratesi
Dipartimento di Fisica
Università di Firenze
Sesto Fiorentino (Fi) - Italia

RS•C
advancing the chemical sciences

ISBN 0-85404-321-7

A catalogue record for this book is available from the British Library

Published by The Royal Society of Chemistry,
Thomas Graham House, Science Park, Milton Road,
Cambridge CB4 0WF, UK

Registered Charity Number 207890

For further information see our web site at www.rsc.org

Typeset by Alden Bookset, Northampton, UK
Printed by Sun Fung Offset Binding Co Ltd, Hong Kong

Preface for the ESP series in photochemical and photobiological sciences

"Its not the substance, it's the dose which makes something poisonous!" When Paracelsius, a German physician of the 14th century made this statement he probably did not think about light as one of the most obvious environmental factors. But his statement applies as well to light. While we need light for example for vitamin D production too much light might cause skin cancer. The dose makes the difference. These diverse findings of light effects have attracted the attention of scientists for centuries. The photosciences represent a dynamic multidisciplinary field which includes such diverse subjects as behavioral responses of single cells, cures for certain types of cancer and the protective potential of tanning lotions. It includes photobiology and photochemistry, photomedicine as well as the technology for light production, filtering and measurement. Light is a common theme in all these areas. In recent decades a more molecular centered approach changed both the depth and the quality of the theoretical as well as the experimental foundation of photosciences.

An example of the relationship between global environment and the biosphere is the recent discovery of ozone depletion and the resulting increase in high energy ultraviolet radiation. The hazardous effects of high energy ultraviolet radiation on all living systems is now well established. This discovery of the result of ozone depletion put photosciences at the center of public interest with the result that, in an unparalleled effort, scientists and politicians worked closely together to come to international agreements to stop the pollution of the atmosphere.

The changed recreational behavior and the correlation with several diseases in which sunlight or artificial light sources play a major role in the causation of clinical conditions (e.g. porphyrias, polymorphic photodermatoses, Xeroderma pigmentosum and skin cancers) have been well documented. As a result, in some countries (e.g. Australia) public services inform people about the potential risk of extended periods of sun exposure every day. The problems are often aggravated by the phototoxic or photoallergic reactions produced by a variety of environmental pollutants, food additives or therapeutic and cosmetic drugs. On the other hand, if properly used, light-stimulated processes can induce important beneficial effects in biological systems, such as the elucidation of several aspects of cell structure and function. Novel developments are centered around photodiagnostic and phototherapeutic modalities for the treatment of cancer, artherosclerosis, several autoimmune diseases, neonatal jaundice and others. In addition, classic research areas such as vision and photosynthesis are still very active. Some of these developments are unique to photobiology, since the peculiar physico-chemical properties of electronically excited biomolecules often lead to the promotion of reactions which are characterized by high levels of selectivity in space and time. Besides the biologically centered areas, technical developments have paved the way for the harnessing of solar energy to produce warm water and electricity or the development of environmentally friendly techniques for addressing problems of large social impact (e.g. the decontamination of polluted waters). While also in use in Western countries, these techniques are of great interest for developing countries.

The European Society for Photobiology (ESP) is an organization for developing and coordinating the very different fields of photosciences in terms of public knowledge and scientific interests. Due to the ever increasing demand for a comprehensive overview of the photosciences the ESP decided to initiate an encyclopedic series, the "Comprehensive Series in Photochemical and Photobiological Sciences". This series is intended to give an in-depth coverage over all the very different fields related to light effects. It will allow investigators, physicians, students, industry and laypersons to obtain an updated record of the state-of-the-art in specific fields, including a ready access to the recent literature. Most importantly, such reviews give a critical evaluation of the directions that the field is taking, outline hotly debated or innovative topics and even suggest a redirection if appropriate. It is our intention to produce the monographs at a sufficiently high rate to generate a timely coverage of both well established and emerging topics. As a rule, the individual volumes are commissioned; however, comments, suggestions or proposals for new subjects are welcome.

Donat-P. Häder and Giulio Jori
Spring 2002

Volume preface

Biology and medicine are increasingly exploiting physical techniques for basic research, laboratory analysis, clinical diagnostics and even therapeutics. In this regard the amount of scientific work on the properties of optical and laser devices is continually expanding, such that it may be difficult for anyone to keep up to date. Hence, a comprehensive view of the state-of-the-art and perspectives of future applications of optical and laser techniques in photobiology and photo-medicine should be widely welcomed and appreciated. Clearly, not all of these issues can be encapsulated in one volume; nevertheless the present endeavour attempts to cover the broad range of topics that hitherto have been scattered throughout various books and specialized journals.

The present volume, the fourth of a series entitled "Comprehensive Series in Photosciences" has been solicited by the European Society for Photobiology and assembled to try bridge the gap between the increasing offering of innovative technologies and the limited demand by the biologists and clinicians who are, with a few exceptions, scarcely conscious, more often entirely unaware, of them. The goal of this endeavour is to offer an extensive and qualified description of the current optical techniques that may stimulate genuine attention (present and future) of many scientists working in the various areas of bio-medicine. This book provides its information through the pen of recognised scientists, each one active in the field.

The authors belong largely to the physics community, but, since the content of the volume (and the aim of the entire series) is mainly dedicated to a life science-oriented auditorium, the basic concepts and essential principles, necessary to understand instrumentation and techniques, are presented in plain language. This, however, is not at expenses of a rigorous approach. Each author gives an initial comprehensive review of laser and/or optical techniques of their interest, goes on to discuss the current, often pilot applications, and concludes by indicating future realistic perspectives in photobiology and photomedicine.

The use of realistic examples has been encouraged especially for those biomedical applications that are currently used and whose knowledge can make other biomedical scientists and clinicians aware of the large and fast growing potentials of optics in medicine. However, the various chapters not only deal with techniques that are already in everyday practice, but also with (a) techniques that have just exceeded the experimental stage and are entering into routine, (b) techniques that are only promising and (c) even those that need much more experimentation before they can be accepted as real breakthroughs.

Each chapter follows a reasonably logical progression, revealing the relevant research for each topical area. This latter feature makes it quite simple for the reader to understand the scientific basis for the subsequent discussion and to retrieve the relevant references from the bibliography.

Indeed, biology and medicine are increasingly attracting the interest of mathematically literate engineers, physicists and computer scientists. Hence, we hope that this book will attract a wide audience and encourage them to apply their most advanced achievements towards optical techniques, not only to benefit basic science, but even to the relief (especially through reliable diagnosis and/or definite prognosis) of patients. In turn, biology and medicine are increasingly exploiting physical techniques and technological products for research, laboratory analysis, clinical diagnostics and therapeutics.

Nevertheless, many biologists and clinicians are terrified by the unfriendly appearance of mathematical expressions customarily employed by engineers and physicists. This is often enough to prevent from reading even a single article. We appreciate the difficulty for the non-mathematical and/or physical researcher to understand some chapters, but we hope the numerous figures that accompany each chapter will provide key signposts so that the biomedical reader can choose to gloss over the mathematics. We expect that the book will give our (hopefully many) readers a strong feel for the capability of these approaches and promote new interdisciplinary interests.

Despite our efforts some important topics are missing. A complete (is it possible?) book would require on unacceptably long gestation period. Hence, the present volume represents a compromise between the availability (present or forthcoming) of other books and/or review papers on specific topics, the acceptance from leading researchers to contribute to the book, and the failure of a few authors to complete their contribution within the agreed dead line.

Finally this volume comes at the right time since it takes advantage of the maturity reached by lasers and, in general, by opto-electronic sciences.

The Volume

As this volume solicited by the ESP, we tried to involve, whenever possible, mostly European authors. The ratio between the number of chapters written by European/ non-European Authors is 3:2, being 16 versus 5 contributions, respectively. Specifically: 2 are from Switzerland, 4 from the UK, 4 from Germany and 6 from Italy, with 4 from the USA, and 1 from Australia.

The Volume is divided into four parts. The apparent dichotomy in the lengths of these parts – some consisting of long, detailed chapters and others consisting of a limited number – is somewhat in accord with the different levels of awareness and development of the individual application/technique described or proposed. The exhaustive list of references at the end of each chapter is of great value to readers and researchers seeking additional information about individual subjects.

Part I

This covers various topics concerning the basic physics and technology of lasers and lamp sources. This is very important since it is unusual to have a comprehensive presentation of the principal laser and lamp sources within the same book. Such information usually appears separately in different and hyper-specialised publications.

The foremost chapters by King (UK), Lüthy & Weber (CH), Unger (D), and Zellmer & Tünnermann (D) deal with the Laser sources (gas, liquid, solid state, semiconductor, diode-pumped, and fibre lasers). The central part of Part I describes efficiently and clearly different types of lamps (Diffey, UK) and solid state lamps (Diehl, D), their nature and principal uses. Of particularly use is the concluding chapter by Nisoli (I) on pulse generation and control, especially in regard to the interesting foreseen perspectives in both basic and applied research (atto-physics).

Part II

Scientists interested in spectroscopic and imaging techniques will find this section valuable. Part II encompasses three exciting chapters, dealing with the most recent progress in this field. This part emphasizes the scientific and technological aspects of the application of advanced spectroscopy. The contributions come from leading laboratories; they review, discuss and illustrate background and advanced biomedical applications of *Autofluorescence spectroscopy* (Bottiroli & Croce, I), *Reflectance and transmittance spectroscopy* (Fantini & Gratton, USA), and *Fluorescence spectroscopy* (Taroni & Valentini, I; Marcu, USA).

Part III

This is dedicated to optical microscopy and includes four chapters, by Schneckenburger (D); Fusi, Monici, & Agati (I). Ascoli, Gottardi & Petracchi (I); and Diaspro (I); Their content spans from *Optical microscopy* to *Scanning probe microscopy* and *Confocal and multiphoton microscopy*. The incredible diffusion and dissemination of techniques and application of microscopy, plus the lack of a recent comparative treatise on the central methodologies presently used in the field, warrants the assembly and eminence of this part.

Part IV

Part IV provides a valuable background that favours a better understanding of the various imaging techniques described throughout. A particularly relevant chapter by Sampson & Hillman (Australia) outlines the theoretical basis and practical applications of the *Optical coherence tomography* (OCT). Indeed, this article fills a large gap in the literature since includes most material and key equations that, even being fundamental to understanding in detail OCT imaging, cannot be found elsewhere. It explains the effects of dispersion, describes the OCT 3-D PSF, noise analysis for balanced detection, highlights the key role of the often overlooked cross-beat noise, provides a good description of the frequency-domain delay line and puts the right emphasis on multiple scattering and speckle, which tend to be glossed over in the literature. Further coverage of the imaging topic is provided by two excellent articles by Jacques and by Jacques & Ramella-Roman (USA), who describe in detail, smoothly explaining and clearly indicating, past, present and future applications of *Laser optoacustic imaging* and *Polarized light imaging*. Finally, a specific chapter by Seeger (CH) is dedicated to *Ultrasensitive fluorescence detection at surfaces systems*. This chapter encompasses a clear explanation of the basic principles, the state-of-art of instrument development and workable applications (present and future) in life science and medicine.

Acknowledgements

We wish to acknowledge our indebtedness to all the contributors for their good work, constructive co-operation, and gracious acceptance of editorial comments. Further we would like to show appreciation to RCS staff for the efficient production and excellent layout. Last but not least we want express our gratitude to our wives Alba and Nadia that, as do all scientists wives, have endless patience.

Giuseppe Palumbo
Riccardo Pratesi

To Alba and Nadia

The editors

Giuseppe Palumbo (GP) was born in Naples, Italy in 1945. He was educated in physical chemistry at the local University, obtaining the "Laurea" in 1969. His education continued at the Clinical Endocrinology Branch of the NIAMDD, National Institutes of Health, Bethesda, Maryland, USA, initially as postdoctoral fellow in the Laboratory of Dr Harold Hedeloch, (1973–1975) and successively as Visiting Scientist (1985–1986 and 1989) in the Laboratory of Dr Jan Wolff. Between 1975 and 1986 he held a permanent position as researcher at the Italian national research Council (CNR). His academic career begun in 1986 when he first became Professor of Physiological Chemistry (1977–1990) at the University of Reggio Calabria (Italy) and then (1990–to date) Professor of General Pathology at the University of Naples Federico II (Italy). The research contributions of GP have been various and embrace numerous studies on of the fluorescent properties of HD and LD lipoproteins (1974–1990), on the structure–function relationship of iodoproteins, on T4 biosynthesis (1980–1990) and, more recently (1990–to date), on the application of photobiology to the study of malignant cells and therapy of human tumors. Professor Palumbo has been Coordinator of a national PhD program in Photobiology (1998–2002), and is currently Coordinator of a multi-centric research project on photodynamic therapy that includes studies at cellular, animal and human levels. His scientific production is testified by more than 75 international peer-reviewed publications in these fields. He is an active member of national and European scientific Societies of Photobiology.

Riccardo Pratesi (RP) was born in Florence, Italy in 1936. He was educated in physics at the local University, obtaining the "Laurea" in 1961 and the "Libera Docenza" in *Quantum Electronics* in 1969. Career: Assistant at the University of Florence (1963–1979); Director of the *Institute of Quantum Electronics* of CNR (National Council of Research) (1970–1990); Director of the SubProject on *Medical applications of lasers* of the CNR, *Finalized Project LASER* (1978–1982); Full Professor of *Structure of Matter* at the University of Florence (1980–to the present); Director of the SubProject on *Biomedical applications of lasers* of the CNR *Finalized Project on Electrooptical Technologies* (1989–1993); President of the *Area of Research of CNR*, Florence, (1990–2000); Director of the *Laser Medical*

Centre (CLAM) of the Optronic Consortium (CEO), Florence (1994–to date); Member of the Board of the Quantum Electronic Division of EPS (1976–1979); Member of the Italian and European Societies for Photobiology (1987–to date). Editor of the volumes *Lasers in Photomedicine and Photobiology* (with C.A. Sacchi, Springer-Verlag (1980), *Lasers in Biology and Medicine* (with F. Hillenkamp and C.A. Sacchi, Plenum Press, 1980), *Optronic Techniques in Diagnostic & Therapeutic Medicine* (Plenum, 1991). The research of RP has focused initially on laser physics and technology, and optical spectroscopy, and on to the photophysics of biomolecules, in particular bilirubin, with application to the phototherapy of neonatal jaundice; and more recently on fluorescence techniques for biomedical applications. His scientific production is testified by about 95 international peer-reviewed publications. RP has greatly contributed, since 1962, to the development of laser technology in Italy and to the birth and fast growth of the local laser industry, with continuous cooperation among research institutions, public administrations and industries. From the very beginning, RP promoted the applications of lasers and optical techniques to biology and medicine.

Contributors

Giovanni Agati
Istituto di Fisica Applicata
"N.Carrara" del C.N.R.
Area della Ricerca di Firenze,
Edificio C Via Madonna del Piano,
50019 Sesto Fiorentino (Firenze),
Italy

Cesare Ascoli
Istituto per i Processi Chimico-Fisici,
IPCF – CNR Area della Ricerca
Via g. Moruzzi 1, 56124 Pisa, Italy

Giovanni Bottiroli
Istituto di Genetica Molecolare –
CNR, Piazza Botta 10, 27100 Pavia,
Italy

Anna Cleta Croce
Istituto di Genetica Molecolare –
CNR, Piazza Botta 10,
27100, Pavia, Italy

Alberto Diaspro
LAMBS-2INFM, e Dipartimento di
Fisica Università di Genova Via
Dodecaneso 33, 16146 Genova, Italy

Roland Diehl
Fraunhofer Institute for Applied
Solid-State Physics Tullastrasse 72,
79108 Freiburg, Germany

Brian L. Diffey
Regional Medical Physics
Department, Newcastle General
Hospital Newcastle upon Tyne NE4
6BE, United Kingdom

Sergio Fantini
Department of Biomedical
Engineering, Bioengineering Center
Tufts University, 4 Colby Street,
Medford, MA 02155, USA

Franco Fusi
Dipartimento di Fisiopatologia
Clinica, Sezione di Fisica Medica
Viale Pieraccini 6, 50139 Firenze,
Italy

Riccardo Gottardi
Istituto per i Processi Chimico-Fisici,
IPCF – CNR, Area della Ricerca di
Pisa Via Moruzzi 1, 56124 Pisa,
Italy

Enrico Gratton
Laboratory for Fluorescence
Dynamics Department of Physics
University of Illinois at Urbana-
Champaign, 1110 West Green Street,
Urbana IL 618, USA

Timothy R. Hillman
Optical + Biomedical Engineering Laboratory, School of Electrical, Electronic & Computer Engineering The University of Western Australia 35 Stirling Hwy, Crawley 6009, Australia

Steven L. Jacques
Oregon Health & Science University, Biomedical Eng. – Vollum 107, 20000 NW Walker Road, Portland, OR 97006, USA

Terry A. King
Department of Physics and Astronomy, University of Manchester, Schuster Laboratory Manchester M13 9PL, United Kingdom

Willy Lüthy
Institute of Applied Physics University of Bern Sidlerstrasse 5, CH-3012 Bern, Switzerland

Laura Marcu
Biophotonics Research and Technology Development, Department of Surgery, Cedars-Sinai Medical Center 8700 Beverly Blvd. Davis G149A, Los Angeles, CA 90048, & Departments of Biomedical and Electrical Engineering University of Southern California Los Angeles, CA 90048, USA

Monica Monici
Centro Laser per Applicazioni Mediche (CLAM) Centro di Eccellenza Optronica (CEO) Largo E.Fermi 6, 50125 Firenze, Italy

Mauro Nisoli
National Laboratory for Ultrafast and Ultraintense Optical Science – INFM, Dipartimento di Fisica, Politecnico di Milano, Piazza L. da Vinci 32, Milano, Italy

Donatella Petracchi
Istituto per i Processi Chimico-Fisici, IPCF – CNR, Area della Ricerca di Pisa, Via Moruzzi 1, 56124 Pisa, Italy

Jessica C. Ramella-Roman
Oregon Health & Science University Oregon Medical Laser Center Providence St. Vincent Medical Center, 9205 SW Barnes Rd., Portland, OR 97225, USA

David Sampson
Optical + Biomedical Engineering Laboratory School of Electrical, Electronic & Computer Engineering The University of Western Australia, 35 Stirling Hwy, Crawley 6009, Australia

Herbert Schneckenburger
Fachhochschule Aalen, Institut für Angewandte Forschung 73428 Aalen, Germany and Institut für Lasertechnologien in der Medizin und Messtechnik an der Universität Ulm, Helmholtzstr. 12, 89081 Ulm, Germany

Stefan Seeger
Physical Chemistry Institute at University of Zürich Winterthurerstrasse 190, 8057 Zürich, Switzerland

Paola Taroni
INFM-Dipartimento di Fisica and IFN-CNR, Politecnico di Milano
Piazza Leonardo da Vinci 32, 20133 Milan, Italy

Andreas Tünnermann
Friedrich Schiller University Jena
Institute of Applied Physics
Max-Wien-Platz 1, 07743 Jena, Germany

Peter Unger
Department of Optoelectronics
University of Ulm
89069 Ulm, Germany

Gianluca Valentini
INFM-Dipartimento di Fisica and IFN-CNR, Politecnico di Milano
Piazza Leonardo da Vinci 32, 20133 Milan, Italy

Heinz P. Weber
Institute of Applied Physics
University of Bern, Sidlerstrasse 5, CH-3012 Bern, Switzerland

Holger Zellmer
Friedrich Schiller University Jena
Institute of Applied Physics
Max-Wien-Platz 1, 07743 Jena, Germany

Contents

Part I: Lasers and Lamps

Part IV: Advancing Imaging Techniques and Novel Ultrasensitive Fluorescence Detection Techniques

Abbreviations

ALA, 5-aminolevulinic acid
APD, avalanche photodiode
BBO, β-barium borate
CCD, charge coupled device
CFD, constant-fraction discriminators
COIL, chemical oxygen iodine laser
COT, cyclooctatetraene
CW (or cw), continuous wave
DBR, distributed Bragg reflector
DFB, distributed feedback
FAD, flavin adenine dinucleotide
FCS, fluorescence correlation spectroscopy
FD, frequency domain
FD-ODL, frequency-domain optical delay line
FISH, fluorescence in situ hybridization
FLIM, fluorescence lifetime imaging
FRAP, fluorescence recovery after photobleaching
FRET, fluorescence resonance energy transfer
FWHM, full-width at half-maximum
GDD, group delay dispersion
GFP, green fluorescent protein
GRINSCH, graded-index separate confinement heterostructure
GVD, group velocity dispersion
HB LED, high-brightness light-emitting diode
Hb, deoxy-hemoglobin
HbO_2, oxy-hemoglobin
HWHM, half-width at half-maximum
LED, light-emitting diode
MBE, molecular beam epitaxy

MOVPE, metal-organic, vapour phase epitaxy
NAD, nicotinamide adenine dinucleotide
NIR, near-infrared
NSOM, near-field scanning fluorescence microscopy
OCM, optical coherence microscopy
OCT, optical coherence tomography
OLED, organic light-emitting diode
OMA, optical multichannel analyzer
OPO, optical parametric oscillators
PCR, polymerase chain reaction
PDT photodynamic therapy
PMT, photomultiplier tube
QCL, quantum cascade laser
S/N, signal-to-noise (ratio)
SAD, seasonal affective disorder
SAM, self-amplitude modulation
SCH, separate confinement heterostructure
SNR, signal-to-noise ratio
SPM, self-phase modulation
TAC, time-to-amplitude converter
TCSPC, time-correlated single-photon counting
TD, time domain
TEA, transversely excited atmosphere
TIRFM, total internal reflectance fluorescence microscopy
TOD, third-order dispersion
VCSEL, vertical cavity surface emitting laser
YAG, yttrium aluminium garnet
ZAP, zero adjustment procedure

Part I
Lasers and Lamps

Chapter 1

Gas state lasers

Terry A. King

Table of contents

Abstract

A brief review of the structures and operation of gas lasers of particular interest in photobiology is given in this chapter along with details of their emission characteristics. Gas lasers provide laser wavelengths over a very broad range from the vacuum UV to the far-IR in continuous wave and pulsed operation and from low to high powers. Many of these wavelengths have found valuable application in photobiology, particularly in the several forms of fluorescence technique, microscopy, imaging and Raman spectroscopy, as well as in such techniques as optical trapping and tweezers, micromanipulation and microdissection. Extensive use has been made of the He-Ne laser (wavelength 632.8 nm), argon ion laser (488.0 and 514.5 nm), krypton ion (647.1 nm), He-Cd laser (441.6 and 325.0 nm), excimer lasers (ArF 193 nm, KrF 248 nm, XeCl 308 nm) and nitrogen laser (337.1 nm). The ion and excimer lasers have application in the pumping of tunable dye and titanium-sapphire lasers. Since the discovery of the first gas laser in 1961 many gas lasers have been devised, a detailed understanding built up of their operation and performance and the commercial technology has reached a highly developed and mature state.

In recent years solid-state alternatives to several of the common gas lasers have been developed based on diode lasers, diode pumped solid-state lasers and fibre lasers which offer advantages of compactness and efficiency and operation from low voltage power supplies. However, solid-state substitutes are not presently available at many of the useful gas laser wavelengths, such as the high power pulsed output from excimer lasers in the UV, high power continuous wave output in the near UV, visible and near IR, and mid-and far-IR wavelengths from molecular gas lasers.

1.1 Introduction

The interaction of laser light with biological samples gives an optical signal which carries information on the composition, structure and function of the sample. The probe laser light may be used in the basic measurements of absorption, fluorescence and phosphorescence emission, and Rayleigh and Raman scattering. In this there is a remarkable variety of techniques which have evolved with important applications in photobiology: spectral microscopy and imaging, scanning confocal microscopy, optical labelling, ultrafast pulse and fluorescence lifetime spectroscopy, fluorescence probe (marker) spectroscopy, fluorescence recovery after photobleaching (FRAP), fluorescence in situ hybridization (FISH), fluorescence resonance energy transfer (FRET) and optical trapping and tweezers [1–4]. Special attention can be drawn to the use of fluorescence in several of these techniques with the use of spectral (continuous), dynamic (pulsed) and imaging methods. These applications include laser scanning confocal fluorescence microscopy, time-correlated single-photon counting lifetime studies, total internal reflection fluorescence, near-field scanning fluorescence optical microscopy (NSOM), fluorescence correlation spectroscopy and multi-photon fluorescence microscopy [4]. The Raman scattering technique in its various forms has widespread application in photobiology through its sensitivity to molecular vibrations. Surface enhanced Raman scattering derives an enhanced Raman signal from molecules attached to surfaces to give molecular structural information. Coherent anti-Stokes Raman scattering microscopy enables the identification of molecular constituents of cells without the use of dye markers.

Gas lasers provide source radiation for many of these techniques. In particular, these include the argon (514.5 and 488 nm) – and krypton (647.1 nm) – ion lasers at medium power levels and which provide UV, visible and near-IR wavelengths; the He-Cd laser (441.6 and 325.0 nm), excimer lasers (ArF 193 nm, KrF 248 nm, XeCl 308 nm), N_2 lasers (337.1 nm) and F_2 lasers (157 nm). The ion lasers and excimer lasers are also used as pump sources for the tunable liquid dye and solid-state titanium-sapphire lasers. In addition there are several gas lasers which have more specialised applications in photobiology.

In this chapter the basic principles of laser operation are briefly reviewed and some of the laser radiation characteristics. Those gas lasers of particular value in photobiology are described to give a survey of their design features and emission properties.

1.2 Basic principles

Before the invention of the laser, available light sources emitted light from thermal excitation such as a tungsten filament lamp with a spectrum corresponding to a quasi-black body emitter at the temperature of the emitter, or by spontaneous emission from atoms or molecules as in a gas discharge arc or fluorescent tube. Their brightness is limited by the temperature of the emitter. In the laser the essential process of stimulated emission was introduced, so that, in conjunction with population inversion, a net optical amplification (or gain) could be created. A brief description of the principles of operation of lasers is given here; extensive discussion can be found in various texts [5–8].

Stimulated emission was first used in 1953 to create a high brightness source from transitions between the two lowest levels of ammonia molecules, giving a very narrow emission line at a wavelength of 12.6 mm; this was termed the maser since the wavelength was in the microwave region.

The basic elements of a laser are an active medium with suitable energy levels, a means of injecting energy into the medium (a process known as pumping) and a resonator in which the amplification process can occur. In a 2-level system interacting with light radiation with which it is in resonance, the processes of absorption, spontaneous emission and stimulated emission can occur. The light wave can be absorbed or be amplified depending on the population densities N_1 and N_2 of the two levels. For a wave travelling in the z direction the change in irradiance of the wave over a length $\mathrm{d}z$ is $\mathrm{d}I = \gamma I \mathrm{d}z$. For an initial irradiance I_0 at $z = 0$, the beam intensity varies along the propagation distance z exponentially as $I(z) = I_0 \mathrm{e}^{\gamma z}$. The unsaturated gain or loss coefficient γ of the medium at frequency ν is

$$\gamma(\nu) = \left(N_2 - \frac{g_2}{g_1}N_1\right)\frac{c^2 A_{21}}{8\pi\nu^2}g(\nu)$$

Here g_1 and g_2 are the degeneracies of the two levels and A_{21} is the Einstein A-coefficient, related to the lifetime τ of the upper level as $A_{21} = 1/\tau$. The quantity $g(\nu)$ is the normalized spectral function (or lineshape function) for the transition, such that

$$\int_{-\infty}^{\infty} g(\nu)\,\mathrm{d}\nu = 1$$

The gain coefficient $\gamma(\nu)$ can be > 0 if $N_2 > g_2/g_1 N_1$. In this case the irradiance grows exponentially. The gain coefficient varies as ν^{-2}, making it easier to achieve higher gain in the infra-red than in the ultra-violet, although some cw UV lasers and pulsed UV lasers are available. Spontaneous emission which scales as ν^3 competes with stimulated emission.

A laser species may be characterized by its stimulated emission cross-section σ_0. This is related to the gain coefficient $\gamma(\nu)$ as

$$\gamma(\nu) = \left(N_2 - \frac{g_2}{g_1}N_1\right)\sigma_0$$

This leads to

$$\sigma_0 = \frac{c^2 A_{21} g(\nu)}{8\pi\nu^2}$$

i.e. the stimulated emission cross-section depends on the characteristics of the emitter.

1.2.1 Line broadening mechanisms in gases

The spectral line of the transition will be broadened by various processes. There is a natural linewidth from the spread in energy for transitions between energy levels in atoms or molecules. For a natural broadened transition the lineshape follows a Lorentzian function, when normalized as

$$g(\nu) = \frac{\Delta\nu_N}{2\pi}\left[(\nu - \nu_0)^2 + \left(\frac{\Delta\nu_N}{2}\right)^2\right]^{-1}$$

Here the linewidth $\Delta\nu_\mathrm{N}$ is related to the lifetime of the transition as $\Delta\nu_N = 1/2\pi\tau$. In the He-Ne laser the natural linewidth is about 20 MHz.

However, in a gas laser, natural broadening is rarely the dominant lineshape determining mechanism. Collisions or electrostatic forces between neighbours induce interactions which equally affects all the emitters. For collisionally broadened transitions the lineshape is also Lorentzian and has a linewidth

$\Delta\nu = \Delta\nu_c$ with

$$\Delta\nu_c = \frac{NQ}{\pi}\left(\frac{16}{\pi}\frac{kT}{m}\right)^{\frac{1}{2}}$$

where N is the number density of atoms or molecules, Q is the collision cross-section and m is the atomic or molecular mass. This also is described by a Lorentzian function and occurs in gases due to collisions of the emitter and may be the predominant broadening mechanism at higher pressure.

The natural and collisional broadening mechanisms are termed homogeneous broadening in which all the emitters are affected equally. Alternatively, when the emitters have a distribution of broadened components this is termed inhomogeneous broadening. In gases the range of thermal velocities of the atoms or molecules leads to Doppler broadening where there is a distribution of Doppler shifts among the emitters. This is usually the dominant broadening mechanism in low density gases.

The Doppler broadened transition has a Gaussian lineshape. For an emitter of mass m with a centre frequency ν_0, the distribution of frequencies is

$$g_{\rm D}(\nu) = \frac{2}{\Delta\nu_{\rm D}}\left(\frac{\ln 2}{\pi}\right)^{\frac{1}{2}}\exp\left[-\left(\frac{2(\nu-\nu_0)}{\Delta\nu_{\rm D}}\right)^2\ln 2\right]$$

The linewidth (full-width at half-maximum) is

$$\Delta\nu_{\rm D} = 2\nu_0\left(\frac{2kT\ln 2}{mc^2}\right)^{\frac{1}{2}}$$

A contribution to line broadening also occurs dependent on the light intensity within the laser cavity (power broadening). For the He-Ne laser a power broadening of up to 100 MHz is found.

The peak value of the normalized lineshape function at line centre $g(\nu_0)$ is inversely proportional to the linewidth $\Delta\nu$ of the transition

$$g(\nu) = \frac{2}{\pi\Delta\nu}$$

Then the stimulated emission cross-section becomes

$$\sigma_0 = \frac{c^2 A_{21}}{4\pi^2\nu^2\Delta\nu}$$

This states that the stimulated emission cross-section for a homogeneously broadened transition is proportional to the ratio $A_{21}/\Delta\nu$ of the spontaneous transition rate to the linewidth.

Line broadening in gas lasers operating at relatively low pressure is mostly dominated by Doppler broadening. Collisional broadening may become dominant, e.g. in the CO_2 laser, at higher pressures.

1.2.2 Saturation irradiance

The gain in the laser is reduced from the unsaturated gain as the irradiance in the cavity increases. For a homogeneously broadened line this leads to the saturated gain coefficient as

$$\gamma_{s,h} = \frac{\gamma}{\left(1 + \frac{I}{I_{s,h}}\right)}$$

Here $I_{s,h}$ is the saturation irradiance

$$I_{s,h} = \frac{4\pi h\nu}{\lambda^2 g(\nu)}(\tau_{nr}/\tau_f)$$

where τ_f is the fluorescence lifetime of the upper laser level, τ_{nr} is the nonradiative decay time and h is Planck's constant. The saturated gain coefficient for an inhomogeneously broadened line is

$$\gamma_{s,i} = \frac{\gamma}{\left(1 + \frac{I}{I_{s,i}}\right)^{1/2}}$$

and the saturation irradiance is

$$I_{s,i} = \frac{2\pi^2 h\nu\Delta\nu}{\lambda^2}\left(\frac{\tau_{nr}}{\tau_f}\right)$$

Here $\Delta\nu$ is the homogeneous linewidth of the components of the inhomogeneously broadened transitions.

1.2.3 Threshold operation

The laser resonator (or cavity) is most usually made up of reflectors either side of the gain medium. For the laser to sustain oscillation the gain of the laser medium must be greater than the losses in the medium. The laser is a threshold device, in which the laser threshold is when the gain of the laser is equal to the losses in the laser cavity. To get a laser beam out of the laser, normally one of the reflectors is given a certain transmission. There will also be other losses in the laser due to diffraction, scattering or absorption.

For a laser cavity with two aligned mirror reflectors of reflectivities R_1 and R_2 spaced by a distance L, the threshold laser gain is

$$\gamma_{th} = k + \frac{1}{2L}\ln\left(\frac{1}{R_1 R_2}\right)$$

Here k accounts for all the losses other than that due to mirror reflectivities.

The threshold population inversion needed to sustain laser oscillation is then,

$$\Delta N_{\text{th}} = \left(N_2 - \frac{g_2}{g_1}N_1\right)_{\text{th}} = \frac{1}{\sigma_0}\left(k + \frac{\ln(R_1R_2)}{2L}\right)$$

1.3 Basic structures of gas state lasers

1.3.1 Gas laser media

Gas lasers often use a mixture of gases arranged to optimize the pumping or emission properties such as discharge characteristics or excited state lifetimes. The emission may be from an electronic transition in neutral atoms (e.g. He-Ne) or ionized atoms, (e.g. Ar^+, Kr^+), electronic transitions in molecules (F_2, N_2), electronic transitions in transient excimer molecules (KrF) or vibrational or rotational transitions in molecules (CO_2, CH_3F) or electronic transitions in molecular ions (N_2^+). A summary of the pulsed or CW operation, wavelengths and energy/power of some of the most common gas lasers is shown in Table 1. The gas lasers are excited by a great variety of pumping methods, including continuous, pulsed or rf electrical discharges, optical pumping, chemical reactions and gas dynamic expansion [9].

The first gas laser, and which was also the first continuous wave (cw) laser, was based on a mixture of helium and neon excited by a radiofrequency electrical discharge [10], producing continuous wave operation at 1.1523 μm. Now the He-Ne laser is more usually used on the 632.8 nm line, which is one of the most common of all lasers. The laser transitions occur in the neon atoms, with the strongest lines occurring at the familiar red line at 632.8 nm, the green line at

Table 1. Examples of some of the most common gas lasers

Laser	λ (nm)	Typical power	Pulsed or CW
(a) Neutral atom			
He-Ne	633	5 mW	CW
Cu	511	20 mW	CW
(b) Ionised atom			
Ar^+	488, 515	10 W	CW
Kr^+	647	0.5 W	CW
He-Cd	441.6, 325.0	50–200 mW	CW
(c) Excimer			
KrF	248	0.5 J	Pulsed
(d) Molecular			
CO_2	10.60×10^3	100–1000 W	CW
N_2	337.1	10 mJ	Pulsed
F_2	157	10 mJ	Pulsed
HCN	336.8×10^3	1 mW	CW
CH_3F	496 μm	1 mW	CW

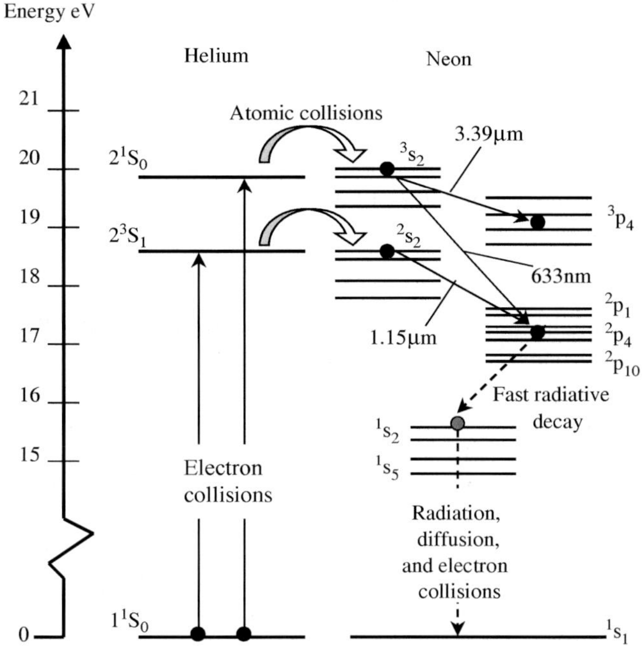

Figure 1. Energy levels of helium and neon and showing laser transitions in neon.

543.5 nm and other visible lines at 594, 612 and 640 nm, and the infra-red lines at 1.15, 1.52 and 3.39 μm.

The operation of the laser can be understood by considering the energy levels of helium and neon, shown in Figure 1. The He 2^3S and 2^1S metastable levels are excited by electron collisions. Transfer of the excitation occurs efficiently in collisions of He atoms with Ne atoms due to the close proximity of the Ne 2s and 3s energy levels to the He metastable levels.

The He-Ne gas mixture is contained in a narrow bore discharge tube, which may be excited by a dc discharge of about 10 mA or a radiofrequency discharge. The gas pressures are usually set for highest gain on the 632.8 nm transition, with partial pressures of about 2.5 mbar (He) and 0.13 mbar (Ne). The discharge tube diameter is kept quite small, since the gain is inversely proportional to the tube diameter, due to collisions of Ne atoms with the tube walls being required to reduce the population of the Ne 1s level to prevent build-up of population in the lower 2p laser levels.

The 632.8 nm and the 3.39 μm laser transitions share a common upper laser level (Figure 1). The 3.39 μm line has a Doppler broadened linewidth about $\frac{1}{5}$ of the 632.8 nm line and consequently has a larger gain. For 632.8 nm laser action, oscillation on the 3.39 μm line is needed to be prevented. This may be accomplished by selecting the laser resonator mirrors to have low reflectivity at 3.39 μm.

1.3.2 Optical resonators

The basic optical resonator of a gas laser is most usually a pair of aligned mirrors, one at each end of the gain medium, which reflect at the laser wavelength. These may be plane or with a selected radius of curvature and with a high degree of flatness to better than $\lambda/20$. The purpose of the optical resonator is to provide optical feedback and retain photons in the laser cavity. For stable laser oscillation with minimal loss in the cavity the resonator mirrors need to be accurately aligned. To obtain laser output one of the mirrors is arranged to have a certain, usually optimized, transmission so that light will be transmitted from the laser cavity, the other mirror normally has high reflectivity. This transmission is often the main optical loss of the laser.

The laser resonator determines the characteristics of the laser output by defining the frequencies of laser oscillation (corresponding to longitudinal modes) and the pattern of the cross-sectional structure of the laser beam (corresponding to transverse modes). The resonance condition for a resonator with two plane mirrors (the Fabry–Perot interferometer arrangement) spaced by a distance L and oscillating at a wavelength λ is $m\lambda/2 = L$, where m is an integer. This defines the resonant frequencies of the longitudinal modes,

$$\nu_{\mathrm{m}} = \frac{mc}{2L}$$

The frequency interval between adjacent modes is then $\Delta\nu = c/2L$.

The transverse mode is an electric and magnetic field configuration at some position in the laser cavity which on propagating one round trip in the cavity returns to that position with the same field pattern. Transverse modes may have circular (polar) symmetry, termed Laguerre–Gaussian modes, or rectangular (Cartesian) symmetry, termed Hermite–Gaussian modes. The fundamental transverse mode, labeled TEM_{00}, has no nodes in its transverse pattern. The fundamental TEM_{00} mode has a Gaussian distribution of irradiance $I(r)$, with peak irradiance on the laser axis. The Laguerre–Gaussian TEM_{01} mode has application in optical trapping. At a distance r from the axis

$$I(r) = I_0 \exp\left(\frac{-2r^2}{w^2}\right)$$

The radial width parameter w is referred to as the spot size of the beam. The spot size is smallest inside the laser cavity where there is a beam waist. For a plane-plane or confocal laser cavity of length L and operating at a wavelength λ, the beam waist is $w_0(\lambda L/2\pi)^{1/2}$. The radial Gaussian dependence of irradiance for laser beams leads to the application of Gaussian optics when dealing with laser beams. The beam may be slightly astigmatic, and therefore not truly Gaussian, induced by the use of Brewster windows or the laser gain medium. Operation in the fundamental Gaussian beam is valuable in laser frequency stabilization or interferometry.

1.3.3 Pumping techniques

Gas lasers are most generally excited by an electric discharge, with continuous, pulsed or radio-frequency discharges [9]. The dominant line broadening mechanisms in gases, at the optimum pressures for laser operation, leads to transition linewidths up to a few GHz. Energy levels can be excited by electron impact excitation or by resonant atom–atom collisions. In contrast optical pumping, commonly used in solid-state lasers, it is rarely used except for the pumping of mid-infrared (8–20 μm) and far infra-red (50 μm to 1 mm) gas lasers. Chemical pumping is effective for particular systems, e.g. in the HF laser and in the photodissociation and chemical oxygen iodine laser (COIL).

1.3.4 Emission characteristics

1.3.4.1 Coherence

A laser beam may be regarded as a stream of coherent photons which are nearly identical in amplitude, phase and direction. This high degree of coherence confers remarkable properties on the laser light. The temporal coherence gives a narrow spectral bandwidth or the ability to create ultrashort pulses. The spatial phase coherence ensures that generally the laser beam has low divergence and may be focused to a spot diameter a few times the wavelength of the light.

The theoretical laser linewidth $\Delta\nu_L$ is determined by the width of the cavity resonance $\Delta\nu_{cav}$ and the power in the laser P

$$\Delta\nu_L = \frac{2\pi h\nu}{P}(\Delta\nu_{cav})^2$$

The actual linewidth of the laser is controlled by thermal and mechanical instabilities.

1.3.4.2 Divergence

The remarkably low divergence of most laser beams is a consequence of their high spatial coherence over relatively large beam cross-sections. The laser beam spreads by diffraction both inside and outside the laser cavity. For the laser beam emerging from the laser cavity originating with a beam waist w_0, the radius of the beam $w(z)$ at a propagation distance z takes the form

$$w(z) = w_0\left[1 + \left(\lambda z/\pi w_0^2\right)^2\right]^{\frac{1}{2}}$$

$$\approx \frac{\lambda z}{\pi w_0} \text{ for } z >> \frac{\pi w_0^2}{\lambda}$$

The divergence of the beam is

$$\theta = \frac{\lambda}{\pi w_0}$$

As expected from diffraction physics, the larger the beam waist the smaller the divergence.

1.3.4.3 Coherence length

The high spatial coherence of the laser beam enables it to be focused to very small spot sizes, a few times the wavelength of the beam.

For a focusing lens of focal length *f*, beam radius w_1, and divergence angle θ, the focused spot size w_2 is

$$w_2 = f\theta = \frac{f\lambda}{\pi w_1}$$

if $2w_1 \approx$ lens aperture, $f/2w_1 \approx$ the *F*-number of the lens, and $w_2 \approx F\lambda$. The spectral bandwidth of the laser determines the temporal coherence, and hence the coherence length. For a laser Gaussian lineshape profile with bandwidth $\Delta\nu_L$ the coherence time is $\tau_c = 1/\Delta\nu_L$. Then the coherence length l_c is

$$l_c = \frac{c}{\Delta\nu_L}$$

A typical He-Ne laser with a practical bandwidth $\Delta\nu_L = 500\,\text{kHz}$ has a coherence length $l_c = 60$ cm, while a stabilized He-Ne with a bandwidth of 1 kHz has $l_c =$ 300 km.

The coherence length is relevant to interferometric applications and in the optical coherence tomography (OCT) technique [11]. There a low coherence light source is used in an interferometric arrangement such that localised structures can be imaged. The axial resolution of OCT, determined by the coherence length of the interferometer source, is dependent on the central wavelength λ_c and bandwidth $\Delta\lambda_c$ as $\lambda^2/(n\Delta\lambda)$, where *n* is the refractive index of the scattering medium.

1.3.5 Pulsewidth control

In some gas lasers, such as the nitrogen and copper vapour lasers, the intrinsic nature of the laser levels limits operation to pulsed or gain-switched output. Gas lasers which nominally operate cw or pulsed may be mode-locked to generate pulses in the picosecond region. In normal laser operation the resonant cavity has a low loss such that the cavity quality factor (*Q*) has a high value. Preventing laser action by mis-aligning one of the resonator reflectors or by the insertion of an in-cavity optical switch, while maintaining pumping of the laser, enables the population inversion to build up to a high value. If the *Q* of the cavity is quickly restored an intense laser pulse builds up. However, in most gas lasers the spontaneous emission lifetime of the laser level is normally very short (nanoseconds) so that little storage of energy is possible. A notable exception is the CO_2 laser with an upper state lifetime of about 4 s. Also in the gaseous atomic iodine laser transition at 1.3 μm, this may be excited by photodissociation from CF_3I or by chemical reaction into the first excited state. Radiative relaxation of the iodine $^2P_{1/2}$ is a magnetic dipole transition, such that the upper laser level lifetime is about 1 ms, thus allowing energy storage and Q-switched operation.

Picosecond and femtosecond pulses may be created by the process of mode-locking. A set of longitudinal modes simultaneously operating on a laser transition may be arranged to maintain a constant relative phase by actively modulating the laser cavity at a frequency that is the inverse of the time for a short pulse to travel round the cavity. In this way coherent oscillation is induced in the set of phase-locked modes. For a laser transition having a bandwidth $\Delta\nu$ and a laser cavity of length L, the mode-locked output is a train of pulses uniformly spaced in time with a period $2L/c$, an individual pulse duration of $\Delta\tau = 1/\Delta\nu$ and a peak irradiance in a pulse of N^2 times (unmode-locked irradiance). For the inhomogeneously broadened argon ion laser, typical parameters are $\Delta\nu = 3.5$ GHz, $L = 1$ m, $N = 20$, leading to mode-locked pulses of 280 ps duration.

Mode-locked lasers on the ps and fs timescales have widespread application in photobiology, in optical coherence tomography (OCT) imaging, in protein dynamics, photosynthesis, femtosecond spectroscopy, primary events in vision on cis–trans isomerization of rhodopsin and DNA/RNA studies [12,13]. These ultrafast lasers enable time-resolved investigations and the observation of transient species. In addition, the high peak power in the pulses, by which multi-photon absorption can be induced, can be used, for example, in the imaging of biological structures and the fragmentation of DNA.

1.3.6 Wavelength selection and frequency control

Many gas lasers can oscillate on more than one transition, these include the He-Ne, Ar ion and CO_2 lasers. Usually the strongest line will oscillate and selection of other lines may be made. In the visible region a tuning prism is frequently used for the He-Ne or argon ion laser. Another common element for wavelength selection in the visible or near infra-red is a birefringent filter. This is simply made up of a birefringent crystal plate with an optic axis parallel to the surface of the plate. When inclined at Brewster's angle and placed between two parallel polarizers it transmits the E-field in the plane of incidence of the plate. The ordinary and extraordinary components of this input beam experience a phase shift in passing through the plate to become elliptically polarized, and which is attenuated by the second polarizer. The thickness of the plate determines the linewidth of the tuning curve and its resolution. For the carbon dioxide laser in the mid-infrared a diffraction grating in the Littrow configuration is used in place of one of the cavity reflectors.

In the single longitudinal mode He-Ne laser, the linewidth of the laser is influenced by mechanical vibrations, air acoustic disturbance and discharge noise. The mode frequency can be stabilized to the gain curve using the Lamb dip, or by reference to absorption of iodine molecules held within a cell for the 632.8 nm He-Ne line, or by Doppler-free spectroscopy [14]. Frequency stabilization to a few parts in 10^{11} can be achieved. The frequency-stabilized laser has application in spectroscopy, metrology, determination of fundamental constants and gravitational wave detection.

1.4 Types of gas lasers

1.4.1 Helium-neon

The 632.8 nm He-Ne laser produces a cw output power ranging from 1 to 100 mW with laser tube lengths ranging from 10 to 100 cm. The He-Ne laser for powers of 1 to 5 mW typically has a tube length of about 15 cm, with resonator mirrors pre-aligned and sealed at each end of the tube. The green line at 543.5 nm is at a useful wavelength for alignment, beam pointing for infra-red lasers and target designation. However, its gain is only about 1/30 of that for the 632.8 nm transition. Consequently, high reflectivity resonator mirrors having low loss are required for the 543.5 nm line to give an output power of about 1mW.

For a typical resonator cavity length (L) of 30 cm the frequency spacing of the longitudinal modes is 500 MHz. The 633 nm transition has a gain bandwidth of about 1.2 GHz such that the number of oscillating longitudinally modes may be 1, 2 or 3. The longitudinal modes oscillate with alternating orthogonal polarization since this minimizes mode competition between adjacent modes.

The He-Ne laser may be operated in an external cavity, rather than with mirrors attached to the laser tube. In an external cavity laser output up to 100 mW can be produced. The ends of the gas discharge tube are then usually sealed with windows set at the Brewster angle such that there is low loss for p-polarization. In this configuration a prism may be placed in the cavity to provide selection and tuning of the laser oscillating wavelengths.

The 1–5 mW He-Ne laser finds extensive general application for optical alignment, optical testing and interferometry. It is also valuable at its red or green wavelengths for target designation. For example, in laser microdissection with a N_2 laser the region to be irradiated has been marked using a He-Ne laser [15]. The He-Ne laser is used in light-scattering, optical trapping, fluorescence correlation spectroscopy and laser scanning confocal fluorescence spectroscopy [16–20].

1.4.2 Ion gas lasers

Pulsed laser action in argon gas was discovered in 1964 and identified as transitions in the argon ion; this was subsequently followed by demonstration of cw operation and laser operation in the ions of other noble element gases [21,22]. These lasers have been and continue to be highly valuable sources of coherent radiation in the ultra-violet (from 275.4 nm), visible (particularly 488.0 and 514.5 nm) and near infra-red (to 1090 nm). Particular advantages of these lasers are the high continuous wave powers in the visible region and the ultra-violet, and the large range of wavelengths in regions not available presently from other sources [23,24]. Typically they provide continuous wave powers on the visible wavelengths of 488.0 and 514.5 nm of 1–20 watts.

In the continuous wave argon ion laser an electrical discharge of high current density is required to provide a sufficient rate of excitation since the laser transitions in the ion are from excited states with lifetimes of a few nanoseconds.

The necessary current density is obtained by constricting a high current discharge in a narrow bore tube. The efficiency of the ion lasers is very low at about 0.1%, this means that powers of kWs are required to be dissipated to maintain temperature equilibrium. This occurs partly since the common ion laser transitions are in singly ionized argon Ar^+ situated at about 35 eV while the lasers produce photons of a few eV. Thus active cooling of the laser tube is required, either by water or air.

The simplified energy level diagram for argon in Figure 2 shows those levels providing the main blue/green laser transitions. Listings of wavelengths for the argon ion and krypton ion lasers are given in Tables 2 and 3, respectively. In the high current electric discharge, electron impact with ground state atoms and metastable atoms and ions creates population in high lying levels of the ions. The upper laser levels are fed by direct excitation, cascade from higher excited states and by recombination from highly ionized species.

Much technological development has been expended on the gas tube for ion lasers over the last 30 years. In one highly developed form of gas plasma tube the envelope of the tube is made of a high thermal conductivity ceramic in which the discharge path is set by a series of tungsten disc apertures, which define the bore of the tube, and held in place by copper supports. Tube lengths range over

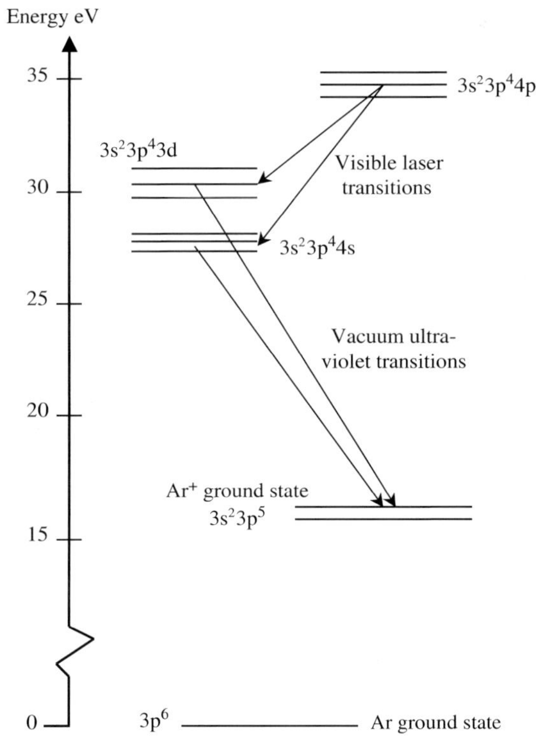

Figure 2. Relevant energy levels for the argon ion laser.

Table 2. Wavelengths and output powers of commercial argon ion lasers (Coherent Inc.)

	Output Power			
Wavelength (nm)	25 W main frame (W)	Frequency-doubled 25 W main frame pump (nm)[a]	(W)	Small-frame laser (mW)
528.7	1.8	264.3	0.10	
514.5	10.0	257.2	1.00	350
501.7	1.8			
496.5	3.0	248.2	0.30	
488.0	8.0	244.0	0.50	200
476.5	3.0			
472.7	1.3			
465.8	0.8			
475.9	1.5	229.0	0.04	80[a]
454.5	0.8			
333.6–363.8	7.0			Multi-line UV 100 mW[a]
350.7				(40)[a]
351.1	1.8		1.0	
363.8	1.7		1.0	
351.1–385.8	4.4			
300.3–335.8	3.0			
275.4–305.5	1.6			
334.5	0.5			
302.4	0.38			
275.4	0.18			

[a] With UV mirror cavity.

50 to 200 cm, in which discharge currents up to 40 A are used. The ends of the gas plasma tube are normally windows held at the Brewster's angle to avoid optical losses. The gas pressure in the tube of typically 1 mbar is maintained by a pressure control system. The operational lifetimes of the ion laser tube is typically up to 6000 h for small lasers and up to 3000 h for larger lasers.

A particular application of the argon ion laser is for the excitation of dye lasers and the solid-state titanium-sapphire laser. This is illustrated in Figure 3(a) for the two configurations and the tuning ranges in Figure 3(b). The full tuning range of the dye lasers requires the use of about 20 dye molecules. Also shown in Figure 3(b) is the tunable ranges of frequency doubled dye lasers and the titanium-sapphire laser. The peak wavelengths of various dye lasers excited by the argon ion lasers are listed in Table 4 for a ring laser configuration, along with peak wavelengths from the titanium-sapphire laser. The table also shows the pumping wavelengths and powers.

In various applications in photobiology laser stability is an important aspect. Feedback control of the laser cavity mirrors enables the alignment of the laser to be maintained, the output power to be stabilized, the modes to be optimized and noise minimized.

Frequency doubling of visible lines to the UV extends the wavelength capabilities of the laser. One configuration is shown in Figure 4 where this is

Table 3. Wavelengths and output powers of commercial krypton ion lasers (Coherent Inc.)

Wavelength (nm)	Output power: Main-frame krypton (W)	Output power: Mixed-gas krypton/argon (W)
793.2–799.3	0.03	–
752.5–799.3	0.25	–
752.5	0.10	0.03
676.4	0.15	–
647.1	0.80	0.25
568.2	0.15	0.15
530.9	0.20	0.13
520.8	0.07	0.13
514.5 (Ar)	–	0.25
488.0 (Ar)	–	0.25
482.5	0.03	–
476.5 (Ar)	–	0.10
476.2	0.05	–
457.9 (Ar)	–	0.03
413.1	0.30	–
	Multi-line Violet 0.60 W	
406.7	0.20	–
356.4	0.12	–
350.7	0.25	–
	Multi-line UV 0.5 W	Multi-line UV 0.05 W

achieved by intracavity critically phased-matched BBO crystals to give wavelengths over 229.0 to 264.3 nm. An advantage of generating UV light by frequency doubling from the visible lines, compared to operating the ion laser on UV lines, is that the lifetime of the laser tube may be almost doubled. There is strong interest in the UV wavelengths since they have application, for example, in resonance Raman spectroscopy to give enhanced Raman signal [25], and in the UV excitation of dyes.

Water-cooled versions of the lasers give output powers up to 25 W and air-cooled versions to 0.5 W. Ion lasers may be operated in a number of configurations. The laser may be operated all-lines for maximum output power. A single wavelength may be selected by use of a tuning prism introduced inside the laser resonator. Commercial lasers with active stabilization systems maintain stable power and frequency and operate in a set transverse mode.

Ion lasers have widespread application in photobiology [1–4], in light-scattering, many fluorescence spectroscopy applications, Raman spectroscopy [26], confocal microscopy, laser scanning confocal fluorescence microscopy [27], DNA micro-array analysis and gene expression, molecular probe spectroscopy and fluorescence correlation spectroscopy. There are extensive applications in biomedical areas such as flow cytometry and microsurgery and for excitation of dye and titanium-sapphire lasers – which may then be applied in optical coherence tomography.

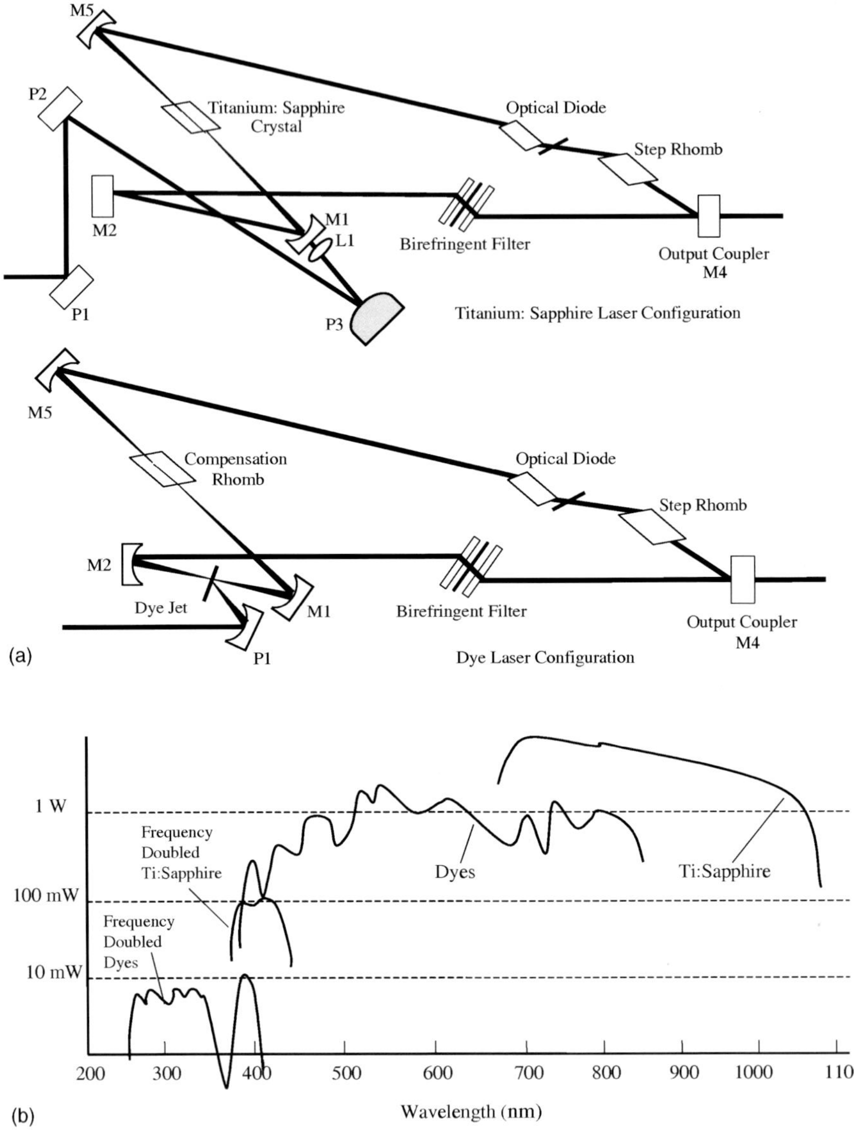

Figure 3. (a) Tunable solid-state and dye lasers pumped by argon ion laser. (b) Tuning ranges for dyes and titanium:sapphire lasers. (Coherent Inc.).

1.4.3 Excimer lasers

The term 'excimer' is short for excited dimer to describe a diatomic molecule which has a bound excited electronic state but which is dissociated (unbound) in the ground state. The first excimer laser demonstrated in 1970 was from liquid Xe_2 molecules excited by an electron beam and in 1972 from xenon gas. This was

Table 4. Wavelengths available from a commercial ring laser with dye or titanium-sapphire pumped by an argon ion laser (Coherent model 899); [B/G-both blue and green argon ion laser pump wavelengths]

Excimer dye	Pumped by	Wavelength (nm) (approx.)
Exalite 392E	5.0 W UV (Ar)	392
Stilbene 1	2.5 W UV (Ar)	416
Stilbene 3	2.5 W UV (Ar)	433
Coumarin 102	3.0 W violet (Kr)	477
Coumarin 30	3.0 W violet (Kr)	518
Coumarin 6	6.0 W 488 nm (Ar)	535
Rhodamine 110	6.0 W 514 nm (Ar)	550
Rhodamine 6G	6.0 W 514 nm (Ar)	575
Rhodamine 6G	4.0 W B/G (Ar)	575
DCM	6.0 W 488 nm (Ar)	640
Pyridine 2	7.5 W 514 nm (Ar)	725
LD700	4.6 W red (Kr)	740
Styryl 8	7.5 W 514 nm (Ar)	764
Styryl 9M	7.5 W 514 nm (Ar)	835
Titanium:sapphire	15.0 W B/G (Ar) SW	760
Titanium:sapphire	15.0 W B/G (Ar) LW	825
Titanium:sapphire	15.0 W B/G (Ar) MW	925

followed by laser action in rare gas halogen excimers XeBr and XeF in 1975 [28,29]. In the excimer lasers an excimer molecule is formed, between two rare gas atoms or a rare gas atom and an halide, into a short-lived bound excited state following an electric discharge. An electronic radiative transition occurs between the excited bound state and an unbound or weakly bound ground state (bound-to-free transition). Transient population inversion following high current pulsed electric excitation leads to pulsed UV laser emission.

The potential energy level diagram for the ArF rare gas-halide excimer is shown in Figure 5, this indicates that a rare gas atom (R) and a halogen atom (Ha) can form a strong ion bound giving RHa in an excited state. An important feature of the

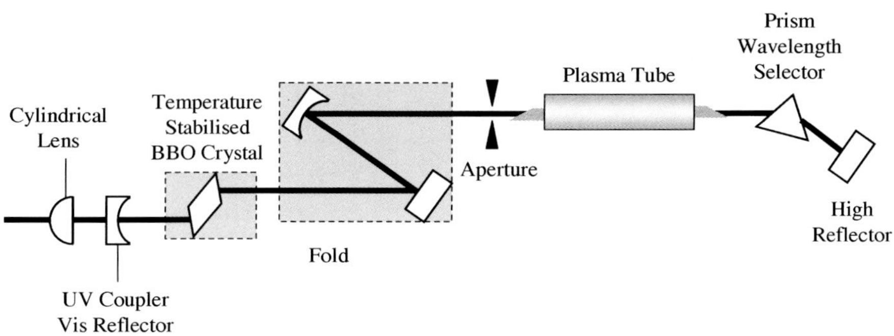

Figure 4. Arrangement for second harmonic generator of argon ion laser lines. (Coherent Inc.).

excimer laser is their UV emission wavelengths in the region where atoms and molecules have strong absorption [30]. Table 5 gives the laser wavelengths of several of the more common rare gas-halide excimer lasers. The laser transitions are $B^2\Sigma$ to $X^2\Sigma$ transitions. The KrF laser may be operated to produce laser pulses of about 300 ns duration at repetition rates up to of 1 kHz at average powers up to 1 kW. The excimer lasers have application in photoexcitation, spectroscopy, photochemistry, photoionization, the pumping of laser dyes and in tissue interaction in laser medicine.

The upper molecular bound state is formed by reacting atomic species and the lower level is fully repulsive or only weakly bound. On radiative emission the molecule splits up within one vibrational period of about 10^{-13} s. The lasers are excited by a pulsed electric discharge in a transverse beam geometry where the electrodes are parallel to the laser axis [31]. To maintain a uniform and stable discharge the medium is pre-ionised. For the KrF excimer laser the gas is a mixture of Kr and F_2 with He or Ne as a buffer gas. Recent research has led to improvements in the understanding of the mechanisms in the excimer laser and on their performance [32,33].

An excimer source operating as a lamp rather than a laser has recently been developed [34] in which the precursor gases are excited by a small electron beam.

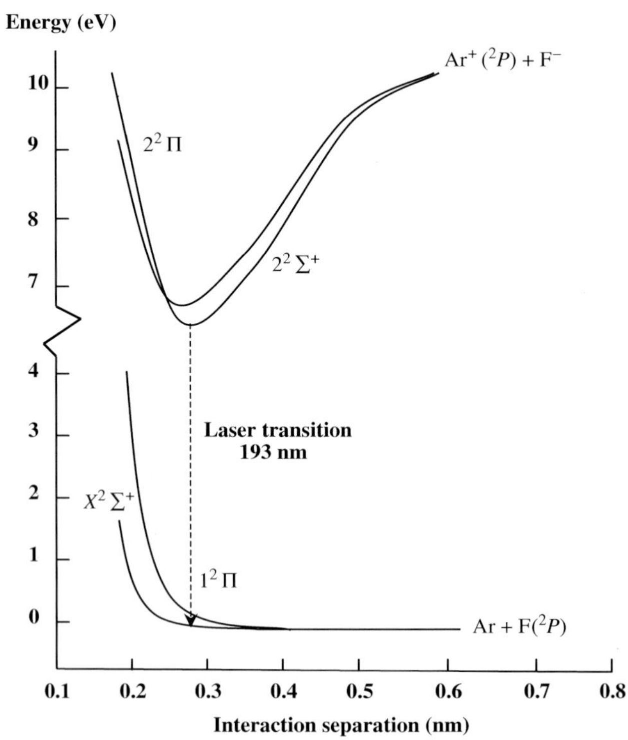

Figure 5. Potential energy level diagram of the ArF excimer laser. The transient bound state of ArF and the laser transition at 193 nm is shown.

Table 5. Laser wavelengths for rare gas-halide excimer, F_2 and N_2 lasers

Laser	Wavelength (nm)
ArCl	169, 175
ArF	193
XeBr	281.8
XeCl	308
XeF	351, 353
XeF ($C \rightarrow A$)	490
KrBr	206.5
KrCl	222
KrF	248
F2	157
N2	337.1

This lamp can produce the excimer wavelengths of 157, 193, 248 and 351 nm wavelengths at total powers of 100 mW and with 1 to 15 nm bandwidth.

1.4.4 Molecular lasers

There are several types of laser based on transitions between electronic, vibrational or rotational molecular energy levels. Laser oscillation on vibrational–rotational transitions of the ground electronic state leads to emission in the mid- to far-IR over 2.5 to 300 μm. Because of the low energy of the levels involved in these transitions, giving a high quantum efficiency and an efficient excitation mechanism, these lasers have high overall efficiency. They are able to give very high cw kW power levels and pulsed energies of 100s J. The most important example of this class is the carbon dioxide laser oscillating at 10.6 μm and other wavelengths in the 9–11 μm region; other examples are the CO laser at 5 to 6.5 μm and the HF chemical laser at 2.7 to 3.3 μm. Transitions in molecules between vibrational levels of different electronic states produce UV laser lines, such as that from the N_2 laser at 337.1 nm. The excimer lasers operating on transient diatomic rare gas or rare gas-halide molecules are considered separately in Section 1.4.3. A third class of molecular lasers involving transition between rotational levels of the same vibrational state leads to laser transitions in the far-IR over 25 μm to 1 mm. Examples are the HCN laser (wavelength 337 μm) and the H_2O laser (wavelengths 28, 78 and 119 μm).

1.4.4.1 Carbon dioxide

Excitation of vibrational levels in the CO_2 molecule can be achieved very efficiently by energy transfer from N_2 molecules, excited in an electric discharge, to give high power CW and pulsed laser output. In fact, the CO_2 laser is both one of the most efficient lasers (about 15–25%) and in special configurations one of the most powerful lasers (kW to 100s kW). The CO_2 laser can be tuned over about 80 lines between 9.1 to 10.9 μm.

The lowest vibrational levels of CO_2 and N_2 are shown in Figure 6. The symmetric stretch, bending and asymmetric stretch modes of vibration correspond to the resonance frequencies ν_1, ν_2 and ν_3. Laser emission may occur on vibrational–rotational transitions of the P ($\Delta J = -1$) or R ($\Delta J = +1$) branches corresponding to the change in J the rotational state quantum number. The most common laser transition in CO_2 at 10.6 μm corresponds to the transition between the 00°1 and 10°0 vibrational levels. The P branch has the lower energy and higher laser gain. For gas pressures greater than about 130 mbar the dominant line broadening mechanism in the CO_2 laser is by collisions, while below that pressure it is Doppler broadening.

In an electric discharge in a mixture of CO_2, N_2 and He excitation to the upper 00°1 laser level is achieved by electron impact or resonant transfer from vibrationally excited ($v = 1$) N_2 molecules. The N_2 ($v = 1$) level is efficiently excited in the discharge and is a long-lived metastable state. The role of He in the gas mixture is to de-activate the lower laser level, to control the electron temperature in the discharge, to stabilize the plasma and to cool the CO_2 molecules by conducting heat to the walls of the discharge tube.

There are a remarkable variety of designs of CO_2 laser: slow axial flow, fast axial flow, diffusion-cooled, sealed-off, transverse flow, waveguide, transversely excited atmospheric (TEA) and gas dynamic lasers. The slow axial flow design is the original conventional form of the laser, in which a gas mixture of CO_2:N_2:He in the approximate ratio of 8:1:1 is slowly flowed along the tube to remove dissociation products and is excited by an electric discharge. These lasers are able to give up to 70 W m^{-1} or about 100 W power for a 1.4 m gain length. For a sealed-off laser giving a reasonable operational lifetime a means of converting CO formed in the discharge back into CO_2 is required. This is achieved by adding H_2 gas which reacts to form OH hydroxyls which in turn react with the CO contaminant. Low power sealed-off lasers are able to operate with up to 10,000 h lifetime.

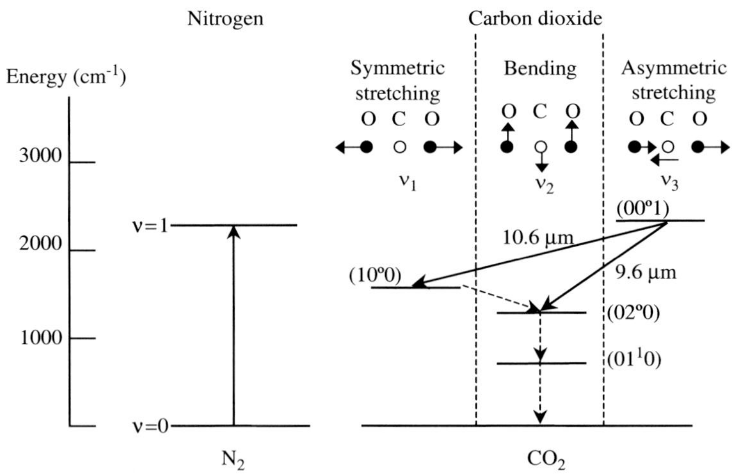

Figure 6. CO_2 vibrational modes and the lowest vibrational levels and laser transitions, energy transfer from N_2 is indicated.

A very compact form of CO_2 laser can be made using a waveguide structure in which the laser radiation is guided by the inside walls of the laser tube, which are typically 2–4 mm apart. The gas pressure can be increased to about 250 mbar and excited by a radiofrequency discharge in a sealed-off structure. For a short laser tube of 50 cm length laser output powers up to 50 W can be generated. Several designs of CO_2 laser have been devised to improve the removal of heat from the laser. The fast axial flow laser overcomes the limitation of the slow axial flow laser by flowing the gas at very high speed, enabling rapid removal of the heat in the tube. The fast axial flow laser may be excited by a longitudinal discharge or a radiofrequency discharge to give output powers of several kW. Other methods of quickly removing heat include the use diffusion-cooling in a slab geometry, in which cooling is effectively achieved by heat flow to the water-cooled electrodes, or to use a transverse flow of gas to remove heat by convection.

In the transversely excited atmospheric (TEA) laser pulsed excitation by transversely mounted electrodes enables the gas pressure to be increased equal to or greater than atmospheric. By pre-ionizing the gas with a pre-pulse and using short pulse excitation instabilities in the discharge are avoided. The transverse flow TEA CO_2 laser produces a pulsed laser output at up to 50 Hz repetition rate and up to 500 W output power. In terms of the cost of the laser per watt of power the CO_2 laser is the leader.

In the gas dynamic laser the gas mixture of CO_2 and N_2 at a high pressure of about 20 atmospheres is heated in a high-pressure vessel up to 2000 K. At that temperature a large fraction of the CO_2 molecules are in the lower or upper laser levels. The gas is then expanded through gas nozzles to a pressure of about 60 mbar such that its temperature drops to less than 400 K. The lifetime of the CO_2 upper laser level is longer than for the lower laser level and relaxation at the lower laser level occurs earlier such that population inversion is obtained over a substantial spatial region. The gas dynamic laser is capable of extremely high powers up to 100 kW.

The CO_2 laser has extensive application in various tissue interaction studies, such as laser surgery. It has been used in the welding of tissues and with a biological solder [35–37]. The addition of a protein solder increases the tensile strength.

1.4.4.2 Carbon monoxide

The CO laser also operates on vibrational–rotational transitions, but at shorter wavelengths over 5 to 6.5 μm compared to the CO_2 laser. Excitation of the CO vibrational levels in an electric discharge can be efficiently achieved. The laser is capable of high efficiency greater than 50% and multi kW power. Similarly as for the CO_2 laser the CO laser designs have included axial flow, transversely excited structures and gas dynamic techniques. To achieve high efficiency and high power the CO laser has been operated at low temperatures down to 77 K. However, room temperature versions of the laser have been produced, operating at 1 kW and 10% efficiency.

1.4.4.3 Nitrogen

The nitrogen laser is an example of a 3-level laser. It produces pulsed output in the UV at 337.1 nm and is a valuable laser source for various applications including laser-induced fluorescence, laser microbeam studies and the pumping of dye lasers. A simplified potential energy level diagram for the N_2 molecule is shown in Figure 7. Electron excitation from the ground state preferentially excites the $C^3\Pi_u$ state. Transient inversion is achieved on the $C^3\Pi_u$–$B^3\Pi_g$ transition. The longer lifetime of the $B^3\Pi_g$ state of 40 μs compared to the lifetime of the $C^3\Pi_u$ state means that the population inversion terminates with laser action and build up of population in the $B^3\Pi_g$ level. This is an example of a "self-terminating laser", as for the metal vapour lasers described in Section 1.4.5.

A common design for the N_2 laser is the transversely excited atmospheric (TEA) configuration, as for the CO_2 laser, in which the laser with a high pressure of N_2 is rapidly excited with a transverse discharge. Output powers at 337.1 nm of greater than 1 MW and a duration of >10 ns and a repetition rate of 100 Hz can be produced. Short duration pulses to 100 ps at 100 kW power can be obtained in very short versions of the laser.

The nitrogen laser pumped dye laser is a simple but valuable source of UV and tunable visible radiation. When used to pump a range of fluorescent dyes a broad

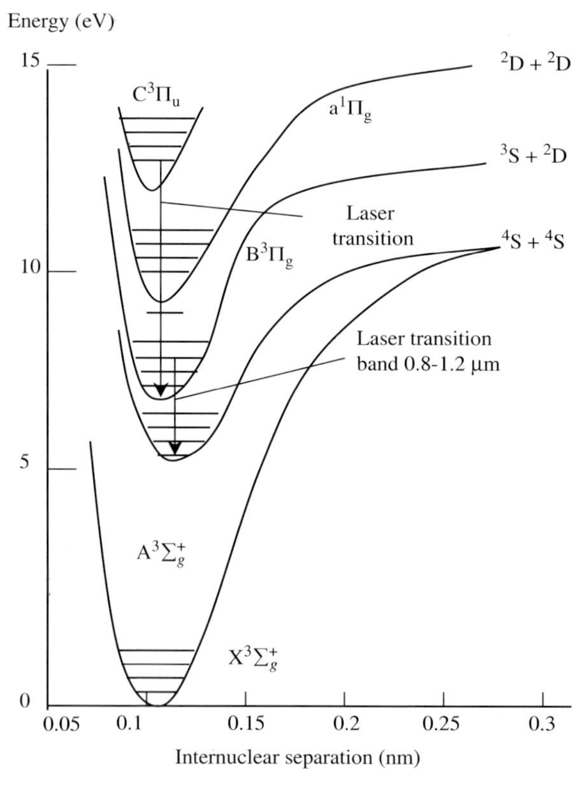

Figure 7. Energy levels of the N_2 molecule showing the two groups of laser transitions.

tuning range can be produced. The 337.1 nm ns/ps pulses from a small N_2 laser are used in the matrix-assisted laser desorption and ionization (MALDI) technique to desorb protein molecules from a solid or liquid matrix for time-of-flight mass spectroscopy.

1.4.4.4 F_2

The molecular F_2 laser operates in the VUV at 157 nm. Although not strictly an excimer laser, since it does not radiate to an unbound ground state, its excitation and laser properties are similar to the excimer lasers. When excited by a fast pulsed discharge it produces output pulses with 10–30 mJ energy and 10–30 ns duration, although pulse durations up to 70 ns have been demonstrated. Atmospheric absorption needs to be taken into account at the F_2 laser 157 nm wavelength.

1.4.5 Metal vapour lasers

The two most used metal vapour lasers are He-Cd, with cw visible output at 441.6 nm and UV output at 354.0 and 325.0 nm, and the pulsed copper vapour laser, with main wavelengths of 511 and 578 nm. In addition there are many other laser transitions in metal vapours [38,39].

1.4.5.1 He-Cd

Laser action occurs in a mixture of He and Cd gases when excited by an electric discharge similar to the He-Ne laser. However, in the He-Cd laser the laser transitions are in single ionized Cd^+[40]. The relevant energy levels of the laser are shown in Figure 8. The Cd upper laser levels are mainly excited by the Penning ionization processes from the He metastable levels, $He^* + Cd \rightarrow He + Cd^+$. The distribution of Cd vapour in the discharge is controlled by the mechanism of cataphoresis. When Cd metal is heated and vapourised at the anode, it is ionized by the discharge and the Cd^+ ions drift towards the cathode end. The Cd condenses at the cooler cathode end. A reservoir of He is connected to the tube to replace He gas lost in the discharge.

The discharge tube length ranges typically from about 25 to 75 cm and is driven by a low current dc discharge of about 100 mA. The 75 cm tube will produce multimode power of about 200 mW at 441.6 nm and 100 mW at 325 nm. These wavelengths are used in many applications, including fluorescence spectroscopy and flow cytometry.

1.4.5.2 Copper and gold vapour

The copper vapour laser (CVL) has pulsed laser lines at 510.5 and 578.2 nm and the gold vapour laser at 312 and 628 nm. These lasers are the most generally used laser types of a set of atomic metal vapours classified as "self-terminating". The lasers work in pulsed mode and can be operated at high repetition rate [41]. The copper vapour laser produces pulse energies of 50 mJ in 50 ns pulses at repetition rates up to 20 kHz such as to generate typically powers of 10 to 100 W. These properties of the CVL give it applications in the pumping of other lasers,

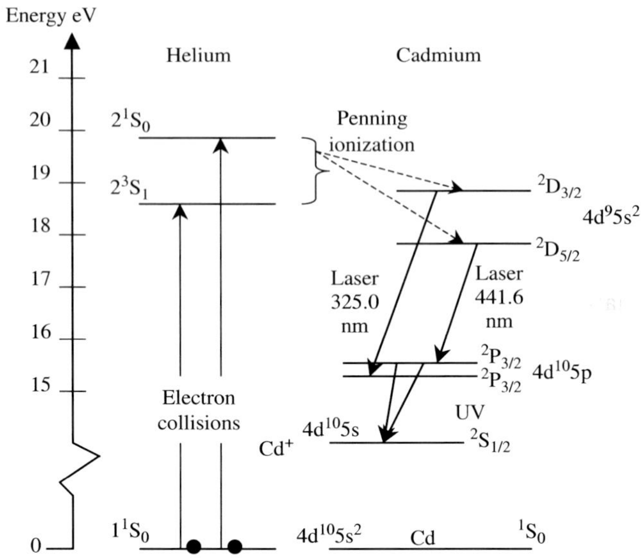

Figure 8. Energy levels and laser transitions relevant to the He-Cd laser.

such as the dye laser and titanium-sapphire laser, and in spectroscopy, high-speed photography and materials processing.

The relevant energy levels of atomic copper and gold are close to the ground state (Figure 9) such that the photon efficiency of the lasers is very high and overall

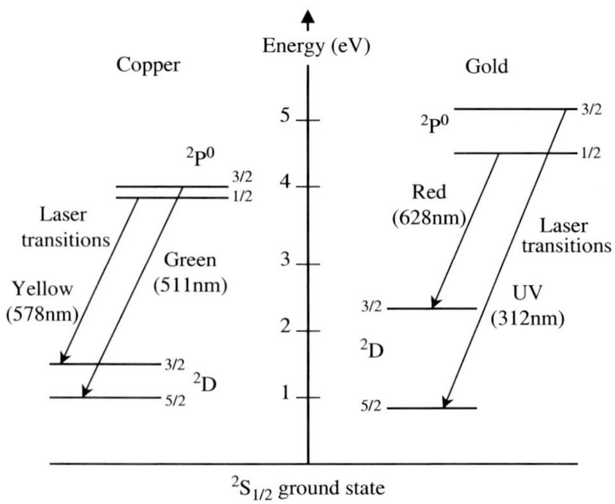

Figure 9. Relevant energy levels of the copper vapour and gold vapour lasers.

laser efficiencies of about 1% can be realized. A large population inversion can be created in the metal vapour when it is excited by a fast pulsed electric current. Copper atoms are excited briefly preferentially into the $3d^{10}4p$ ($^2P_0$) configuration rather than into the $3d^94p^2(^2D)$ configuration. The lower laser levels are metastable and the relatively longer lifetime of the lower laser level prevents cw oscillation, hence the "self-terminating" nature of the laser operation.

The most common copper vapour laser operates on elemental copper in which the source of Cu vapour is by evaporation of copper pellets in an alumina discharge tube operating at a temperature of about 1500°C. This is thermally insulated from a surrounding glass envelope and also contains a Ne buffer gas and a trace of H_2. A pulse discharge circuit is triggered by a thyratron or a solid-state-switch to deliver pulsed current to the laser at 10s of kHz repetition rates. Improvement of performance over the elemental copper vapour lasers has been obtained by adding H_2 and HCl to the Ne buffer gas as in the kinetically-enhanced CVL [42]. Copper vapour lasers in which the copper is introduced in the form of copper halide such as CuBr minimise the warm-up time of the laser. A further extension is to generate the copper halide inside the discharge tube by reacting HBr gas with metallic copper, this type has been termed the copper HyBrID laser.

1.5 Perspectives

The gaseous lasers discussed in this chapter provide many wavelengths from the UV to the far-IR in both cw and pulsed outputs. They are based on highly developed technology and offer reliable operation with good performance. However, since they operate in a low density gas medium they need a relatively large physical size, e.g. compared to semiconductor lasers or diode pumped solid-state lasers, to achieve sufficient gain and output power or pulse energy. In addition, excitation by cw or pulsed discharges requires high voltage power supplies. In comparison the emergence and development of diode pumped solid-state lasers, semiconductor lasers, solid-state lasers combined with frequency conversion and fibre lasers provide in many cases superior alternative sources, which may be smaller, more efficient and convenient or operate from low voltage supplies. An alternative to the He-Ne laser is the red-emitting semiconductor AlGaInP laser. Further developments expected in semiconductor diode lasers, in the red and blue/violet regions, and advances in vertical cavity surface emitting lasers (VCSELs), will bring alternatives to red and blue gas lasers. Improvement has been made in performance of the red-emitting VCSEL, combining an AlGaInP quantum well with AlGaAs multiplayer mirrors [43]. The pre-eminence of the argon ion laser in providing medium power visible cw radiation is being challenged by the frequency doubled diode pumped $Nd{:}VO_4$ solid-state laser.

However, solid-state alternatives to many gas lasers wavelengths are not currently available. This applies to the CO_2 and CO lasers in the mid-IR region, far-infrared gas lasers and to the ion lasers and excimer lasers in the UV region.

Gas lasers can also often offer power scaling in cw or pulsed systems which are less readily achievable with solid-state lasers.

There is extensive use of lasers in photobiology. Gas lasers are an essential source in many of the established techniques, such as microscopy, imaging, Raman spectroscopy and fluorescence spectroscopy, and in the emerging techniques of optical trapping, micromanipulation, microbeam dissection, DNA micro-arrays and genomics.

References

1. T. Vo-Dihn (Ed) (2003). *Biomedical Photonics Handbook.* Chemical Rubber Company, Boca Raton, FL.
2. V.V. Tuchin (Ed) (2002). *Handbook of Optical Biomedical Diagnosis.* SPIE, Bellingham WA.
3. J.R. Lakowicz (1999). *Principles of Fluorescence Spectroscopy*, Plenum, New York.
4. B. Pogue, M. Mycek (2003). *Fluorescence in Biomedicine.* Marcel Dekker, New York.
5. P.W. Milonni, J.H. Eberly (1998). *Lasers.* Wiley, New York.
6. W.T. Silfvast (1996). *Laser Fundamentals.* Cambridge University Press, Cambridge.
7. C.C. Davis (1996). *Lasers and Electro-Optics.* Cambridge University Press, Cambridge.
8. O. Svelto (1998). *Principles of Lasers.* 3rd edn. Plenum Press, New York.
9. C.S. Willett, (1974). *Introduction to Gas Lasers: Population Inversion Mechanisms*, Pergamon Press, Oxford.
10. A. Javan, W.R. Bennett, D.R. Herriott (1961). Population inversion and continuous optical maser oscillation in a gas discharge containing a helium-neon mixture. *Phys. Rev. Lett.* **6**, 106–110.
11. B.E. Bouma, G.J. Tearney (2001). *Handbook of Optical Coherence Tomography.* Marcel Dekker, New York.
12. J.C. Diels, W. Rudolph (1996). *Ultrashort Laser Pulse Phenomena: Fundamentals, Techniques and Applications.* Academic, San Diego.
13. A.H. Zewail (1994). *Femtochemistry, Ultrafast Dynamics of the Chemical Bond.* World Scientific, Singapore.
14. T.J. Quinn (1996). Results of recent international comparison of national measurement standards carried out by BIPM. *Metrologica* **33**, 271–287.
15. W. Meier-Ruge, W. Bielser, E. Remy, F. Hillenkamp, R. Nitsche, E. Ünsold (1976). The laser in the Lowry technique for microdissection of freeze-dried tissue slices. *Histochem. J.* **8**, 387–401.
16. A. Ashkin, J.M. Dziedzic (1987). Optical trapping and manipulation of viruses and bacteria. *Science* **235**, 1517–1520.
17. A. Ashkin, J.M. Dziedzic, T. Yamane (1987). Optical trapping and manipulation of single cells using infrared laser beams. *Nature* **330**, 769–771.
18. S.B. Smith, Y.J. Cui, C. Bustamante (1996). Overstretching β-DNA: The elastic response of individual double-stranded and high-stranded DNA molecules. *Science* **271**, 795–799.
19. Y. Ishii, A. Ishijima, T. Yanagida (2001). Single molecule nanomanipulation of biomolecules. *Trends Biotechnol.* **19**, 211–216.
20. K.O. Greulich (1999). *Micromanipulation by Light in Biology and Medicine.* Birkhauser Verlag, Basel.
21. W.B. Bridges (1964). Laser oscillation in singly ionized argon in the visible spectrum. *Appl. Phys. Lett.* **4**, 128–130.

22. G. Convert, M. Armand, P. Martinot-Lagarde (1964). Transitions lasers visibles dans l'argon ionise. *C.R. Acad. Sci.* **258**, 4467–9.
23. W.B. Bridges (1982). Ionized gas lasers. In: M.J. Weber (Ed.), *Handbook of Laser Science and Technology* (Volume II). Chemical Rubber Company, Boca Raton, FL.
24. C.C. Davis, T.A. King (1975). Gaseous ion lasers. *Adv. in Quantum Electron.* **3**, 169–454.
25. S.A. Asher, R.W. Bormett, X.G. Chen, D.H. Lemmon, N. Cho, P. Peterson, M. Arrigoni, L. Spinelli, J. Cannon (1993). UV resonance Raman spectroscopy using a new cw laser source: convenience and experimental simplicity. *Appl. Spectrosc.* **47**, 628–633.
26. I.R. Lewis, H.G.M. Edwards (Eds) (2001). *Handbook of Raman Spectroscopy*. Marcel Dekker, New York.
27. J.B. Pawley (Ed) (1995). *Handbook of Biological Confocal Microscopy*. Plenum, New York.
28. S.K. Searles, G.A. Hart (1975). Stimulated emission at 281.8 nm from XeBr. *Appl. Phys. Lett.* **27**, 243–245.
29. C.A. Brau, J.J. Ewing (1975). 354 nm laser action on XeF. *Appl. Phys. Lett.* **27**, 435–457.
30. D.J. Elliott (1995). *Ultraviolet Laser Technology and Applications*. Academic, New York.
31. W.J. Witteman, G.B. Ekelmans, M. Trenkelman, F.A. von Goor (1990). Discharge technology for excimer lasers of high average power. In: J.M. Orza, C. Domingo (Eds), *Eighth Int. Symp. On Gas Flow and Chemical Lasers* (SPIE, Volume 1397). SPIE, Bellingham.
32. T. Saito, S. Ito, A. Tada (1996). Long lifetime operation of an ArF excimer laser. *Appl. Phys., B*, **63**, 229–235.
33. S. Nagai, M. Sakai, H. Furuhashi, A. Kono, T. Goto, Y. Uchida (1998). Effects of F^- and F_2 molecules on the oscillation process of a discharge pumped ArF excimer laser. *IEEE J. Quantum Electron* **34**, 40–46.
34. J. Wallace (2003). E-lamp emits ultraviolet simply. *Laser Focus World*, May, 20–21.
35. L.S. Bass, M.R. Treat, (1995). Laser tissue welding: a comprehensive review of current and future clinical applications. *Lasers Surg. Med.* **17**, 315–349.
36. A. Lauto (1998). Repair strength dependence on solder protein concentration: A study in laser tissue welding. *Lasers Surg. Med.* **22**, 120–125.
37. E. Strassmann, N. Loya, D. Gaton, A. Ravid, N. Kariv, D. Weinberger, A. Katzir (2001). Laser soldering of the cornea in a rabbit model using a control-temperature CO_2 laser system. *Proc. SPIE* **4244**, 253–265.
38. C.E. Little, N.V. Sabotinov (Eds) (1996). *Pulsed Metal Vapour Lasers – Physics and Emerging Applications in Industry, Medicine and Science*, (NATO ASI Series 1, Volume 5). Kluwer Academic, Dordrecht.
39. C.E. Little (1999). *Metal Vapour Lasers: Physics, Engineering and Applications*. Wiley, Chichester.
40. W.T. Silfvast, G.R. Fowles, B.D. Hopkins (1966). Laser action in singly ionized Ge, Sn, Pb, In, Cd and Zn. *Appl. Phys. Lett.* **8**, 318–319.
41. A.A. Isaev, M.A. Karayan, G.G. Petrash (1972). Effective pulsed copper-vapour laser with high average generation power. *JETP Lett.* **16**, 27–29.
42. M.J. Withford, D.J.W. Brown, R.J. Carman, J.A. Piper (1998). Enhanced performance of elemental copper-vapor lasers by use of H_2-HCl-Ne buffer gas mixtures. *Opt. Lett.* **23**, 706–708.
43. R.P. Schneider, C.M. Hagerott, K.D. Choquette, K.L. Lear, S.P. Kilcoyne, J.J. Figiel (1995). Improved AlGaInP-based red (670-690 nm) surface emitting lasers with novel C-doped short cavity epitaxial design. *Appl.Phys.Lett.* **67**, 329–331.

Chapter 2

Liquid state lasers

Terry A. King

Table of contents

Abstract

Liquid state lasers mainly use organic dye molecules in solution as the gain medium, which absorb and emit over broad wavelength bands that span from the UV to IR regions. The dye molecules may be excited by broadband incoherent sources, such as flashlamps, or by other pulsed or continuous lasers, with the most common including argon and krypton ion, Nd:YAG, excimer, nitrogen and copper vapour lasers.

Dye lasers show great versatility in their output providing tunable radiation from 320 to 1500 nm using a range of dyes, continuous tunability over about 30 nm from an individual dye and frequency stabilization in CW dye lasers to <1 kHz. The lasers can be scaled to provide high peak powers and high pulse energies. The large fluorescence bandwidth enables the mode-locking technique to be used to generate ultrafast pulses in the picosecond and femtosecond range. With colliding pulse mode-locking and time compression pulses of only several cycles down to a few femtoseconds can be produced.

These properties have prompted the dye laser to be used in a large diversity of applications in science, medicine and engineering. For several of these applications the dye laser has been replaced by diode pumped solid-state lasers, semiconductor diode lasers or optical parametric oscillators, while the dye laser continues to be used for wavelength conversion, tunable selective excitation, power applications and ultrashort pulse systems.

2.1 Introduction

Laser action in a solution of an organic dye molecule chloroaluminium phthalocyanine was demonstrated in 1966 by Sorokin and Lankard [1] and independently in 3,3^1-diethylthiatricarbocyanine bromide in methanol by Schäfer et al. [2]. These were the first dye liquid lasers and were excited by a pulsed ruby laser at 694 nm and gave laser emission near 755 nm. These discoveries were quickly followed over the next few years by demonstration of laser operation in many organic dye molecules using alternative excitation sources. Laser action was also demonstrated in rare earth chelates and trivalent rare earth ions in aprotic inorganic solvents. Eu-chelate dissolved in alcohol was the first liquid medium to exhibit laser action, in 1963 [3].

Continuous operation of the dye laser was not achieved until 1970 [4] when an argon ion laser operating at 515 nm was used to excite rhodamine 6G in water solution. The availability of shorter wavelength lasers as pump sources, such as the 337.1 nm nitrogen laser, the excimer lasers with UV output or the higher harmonics of solid-state lasers, enabled dye lasers operating at shorter wavelengths to be devised. The strong absorption and emission cross-sections of the dye molecules of about 10^{-16} cm^2 and the corresponding lifetimes of the laser emitting state of a few nanoseconds means that a high pumping intensity is required. This is achieved optically by intense flashlamps or by another laser. The development of the dye laser is closely linked generally to the evolution of laser technology over the last forty years [4–12].

Dye lasers offer a great versatility in their output resulting from the broad energy bands characteristic of the electronic energy states of the organic molecules in solution. Particular features are the wide range of wavelengths which can be produced from the many organic molecules which have efficient fluorescence emission. Output wavelengths can be produced from the UV (~320 nm) to the near-IR (~1500 nm) and with frequency conversion the range of 200–320 nm is achieved. Continuous tunability of the output by prism, grating or birefringent plate is attainable over about 30 nm for an individual dye molecule. The large number of organic dyes provide overlapping tuning ranges. Narrow linewidth emission for high resolution spectroscopy provides linewidths in pulsed dye lasers to 100 MHz and in CW lasers to 1 kHz. The dye laser can operate in continuous wave (CW) at power levels up to 100s W and in pulsed output generating pulse energies of mJ to 100s J.

The large fluorescence bandwidths of the spontaneous emission of the dye molecules enables ultrafast pulses as short as 100 fs to be produced by mode-locking, and which may be further shortened to a few femtoseconds by pulse compression techniques. The first active mode-locked dye laser was by synchronous pumping with a mode-locked Nd:glass laser; alternatively passive mode-locking was achieved by a saturable absorber cell in the laser cavity.

These versatile properties of dye lasers have prompted their use in many scientific, industrial and medical applications. In photobiology the dye laser has extensive application in fluorescence-based techniques of endogenous and exogenous species. Ultrafast laser techniques have been applied [13] to the study of conversion of light energy into chemical energy, photosynthesis, haem protein dynamics and in vision in the function of rhodopsin and in bacteriorhodopsin, and spectral imaging. Other applications have included high resolution spectroscopy using single mode, narrow bandwidth tunable lasers; ultrafast time-resolved dynamical studies; fluorescence microscopy, selective excitation in photochemistry [14–21]; state selection of nuclear isotopes [22–24].

The dye laser has found widespread application in medical phototherapy [25–27]. This has included selective absorption in dermatology, treatment of cutaneous vascular lesions, fragmentation of stones in lithotripsy, microdissection, photodynamic therapy in the detection and treatment of cancer, aethestics surgery in the removal of tattoos and in diagnostics and flow cytometry. Several commercial medical dye lasers are available based on flashlamp pumped systems or pumped by argon ion, Nd:YAG, excimer or copper vapour lasers.

2.2 Basic principles

The most common liquid state laser contains organic dye molecules in a solvent, typically ethanol or methanol or water. Organic dyes possess a backbone of carbon atoms with conjugated double bonds and delocalised π electrons that confer their characteristic absorption and emission properties.

2.2.1 Absorption and emission characteristics of dye molecules

The electronic energy levels of a dye molecule are shown in Figure 1. The outer $2s^22p^2$ electrons of the carbon atom are hybridized into three sp^2 trigonal orbitals to form planar and structural bonds. The fourth p electron is in a delocalized π molecular orbital; in a carbon chain these form two planar distributions, above and below the molecular plane, that can move freely within the chain. The π electrons determine the electronic states for the outer electrons. The states are simply arranged into a set of singlet states (S_o, S_1, S_2...) having a total spin equal to zero, and a set of triplet states having a total spin of one. To each electronic state there corresponds molecular vibrational modes (indicated in Figure 1) and also rotational modes.

From the selection rule for electric-dipole radiation, $\Delta S = 0$ where S is the total electron spin, singlet–singlet and triplet–triplet transitions are allowed, while singlet–triplet transitions are forbidden. These transitions are indicated on Figure 1. It is a characteristic feature of dye molecules that excitation of molecules to the levels S_2, S_3 or to higher vibrational levels of S_1 is rapidly followed on a picosecond timescale by non-radiative collisional internal conversion into the lowest vibrational level of S_1. From the S_1 level the molecule may radiatively decay to S_0 emitting fluorescence, or by intersystem crossing convert into T_1 assisted by collisions. From the energy level structure it is seen that the absorption and emission spectra are approximately mirror images of each other, as indicated in Figure 2 for rhodamine 6G in ethanol. When in the T_1 level the molecule may absorb further radiation in T_1–T_n transitions.

The decay rate k_s of the S_1 level (lifetime $\tau = 1/k_s$) is determined by the fluorescence radiative rate k_f (lifetime $\tau_f = 1/k_f$), and the rate of intersystem crossing, k_{ST}.

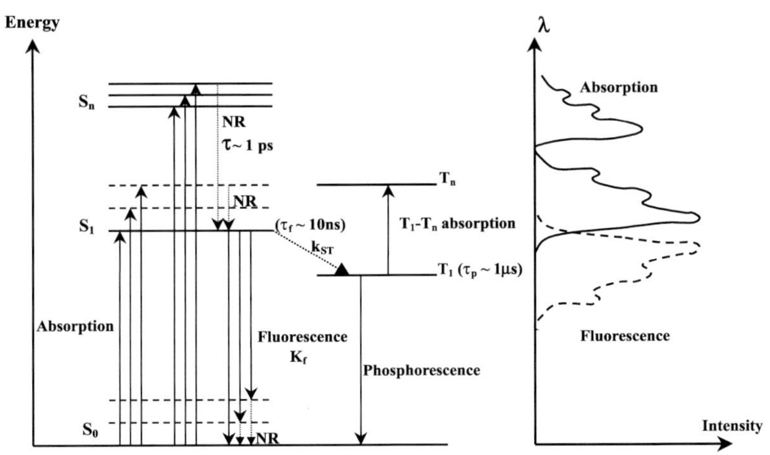

Figure 1. Singlet and triplet electronic energy levels for a dye molecule with some vibrational levels indicated, absorption and fluorescence spectra, and radiative and nonradiative transitions.

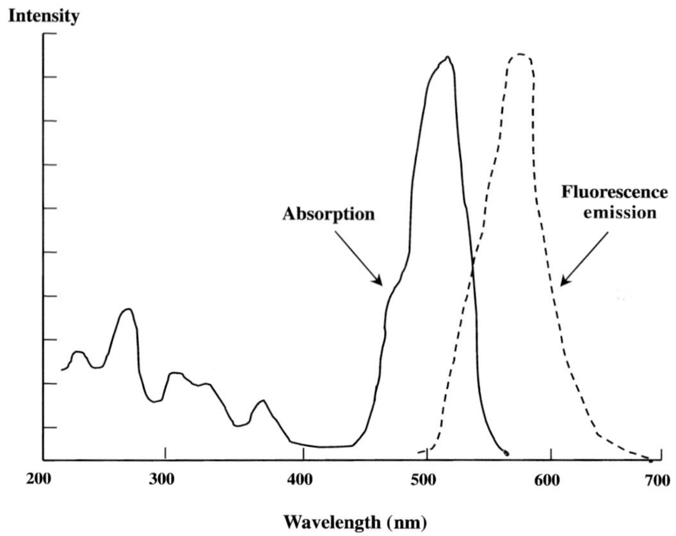

Figure 2. Absorption and fluorescence normalized spectra of rhodamine 6G in ethanol.

$$k_s = \frac{1}{\tau} = k_f + k_{ST}$$

The number of photons emitted by fluorescence compared with the total number of molecules excited to the S_1 level is the fluorescence quantum yield φ,

$$\varphi = \frac{k_f}{k_s} = \frac{\tau}{\tau_f}$$

The radiative lifetimes of laser dye molecules are in the range of a few nanoseconds and the fluorescence quantum yield falls typically in the range of 0.5 to near 1. The formation of molecules in the T_1 level has important implications for dye lasers, as the allowed T_1–T_n absorption can overlap the S_1–S_0 emission, leading to high loss in the laser. The T_1–S_1 transition can be induced by collisions which conserve the total spin of the colliding species, e.g. with oxygen. The lifetime of the T_1 metastable level for deoxygenated solution is typically 10^{-3} s, and reducing to $<10^{-6}$ s for solutions containing oxygen. The T_1 population may also be reduced by the addition of triplet quenchers such as cyclooctatetraene (COT), or by heavy atom molecules such as iodomethane. A condition for CW laser action is that the triplet lifetime τ_T satisfies the condition $\tau_T < (\sigma_e/\sigma_T k_{ST})$ where σ_T is the triplet–triplet absorption cross-section. Pulsed-dye laser operation can readily be achieved momentarily by rapid excitation before the population builds up in T_1.

2.2.2 Gain and threshold operation in dye lasers

The internal conversion and thermalization processes within S_2, S_1 and S_0 act such that the dye laser works as a 4-level system. The threshold pump power

P_{th} for a 4-level laser is related to the lifetime τ and stimulated emission cross-section σ_e as,

$$P_{th} \propto \frac{1}{\sigma_e \tau}$$

Although σ_e is $\sim 10^{-16}$ cm^2 the nanosecond lifetime of the S_1 level means that P_{th} is still high, e.g. compared with many solid state lasers, and a relatively high threshold pump power is required.

The small signal gain of the laser may be expressed in terms of the total molecular number N, gain length L, singlet absorption coefficient at the laser wavelength λ_1 and pump intensity I_p as [10],

$$k_0 L = NL\left[\frac{\sigma_e \sigma_p I_p - \sigma_a\left(\frac{hc}{\lambda_p}\right)\left(\frac{1}{\tau\varphi}\right)}{\sigma_p I_p + \left(\frac{hc}{\lambda_p}\right)\left(\frac{1}{\tau\varphi}\right)}\right]$$

where σ_p is the absorption cross-section and σ_a is the singlet absorption cross-section at the laser wavelength λ_1.

The dye laser acts predominantly as a homogeneously broadened laser. The saturated gain for laser intensity I is then,

$$k(I) = \frac{k_o}{1 + \frac{I}{I_s}}$$

The intensity I at which the gain falls to ½ of its small signal value is the saturation intensity, I_s. This is then,

$$I_s = \frac{\lambda_p}{\lambda_1}\left[\frac{\sigma_p I_p + \left(\frac{hc}{\lambda_p}\right)\left(\frac{1}{\tau\varphi}\right)}{(\sigma_e + \sigma_a)}\right]$$

The saturation intensity is seen to be inversely proportional to $(\sigma_e + \sigma_a)$. The minimum pump intensity required for a dye laser to reach threshold for a cavity with absorption losses a and for scattering and output transmission losses t is

$$I_p^{th} = \left(\frac{1}{\sigma_p}\right)\left(\frac{hc}{\lambda_p}\right)\left(\frac{1}{\tau\varphi}\right)\left[\frac{NL\sigma_e}{(a+t)} - 1\right]^{-1}$$

Typical values of the parameters are $\sigma_p = \sigma_e = 10^{-16}$ cm^2, $\lambda_p = 500$ nm, $\tau\varphi = 5 \times 10^{-9}$ s, $N = 10^{18}$ cm^{-3}, $(a+t)/L = 10$ cm^{-1}. This leads to a relatively high value of threshold pump intensity of 90 kW cm^{-2}. As described in Section 2.3, this high value of threshold pump intensity means that particular pumping conditions for pulsed and CW operation are required. The output power for when the laser is pumped well above threshold becomes

$$P_{out} = \left(\frac{t}{a+t}\right)\left(\frac{\lambda_p}{\lambda_1}\right)\left(NL\sigma_p\right)\rho P_{in}$$

under the assumption that $k_0 L/(a+t) >> 1$, $\sigma_a = 0$ and $\rho = \omega_0^2/\omega_p^2$ is the ratio of dye beam area to pumped area. Interestingly, there is a difference in alignment dependence of the dye laser at opposite ends of the tuning range. In the shorter wavelength region the laser is tolerant to misalignment while at longer wavelength region it is less tolerant.

2.3 Structures of liquid state lasers

2.3.1 Liquid state laser media

Laser dye molecules can be classified into several categories, these include (a) xanthene dyes, mostly operating in the visible region, e.g. rhodamine 6G; (b) polymethine dyes, operating in the red or infrared, e.g. pyrromethene 567; (c) coumarin dyes operating in the blue-green region, e.g. coumarin [5–11,28,29].

Since the dye solution is readily prepared, the concentration of the dye, typically in the range 10^{-5} to 10^{-3} M, can be specifically prepared and optimized for a particular application. The liquid host means that heat can be dispersed by a flowing system for cooling and also to provide a reservoir of active molecules. The common solvents for dye lasers are alcohols, ethylene glycol or water with added surfactant to reduce dimerization.

Photodegradation of the dye occurs particularly for short wavelength excitation and primarily through photooxidation reactions [30,31]. These effects can be reduced by removing oxygen from the solution and the addition of triplet quenching agents. In the flashlamp pumped dye laser the exciting light can be filtered to remove the UV content and fluorescence filters used to convert the UV into visible light, and thereby increase the pumping efficiency.

2.3.1.1 Rare earth ion lasers in inorganic solvents

Liquid lasers based on trivalent rare earth ions Nd^{3+} in a suitable solvent have been demonstrated [3]. The solvent is chosen to minimise non-radiative deactivation, – solvents selenium oxychloride and phosphorous oxychloride have been used. The Nd^{3+} laser emission is near the common Nd wavelength of 1.06 μm and the fluorescence emission band of the ion in the liquid is narrower than for a glass host. Pulsed output of a few hundred joules and GW peak power has been produced. A disadvantage of this liquid laser is that thermal distortion due to refractive index changes is greater than for the solid state.

2.3.2 Pulsed dye lasers

The pulsed dye lasers produce tunable, narrow bandwidth, high energy pulses [7–9]. Pulsed dye lasers at up to a few tens of watts average power may be excited by several pump lasers, including nitrogen, Nd:YAG, excimer and copper vapour lasers. These provide the necessary high pump intensity around 100 kW cm^{-2} while operating on a time scale in which triplet–triplet absorption and thermal distortion of the laser medium is minimal. The dye may be excited at high pulse repetition rate with the solution circulated to provide a uniform medium for each excitation pulse. Ranges of wavelengths for nitrogen, excimer and Nd:YAG pump sources are illustrated in Figure 3. The output of pulsed dye lasers is summarised in Table 1. The short fluorescent lifetimes of dyes of a few nanoseconds mean that energy storage is short lived and precludes the Q-switching of the lasers.

The nitrogen laser provides an effective pump source for dye lasers with an excitation wavelength of 337.1 nm, which is within the absorption band of many dye molecules with fluorescence in the visible or near-UV regions. It provides pump pulses of 4–10 ns duration at repetition rates of up to few 100 Hz. With this pump the dye laser operates as a gain-switched system, giving output pulses of about 5 ns duration. In the Hänsch design of a Littrow cavity [32] (see Figure 5(a), the nitrogen laser pumped dye cell is within a short laser cavity which is tuned

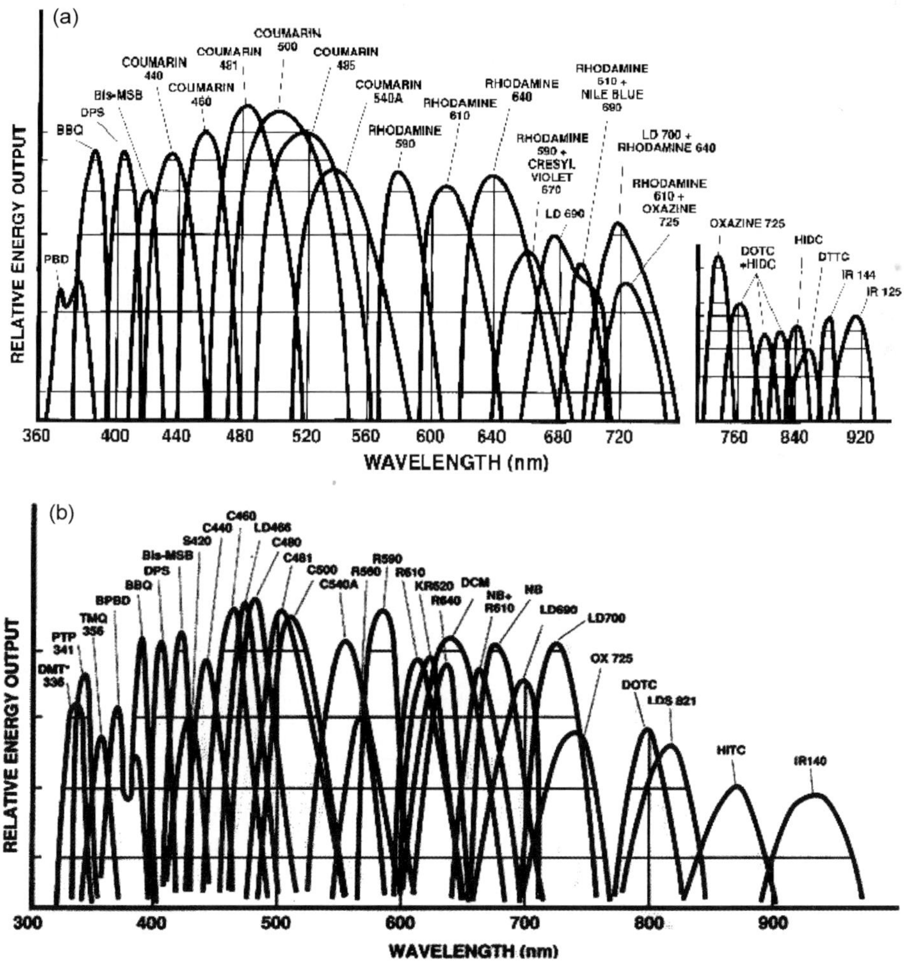

Figure 3. Tuning performance of dye lasers with pumping by (a) nitrogen lasers, (b) krypton fluoride or xenon chloride excimer lasers, (c) Nd:YAG laser (Exciton Inc.).

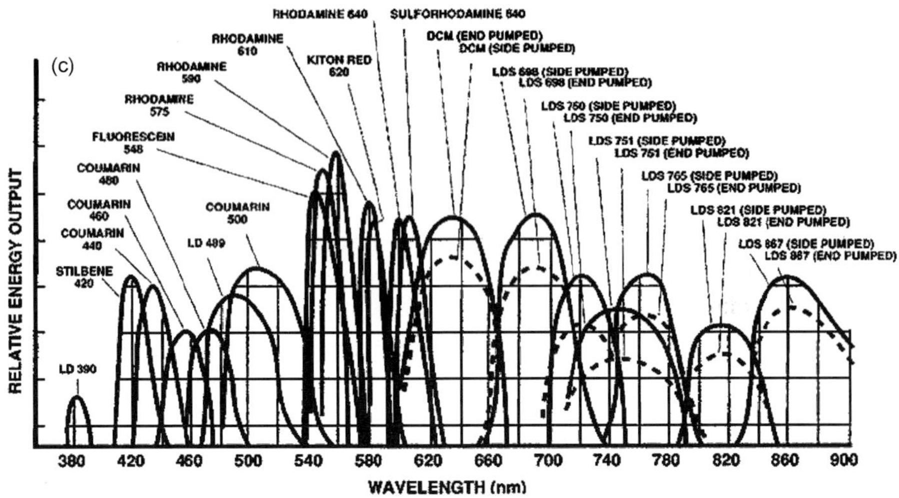

Figure 3. Continued.

by an echelle grating acting as one of the reflectors. An intra-cavity telescope is used to expand the beam onto the grating in order to increase the resolution. An etalon may be added to give narrow frequency operation with a free spectral range sufficiently large that only one mode lays within the grating profile. The in-cavity telescope has subsequently been replaced by a set of four prisms acting as the beam expander [33]. The prism set polarizes the laser output and gives a large beam expansion in a short laser cavity. To use a large surface area on the grating and hence to accommodate high angles of incidence a coarse grating (600 lines mm^{-1}) is used. The original design worked on several dyes tunable over the visible region and produced 0.1 mJ, 5 ns pulses with a spectral width of 0.1 cm^{-1} without the etalon and 0.01 mJ, 0.01 cm^{-1} pulses with the etalon. Single longitudinal mode operation of a dye laser using a four-prism beam expander and pumped at 511 and 578 nm by a copper vapour laser was obtained with a 60 MHz linewidth tunable over 16GHz [34].

Table 1. Typical output characteristics of dye lasers for various pump sources (Modified from Ref. 80)

Pump source	Tuning range (nm)	Pulse duration (ns)	Peak power (W)	Pulse energy (mJ)	Repetition rate (Hz)	Average power (W)
Flashlamp	300–800	300–10^4	10^2–10^4	<5000	1–100	0.1–200
Ar^+ or Kr^+	400–1100	CW	CW	–	CW	0.1–5
N_2	370–1020	1–10	$<10^5$	<1	$<10^3$	0.01–0.1
Excimer	370–985	10–200	$<10^7$	≤ 300	20–2000	0.1–10
Nd:YAG	400–920	10–20	10^5-10^7	10–100	10–30	0.1–1
CVL	530–980	30–50	10^4–10^5	≈ 1	$\approx 10^4$	≤ 10

Excimer lasers (e.g. XeCl, 308 nm, XeF 351 nm, KrF 248 nm, ArF 193 nm) and Nd:YAG lasers with second (532 nm) and third (355 nm) harmonic conversion can each produce intense short wavelength light suitable for the pumping of dye lasers. Excimer lasers give higher powers and shorter wavelengths but with lower spatial beam quality than the Nd:YAG laser. The Nd:YAG laser also provides higher repetition rates. A commercial dye laser (Lambda Physik Scanmate 1) provides for pumping by the XeCl or Nd:YAG at an optical conversion efficiency of 11% to give a tuning range of 320–860 nm and a bandwidth of 0.15 cm^{-1} without an etalon and 0.03 cm^{-1} with an etalon. Pumping with the harmonics of the 5–10 ns Q-switched Nd:YAG laser the dye laser gives the highest peak power but at lower average power and pulse repetition rate. The tuning curves of the dye lasers are shown in Figure 3. The two systems are complementary in that the excimer laser gives greater output in the UV/blue and the Nd:YAG laser in the red region.

With the use of stimulated Raman scattering in a gas cell the long wavelength outputs can be extended into the near-IR to about 4.5 μm. With a H_2 Raman shift cell, coherent stimulated scattering shifts the pump wavelength by the first Stokes frequency from the H_2 vibrational level spacing of 4155 cm^{-1} at a conversion efficiency of about 20%. In pulsed dye lasers a combination of an oscillator with one or more amplifiers (termed master oscillator pump amplifier, MOPA) enables high pulse energy systems to be operated with average powers of up to several kW.

2.3.2.1 Flashlamp pumped dye lasers

Many types of flashlamp pumped dye laser systems have been developed and engineered to provide the high threshold pump intensity indicated in Section 2.2.2 [6–9,35–41]. Research lasers based on arc lamp and surface discharges have also been developed. Pulsed flashlamp pumped dye lasers can provide tunable pulses with joules of energy in microsecond pulses.

To provide the high pump intensity the flashlamp is required to use a rapid discharge circuit with a pump pulse length of about 10–50 μs to give black body emission temperature around 6000 K. The fast excitation creates an inverted population and laser action before significant population can build up in the triplet state and before thermally generated medium inhomogeneities degrade the optical quality of the medium. The light from the flashlamp is filtered to remove short wavelength UV radiation which would induce dye photodegradation. Triplet state quenchers such as cyclooctatetraene (COT) are added to reduce loss from triplet–triplet absorption and an additive such as ammonyx LO is used to reduce dimer formation and thermally generated optical distortion of the gain medium.

Operation of flashlamp pumped dye lasers to give long output pulses to a few hundred microseconds is possible by extending the lamp excitation pulse using a pulse forming network and to reduce the triplet state population by the use of quenching agents such as COT [41]. This molecule has an acceptor state, which is below the first triplet state of many dye molecules.

The common form of flashlamp excitation is with one or more lamps mounted in a linear elliptical or close-coupled arrangements in a pumping chamber. An alternative is a coaxial lamp which is concentric with the dye medium and which gives very rapid excitation of a few microseconds.

A small-scale flashlamp pumped dye laser emits an average power of about 1 W at an efficiency of about 1% for the most efficient. This may be scaled to larger powers such that average powers of greater than 1 kW can be produced, at repetition rates of 1 kHz or 1 kJ in a single pulse. The operational lifetime of the dye may be extended with a recirculation system containing a dye solution reservoir to compensate for photodegradation of the dye.

2.3.3 CW dye lasers

The minimum pump intensity for a dye laser to reach threshold is rather high at ~90 kW cm^{-2}, as indicated in Section 2.2.2. To achieve this intensity for CW laser operation, the pump source is focused into a small volume of dye of about 50 μl. The dye solution is flowed in a recirculation arrangement to maintain a constant dye temperature and to remove triplet states from the active region. Since the pump light is focused to a small spot ~20 μm diameter, the beam waist in the dye laser cavity is required to match this size and be accurately located at the pump focused region. In the jet stream laser the active region of the flowing dye is formed by spraying a thin, planar jet from a nozzle which is then collected and returned to the recirculating system [7–9,42,43]. The angle of the dye jet is set to be at the Brewster angle to minimise optical losses, so that the laser output is linearly polarized with its electric field in the plane of incidence. The dye solvent is usually ethylene glycol as this has suitable viscosity and thermal properties. The pump laser for CW operation is the argon or krypton ion lasers operating on their visible or UV lines, the frequency doubled Nd:YAG lasers or diode lasers. With the use of various dye and solvent combinations output wavelengths over the 360–1000 nm range can be produced (Figure 4) [43]. Tuning of the CW laser may be achieved by an intracavity prism, birefringent plate or a reflective grating.

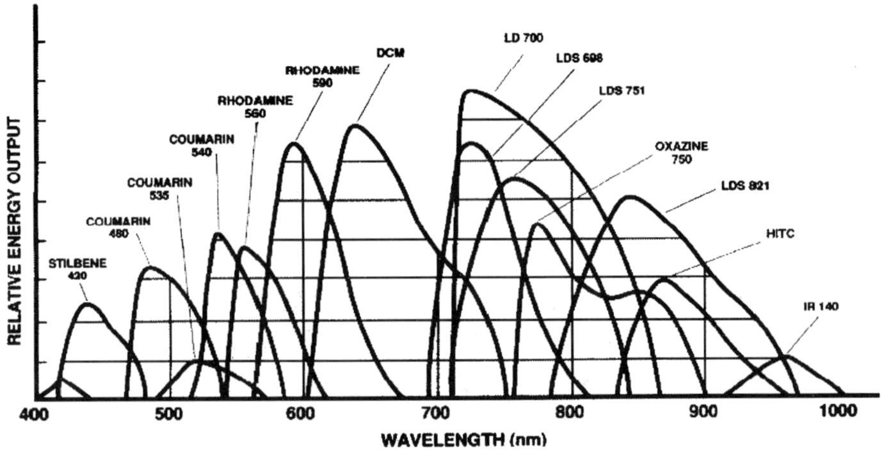

Figure 4. Tuning performance for argon-ion and krypton-ion pumped dye lasers (Exciton Inc.).

2.3.4 Solid-state dye lasers

Laser dyes used in the liquid state can also be doped into glasses and polymer hosts to form a solid-state dye laser [30,44–53]. This makes for a convenient laser system, avoiding the liquid handling and offering a low cost gain medium with wavelength range, tunability and power capabilities similar to the liquid state laser. The first solid-state laser, in 1967, was rhodamine 590 in poly(methyl methacrylate) excited by a laser [44], and was followed by flashlamp excitation in 1968 [45]. Solid-state dye lasers using various polymer hosts and co-polymers have been developed [44,45,47,49–51]. Other solid-state media has been sodium borosilicate glasses leached to make a porous medium and doped polyacrilamide gels. Sol-gel glasses have been shown to be effective hosts for solid-state lasers. Sol-gel glass can be prepared from solution and room temperature into a hard porous medium with pore sizes over 5–20 nm and with good pore interconnectivity. Laser dyes may be post-doped into the porous glass such that the dye concentration can be controlled [48]. A method to fill the residual pore free volume with monomer followed by polymerization has been developed to reduce the optical scatter and enhance the optical quality [49]. The incorporation of organic functional groups into the glass to form an organic–inorganic composite gives a solid-state medium (Ormosil) which can be pre-doped at the fabrication stage. New laser dyes based on pyrromethene and perylene have been introduced to give increased photostability and efficiency.

The presence of oxygen accelerates the photodegradation of pyrromethene 567 in both liquids and solid media [30,31]. Pulse radiolysis and laser flash photolysis have been used to characterize the triplet state of pyrromethene and the production of singlet oxygen [31]. Self-sensitised photooxidation was identified as the main photodegradation mechanism, and the use of a singlet oxygen quencher enhanced photostability. De-oxygenated pyrromethene 567 doped modified poly(methyl methacrylate) is five times more photostable than oxygenated solutions.

There has been recent investigation of solid-state lasers based on semiconducting conjugated polymers [54,55]. These are relatively easily processed from solution and may be fabricated into waveguide resonator structures and distributed feedback lasers. The finite semiconductivity suggests the possibility of direct electrical excitation. Nitrogen-and-excimer-pumped dye lasers have been used in some of these studies [56,57]. Semiconducting polymers offer exciting prospects as compact, efficient and flexible laser sources.

2.4 Manipulation of dye laser output

2.4.1 Laser tuning

The output linewidth of a dye laser without a tuning element is several nanometres. The broad laser bandwidth from dye lasers provides wide range tuning, while the partial homogeneous nature of the transition enables efficient spectral condensation into a narrow linewidth for many applications of the tunable dye laser [6–9,13,14].

In addition to narrow linewidth, good beam quality and pointing stability are often required, and a high repetition rate for pulsed lasers [58].

For a pulsed laser the theoretical ultimate limit of the linewidth is determined by the uncertainty principle. A Gaussian-shaped laser pulse of duration Δt has a minimum linewidth $\Delta\nu$ of

$$\Delta\nu = \frac{0.441}{\Delta t}$$

The most common tuning element is a diffraction grating that is mounted to replace one of the laser mirrors; three versions for pulse pumping are shown in Figure 5. The limiting resolution of the grating depends on the total number of lines N which are illuminated on the grating, or equivalently on the length of the grating illuminated.

In the Littrow configuration a particular wavelength λ diffracted in a particular order is reflected back along its incident path such that $\theta_i = \theta_d$. In that instance,

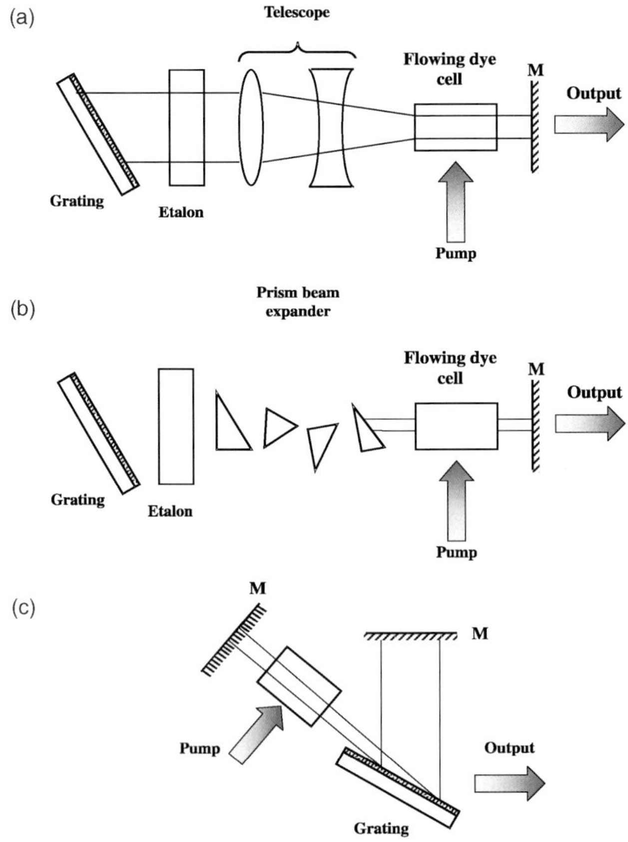

Figure 5. Tunable pulsed dye laser cavities: (a) Hänsch-type, (b) multiple prism beam expander, (c) grazing-incidence.

for a grating period d, $2d\ \sin\theta_i = m\lambda$ where θ is the incidence angle and m is the order of diffraction. Those wavelengths unable to satisfy this equation are not fed back into the cavity. The angular dispersion of the grating is

$$\frac{\mathrm{d}\theta}{\mathrm{d}\lambda} = m(d\cos\theta)^{-1}$$

The laser linewidth is determined by the range of wavelengths for which their angular spread is within the divergence of the beam. For a half-angle divergence of amplified spontaneous emission $\delta\alpha$ and a single pass through the cavity, the linewidth $\delta\lambda$ is

$$\delta\lambda = 2\delta\alpha\left(\frac{\mathrm{d}\theta}{\mathrm{d}\lambda}\right)^{-1}$$

For short pulse dye lasers of $\sim$10 ns duration only a few round trips are completed in the laser. For a grating with a total number of lines N in the Littrow configuration the resolving power is $R = (\lambda/\Delta\lambda) = (2Nd\sin\theta_l/\lambda)$. To achieve the largest resolution and the narrowest laser linewidth the condition that N should be as large as possible is met by expanding the beam with a telescope or prism (Figure 5).

In the grazing incidence grating cavity the length of the grating that is used is increased by increasing the angle θ without the use of beam expansion [59,60]. In addition, the grating is used twice in each round trip of the cavity, thereby further narrowing the linewidth. A cavity of this type, as in Figure 6, when pumped with a 6 kHz copper vapour laser has produced a single mode near diffraction-limited beam at over 1% efficiency [61].

Linewidths in pulsed dye lasers with a prism tuning element are typically 100 GHz ($\equiv$3.3 cm^{-1}). This may be narrowed to about 1 GHz (0.03 cm^{-1}) with dual tuning elements. A pulsed laser with a distributed feedback (DFDL) cavity can also produce narrow linewidth, broadly tunable output. In one configuration [62] (Figure 7) operating on the dye DCM in methanol, the DFDL gave a time-averaged linewidth of 5 GHz over the tuning range of 610–670 nm when pumped by a frequency doubled Nd:YAG laser. Linewidths in pulsed lasers down to 100 MHz have been achieved, which can be further narrowed by additional techniques to about 1 MHz.

For high-resolution spectroscopy the CW dye laser should operate on a single longitudinal mode and a single transverse mode. A feature of the standing wave cavity is that more than one longitudinal mode may operate since their standing wave patterns can derive gain from different spatial regions of the gain volume. To overcome this the standing wave cavity has been replaced by the traveling wave or ring cavity. The direction of the traveling wave is set by an optical device to limit light propagation in only one direction in the ring [63,64].

A common configuration is to add a birefringent filter and two etalons to the cavity to achieve narrow bandwidth and a rhomb prism to compensate for astigmatism [65]. In this case the diffraction grating is inappropriate because of its losses and the low gain of the CW laser. A birefringent plate acts as a retardation waveplate and is mounted at Brewster's angle in the laser cavity. The retardation of the plate at the resonant wavelength is an integer number of wavelengths such that

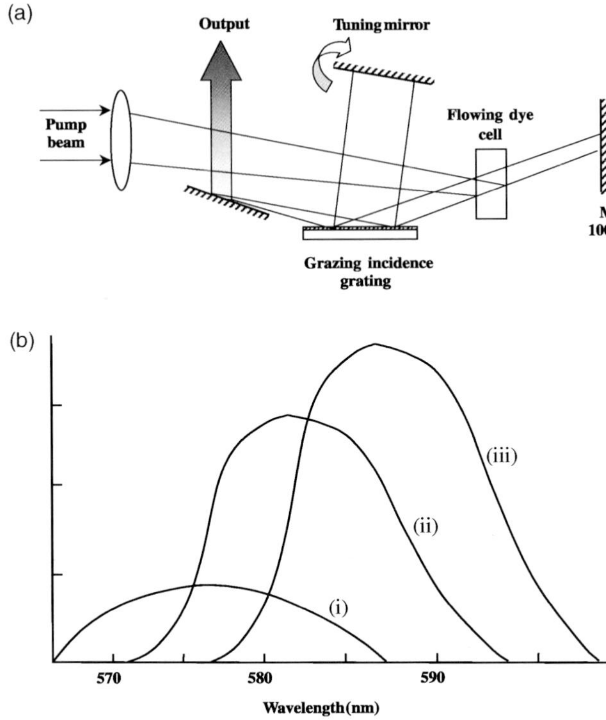

Figure 6. (a) Tunable dye laser using a grazing-incidence grating and longitudinal pumping. (b) Tuning ranges for rhodamine B at concentrations of (i) 10^{-4}, (ii) 2×10^{-4}, (iii) 4×10^{-4} M.

the polarization is unchanged giving lossless insertion of the plate. At other wavelengths, the polarization of transmitted light is changed and is partially reflected out of the cavity. Tuning is achieved by rotation of the birefringent plate [66,67]. The linewidth of the ring laser is determined by instabilities of the cavity mode frequency, where the path round the cavity is perturbed by random environmental fluctuations from vibrations, noise and the jet stream. Under these circumstances the effective linewidth is about 15 MHz (0.0005 cm^{-1}).

The linewidth is further reduced by use of a frequency stabilization servo that locks the dye cavity mode frequency to a stable reference interferometer. With these systems a laser linewidth down to 200 kHz may be obtained. With further stabilization technique linewidths to 100 Hz have been achieved.

2.4.2 Generation of ultrafast pulses

Ultrafast lasers allow transient species to be observed in time-resolved experiments, vision and photosynthesis as well as providing high peak powers for molecular fragmentation, multiphoton absorption and study of biochemical reactions. The very large fluorescence gain bandwidth $\Delta\nu$ of the dye molecules and the partial homogeneous nature of the broadening of the transition indicates that very short

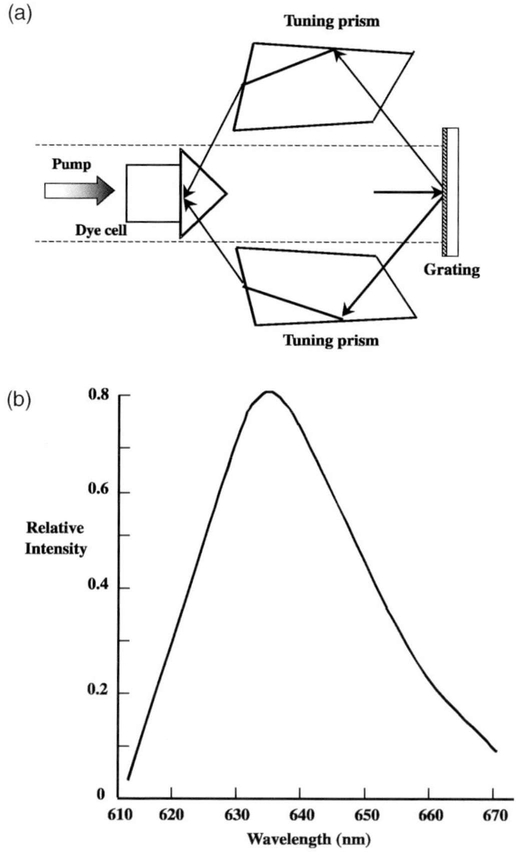

Figure 7. Distributed feedback dye laser. (a) Configuration: the pump beam passes above the dye cell and is reflected by the grating. (b) Tuning response for laser dye 10^{-4} M DCM in ethanol.

duration mode-locked pulses can be produced. According to the frequency bandwidth-time relation, $\Delta\nu \cdot \Delta t \approx 1$, for a typical bandwidth $\Delta\nu \sim 10$ nm ($\equiv$120 GHz), a corresponding mode-locked pulse duration is 8×10^{-12} s. In the mode-locking technique the oscillating longitudinal modes of the laser are synchronized in phase with either active mode-locking (by the use of an acousto-optic or electro-optic modulator) or with a passive saturable absorption element [5,7,68–71]. Techniques and performance for mode-locked dye lasers are summarized in Table 2.

In *passive mode-locking* a medium is inserted in the cavity which is selected to have an absorption that can be saturated at low intensity and which has a very fast relaxation time [71]. The common saturable absorbers are cyanine dyes or semiconductor materials. The saturable absorber permits gain in the laser cavity when it is switched from the unsaturated to the saturated state.

Mode-locking can be induced by varying periodically the gain of the laser. In the CW dye laser pumped by an argon ion laser, if the argon ion laser is itself mode-locked this will directly induce periodic variation in the dye laser gain.

Table 2. Ultrashort pulse generation in dye lasers by mode-locking and associated techniques

Mode-locking technique	Mode-locking element	Pulse duration (ps)	Pulse energy (nJ)
Passive	Saturable	1	1
Synchronous pumped	Mode-locked pump with matched dye resonator	1	10
Colliding pulse (CPM)	Passive mode-locking and synchronous pumping in ring	0.05–0.1	1
CPM and pulse compression	CPM and fibre/grating pulse compression	0.005	0.5

Because of the much greater gain bandwidth of the dye laser compared to the argon ion laser (typically 45 THz compared to 3.5 GHz) much shorter mode-locked pulses can be generated by the dye laser. In this technique of *synchronous pumping* the cavities of the argon ion pump and the dye laser need to be matched so that the pump and dye pulses are synchronous. The laser beam inside the cavity is focused on an absorbing dye so that saturation of the dye provides phase-locking of the cavity modes. The synchronously pumped dye laser gives tunable ultrafast pulses in the shorter wavelength visible region that are not available from mode-locked titanium-sapphire lasers.

In the ring dye laser [72] the technique of *colliding pulse mode-locking* gives very short pulse durations [73,74]. The dye laser is pumped by a mode-locked CW argon ion laser or a diode pumped solid state laser, this generates two pulses traveling in opposite directions in the ring. The laser is passively mode-locked by a slow saturable absorber (e.g. 3,3-diethyloxadicarbocyannine iodide, DODCI). The two counter propagating pulses are arranged to meet at the position of the saturable absorber; the saturable absorber has lowest loss when the two pulses coincide at its position, the peak gain is increased and mode synchronization occurs. The mode-locked pulse duration is usually much shorter than the lifetime of the gain medium or the saturable absorber recovery time. Mode-locked pulses using the colliding pulse technique of 27 fs have been obtained for the rhodamine 6G dye laser operating at 570 nm [75]; the bandwidth gain of that laser of about 45 THz indicates an ultimate pulse duration of 10 fs.

Ultrashort laser pulses may be further shortened by the technique of *pulse compression* [76]. A mode-locked pulse passing through a single mode fibre undergoes group-velocity dispersion, which imposes a frequency chirp on the pulse, with the longer wavelengths emerging first. The pulse is then passed through a combination of two diffraction gratings, this is arranged to act as a dispersive delay line with a path length proportional to wavelength and which compensates the chirp, thereby giving a pulse compressed in time. In one system a 50 fs pulse from a colliding pulse mode-locked dye laser has been compressed to 6 fs, a record shortest pulse duration [76].

The alternative technique to mode-locking of *cavity-dumping* can give pulses of a few nanoseconds. This may be produced in a picosecond dye laser synchronously pumped by a mode-locked argon ion laser or a Nd:YAG laser. In this method an optical switch in the cavity is used rapidly to switch out the in-cavity pulse.

2.4.3 Short wavelength and UV dye lasers

Several developments have led to an increase in the performance of short wavelength and UV dye lasers [77,78]. These include the availability of improved nonlinear optical doubling crystals with UV transmission, the increased powers of short wavelength laser lines from ion lasers for the pumping of dye lasers and the development of new laser dyes with short wavelength emission. Argon ion lasers with output powers on UV lines of greater that 7W and the generation of lines in the 300 to 335 nm band form powerful pump sources for tunable short wavelength dye lasers. The tuning ranges are shown in Figure 4. The scintillation molecules stilbene 1 and stilbene 3 produce 0.42 W and 1.0 W in the band 400–480 nm. The dye polyphenyl 2 produces laser output over 364–408 nm with 0.25 W peak power at 383 nm.

Generation of the higher harmonics of dye lasers in nonlinear crystals valuably extends the short wavelength range. Frequency doubling of the laser dyes R110, R6G, kiton red, DCM and LD700 in KDP or $LiIO_3$ nonlinear crystals produces wavelengths in the 270 to 410 nm range (Coherent Inc). The doubling crystal β-barium borate (BBO) has transmission to just below 200 nm and has high optical quality with low loss and a high damage threshold. When used with stilbene 3 and coumarin 102 tuning over the range of 215 to 260 nm can be obtained. A dye laser pumped by a copper vapour laser to produce 468 nm light has been used to produce 196 nm radiation, using a combination of doubling in BBO (to give 234 nm) followed by a H_2 Raman shift cell (to give 196 nm).

2.5 Perspectives

The characteristic features of liquid lasers based on solutions of dye molecules are the availability of a very broad range of laser wavelengths from the UV to the IR, wide range continuous tuning both for an individual dye and across the range, pulsed and CW operation up to high powers and pulse energies, very narrow stabilized linewidths in CW and pulsed laser tunable lasers and the ability to generate ultrafast pulses down to a few femtoseconds.

The emergence of the tunable solid-state lasers, such as the titanium-sapphire, alexandrite and Cr-forsterite-Mg_2SiO_4 lasers, semiconductor diode lasers and the optical parametric oscillators (OPO) offers alternatives to the dye-laser for the generation of coherent radiation in certain wavelength ranges. As an example, the titanium:sapphire laser can operate over 670–1070 nm and its second harmonic can be scanned from 330 to 580 nm, leaving a gap of about 100 nm in the visible region. A commercial laser system is available (Coherent Inc 899-01 ring laser) in which a common pump laser can be used to excite either a dye cell or a titanium-sapphire crystal (with tuning over 700 to 1100 nm). The OPO also has a wide tuning range, relatively high pulse energy, with minimal degradation of the active medium and at a cost comparable to a dye laser [79]. Where comparable or better performance is available the solid-state laser or OPO systems are preferable since they avoid the degradation and liquid handling involved with the dye laser. As a consequence there has been a significant move to the use of solid-state laser systems and frequency conversion techniques in the last decade, often substituting for dye lasers. This trend can be expected to continue as the solid-state systems undergo further development.

The various forms of dye laser are a mature and reliable technology [80] that offers specialist systems not yet available from solid-state lasers and OPOs, such as lasers operating at particular wavelengths in the UV, high pulse energy flashlamp pumped lasers across the visible region, wavelength conversion from existing pump lasers such as Nd:YAG, excimer, copper vapour or nitrogen lasers, and as a low cost tunable source pumped by a nitrogen laser. The development of the solid-state dye laser brings some of the advantages of solid-state technology to the dye lasers.

References

1. P.P. Sorokin, J.R. Lankard (1966). Stimulated emission observed from an organic dye, chloro-aluminium phthalocyanine. *IBM J. Res. Dev.* **10**, 162–163.
2. F.P. Schäfer, W. Schmidt, J. Volze (1966). Organic dye solution laser, *Appl. Phys. Lett.* **9**, 306–309.
3. (a) A. Lempicki, H. Samelson (1966). Organic laser systems. In: Lasers, (Volume 1, pp. 181–252). Marcel Dekker, New York. (b) A. Lempicki (1971). Rare earth liquid lasers. In: W.L. Weber (Ed), *Handbook of Lasers*, (Volume 1, pp. 355–359). Chemical Rubber Co., Boca Raton, FL.
4. O.B. Peterson (1970). CW operation of an organic dye solution laser. *Appl. Phys. Lett.* **17**, 245–247.
5. C.V. Shank (1975). Physics of dye lasers. *Rev. Mod. Phys.* **47**, 649–657.
6. B.B. Snavely (1977). Continuous-wave dye lasers. In: F.P. Schäfer (Ed), *Dye Lasers*, 2nd revised edition, Topics in Applied Physics, (Volume 1). Springer-Verlag, Berlin.
7. F.J. Duarte, L.W. Hillman (Eds) (1990). *Dye Laser Principles with Applications*. Academic Press, New York.
8. F.P. Schäfer (1990). *Dye Lasers*, (Volume 1 of Topics in Applied Physics, 3rd Edn.). Springer-Verlag, Berlin.
9. F.J. Duarte (Ed) (1991). *High-Power Dye Lasers*. Springer, Berlin.
10. T.F. Johnston (1991). Tunable dye lasers. In: *Encyclopaedia of Lasers and Optical Technology*. Academic Press, San Diego.
11. M. Stuke (Ed.) (1992). *25 Years Dye Laser*. (Topics in Applied Physics Volume 70). Springer, Berlin.
12. X. Hou, J.X. Zhou, K.X. Kang, P. Stchur, R.G. Michel (1999). New types of tunable lasers. In: J. Sneddon (Ed), *Advances in Atomic Spectroscopy*, Jai Press, Stanford.
13. R.M. Hochstrasser, C.K. Johnson (1993). Biological processes studied by ultrafast laser techniques. In: *Ultrashort Laser Pulses, Generation and Applications*, (Topics in Applied Physics, Volume 60, 2nd edn). Springer, Berlin.

14. J.P. Hohimer, P.J. Hargis (1977). Picogram detection of cesium in aqueous solution by nonflame atomic fluorescence spectroscopy with dye laser excitation. *Appl. Phys. Lett.* **30**, 344–346.
15. M. Alden et al. (1984). Application of a two colour dye laser in CARS experiment. *Appl. Opt.* **23**, 2053–2055.
16. W. Sdorra, A. Quentmeier, K. Niemax (1989). Basic investigations for laser microanalysis: II Laser-induced fluorescence in laser-produced sample plumes. *Microchem. Acta II* **2**, 201–218.
17. M. Schutz, U. Heitmann, A. Hess (1995). Development of a dual-wavelength dye-laser system for the UV and its application to simultaneous multi-element detection. *Appl. Phys. B* **61** 339–343.
18. I.B. Gornushkin, J.E. Kim, B.W. Smith, S.A. Baker, J.D. Winefordner (1997). Determination of cobalt in soil, steel and graphite using excited state laser fluorescence induced in a laser spark. *Appl. Spectrosc.* **51**, 1055–1059.
19. R.Q. Aucélio, V.M. Rubin, B.W. Smith, J.D. Winefordner (1998). Ultratrace determination of platinum in environmental and biological samples by electrothermal atomization laser-excited atomic fluorescence using a copper vapour laser pumped dye laser *J. Anal. At. Spectrom.* **13**, 49–54.
20. S.J. Kok, I.C.K. Isberg, C. Groijer, U.A. Th. Brinkman, N.H. Velthorst (1998). Ultraviolet laser-induced fluorescence detection strategies in capillary electrophoresis: determination of naphthalene sulphonates in river water. *Anal. Chem. Acta* **360**, 109–118.
21. P. Stchur, K.X. Yang, X. Hon, T. Sun, R.E. Michel (2001). Laser excited atomic fluorescence spectrometry. *Spectrosc. Acta. B* **56** 1565–1592.
22. D.A. Ackerman (1990). Dye laser isotope separation. In: F.J. Duarte and L.W. Hillman (Eds), *Dye Laser Principles with Application*. Academic, Boston.
23. J. Billowes, P. Campbell (1995). High-resolution laser spectroscopy for the study of nuclear sizes and shapes. *J. Phys. G Nucl. Part. Phys.* **21**, 707–739.
24. P. Campbell, H.L. Thayer, J. Billowes, P. Dendooven, K.T. Flanagan, D.H. Forest, J.A.R. Griffith, J. Huikari, A. Jokinen, R. Moore, A. Nieminen, G. Tungate, S. Zemlyanoi, J. Āystö (2002). *Phys. Rev. Lett.* **89**, 82501–82504.
25. C.A. Puliafito (Ed) (1996). *Laser Surgery and Medicine: Principles and Practice*. Wiley Liss Inc., New York.
26. T. Vo-Dinh (Ed) (2002). *Biomedical Photonics Handbook*. CRC Press, Boca Raton, FL.
27. B. Pogue, M. Mycek (Eds) (2003). *Fluorescence in Biomedicine*. Marcel Dekker, New York.
28. (a) M. Maeda (1984). *Laser Dyes: Properties of Organic Compounds for Dye Lasers*. Academic Press, Orlando, FL. (b) T.G. Pavlopoulos (2000). Laser dyes. In: *Encyclopaedia of Materials Science and Engineering* (Chapter 7). Wiley, New York.
29. R. Scheps (1995). Near-IR dye laser for diode-pumped operation. *IEEE J. Quantum Electron.* **32**, 126–34.
30. M.D. Rahn, T.A. King, A.A. Gorman, I. Hamblett (1997). Photostability enhancement of pyrromethene 567 and perylene orange in oxygen-free liquid and solid dye lasers. *Appl. Opt.* **36**, 5862–71.
31. A.A. Gorman, I. Hamblett, T.A. King, M.D. Rahn (2000). A pulse radiolysis and pulsed laser study of the pyrromethene 567 triplet state. *J. Photochem. Photobiol.* **A 130**, 127–32.
32. T.W. Hansch (1972). Repetitively pulsed tunable dye laser for high resolution spectroscopy. *Appl. Opt.* 11, 895–898.
33. G.K. Klauminzer (1977). New high-performance short-cavity dye laser design. *IEEE J. Quan. Electron.* **QE-13**, 904–905.

34. A.F. Bernhardt, P. Rasmussen (1981). Design criteria and operating characteristics of a single-mode pulsed dye laser. *Appl. Phys. B* **26**, 141–146.
35. J. Jethwa, F.P. Schafer, J. Jasney (1978). A reliable high average power dye laser. *IEEE J. Quantum Electron.* **QE-14**, 119–121.
36. R.G. Morton, V.G. Draggoo (1981). Reliable high average power high pulse energy dye laser. *J. Quantum Electron* **17**, 2576–2577.
37. L.K. Danisov, A.F. Loshin, N.A. Kozlov, V.G. Nikiforov (1985). Determination of sodium and barium by the atomic fluorescence method with the excitation by a lamp-pumped pulsed dye laser. *Zh. Prikl. Spektrosk.* **43**, 566–570.
38. P.N. Everett, et al. (1986). Efficient 7J flashlamp pumped dye laser at 500 nm wavelength. *Appl. Opt.* **25**, 2142–7.
39. F.J. Duarte, R.W. Conrad (1987). Diffraction-limited single-longitudinal-mode multiple-prism flashlamp-pumped dye laser oscillator. *Appl. Opt.* **26**, 2567.
40. D.E. Klimek, H.R. Aldag (1994). 10 Hz kilowatt-class dye laser system, *SPIE Conf. On Intense Laser Beams and Applications*, (Volume 1871, p.11). SPIE, Bellingham, WA.
41. A. Hirth, H. Fagot (1977). High average power from long pulse dye laser. *Opt. Comm.* **21**, 318–320.
42. W. Jitschin, G. Meisel (1979). Fast frequency control of a CW dye jet laser. *Appl. Phys. B* **19**, 181–184.
43. B. Pense (1985). New developments in CW dye lasers. In: N.B. Abraham, F.T. Arecchi, A. Mooradian, A. Sona (Eds), *Physics of New Laser Sources*. Plenum, New York.
44. B.H. Soffer, B.B. McFarland (1967). Continuously tunable, narrow band organic dye lasers. *Appl. Phys. Lett.* **10**, 266–267.
45. O.G. Peterson, B.B. Snavely (1968). Stimulated emission from flashlamp-excited organic dyes in poly(methylmethacrylate). *Appl. Phys. Lett.* **12**, 238–240.
46. A. Charlton, I.T. McKinnie, M.A. Meneses-Nava, T.A. King (1992). A tunable visible solid state laser. *J. Mod. Opt.* **39**, 1517–1523.
47. K.M. Dyumaev, A.A. Manenkov, A.P. Maslynkov, G.A. Matyushin, V.S. Nechitailo, A.M. Prokhorov (1992). Dyes in modified polymers: problems of photostability and conversion efficiency at high intensities. *J. Opt. Soc. Am. B* **9**, 143–151.
48. M.D. Rahn, T.A. King (1998). High performance solid-state dye laser based on perylene-orange-doped polycom glass. *J. Mod. Opt.* **45**, 1259–1267.
49. M.D. Rahn, T.A. King (1995). Comparison of laser performance of dye molecules in solgel, polycom, ormosil and poly(methylmethacrylate). *Appl. Opt.* **34**, 8260–8271.
50. S.M. Griffin, I.T. McKinnie, W.J. Wadsworth, A.D. Woolhouse, G.J. Smith, T.G. Haskell (1999). Solid state dye lasers based on 2-hydroxyethyl methacrylate and methyl methacrylate co-polymers. *Opt. Commun.* **161**, 163–170.
51. G. Somasunderam, A. Ramalingam (2000). Gain studies of coumarin 1 dye-doped polymer laser. *J. Lumin.* **90**, 1–5.
52. D. Lo, S.K. Lam, C. Ye, K.S. Lam (1998). Narrow linewidth operation of solid state dye laser based on sol-gel silica. *Opt. Commun.* **156**, 316–320.
53. K.M. Abedin, M. Alvarez, A. Costela, I. Garcia-Moreno, O. Garcia, R. Sastre, D.W. Coutts, C.F. Webb (2003). 10 kHz repetition rate solid-state dye laser pumped by diode-pumped solid-state laser. *Opt. Commun.* **218**, 359–363.
54. M.D. McGehee, A.J. Aeeger (2000). Semiconducting (conjugated) polymers as materials for solid-state lasers. *Adv. Mater.* **12**, 1655–1668.
55. W. Holzer, A. Penzkofer, T. Pertsch, N. Danz, A. Braner, E.B. Kley, H. Tillmann, C. Bader, H.H. Horhold (2002). Corrugated neat thin-film conjugated polymer distributed feedback laser. *Appl. Phys. B* **74**, 333–342.

56. G.A. Turnbull, T.F. Krauss, W.L. Barnes, I.D. W. Samuel (2001). Tuneable distributed feedback lasing in MEH-PPV films. *Synth. Met.* **121**, 1757–1758.
57. C. Bauer, H. Griessen, B. Schnabel, E.B. Kley, C. Schmitt, U. Scherf, R. F. Mahrt (2001). *Adv. Mater.* **13**, 1161.
58. J. Hough, D. Hills, M.D. Rayman, L.S. Ma, L. Hollberg, J.L. Hall (1984). Dye-laser frequency stabilization using optical resonators. *Appl. Phys. B* **33**, 179.
59. M.G. Littman (1978). Single-mode operation of grazing-incidence pulsed dye laser. *Opt. Lett.* **3**, 138.
60. M.G. Littman, H.J. Metcalf (1978). Spectrally narrow dye laser without beam expander. *Appl. Opt.* **17**, 2224–2227.
61. A.J. Berry, I.T. McKinnie, T.A. King (1990). Narrow linewidth operation of a pulsed grazing incidence dye laser. *J. Mod. Opt.* **37**, 463–471.
62. I.T. McKinnie, A.J. Berry, M.D.L. Stonefield, T.A. King (1990). A widely tunable narrow-linewidth distributed-feedback dye laser. *J. Mod. Opt.* **37**, 483–491.
63. T.F. Johnson, Jr., R.H. Brady, W. Proffitt (1982). Powerful single-frequency ring dye laser spanning the visible spectrum. *Appl. Opt.* **21**, 2307–2316.
64. L.E. Jusinski, C.A. Taatjes (2001). Efficient and stable operation of an Ar^{+}-pumped CW ring laser from 506–560 nm using a coumarin laser dye. *Rev. Sci. Instum*, **72**, 2837.
65. H.W. Kogelnik, E.P. Ippen, A. Dienes, C.V. Shank (1972). Astigmatically compensated cavities for CW dye lasers. *IEEE J. Quan. Electron* **QE-8**, 373–379.
66. M. Okada, S. Ieiri (1975). Electronic tuning of dye lasers by an electro-optical birefringent Fabry-Perot etalon. *Opt. Commun.* **14**, 4–7.
67. M. Okada, K. Takizawa, S. Ieiri (1976). Tilted birefringent Fabry-Perot etalon for tuning of dye lasers. *Appl. Opt.* **15**, 472.
68. C.V. Shank, E.P. Ippen (1974). Subpicosecond kilowatt pulses from a mode-locked CW dye laser. *Appl. Phys. Lett.* **24**, 373–375.
69. K. Smith, N. Langford, W. Sibbett, J.R. Taylor (1985). Passive mode-locking of a continuous-wave dye laser in the red near-infrared spectral region. *Opt. Lett.* **10**, 559–561.
70. P.M.W. French, J.R. Taylor (1986). In: *Ultrafast Phenomena V*, p. 11. Springer-Verlag, Berlin.
71. E.P. Ippen (1972). Passive mode locking of the CW dye laser. *Appl. Phys. Lett.* **21**, 348–350.
72. H.W. Schroder, L. Stein, D. Fröhlich, F. Fugger, H. Welling (1977). A high-power single-mode CW dye ring laser. *Appl. Phys.* **14**, 377–380.
73. R.L. Fork, B.I. Greene, C.V. Shank (1981). Generation of optical pulses shorter than 0.1 psec by colliding pulse mode-locking. *Appl. Phys. Lett.* **38**, 671–672.
74. P.M.W. French, J.R. Taylor (1988). Generation of sub-100 fsec pulses tunable near 497 nm from a colliding-pulse mode-locked ring dye laser. *Opt. Lett.* **13**, 470–472.
75. J.A. Valdamis, R.L. Fork, J.P. Gordon (1985). Generation of optical pulses as short as 27 femtoseconds directly from a laser balancing self-phase modulation, group-velocity dispersion, saturable absorption and gain saturation. *Opt. Lett.* **10**, 131–133.
76. R.L. Fork, C.H. Brito Cruz, P.C. Becker, C.V. Shank (1987). Compression of optical pulses to six femtoseconds using cubic phase compression. *Opt. Lett.* **12**, 483–5.
77. N. Wang, V. Gaubatz (1986). Optical frequency doubling of a single-mode dye laser in an external resonator. *Appl. Phys. B* **40**, 43.
78. U. Heitmann, M. Kotteritzsch, S. Heitz, A. Hese (1992). Efficient generation of tunable VUV laser radiation below 205 nm by SFM in BBO. *Appl. Phys. B* **55**, 419–423.
79. J.X. Zhou, X. Hou, K.X. Kang, S.J. Tsai, R.G. Michel (1998). Lasers based on optical parametric devices: wavelength tunability empowers laser-based techniques in the UV, Vis, and near-IR, *Appl. Spectrosc.* **52A**, 176–189.
80. W. Demtroder (1996). *Laser Spectroscopy*, (2nd edn.), Springer, Berlin.

Chapter 3

Solid state lasers

Willy Lüthy and Heinz P. Weber

Table of contents

Abstract

Lasers are described that are based on solid material as amplifying medium in contrast to gases or liquid solutions. This solid material can be a dielectric or a semiconductor. Since the excitation mechanisms of a semiconductor are completely different with respect to a wide-bandgap dielectric material, semiconductor-based lasers are described separately in Chapter 4. Lasers based on dielectrics consist of active ions incorporated in a crystalline or glassy host. The combination of many suitable ions with many possible hosts opens the possibility of an enormous number of laser materials although, actually, only a few dozen are of practical interest.

3.1 Introduction

The basic principle of laser action was first described by Schawlow and Townes in 1958. The practical demonstration was realized by Maiman with the first Ruby laser. Ruby or Cr^{3+}:Al_2O_3 is the first laser medium based on transition element doped crystals. Sorokin and Stevenson realized lasers with Sm^{3+}:CaF_2 and U^{3+}:CaF_2 demonstrating the importance of trivalent lanthanides (rare earths) and actinides [1].

3.2 Basic principles and theoretical background

Laser action is strongly enhanced if all the ions in the host have identical energy level schemes with very narrow bandwidths of the transitions. This is not the case if the relevant electrons of the ions are involved in chemical bonds or if they are strongly influenced by Stark effect or magnetic interaction with the ligands of the host. It is therefore favourable to use ions where the active electrons are not involved in chemical bonding because they are buried inside a protective shell of noble-gas-like electrons. This is the case for the 4f electrons of the rare earths. Table 1 shows the electron configuration of the trivalent rare earths.

Similarly, the actinides also have suitable electrons in the partly filled 5f states. Since the actinides are generally unstable, only U^{3+} has been used for laser action.

3.2.1 Basic spectroscopy

Amongst the trivalent ions, Ce^{3+} has only one electron in the 4f state and Yb^{3+} has 13. Since a maximum of 14 electrons can be arranged in the 4f state, the Yb^{3+} 4f state can be considered as an element with one electron missing, a so-called hole. Spectroscopically, this hole can be treated like an electron. The energy level schemes of Ce^{3+} and Yb^{3+} are, therefore, very similar. The same holds for two electrons/holes, as in Pr^{3+} and Tm^{3+}, etc.

Table 1. Electron configuration for the trivalent ions from Ce^{3+} to Yb^{3+}; the electrons in the incompletely filled 4f state are indicated in bold; these 4f electrons are buried below a filled shell of 5s and 5p electrons similar to the L-shell of neon or the M-shell of argon and thereby strongly protected from external influences

Shell	K	L		M			N				O	
Principal quantum number (*n*)	1	2		3			4				5	
Orbital quantum number	s	s	p	s	p	d	s	p	d	**f**	s	p
Ce^{3+}	2	2	6	2	6	10	2	6	10	**1**	2	6
Pr^{3+}	2	2	6	2	6	10	2	6	10	**2**	2	6
Nd^{3+}	2	2	6	2	6	10	2	6	10	**3**	2	6
Pm^{3+}	2	2	6	2	6	10	2	6	10	**4**	2	6
Sm^{3+}	2	2	6	2	6	10	2	6	10	**5**	2	6
Eu^{3+}	2	2	6	2	6	10	2	6	10	**6**	2	6
Gd^{3+}	2	2	6	2	6	10	2	6	10	**7**	2	6
Tb^{3+}	2	2	6	2	6	10	2	6	10	**8**	2	6
Dy^{3+}	2	2	6	2	6	10	2	6	10	**9**	2	6
Ho^{3+}	2	2	6	2	6	10	2	6	10	**10**	2	6
Er^{3+}	2	2	6	2	6	10	2	6	10	**11**	2	6
Tm^{3+}	2	2	6	2	6	10	2	6	10	**12**	2	6
Yb^{3+}	2	2	6	2	6	10	2	6	10	**13**	2	6

The spectroscopic terms of the ground state can easily be estimated by considering Hund's rules:

- The spins of equivalent electrons are parallel.
- The lowest state has the largest possible orbital momentum quantum number *L* considering the Pauli exclusion principle

Further, if the 4f shell is less than half-filled, the smallest *J*-value is lowest, with more than half-filled 4f shells the highest *J*-value is lowest. This leads to the scheme of Table 2.

3.2.2 Transition elements

In the ions of the transition elements the active 3d electrons are not as well shielded as the 4f electrons of the rare earth ions. Therefore, they suffer a strong interaction with the surrounding ligands. The resulting spectra are not similar to those of the free ions. Their energy level schemes are, therefore, not given by regular spectroscopic terms but are described by terms taken from the theory of symmetry groups. A and B indicate one-dimensional representations, E indicates a two-dimensional representation and F, or T, a three-dimensional representation of a point group.

As an example the ruby laser shows broad $^4T_1$ or $^4F_1$ (blue band) and $^4T_2$ or $^4F_2$ (green band) absorption bands that make it very suitable for flash lamp pumping. The upper laser level is ^2E with the laser transitions to the $^4A_2$ ground state.

Table 2. Electron quantum numbers and resulting ground states of the trivalent rare earths

Number of 4f electrons	Total spin S	$2S$+1	Maximum L	Ground state	Ions		Ions	
1 or 13	½	2	3	^{2}F	Ce^{3+}	$^2F_{5/2}$	Yb^{3+}	$^2F_{7/2}$
2 or 12	1	3	5	^{3}H	Pr^{3+}	$^3H_4$	Tm^{3+}	$^3H_6$
3 or 11	3/2	4	6	^{4}I	Nd^{3+}	$^4I_{9/2}$	Er^{3+}	$^4I_{15/2}$
4 or 10	2	5	6	^{5}I	Pm^{3+}	$^5I_4$	Ho^{3+}	$^5I_8$
5 or 9	5/2	6	5	^{6}H	Sm^{3+}	$^6H_{5/2}$	Dy^{3+}	$^6H_{15/2}$
6 or 8	3	7	3	^{7}F	Eu^{3+}	$^7F_0$	Tb^{3+}	$^7F_6$
7	7/2	8	0	^{8}S	Gd^{3+}	$^8S_{7/2}$		

The energy of the ^{4}T levels strongly depends on the magnetic field of the ligands at the position of the transition element ion. With a suitable choice of the host crystal the spacing of the ^{2}E levels and the $^4T_2$ band can be varied. In the case of alexandrite, Cr^{3+}:Be Al_2O_4, the $^4T_2$ band is very close (800 cm^{-1}) to the ^{2}E multiplet. Interaction of the states allows a broad band laser transition with emission wavelengths ranging from 700 to 800 nm.

3.3 Basic structure of solid state lasers

Let us assume a system with two levels, an upper level with energy E_2 and a lower level with energy E_1 associated with an optical transition with frequency v so that

$$hv = E_2 - E_1 \tag{1}$$

where h is Planck's constant. The two levels E_1 and E_2 may be populated with population densities of N_1 and N_2 respectively. Light of frequency v travelling through this medium can suffer absorption if $N_1 > N_2$, leading to a transition from E_1 to E_2 equivalent to the loss of a photon. It can also lead to stimulated emission if $N_2 > N_1$, leading to a transition from E_2 to E_1 with generation of a photon. Thus for $N_1 > N_2$ the light is weakened, for $N_2 > N_1$ it is amplified.

Normally N_2 is zero or very small due to thermal excitation and stimulated emission is rarely observed in daily life, whereas absorption is quite normal. With suitable excitation, however, it is possible to invert this normal population so that $N_2 > N_1$. In this case an amplifier with amplification G is generated.

$$G = \mathrm{e}^{(N_2 - N_1)\sigma L} \tag{2}$$

where σ is the emission cross-section and L is the length of the amplifying medium. Such an amplifier can be transformed into an emitter (laser) by using feedback.

3.3.1 Optical resonator

Feedback is reached if the amplifier is placed between two mirrors that repeatedly send back and forth the amplified lightwave through the amplifier (Figure 1).

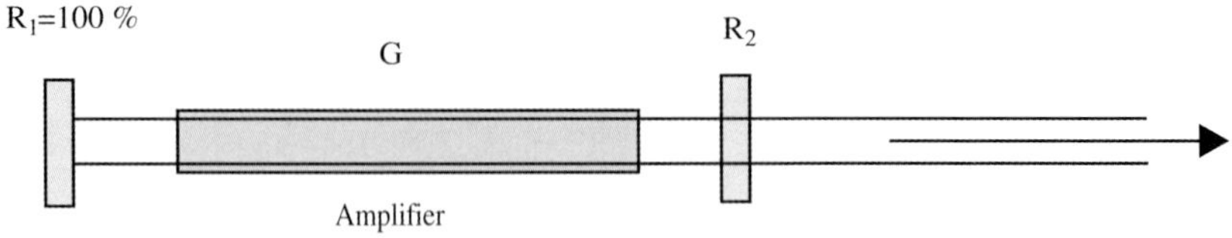

Figure 1. Laser resonator with flat mirrors of reflectance $R_1 = 100\%$ and R_2.

To avoid excessive losses, these mirrors have to be aligned parallel, within about one minute of arc.

With a partially transmitting mirror of reflectance R_2, the fraction $1-R_2$ of the light intensity is transmitted and can be measured outside the resonator. The threshold for laser emission is reached if the gain G is sufficient to compensate the losses $1-R_2$

$$G^2 R_1 R_2 = 1 \tag{3}$$

3.3.2 Laser medium

The laser medium can be used in different geometries. The shape is determined in a rather complex way by the required amount of active ions given by dopant concentration and volume, the way of pumping, the absorption length of pump radiation and the required laser gain in the resonator.

3.3.3 Rod geometry

Most typical is the cylindrical rod as it was used in the first ruby lasers. The rather large volume allows high output power and the geometry is very suitable for strong absorption of side-launched pump light in the relatively small diameter of the rod. Through this absorption a high gain is built up along the long axis of the cylinder.

To reduce the Fresnel reflections at the end faces, the crystal ends are often cut under the Brewster angle (Figure 2). At the Brewster angle only s-polarization suffers reflection losses, p-polarization will therefore be favoured to reach threshold at a lower pump energy than s-polarization. Brewster windows lead to lower threshold and to p-polarized laser emission. In birefringent laser crystals, e.g. 90° or 60°. Ruby, the cross-sections for stimulated emission are different for the ordinary wave and the extraordinary wave. In this case the Brewster window has to be cut so that the wave with the higher gain corresponds to the p-polarized emission.

Besides the laser radiation that builds up from the reflection of the mirrors to well-controlled radiation as described later, undesired disturbing emission can also occur, e.g. in polished laser rods the fundamental mode might have to compete with

Figure 2. Laser rod with end faces cut under the Brewster angle.

wave forms (whispering modes) that undergo total internal reflection at the cylinder wall, propagating zig-zag through the laser medium. The zig-zag trace with its higher length leads to a higher gain that can eventually compensate losses at the rod windows. These whispering modes can be avoided by a frosted or corrugated cylinder surface of the laser rod.

3.3.4 Slab geometry

It is, however, also possible to profit from the high gain in zig-zag waves. Here, the cylinder symmetrical rod is replaced by a slab with a rhomboedric shape combining Brewster windows and the desired total internal reflections (Figure 3).

In extremely powerful solid state lasers as they are used in nuclear fusion experiments the intensity in the laser becomes so large that the material may be destroyed by dielectric breakdown in the material or by thermal damage in adlayers or dust on the windows. In this case the cross-sectional area of the-rod has to be enhanced to bring the intensity below the damage threshold. With the enhancement of the diameter the rod-shaped material then becomes a disc (Figure 4). The disc as a whole can now be aligned under the Brewster angle. Discs with stepwise growing diameter are used in amplifier stages of GW and TW lasers.

Disc geometry is, however, also very useful at the lower end of the power scale. Laser materials that require a very high pump intensity can be excited longitudinally with focused light of a pump laser, often a diode laser. Since the high intensity region in the focal depth is very limited a thin disc is the optimum geometry for this purpose. Numerous miniature lasers have been realised with this geometry.

3.3.5 Waveguide and fibre geometry

Maintenance of the high pump intensity in the focal region of the pump optics can be achieved if the laser material is a wave guide for the pump light as well as for the

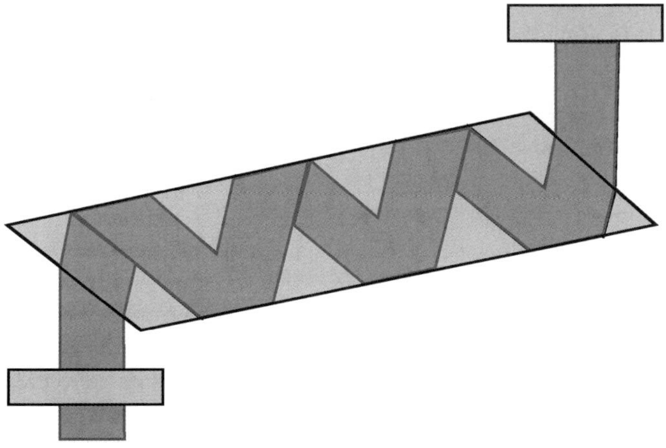

Figure 3. Slab laser with Brewster windows and six internal reflections.

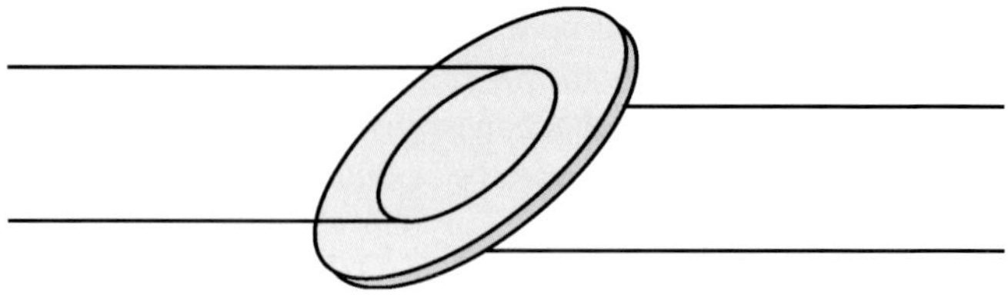

Figure 4. Laser material in the shape of a disc.

laser light. Maintaining a high pump intensity over a long distance in a waveguide-laser allows strong excitation and very low threshold. The most important representative of this type of laser is the fibre laser.

3.4 Special properties of laser media

Besides geometrical consideration the host also has a strong influence on the spectral properties of the laser. The strong influence of magnetic fields of the ligands on the energy levels of 3d electrons has already been mentioned. In addition, the well-shielded 4f electrons suffer from the influence of the electrostatic and magnetic fields of their nearest neighbours. In a similar way as in Zeeman effect, the electronic levels are split in $2J + 1$ sublevels. Each state of the electronic multiplets is thereby transformed into a manifold of sublevels. Since the interaction with 4f electrons is very weak the sublevels are spaced only about few tens to few hundred cm^{-1}. In the rare earth ions the interaction is dominated by Stark effect due to the local field. In this case, with an odd number of 4f electrons and with less than cubic symmetry the Stark sublevels are simply degenerate. The number of Stark sublevels is then $J + ½$. This degeneration is known as Kramers degeneration and the lines in the manifold are known as Kramers doublets.

The energy levels of these Stark manifolds vary for different host crystals. This can have a great influence on the thermal population of the sublevels or for the spectral overlap of transitions involved in cross-relaxation processes. Consequently, the level lifetimes depend on the specific host material.

3.4.1 Phonon energy

The lifetimes of the states in the laser medium strongly depend on the possibility of phonon relaxation. These radiationless transitions reduce the lifetimes of excited states. If the energy difference between two adjacent levels exceeds the maximum phonon energy of the material, then there is still some probability for multiphonon relaxation. However, the probability of multiphonon processes rapidly decreases with the number of phonons involved. For the transition of the pump level to the upper laser state and for the depopulation of the lower laser state these transition need to be efficient. As competing processes for the laser transition they may be fatal. Materials with low phonon energies are in most situations favourable since a given energy difference requires a larger number of phonons.

Table 3. Maximum phonon energies of some laser hosts

Medium	**Highest phonon energies (cm^{-1})**
Phosphate glass	1300
Fused silica (SiO_2)	1100
YPO_4	1050
Tungsten-telluride (TWL) glass	990
Sapphire Al_2O_3	950
Germanate glass (GeO_2)	800
Yttrium aluminium garnet YAG ($Y_3Al_5O_{12}$)	700
Sc_2O_3	625
YAP ($YAlO_3$)	600
ZBLAN	575
YLF ($YLiF_4$)	500
Heavy metal fluoride glass	500
Y_2O_3	460
BaY_2F_8	415
ZnS	323
Fluorozincate ZnABSY	300
$Cs_3Er_2Br_9$	190

The highest phonon energies result from longitudinal vibrations of light particles. In quartz this is the vibration between silicon and oxygen. The vibration energy is lower the heavier the involved atoms are. Therefore oxides will have a higher maximum phonon energy than halides. The highest phonon energy is found in O–H bonds. Therefore, OH content of glasses or crystals should be avoided. Table 3 shows the highest phonon energies of some laser host materials.

3.4.2 Heat conduction

Excitation of the laser medium with light from a flash lamp or from laser diodes generally leads to a number of radiationless transitions that result in heating of the material. If the material is cooled at the surface, a temperature gradient between bulk and surface is established. For a rod of infinite length that is homogeneously heated, a parabolic temperature distribution results, with the highest temperature in the centre of the rod [2].

The hot core of the rod leads to thermal expansion with respect to the cooler outer zone, resulting in mechanical stress. The stress is also parabolic with respect to the radius. This leads to compression in the centre and to stress for the nonradial components at the surface. Exceeding the tensile strength of, e.g. Nd:YAG of about 20,000 Nt cm^{-2}, leads to fracture. Consequently, a good laser material should have good heat conduction and a high fracture limit.

3.5 Pump sources

The optical excitation of the laser material is most easily performed by discharge of electric flash lamps typically filled with a gas mixture containing xenon. They are operated with a triggered discharge of a capacitor C charged with voltage U. The excitation energy E is then proportional to the electric energy of the discharge

$$E = \frac{CU^2}{2} \tag{4}$$

The spectral emission of the flash lamp depends on the discharge energy and duration. In any case, however, it is broad band, ranging from UV to infrared. It is therefore unsuitable for direct pumping of a single narrow energy level. It is necessary, rather, to absorb the pump light in a multitude of levels or in broad absorption bands with a rapid radiationless transfer to the upper laser level. Example broad absorption bands are the $^4F_2$ (around 550 nm) and $^4F_1$ (around 400 nm) states in ruby (cf. Figures 5 and 6).

A multitude of narrow absorbing levels is more typical for trivalent rare earth ions such as Nd^{3+} or Er^{3+} (cf. Figure 6). In both cases numerous high-lying levels are excited with the emission of the flash lamp, and from these levels the excitation rapidly decays to the upper laser level via phonon and multiphonon transitions.

The absorption can even be enhanced by co-doping the laser material with a suitable absorber. A typical example is the erbium laser co-doped with chromium, Cr^{3+}:Er^{3+}:YAG.

For CW lasers, incandescent lamps (quartz-halogen lamp) or a krypton arc lamp can be used. The lamps are usually mounted with the rod inside an elliptical pump-cavity, which is water cooled. The elliptical surface may be coated to enhance the reflectance. Besides polished aluminum, gold is the best reflector for the pump light. Chromium plating is also often used. Single elliptical and double elliptical cavities with two lamps have been employed; however, close-coupled cavities with a diffuse white reflector are also often used.

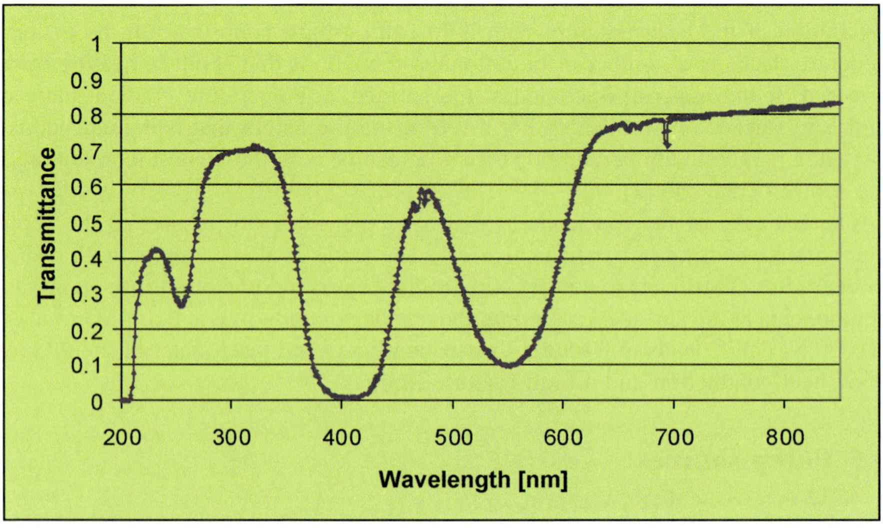

Figure 5. Transmittance of ruby in the wavelength region 200–850 nm.

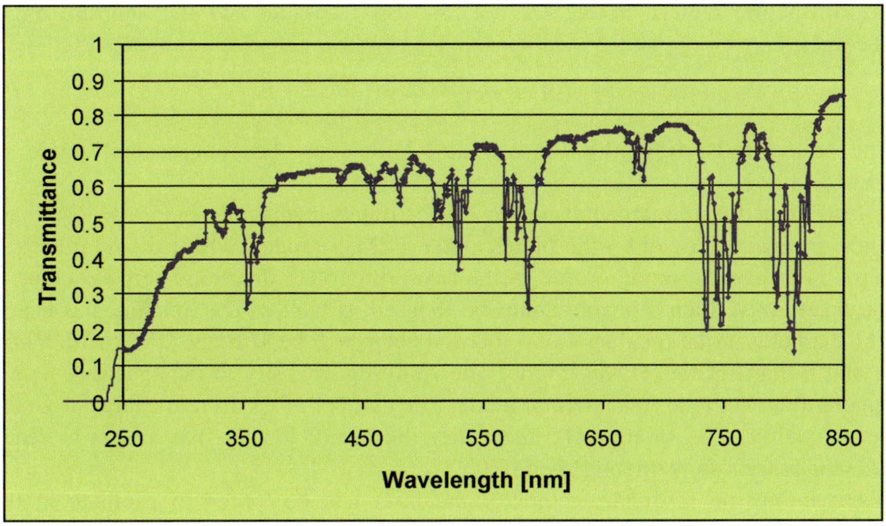

Figure 6. Transmittance of Nd:YAG in the wavelength range 250–850 nm.

Even solar power can be concentrated and used for laser pumping. This procedure has much potential in countries favoured by the sun or for space applications.

The greatest potential for future solid state laser development is pumping with laser diodes. It has the advantage of relatively low thermal load, a very high efficiency and the possibility to design all solid state laser systems without vacuum components and high voltage. It promises very high reliability and lifetime. In view of the importance of this subject, diode laser excitation is treated separately in Chapter 5.

3.6 Emission characteristics

3.6.1 Aperture angle

The highest amplification in a laser is reached when the beam in the resonator can be reflected forth and back as many times as possible through the amplifying medium. This is possible only for a light wave propagating perpendicularly to the mirrors. Deviation even by a small angle will eventually lead to a lateral walk out of the beam, resulting in a loss. As a consequence laser emission occurs only within a very small angle perpendicular to the resonator mirrors.

The intensity distribution across the beam is, however, not homogeneous but has an intensity maximum in the centre and a smaller intensity off-axis. Such an intensity distribution can be approximated by a Gaussian distribution. The smallest lateral radius at $1/e^2$ of the maximum intensity is w_0, the beam waist. In this case the diffraction pattern in the far field is also Gaussian and the aperture angle becomes

$$\vartheta = \operatorname{arctg}\frac{\lambda}{\pi w_0} \approx \frac{\lambda}{\pi w_0} \quad (5)$$

The emission is diffraction limited and known as the fundamental mode or TEM_{00} mode.

Instead of the fundamental mode, higher transversal modes can also oscillate. They are characterized by the higher order TEM_{mn} modes, where m and n are the zeros in transverse mode profile in the two orthogonal directions perpendicular to the beam axis. Such a beam cannot be focused as well as the fundamental beam. The diameter of the focal spot of a focused beam will be M times larger with larger m and n than for the TEM_{00} mode. The resulting intensity in the spot will be M^2 smaller than with the fundamental mode. The number M^2 is used to characterize the beam quality. The smaller M^2, the better the beam. Ideally, the aim is to reach $M^2 = 1$ as in a fundamental mode.

Depending on resonator geometry the laser can be forced to oscillate in the fundamental mode by applying an additional aperture that favours TEM_{00} oscillation but not, however, without losing some of the output power. With high pump power and thermal load, however, the laser medium itself will form a lens due to thermal expansion and the thermal dependence of the refractive index. Such a lens in the resonator destroys the beam quality again and leads to larger M^2. For a given load the thermal lens can be compensated with an optical element such as a dispersive lens or a modification of the laser resonator. But for a different thermal load the beam is distorted again. Compensating optics that automatically adapt to the thermal load are described in Refs. 3 and 4.

3.6.2 Pulse-width control

A constant excitation of a laser material, especially with laser diodes can lead to CW laser emission. In pulsed operation a novel phenomenon occurs. When the pump light is switched on, a series of relaxation oscillations may appear that are then damped when CW emission is reached. Relaxation oscillations will also appear as a reaction of rapid pump light fluctuations, thermal distortions or spectral hole burning. In free running mode these relaxation oscillations lead to spikes in the output power of the laser. Typical spike duration is 1 μs with about 10 μs between spikes. Spikes can be regular, with a rather constant frequency or irregular with chaotic character. They further can be superimposed to a CW part of the emission with different modulation depth.

Well-controlled pulses of high intensity can be generated by artificially changing quickly the reflectivity of one laser mirror from a very low to a high value. The quality Q of the resonator is switched. Before switching Q is kept low, so that threshold is not reached. When excitation has reached its maximum Q is switched to very high values so that the laser process can start. With this technique laser pulses with a duration in the order of 10 ns and 100 MW peak power can be generated.

Active Q-switching can be achieved mechanically with a rotating mirror or with a piezo-driven prism pair using the optical tunnel-effect. It can also be performed with an acousto-optic element or with electro optical components based on Pockels-effect or Kerr-effect. Active switching requires synchronisation between excitation and switching as well as driving electronics for the switch. All this is not necessary in the case of passive switching. Passive switching is based on bleaching of saturable absorbers or exploding thin films.

The generation of ultrashort pulses in the ps and fs range has to consider uncertainty relations between energy, $E = h\nu$, and time t (time-bandwidth product):

$$\Delta E \Delta t \geqslant \tfrac{1}{2}\hbar \quad \text{or} \quad \Delta\nu\Delta t \geqslant \frac{1}{4\pi} \tag{6}$$

where Δ is the variance of the observables. The broader the spectral width of the gain medium, shorter and shorter pulse generation becomes possible. Technically this is realized by coupling a large number of resonator modes in such a way that at one instant in time a constructive interference is reached. This process to generate ultrashort pulses is called mode locking. Practically this locking is performed with a modulation of the laser beam synchronously to the resonator round-trip time. The technical equipment to achieve this modulation is about the same as in Q-switching.

3.6.3 Frequency selection control

The spectral distribution of the fluorescence spectrum of a laser material is generally described by a Voigt profile [5], a bell-shaped distribution. This is a combination of both homogeneous broadening with Lorentzian shape and inhomogeneous broadening with Gaussian shape. In a crystal the spectral broadening is mainly due to phonon interaction and therefore homogeneous. In this case Lorentzian shape dominates. In a glass, however, with varying site symmetry of the active ions, inhomogeneous broadening, Gaussian in shape, becomes more important. Typical bandwidth FWHM of fluorescence is often of the order of a few 100 GHz (ruby 340 GHz [6], (Nd:YAG 0.45 nm [7] corresponding to 120 GHz). Much higher values are also possible, as with Yb:YAG or Ti:sapphire.

If fluorescence is amplified in an amplifier with a Voigt-shaped gain profile, the spectrum of the wave is narrowed with the number of transitions through the amplifier. As a consequence a Q-switch pulse with about 50 transitions through the crystal will show a bandwidth of about 10 GHz, a single spike of a free-running laser with about 10,000 transitions will have a spectral width below 100 MHz.

With frequency selective elements in the laser resonator the emission can be forced to oscillate between a chosen pair of Stark sublevels or it can be tuned within the fluorescence line of a given transition. For coarse tuning prisms, gratings or Lyot filters can be used, for fine tuning Fabry–Perot etalons are suitable. Especially in fibre lasers, fibre gratings are also a powerful tool.

3.7 Types of lasers

3.7.1 Transition metal lasers (in crystals)

The weak shielding of the 3d electrons in transition metal ions leads to strong interaction with ligands and can therefore give rise to very broad bandwidth. This allows ultrashort pulse generation and wide tunability. Examples are listed in Table 4.

3.7.1.1 Ruby laser

The energy level scheme of the ruby laser is shown in Figure 7. Since the lower laser level of ruby is identical with the ground state, ruby belongs to the family of three-level lasers, characterized by strong reabsorption losses and high excitation energy required. Therefore, since about 1970 ruby has been replaced in most applications by Nd:YAG. It is, however, still important if the specific wavelength is required for a given application, e.g. in tattoo removal.

3.7.2 Rare earth lasers

The are numerous rare earth lasers, with thirteen potential ions of the lanthanides and uranium from the actinides that can be incorporated in numerous crystal and glass hosts. Most of these rare earths show several laser transitions. Due to the good concentration of pump light in a guiding structure fibre lasers, especially, can be operated with low laser thresholds. This subject is treated in Chapter 8. A detailed treatment of crystal lasers is outside the scope of this chapter. It can, however, be found in Ref. 14. In the following we concentrate on three of the most important laser systems.

3.7.2.1 Nd:YAG laser

The Nd:YAG laser is the work horse of the crystal lasers. KW output power is reached so that Nd:YAG can be used for many industrial applications like drilling, welding and cutting, in competition with the CO_2 laser.

Table 4. Some transition metal lasers and their emission wavelengths

Laser	Crystal	Wavelength (nm)	Reference
Ruby	Cr^{3+}:Al_2O_3	694.4 692.9	
Alexandrite	Cr^{3+}:$BeAl_2O_4$	700–800	8
Titanium–sapphire	Ti^{3+}:Al_2O_3	660–1180	9
Forsterite	Cr^{3+}:Mg_2SiO_4	1167–1345	10,11
Nickel–magnesium	Ni:MgF_2	1610–1730	12
Cobalt–magnesium	Co:MgF_2	1650–2010	12,13

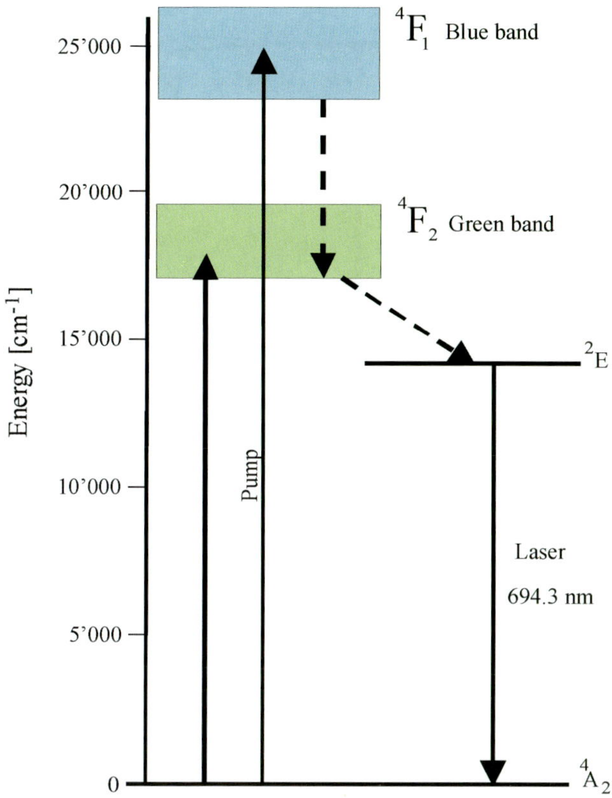

Figure 7. Simplified energy level scheme of the ruby laser. Pump light leads to excitation of the broad green and blue bands. From there the upper laser level is populated by radiationless transitions. The lifetime of the upper laser level is very long ($\tau = 3$ ms at 300 K). A population can easily be accumulated with flash lamp pumping. Due to strong reabsorption from the ground state, more than half of all chromium ions have to be excited to the upper laser state to reach the threshold.

A simplified energy level scheme is shown in Figure 8. Nd:YAG is an ideal four-level laser material. It has a relatively long lifetime of the upper laser level of about 250 μs and the laser transition at 1.064 μm ends about 2000 cm^{-1} above the ground state. Thus the thermal population of the lower laser level at 300 K is less than 10^{-4} and can be neglected. Excitation at 809 nm can ideally be achieved with powerful AlGaAs laser diodes.

3.7.2.2 Erbium laser

Erbium lasers are used with different laser wavelengths. The energy level scheme is shown in Figure 9. The transition from $^4I_{13/2}$ to $^4I_{15/2}$ at a wavelength of 1.56 μm is used for fibre communication with minimum fibre losses in the "third window". The transition from $^4I_{11/2}$ to $^4I_{13/2}$ occurs at 2.94 μm in Er:YAG. This wavelength

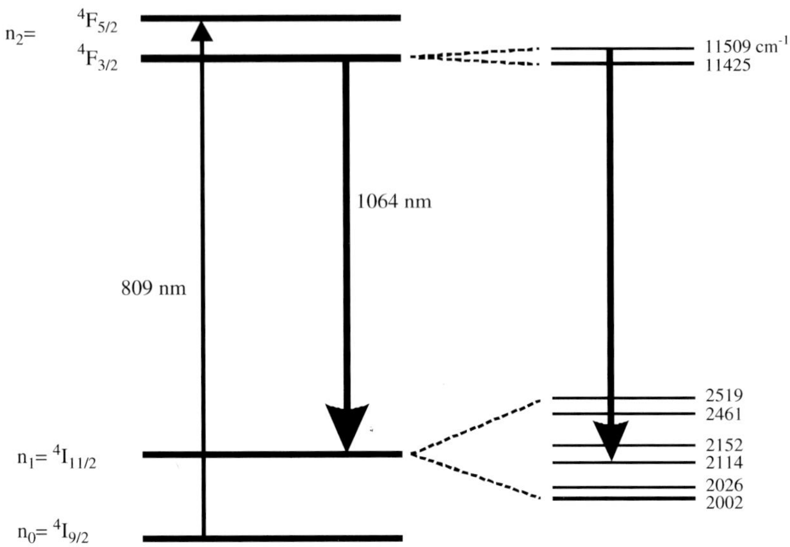

Figure 8. Simplified energy level scheme of Nd:YAG with the main levels (left) and the Stark-split levels (right).

corresponds with the maximum absorption of water in the infrared region. The Er:YAG laser therefore finds many applications in medicine, especially as a surgical knife for cutting of tissue.

Excitation of Er:YAG or Er:$YAlO_3$ is performed with conventional flash lamps. Diode-laser pumping has been demonstrated with Er:YLF, Er:BYF, Er:YSGG and Er:ZBLAN [15–20].

3.7.2.3 Holmium laser

Lasers emitting in the 2 μm eye-safe spectral region are of interest due to their absorption properties in water, atmospheric water vapour and trace gases such as CO_2. Applications range from LIDAR to medicine. Medical applications involve coagulative cutting and tissue welding [21–25]. The wavelength of 2 μm can be emitted from the ${}^5I_7 \rightarrow {}^5I_8$ transition of Ho^{3+} [26]. The ${}^5I_7$ level lies approximately 5000 cm^{-1} above the ${}^5I_8$ ground state (Figure 10). Direct excitation can be performed with laser diodes emitting 2 μm radiation. Powerful AlGaAs laser diodes emitting in the range of about 810 nm or standard InGaAs diodes emitting at 900 to 1100 nm are, however, more easily available. It is therefore interesting to evaluate the possibilities to excite Ho^{3+} by sensitisation of the crystal with other elements that absorb the pump light and transfer the energy into the upper Ho^{3+} laser level. Excitation of Ho^{3+} must be rather strong because the gain must overcome the unavoidable resonant absorption of the quasi-three level Ho^{3+} system. The most common sensitizer for Ho^{3+} -doped crystals is Tm^{3+}. The great advantage of sensitization with Tm^{3+} is the quantum efficiency of 2 [26] that leads to a theoretical maximum achievable efficiency of 75% under pumping at 785 nm and lasing at 2.1 μm [26] (cf. Figure 10).

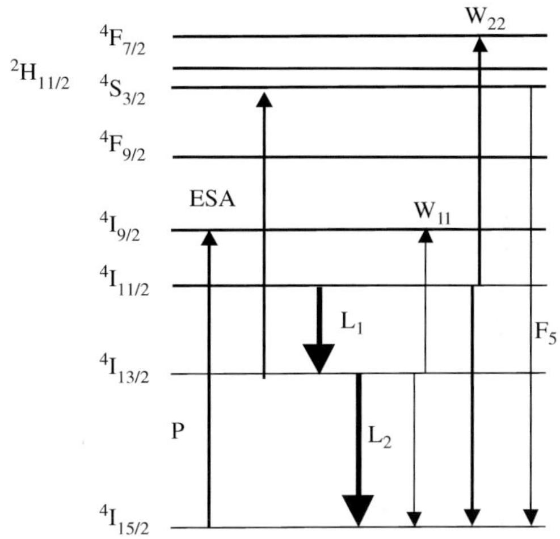

Figure 9. Simplified energy level scheme of Er:YLF, showing: P, Pump transition; ESA, excited state absorption; L_1, the 3 μm laser transition; L_2, the 1.56 μm laser transition; W_{ij}, energy transfer up conversion and F_5, the green fluorescence transition.

3.7.3 New solid state materials

New laser materials can be developed by combining different dopants with the aim of sensitizing the material for optimum pump light absorption and to use energy transfer mechanisms to excite the desired energy levels. This also allows up conversion into shorter wavelengths. Conversely, new host materials have to be developed with properties such as:

- Good mechanical properties with high breaking strengths
- Good heat conduction
- Small thermal lens formation for high power applications
- The ability to incorporate dopant ions in the desired sites without clustering
- Low phonon energy to reduce multiphonon deexcitation to obtain longer fluorescence lifetimes
- Suitable Stark splitting leading to optimum resonance in desired energy transfer up conversion processes or cross-relaxations. Depending on the goals such resonances can also be minimized to avoid undesired processes.
- Depending on the application either birefringent or isotropic materials can be favourable.
- Finally, the laser host should be easy to manufacture at low costs.

3.8 Perspectives

The future solid state laser will be excited with laser diodes, profiting from the robustness and reliability of an all-solid-state system that can be operated at low voltage. With a wider spectrum of the excitation wavelengths the pump light

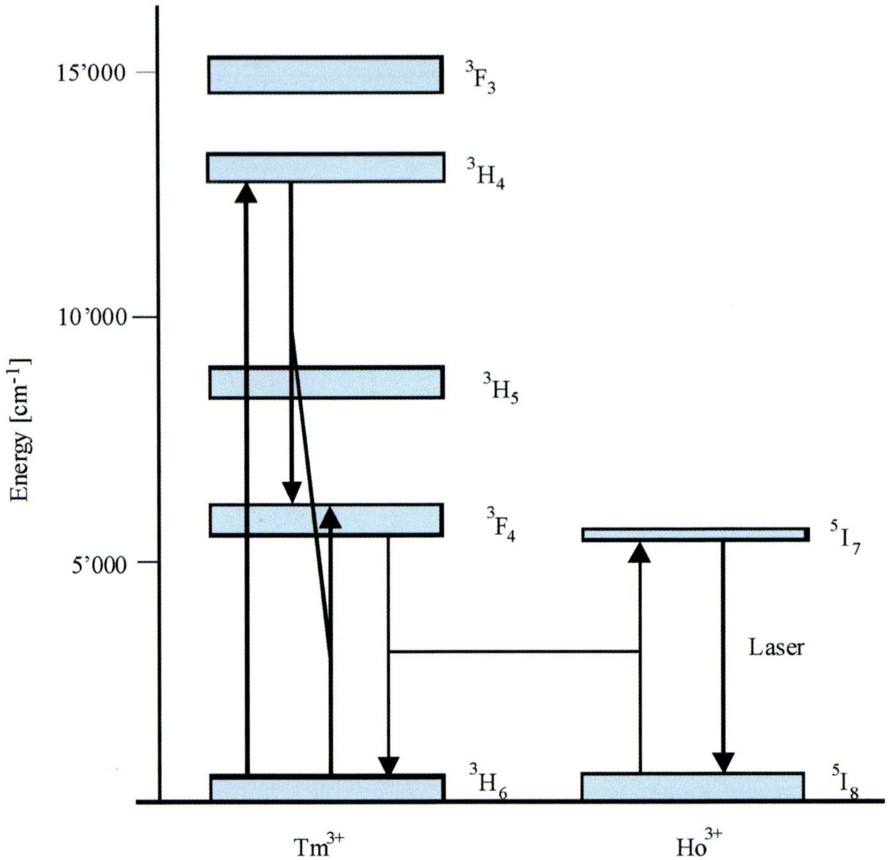

Figure 10. Energy level scheme of Ho^{3+}:YAG co-doped with Tm^{3+}. Pumping with a wavelength of about 785 nm leads to excitation of the Tm^{3+} $^3H_4$ level. Excited Tm^{3+} $^3H_4$ and ground state $^3H_6$ ions can exchange energy, leading to a two-fold population of $^3F_4$. From there a second energy transfer excites the Ho^{3+} upper laser level.

absorption of the laser crystal can better be matched and excess heating of the laser material can be reduced. With further good control of unavoidable thermal distortions the way is open to small high-power systems that can be applied in surgery as well as on an industrial robot or in metrology. In addition to crystal lasers fibre lasers also seem to be extremely promising for reaching kW output power at extremely good beam quality with M^2 close to 1.

References

1. D.W. Goodwin (1965). Crystalline solid state lasers. *Z. Mathematik Phys. ZAMP*, **16**, 35–48.
2. Walter Koechner (1996). *Solid-State Laser Engineering* (4th edn, Chapter 7, p. 393 ff). Springer Verlag.

3. Th. Graf, E. Wyss, M. Roth, H.P. Weber (2001). Laser resonator with balanced thermal lenses. *Opt. Commun.* **190**, 327–331.
4. E. Wyss, M. Roth, Th. Graf, H.P. Weber (2002). Thermooptical compensation methods for high-power lasers, *IEEE J. Quantum Electron.* **38**(12), 1620–1628.
5. A.E. Siegmann (1986). *Lasers* (p. 165 ff). University Science Books Sausalito, CA.
6. W. Koechner (1996). *Solid-State Laser Engineering* (4th edn, Chapter 1, p. 15). Springer Verlag.
7. Walter Koechner (1996). *Solid-State Laser Engineering* (4th edn, Chapter 2, p. 49). Springer Verlag.
8. J.C. Walling, D.F. Heller, H. Samelson, D.J. Harter, J.A. Pete, R.C. Morris (1985). *Tunable alexandrite lasers: development and performance. IEEE J. Quantum Electron.*, *QE*-21(10), 1568–1581.
9. P. Albers (1985). *Ti^{3+} dotierter Saphir und YAG, Elektron-Phononkopplung und Lasereigenschaften eines $3d^1$-Elektronensystems*, Thesis, University of Hamburg.
10. V. Petricevic, S.K. Gayen, R.R. Alfano (1988). Laser action in chromium-doped forsterite. *Appl. Phys. Lett.* **52**(13), 1040–1042.
11. V. Petricevic, S.K. Gayen, R.R. Alfano (1989). Near infrared tunable operation of chromium doped forsterite laser. *Appl. Opt.* **28**(9), 1609–1611.
12. B.C. Johnson, P.F. Moulton, A. Mooradian (1984). Mode-locked operation of Co:MgF_2 and Ni:MgF_2 lasers. *Opt. Lett.* **10**(4), 116–118.
13. P.F. Moulton (1985). An investigation of the Co:MgF_2 laser system. *IEEE J. Quantum Electron. QE*-21(10), 1582–1595.
14. A.A. Kaminskii (1981). *Laser Crystals*. Springer Verlag, Berlin, Heidelberg, New York.
15. Chr. Wyss, W. Lüthy, H.P. Weber, P. Rogin, J. Hulliger (1997). Emission properties of an optimised 2.8 μm Er^{3+}:YLF laser. *Opt. Commun.* **139**, 215–218.
16. M. Pollnau, W. Lüthy, H.P. Weber, T. Jensen, G. Huber, A. Cassanho, H.P. Jenssen, R.A. McFarlane (1996). *Investigation of diode-pumped 2.8-μ*m laser performance in Er:BaY_2F_8. *Opt. Lett.* **21**(1), 48–50.
17. M. Tempus, W. Lüthy, H.P. Weber, V.G. Ostroumov, I.A. Shcherbakov (1994). 2.79 μm YSGG:Cr:Er laser pumped at 790 nm. *IEEE J. Quantum Electron.* **30**(11), 2608–2611.
18. B.J. Dinerman, P.F. Moulton (1994). 3-μm cw laser operations in erbium-doped YSGG, GGG, and YAG. *Opt. Lett.* **19**, 1143–1145.
19. M. Pollnau, C. Ghisler, W. Lüthy, H.P. Weber, J. Schneider, B. Unrauh (1997). Three-transition cascade erbium laser at 1.7, 2.7, and 1.6 μm. *Opt. Lett.* **22**(9), 612–614.
20. T. Huber, W. Lüthy, H.P. Weber, D.F. Hochstrasser (1999). Q-switching of a diode cladding-pumped erbium fibre laser at 2.7 μm. *Opt. Quantum Electron.* **31**, 1171–1177.
21. B. Ott, B. Züger, D. Erni, A. Banic, T. Schaffner, M. Frenz, H.P. Weber (2001). Comparative in-vitro study of tissue welding using a 808 nm diode laser and a Ho:YAG laser. *Lasers Med. Sci.* **16**, 260–266.
22. M. Ith, M. Frenz, H.P. Weber (2001). Scattering and thermal lensing of 2.12-μm laser radiation in biological tissue. *Appl. Opt.* **40**, 2216–2223.
23. I.A. Genyk, M. Frenz, B. Ott, B. Walpoth, T. Schaffner, T. Carrel (2000). Acute and chronic effects of transmyocardial laser revascularization in the non-ischemic pig myocardium by using three laser systems. *Lasers Surg. Med.* **27**, 438–450.
24. M. Frenz, H. Pratisto, F. Könz, E.D. Jansen, A.J. Welch, H.P. Weber (1996). Comparison of the effects of absorption coefficient and pulse duration of 2.12 μm and 2.79 μm radiation on laser ablation of tissue. *IEEE J. Quantum Electron.* **32**, 2025–2036.

25. E.D. Jansen, T. Asshauer, M. Frenz, M. Motamedi, G. Delacrétaz, A.J. Welch (1996). Effect of pulse duration on bubble formation and laser-induced pressure waves during holmium laser ablation. *Lasers Surg. Med.* **18**, 278–293.
26. Th. Rothacher, W. Lüthy, H.P. Weber (1998). Spectral properties of a Tm:Ho:YAG laser in active mirror configuration. *Appl. Phys. B* **66**, 543–546.

Chapter 4

Semiconductor lasers

Peter Unger

Table of contents

Abstract

A synopsis is presented on the physics and device properties of semiconductor lasers. After a brief introduction to the basic principles and design aspects, the different epitaxial material systems together with the appropriate wavelength ranges are overviewed. The epitaxial layer designs for quantum-well diode lasers and quantum cascade lasers are explained and examples for typical edge- and surface-emitting device implementations given.

4.1 Introduction

Semiconductor lasers are gaining increasing interest because of their small size, high electrical-to-optical conversion efficiency, and long device lifetime. Moreover, semiconductor lasers are easy to modulate with electrical signals up to frequencies of severals GHz, which makes them the ideal optical emitter devices in fiber communication systems. Other applications include optical data storage, optical pumping of solid-state lasers and fiber amplifiers, medical applications, material treatment, laser pointers, and illumination. In this chapter, a basic understanding of semiconductor devices and typical device implementations are provided. More information on semiconductor lasers can be found in Refs. 1 and 2, applications for semiconductor lasers in medicine are summarized in Refs. 3 and 4.

Most semiconductor lasers consist of III–V compound materials that are based on alloys of elements in the third group of the periodic table of elements, such as Ga, Al, In, and the fifth group, such as As, P, N, Sb. Examples include GaAs, AlGaAs, InGaAs, InP, GaInAsP, GaInP, and GaN. The emission wavelengths range from 400 nm for nitride-based laser diodes in the blue and ultraviolet regime to more than 10 μm for quantum-cascade lasers in the mid-infrared regime. Continuous optical output powers from several mW for data communication applications up to more than 100 W for laser bars used for pumping of solid-state lasers are commercially available.

The first semiconductor lasers were realized in 1962 almost simultaneously by research groups from General Electric (GE) Laboratories, IBM Research Division, and Massachusetts Institute of Technology (MIT) Lincoln Laboratories. These early devices operated only at cryogenic temperatures or under pulsed conditions. They were homojunction lasers, which means that the same material (GaAs) was used in the active region as well as for the p- and n-doped side of the laser diode junction. In 1963, Herbert Kroemer from the University of Colorado came up with the idea of heterostructure lasers, where a thin active layer is sandwiched between two slabs of different material having a higher band-gap energy to confine the carriers, but it was not until 1970 that this idea was realized in the AlGaAs/GaAs material system almost simultaneously by Zhores Alferov at the Ioffe Physical Institute in St. Petersburg and by Morton Panish and Izuo Hayashi at Bell Laboratories, leading to the first semiconductor lasers that operated continuously at room temperature. The following two decades saw a dramatic development in the device properties and an extension of the emission-wavelength range that was mainly driven by the development of fiber communication systems. Prominent achievements during this time period were substantial improvements of epitaxial growth techniques, the establishment of lasers in the GaInAsP/InP material system allowing emission wavelengths of 1.3 and 1.55 μm for fiber communication, the use of quantum wells in the active region of the devices, the introduction of Distributed FeedBack (DFB) and Distributed Bragg Reflector (DBR) gratings to stabilize the emission wavelength, the implementation of strained quantum wells, the development of red-emitting lasers in the AlGaInP/GaAs material system, and the invention of Vertical-Cavity Surface-Emitting Lasers (VCSELs). Even today, tremendous progress is observed and novel semiconductor laser devices appear in the literature or even on the market. Examples are the quantum cascade laser, published by Frederico Capasso's group at Bell Laboratories in 1994, and the blue-light-emitting GaN laser diode, reported by Shuji Nakamura of Nichia Chemical Industries, Japan, in 1996.

4.2 Basic principles and theoretical background

Gas and solid-state lasers have electronic energy levels that are nearly as sharp as the energy levels of isolated atoms. In semiconductors, these energy levels are broadened into energy bands due the overlap of the atomic orbitals. In an undoped semiconductor with no external excitation at low temperature, the uppermost energy band, called the conduction band, is completely empty and the energy band below the conduction band, called the valence band, is completely filled with electrons. Conduction and valence bands are separated by an energy gap (band gap), which has a value of $E_g = 0.5$–2.5 eV for the typical materials which semiconductor lasers are made of. Two types of carriers contribute to electronic conduction, these are electrons in the conduction band and holes (missing electrons) in the valence band. Holes can be regarded as particles having a positive charge.

The semiconductor material, which is used in the active region of semiconductor lasers, must have a direct band gap, which means that in a diagram showing the energy of the electronic carriers versus their momentum (the momentum of a quantum-mechanical particle is proportional to its wavenumber) the energy minimum for electrons in the conduction band and the energy maximum for holes in the valence band are both located at the point of zero momentum. Direct semiconductor material allows radiative band-to-band transitions, which are the generation and recombination of electron–hole pairs associated with absorption or emission of photons. In indirect semiconductors like silicon and germanium, band-to-band recombinations can only occur with the contribution of phonons or traps. These transitions are unsuitable for laser activity, because the density of traps is very low and the transitions are mostly non-radiative or rather unlikely.

In thermal equilibrium, the occupation of the electronic states is characterized by the Fermi-level energy E_F. At low temperatures, all electronic states are filled below the Fermi-level energy, above it, all states are empty. Laser activity can only occur when stimulated emission is larger than absorption, which is not possible if the electronic carriers are in thermal equilibrium. Laser activity requires a process called pumping, which builds up and maintains a nonequilibrium carrier distribution (inversion) in the semiconductor material. In semiconductor lasers, this is provided by the generation of electron–hole pairs either caused by absorption of photons (optical pumping) or by the electrical injection of electron–hole pairs in a forward-biased p-n junction (laser diode). The nonequilibrium carrier distribution can be described using separate Fermi-level energies E_{Fc} and E_{Fv} for electrons in the conduction band and holes in the valence band, respectively. Inversion in semiconductors is achieved when the difference of these quasi-Fermi levels is larger than the band-gap energy E_g.

All state-of-the-art laser diodes use forward-biased double-hetero p-i-n structures to achieve carrier inversion. In this type of structure, an undoped (intrinsic) semiconductor layer with a direct band gap is sandwiched between p-doped and n-doped material (cladding layers) with a higher band gap. When the junction is forward biased, the quasi-Fermi levels E_{Fc} and E_{Fv} in the intrinsic layer are located inside the conduction and valence bands as illustrated in Figure 1. Thus, this region acts as a laser-active layer which amplifies optical radiation by stimulated emission. Furthermore, the double heterostructure has two additional advantages. First, the carriers (electrons and holes) are confined between the double heterobarriers in the conduction and the valence bands and are therefore forced to recombine inside the intrinsic layer of direct semiconductor material. Second, this layer sequence works like an optical waveguide since, for most semiconductor-material systems, the low-band-gap layer in the middle of the structure has a higher refractive index. The conventional resonator design for semiconductor lasers is a Fabry–Pérot resonator consisting of two flat mirror facets, which are obtained by cleaving the wafer along crystal planes. This type of laser is called edge-emitting laser.

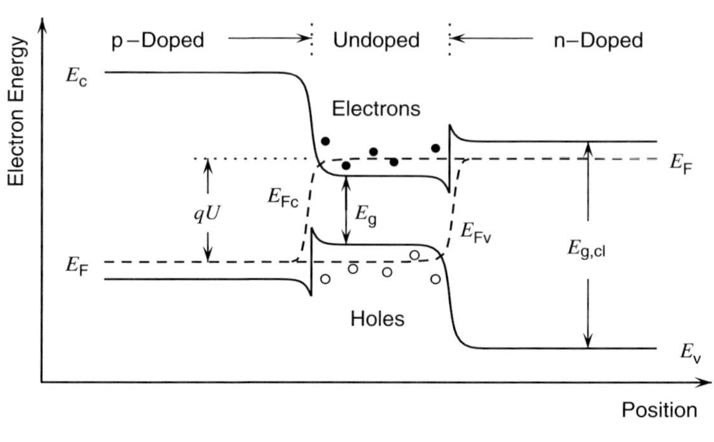

Figure 1. Forward-biased double-heterostructure p-i-n junction. Conduction E_c and valence band edges E_v are plotted as solid lines. The Fermi-level energy E_F, which is represented by dashed lines, splits into quasi-Fermi levels E_{Fc} and E_{Fv} in the undoped transition region, where holes and electrons coexist. In this region, inversion is achieved since the quasi-Fermi levels are inside the bands. E_g is the band-gap energy of the active region, $E_{g,cl}$ is the band-gap energy of the cladding layers. U is the electrical voltage applied to the diode, q the elementary charge.

4.3 Optical gain in semiconductor laser structures

4.3.1 Gain in edge-emitting diode-laser structures

The left-hand side of Figure 2 shows a material gain spectrum of a GaAs layer at room temperature plotted for different carrier densities N. The maximum gain is observed at photon energies which are slightly higher than the band-gap energy. For photon energies below the band-gap energy, the semiconductor is transparent, if the photon energy is significantly higher than the band-gap energy, absorption is observed.

In double heterostructures, the typical thickness of the active layer is 50–300 nm. If the thickness of the active layer is shrunk to 5–10 nm, the electronic wavefunctions in the quantum well show quantization, resulting in discrete energy levels. In this case, the density of states close to the lowest-energy level in a quantum well is much higher than the density of states at the band edge in bulk material. This results in a gain spectrum for a quantum-well laser which has a higher maximum value and a smaller energetic width.

Due to the small active volume of a quantum-well laser, low threshold currents can be obtained. Additionally, the material gain is higher and the spectral shift of the gain curve is lower due to the smaller band-filling, because of the higher carrier density and its narrower energetic distribution. The main advantage of quantum-well structures, however, is the possibility to introduce compressive or tensile

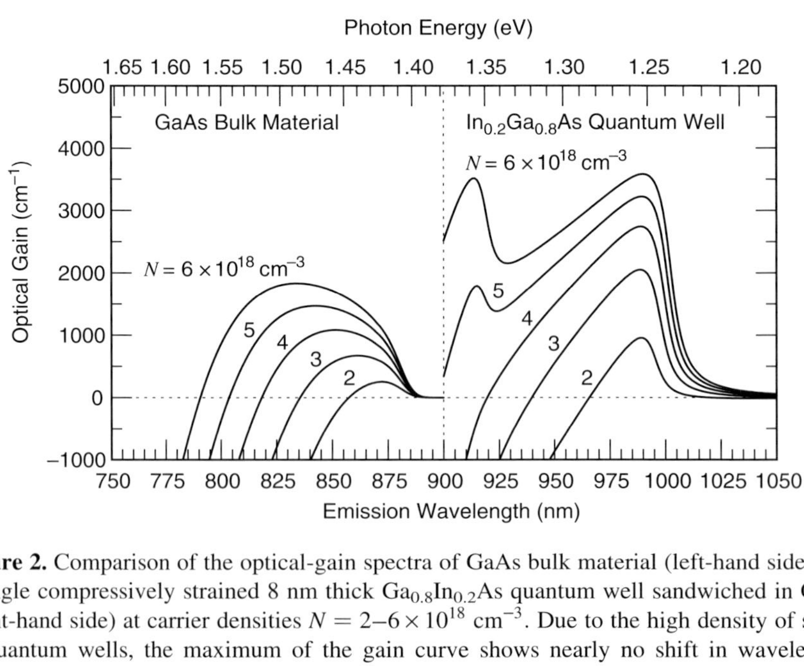

Figure 2. Comparison of the optical-gain spectra of GaAs bulk material (left-hand side) and a single compressively strained 8 nm thick $Ga_{0.8}In_{0.2}As$ quantum well sandwiched in GaAs (right-hand side) at carrier densities $N = 2–6 \times 10^{18}$ cm^{-3}. Due to the high density of states in quantum wells, the maximum of the gain curve shows nearly no shift in wavelength. At higher carrier densities, transitions to the second subband of the quantum well contribute to the gain.

mechanical biaxial strain. In this way, the useable wavelength range of a particular material system can be extended, e.g. the incorporation of In instead of Ga into a thin GaAs quantum well-layer results in a compressively strained quantum well and the accessible wavelength now ranges from 870 nm for bulk GaAs into the long-wavelength region up to approximately 1100 nm. This is illustrated by the spectral gain of an 8 nm thick $Ga_{0.8}In_{0.2}As$ quantum well embedded in GaAs, which is plotted at the right-hand side of Figure 2. The product of quantum-well film thickness and strain must be below a critical value. Above this value, the film experiences relaxation, which is associated with a high number of traps, leading to nonradiative recombination. In compressively strained quantum wells, the light and heavy hole bands are split and the effective mass of the holes in the valence band is reduced. The effective masses of electrons and holes are now comparable, resulting in a more-efficient population inversion in the quantum well. In the long-wavelength range, any kind of strain is beneficial due to the reduced inter-valence-band absorption and Auger recombination. Especially for compressively strained quantum-well lasers, a significantly improved reliability has been observed. For these reasons, strained quantum wells are the rule rather than the exception in state-of-the-art diode lasers.

Since quantum wells are very thin, the confinement of the optical mode is poor. This can be overcome by a Separate-Confinement Heterostructure (SCH) where the confinement of the optical mode is provided by a separate waveguide structure. Two examples of such vertical structures are shown in Figure 3. If the waveguide includes a graded refractive-index profile, the structure is called a GRaded-INdex Separate-Confinement Heterostructure (GRINSCH).

4.3.2 Semiconductor laser material systems

The multilayer structures of diode lasers are fabricated using epitaxial growth techniques like Metal-Organic Vapor-Phase Epitaxy (MOVPE) or Molecular Beam Epitaxy (MBE). In these processes, single-crystal lattice-matched layers with precisely controlled thickness, material composition, and doping profiles are deposited onto substrate wafers. The layer sequence consists of a buffer layer, p- and n-doped cladding layers, a layer for the ohmic contact, and the active region, which may be simple bulk material or a sophisticated structure containing one or more quantum wells with separate optical confinement.

Figure 4 shows suitable III–V semiconductor compounds that can be epitaxially grown on GaAs and InP substrates. The following list gives an overview of material systems commonly used for semiconductor lasers.

- $Al_xGa_{1-x}As$ grown on GaAs is the classical material for semiconductor lasers. Since the radii of gallium and aluminium ions are nearly equal, $Al_xGa_{1-x}As$ can be grown lattice-matched for any composition x, and laser emission in the wavelength range 700–870 nm can be realized.
- Using an InGaAs quantum well, the emission range of the AlGaAs/GaAs system can be extended to longer wavelengths. Since an indium ion has a larger radius

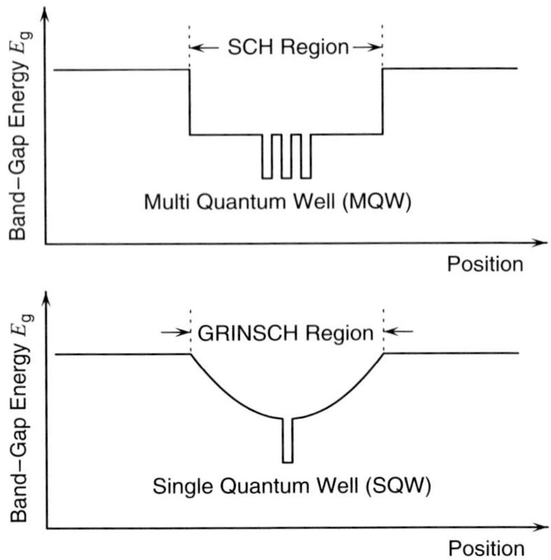

Figure 3. Different vertical structures for separate confinement of electronic carriers and the optical mode. Plots are of the band-gap energies E_g versus position. In the upper diagram, a Multi-Quantum-Well Separate-Confinement Heterostructure (MQW-SCH) with three quantum wells is sketched. The lower part of the figure shows a Single-Quantum-Well GRaded-INdex Separate-Confinement Heterostructure (SQW-GRINSCH).

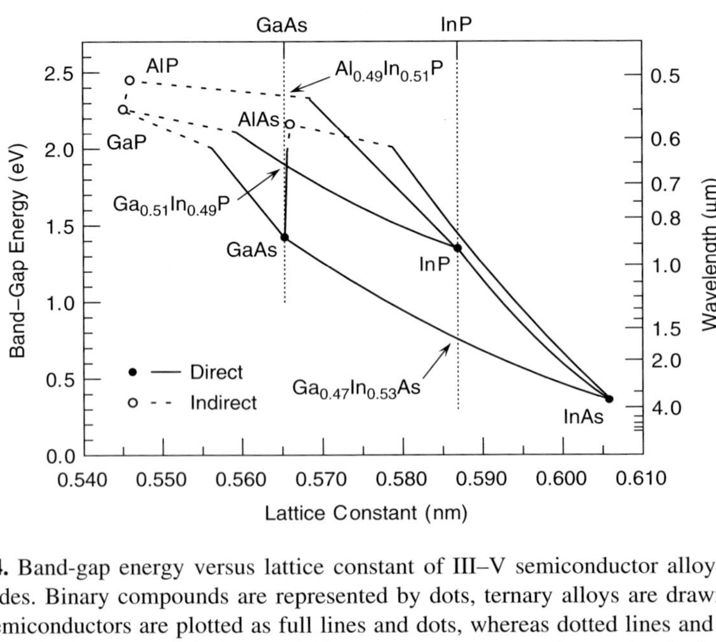

Figure 4. Band-gap energy versus lattice constant of III–V semiconductor alloys used for laser diodes. Binary compounds are represented by dots, ternary alloys are drawn as lines. Direct semiconductors are plotted as full lines and dots, whereas dotted lines and open dots are used for indirect material. Arrows indicate ternary compounds growing lattice-matched on the common substrate materials GaAs and InP. Data derived from [5].

than a gallium ion, the lattice parameter of GaInAs is higher than the lattice constant of GaAs. Therefore, only thin GaInAs quantum-well layers containing compressive mechanical strain can be grown keeping lattice matching. The emission wavelength can be adjusted in the range 800–1100 nm by varying the thickness and indium content of the strained quantum well [6]. Even longer wavelengths can be achieved by compensating the strain using InGaAsN quantum wells [7].

- The same wavelength range can be covered using the strained GaInAs quantum well in combination with GaInAsP separate-confinement layers and $Ga_{0.51}In_{0.49}P$ cladding layers. Like GaInAs-AlGaAs/GaAs, this system is also lattice-matched to GaAs but is completely aluminium-free [8].
- The visible-red short-wavelength range 600–700 nm can be accessed using $(Al_xGa_{1-x})_{0.5}In_{0.5}P$. Again, this material is lattice-matched to GaAs for any aluminium concentration x since the ion radii of aluminium and gallium are approximately equal [9].
- $Ga_xIn_{1-x}As_yP_{1-y}$ grows lattice-matched on InP substrates if the equation $x = 0.4y + 0.067y^2$ is fulfilled. The material has a direct band gap ranging from $E_g = 0.75$ eV for $Ga_{0.47}In_{0.53}As$ to $E_g = 1.35$ eV for InP. Lasers of this type are implemented in fiber communication systems at the wavelengths 1.3 and 1.55 μm [10].
- Blue-light-emitting GaN-based semiconductor lasers have been available since 1996. In these devices, the light-emitting quantum wells consist of InGaN embedded in (Al)GaN layers. The epitaxial structure is grown by MOVPE on sapphire (Al_2O_3) or silicon carbide (SiC) substrates [11]. The main applications of blue diode lasers are optical data storage and spectroscopy.

4.3.3 Quantum cascade lasers

Quantum Cascade Lasers (QCLs) involve only one type of carriers. The laser is based on two fundamental phenomena of quantum mechanics, namely tunnelling and quantum confinement [12]. In conventional heterostructure laser diodes, the light originates from the recombination of negative and positive charges (electrons and holes) across the energy gap existing between the conduction and valence band of the semiconductor crystal. The band-gap energy thus determines the lasing wavelength. The quantum cascade laser, however, is based on a completely different approach. Electrons are making transitions between bound states created by quantum confinement in quantum wells. Since these quantum wells have a size comparable to the de Broglie wavelength of the electron, they restrict the electron motion perpendicular to the quantum-well plane. This effect is called quantum confinement. The electron can only jump from one state to another by discrete steps, thereby emitting or absorbing photons. The spacing between these energy steps depends on the width of the quantum well. The emission wavelength now depends on the layer thickness and not on the band gap of the semiconductor material. This allows the manufacture of lasers with an emission wavelength in the range of 3–10 μm using conventional semiconductor materials like InGaAs/AlInAs on InP substrate. Recent results have presented QCLs operating continuous wave at room temperature [13].

The right-hand side of Figure 5 illustrates the active zone of a QCL. The electrons are injected (from the left) into a narrow quantum well by tunneling. Due to the overlap of the wavefunctions there is a lasing transition between this energy level and the energy levels of a neighboring pair of wider quantum wells. This pair of quantum wells has two energy levels. The relaxation between these energy levels is rather fast; therefore, the upper energy level is always empty, resulting in a population inversion. The electrons leave the lowest energy level by a tunneling process (to the right). This active region can by repeated in a cascade of typically 20–50 identical stages, allowing one electron to emit many photons.

4.4 Edge-emitting lasers

The mirror facets of edge-emitting lasers are obtained by cleaving the wafer along crystal planes. The mirror reflectivities are approximately 30% if the facets are uncoated. Mirror coatings can be applied to change these reflectivities and to passivate the surfaces. The propagation direction of the optical mode in the resonator is in plane with the substrate surface and is referred to as the axial direction. A planar optical waveguide and the laser-active region are formed by depositing a layer sequence onto the substrate surface using epitaxial growth techniques where the deposited single-crystal layers are lattice-matched to the substrate. The growth direction, which is perpendicular to the substrate surface, is called the transverse or vertical direction. The lateral direction is in the substrate plane normal to the axial direction.

The active region has a lateral width *W*, a vertical height given by the thickness of the epitaxially grown active layer, and an axial length *L* that is identical to the cavity length. Since edge-emitting lasers have typical cavity lengths *L* in the range 300–2000 μm, the order number of longitudinal optical modes is very large

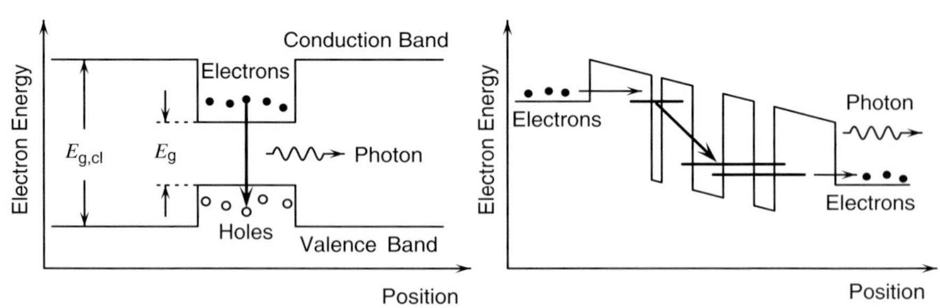

Figure 5. Comparison of a conventional double-heterostructure laser and a quantum cascade laser (QCL). The conventional laser makes use of an electronic transition between conduction and valence band (interband transition) by electron–hole recombination. In a QCL, only one type of carrier is involved (electrons) and the lasing transition is an intersubband transition in quantum wells.

(1000–20,000), the spectral density of the longitudinal modes is very high, and a lot of possible modes can exist within the bandwidth of the spectral gain.

4.4.1 Ridge-waveguide lasers

The left-hand side of Figure 6 shows a narrow-stripe laser. If the stripe width W is small enough, only a single lateral mode can propagate under the stripe, which acts as an optical waveguide. There are in principle two methods to achieve lateral guiding in semiconductor lasers, gain guiding and index guiding. In a gain-guided laser, the lateral width of the active layer is defined by the top ohmic contact. Since the current through the device is restricted to a narrow stripe, inversion and thus optical gain is only possible in the region below the top contact. Outside this area, the optical wave is absorbed. A common way to implement an index-guided waveguide is the ridge-waveguide laser. In this case, the waveguide is defined by etching a ridge into the top cladding layer of the laser. An optical wave experiences a higher effective refractive index when traveling directly below the ridge profile compared to a wave traveling beside the ridge. The main advantage of an index-guided laser are the planar wavefronts of the emitted light, resulting in a low astigmatism. Ridge-waveguide lasers have output power levels in the range of 5–200 mW, exhibit excellent beam profiles, and are used for optical data storage and as pump lasers for fiber amplifiers.

4.4.2 Broad-area lasers and laser bars

As illustrated in the right-hand side of Figure 6, a rather simple approach to achieve high output powers are broad-area laser diodes having a broad stripe design, which allows the propagation of numerous different lateral modes. These devices are widely employed for applications where single-mode beam profiles are not required,

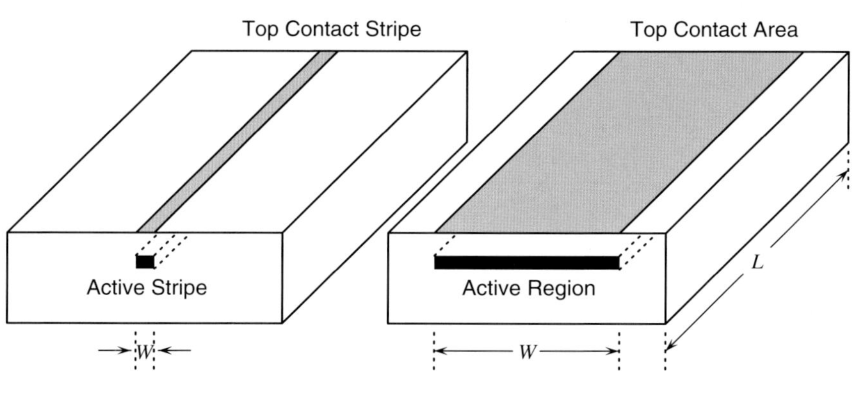

Figure 6. Comparison of a narrow-stripe laser having single-lateral-mode emission and a broad-area laser showing high optical output power.

like infrared illumination, material treatment and solid-state laser pumping. Electrical-to-optical conversion efficiencies of more than 50% can be achieved making broad-area lasers the most efficient technical light sources. Record output powers around 10 W for lasers with cavity lengths of 2 mm and lateral stripe widths of 100 μm have been reported for the InGaAs/AlGaAs material system as well as for aluminium-free InGaAs/GaInAsP/GaInP lasers. An example for the characteristics of a broad-area InGaAs/AlGaAs laser diode is shown in Figure 7. For pumping of solid-state lasers, several broad-area lasers are usually combined to a monolithically integrated bar having a typical width of 1 cm. When properly cooled, such bars yield continuous output power levels in the 50–200 W range.

4.4.3 Laser arrays

Monolithic integration of laser diodes can be easily performed by arranging the individual devices onto the chip in form of an one-dimensional array. If the distance between the individual emitters is large enough, no optical coupling is observed and the emission of each laser is independent of the emission of its neighbors. Typical array devices of this type are the 1 cm wide bars used for the pumping of solid-state lasers. Common emission wavelengths are 808 nm and 940 nm for pumping of Nd:YAG and Yb:YAG, respectively. Commercially available 1 cm wide laser bars with a cavity length of 900 μm are specified for

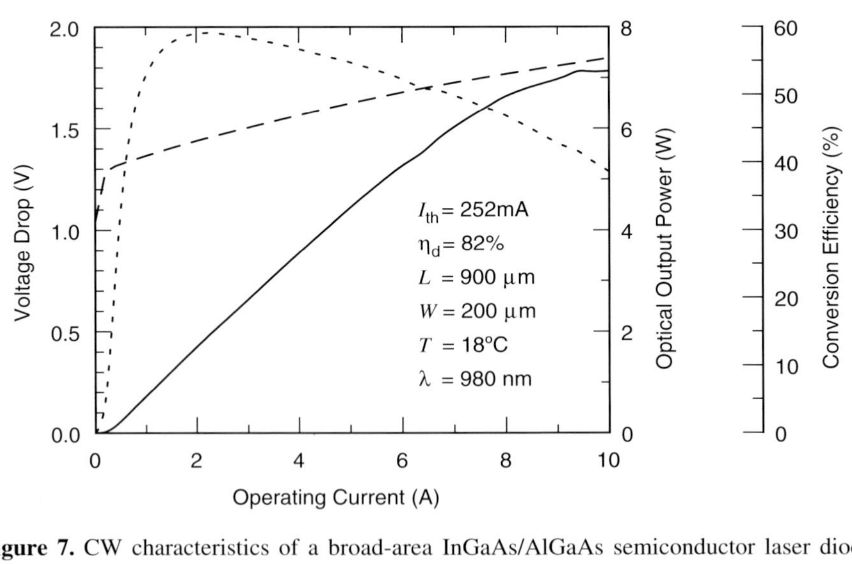

Figure 7. CW characteristics of a broad-area InGaAs/AlGaAs semiconductor laser diode. The optical output power (full curve), voltage drop (dashes) and electrical-to-optical power conversion efficiency (hyphens) are plotted. Listed are the threshold current I_{th}, the differential quantum efficiency η_d determined from the slope of the output power characteristic, the cavity length L, the lateral width of the active region W, the temperature of the heat sink T and the emission wavelength λ (after [14]).

continuous optical output powers of 50 W at operating currents of approximately 70 A. Although such devices can be driven to output powers of more than 100 W, the lifetime is drastically reduced under such conditions. When operated under specified conditions, lifetimes of up to 10,000 h have been reported. Another example of broad-area laser arrays are individually addressable arrays of single-mode lasers used for laser printers.

If the single emitters are arranged in close proximity, as illustrated in Figure 8, the optical field generated by each element of the array is coupled to other elements. Typical arrays of this type consist of single-mode index- or gain-guided waveguides which are closely spaced. These devices are called phase-locked laser arrays, because there is a fixed phase correlation between the individual emitters. Optical coupling can occur either via evanescent waves or leaky modes (radiation modes). The nature of the coupling strongly depends on the geometry of the waveguides and the effective refractive index step from the waveguide regions to the regions that mediate the optical coupling between the waveguides. For evanescent-type array modes, the fields are peaked in the high-effective-index regions, whereas for leaky-type array modes, the fields are peaked in the low-index array regions. Evanescent-wave arrays have propagation constants between the low and high refractive-index level, while leaky modes have values for the propagation constant below the low effective refractive index. The modes are said to be 'in-phase' when there is no phase shift for the fields of each waveguide and are called 'out-of-phase' when the fields of adjacent waveguides show a phase shift of π.

Generally, evanescent-wave arrays tend to operate in the out-of-phase mode, resulting in a double-lobed far field. The reason for this behavior is the better field overlap of the out-of-phase mode with the gain regions because this mode has zero intensity in the regions between the waveguides. Another disadvantage of evanescent-wave arrays is the limitation in the effective refractive-index step between the waveguides and the inter-element regions since a high index step will allow the propagation of higher-order modes in the individual waveguide elements. This low index step makes the device sensitive to spatial hole burning, resulting in deteriorated beam properties at high output-power levels.

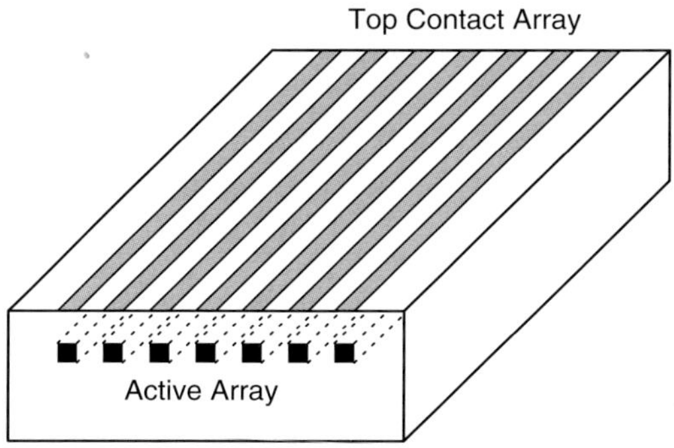

Figure 8. Semiconductor laser array consisting of an array of coupled narrow-stripe waveguides.

In leaky-mode arrays, the optical gain is provided in the low-index regions. Unlike evanescent-wave arrays, there is no limitation in the refractive index step, resulting in a stable operation at high output power. When properly designed, leaky-mode arrays operate in-phase and exhibit a single-mode beam profile at output power levels in the 1–2 W range. More information of this topic can by found in Ref. 15.

4.4.4 High-brightness lasers

With tapered laser devices, the beam quality of wide-aperture laser diodes at high output powers can be significantly improved in comparison to broad-area devices. Figure 9, presents three examples for tapered lateral designs.

A tapered traveling-wave laser amplifier has a narrow-aperture input facet and a wide-aperture output facet [16]. Both mirror facets are covered with antireflection coatings to suppress any optical resonances. The basic function of the device is the amplification of a low-power beam profile from an external master-oscillator laser while maintaining its beam properties. The master oscillator can be any type of laser having an excellent beam profile, usually another low-power single-mode laser diode. During amplification inside the tapered traveling-wave amplifier area, the beam propagates freely in lateral direction. Therefore, the taper angle must be properly adjusted to the diffraction angle of the propagating beam. The mode pattern of the output beam is very sensitive to the coupling accuracy of the input beam, the quality of the antireflection coatings and the taper geometry. Tapered laser oscillators with wide output apertures consist of a tapered gain region and a lateral waveguide section at the narrow-aperture mirror facet. The output facet is typically antireflection coated to a reflectivity of less than 1%, therefore only a small part of the output light is back reflected towards the waveguide section, which works like a spatial mode filter. Cavity-spoiling grooves around the

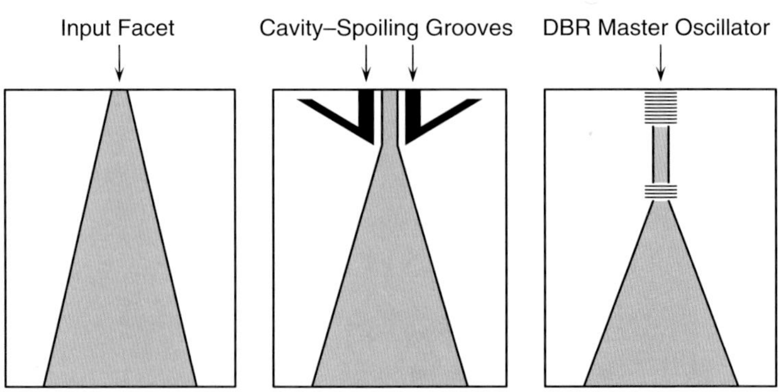

Figure 9. Lateral design of the active area for different types of tapered laser diodes with wide-aperture output facets. Left-hand side, a tapered traveling-wave amplifier. Center, a tapered laser oscillator having cavity-spoiling grooves at the narrow end of the taper. Right-hand side, a monolithically integrated master-oscillator power amplifier (MOPA).

waveguide, which are tilted against the output facet, suppress any further Fabry–Pérot oscillations.

The third example for a tapered laser design shown at the right-hand side of Figure 9 is a monolithically integrated master-oscillator power amplifier (MOPA) [17]. Again, there is a tapered gain region with an antireflection coating at the wide-aperture output facet. At the narrow end of the taper, a master laser with a single-mode waveguide is monolithically integrated into the chip. The resonator mirrors of the master laser are realized using distributed Bragg reflectors (DBRs).

In recent years, considerable interest has been generated in semiconductor device research to realize unstable-resonator designs. The most promising approaches use curved mirror facets [14]. To produce these devices, a reliable dry-etching process for vertical, flat and smooth mirror facets is required. The curved mirrors create a high-loss resonator through the introduction of lateral divergences in the oppositely propagating wave fronts reflected by the curved mirror surfaces. The design of the mirror curvature has to be optimized so that the lowest-loss mode will produce a single-lobed output beam and all higher-order modes are strongly suppressed because of their substantially higher losses.

4.5 Surface-emitting lasers

4.5.1 Vertical-cavity surface-emitting lasers

In vertical-cavity lasers, the optical propagation direction is normal to the substrate surface and the effective cavity length is very short (typically 1–3 μm), allowing the existence of only a single longitudinal mode within the spectral-gain range. To avoid extremely high mirror losses in the short cavity, the reflectivity of the Fabry–Pérot mirrors must be close to 100% according. This can be achieved by using Bragg reflectors that consist of typically 20 pairs of epitaxially grown GaAs–AlAs layers having alternating high and low refractive index and a thickness of a quarter wavelength. Between the mirrors, a set of quantum wells is sandwiched, providing the optical gain in the active region. In the case of strained InGaAs quantum wells, all semiconductor material including the substrate is transparent for light in the wavelength range 870–1100 nm generated in these quantum wells. The choice of the appropriate mirror reflectivities allows the VCSEL to be operated as a top or a bottom emitter. In the schematic illustration of a VCSEL shown in Figure 10, the electrical current is supplied through the p- and n-doped mirrors.

The emitting lateral aperture normally has a circular geometry with a diameter of a few microns, allowing single-lateral-mode operation and a highly effective coupling to optical fibers. Additional advantages are the ultra-low threshold current (<1 mA), excellent dynamic properties, a high electrical-to-optical power-conversion efficiency, insensitivity to optical feedback, and the absence of sudden device failures attributed to mirror damage. Although the epitaxial growth of VCSEL structures is rather sophisticated, the fabrication process is similar to the manufacturing of LEDs, which allows wafer-scale processing and on-wafer device testing. VCSELs can be easily arranged in two-dimensional arrays and coupled to

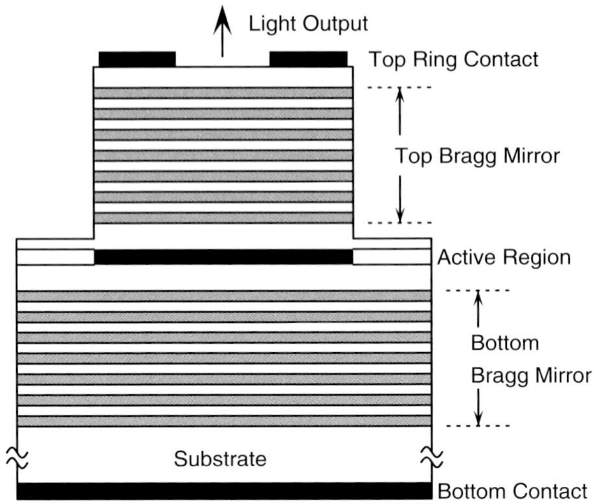

Figure 10. Schematic cross-sectional drawing of the Vertical-Cavity Surface-Emitting Laser (VCSEL). The laser cavity of the top-emitting VCSEL is perpendicular to the substrate plane. The mirrors of the Fabry–Pérot resonator consist of Bragg reflectors with reflectivities close to 100%. The total length of the device in the vertical direction is about 7 μm and the effective cavity length is in the range 1–3 μm.

parallel optical-fiber bundles. The single-mode output power of a VCSEL is in the mW range. Higher power levels can be achieved by enlarging the diameter of the emitting aperture or by a densely packed arrangement. Since VCSELs are rather novel devices, their full potential has not yet been exploited. Certainly, there will be an increasing range of applications for VCSELs in the high-power regime. More comprehensive information on the properties and applications of VCSELs can be found in Ref. 18.

4.5.2 Optically pumped semiconductor disk lasers

An example of a new promising device concept is the optically pumped semiconductor disk laser, which is also referred to as Vertical-External-Cavity Surface-Emitting Laser (VECSEL) [19]. As shown in Figure 11, the semiconductor chip is mounted on a heat sink and forms a resonator together with an external concave mirror. The pump laser beam from a broad-area laser diode is focused onto the semiconductor disk. The laser cavity is formed by an AlAs–GaAs Bragg mirror which is grown directly onto the GaAs substrate and an external concave dielectric mirror, resulting in a stable concentric (hemispheric) resonator configuration. The active laser medium is a resonant periodic gain structure with quantum wells located in antinodes of the standing wave pattern of the longitudinal optical mode. Around each quantum well are AlGaAs pump light absorbing lasers. The pump light is absorbed in the AlGaAs layers and the carriers (electrons and hole) relax

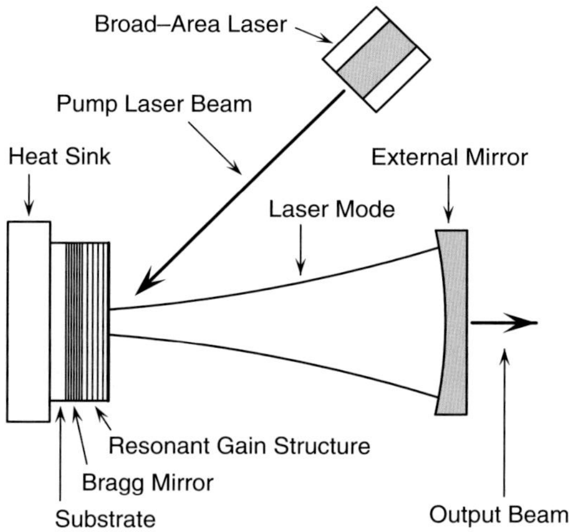

Figure 11. An optically pumped semiconductor disk laser. The semiconductor disk consists of an epitaxially grown Bragg mirror and a resonant quantum-well gain structure. Together with an external concave mirror, a fundamental transversal mode develops in the laser resonator. The laser disk is optically pumped by a focused beam from a high-power broad-area laser diode.

into the quantum wells, where they provide gain to the lasing mode. The device is similar to the solid-state disk laser (thin disk laser) besides the fact that the emission wavelength can be chosen by the design of the epitaxially grown quantum wells. Also the optical pumping of the semiconductor disk is easier, because the pump light can be absorbed in a single pass. The optically pumped semiconductor disk laser has excellent beam properties, is scalable in output power, and allows intracavity frequency doubling [20].

4.6 Perspectives

Semiconductor diode lasers providing high optical output powers are attractive devices for applications where a small size, a high electrical-to-optical power conversion efficiency, reliable operation and a low price is desired. Broad-area laser devices and laser diode bars exhibit optical output powers up to 100 W and are widely used for pumping of solid-state lasers, material treatment, medical applications and infrared illumination. Due to filamentation effects, the beam quality of these devices is rather poor. State-of-the-art single-mode devices like ridge-waveguide and narrow-stripe lasers are limited to output power levels of several 100 mW.

A considerable part of the ongoing reseach activities on high-power lasers is attributed to approaches that allow high optical output powers and diffraction-limited beam profiles at the same time. Devices of this kind have tapered lateral designs, implement unstable resonator concepts, stabilize the mode by lateral angled gratings or use coupled arrays of single-mode waveguides.

In future, the wavelength range of 400–1100 nm for commercially available semiconductor laser diodes will be extendend towards longer wavelengths, e.g. using InGaAsN quantum wells and quantum cascade lasers. Laser devices with monolithically integrated functionality like wavelength tunability or integrated monitor photodiodes may enter the market. Surface-emitting semiconductor lasers like optically pumped semiconductor disk lasers will have an interesting perspective.

References

1. L.A. Coldren, S.W. Corzine (1995). *Diode Lasers and Photonic Integrated Circuits*, Wiley, New York.
2. P. Unger (2000). Introduction to power diode lasers. In: R. Diehl (Ed), *High-Power Diode Lasers* (Topics in Applied Physics, Volume 78, pp. 1–53). Springer-Verlag, Berlin, Heidelberg.
3. R. Pratesi (1984). Diode lasers in photomedicine. *IEEE J. Quantum. Electron.* **20**, 1433–1439.
4. R. Pratesi (1991). Initial applications and potential of miniature lasers in medicine. In: R. Pratesi (Ed), *Optronic Techniques in Diagnostics and Therapeutic Medicine,* (pp. 271–285). Plenum Press, New York.
5. O. Madelung, R. Poerschke (Eds) (1991). *Data in Science and Technology—Semiconductors: Group IV Elements and III–V Compounds.* Springer-Verlag, Berlin, Heidelberg.
6. R. Jäger, J. Heerlein, E. Deichsel, P. Unger (1999). 63% wallplug efficiency MBE grown InGaAs/AlGaAs broad-area laser diodes and arrays with carbon p-type doping using CBr_4. *J. Crystal Growth* **201/202**, 882–885.
7. A.Y. Egorov, D. Bernklau, D. Livshits, V. Ustinov, Z.I. Alferov, H. Riechert (1999). High-power cw operation of InGaAsN lasers at 1.3 μm. *Electron. Lett.* **35**, 1643–1644.
8. A. Al-Muhanna, L.J. Mawst, D. Botez, D.Z. Garbuzov, R.U. Martinelli, J.C. Connolly (1998). High-power (>10 W) continuous-wave operation from 100 μm-aperture 0.97 μm-emitting Al-free diode lasers. *Appl. Phys. Lett.* **73**, 1182—1184.
9. P. Unger, G.-L. Bona, R. Germann, P. Roentgen, D.J. Webb (1993). Low-threshold strained GaInP quantum-well ridge lasers with AlGaAs cladding layers. *IEEE J. Quantum Electron.* **29**, 1880–1884.
10. M.J. Adams, A.G. Steventon, W.J. Devlin, I.D. Henning (1987). *Semiconductor Lasers for Long-Wavelength Optical-Fibre Communications Systems* (IEE Materials and Devices Series). Peter Peregrinus, London.
11. S. Nakamura, S. Pearton, G. Fasol (2000). *The Blue Laser Diode—The Complete Story.* Springer-Verlag, Berlin, Heidelberg.
12. J. Faist, F. Capasso, D.L. Sivco, C. Sirtori, A.L. Hutchinson, A.Y. Cho (1994). Quantum cascade laser. *Science* **264**, 553–556.
13. M. Beck, D. Hofstetter, T. Aellen, J. Faist, U. Oesterle, M. Ilegems, E. Gini, H. Melchior (2002). Continuous wave operation of a mid-infrared semiconductor laser at room temperature. *Science* **295**, 301–305.
14. E. Deichsel, R. Jäger, P. Unger (2002). High-brightness unstable-resonator lasers fabricated with improved dry-etching technology for ultra-smooth laser facets. *Jpn. J. Appl. Phys. Part I* **41**, 4279–4282.
15. M.W. Carlson (1994). *Monolithic Diode-Laser Arrays.* Springer-Verlag, Berlin, Heidelberg.

16. M. Mikulla, P. Chazan, A. Schmitt, S. Morgott, A. Wetzel, M. Walther, R. Kiefer, W. Pletschen, J. Braunstein, G. Weimann (1998). High-brightness tapered semiconductor laser oscillators and amplifiers with low-modal gain epilayer-structures. *IEEE Photon. Technol. Lett.* **10**, 654–656.
17. S. O'Brien, R. Lang, R. Parke, J. Major, D.F. Welch, D. Mehuys (1997). 2.2-W continuous-wave diffraction-limited monolithically integrated master oscillator power amplifier at 854 nm. *IEEE Photon. Technol. Lett.* **9**, 440–442.
18. K. Iga, F. Koyama, S. Kinoshita (1988). Surface emitting semiconductor lasers. *IEEE J. Quantum Electron.* **24**, 1845–1855.
19. M. Kuznetsov, F. Hakimi, R. Sprague, A. Mooradian (1999). Design and characteristics of high-power (>0.5 W cw) diode-pumped vertical-external-cavity surface emitting semiconductor lasers with circular TEM_{00} beams. *IEEE J. Select. Topics Quantum Electron.* **5**, 561–573.
20. E. Schiehlen, M. Golling, P. Unger (2002). Diode-pumped semiconductor disk laser with intracavity frequency doubling using lithium triborate (LBO). *IEEE Photon. Technol. Lett.* **14**, 777–779.

Chapter 5

Diode-pumped solid state lasers

Holger Zellmer and Andreas Tünnermann

Table of contents

Abstract

By the application of diode lasers as pump source for solid state lasers the advantages of the semiconductor laser material regarding compactness, reliability and efficiency can be transferred to the whole solid state laser system. In addition, it allows the realization of novel laser concepts that have output parameters which cannot be achieved by conventional arc lamp excited solid state lasers.

5.1 Introduction

After more than forty years of development solid state lasers have become attractive sources for research and industry. The scope of their applications ranges from basic research over life science and metrology to manufacturing in industry. Today, the different fields of application require powerful laser sources with excellent beam quality and long maintenance intervals. These requirements can only be fulfilled by applying diode lasers as pump source for solid state lasers.

The most common active material for diode-pumped solid state lasers are neodymium-doped garnets such as Nd:YAG operated at a laser wavelength of 1064 nm. In addition, several different laser materials have been developed for various applications in material processing, metrology and medicine. Their emission wavelengths are located in the near- and middle-infrared (Table 1).

In general, solid state lasers consist of an active medium in the shape of a rod. This geometry can be easily manufactured and gives the output beam a rotational symmetry which is preferred for most applications. Other geometries of the active medium, like rectangular slabs, are used only if special laser parameters have to be realized [1].

In conventional solid state lasers the excitation of the laser process is performed by noble gas arc lamps. However, the spectral overlap of the broad emission spectrum of the arc lamp and the narrow absorption lines of the active ions in the laser material is small, resulting in a low efficiency. Typically, less than 3% of the electrical input power of a conventional solid state laser is converted into laser radiation [2].

In addition, due to the Stokes shift between the pump radiation and the laser emission, heat is deposited in the laser material. This heat load usually limits the performance of solid state lasers in terms of output power and beam quality. The heat is generated in the total volume of the active material. The cooling, however,

Table 1. Emission and excitation wavelengths of selected laser materials

Laser material	Laser wavelength	Pump wavelength (nm)
Nd:YAG, Nd:YVO	1064 nm	808
Yb:YAG	1030 nm	940
Tm:YAG	2.1 μm	780
Er:YAG	3.0 μm	795, 975

is performed only at the edges of the laser crystal, either by liquid cooling or by conduction cooling to a heat sink. This causes a temperature gradient in the laser material. This temperature gradient leads to a refractive index gradient in the laser crystal because of the temperature dependence of the refractive index. The temperature gradient also creates mechanical stress that causes a refractive index variation by the photo-elastic effect. In addition, the mechanical stress induces birefringence, which causes depolarization of the generated laser radiation.

Compared with arc lamps, diode lasers are an ideal pump source for solid state lasers with their high electrical-to-optical efficiency of up to 50%. Their emission spectrum can be tuned to perfectly match the excitation spectrum of the active material resulting in a high efficiency and a minimized heat deposition inside the laser material. Further, the lifetime of diode lasers of more than 10,000 h exceeds that of arc lamps by at least a factor of 10.

Becoming available in large quantities during the last years, powerful diode lasers triggered a wide spectrum of different new concepts and geometries for diode-pumped solid state lasers.

5.2 Concepts for diode-pumped solid state lasers

For the design of diode-pumped solid state lasers two basic configurations are possible: transversally and longitudinally pumped laser systems.

In longitudinally pumped lasers the directionality of the diode laser radiation is exploited (Figure 1). In contrast to lamps the radiation of diode lasers can be focused to small spots, enabling very high pump intensities. This allows the operation of new laser transitions in small and efficient laser devices. Further, by launching the pump light through one of the resonator mirrors into the laser crystal it is possible to match the volume of the optically excited material exactly to the volume of the fundamental laser mode of the laser resonator, allowing a very high optical efficiency [3–5]. In addition, by careful design of the laser system, single transverse mode operation can be easily achieved, ensuring an optimum beam quality [6].

However, the amount of pump power that can be coupled into the active laser medium through the end facet of the laser rod is limited. Due to thermal stress at

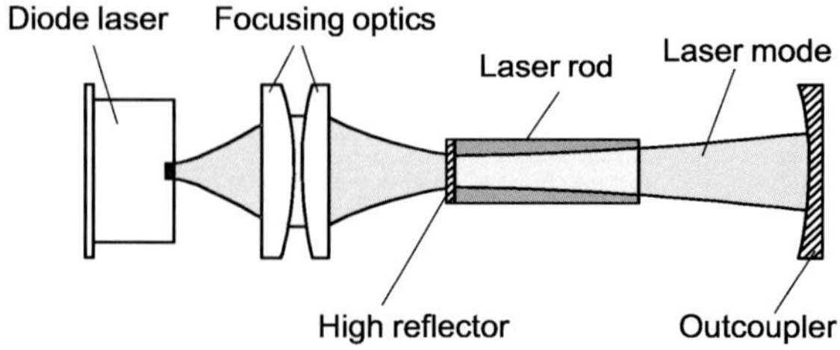

Figure 1. Longitudinally diode pumped solid state laser. The pump light is launched through one of the resonator mirrors into the end face of the laser material. By matching the excited volume to the mode volume of the fundamental mode an excellent beam quality is obtained.

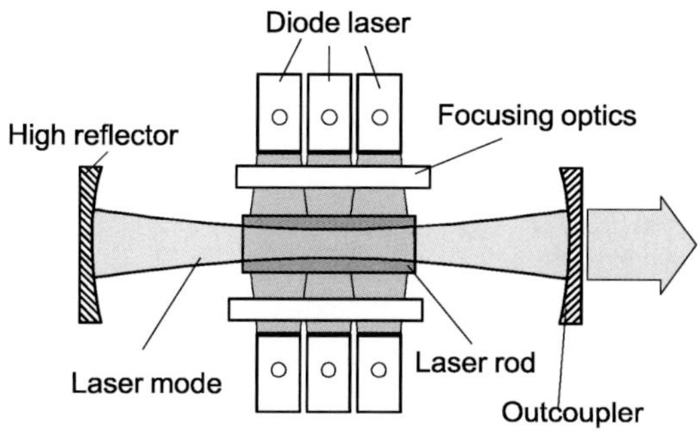

Figure 2. Transversally diode pumped solid state laser. The pump diodes, usually diode laser bars or stacked arrays, are mounted on the side of the active material and their radiation is coupled through the barrel face into the laser rod or slab. Power scaling is possible by increasing the rod length while keeping the pump power density constant.

the end face of the laser rod the output power that can be achieved from an end-pumped Nd:YAG laser is limited to approximately 60 W [7]. Hence, for high power applications the pump light is launched through the barrel faces of the laser crystal. Within certain limits the output power of such side pumped systems can be scaled by increasing the length of the laser crystal, thus enabling more pump power to be launched to the laser crystal (Figure 2).

The optical excitation process in a solid state laser generates heat in the active material, due to the energy difference of the pump and laser photon, which is deposited in the host material and due to non-radiative transitions. Although the heat deposited in the active material in diode-pumped solid state lasers is up to four times less than in lamp pumped systems operating at the same power level, the heat load at high power operation is not negligible [8]. Nd:YAG rods can tolerate only pump powers in the order of 300 W cm^{-1} in safe operation [9,10]. More pump power probably leads to crystal damage by tensile stress inside the rod.

5.3 New concepts

Novel concepts for diode-pumped solid state lasers try to circumvent the consequences of thermal effects. The most promising ideas are the thin disk and the fiber laser [11].

Thin disk lasers (Figure 3) are based on a disk of laser material with a height of typically 200 μm and a diameter of up to more than 10 mm. One end face of the disk is coated with a dielectric high-reflecting mirror and mounted to a heat sink. The pump light is launched into the disk through the opposite end face. This concept leads to a homogeneous, one-dimensional temperature gradient along the axis of the crystal, i.e. parallel to the generated laser beam. Hence, the thermally induced distortions of the laser mode are very small and an excellent beam quality can be achieved. The drawback of the thin active material is the short absorption

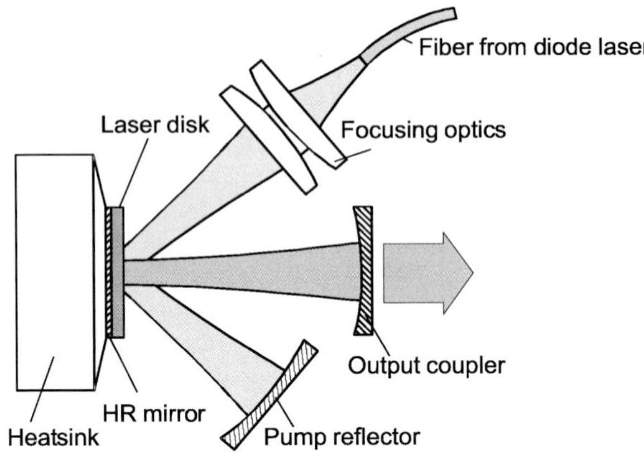

Figure 3. Thin disk laser. The laser crystal, coated with the high reflecting mirror on one side, is mounted on a heat sink. The pump light is coupled into the crystal from the opposite side.

length of the pump light. High doping concentrations in the laser crystal and multi-pass configurations for the pump light solve this problem and allow high optical efficiencies [12,13].

Whereas thin disk lasers use a short active medium with large diameter to avoid thermal effects, fiber lasers (Figure 4) apply an active medium with a diameter of only a few micrometres but with a length of several metres. Usually, neodymium-, ytterbium- or erbium-doped silica fibers are used for high power fiber lasers. The beam quality of a fiber laser is solely determined by the refractive index profile of the active fiber. Refractive index variations due to thermal effects are very small in comparison to the fiber's refractive index profile and will not disturb the laser mode. Furthermore, the thermal load caused by the pumping process is spread over a long region and, due to the good ratio between surface and active volume, heat is efficiently removed. Even at high power operation no active cooling is required.

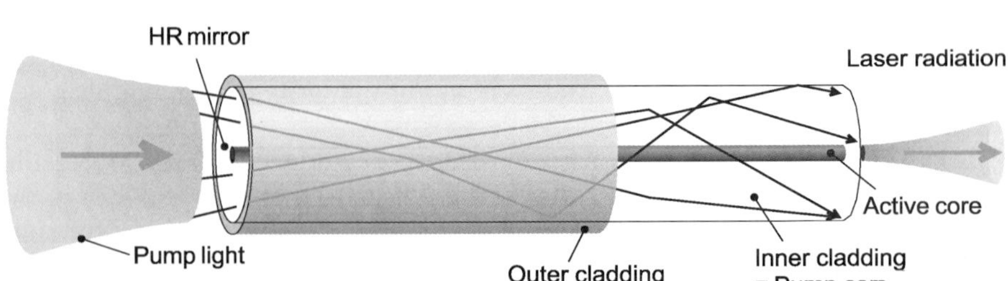

Figure 4. Fiber laser. A double clad structure with a single mode core for the generated laser radiation and a surrounding multimode core for the pump light allows single transverse mode operation and the launching of very high power into the fiber.

For high power operation a double clad fiber design has to be applied [14]. The fiber core, which is doped with the active material, is surrounded by a multimode core for the pump light. The pump core has a diameter of the order of 400 μm and a NA of 0.4, typically. High power diode lasers can be efficiently launched to it. During the propagation in the pump core the pump light is absorbed in the active core and excites the laser process.

Due to the waveguide structure with a diameter of 10 μm, typically, the generated laser power is concentrated on a very small area, resulting in extreme power densities. To reduce the power density special large mode area fibers have been designed [15]. With such fibers a single transverse mode output power of cw 0.5 kW has been demonstrated [16].

The length of the active fibers and the small diameter of the active medium results in a high single pass gain that can be exploited in the amplification of cw [17] and pulsed [18] signals. Further, the high gain in fiber amplifiers offers a simple solution for high energy, high repetition rate, short-pulse laser systems [19,20].

5.4 Conclusion

Diode-pumped solid state lasers offer the possibility to achieve high output powers up to the kilowatt range with high electric-to-optic conversion efficiency. Novel designs like fiber and thin disk lasers reduce thermal effects in the laser medium and allow excellent beam quality even at high output power.

References

1. R.J. Shine, Jr., A.J. Alfrey, R.L. Byer (1995). 40-W cw, TEM00-mode, diode-laser-pumped, Nd:YAG miniature-slab laser, *Opt. Lett.* **20**, 459–461.
2. W. Koechner (1999). *Solid State Laser Engineering*, Springer Verlag, New York.
3. T.W. Fan, A. Sanchez (1990). Pump source requirements for end-pumped lasers. *IEEE J. Quantum Electron.* **26**, 311.
4. F. Salin, F. Squier (1991). Geometrical optimization of longitudinally pumped solid state lasers. *Opt Commun.* **86**, 397.
5. Y.F. Chen, C.F. Kao, T.M. Huang, C.L. Wang, S.C. Wang (1997). Influence of thermal effects on power optimization in fiber-coupled laser-diode end-pumped lasers. *IEEE J. Selected Top Quantum Electron.* **3**, 29.
6. P. Laporta, M. Brussard (1991). Design criteria for mode size optimization in diode pumped solid state lasers. *IEEE J. Quantum Electron.* **27**, 2319.
7. S.C. Tidwell, J.F. Seamans, M.S. Bowers, A.K. Cousins (1992). Scaling cw diode-end-pumped Nd:YAG lasers to high average powers. *IEEE J Quantum Electron.* **28**, 997 ff.
8. U. Brauch, M. Schubert (1995). Comparison of lamp and diode pumped cwNd:YAG slab lasers. *Opt. Commun.* **117**, 116 ff.
9. D. Golla, M. Bode, S. Knoke, W. Schöne, F. von Alvensleben, A. Tünnermann (1996). High power operation of Nd:YAG rod lasers pumped by fiber-coupled diode lasers. *OSA Trends Opt. Photonics Series TOPs* **1**, p. 198 ff.

10. A. Takada, Y. Akiyama, T. Takase, H. Yuasa, A. Ono (1999). Diode laser-pumped cw Nd:YAG lasers with more than 1 kW output power, *Advanced Solid State Lasers 1999, OSA Tech. Digest Ser.* paper MB 18, p. 69.
11. A. Tünnermann, H. Zellmer, W. Schöne, A. Giesen, K. Contag (2000). New concepts for diode pumped solid state lasers. In: R. Diehl (Ed), *High-Power Diode Lasers* (Topics in Applied Physics Volume 78, pp. 369–408). Springer Verlag, Heidelberg.
12. M. Karszewski, U. Brauch, A. Giesen, I. Johannson, C. Stewen, A. Voss (1998). 100 W TEM_{00} operation of Yb:YAG thin disk laser with high efficiency. *OSA Trends Opt. Photonics Ser. TOPs* **16**, 296 ff.
13. S. Erhard, A. Giesen, M. Karszewski, T. Rupp, C. Stewen, I. Johannsen, K. Contag (1999). Novel pump design of Yb:YAG thin disk laser for operation at room temperature with improved efficiency. *OSA Trends Opt. Photonics Ser. TOPs* **26**, 38 ff.
14. E. Snitzer, H. Po, F. Hakimi, R. Tumminelli, B.C. McCollum (1988). Double clad, offset coreNd fiber laser, *Optical Fiber Sensors, New Orleans*, January 27–29, Postdeadline Paper PD5.
15. J.A. Alvarez-Chavez, H.L. Offerhaus, J. Nilsson, P.W. Turner, W.A. Clarkson, D.J. Richardson (2000). High-energy, high-power ytterbium-doped Q-switched fiber laser. *Opt. Lett.* **25**(1), 37–39.
16. J. Limpert, A. Liem, H. Zellmer, A. Tünnermann (2003). Continuous-wave ultra high brightness fiber laser systems, *Advanced Solid State Photonics*, San Antonio, February 2–5, Post Deadline Paper PD1.
17. S. Höfer, A. Liem, J. Limpert, H. Zellmer, A. Tünnermann, S. Unger, S. Jetschke, H.-R. Müller, I. Freitag (2001). Single-frequency master-oscillator fiber power amplifier system emitting 20 W of power. *Opt. Lett.* **26**, 1326–1328.
18. J. Limpert, S. Höfer, A. Liem, H. Zellmer, A. Tünnermann, S. Knoke, H. Voelckel (2002). 100 W average power high energy nanosecond fiber amplifier. *Appl. Phys. B* **75**, 477–479.
19. A. Liem, D. Nickel, J. Limpert, H. Zellmer, U. Griebner, S. Unger, A. Tünnermann, G. Korn (2000). High avarage power ultrafast fiber CPA system. *Appl. Phys. B* **71**(6), 889–891.
20. A. Galvanauskas, G.C. Cho, A. Hariharan, M.E. Fermann, D. Harter (2001). Generation of high-energy femtosecond pulses in multimode-core Yb-fiber chirped-pulse amplification systems. *Opt. Lett.* **26**, 12, 935–937.

Chapter 6

Incoherent light sources

Brian L. Diffey

Table of contents

Abstract

Artificial sources of incoherent optical radiation may be produced either by heating a body to an incandescent temperature or by the excitation of a gas discharge. The most common source of incandescent optical radiation is the sun, and although there are artificial incandescent sources, in general, they are not efficient emitters of the ultraviolet component of optical radiation. For this reason, the most usual way to produce ultraviolet radiation artificially is by the passage of an electric current through a gas, usually vaporized mercury. Examples of lamps that emit incoherent ultraviolet radiation include low and high-pressure mercury arc lamps, fluorescent lamps, metal halide lamps and xenon arc lamps. All of these lamps find applications in medicine, principally in the treatment of disease.

6.1 Introduction

Artificial sources of incoherent (non-laser) optical radiation may be produced either by heating a body to an incandescent temperature or by the excitation of a gas discharge [1].

6.2 Incandescence

A body heated to a high temperature radiates as a result of its constituent particles becoming excited by numerous interactions and collisions. For a perfect black body the power radiated at any wavelength from a unit surface area is determined by its temperature, in accordance with Planck's law:

$$M_{e\lambda} = A/\{\lambda^5[\exp(B/\lambda T) - 1]\} \tag{1}$$

where $M_{e\lambda}$ is the spectral radiant excitance at wavelength λ (nm), $T\,(K)$ is the absolute temperature of the radiator, and A and B are constants. As the temperature is raised, not only does the maximum power radiated increase rapidly, but also the peak of the emission curve λ_{max} (nm) moves to a shorter wavelength, given by Wien's displacement law as:

$$\lambda_{max} = 2.898 \times 10^6/T\,(K) \tag{2}$$

The sun is the most celebrated source of incandescent ultraviolet radiation (UVR); artificial incandescent sources are not efficient emitters of UVR. The ultraviolet emission from a general-purpose tungsten filament lamp is only 0.08% of the rated power for a 40 W lamp, rising to 0.1% for a 100 W lamp and 0.17% for a 1 kW lamp [2].

6.3 Gas discharges

The most usual way to produce UVR artificially is by the passage of an electric current through a gas, usually vaporized mercury. The mercury atoms become excited by collisions with the electrons flowing between the lamp electrodes. The excited electrons return to particular electronic states in the mercury atom and in doing so release some of the energy they have absorbed in the form of optical radiation – that is, ultraviolet, visible and infrared radiation. The spectrum of the radiation emitted consists of a limited number of discrete wavelengths (so-called 'spectral lines') corresponding to electron transitions characteristic of the mercury atom, and the relative intensity of the different wavelengths in the spectrum depends upon the pressure of the mercury vapour. For low-pressure discharge tubes containing mercury vapour at about 1 Nm^{-2}, more than 90% of the radiated energy is at a wavelength of 253.7 nm.

As the pressure in a discharge tube is raised to a few atmospheres (1 atm = $1.013 \times 10^5 Nm^{-2}$), two principal changes occur:

- The gas temperature increases due to the increasing number of collisions (mainly elastic collisions) with the energetic electrons;
- the high temperature becomes localized at the centre of the discharge, there now being a temperature gradient towards the walls, which are much cooler.

The wall becomes much less important at high pressures, and not altogether essential: discharges can operate between two electrodes without any restraining wall, and are then referred to as arcs. At high pressures the characteristic lines present in the low-pressure discharge spectrum broaden and are accompanied by a low-amplitude continuous spectrum. By doping mercury vapour lamps with traces of metal halides it is possible to enhance both the power and the width of the spectrum emitted, particularly in the UVA (315–400 nm) and visible regions.

6.4 Lamps

6.4.1 Incandescent lamps

Incandescent lamps emit optical radiation as a result of the heating of a filament, made almost exclusively from tungsten filaments. As the melting temperature of tungsten is 3653 K, the theoretical maximum possible luminous efficacy of an incandescent filament lamp corresponds to that of a perfect black-body radiator at that temperature. The luminous efficacies of incandescent lamps are generally much less for two main reasons:

- With the exception of photoflood lamps, lamp filaments are generally not heated to temperatures in excess of 3000°C – this prolongs their useful life;
- the spectral emissivity of tungsten varies with temperature.

Incandescent lamps are made in a wide range of shapes and sizes, including the simple domestic light bulb, reflective display lamps, car headlamps, panel bulbs and photoflood lamps. The envelopes of low-power incandescent lamps are normally made from lead or soda glass and those of higher-power ones from silica glass or quartz. However, even with quartz bulbs, the levels of emission of UVR, and in particular actinic ultraviolet, from standard incandescent lamps are small enough not to constitute an ultraviolet exposure hazard.

Conversely, tungsten halogen lamps usually operate at a colour temperature between 2900 and 3450 K and are much higher-power devices (up to 5 kW) than standard vacuum or gas-filled incandescent lamps. Because of their higher black-body equivalent temperature and the use of quartz as an envelope material they may emit UVR in excess of the recommended maximum permissible exposure [3].

The principal applications of tungsten halogen lamps are flood projector lamps, studio and theatre lighting, car headlamps, photocopier lamps, and a wide variety optical instrument lamps.

6.4.2 Germicidal lamps

The principal radiation emitted by low-pressure mercury vapour discharge lamps is at 253.7 nm. This radiation is particularly efficacious for preventing the growth of moulds and bacteria. In germicidal lamps there is no phosphor present and a fused silica (quartz) envelope is used. Therefore, such lamps are very efficient sources of UVC (200–280 nm) radiation. Approximately 50% of the electrical power is converted into radiation of which up to 95% is emitted in the 253.7 nm line. Germicidal lamps are available in a range of sizes, shapes and powers. The tubular types are the most common and they fit into standard domestic and commercial fluorescent lamp fittings (bi-pin) and operate on standard fluorescent lamp control gear. Small low-wattage germicidal type lamps are often used as fluorescence lamps for chromatographic analysis, and identification of minerals. A germicidal type lamp is often used in combination with a UVA "blacklight" fluorescent lamp for such purposes.

6.4.3 Fluorescent lamps

A fluorescent lamp is a low-pressure mercury discharge lamp that has a phosphor coating applied to the inside of the envelope. Fluorescent radiation is produced by the excitation of the phosphor by radiation at 253.7 nm. The spectral power distribution of the fluorescent radiation will be a property of the chemical nature of the phosphor material. The exact chemical composition of a phosphor will depend upon the desired spectral power distribution; in general, phosphors consist of mixtures of silicates, borates and phosphates of alkaline earth metals.

Since their introduction around 1940, the use of fluorescent lamps for domestic, commercial, and industrial general lighting has grown enormously. This has been due to several factors:

- The introduction of mass-production techniques for their manufacture and testing, hence low production costs,
- long life and high conversion efficiency of electrical power to light, hence low running costs,
- flexibility of size, from a few cm to 2 m in length,
- low surface brightness, hence reduced glare,
- good colour rendering properties.

There are few public, commercial, or industrial premises that do not use fluorescent lamps for general lighting purposes. Ultraviolet fluorescent lamps have also found many specialized applications, including the treatment of skin diseases, cosmetic tanning and in the printing industry for drying inks. The ultraviolet spectral emissions from three fluorescent lamps commonly used in the phototherapy of skin diseases, notably psoriasis, are shown in Figure 1. The phototherapy of neonatal jaundice is achieved with visible, and not ultraviolet, radiation.

Fluorescent lamps used for this purpose have normally emitted white light or blue light. However, a recent development [4] has shown that fluorescent lamps emitting turquoise light are equally effective, and probably safer, in this treatment than the commonly-used blue lamps. The spectral emission of these two lamps are compared in Figure 2.

A fluorescent lamp normally fails when the emissive material on the cathodes ceases to produce sufficient electrons to permit the lamp to strike. A sign of a lamp nearing the end of its useful life is severe blackening at both ends of the tube. The light output of a fluorescent tube designed for general lighting purposes falls by about 2–4% during its first 100 h of life and thereafter at a slower rate until at 2000 h it has fallen by only a further 5–10%. However, ultraviolet fluorescent lamps have phosphors that are less stable than those used in general lighting lamps and the radiation output shows a more rapid degradation in the first 100 h, with the lamp restricted to a useful life of about 1000 h.

6.4.4 Mercury arc lamps

The radiation emitted from a mercury-vapour arc lamp arises from two mechanisms. Line or characteristic radiation is produced as a result of excitation of the constituent atoms together with a spectral continuum chiefly due to ion and electron recombination. When a mercury arc lamp is switched on the vapour pressure of mercury is low and a discharge fills the lamp and appears blue, with a relatively high proportion of radiated energy in the ultraviolet. As the temperature of the

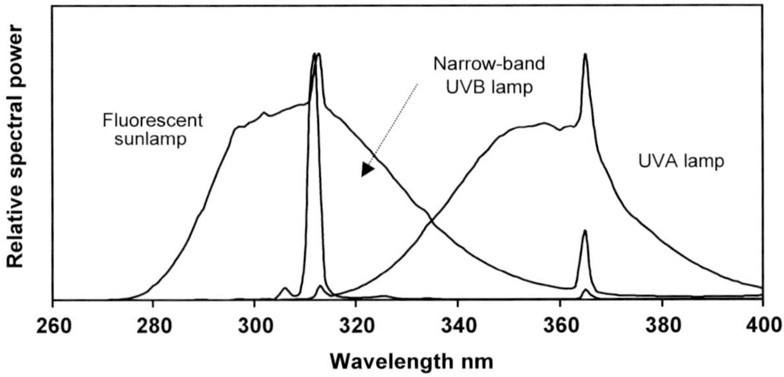

Figure 1. Spectral power distribution of three common ultraviolet fluorescent lamps used in the phototherapy of skin diseases.

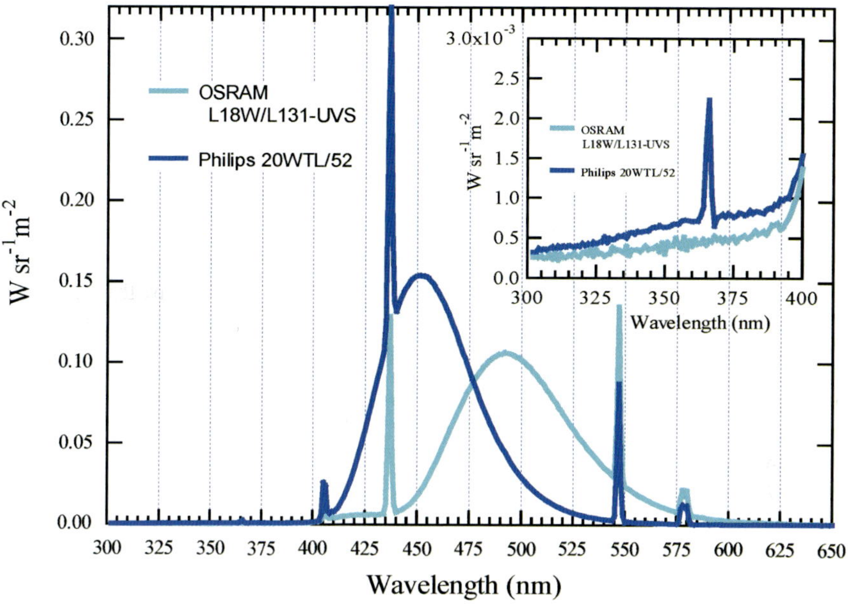

Figure 2. Spectral power distribution of two visible fluorescent lamps used in the phototherapy of neonatal jaundice.

discharge rises, and with it the mercury vapour pressure, the radiated energy is concentrated progressively in the spectral lines of the visible region, and this, together with the introduction of a small proportion of continuous radiation, results in the discharge becoming whiter. After about 5–6 min the lamp is fully run-up and in the medium-pressure mercury arc lamp the mercury vapour pressure is in the range 2–10 atm depending on the lamp rating. High-pressure mercury arc lamps operate at a pressure of 10–100 atm, which results in broadening of the spectral lines and an increase in the continuum relative to the characteristic radiation. The spectral power distribution of a medium-pressure arc lamp in a quartz envelope is shown in Figure 3, consisting of the line spectrum of mercury with some low-level continuous radiation.

6.4.5 Metal halide lamps

In high-pressure mercury lamps a considerable volume of the arc tube is not effectively used to emit radiation. This non-productive space is required to reduce energy dissipation at the wall of the arc tube to maintain a long lamp life. If, however, another element of low excitation potential can be introduced without interfering unduly with the mercury discharge, then this element can be excited in the otherwise non-productive space, resulting in additional output and spectral content. The most common added elements are the alkali metals, used in the alkali metal iodide form to eliminate chemical attack on the silica envelope.

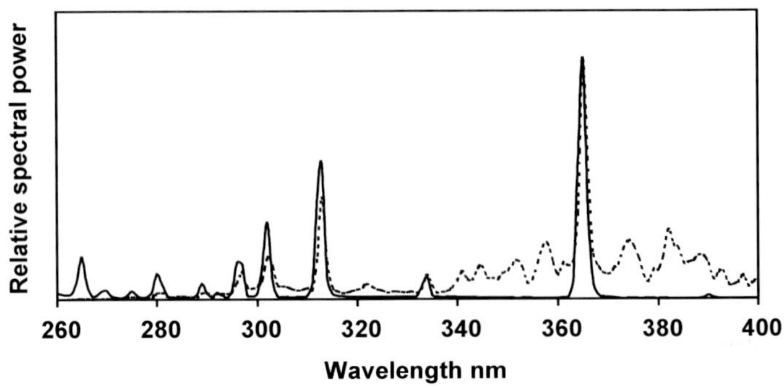

Figure 3. Spectral power distribution of a medium-pressure mercury arc lamp (solid line) and an iron iodide lamp (broken line).

The inclusion of metal halides in discharge lamps provides a significant gain in UV output compared with ordinary mercury arc lamps; a metal halide lamp of 1800 W emits about 22% of its radiation at wavelengths less than 400 nm. The spectral power distribution depends, of course, on the particular metal halide. Lamps that contain halides of gallium, iron and dysprosium, as well as mercury, have been employed as UV sources for irradiation of patients with skin disease. The spectral emission of an iron iodide lamp is compared in Figure 3 with a medium-pressure mercury lamp.

The use of long-wave UV radiation (UVA1; 340–400 nm) in the treatment of various skin disorders is a relatively new modality [5]. There are various UVA1 sources that have been employed, ranging from arrays of fluorescent lamps with irradiances at the patients' surface of around 20 mW cm^{-2}, permitting exposures of 10–30 J cm^{-2}, in a single exposure to high-output metal halide lamps (irradiance at the patients' surface of around 50 mW cm^{-2}), allowing the administration of single doses of 100 J cm^{-2} or higher.

6.4.6 Xenon arc lamps

In the xenon arc the radiation is emitted primarily as a continuum, unlike the mercury arc, which essentially emits a line spectrum. The production of the continuum is optimum under conditions of high specific power, high current density and high internal pressure, leading to compact, bright sources. Because of high operating temperatures the lamp envelope is normally constructed from fused silica. Xenon lamps consist of an arc burning between solid tungsten electrodes in a pressure of pure xenon and may be designed to operate from AC or DC. Cold-filling pressures up to 12 atm are commonly used and as a result there exists a potential hazard from explosive failure of the lamp, although in practice lamp explosions are rare. The lamps can be of the compact form, when the bulb is nearly spherical, or the linear form, which utilises a cylindrical lamp envelope. Because xenon lamps contain a permanent gas filling, the full radiation output is available immediately after switching on; there is no run-up period as in arc lamps containing mercury that has to vaporise.

6.5. Simulated sources of sunlight

6.5.1 Xenon arc lamps

For many experimental studies in photobiology it is simply not practicable to use natural sunlight and so artificial sources of UV radiation designed to simulate the UV component of sunlight are employed. No such source will match exactly the spectral power distribution of sunlight and as the shorter UV wavelengths (less than around 340 nm) are generally more photobiologically active than longer UV wavelengths, the usual goal is to match as closely as possible the UVB (280–315 nm) and UVAII (315–340 nm) regions.

The classical so-called *solar simulator* consists of an optically filtered xenon arc lamp. This lamp has a smooth continuous spectrum in the UV region and various models of solar simulator are available with input power in the range 75 W to 6 kW and above, from companies that include Oriel Corporation, Solar Light Company, Spectral Energy Corporation and Schoeffel Optical [6]. Optical filters and dichroic mirrors are used to shape the spectrum. In most cases a 1 mm thick Schott type WG320 filter is used to control the short wavelength end of the spectrum. By varying the thickness of the filter from 1 to 1.5 or 2 mm, spectra are obtained that approximate varying solar altitudes. The simulator normally also incorporates an UV-transmitting, visible-light absorbing filter (e.g. Schott UG5 or UG11) or other filters or multiple dichroic mirrors to remove visible and infrared wavelengths. The spectrum of a solar simulator is compared with natural sunlight in Figure 4.

6.5.2 Fluorescent lamps

A drawback of arc lamp solar simulators is that the irradiation field is generally limited to less than around 15 × 15 cm, although it is possible to achieve a uniform flux over a larger area, albeit with a reduction in irradiance. This may pose little

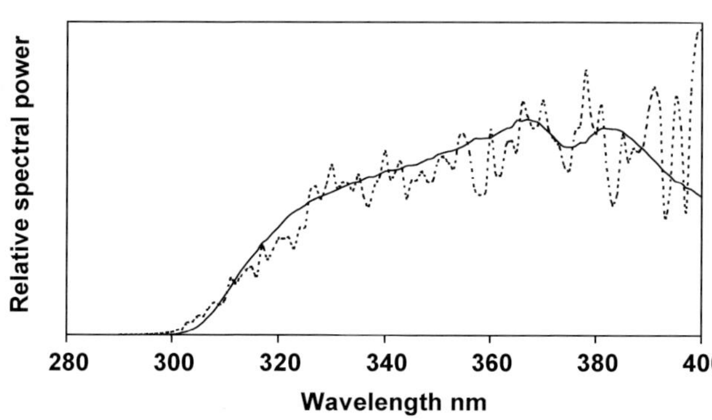

Figure 4. Spectral power distribution of clear sky, terrestrial UV radiation measured at around noon in summer at a latitude of 38° S (broken line) and a xenon arc filtered with a WG320 (2 mm thick) and UG5 (1 mm thick) filter (solid line).

problem if the object is to irradiate small areas of skin but for studies where large numbers of experimental animals or plants are to be irradiated the limited irradiation field is a real problem. Because of this, attention has turned to fluorescent lamps as sources of simulated UV [7]. One way to evaluate candidate lamps and decide which is the most appropriate approximation to sunlight is to calculate the % relative cumulative erythemal effectiveness (%RCEE) for several wavebands and to compare these values with the %RCEE values of a "standard sun" [8]. The %RCEE for the spectral range 290 to λ_c is the erythemally-effective UV radiation within this waveband expressed as a percentage of the total erythemally-effective radiation from 290 to 400 nm. This is calculated mathematically as:

$$\%\,\mathrm{RCCE} = 100 \times \sum_{290}^{\lambda_c} E(\lambda)\varepsilon(\lambda)\Delta(\lambda) \Big/ \sum_{290}^{400} E(\lambda)\varepsilon(\lambda)\,\Delta(\lambda) \tag{3}$$

where $E(\lambda)$ is the relative spectral power distribution of the UV source and $\varepsilon(\lambda)$ is the effectiveness of radiation of wavelength λ (nm) in producing erythema in human skin [9]. Table 1 compares %RCEE values from a number of fluorescent lamps with lower and upper acceptance limits for a "standard sun" given by the European Cosmetic, Toiletry and Perfumery Association (COLIPA). The Arimed B lamp is perhaps the best choice as a source of simulated solar UV radiation from those given, although there is little to chose between this lamp and some of the others. The TL-12 (equivalent spectrum to the Westinghouse sunlamp), a mainstay of photobiological research for many years, is a poor surrogate for solar UV radiation [7].

Table 1. UVB & UVA components (%) and the percentage relative cumulative erythemal effectiveness (%RCEE) for the summer sun[a] and several fluorescent lamps[b]

	Lamp					
	Sun[a]	**A**	**B**	**C**	**D**	**E**
UVB (290–315 nm)	3.35	55.64	2.58	4.54	4.30	3.43
UVA (315–400 nm)	96.65	44.36	97.42	95.46	95.70	96.57
Lower & upper limits of the %RCEE according to COLIPA (5)						
<290 nm (<1.0%)	0.047	19.6	0.087	0.095	0.000	0.089
290–310 nm (46.0–67.0%)	62.3	77.6	51.4	60.7	42.8	53.4
290–320 nm (80.0–91.0%)	86.4	80.2	79.2	86.7	80.9	81.9
290–330 nm (86.5–95.0%)	91.7	80.4	86.5	92.4	88.8	89.0
290–340 nm (90.5–97.0%)	94.0	80.4	91.0	95.1	93.0	92.8
290–350 nm (93.5–99.0%)	95.8	80.4	94.5	97.1	96.4	95.9

[a]Measured in Melbourne (38° S) at solar noon on 17 January 1990. Measurements were made at the Australian Radiation Laboratory with a Spex 1680B double monochromator with a resolution of 1 nm.

[b]Lamp A, TL-12 ("fluorescent sunlamp"); Philips Lighting, The Netherlands; Lamp B, Bellarium S; Wolff System, Germany; Lamp C, Arimed B; Cosmedico, Germany; Lamp D, CLEO Natural; Philips Lighting, The Netherlands; Lamp E, UVA-340; Q-Panel Lab Products, Cleveland OH, USA.

References

1. R. Phillips (1983). *Sources and Applications of Ultraviolet Radiation*. Academic Press, London.
2. B.L. Diffey (1982). *Ultraviolet Radiation in Medicine*. IoP Publishing, Bristol.
3. A.F. McKinlay (1992). Artificial sources of UVA radiation: uses and emission characteristics. In: F. Urbach (Ed), *Biological Responses to Ultraviolet A Radiation* (pp. 19–38). Valdenmar Publishing Company, Overland Park, KS.
4. F. Ebbesen, G. Agati, R. Pratesi (in press). Phototherapy with turquoise versus blue light. *Arch. Dis. Child Fetal Neonatal Ed.*, **88**, F430–431.
5. R. Dawe (2003). Ultraviolet-A1 phototherapy. *Br. J. Dermatol.* **148**, 626–637.
6. F. Wilkinson (1998). Solar simulators for sunscreen testing. In: R. Matthes, D. Sliney (Eds), *Measurements of Optical Radiation Hazards* (pp. 653–684). International Commission on Non-Ionizing Radiation Protection.
7. D.B. Brown, A.E. Peritz, D.L. Mitchell, S. Chiarello, J. Uitto, F.P. Gasparro (2000). Common fluorescent sunlamps are an inappropriate substitute for sunlight. *Photochem. Photobiol.* **72**, 340–344.
8. COLIPA *Sun Protection Factor Method* (1994). Published by the European Cosmetic Toiletry, and Perfumery Association (COLIPA), Brussels.
9. CIE Standard (1998). *Erythema Reference Action Spectrum and Standard Erythema Dose*. CIE S 007/E-1998. Commission Internationale de l'Éclairage, Vienna.

Chapter 7

Solid state lamps

Roland Diehl

Table of contents

Abstract

Light-emitting diodes (LEDs) as solid state lamps are going to make dramatic inroads into mainstream lighting applications such as general illumination, traffic lights, automotive lighting and large outdoor information screens. LEDs are small, efficient and long-living photonic devices that exploit the principle of injection luminescence to generate light in an active zone where p- and n-type semiconductor material sections meet to form a p-n junction. Based on the III–V compound semiconductor alloys AlGaAs, AlGaInP and AlGaInN, their emission wavelengths cover the visible as well as the near-IR and UV spectrum with high-brightness performance. Advanced high-brightness LEDs are designed as double heterostructures with the active zone sandwiched between carrier confining regions, which allows for efficient carrier injection and high radiative recombination rates. The crucial process step for the realisation of double heterostructures is metal-organic vapour phase epitaxy (MOVPE) which is the method of choice for cost-effective and reliable large-scale production of high-brightness LEDs. Meanwhile, these solid state lamps have also found application potential in the life sciences such as photomedicine and photobiology.

7.1 Introduction

At the beginning of this millennium tungsten incandescent lamps were still the number one provider of residential lighting, although they are challenged by compact fluorescent lamps due to their higher efficiency. Lighting in work and industrial environments is dominated by fluorescent tubes, and street lighting often employs sodium lamps. An alternative to all is solid state lighting based on photonic devices called light-emitting diodes (LEDs), which have already conquered niches such as signalling lamps and advertising displays. The science and commercialisation of LEDs have advanced so rapidly towards dramatic improvements in brightness, power and cost that these solid state lamps are now penetrating mainstream applications such as general illumination, traffic lights, automotive lighting and large outdoor information screens. LEDs deliver high brightness, small space, consume little energy from a direct voltage source, generate little waste heat, are reliable and shine with a longevity of up to 100,000 h. These inherent advantages have provoked tough global competition for market share in the leading industrialized countries with further rapid progress to be expected in terms of even higher performance and lower cost.

LEDs exhibit an efficient kind of electroluminescence caused by the injection of an electrical current into semiconductor materials. This injection luminescence was discovered in 1907 in silicon carbide, SiC, when carriers (electrons, holes) were injected from metal contacts and a yellowish light was observed [1]. The first LED appeared in 1962 when strong red electroluminescence radiation from a GaAs-based device structure was demonstrated [2]. Since then injection electroluminescence has received considerable attention that cumulated in the most recent developments of high-brightness and high-power LEDs (for details see reference books, e.g. [3–5]).

7.2 Fundamentals and characteristics of solid state lamps

Solid state lamps are made from semiconductor materials. Such lamps are diodes that emit light if properly biased. Basically, light-emitting diodes are composed of two sections, an n-type one having a surplus of electrons and a p-type one having a surplus of holes. The boundary between both is called a p-n junction (Figure 1). Excess charge carriers such as electrons and holes that can diffuse through the semiconductor crystal lattice are generated by doping. As an example, the semiconductor material gallium arsenide, GaAs – a member of the family of III–V compound semiconductors composed of a group III and a group V element of the periodic system – has a crystal lattice in which each Ga atom has four As atoms surrounding it in the form of a tetrahedron, and vice versa (zinc blende (ZnS)-type lattice). The Ga atom extends three valence electrons to three of its As neighbours, the As atom extends four of its five valence electrons to its four Ga neighbours, the fifth one "being lent" to the Ga atom to allow fourfold coordination in space.

High-purity GaAs is an intrinsic semiconductor with electrons tightly bound to the atoms. Its conductivity at room temperature is very low ($\sim 10^8\,\Omega\,\mathrm{cm}$). If a Ga atom in the GaAs crystal lattice is substituted by, e.g., a Si atom that has four valence electrons, one electron is obsolete for chemical bonding and is given off to the lattice where it can move freely. As the electron is a negatively charged carrier, the GaAs is doped n-type. Conversely, if a Ga atom is replaced, e.g., by a Zn atom that has two valence electrons, one electron is missing in one of the Zn – As bonds. At the position of the electron a hole is localized that behaves like a positively charged carrier and the GaAs is doped p-type. In thermal motion electrons can jump to the places of holes, leaving holes behind. Through this mechanism holes can also move through the lattice. Doping levels in III–V semiconductors are usually of the order $10^{18}\,\mathrm{cm}^{-3}$. If forwardly biased, i.e. with the n-type section of the LED connected to the cathode of a battery and the p-type section connected to the anode, electrons and holes flow from the contacts into the device both being pushed by the voltage towards the p-n junction where they recombine, annihilating each other (Figure 2). The area where recombination takes place is called the active zone.

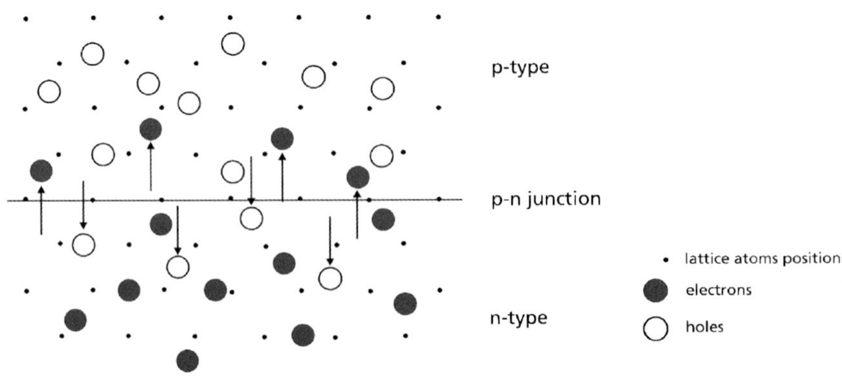

Figure 1. Principle of a p-n junction in a doped semiconductor.

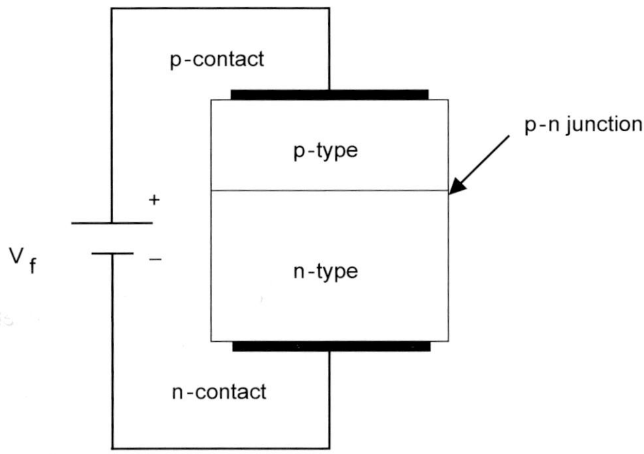

Figure 2. Schematic of a light emitting diode with forward bias voltage applied.

Electrons in the p-type region and holes in the n-type region are termed minority carriers. Diffusion of the minority carriers through the p-n junction is due to the forward bias voltage. The increase in minority carrier density is called injection. Injected electrons and holes can recombine both radiatively and nonradiatively. Radiative recombination yields photons that are emitted to the outside of the semiconductor. The light emission is spontaneous and its phase random (incoherent radiation). The photon energy (wavelength, colour) is equal to the distance of the energy levels occupied by the electron and the hole, respectively, and is specific to the semiconductor material from which the LED is made. Using the III–V semiconductors not only the entire visible spectrum can be covered but also wavelengths in the infrared (IR) and the ultraviolet (UV). Nonradiative recombination gives rise to vibration of the semiconductor lattice thus not creating photons but phonons and, finally, heat.

Radiative and nonradiative recombination of electrons and holes compete, by which the internal quantum efficiency n_{int} of the LED is determined:

$$n_{\text{int}} = n_{\text{rre}}/n_{\text{are}} \tag{1}$$

where n_{rre} is number of radiative recombination events and n_{are} the number of all recombination events. In modern III–V semiconductor LEDs n_{int} is usually greater than 90%.

Another important characteristic of LED performance is the external quantum efficiency, n_{ext}, which is essentially the product of n_{int}, and n_{lee}, the light-extraction efficiency:

$$n_{\text{ext}} = n_{\text{int}} \times n_{\text{lee}} \tag{2}$$

Light-extraction efficiency is the fraction of generated photons that escape from the device; n_{ext} greatly depends on the specific design of the LED and may vary between a few percent to more than 50%.

An LED is typically a tiny cube with edge dimensions of the order of a few 100 μ. Its internal layout can be rather complicated depending on whether it is a homostructure or a heterostructure. A homostructure has a p-n junction formed by p- and n-type regions of the same material whereas a heterostructure has a p-n junction where p- and n-type regions of different materials are in contact. The simple design of a homostructure is no longer used in state-of-the-art high-brightness (HB) LEDs because of considerable shortcomings due the partial re-absorption of photons generated in the active zone by the p- and n-regions. This reduces the efficiency of light extraction. Furthermore, injection of electrons is preferred in the p-type section due to the higher mobility of electrons compared with holes which reduces the internal quantum efficiency.

These drawbacks can be overcome by changing the material composition around the active zone. Nearly all state-of-the-art HB LEDs are composed of double heterostructures (DH). In such a structure, a thin active layer of a high electron (or hole) affinity material is sandwiched between p- and n-conducting regions of material layers of lower electron (or hole) affinity (confinement regions). This allows for very efficient bi-directional injection into the active zone where electrons and holes recombine with a high radiative recombination rate. The narrow active zone and a potential barrier on either side greatly enhance carrier confinement and hence the internal quantum efficiency. Confinement regions are also called cladding regions.

LED heterostructures which might be composed of sophisticated semiconducting layer sequences require materials with good matching of their respective lattice constants. Lattice mismatch causes lattice defects to be generated in the regions around the heterointerfaces, which give rise to nonradiative recombination. Hence, lattice matching is an important prerequisite for highly efficient and reliable heterostructure LEDs, and this limits the choice of materials systems suitable to manufacture high-performance solid state lamps.

7.3 Material systems for LEDs

Group III–V compound semiconductors are the materials of choice for LED production. Ternary and quaternary alloys containing Al, Ga, In as the group III members and N, P, As as the group V members cover the visible spectrum. At present, the LED industry is based on the three systems AlGaAs, AlGaInP and AlGaInN. The ternary alloy AlGaAs should be written as $Al_xGa_{1-x}As$ to indicate a solid solution between AlAs and GaAs having an As sub-lattice with Al and Ga distributed randomly over the metal sub-lattice. The same holds for the quaternary alloys with random distribution of three kinds of metal atoms. Usually, the sub-indices are omitted for simplicity. Mature techniques to deposit layers of excellent crystalline quality within wide ranges of compositions and both n-type and p-type doping as well as lattice robustness have allowed the fabrication of heterostructures with highly efficient injection and radiative recombination.

Alternatively, group II–VI compound semiconductors such as ZnS, ZnSe, CdO, etc. are also candidates to realize visible LEDs. Ease of defect formation due to low lattice rigidity has rendered this material family less suitable for LED fabrication [6]. Organic molecules and polymers are also used as materials for LEDs. These organic LEDs (OLEDs) have great application potential with the focus presently on large and cost-efficient emissive displays as possible substitutes for liquid crystal displays (see, e.g., Ref. [7]). Although OLEDs have potential as solid state lamps, they are not yet considered a competition to the versatile application spectrum of HB LEDs.

Returning to the III–V semiconductors, the ternary alloy $Al_xGa_{1-x}As$ has a lattice constant as a function of x which is expressed by

$$a(\text{nm}) = 0.56533 + 0.00078x \quad [8] \tag{3}$$

with 0.56533 nm being the lattice constant of GaAs. This means excellent lattice matching of AlGaAs layers of any composition or layer sequence to GaAs, a substrate material available from large-scale production in the form of round wafers with diameters of up to 150 mm. Epitaxially grown AlGaAs/GaAs heterostructures were the first red LEDs with excellent brightness and lifetime. With AlGaAs the wavelength range from 780 to about 650 nm is most effectively covered. Pure GaAs LEDs shine in the near-infrared around 850 nm and are used, e.g., as TV teleswitches.

The energy levels of electrons and holes in AlGaInP depend on the composition of this quaternary alloy. This material system is written as $(Al_xGa_{1-x})_yIn_{1-y}P$. For $y \sim 0.5$ the alloy shows a nearly perfect lattice match to GaAs [9], which is used as substrate for the relevant heterostructures. Through the parameter x the wavelength of the emitted light can be tuned over a spectrum, ranging from 660 to about 550 nm, which covers the colours red, orange, amber, yellow and green.

With the solution of problems due to p-type doping and crystal growth in the last decade of the 20th century, the AlGaInN alloy system saw a breakthrough as the material base for HB LEDs in the yellow, green, blue and near-UV spectral range [10]. Despite the non-availability of a lattice-matched substrate, the progress in group III nitride technology was remarkable. The substrate material of choice still is sapphire (Al_2O_3), which has a lattice mismatch of 16% compared with GaN. SiC, with a lattice mismatch of 3.5%, is also used. Both are transparent to the emitted light which allows easy light extraction from the LED through the substrate. Introducing a GaN buffer layer grown at low temperatures between the substrate and the heterostructure reduced the defect density at the interfaces giving rise to high internal quantum efficiency [11]. This should be further improved when GaN substrates are employed, which have become commercially available only relatively recently [12]. Figure 3 shows the typical electroluminescence spectra of a few real HB LEDs.

Modern HB LEDs are fabricated by depositing the desired heterostructures as a stack of single-crystalline III–V semiconductor layers on suitable substrates. Subsequently, the coated wafers are diced into numerous small cubes (photonic "chips"), contacted with appropriate metal alloys and mounted into various types of packages such as, e.g., into a reflecting cup under a transparent plastic dome (Figure 4).

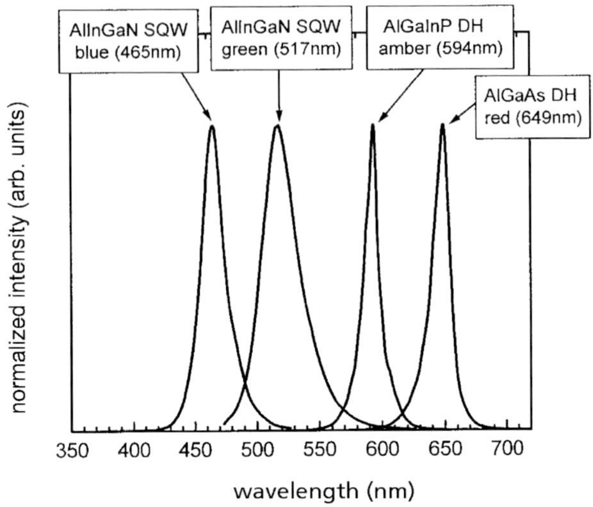

Figure 3. Typical electroluminescence spectra of HB LEDs for various alloys and their respective peak wavelengths (forward current 20 mA, intensity normalized for clarity; after [5]).

The deposition of single-crystalline layers on a lattice-matched substrate is called epitaxy, the layers are epilayers, the coated wafers are epiwafers. The method of choice for the fabrication of HB LEDs is MOVPE [13]. In a reaction chamber heated to between 500 and 1000°C, depending on the material system to be deposited, gaseous reactants such as, e.g., trimethylgallium TMGa ($Ga(CH_3)_3$) and arsine (AsH_3) are flushed by a carrier gas such as hydrogen (H_2) where they react at the surface of the heated substrate to form a GaAs layer (Figure 5).

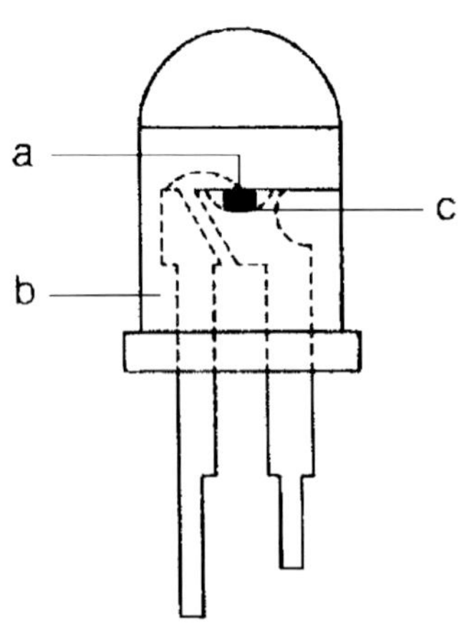

a) LED chip
b) epoxy resin
c) reflector cup

Figure 4. Section through a conventional LED package with the chip mounted into a reflector cup and contacted to the DC leads.

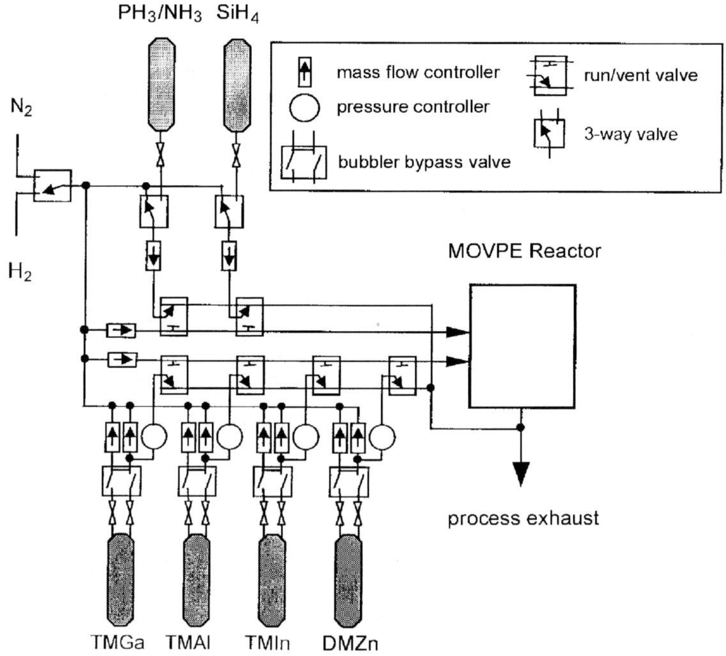

Figure 5. Schematic of the constituent and doping gas delivery system for a MOVPE reactor (after [14]).

Other group III precursors are the alkyls aluminium- or indium-trimethyl or -triethyl, other group V precursors are the hydrides PH_3 and NH_3. Dopant precursors are, e.g., silane (SiH_4) and dimethylzinc DMZn ($Zn(CH_3)_2$). MOVPE allows tight control of layer thickness, composition and doping profile. Several wafers can be coated simultaneously, which renders MOVPE a very economic technique for large-scale LED production with high yields. A disadvantage is that nearly all the precursors used are hazardous gases (toxic, inflammable), which requires safety and environmental issues to be properly addressed.

7.4 Advanced solid state lamps

Advanced solid state lamps are LEDs designed for high-brightness and high-power performance. A prerequisite for both are high values for injection and internal quantum efficiencies. Heterostructures with the active region sandwiched between two cladding (or confining) layers that act as potential barriers minimize carrier leakage at high injection currents, thus allowing very high internal quantum efficiencies. The other important parameter that determines LED performance is the extraction efficiency, and special device designs have been developed to allow a maximum of light to escape from the device. An important figure of merit used to evaluate the performance of visible LEDs is the luminous efficiency, expressed in

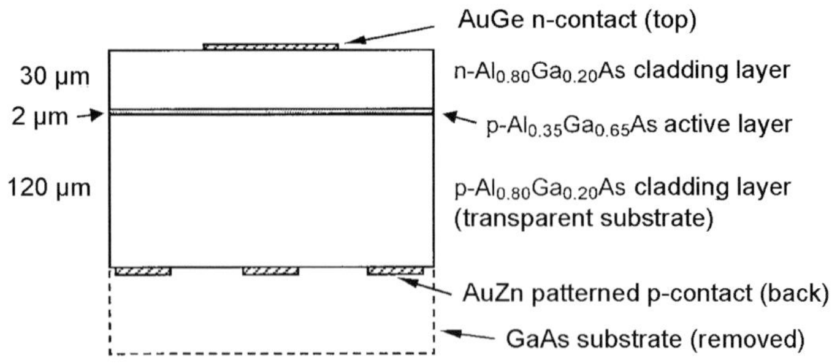

Figure 6. Typical chip structures designed for high brightness: AlGaAs double heterostructure with transparent substrate (after [15]).

lumens per watt ($lm\,W^{-1}$). This is primarily determined by the external efficiency and the operating voltage.

A HB LED design consists of the heterostructure sandwich with top and bottom confining layers followed by thick layers called window layers to improve the extraction of light directed towards the sides of the chip. In addition, the optimum for light extraction is a non-absorbing substrate. Figures 6 and 7 shows two typical designs for commercial HB AlGaAs and AlGaInP LEDs. The AlGaAs DH device contains an active zone with 35% Al, yielding a peak wavelength at 660 nm, between two thick cladding layers with 80% Al that also act as window layers (Figure 6) [15]. With the absorbing GaAs substrate removed, an external efficiency around 20% and a luminous efficiency of about 10–15 $lm\,W^{-1}$ have been achieved [16]. Emitting at a wavelength around 600 nm (amber) the AlGaInP device has a 0.5 μm thick active layer composed of $(Al_{0.2}Ga_{0.8})_{0.5}In_{0.5}P$ that is unintentionally or slightly doped, sandwiched between n- and p-type cladding layers of composition $(Al_{0.7}Ga_{0.3})_{0.5}In_{0.5}P$ that are about 1 μm thick, both in contact with thick window layers of transparent GaP (Figure 7). The lower window layer has been realized by removing the absorbing GaAs substrate by selective etching to be replaced by a GaP substrate by solid-state waver bonding [17]. AlGaInP-based solid state lamps show the highest luminous efficiencies so far demonstrated with LEDs in any material system. For conventional lamp designs more than 70 $lm\,W^{-1}$ have been

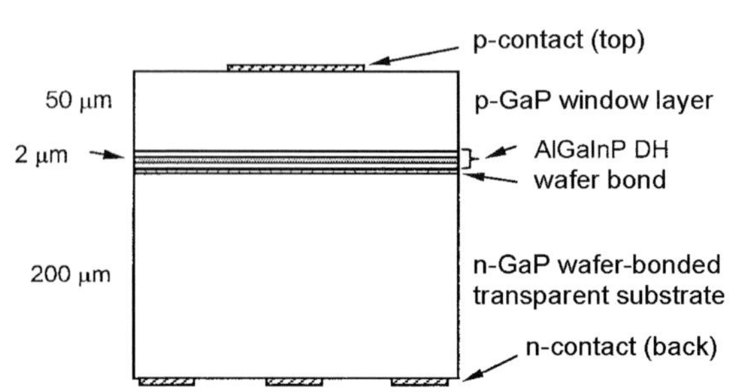

Figure 7. AlGaInP double heterostructure with a wafer-bonded transparent GaP substrate (after [17]).

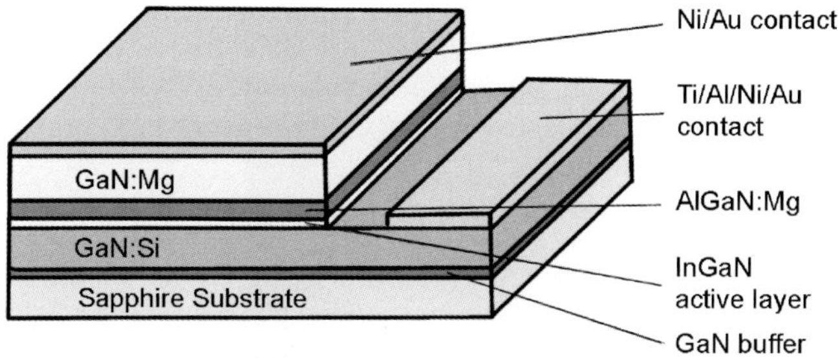

Figure 8. Schematic drawing of a GaN/InGaN/AlGaN double heterostructure LED (after [19]).

achieved [18]. This was surpassed by special lamps designed for high power, to increase the luminous flux (measured in lumens) per LED, by making the chip area larger. This requires a high-power package to dissipate the generated waste heat. The highest luminous efficiency of $102\,\mathrm{lm\,W^{-1}}$ was achieved for 611 nm lamps [18].

The AlGaInN material system is ideal for high-brightness/high-power performance with its a priori transparent substrate that allows for very high extraction efficiencies. A schematic drawing of a mesa diode prepared from a GaN/InGaN/AlGaN layer stack is shown in Figure 8. The layer structure of a blue-emitting DH LED consists of a 3 μm thick n-GaN layer followed by a 10 nm thick $In_{0.1}Ga_{0.9}N$, a 15 nm thick $Al_{0.12}Ga_{0.88}N$ and a 0.5 μm thick p-GaN layer. This or similar designs show luminous efficiencies from 10 to $60\,\mathrm{lm\,W^{-1}}$ from blue to green. High-power designs allowing higher current, a larger light-emitting surface and proportionally higher output are being developed. Mounted p-side-down on a heat sink in a special package, blue-green power LED chips deliver up to 1W of light output power, generating 25 lm [20].

AlGaInN-based LEDs emitting in the blue, violet and near-UV form the basis of luminescence converting LEDs (LucoLEDs) applying radiation down-conversion in phosphors. In a Luco-LED (Figure 9) the LED chip mounted in a reflector cup is covered with a luminescent dye, the rest is standard LED technology [19]. The dye

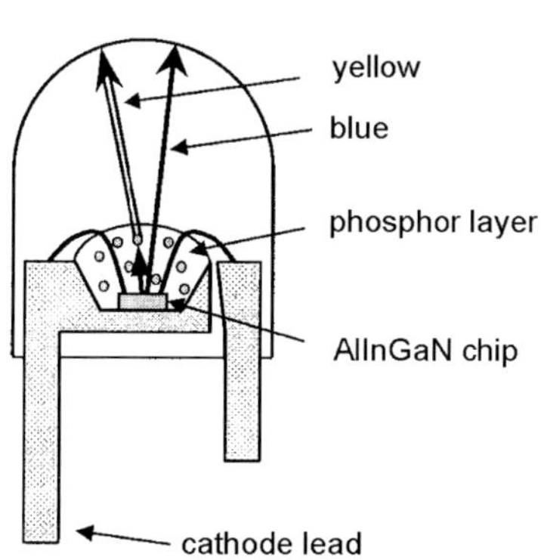

Figure 9. Schematic design of an AlGaInN-based luminescence-conversion white LED (LucoLED) (after [19]).

Figure 10. Photograph of a white-emitting luminescence-conversion LED based on an inorganic phosphor (after [19]).

absorbs the blue light and emits at longer wavelengths. Various mixed colours have been demonstrated by additive colour mixing of the primary LED-blue with a secondary luminescence light. White-emitting diodes can also be realized when adding a green- and red-emitting dye simultaneously [21]. For yellow light emission under blue photo-excitation, only one converter dye species is necessary for white light generation, since the complementary colours blue and yellow result in white light after appropriate additive mixing (Figure 10). Luminous converters for white can be organic dyes or Ce-doped inorganic oxides [10]. LucoLEDs with a luminous efficiency of $15\,\mathrm{lm\,W^{-1}}$ exceeding that of incandescent lamps, have become available. Further developments will make $50\,\mathrm{lm\,W^{-1}}$ feasible, with an upper theoretical limit of $270\,\mathrm{lmW^{-1}}$ [5,22]. This holds great promise for general solid state lighting application when the LED luminous efficiency approaches $100\,\mathrm{lm\,W^{-1}}$ (for comparison: fluorescent tube, $85\,\mathrm{lm\,W^{-1}}$) and when the system costs per metre come down further.

7.5 Application perspectives of LEDs in photomedicine and photobiology

Apart from low-power signalling lamps of various colours, which have long been in use, numerous new fields of application have opened up to solid state lamps as they became available with high-brightness and high-power performance. Traffic lights, automotive interior lighting and taillights, even headlamps, safety signs, full-colour (video) displays, and illumination are among the most important [5]. Photonic technologies are finding uses in medicine and biology [23–25]. Minimal invasive surgery, ophthalmologic and dental treatments as well as hair and wrinkle removal using lasers are major applications. Conventional LEDs are found in various appliances where small reliable light sources are an advantage. New generations of HB LEDs offering considerable luminous flux enable new applications in photomedicine and photobiology [5]. Photodynamic therapy, phototherapy of neonatal jaundice and phototherapy of seasonal affective disorder (SAD) are opportunities for solid state lighting in medicine. Photostimulation for wound healing [26] is another example where the cool light beam of LEDs offer advantages over halogen lamps. Photodynamic therapy is applied to treat some types of cancer. A photosensitising agent that is a phototoxic substance is fed into the bloodstream, which carries it to the cancer tissue. Most agents are activated by red light under the irradiation of which they produce active oxygen that preferentially kills the tumour cells [27]. Another application is photodynamic inactivation of viruses [28]. Red-emitting diode lasers as well as arrays of red AlGaAs LEDs have been successfully applied for these therapeutic purposes [29].

Statistically, more than 50% of newborn babies suffer from neonatal jaundice, which is recognized by a yellow colouration of the skin and the white of the eye. Jaundice originates from a high concentration of bilirubin in the bloodstream. Neonatal jaundice can be controlled by phototherapy applying blue light around a wavelength of 450 nm. HB AlGaInN LEDs have been proposed as light sources for the phototherapy of hyperbilirubinemic newborns to replace fluorescent tubes which generate UV radiation and ozone [30].

SAD is a kind of depression experienced by some people during the winter. This mood disorder is successfully treated when the patients are flooded with bright light [31]. Blue, green, yellow and white high-brightness LED arrays seem to have the best therapeutic potential for SAD phototherapy [32].

Also discussed are dental hardening systems using blue and violet HB LEDs to replace halogen lamps [33]. This will greatly reduce the curing times of light-activated dental polymer composites while consuming only a fraction of the power. Another advantage is a lightweight handheld lamp design.

The green pigment chlorophyll widely absorbs light of the visible spectrum, converting the photon energy into photosynthetic reactions that produce organic compounds and oxygen from carbon dioxide, water and minerals. LED-based photosynthetic systems with a spectrum close to the absorption spectrum of chlorophyll are under development to improve photosynthetic efficiency [34]. Mixtures of red and blue HB LED irradiation sources yielded the best results in a model system for lettuce growth [35].

Arrays of HB AlGaAs LEDs have been utilized to realize a photobioreactor to exploit efficient algae production [36]. Algae cultures are suitable, e.g., for wastewater treatment, oxygen regeneration and agricultural applications. Another LED-based photobioreactor has been devised for bacteria cultivation [37]. LED arrays are ideally suited for such experiments as they can easily be modulated in time, brightness and wavelength, thus taking care of the peculiarities of photosynthetic reaction cycles.

References

1. H.J. Round (1997). A note on carborundum. *Electrical World* **49**, 309.
2. N. Holonyak Jr., S.F. Bevacqua (1962). Coherent (visible) light emission from $Ga(As_{1-x}P_x)$ junctions, *Appl. Phys. Lett.* **75**, 82–83.

3. G.B. Stringfellow, M.G. Craford (Eds) (1997). R.K. Willardson, E.R. Weber (series eds), High-brightness light emitting diodes. *Semiconductors and Semimetals* (Volume 48, 469 pp). Academic Press, New York.
4. G. Müller (Ed) (2000). *Electroluminescence I, Semiconductors and Semimetals*, R.K. Willardson, E.R. Weber (series Eds), (Volume 64, 331 pp.) Academic Press, New York.
5. A. Zukauskas, M.S. Shur, R. Gaska (2002). *Introduction to Solid-state Lighting* (207 pp.). John Wiley & Sons, Inc., New York.
6. N. Nakayama, S. Itoh, A. Ishibashi, Y. Mori (1996). High-efficiency InCdSe/ZnMgSSe green and blue light-emitting diodes. *Proc. SPIE* **2693**, 36–42.
7. Y. Sato (2000). Organic LED system considerations. In: R.K. Willardson, E.R. Weber (series Eds), *Electroluminescence I, Semiconductors and Semimetals* (Volume 64, pp. 209–254). Academic Press, New York.
8. M. Levinshtein, S. Rumyantsev, M. Shur (1999). *Handbook Series on Semiconductor Parameters*, Volume 2: *Ternary and Quaternary III-V Compounds* (205 pp.). World Scientific, Singapore.
9. H.C. Casey, Jr., M.B. Panish (1978). *Heterostruture Lasers*, Part B: *Materials and Operating Characteristics* (330 pp.). Academic Press, New York.
10. S. Nakamura, G. Fasol (1997). *The Blue Laser Diode: GaN Based Light Emitters and Lasers* (343 pp.). Springer, Berlin.
11. S. Nakamura (1991). GaN growth using GaN buffer layer. *Jpn. J. Appl. Phys.* **30**, L1705–L1707.
12. J. Newey (2000). Perfect substrate within reach for wide-bandgap materials. *Compound Semicond.* **8**(6), 45–48.
13. H.M. Manasevit (1968). Single-crystal gallium arsenide on insulating substrates. *Appl. Phys. Lett.* **12**, 156–159.
14. C.H. Chen, S.A. Stockman, M.J. Peansky, C.P. Kno (1997). OMVPE growth of AlInGaP for high-efficiency visible light-emitting diodes. In: G.B. Stringfellow, M.G. Craford (Eds), *High-Brightness Light Emitting Diodes*, R.K. Willardson, E.R. Weber (series Eds), *Semiconductors and Semimetals* (Volume 48, pp. 97–148). Academic Press, New York.
15. F.M. Steranka (1997). AlGaAs red light emitting diodes. In: G.B. Stringfellow, M.G. Craford (Eds), *High-Brightness Light Emitting Diodes*, R.K. Willardson, E.R. Weber (series Eds), *Semiconductors and Semimetals* (Volume 48, pp. 65–96). Academic Press, New York.
16. M.G. Craford (1997). Overview of device issues in high-brightness light emitting diodes. In: G.B. Stringfellow, M.G. Craford (Eds), *High-Brightness Light Emitting Diodes*, R.K. Willardson, E.R. Weber (series Eds), *Semiconductors and Semimetals* (Volume 48, pp. 47–63). Academic Press, New York.
17. F.A. Kish, R.M. Fletcher (1997). AlGaInP light emitting diodes. In: G.B. Stringfellow, M.G. Craford (Eds), *High-Brightness Light Emitting Diodes*, R.K. Willardson, E.R. Weber (series Eds), *Semiconductors and Semimetals* (Volume 48, pp. 149–226). Academic Press, New York.
18. M. Holcomb, P. Grillot, G. Höfler, M. Kramer, S. Stockman (2001). AlGaInP LEDs break performance barriers. *Compound Semicond.* **7**(3), 59–64.
19. P. Schlotter, J. Baur, Ch. Hielscher, M. Kunzer, H. Obloh, R. Schmidt, J. Schneider (1999). Fabrication and characterization of GaN/InGaN/AlGaN double heterostructure LEDs and their application in luminescence conversion LEDs (LucoLEDs), *Mat. Sci. Eng.* **B59**, 390–394.
20. D. Silkwood (2003). Application surge for high-power LEDs. *Photonics Spectra* (Jan. 2003), 92–93.

21. P. Schlotter, R. Schmidt, J. Schneider (1997). Luminescence conversion of blue light emitting diodes. *Appl. Physics A* **64**, 417–418.
22. R. Müller-Mach, G.O. Müller (2000). White light emitting diodes for illumination. *Proc. SPIE* **3928**, 30–41.
23. R. Pratesi (1984). Diode lasers in photomedicine. *IEEE J. Quant. Electron.* **QE-20**, 1433–1439.
24. R. Pratesi (1985). Potential use of incoherent and coherent light emitting diodes (CLEDs) in photomedicine. In: A.N. Chester, S. Martellucci (Eds), *Laser Photobiology and Photomedicine* (pp. 293–308). Plenum Press, New York.
25. R. Pratesi (1991). Initial applications and potential of miniature lasers in medicine. In: R. Pratesi (Eds), *Optronic Techniques in Diagnostic and Therapeutic Medicine* (pp. 271–285). Plenum Press, New York.
26. A.P. Sommer, A.L.B. Pinheiro, A.R. Mester, R.P. Franke, H.T. Whelan (2001). Biostimulatory windows in low-intensity laser activation: Lasers, scanners and NASA's light-emitting diode array system. *J. Clin. Laser Med. Surg.* **19**, 29–33.
27. M.H. Schmidt, D.M. Bajic, K.W. Reichert II, T.S. Martin, G.A. Meyer, H.T. Whelan (1996). Light-emitting diodes as a light source for intraoperative photodynamic therapy. *Neurosurgery* **38**, 552–557.
28. F. Ben-Hur, W.S. Chan, Z. Yim, M.M. Zuk, V. Dayal, N. Roth, E. Heldman, A. Lazo, C.R. Valeri, B. Horowitz (1999). Photochemical decontamination of red blood cell concentrates with the silicon phthalocyanine PC 4 and red light. *Dev. Biol. Stand* **102**, 149–155.
29. A. Colosanti, A. Kisslinger, D. Kusch, R. Liuzzi, M. Mastrocinque, F.P. Mintforts, M. Quarto, P. Riccio, G. Roberti, F. Villani (1957). *In-vitro* photo-activation of newly synthesized chlorin derivatives with red-light-emitting diodes. *J. Photochem. Photobiol.* **338**, 54–60.
30. H.J. Vreman, R.J. Wong, D.K. Stevenson, R.K. Route, S.D. Leader, M.M. Fejer, R. Gale, D.S. Seidman (1998). Light-emitting diodes: A novel light source for phototherapy. *Pediatr. Res.* **44**, 804–809.
31. N.E. Rosenthal, D.A. Sack, R.G. Skwerer, F.M. Jacobsen, T.A. Wehr (1989). Phototherapy for seasonal affective disorder. In: N.E. Rosenthal, M.C. Blehar (Eds), *Seasonal Affective Disorders and Phototherapy* (pp. 273–294). Guildford Press, New York.
32. T.M.C. Lee, C.C.H. Chan, J.G. Paterson, H.C. Janzen, C.A. Blashko (1997). Spectral properties of phototherapy for seasonal affective disorder: A metaanalysis. *Acta Psychiatr. Scand.* **96**, 117–121.
33. R.W. Mills (1995). Blue light emitting diodes – another method of light curing? *Br. Dent. J.* **178**, 169.
34. J.R.J. Bula, R.C. Morrow, T.W. Tibbitts, D.J. Barta, R.W. Ignatius, T.S. Martin (1991). Light-emitting diodes as a radiant source for plants. *HortScience* **26**, 203–205.
35. M.E. Hoenecke, R.J. Bula, T.W. Tibbitts (1992). Importance of 'blue' photon levels for lettuce seedlings grown under red-light-emitting diodes. *HortScience* **27**, 427–430.
36. C.G. Lee, B. Palsson (1954). High-density algal bioreactors using light-emitting diodes. *Biotechnol. Bioeng.* **44**, 1161–1167.
37. H. Takano, T. Arai, M. Hirano, T. Matsunaga (1995). Effects of intensity and quality of light on phytocyanin production by a marine cyanobacterium. *Appl. Microbiol. Biotechnol.* **43**, 1014–1018.

Chapter 8

Fibre lasers

Terry A. King

Table of contents

Abstract

Rare-earth-doped glass fibre lasers provide many efficient wavelengths from the UV to the mid-IR. The broad bandwidths of the laser transitions provide for tuning of the laser and also for mode-locked operation to give picosecond-femtosecond pulses. Developments in double-clad side-pumped fibre structures have enabled efficient high power CW operation with pumping by semiconductor diode arrays. This has led to powers in excess of 100 W, and with potential for power levels greater than 1 kW. Fibre laser technology continues to develop at a rapid pace with new materials and techniques such that the output characteristics continue to be enhanced and to cover a broad range. More recent developments in improved fibre materials, microstructured photonic crystal fibres, oscillator-amplifier configurations and the use of nonlinear optical processes have expanded the performance and capabilities of fibre lasers. The range of wavelengths provided by fibre lasers, and other valuable characteristics such as high power and ultrashort pulse operation gives the fibre laser a wide range of application. Recent developments of high power fibre lasers with wavelengths in the 1 to 3 μm region, which match water and biotissue absorption, are opening up new applications in photobiology and medicine.

8.1 Introduction

Fibre lasers take the form of that of a basic optical fibre, a long sub-millimetre diameter glass filament consisting of a central core surrounded by a cladding of lower refractive index, but with the core (and sometimes the cladding) doped with ions to give optical gain. The rare-earth dopant is typically Er, Nd, Tm or Yb and the core is most commonly silica or a fluoride glass [1–3]. With $n_{\mathrm{core}} > n_{\mathrm{cladding}}$ total internal reflection occurs at the core-cladding interface, leading to waveguiding and trapping of pump and fibre laser light in the core. Attaching mirrors to each end of the fibre or by the natural Fresnel reflection at each facet of the fibre creates the optical resonator for the fibre laser.

Excitation of the fibre laser is optically, usually by end-pumping through one of the mirrors (longitudinal pumping) or by side-pumping in special optical arrangements. The beam quality of the fibre laser output radiation is determined by the refractive index profile of the core-cladding materials, in contrast to other conventional solid-state lasers. The confined waveguided light and long interaction lengths provided by the fibre structure gives the fibre laser valuable properties of high gain, low threshold, high efficiency and compactness. The broadening of the linewidth of the laser transition due to the host glass matrix enables some degree of tuning of the laser output. This also provides the capability for ultrashort femtosecond pulse generation by the technique of mode-locking. Fibre laser wavelengths are available from the UV to the mid-IR and with the capability of tuning in many systems.

The first fibre laser, in 1961 [4], was based on the trivalent neodymium ion, Nd^{3+}, in an alkali – alkaline earth silicate with a core of refractive index 1.54 surrounded by a cladding of refractive index 1.52. The core diameter was 0.3 mm diameter, which supported multiple radiation modes, and was side-pumped by a pulsed flashlamp. Later, in 1964, reduction of the core diameter led to single mode operation [5] and power amplification, in 1969, in which a pulsed He-Ne laser beam of 0.230 mW at 1.0621 μm was amplified to 0.6 W in a Nd^{3+}-doped fibre with a 15 μm core diameter [6]. Continuous wave (CW) oscillation in a Nd^{3+}-doped fibre was demonstrated in 1973 [7] and, significantly, the development of a doped monomode silica fibre pumped by a semiconductor diode laser [8] and, most importantly, the erbium-doped fibre amplifier (EDFA) [9,10]. These developments opened up the application of optical fibre communications in the telecommunications windows at 850, 1300 and 1550 nm [11,12]. The 1.55 μm EDFA has ideal properties for communications of high gain, low noise and low pump power.

Over the intervening time, several thousand different fibre glass compositions have been made to show optical gain as amplifiers or lasers. Developments in fibre glasses in addition to silica fibres, including fluoride, tellurite, heavy metal oxide and chalcogenide glasses, have enabled fibre laser transitions in both the visible and infra-red regions. Fluoride glasses in particular have low phonon energies, which confer a low nonradiative decay rate for excited states, so that the upper laser level lifetimes are increased and laser action on new transitions become possible. A wide range of fibre designs, including linear and ring structures, with a remarkable variety of output properties have now been produced, providing a great versatility in their applications. Fibre lasers also offer some potential advantages over semiconductor diode lasers in terms of a wide range of wavelengths not presently available from diode lasers, high brightness, good temperature stability, high beam optical quality and efficient coupling into delivery optical fibres.

The application of fibre lasers has extended beyond optical communications to many other areas [12–17]. New fibre configurations have enabled the earlier low power devices to be extended so that high power fibre lasers, with an output of over 1 kW, are now available for use in materials processing and other industrial applications. Excitation by efficient semiconductor laser diodes and diode-pumped solid-state lasers provide efficient and convenient pumping sources [18]. Q-switching and mode-locking techniques have been used to produce nanosecond – femtosecond and soliton laser pulses. The incorporation of Bragg gratings by optically writing into the optical fibre enables tunable and narrow bandwidth operation. These lasers have widespread application, in addition to optical communications, in metrology, sensing, fibre-optic gyroscopes, spectroscopy, imaging, biology, medicine, optical data storage, laser projection displays and laser range-finding.

There are significant photobiological and medical applications for infra-red lasers operating in the 1 to 3 μm region. These applications are partly based on the matching of the laser wavelengths with peaks in the absorption spectrum of water. The fundamental OH absorption peak near 3 μm and the weaker harmonic absorptions at 1.4 and 1.9 μm can be probed by wavelengths produced by Er, Tm and Ho fibre lasers [17,19–23,60]. The ability of these fibre lasers to operate at power levels greater than 1 W at the peaks in water and tissue absorptions opens up many biomedical applications.

8.2 Elements of fibre lasers

The ground-state electron configuration of the rare earth trivalent ions has an electron configuration of the xenon core together with outer 4f electrons. In a glass host the optically active electrons in the partially filled 4f shell are influenced by the crystal field from neighbouring ions. The degenerate levels are Stark split into a number of sub-levels, and electric dipole transitions between Stark levels in different multiplets become allowed. Broadening of the absorption and emission bands of the rare earth ions in the fibre glass occurs with a random variation of the degree of the Stark splitting. At room temperature the absorption and emission bands are relatively narrow, primarily broadened uniformly over the population of ions (homogeneously broadened) and are somewhat insensitive to the host medium. Individual transitions between Stark levels of different energy levels are rarely resolved except at very low temperatures.

Fibre laser action has been observed in rare-earth doped glass hosts, and also less usually in some crystalline fibre host media. The trivalent rare-earth dopants Nd^{3+}, Er^{3+}, Yb^{3+}, Tm^{3+}, Ho^{3+}, Sm^{3+} and Pr^{3+} are commonly doped into silica or fluoride host glasses. The wavelengths of fibre lasers in silica and ZBLAN fluoride glasses are shown in Figure 1 [3] and Tables 1 and 2. A greater range of laser wavelengths are seen in the ZBLAN glass than in silica since ZBLAN has a lower phonon energy ($650\,cm^{-1}$) and a transparency that extends further into the IR. ZBLAN glass is made up of the fluorides $ZrF_4 + BaF_2 + LaF_3 + AlF_3 + NaF$. The higher phonon energy in silica ($1160\,cm^{-1}$) increases the multiphonon nonradiative decay rate, decreases the upper level lifetimes and increases the laser threshold.

8.2.1 Fibre fabrication

Rare-earth-doped fibres can be fabricated by several different methods [1,2,24]. In vapour deposition, halides of Si, Ge, P and B are reacted in a hydrogen flame and the product soot collected on a mandrel, which is subsequently sintered. In an alternative method of modified chemical vapour deposition (MCVD) the gaseous chlorides are reacted inside a substrate tube which becomes part of the cladding. The rare earths may be introduced as a gaseous compound into the reaction. These methods make a pre-form from which, when one end is heated, may be drawn into a fibre. Dopant ions to raise the refractive index (Ge, P, Al, Ti) or to lower the refractive index (B, F) are added to the reacting vapour stream. A further method is the solution-doping technique in which the core produced by MCVD is deposited as a porous soot and then immersed in an aqueous solution containing the rare earth salts. When the water is evaporated the rare earth deposits in the pores, which is then compacted and the core collapses. Typical dopant concentrations are 50 to 1000 ppm.

In fibre lasers one notable advantage is that heat is efficiently dissipated from the core of the fibre due to the large surface area:volume ratio. Then thermal stress and thermal lensing are less evident than in bulk solid-state lasers. Heat dissipation in the fibre can also be controlled to some extent by control of the dopant concentration.

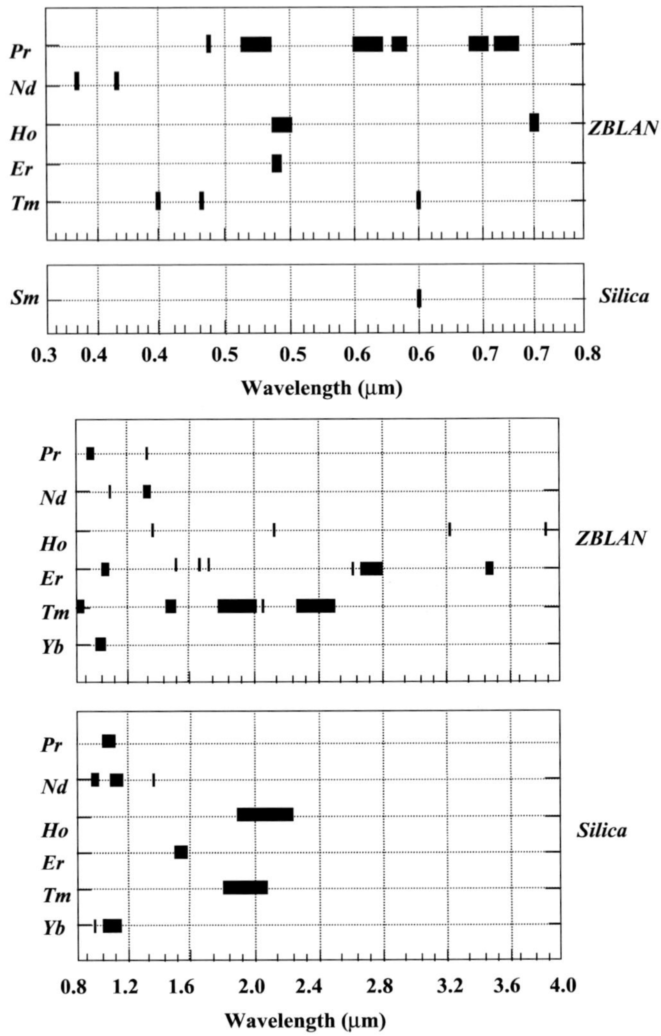

Figure 1. Wavelengths of rare earth-doped silica and ZBLAN fibre lasers [3].

Ion – ion interactions between the dopants lead to the mechanisms of energy transfer, cross-relaxation, up-conversion and concentration quenching. These influence the design and detailed operation of the fibre laser.

8.3 Basic principles

A common method to pump the fibre is optically through an end facet of the fibre. In this case the overall efficiency of the laser depends in part on the coupling of

Table 1. Operating wavelengths below 1 μm of fibre lasers and amplifiers. UC, up-conversion; ST, apparent self-terminating; 3L, three level; 4L, four level [2]

Operating range (nm)	Dopant ion	Transition	Type of host: Oxide	Type of host: Fluoride	Type of transition
≈455	Tm^{3+}	$^1D_2 \rightarrow {}^3F_4$		Yes	UC, ST
≈480	Tm^{3+}	$^1G_4 \rightarrow {}^3H_6$	Yes		UC, 3L
≈490	Pr^{3+}	$^3P_0 \rightarrow {}^3H_4$		Yes	UC, 3L
≈520	Pr^{3+}	$^3P_1 \rightarrow {}^3H_5$		Yes	UC, 4L
≈550	Ho^{3+}	$^5S_2, {}^3F_4 \rightarrow {}^5I_8$	No	Yes	UC, 3L
≈550	Er^{3+}	$^4S_{3/2} \rightarrow {}^4I_{15/2}$	No	Yes	UC, 3L
601–618	Pr^{3+}	$^3P_0 \rightarrow {}^3H_6$		Yes	UC, 4L
631–641	Pr^{3+}	$^3P_0 \rightarrow {}^3F_2$		Yes	UC, 4L
≈651	Sm^{3+}	$^4G_{5/2} \rightarrow {}^6H_{9/2}$	Yes		4L
707–725	Pr^{3+}	$^3P_0 \rightarrow {}^3F_4$		Yes	UC, 4L
≈753	Ho^{3+}	$^5S_2, {}^3F_4 \rightarrow {}^5I_7$	No	Yes	UC, ST?
803–825	Tm^{3+}	$^3H_4 \rightarrow {}^3H_6$	No	Yes	3L
≈850	Er^{3+}	$^4S_{3/2} \rightarrow {}^4I_{13/2}$	No	Yes	4L
880–886	Pr^{3+}	$^3P_1 \rightarrow {}^1G_4$		Yes	4L
902–916	Pr^{3+}	$^3P_1 \rightarrow {}^1G_4$		Yes	4L
900–950	Nd^{3+}	$^4F_{3/2} \rightarrow {}^4I_{9/2}$	Yes		3L
970–1040	Yb^{3+}	$^5F_{3/2} \rightarrow {}^5F_{7/2}$	Yes		3L
980–1000	Er^{3+}	$^4I_{11/2} \rightarrow {}^4I_{15/2}$	No	Yes	3L
1000–1150	Nd^{3+}	$^4F_{3/2} \rightarrow {}^4I_{11/2}$	Yes	Yes	4L

Table 2. Operating wavelengths above 1 μm of fibre lasers and amplifiers. UC, up-conversion; ST, apparent self-terminating; 3L, three level; 4L, four level [2]

Operating range (nm)	Dopant ion	Transition	Type of host: Oxide	Type of host: Fluoride	Type of transition
1060–1110	Pr^{3+}	$^1D_2 \rightarrow {}^3F_4$	Yes		4L
1260–1350	Pr^{3+}	$^1G_4 \rightarrow {}^3H_5$	No	Yes	4L
1320–1400	Nd^{3+}	$^4F_{3/2} \rightarrow {}^4I_{13/2}$	Yes	Yes	4L
≈1380	Ho^{3+}	$^5S_2, {}^5F_4 \rightarrow {}^5I_5$	?	Yes	4L
1460–1510	Tm^{3+}	$^3H_4 \rightarrow {}^3F_4$	No	Yes	ST
≈1510	Tm^{3+}	$^1D_2 \rightarrow {}^1G_4$		Yes	UC, 4L
1500–1600	Er^{3+}	$^4I_{13/2} \rightarrow {}^4I_{15/2}$	Yes	Yes	3L
≈1660	Er^{3+}	$^2H_{11/2} \rightarrow {}^4I_{9/2}$	No	Yes	4L
≈1720	Er^{3+}	$^4S_{3/2} \rightarrow {}^4I_{9/2}$	No	Yes	4L
1700–2015	Tm^{3+}	$^3F_4 \rightarrow {}^3H_6$	Yes	Yes	3L
2040–2080	Ho^{3+}	$^5I_7 \rightarrow {}^5I_8$	Yes	Yes	3L
2250–2400	Tm^{3+}	$^3H_4 \rightarrow {}^3H_5$	No	Yes	4L
≈2700	Er^{3+}	$^4I_{11/2} \rightarrow {}^4I_{13/2}$	No	Yes	ST
≈2900	Ho^{3+}	$^5I_6 \rightarrow {}^5I_7$	No	Yes	ST

pump light into the fibre. The maximum angle of incidence onto the entry face of the fibre in which light is accepted into the fibre and transmitted in the core, θ_m, is defined by the numerical aperture (NA)

$$NA = \sin\theta_m = (n_1^2 - n_2^2)^{\frac{1}{2}}$$

where n_1 is the refractive index of the core and n_2 the refractive index of the cladding.

Fibre lasers have operated with efficiencies greater than 80%. These high efficiencies result mainly from the overlap of the pump and laser beams through the length of the fibre. The important factor influencing the efficiency of the fibre laser is the coupling efficiency for the pump into the fibre. This depends on the optical or spatial quality of the pump source and the diameter and NA of the fibre, this also determines the number of transverse modes that the fibre will support. Typical coupling efficiencies into a single mode fibre from a TEM_{00} pump beam may be up to 70%, while greater efficiencies are achievable for coupling into a multi-mode fibre. For the most common pump source of a semiconductor diode laser, which is multi-transverse mode, the coupling efficiency is typically 10%.

The fibre may be classified as single mode or multimode depending on V, a parameter that depends on the wavelength of the light and the radius a of the core,

$$V = \frac{2\pi a}{\lambda}(NA)$$

For single mode LP_{01} operation, the V should be <2.4. A wavelength of 1 μm, $NA = 0.1$ and $V < 2.4$, for single mode operation, sets $2a = 7.6$ μm.

As indicated in Tables 1 and 2 and Figure 2, fibre lasers may be classified as 3-level or 4-level, depending on the arrangement of energy levels for the lasers

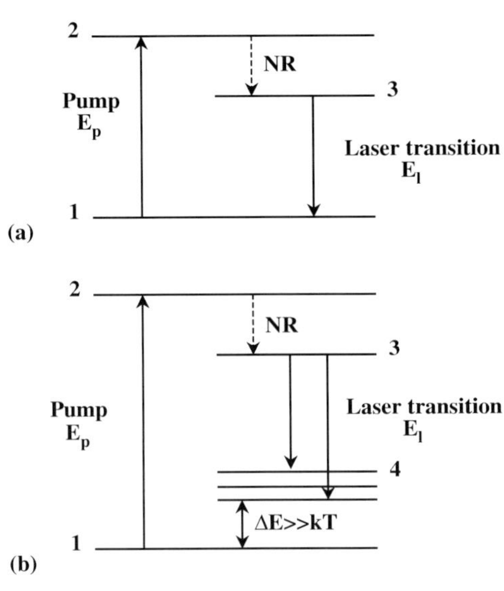

Figure 2. (a) Three-level laser operation, (b) four-level laser operation. Pump photon energy E_p, laser photon energy E_l, NR-nonradiative transition.

(Figure 2). Where the lower laser is also the ground level, or where the lower laser level is close to the ground level, $\Delta E \ll kT$, then the laser operates as a 3-level system. For $\Delta E \gg kT$ the behaviour is 4-level. This classification is not rigorous as a fibre laser may exhibit both 3-level and 4-level properties. For example, the Er^{3+} laser at 1.55 μm is 3-level and quasi-4-level at $\lambda > 1.6$ μm. Similarly the Yb and Tm lasers show 3-level characteristics at shorter wavelengths and 4-level at longer wavelengths.

We have previously noted that the fibre laser may exhibit both homogeneous and inhomogeneous broadening characteristics. The gain in a fibre laser operating on a homogeneously broadened transition depends on the laser intensity in the fibre in which, at higher intensities, saturation of the gain occurs. For a laser operating at frequency v with an in-cavity intensity I the small gain $\gamma_0(v)$ is modified to $\gamma(v)$ where

$$\gamma(v, I) = \frac{\gamma_0(v)}{1 + \frac{I}{I_s}}$$

where I_s is the saturation intensity, $I_s = hv_p/\sigma_p\tau$, σ_p is the pump absorption cross-section and τ is the lifetime of the upper laser level.

For a linear fibre laser supporting standing waves the intensity takes the form $I = I_1 + I_2 + 2I_1I_2\cos(\boldsymbol{k}z)$ where $I_{1,2}$ are the laser intensities travelling in the two axial directions, z is the distance along the fibre and $\boldsymbol{k}$ the wave-vector. This indicates that there is a spatial modulation of the gain along the length of the fibre.

For an internal cavity field $E(v)$ at frequency v, the cavity supports longitudinal (axial or frequency) modes with resonances at frequencies v_n. The cavity resonances are described by,

$$E_c(v) = \frac{E(v)}{1 - r_1r_2\exp(i\varphi - 2\gamma l)}$$

where the in-cavity phase shift is

$$\varphi = \frac{4\pi vl}{c}$$

The frequency spacing of the longitudinal modes is $\Delta v = n(c/2l)$. For a typical fibre laser length of $l = 1$ m, the longitudinal mode spacing $\Delta v_n = 150$ MHz.

For a fibre laser in which the pump and fibre laser signal is confined in the fibre core the launched pump threshold power P_{th} is a measure of the figure-of-merit of the fibre laser. P_{th} is related to the effective core area A_c, the fraction of pump power absorbed ε_p, the stimulated emission cross-section σ_e and the fluorescence lifetime τ_f as [25],

$$P_{th} \propto \frac{A_c}{\sigma_e\tau_f\varepsilon_p}$$

As expected, for a given fluorescence lifetime and stimulated emission cross-section, the pump threshold is lowered by a reduction in the core area and an increase in the absorbed power.

The pump threshold to reach laser oscillation in a 3- or 4-level fibre laser is usually readily achieved since the confined pump light in the small diameter core reaches a high intensity, even at low pump powers. For example, for a fibre with a core diameter of 7.5 μm being pumped with a power of 10 mW, the intensity in the fibre core is $\sim 2\,\mathrm{kW\,cm^{-2}}$.

In fibre lasers the transverse mode structure is determined by the fibre cylindrical optical waveguide. The laser intensity distribution is in a set of low order LP_{lm} transverse modes, in which l is the azimuthal mode number and m is the radial mode number. These correspond to the number of minima in the intensity distribution in the azimuthal and radial directions. The fundamental transverse mode is LP_{01} with the maximum intensity on the axis of the fibre and has the highest gain. Fibre lasers that support one transverse mode provide an output beam that is diffraction-limited.

8.4 Fibre laser structures

Some basic fibre laser resonator structures are shown in Figure 3. The pump source is most often a semiconductor diode laser whose output is conditioned for

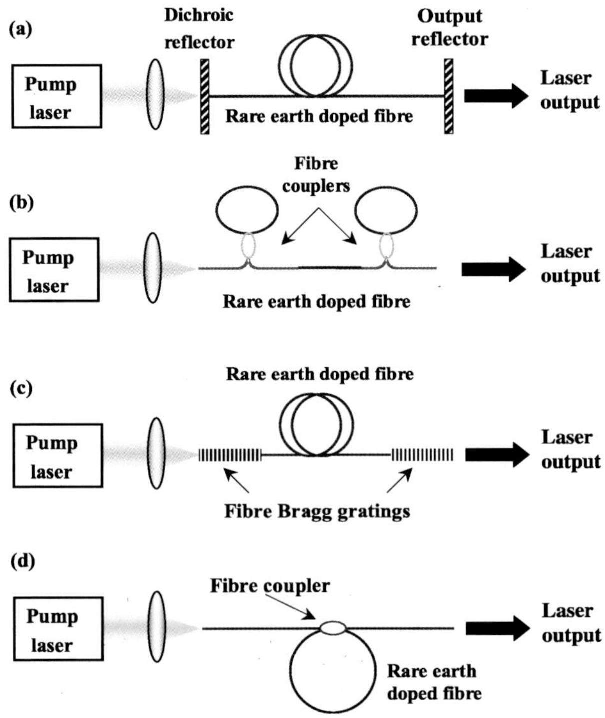

Figure 3. Schematic of various fibre resonators: (a) Fabry–Perot with dielectric reflectors; (b) Fabry–Perot with all-fibre reflectors; (c) Fabry–Perot with fibre Bragg gratings; (d) ring laser [1].

suitability for coupling into the fibre by one of several optical systems. The optimum length of the fibre laser is determined by several factors, including the dopant concentration, absorption coefficient at the pump wavelength, resonator mirror transmissions and optical losses.

The optical resonator may be formed in several ways. This may be simply by the Fresnel reflection at each end of the fibre, by dielectric mirrors butt-jointed to the fibre or coated onto the fibre, or by Bragg reflection gratings optically written onto the fibre.

The fibre laser normally operates without cooling since the cross-sectional area of the core and cladding is very small, so that heat is efficiently transferred from the core or cladding to the surrounds.

The double-clad structure for pumping high power fibre lasers is illustrated in Figures 4 and 5. In these illustrations the core of the fibre contains the rare earth dopant and the inner cladding is undoped [26–29]. Pump light is directed into the inner

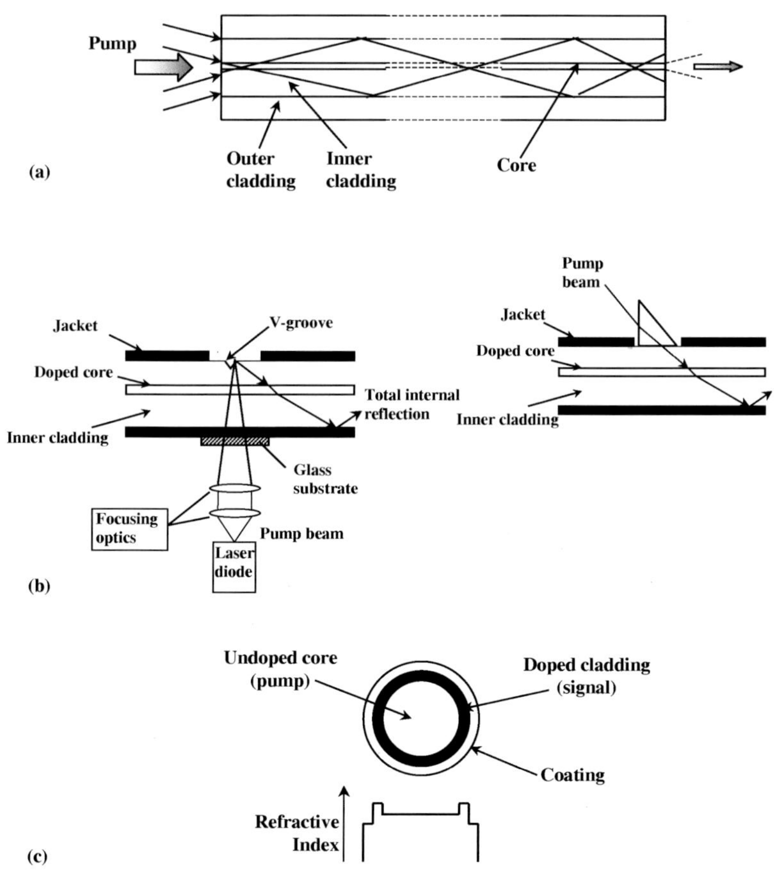

Figure 4. (a) Double-clad pumping fibre arrangement; (b) side-pumping into double-clad fibre with either V-groove coupling or prism coupling; (c) M-profile, cladding-doped fibre.

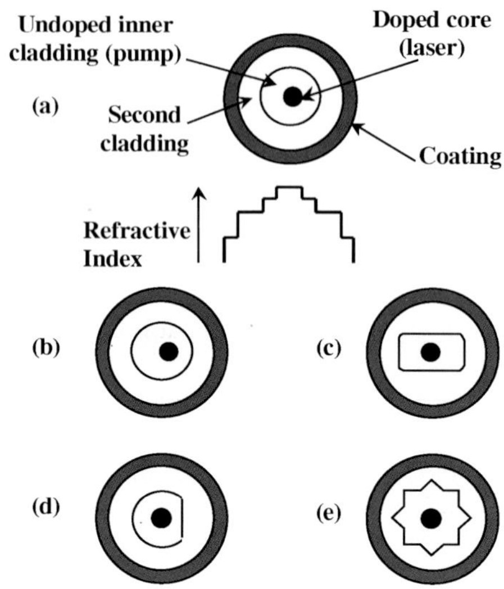

Figure 5. Double-clad fibre configurations: (a) circular, symmetric, (b) circular, offset core, (c) rectangular cladding, (d) D-shaped cladding, and (e) crenellated cladding.

cladding, and as it propagates down the fibre it transits the doped core, where some light is absorbed. Much greater coupling efficiency into the inner cladding is achieved compared with pumping directly into the small diameter core. In addition high pumping powers may be employed. Figure 4(b) illustrates two side-pumping techniques [30,31]. These techniques enable light from divergent pump sources such as from high power semiconductor diode lasers to be efficiently coupled into the fibre. The M-profile fibre enables both a greater laser mode area than can be achieved with double-clad fibre and, hence, potentially higher output powers. The ability of fibre lasers to operate at high peak powers is limited by optical nonlinearities. A fibre design for high power operation is to increase the mode area for single transverse mode operation [32,33]. The large-mode-area fibre allows reduced intensity in the fibre and greater stored energy for short pulse operation.

The use of Bragg gratings in the fibre structure, to act as a laser mirror or a wavelength filter, enables wavelength selection from the fibre laser emission bands and also for single frequency oscillation [34]. To optically write the Bragg grating into the fibre it is necessary for the fibre to be photosensitive at the wavelength of the writing beam [34,35].

8.5 Main fibre lasers

8.5.1 Erbium

The erbium-doped fibre amplifier (EDFA) is a key component in optical communication systems for operation in the third communication window

[11,12,36–38]. The lower energy levels of Er^{3+} are shown in Figure 6. The lifetime of the upper laser level is about 10 ms. Pumping of the laser is at 980 nm or at 810 and 1480 nm with diode lasers and with laser emission in a band around 1.5 μm over 1.5 to 1.62 μm. At low pump intensities the laser operates at longer wavelengths appropriate to 4-level oscillation; it changes to shorter wavelengths at higher pump intensities, then corresponding to the normal 3-level operation. The laser also exhibits excited state absorption (ESA) when pumped at 800 nm.

8.5.2 Thulium

The thulium fibre laser at 1.9 μm has assumed increasing importance due to its high power capability, its tunability and potential applications, particularly in bioscience and medicine [39]. The energy levels of Tm^{3+} are shown in Figure 7(a), with the upper laser level lifetime being about 0.2 ms. The absorption spectrum of Tm^{3+} is shown in Figure 7(b), each of the main absorption bands centred on 0.680, 0.790, 1.22 and 1.62 μm have been used for excitation and laser emission at ~1.9 μm. The thulium-silica laser has shown high power (>40 W) and high efficiency (>70%) with continuous tuning over 1.75 to 2.1 μm in silica and ZBLAN and 2.25–2.40 μm in a ZBLAN host.

8.5.3 Neodymium

A partial energy level diagram for Nd^{3+} is shown in Figure 8 with pumping at 800 nm, which is conveniently available from high power laser diodes. The laser

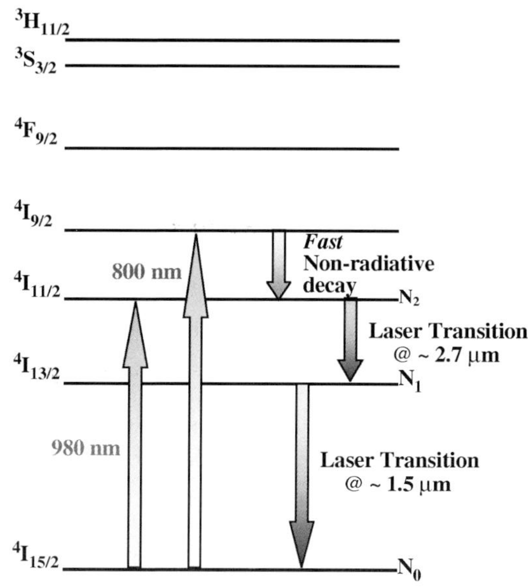

Figure 6. Energy levels of Er^{3+}, showing the possible pump wavelengths for the 1.5(5) and 2.7 μm transitions and the fast non-radiative decay.

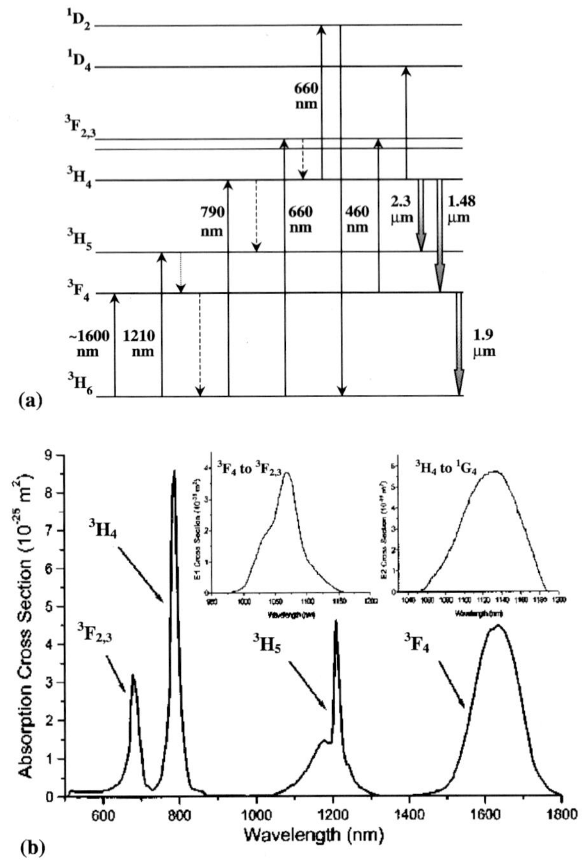

Figure 7. (a) Energy levels of Tm^{3+} with peak pumping bands identified and the up-conversion transition at 460 nm. Laser transitions at 1.48, 1.9 and 2.3 μm are shown. Note that the notation of the $^3F_4$ and $^3H_4$ levels are often interchanged in the literature. (b) Absorption spectrum of Tm^{3+}. Insets show the excited state absorption for the $^3F_4$ and $^3H_4$ levels.

transitions in Nd^{3+}-silica occur at 900–950, 1000–1150 and 1320–1400 nm, with the most common at 1060 nm, which is limited by amplified spontaneous emission [40]. Simultaneous wavelength oscillation is also possible in Nd^{3+}. The transition at 1300 nm is subject to excited state absorption (ESA) but not that at 1060 nm, which acts as a 4-level laser. Convenient and efficient pumping of this laser is available at around 810 nm by AlGaAs semiconductor diode lasers. Concentration quenching occurs at high Nd^{3+} concentrations.

8.5.4 Ytterbium

The Yb fibre laser is highly efficient and operates readily at high power [41–44]. Specialist Yb^{3+} lasers up to 6 kW power have been produced. As shown in

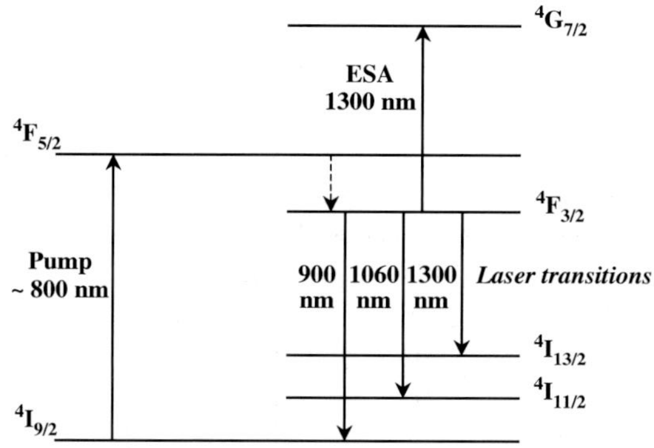

Figure 8. Partial energy level diagram for Nd^{3+}. The main laser transitions are identified, and the transition for excited state absorption (ESA) at 1.3 μm.

Figure 9, Yb^{3+} has a simple energy level structure with higher energy levels occurring in the UV region. The absorption of Yb^{3+} is particularly broad, ranging from 850 to 1070 nm, and hence the laser may be excited by several laser pump sources, including AlGaAs (800–920 nm) and InGaAs (980 nm) laser diodes.

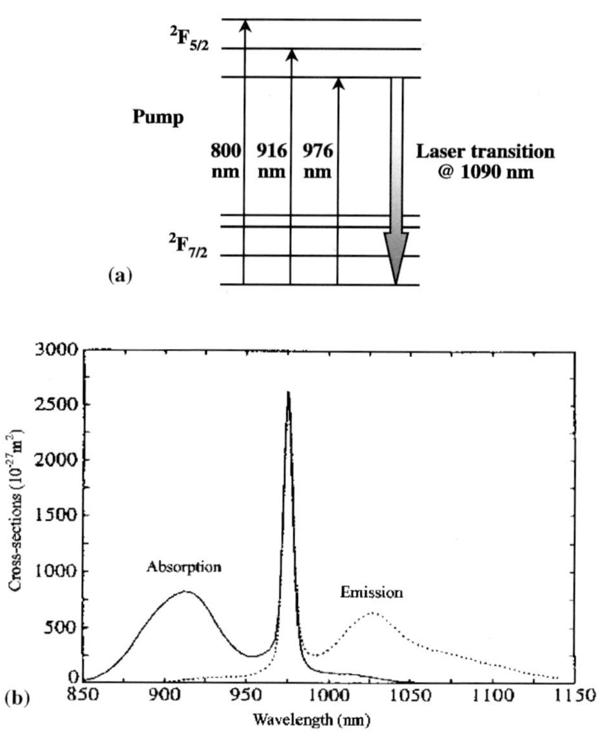

Figure 9. (a) Lowest energy levels of Yb^{3+}. (b) Absorption and emission cross-sections and spectra for Yb^{3+} in germanosilicate glass [41].

Because of the broad absorption the pump laser diodes also do not require temperature control. The lifetime of the upper laser level is within 700–1400 μs. This simple energy level structure ensures that there is no ESA and ion–ion interactions only transfer excitation energy from one ion to a neighbour. Laser emission is possible over a broad range, from 970 to 1200 nm, which is valuable for spectroscopic applications.

Yb-Er (Figure 10(a)) and Yb-Pr co-doped lasers operate with absorption into Yb^{3+} and then transfer to Er^{3+} or Pr^{3+}, and hence provide an alternative pumping route for Er^{3+} or Pr^{3+} with a larger absorption cross-section than for direct excitation into Er^{3+} or Pr^{3+}. A similar mechanism operates in the Tm^{3+}-Ho^{3+} system (Figure 10(b)).

8.5.5 Holmium

Ho^{3+} has absorption bands near 450, 540 and 650 nm with laser emission near 2 μm on the $^5I_7$ to $^5I_8$ transition and an emission bandwidth up to 200 nm. This emission band is similar to that of Tm^{3+} and the transition may be sensitized by energy transfer from Tm in a co-doped Tm-Ho silica or ZBLAN glass (Figure 10(b)). The Ho^{3+}BLAN laser can also operate as an up-conversion laser, giving green emission over 545 to 550 nm.

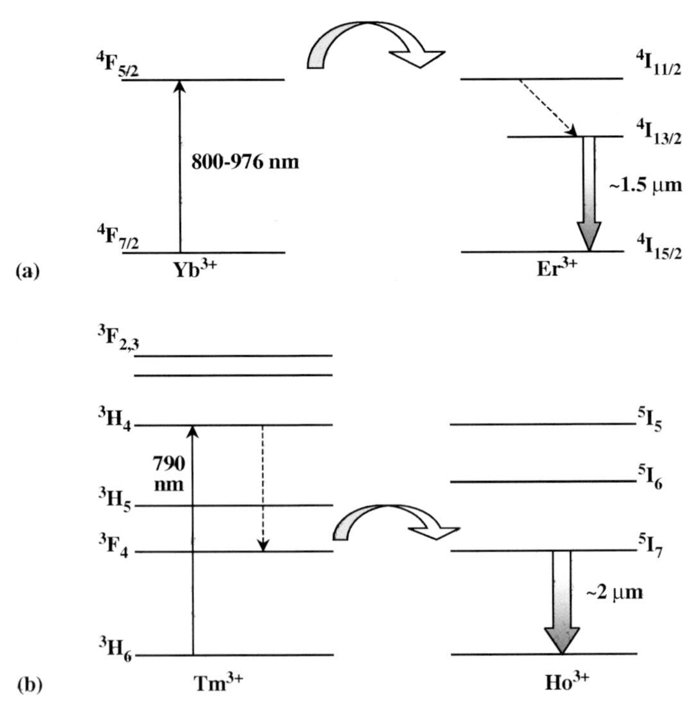

Figure 10. Examples of sensitisation in co-doped laser systems: (a) energy transfer from Yb^{3+} to Er^{3+}, (b) Ho^{3+} 2.0 μm transition sensitised by energy transfer from Tm^{3+}.

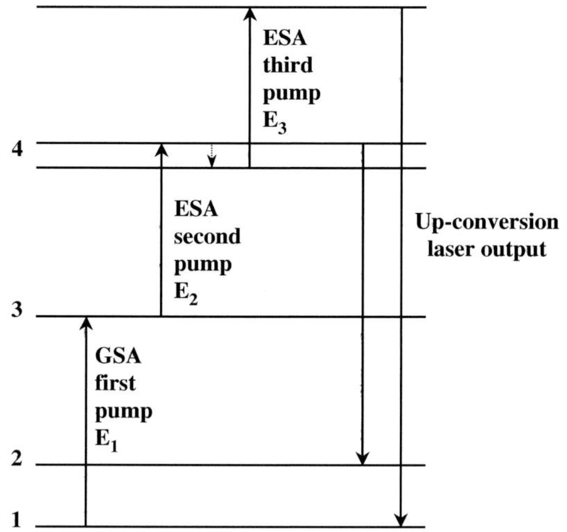

Figure 11. Arrangement of energy levels of up-conversion lasers for two- or three-pumping transitions; GSA – ground state absorption, ESA – excited state absorption.

8.5.6 Up-conversion fibre lasers

Levels at energies greater than the pump photon energy may be reached by sequential excitation [1,45]. These levels may subsequently radiate to produce laser radiation at wavelengths shorter than the pump wavelength (Figure 11). Many up-conversion fibre lasers have been reported. This is a method to obtain short wavelength UV or blue laser emission. The process is most efficient in host glasses having a low phonon energy such as the ZBLAN glasses, whereas in the silica glass rapid multiphoton decay depletes the upper levels. The first up-conversion laser in Tm^{3+}-ZBLAN glass, when excited at 647 and 676 nm, produced laser emission at 455 and 480 nm. A Nd^{3+}-ZBLAN UV fibre laser gave emission at 381 nm when pumped at 590 nm. Up-conversion from the IR to the visible has been achieved in Pr-ZBLAN in which pumping at 1010 and 835 nm produced laser emission at 491, 520, 605 and 635 nm. Efficient up-conversion lasers operating at room temperature are now practicable sources.

8.6 Fibre laser performance

Tables 1 and 2 [2] have fibre lasers based on oxide and fluoride hosts separately identified. Laser action has been observed in seven of the fifteen rare earth elements. In Figure 1 the main emission wavelengths are also shown for fluoride and silica host media. Notably, there are more transitions in the ZBLAN fluoride host than in silica and also these extend further into the IR due to the extended optical transmission of the fluoride glass. This is due to the higher phonon energy in the silica glass ($1160\,cm^{-1}$) compared with the fluoride glass ($650\,cm^{-1}$); multiphonon decay transitions then more readily reduce the laser level lifetime in silica glass.

In addition to the silica and fluoride glasses other glass media that have been used for fibre lasers include borates, phosphates, germinates, tellurite, sulfide and LaF_3 crystal fibres.

8.6.1 *High power fibre lasers*

The difficulty of coupling light from powerful but low quality, low brightness sources, such as the semiconductor diode lasers and arrays, has been largely solved by the double cladding fibre (Figures 4 and 5). In the double-clad fibre, pumping into the larger inner cladding is significantly more efficient than pumping into the core. As the pump light propagates down the inner cladding, that part which crosses the doped core will be partially absorbed. The diameter of the inner cladding may be rather large, up to a few hundred micrometres and have a numerical aperture of 0.4 to 0.7, such that a high coupling efficiency from high power diode arrays is achieved. Double-clad Yb^{3+} fibre lasers with CW output powers in excess of 100 W and with optical efficiencies up to 80% have been demonstrated. Current developments using multiple side-pumping of double-clad fibres have achieved, at CW, fibre laser powers greater than 1 kW. Scaling of the power in fibre lasers by further amplification, in the master oscillator-power amplifier arrangement, is an effective method to achieve very high power levels. Techniques can be employed to reduce the power limitations intrinsic to a small core diameter, e.g. by using large mode area fibres.

8.6.2 *Pulsed fibre lasers*

Many applications require short duration high peak power laser pulses, e.g. as sources for remote or distributed gas sensing, nonlinear frequency conversion, range-finding, photomedicine and long distance, low loss communications. Pulsed fibre lasers can provide pulse energies, peak powers and average powers comparable to bulk diode-pumped solid-state lasers. Q-switching of fibre lasers can be induced by acousto-optic, electro-optic or mechanical cavity modulation techniques. Q-switching of the Nd^{3+}, Yb^{3+}, Er^{3+} and Tm^{3+} fibre lasers has been demonstrated with pulse energies of up to 1 mJ and peak powers of up to 4 kW in pulses with durations as short as 2 ns and up to a few hundred nanoseconds [46–48].

Mode-locked fibre lasers are important sources of picosecond – femtosecond pulses [49–53]. The mode-locked fibre may operate in either single-mode or multimode forms and operate as oscillators or fibre amplifiers. Modulation of the lasers to phase-lock the longitudinal modes can be achieved by active or passive modulation. Passively mode-locked Er^{3+} single-mode fibre lasers have produced 3 nJ, 100 fs pulses and up to 20 mJ for multimode passively mode-locked. Actively mode-locked fibre lasers can produce pulses of about 1 ps with pJ pulse energy and at GHz repetition rates. Combined oscillator – amplifier fibre lasers are competitive with conventional solid-state lasers in being able to provide femtosecond pulses at microjoule energies.

Fibre lasers that are excited by pulsed lasers produce a gain-switched output [54,55]. Pulse-pumped fibre lasers are useful in providing high energy pulses at high efficiency. In an early demonstration a flashlamp-pumped titanium-sapphire laser was used to excite, at 791 nm, an Er^{3+}-ZBLAN fibre. This produced 1.9 mJ output pulses at 2.7 μm with a duration of a few microseconds in a near single transverse mode. With the same excitation source a Tm^{3+}-silica double-clad fibre laser gave 10.1 mJ pulses at 1.9 μm.

8.6.3 Superfluorescence fibre sources

Incoherent broadband light sources have various optical applications. One particular application is in optical coherence tomography (OCT), where a low coherence source is used in a Michelson interferometer arrangement to derive spatially-resolved images from a probed medium, such as biomedical tissue. The source requirements are a broad optical bandwidth of about 50 nm, small focused spot size and power levels greater than 1 mW, together with good penetration in biological tissue. Doped fibres have suitable characteristics for this application, producing amplified spontaneous emission (superfluorescence) but without the extreme spectral narrowing characteristic of laser cavities. Recently, a Tm^{3+}: Ho^{3+} co-doped silica fibre laser at near 1.8 μm has been demonstrated to operate as a superfluorescent source with a bandwidth of 70 nm and a power of 40 mW when pumped at 1.6 μm by a Yb:Er fibre laser [56]. In addition to OCT and its use, for example, in ophthalmology these sources have various applications, including in the fibre-optic gyroscope.

8.6.4 Raman-fibre and Brillouin-fibre wavelength conversion

The nonlinear optical effects of stimulated-Brillouin (SBS) and stimulated-Raman scattering (SRS) occur in optical fibres above a certain threshold intensity [57,58]. These act as power-limiting mechanisms, but also can be used for frequency conversion. Stimulated Raman scattering is scattering by vibrational modes (optical phonons) of the glass and in silica glass the Raman-active vibrational mode at 440 cm^{-1} has the greatest strength. This may be used for a fibre Raman laser acting as a wavelength converter, with the dominant scattering being in the forward direction. Fibre-Raman lasers have operated in the 1.0 to 1.6 μm region when pulsed pumped by a Nd:YAG laser and in the UV when pumped by an excimer laser. Stimulated Brillouin scattering occurs by coherent scattering off acoustic phonons. The SBS shows a frequency shift of typically 10 GHz and the SBS gain coefficient is about 100 × greater than the SRS gain.

8.7 Perspectives

Fibre lasers based on rare-earth-doped silica and fluoride glasses provide a broad range of wavelengths from the UV to the mid-IR. The broadened transitions operating in the glass host enable a certain degree of tuning, such as for the Tm^{3+}-silica fibre laser over 1.75 to 2.07 μm and Ho-silica over 1.85–2.3 μm. The recent development of the double-clad fibre, and with diode laser side-pumping, has opened up the fibre laser to high power operation. Here the ability to pump the fibre by high power semiconductor diode arrays gives an efficient all-solid-state laser with high beam quality. The double-clad structure has also led to significant development in other areas of fibre laser, including Q-switched pulsed operation. The mode-locked femtosecond fibre lasers have developed into practical devices with advantages over conventional solid-state lasers such as titanium-sapphire in offering alternative wavelengths and higher pulse energy.

The use of photosensitive fibres and the writing of fibre Bragg gratings in distributed feedback lasers provides a means of wavelength selection and tuning. Another important recent development has been the photonic crystal micro-structured fibre (holey fibre) [59], which offers significant potential application to fibre lasers in refractive index control, and high power application. In these structures a high numerical aperture can be provided along with a new versatility in fibre design.

It unsurprisingly, the remarkable range of properties of the fibre laser is finding widespread application, not only as lasers and amplifiers in optical communications and information technology, where it is pre-eminent, but in a great variety of fields. These include biology and medicine [60], materials processing, semiconductor fabrication, laser marking, optical data storage, displays, metrology, range-finding and sensing. The selection of lasers for these applications also includes the choice from the semiconductor diode lasers and arrays, being developed to give new wavelengths in the visible region and high power levels, and the diode pumped solid-state crystal lasers, in which there has been substantial progress. However, the range of wavelengths available from fibre lasers, their high power and short pulse capability, intrinsic versatility and relative simplicity confer particular advantages.

References

1. M.J.F. Digonnet (Ed) (2001). *Rare-Earth-Doped Fiber Lasers and Amplifiers* (2nd edn.). Marcel Dekker, New York.
2. W.J. Miniscalco (2001). Optical and electronic properties of rare earth ions in glasses. In: M.J.F. Digonnet (Ed), *Rare-Earth-Doped Fiber Lasers and Amplifiers* (2nd edn. pp. 17–112). Marcel Dekker, New York.
3. D.S. Funk, J.G. Eden (2001). Visible fluoride fiber lasers. In: M.J.F. Digonnet (Ed), *Rare-Earth-Doped Fiber Lasers and Amplifiers* (pp. 171–242). Marcel Dekker, New York.
4. E. Snitzer (1961). Optical maser action in Nd^{3+} in barium crown glass. *Phys. Rev. Lett.* **7**, 444–446.
5. C.J. Koester, E. Snitzer (1964). Amplification in fibre laser. *Appl. Opt.* **3**, 1182–1186.
6. G.C. Holst, E. Snitzer, R. Wallace (1969). High-coherence high-power laser system at 1.0621 μm. *IEEE J. Quantum Electron.* **QE-5**, 342.
7. J. Stone, C.A. Burns (1973). Neodymium-doped silica lasers in end pumped fibre geometry. *Appl. Phys. Lett.* **223**, 388–389.

8. R.J. Mears, L. Reekie, S.B. Poole, D.N. Payne (1985). Neodymium-doped silica single-mode fibre lasers. *Electron. Lett.* **21**, 738–740.
9. R.J. Mears, L. Reekie, S.B. Poole, D.N. Payne (1986). Low-threshold-tunable CW and Q-switched fibre laser operating at 1.55 μm. *Electron. Lett.* **22**, 159–160.
10. R.J. Mears, L. Reekie, I.M. Jauncey, D.N. Payne (1987). Low-noise, erbium-doped fiber amplifier operation at 1.55 μm. *Electron. Lett.* **23**, 1026–1028.
11. S. Sudo (Ed) (1997). *Optical Fiber Amplifiers: Materials, Devices and Applications.* Artech House, Boston MA.
12. A. Othonos, K. Kalli (1999). *Fiber Bragg Gratings: Fundamentals and Applications in Telecommunications and Sensing.* Artech House, Boston MA.
13. E. Maurice, S.A. Wade, S.F. Collins, G. Monnom, G.W. Baxter (1997). Self-referenced point temperature sensor based on a fluorescence intensity ratio in Yb^{3+}-doped silica fiber. *Appl. Opt.* **36**, 8264–8269.
14. D.G. Lancaster, D. Richter, R.F. Curl, F.K. Tittel, L. Goldberg, J. Koplow (1999). High power continuous-wave mid-infrared radiation generated by difference frequency mixing of diode-laser-seeded fiber amplifiers and its application to dual-beam spectroscopy. *Opt. Lett.* **24**, 1744–1746.
15. H.J. Lee, S.J.B. Yoo, V.K. Tsui, S.K.H. Fong (2001). A simple all-optical label detection and swapping technique incorporating a fiber Bragg grating filter. *IEEE Phot. Technol. Lett.* **13**, 635–637.
16. O. Hadeler, M. Ibsen, M.N. Zervas (2001). Distributed-feedback fiber laser sensor for simultaneous strain and temperature measurements operating in the radio frequency domain. *Appl. Opt.* **40**, 3169–3175.
17. M. Pierce, S. Jackson, P. Golding, B. Dickinson, M. Dickinson, T. King, P. Sloan (2001). Development and Application of Fibre Lasers for Medical Applications. SPIE Volume 4253, pp. 144–154. SPIE, Bellingham, WA.
18. R. Diehl (2000). *High Power Diode Lasers.* Springer, Berlin.
19. M.C. Pierce, S.D. Jackson, M.R. Dickinson, T.A. King (1999). Laser-tissue interaction with a high power 2-μm fiber laser. *Lasers Surg. Med.* **25**, 407–413.
20. T. Sumiyoshi, H. Sekita, T. Arai, S. Sato, M. Ishihara, M. Kikuchi (1999). High power continuous wave 3- and 2-μm cascade Ho^{3+}:ZBLAN fiber laser and its medical applications. *IEEE J. Quantum Electron.* **5**, 936–943.
21. M.C. Pierce, S.D. Jackson, M.R. Dickinson, T.A. King, P. Sloan (2000). Laser-tissue interaction with a continuous wave 3-μm fibre laser: Preliminary studies with soft tissue. *Lasers Surg. Med.*, **26**, 491–495.
22. K. Naruse, T. Arai, S. Kawanchi, T. Sumiyoshi, S. Kiyoshima, M. Ishihara, S. Sato, M. Kikuchi, H. Sekita, M. Obara (2000). Development for medical surgical system with function variability and flexible delivery using dual-wavelength infrared fiber laser (SPIE Volume 4617, pp. 118–124). SPIE, Bellingham, WA.
23. T. Arai, T. Sumiyoshi, K. Naruse, M. Ishihara, S. Sato, M. Kikuchi, T. Kasamatsu, H. Sekita, M. Obara (2000). Laser tissue interaction of a continuous wave 2 μm, 3 μm cascade oscillation fiber laser: sharp incision with controlled coagulation layer thickness. SPIE Volume 3914, pp. 252–259. SPIE, Bellingham, WA.
24. S.B. Poole, D.N. Payne, M.E. Fermann (1985). Fabrication of low-loss optical fibers containing rare earth ions. *Electron. Lett.* **21**, 737–738.
25. M.J.F. Digonnet, C.J. Gaeta (1985). Theoretical analysis of optical fiber laser amplifiers and oscillators. *Appl. Opt.* **24**, 333–342.
26. L. Zenteno (1993). High power double-clad fiber laser. *J. Lightwave Technol.* **11**, 1435–1446.

27. H. Po, J.D. Cao, B.M. Laliberte, R.A. Minns, R.F. Robinson, B.H. Rockney, R.R. Tricca, Y.H. Zhang (1993). High power neodymium-doped single transverse mode fibre laser. *Electron. Lett.* **29**, 1500–1501.
28. H. Zellmer, A. Tunnermann, H. Welling, V. Reichel (1997). Double-clad fibre laser with 30 W output power. In: A. Willner, M. Zervas, S. Sasaki (Eds), *Optical Amplifiers and their Applications, OSA Trends in Optics and Photonics* (Volume 6, p. 137). Cambridge University Press, Cambridge.
29. A.S. Kurkov, A. Yu Laptev, E.M. Dianov, A.N. Guryanov, V.I. Karpov, V.M. Paramonov, O.I. Mednedkov, A.A. Umnikov, V.N. Protopopov, N.N. Vechkanov, S.A. Vasiliev, E.V. Pershina (2000). Yb^{3+}-doped double-clad fibers and lasers. In: M. Dianov (Ed), *Advances in Fiber Optics* (SPIE Volume 4083, pp. 118–126). SPIE, Bellingham, WA.
30. L. Goldberg, B. Cole, E. Snitzer, V-groove side-pumped 1.5 μm fiber amplifier. *Electron. Lett.* **33**, 2127–2129.
31. L. Goldberg, J.P. Koplow, D.A.V. Kliner (1999). Highly efficient 4-W Yb-doped fiber amplifier pumped by a broad-stripe laser diode. *Opt. Lett.* **24**, 673–675.
32. N.G.R. Broderick, H.L. Offerhaus, D.J. Richardson, R.A. Sammut (1998). Power scaling in passively mode-locked large mode area fiber lasers. *IEE Phot. Tech. Lett.* **10**, 1718–1720.
33. N.G.R. Broderick, H.L. Offerhaus, D.J. Richardson, R.A. Sammut, J. Caplen, L. Dong (1999). Large mode area fibers for high power applications. *Opt. Fiber Tech.* **5**, 185–196.
34. R. Kashyap (1999). *Fiber Bragg Gratings*. Academic, San Diego, CA.
35. K.O. Hill, G. Meltz (1997). Fibre Bragg grating technology fundamentals and overview. *J. Lightwave Technol.* **15**, 1263–1276.
36. E. Desurvire (1994). *Erbium-Doped Fibre Amplifers*. Wiley, New York.
37. P.C. Becker, N.A. Olson, J.R. Simpson (1999). *Erbium-doped Fiber Amplifiers*. Academic, San Diego CA.
38. E. Desurvire (2001). Erbium-doped fiber amplifers: basic physics and characteristics. In: M.J.F. Digonnet (Ed), *Rare-Earth-Doped Fiber Lasers and Amplifiers* (pp. 531–582). Marcel Dekker, New York.
39. S.D. Jackson, T.A. King (1998). High power diode-cladding-pumped Tm-doped silica fiber lasers. *Opt. Lett.* **23**, 462–464.
40. H. Zellmer, U. Williamowski, A. Tunnermann, H. Welling, S. Unger, V. Reichel, H-R. Müller, J. Kirchhof, P. Albers (1995). High power cw neodymium-doped fiber laser operating at 9.2 W with high beam quality. *Opt. Lett.* **20**, 578.
41. H.M. Pask, R.J. Carman, D.C. Hanna, A.C. Tropper, C.J. Mackechnie, P.R. Barber, J.M. Dawes (1995). Ytterbium-doped silica fiber lasers: versatile sources for the 1–1.2 μm region. *IEEE J. Sel. Top. Quantum Electron.* **1**, 2–13.
42. J.P. Koplow, D.A.V. Klinv, L. Goldberg (1998). UV generation by frequency quadrupling of a Yb-doped fiber amplifier. *IEEE Photonics Tech. Lett.* **10**, 75–77.
43. J. Nilsson, J.A. Alvarez-Chavez, P.W. Turner, W.A. Clarkson, C.C. Renaud, A.B. Grudinin (1999). Widely tunable high power diode-pumped double clad Yb^{3+}-doped fiber laser, in *Adv. Solid-State Lasers, OSA* Technical Digest (Optical Society of America, Washington DC) 147–149.
44. V. Dominic, S. MacCormack, R. Waarts, S. Sanders, S. Bicknese, R. Dohle, E. Wolak, P.S. Yeh, E. Zucker (1999). 110 W fibre laser. *Electron. Lett.* **35**, 1158–1160.
45. T. Sandrock, H. Scheife, E. Henmann, G. Huber (1997). High-power continuous-wave upconversion fiber laser at room temperature. *Opt. Lett.* **22**, 808–810.
46. J.A. Alvarez-Chavez, H.L. Offerhaus, J. Nilsson, P.W. Turner, W.A. Clarkson, D.J. Richardson (2000). High energy, high power ytterbium-doped Q-switched fiber lasers. *Opt. Lett.* **25**, 37–39.

47. C.C. Renaud, H.L. Offerhaus, J.A. Alvarez-Chavez, J. Nilsson, W.A. Clarkson, P.W. Turner, D.J. Richardson, A.B. Grudinin (2001). Charactersitics of Q-switched cladding-pumped ytterbium-doped fiber lasers with different high-energy fiber designs. *IEEE J. Quantum Electron.* **37**, 199–205.
48. A.F. El-Sherif, T.A. King (2003). Analysis and optimization of Q-switched operation of a Tm^{3+}-doped silica fiber laser operating at 2 μm. *IEEE J. Quantum Electron.* **39**, 759–765.
49. J. Porta, A.B. Grudinin, Z.J. Chen, J.D. Minelly, N.J. Traynor (1998). Environmentally stable picosecond ytterbium fiber laser with a broad tuning range. *Opt. Lett.* 23, 615–617.
50. A. Liem, D. Nickel, J. Limpert, H. Zellmer, U. Griebner, S. Unger A. Tunnermann, G. Korn (2000). High average power ultrafast fiber chirped pulse amplification system. *Appl. Phys. B* **71**, 889.
51. A. Galvanauskas, G.C. Cho, A. Hariharan, M.E. Fermann, D. Harter (2001). Generation of high-energy femtosecond pulses in multimode-core Yb-fiber chirped-pulse amplification systems. *Opt. Lett.* **26**, 935–937.
52. M.E. Fermann, M. Höfer (2001). Mode-locked fibre lasers. In: M.J.F. Digonnet (Ed), *Rare-Earth-Doped Fibre Lasers and Amplifiers* (2nd edn, pp. 395–447). Marcel Dekker, New York.
53. J.H.V. Price, K. Furnsawa, T.M. Monro, L. Lefort, D.J. Richardson (2002). Tunable, femtosecond pulse source operating in the range 1.06–1.33 μm based on an Yb^{3+}-doped holey fiber amplifer. *J. Opt. Soc. Am.* **19**, 1286–1294.
54. B.C. Dickinson, S.D. Jackson, T.A. King (2000). 10 mJ total output from a gain-switched Tm-doped fibre laser. *Opt. Common.* **182**, 199–203.
55. B.C. Dickinson, P.S. Golding, M. Pollnau, T.A. King, S.D. Jackson (2001). Investigation of a 791 nm pulse pumped 2.7 μm Er-doped ZBLAN fibre laser. *Opt. Commun.* **191**, 315–321.
56. Y.H. Tsang, A.F. El-Sherif, T.A. King (2004). Broadband amplified spontaneous emission fibre source near 2 microns using resonant in-band pumping. *J. Mod. Opt.*, in press.
57. G.P. Agrawal (1995). *Nonlinear Fiber Optics* (2nd edn). Academic, San Diego CA.
58. G.P. Agrawal (2001). *Applications of Nonlinear Fiber Optics*. Academic, San Diego CA.
59. J.C. Knight, T.A. Birks, P.St.J. Russell, D.M. Atkin (1996). All-silica single mode optical fibre with photonic crystal cladding. *Opt. Lett.* **21**, 1547–1549.
60. A.F. El-Sherif, T.A. King (2003). Soft and hard tissue ablation with short-pulse high peak power and continuous thulium-silica fibre lasers. *Lasers Med. Sci.* **18**, 139–147.

Chapter 9

Methods for the generation of light pulses: from nanoseconds to attoseconds

Mauro Nisoli

Table of contents

Abstract

With the advent of the laser, the race towards the generation of short light pulses has been characterized by a sudden dramatic progress. The duration of the shortest light pulses has been reduced by six orders of magnitude in the last forty years. Pulsed lasers, with pulse durations ranging from a few tens of nanoseconds to a few tens of femtoseconds, are now commonly used in various technological applications. In this chapter the techniques for the generation of light pulses are reviewed, and the basic concepts of the Q-switching and mode-locking methods are discussed. The present state of the art of the generation of ultrashort pulses, down to the few-optical-cycle regime, is presented.

9.1 Introduction

Since the invention, in the 19th century, of flash photography, short light pulses have been employed to *freeze* the temporal evolution of short events. In 1851 Talbot used microsecond sparks to obtain photographs of a fast revolving newspaper page. In 1878, using spark photography, Muybridge obtained high-speed photographs of a galloping horse. In 1866 Töpler was able to take pictures of rapidly vibrating flames and to freeze the variation of atmospheric pressure in acoustic waves. He used a microsecond light spark to generate a sound wave, and a second time-delayed identical spark, triggered by the first one, to photograph the sound wave. The temporal evolution of the sound wave could be obtained by taking a series of snap-shots of the event as a function of the temporal delay between the two light sparks. This pioneering experiment can be regarded as the precursor of the modern time-resolved spectroscopy. To measure a fast physical event, a source is needed, whose temporal duration is comparable with the event itself. Since, in Töpler's experiment, the oscillation cycle of the sound wave is of the order of a millisecond, the light spark with a microsecond-duration was more than enough to *freeze* the sound wave evolution at precise instants. Between the beginning of the 20th century and the invention of laser, in 1960, the duration of the shortest light pulses was of the order of one nanosecond (1 ns $= 10^{-9}$ s). With the advent of laser the race towards the generation of extremely short light pulses has become very exciting, with important achievements. Light pulses with duration from a few nanoseconds to a few tens of nanoseconds with high-peak power (in the megawatt range) were generated in the early period of the laser development, using the technique of *Q-switching*. The most widely used technique for the generation of short light pulses, in the picosecond and femtosecond range, named *mode-locking*, was introduced in 1964 [1,2]. The first generation of mode-locked lasers, producing pulses with a duration shorter than 100 ps, was based on solid-state active media such as ruby, Nd:glass or Nd:YAG. For several years the standard sources of picosecond pulses were the actively mode-locked continuous wave (cw) Nd:YAG laser and the passively mode-locked, flashlamp-pumped Nd:glass laser. A significant step forward in the generation of even shorter pulses, down to the femtosecond regime, was obtained with the development of a second-generation mode-locked lasers, based on the discovery of the cw organic dye laser [3]. Pulses as short as 100 fs were generated in 1981 with the development of the colliding pulse mode-locked (CPM) dye laser [4]. Pulses as short as 27 fs were generated in 1985 using a prism-controlled CPM laser [5]. The evolution of the femtosecond dye laser technology during the 1980s revolutionized molecular and condensed-matter spectroscopy. An extremely important technological breakthrough in femtosecond lasers was achieved in 1991 with the first demonstration of the self-mode-locked Ti:sapphire laser by Spence et al. [6]. This triggered a renaissance of solid state lasers. Since then, a dramatic reduction in achievable pulse duration has been obtained. Sub-6-fs pulses have been generated directly from Kerr-lens mode-locked Ti:sapphire lasers [7,8]. Because the period of the optical cycle in the visible and near-infrared is 2–3 fs, this pulse duration is approaching the physical limit of devices operating in this wavelength range. In parallel with this progress in femtosecond pulse generation, the introduction of the technique of chirped-pulse amplification (CPA) [9] has made possible the amplification of ultrashort pulses to unprecedented power levels. Pulses as short as 20 fs have now become available with terawatt peak powers at repetition rates of 10–50 Hz, and with multigigawatt peak power at kilohertz rates.

Sub-10-fs light pulses can also be generated by external compression. In 1981 Nakatsuka et al. [10] introduced a method for optical pulse compression based on the interplay between self-phase modulation (SPM) and group velocity dispersion (GVD) that arises during the propagation of short light pulses in single-mode optical fibers. Using this technique pulses as short as 6 fs at 620 nm were obtained in 1987 [11] and, employing an improved ultrabroad-band dispersion compensation, pulses as short as 4.5 fs at 800 nm were generated in 1997 [12]. The use of single-mode optical fibers limits the pulse energy to a few nanojoules. In 1996 a powerful pulse compression technique based on SPM-induced spectral broadening in a hollow fiber filled with noble gases demonstrated the capability of handling high energy pulses (mJ range) [13]. The implementation of the hollow-fiber technique using 20 fs seed pulses from a Ti:sapphire system and a high-throughput broadband compressor has led to the generation of pulses with durations down to 4.5 fs [14] and energy up to 0.55 mJ [15]. More recently, pulses as short as 3.8 fs have been generated using two hollow-fibers and a spatial light modulator [16]. A remarkable technique for the generation of frequency tunable ultrashort pulses is based on *optical parametric amplification*. Sub-10 fs light pulses tunable in the visible have been generated using optical parametric amplifiers (OPA) pumped by the second harmonic of a Ti:sapphire laser. Ultrashort pulses widely tunable in the near-infrared have been generated using OPAs pumped by Ti:sapphire lasers.

The chapter is organized as follows. Section 9.2 presents the Q-switching technique for the generation of nanosecond high-peak power pulses. Wavelength control methods in pulsed lasers are briefly reviewed in Section 9.3. In Section 9.4 the mode-locking technique is described. The role of cavity dispersion and the pulse shortening mechanism, which leads to the generation of femtosecond light pulses, are analyzed in Section 9.5. Section 9.6 presents the general scheme of pulse compression, which allows the generation of light pulses approaching the single-cycle regime in the visible spectral region. Section 9.7 describes the generation of few-optical-cycle parametric pulses tunable in the visible and in the near-infrared. Applications and perspectives are discussed in the last section, with particular emphasis on the generation of attosecond XUV light pulses.

9.2 Generation of nanosecond light pulses: Q-switching

A simple and direct way to generate light pulses is based on the use of a cw laser with an external modulator that transmits light only during particular temporal intervals. The disadvantages of this scheme are evident: (i) the system is greatly inefficient; (ii) the pulse peak power is limited by the steady power of the cw laser; (iii) the shortest achievable pulse duration is limited by the characteristics of the modulator. A more efficient solution can be obtained by inserting the modulator inside the laser resonator. In this way the pumping energy can be stored either in the gain medium, in the form of population inversion (Q-switching), or in the resonator, in the form of light that periodically escapes from the laser cavity (mode-locking). We will concentrate here on the first technique.

Under cw operation the steady-state population inversion (defined as the difference between the population density, N_2, of the upper laser level and that of the lower laser level, N_1) always equals the critical (threshold) inversion, which can be calculated by equating gain and losses. Therefore, below threshold the pump rate increases the energy stored in the material in the form of population inversion; above threshold the pump rate increases the number of photons, that is the electromagnetic energy stored in the cavity. If a shutter, assumed to be inserted inside the laser cavity, is closed, laser action is prevented since the gain is always less than the (very high) losses. In this situation the pumping process leads to a continuous increase of the population inversion in the gain medium, which may far exceed the threshold population. Suddenly, the shutter is opened. Since the cavity losses are suddenly reduced the laser gain largely exceeds losses and the stored energy may be released as a short, intense light pulse. This technique for the generation of short light pulses is called *Q-switching* because it is based on the switching of the cavity Q-factor from a low to a high value. We recall that, for any resonant system (in particular for a resonant optical cavity), the cavity Q-factor is defined as $Q = 2\pi\times$ (energy stored)/(energy lost in one oscillation cycle). Therefore a high-Q cavity is characterized by low losses, whereas a lossy cavity has a low Q. Q-switching allows the generation of light pulses with a duration comparable to the photon decay time (from a few nanoseconds to a few tens of nanoseconds) and high peak power (in the megawatt range). In the following we will qualitatively describe the dynamic behavior leading to the generation of the Q-switching pulses.

We assume that at time $t = 0$ the pump is turned on and it is maintained at a constant value; the shutter inside the cavity is closed. The population inversion increases, with a time constant given by the lifetime of the laser upper state, τ, up to an asymptotic value given by $N_\infty = R_p\tau$, where R_p is the constant pump rate. To achieve a sufficiently large inversion (i.e. a large stored energy) a long lifetime

τ is required, in the millisecond range, typical of an electric-dipole forbidden laser transition. Therefore, efficient Q-switching can be obtained in the case of most solid-state lasers (e.g. Nd in different host materials; ruby; Cr-doped materials) and some gas lasers (e.g. CO_2 or iodine). For dye lasers and several gas lasers (e.g. He-Ne or Ar), the laser lifetime is of the order of a few to a few tens of nanoseconds, therefore the Q-switching technique is not effective since there is not enough time to accumulate a high population inversion. At $t = t_1$ the shutter is suddenly opened, so that losses are switched to a low value (*fast switching*). At this point gain is larger than losses and laser action can start. The qualitative temporal evolution of the population inversion, $N(t)$, and of the photon density inside the laser cavity, $\phi(t)$, are shown in Figure 1. Laser oscillation begins and the cavity photon density rises sharply, thus depleting the population inversion, so that $N(t)$ begins to decrease. When $N(t)$ eventually falls to the threshold inversion, N_c, the photon density reaches its peak value. From this point the losses again exceed the gain, so that the photon density decreases to zero, with a time constant of the order of the cavity photon decay time (usually 5–50 ns). The described process is repeated periodically in such a way that a periodic pulse train is generated.

9.2.1 *Methods of Q-switching*

There are several methods to obtain Q-switching [17,18]. Four common methods are the following: (i) rotating-mirror method; (ii) electrooptical Q-switching; (iii) acousto-optical Q-switching; (iv) passive Q-switching. In the following we will briefly describe such methods.

(i) Rotating-mirror method: This was the first developed method to achieve Q-switching. In this case one of the cavity end mirror rotates at a very high angular velocity about an axis perpendicular to the cavity axis. The losses are very large except when the rotating mirror passes through a position parallel to the other cavity mirror. This method is simple and can be used at

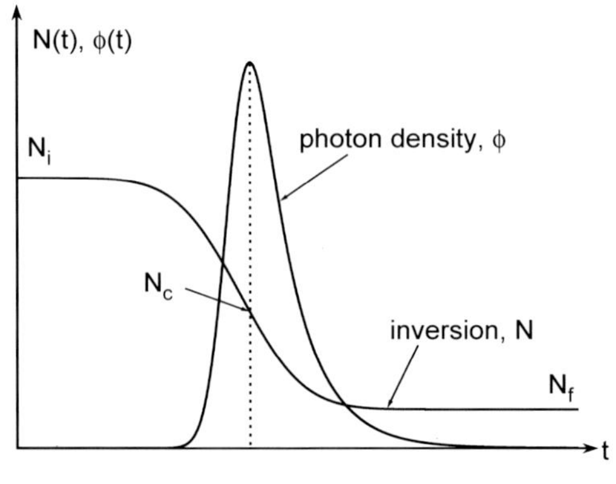

Figure 1. Temporal evolution of the population inversion, N, and of the photon density, ϕ, in a Q-switched laser, after the (fast) opening of the intracavity shutter.

any wavelength, but it is basically a slow switching device, due to the limited speed of the rotating motor, which sometimes results in the generation of multiple pulses. Moreover it leads to lower peak powers than other methods.

(ii) Electro-optical Q-switching: The basic element used in this method is an electrooptic medium, which becomes birefringent when an electric field is applied. The induced birefringence, defined as the difference between the refractive index for light polarized parallel to the direction of the inducing field and the refractive index for light polarized at right angles, can be proportional either to the applied electric field (*Pockels effect*) or to the square of the applied field (*Kerr effect*). Pockels cell Q-switches are generally preferred because of the lower voltage needed to produce the effect. Depending on the particular electrooptic crystal, the crystal dimension and the operating wavelength, typical voltages are in the 1–5 kV range. Q-switching can be obtain using a Pockels cell in combination with a polarizer located inside the laser cavity, as schematically shown in Figure 2. When the voltage applied to the cell is turned on, and the cell is properly aligned, the linearly polarized light passing through the polarizer is converted into circularly polarized light. The beam is then reflected back by the end mirror and passes a second time in the Pockels cell. At the output of the cell the beam is linearly polarized orthogonally to the polarization axis. Therefore when the voltage is on, optical feedback in the cavity is prevented. When the voltage is switched off, the cell is no longer birefringent so that the light polarization is no longer rotated. Optical feedback is active, resulting in a sudden increase of the cavity Q, leading to the generation of the Q-switched pulse.

(iii) Acousto-optical Q-switching: The acousto-optic effect is the change in the refractive index of a medium caused by the presence of sound. An acoustic wave creates a perturbation of the refractive index, leading to the generation of an optical phase grating, with a period equal to the acoustic wavelength and amplitude proportional to the sound amplitude, which can be used to deflect a fraction of the incident light beam. Therefore, if an acousto-optical cell is inserted in a laser cavity, an additional loss is produced when an acoustic wave is launched in the cell, by means of a suitable transducer. If the amplitude of the acoustic wave is sufficiently large, this additional loss is sufficient to prevent laser oscillation. When

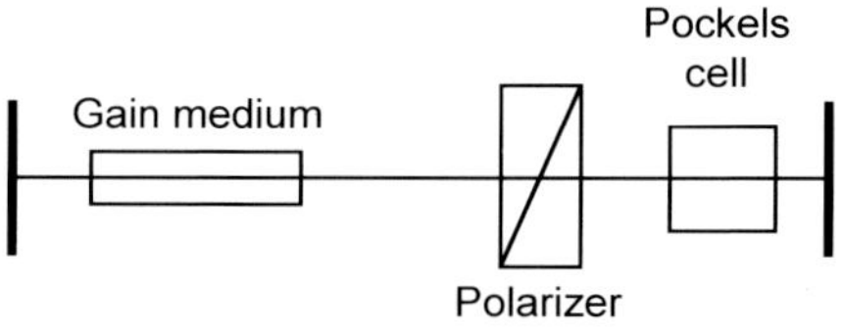

Figure 2. Schematic set-up of a laser cavity for electrooptical Q-switching with a Pockels cell.

the acoustic wave is turned off the light beam is no longer deflected, and the laser returns to its high-Q condition. Acousto-optic modulators have the advantage of low optical insertion losses.

(iv) Passive Q-switching: Passive Q-switching is based on the use of *saturable absorbers*, which are materials whose absorption can be saturated (or bleached) by laser radiation. Therefore, the absorption coefficient decreases upon increasing the intensity of the *resonant* incident radiation. Typically used saturable absorbers are solutions of a dye in a suitable solvent. Solid state and gaseous saturable absorbers are also used. This method is called *passive* since it is not driven by an external source, in contrast to the *active* methods described above. The effect of the saturable absorber in inducing Q-switching can be qualitatively understood as follows. Assume that the absorber, with peak absorption wavelength coincident with laser wavelength, is inserted in the laser cavity. Initially, the cavity losses introduced by the unsaturated absorber prevent laser oscillation. Therefore the population inversion can be significantly increased by pumping, without a significant increase of light intensity inside laser cavity. When laser action eventually starts, the beam intensity inside the cavity grows rapidly. This rapid intensity increase saturates the absorber, thus decreasing abruptly the cavity losses. This leads to the generation of a Q-switched pulse. Passive Q-switching is the simplest method of Q-switching, moreover it leads rather easily to emission in a narrow linewidth (single-mode operation).

9.2.2 Operating regimes

A widely used operating regime is the *continuously pumped, repetitively Q-switched* operation. In this case a cw pump is employed and cavity losses are periodically switched from a high to a low value, thus producing a continuous train of giant pulses. In this operating regime, the population inversion undergoes a periodic variation from its initial value N_i (before Q-switching) to the final value N_f (after the Q-switched pulse). After the Q-switched pulse, the continuous pumping restores the initial population inversion N_i, with a time constant equal to the upper-state lifetime τ, as described in Section 9.2. For this reason the temporal separation between two consecutive switching events (and therefore between two consecutive pulses) must be of the order of the lifetime τ. In fact, much longer temporal separation does not produce a significant increase of the population inversion and pump power is wasted through spontaneous decay. Repetition rates of cw-pumped Q-switched lasers are therefore typically from a few kilohertz to a few tens of kilohertz. In this case mechanical shutters or, more commonly, acousto-optic devices are used.

A Q-switched laser can also operate in a *pulsed regime*. In this case a pump pulse is used and the cavity Q is switched, mainly using electro-optical devices, when the population inversion reaches its maximum value. Repetitive operation with a typical repetition rate ranging from a few to a few tens of hertz can be achieved.

9.2.3 Gain switching

Like Q-switching, gain switching is a technique allowing the generation of laser high-peak-power pulses of short duration, typically from a few tens to a few hundreds of nanoseconds. Based on a fast switching of the laser gain, it is achieved by pumping the laser at such a fast pump rate that the population inversion, and therefore the laser gain, reach a value considerably above threshold before the laser oscillation has time to build up in the resonator. Note that any laser can in principle be gain-switched if a sufficiently short and intense pump pulse is available, even if the lifetime of the upper laser level falls in the nanosecond range. In this case, the pump pulse and the cavity photon decay time must be appreciably shorter than this lifetime, and very short gain-switched pulses, shorter than 1 ns in duration, can be generated.

9.3 Wavelength control methods in pulsed lasers

Several applications require the laser to operate on a narrow and tunable spectral bandwidth. Most applications of transition metal solid-state lasers (e.g. Ti:sapphire, $Cr:LiCaAlF_6$, $Cr:LiSrAlF_6$, etc.) take advantage of the broad tunability of these gain media. In this section the most widely used wavelength-control devices will be briefly described. For broad tuning, a coarse wavelength control can be obtained using gratings, prisms or birefringent filters. Gratings can produce narrow spectral bandwidths at the expense of high insertion losses. They can be used when the gain of the laser medium is high and when the pulselength of the laser is long, as in the case of dye lasers. Gratings can be used in either a Littrow or a grazing-incidence configuration. In a Littrow configuration the selected spectral components of the incident radiation are retroreflected. In this case the grating can be used as one of the cavity end mirrors. With a grazing-incidence configuration the grating is used as an internal mirror.

Birefringent filters consist of a plate of a suitable birefringent crystal (e.g. quartz or KDP) inclined at Brewster's angle to the beam direction, placed inside the laser cavity. In this configuration, the Brewster's angle surfaces act as a polarizer. Provided the optic axis of the birefringent crystal is neither perpendicular nor parallel to the plane of incidence, the input linearly polarized beam contains both ordinary and extraordinary components, which will experience a relative phase shift. For normal incidence such phase shift is given by the following expression:

$$\Delta\phi = 2\pi(n_e - n_o)L/\lambda \tag{1}$$

where: n_o and n_e are refractive indices for ordinary and extraordinary beams, respectively; L is the plate thickness. After passing through the plate, the two beam components will combine to form a resultant beam with elliptical polarization, unless $\Delta\phi$ is an integer number of 2π. In this last case beam polarization remains unchanged. Therefore, if a polarizer is used after the wave plate, only the wavelengths with the correct polarization will suffer no loss. The disadvantage of this wavelength control technique is due to unwanted secondary and tertiary transmission peaks, which can cause erratic wavelength tuning characteristics, particularly in ultrabroadband lasers, such as Ti:sapphire.

Prisms present low insertion loss when set at Brewster's angle; they have high damage threshold and do not present the problems associated with secondary transmission peaks. When a prism is inserted in a laser cavity, the resonator will be aligned correctly only for one wavelength. Wavelength control can be achieved by changing the orientation of the resonator mirrors.

For narrow and fine tuning, a narrow wavelength control device has to be used in addition to the broad wavelength control devices described above: etalons are virtually the only choice. An etalon consists of a plane-parallel plate of transparent material (fused silica or glass for visible or near-infrared radiation), with refractive index n, whose two plane surfaces are coated to a suitably high reflectivity value R. An etalon is hence a resonant structure: the resonance condition is obtained for those wavelengths that fill the distance between the reflecting planes with an integer number of half-wavelengths. Single-pass spectral resolution is given by Equation (2)

$$\Delta\nu_1 = c/(2ndF\cos\theta) \tag{2}$$

where d is the etalon thickness; θ is the angle of propagation; F is the etalon finesse defined as $F = \pi R^{1/2}/(1-R)$. When used in a laser resonator, the resolution of any wavelength control device is higher than in the single-pass configuration. In a pulsed laser the pulse propagates through the device several times during its evolution in the laser cavity. If p passes are made through the wavelength control device, the achievable spectral bandwidth [19] is reduced by a factor $p^{1/2}$. Therefore, spectral bandwidth depends on the temporal evolution of the laser pulse inside the cavity. The pulse evolution time interval τ_e and the number of passes are related simply as $p = \tau_e c/L$ [19], where L is the length of the resonator. Consequently, the spectral bandwidth is inversely proportional to the square root of the pulse-evolution time interval. Such a temporal interval can be significantly increased using an injection locked scheme, which gives narrower spectral bandwidths [19]. In this configuration, a low-power laser (*seed oscillator*) is operated with a narrow spectral bandwidth. The seed beam is then injected into a second high-power oscillator (*power oscillator*), which is induced to operate with a narrow spectral bandwidth. Since the power oscillator does not contain line-narrowing elements, the associated losses and the possibility of optical damage are both reduced.

9.4 Mode locking

As seen in Section 9.2, the minimum pulse duration that can be obtained using the Q-switching technique is of the order of the nanosecond. For a number of applications, even shorter pulses are required. As mentioned in the Introduction, the most widely used technique to generate short light pulses, in the picosecond and femtosecond range, is *mode-locking*. A typical laser consists of an optical resonator, made up of either plane or curved mirrors, enclosing the laser gain medium. The electromagnetic field inside the laser cavity can be described in terms of cavity modes (longitudinal modes), characterized by particular field distributions and regularly spaced resonance frequencies. The frequency separation between consecutive modes, Δv, is given by the following expression: $\Delta v = c/2L$, where c is the speed of light and L is the optical path length between the cavity end mirrors. Oscillation of the modes is sustained within the laser cavity when the gain exceeds the total losses. Therefore, the number of oscillating longitudinal modes depends upon the spectral width of the gain medium. If the cavity does not contain mode selecting elements, the output beam consists of a sum of the various frequency components, corresponding to the oscillating modes. The total electric field $E(t)$ of the output beam can be written as:

$$E(t) = \sum_n E_n \exp \mathrm{i}[(\omega_0 t + n\Delta\omega) + \phi_n] \tag{3}$$

where: E_n is the field amplitude of the nth mode; $\Delta\omega = 2\pi\Delta v$ is the mode angular frequency spacing and ϕ_n is the phase of the nth mode. In general, the phases of the modes have random values. In this case the intensity of the output beam shows random time behavior. Figure 3 shows the output intensity in the case of 20 oscillating modes, all with the same amplitude, E_0, and with random phases. The output beam consists of a random sequence of light pulses. It is worth pointing out a few characteristics of the temporal evolution of the output intensity shown in Figure 3. The waveform is periodic with a period $T = 1/\Delta v$, each intensity spike is characterized by a duration roughly given by $\tau = 1/(N\Delta v)$, where N is the total number of oscillating modes (so that $N\Delta v$ is the total oscillating bandwidth). Finally, the average power is the sum of powers in the oscillating modes, hence it is proportional to NE_0^2.

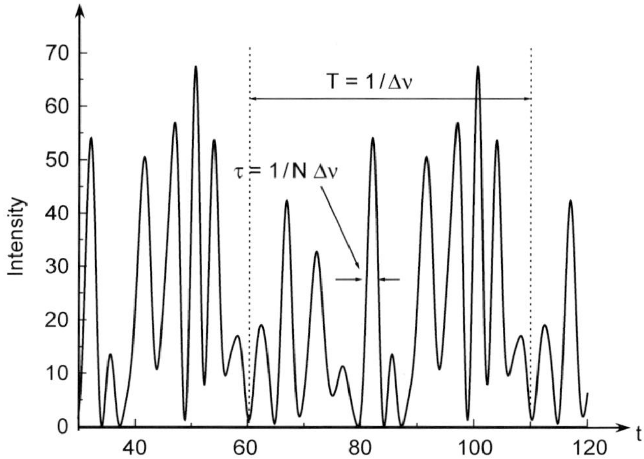

Figure 3. Temporal evolution of the laser output intensity in the case of 20 oscillating longitudinal modes, with the same amplitude and with random phases.

If now the phases of the oscillating modes are *locked*, that is if the modes are made to oscillate with some definite relation among their phases, the various modes will constructively interfere at a certain instant around a given point in the resonator. This is qualitatively shown in Figure 4, which, in the upper panel, displays the temporal evolution of the electric field of five modes with a constant phase relation ($\phi_n = n\phi$), and the same amplitude E_0. In the lower panel the resultant temporal evolution of the output intensity is shown: a pulse is generated, which travels back and forth inside the laser cavity. The phase locking of numerous longitudinal modes produces a total electric field equal to zero most of the time, except for very short intervals. The temporal separation between consecutive pulses is $T = 1/\Delta\nu = 2L/c$, which corresponds to the resonator round-trip time. The pulse duration, τ, is inversely proportional to the number of oscillating modes and therefore to the gain bandwidth of the active medium, $\tau = 1/(N\Delta\nu)$: the broader the oscillating bandwidth, the shorter is the obtainable pulse duration. The energy of the laser is concentrated within these short pulses as a result of the constructive interference between the oscillating modes. Note that the peak power of the pulse is proportional to $N^2E_0^2$. Therefore, for the same number of oscillating modes and for the same field amplitudes E_0, the ratio between the pulse peak power in the mode-locked case and the average power in the non-mode-locked case is equal to the number, N, of oscillating modes, which for solid-state or liquid lasers, can be very high (10^3–10^4). Mode-locking is thus useful not only for the generation of very short pulses, but also for the generation of high peak powers. Actual pulse shapes depend upon the method of mode-locking and a variety of material properties. Incomplete locking of the modes results in an envelope larger than predicted by the oscillating spectrum.

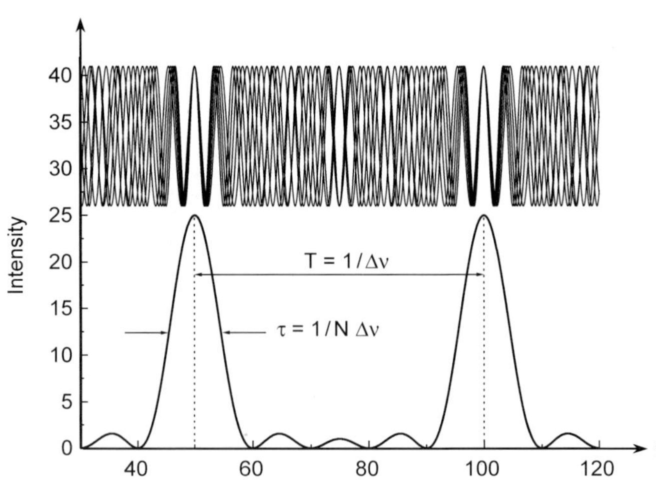

Figure 4. Upper panel: temporal evolution of the electric field of five oscillating longitudinal modes, with the same amplitude and locked phases; lower panel: temporal evolution of the corresponding laser output intensity.

The minimum pulse width obtainable from a given spectrum is referred to as the Fourier-transform-limited pulse-width.

Notably the spatial extension of a pulse in a typical mode-locked laser, $\Delta z = c\tau$, is usually much shorter than the cavity length. Therefore the temporal behavior of a mode-locked laser can be visualized considering a single ultrashort pulse propagating back and forth within the cavity. Each time the pulse is reflected onto the laser output coupler, a small fraction of its energy is coupled out of the resonator, thus leading to a train of ultrashort pulses at the output of the laser. In the framework of this time-domain picture of mode locking, the formation of the ultrashort pulse can be seen as produced by a suitably fast optical shutter placed at one end of the cavity, periodically opened, with a period equal to the cavity round-trip time ($2L/c$), for a short temporal interval, during which it is crossed by the pulse. Suppose that a non-mode-locked beam is present within the laser cavity, whose temporal intensity evolution can be represented as in Figure 3, and that the shutter is opened when the most intense noise pulse in Figure 3 reaches the shutter. If the opening time is comparable with the duration of the pulse, then only this pulse will grow in intensity, thus leading to the mode-locking condition. Such fast optical shutters can be implemented using several techniques, which fall into two main categories: (i) *active* mode-locking, in which the mode-locking element is driven by an external source; (ii) *passive* mode-locking, in which the pulses provide their own modulation, exploiting particular nonlinear optical effects. In the following such techniques will be briefly discussed.

9.4.1 Active mode locking

Active mode locking, first demonstrated in 1964 [2], is achieved when a modulation of the losses and/or gain is introduced into the laser resonator. In the case of gain modulation, the active medium is repetitively pumped by a train of regularly spaced pump pulses. If the pump pulse repetition frequency is equal to the round-trip frequency, the gain is modulated at a frequency corresponding to the mode spacing, thus resulting in efficient mode locking. This technique is also called *synchronous pumping*. This mode-locking technique, now less widely used, requires active media (typically dye solutions) with a gain relaxation time shorter than the pulse round-trip time in the cavity (nanosecond range). Moreover the repetition rate of the pump pulses must be equal (with high precision in the case of short pulses) to the repetition rate of the laser cavity, and therefore the cavity lengths of the two lasers must closely match. Pulse durations shorter than 1 ps are difficult to achieve from a synchronously pump dye laser.

With amplitude modulation (*AM mode locking*), a suitable amplitude modulator inside the laser cavity produces a window of net gain at the cavity round-trip frequency. A pulse arriving in the modulator at a time of minimum loss will return to the modulator after a time, $2L/c$, where the loss is again at a minimum. All the radiation in the laser resonator experiences loss except that radiation which passes through the modulator when loss is at the minimum value. Figure 5 shows how the amplitude modulator generates a window of gain for the peak of the pulse,

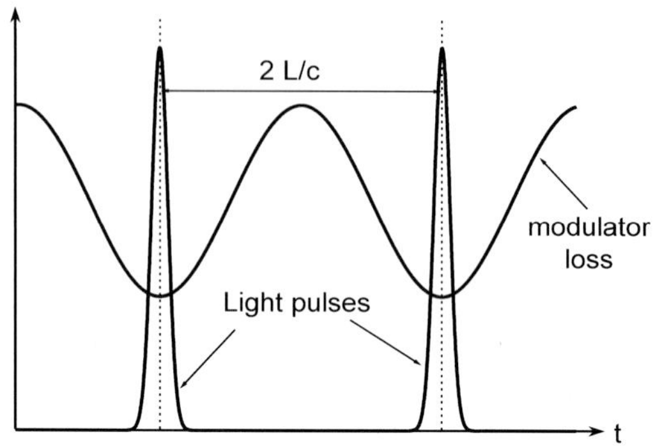

Figure 5. Time-domain description of active mode locking.

considering a gain medium with a long relaxation time, as in the case of solid-state lasers, so that time variation of the gain during the pulse can be neglected. Since the amplitude modulation of the pulse is the same for each pass, the effectiveness of the modulator decreases as the pulse gets shorter. Moreover, the slow variation of the imposed amplitude modulation provides only a weak and slow mode-locking effect. For this reason this technique is not suitable for the generation of ultrashort (sub-ps) pulses.

Amplitude modulation is commonly achieved using a Pockels cell modulator, for a pulsed and generally high-gain laser. For a cw-pumped and generally low-gain laser, amplitude modulation is achieved using an acoustooptic modulator, which has lower insertion losses than a Pockels cell modulator. In this case an acoustic standing wave is produced in the modulator. If the acoustic wave is oscillating at frequency ω_s, the diffraction loss will be modulated at frequency $2\omega_s$. If $2\omega_s = \pi c/L$, the cavity loss is modulated at the mode frequency separation, as required for mode locking.

Active mode locking can also be induced by a phase modulator (*FM mode locking*). Here the refractive index, n, of the modulator placed inside the laser cavity is sinusoidally modulated. This determines a corresponding modulation of the phases of the oscillating longitudinal modes: $\phi = (2\pi L'/\lambda)n(t)$, where L' is the modulator length. It is possible to demonstrate that such a phase modulation gives rise to two stable mode-locking states, corresponding to a light pulse passing through the modulator either at each minimum of $n(t)$ or at each maximum. The FM mode-locking technique is less widely used for two main reasons: (i) the generated pulses are frequency modulated; (ii) mode locking tends to be unstable because switching between the two stable mode-locking states mentioned above often occurs in practice.

9.4.2 Passive mode locking

Passive mode locking involves the use of a system which transmits or reflects intense light pulses with less losses than those experienced by a constant low level of light. Under proper conditions this process leads to the generation of a train of pulses. Since its first demonstration, in 1966 [20], passive mode locking has led to the generation of the first sub-picosecond pulses, in 1974 [21], and of the first sub-100 fs pulses in 1981 [22]. Passive mode locking is now used to generate sub-10 fs light pulses. The key characteristic of this mode-locking technique is that the pulse inside the laser cavity self-modulates itself, more rapidly than would be possible with any active modulation. The most common technique of passive mode-locking makes use of a saturable absorber in the laser cavity. This passive device is characterized by a transmission that increases nonlinearly with increasing light intensity. There are two main classes of passive mode locking, depending on the nature of the employed saturable absorbers: (i) *slow saturable absorber* mode locking, in which the absorber cannot recover its original absorption on the time-scale of the pulse; (ii) *fast saturable absorber* mode locking, in which the absorber recovers its initial absorption on a time-scale shorter than the pulse duration. In the following we will briefly discuss these two classes.

9.4.2.1 Mode locking with a slow saturable absorber

The use of a slow saturable absorber can lead to the generation of ultrashort pulses when saturation of the gain medium also occurs. The combined action of absorber and gain can generate a very short window of net gain, thus giving rise to pulse formation. The relevant parameters for the analysis of this technique are the saturation fluences of both gain medium, $\Gamma_{sg} = h\nu/\sigma_g$, and saturable absorber, $\Gamma_{sa} = h\nu/2\sigma_a$, where σ_g and σ_a are the effective gain and absorber cross-sections [17]. When the pulse is much shorter than the recovery time of either the absorber or the gain, the loss term (ℓ) and the gain (g) experienced by the pulse can be written as

$$\ell(t) = \ell_0 \exp[-\Gamma(t)/\Gamma_{sa}] \tag{4}$$

$$g(t) = g_0 \exp[-\Gamma(t)/\Gamma_{sg}] \tag{5}$$

where, indicating by $I(t)$ the pulse intensity, $\Gamma(t) = \int_{-\infty}^{t} I(t)\mathrm{d}t$ is the pulse energy fluence. We assume that both gain medium and absorber recover on a time scale comparable with the cavity round-trip time. If the saturation fluence of the gain medium is larger than that of the saturable absorber, a net gain window can be produced. With the help of Figure 6, this mode locking mechanism can be easily explained. Before the arrival of the pulse, the gain must be smaller than the losses, in order to have a loss for the pulse leading edge. At some time during the leading edge of the pulse, when the pulse fluence becomes comparable to Γ_{sa}, saturation of the absorber begins and the loss can thus become smaller than the gain. Starting from this time, the pulse will be amplified rather than attenuated. Subsequently, when the pulse fluence becomes comparable to Γ_{sg}, saturation of the gain medium will occur. The gain can thus become smaller than the loss on the trailing edge of the pulse. Under the above conditions, the pulse will experience a net gain in its central part and net loss in its wings. A steady-state condition is obtained when the pulse-shortening mechanisms are balanced by the pulse broadening processes present in the laser cavity.

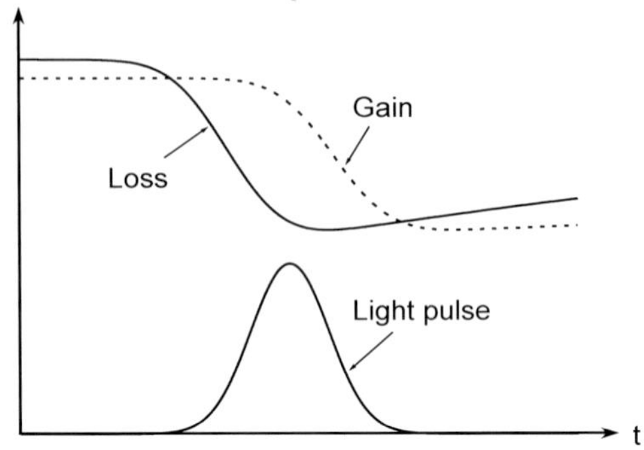

Figure 6. Time-domain description of passive mode locking with a slow saturable absorber.

Notably, according to Equations (4) and (5), the amplitude of the net-gain window depends only on the pulse fluence and is independent of pulse duration, whilst the speed of the modulation is determined by the pulse duration, since the net-gain window narrows upon decreasing the pulse width. Therefore, in this case the effectiveness of the saturable absorber modulator does not decrease as the pulse gets shorter, as for active mode locking; indeed, the shortening mechanism is independent of pulse duration [23].

As mentioned above, the relaxation time of both absorber and gain medium must be comparable to the cavity round-trip time. Moreover, the saturation fluences of both gain medium and absorber must be sufficiently low to allow the two media to be saturated by the laser pulse. This requires the use of a short-lifetime (a few nanosecond) high cross-section ($\sim 10^{-16}\,\mathrm{cm}^2$) gain media, such as dyes or semiconductors. This type of mode locking cannot occur with long-relaxation-time (hundreds of microsecond) solid-state gain media, where dynamic gain saturation cannot occur.

9.4.2.2 Mode locking with a fast saturable absorber

A fast saturable absorber is a medium that recovers from absorption saturation on a time scale shorter than the pulse duration. Such absorber can be employed to achieve mode locking in combination with long-relaxation-time gain media. Figure 7 shows the physical picture of this type of mode locking. The gain is assumed constant during the pulse, and equal to its saturated value established by the average intracavity laser power. The cavity power loss can be written as:

$$\ell(t) = \ell_0 - \gamma I(t) \tag{6}$$

where $\gamma I(t) \ll \ell_0$. As it is evident from Figure 7, the window of net gain is completely determined by the temporal dynamics of the absorber. The pulse tails experience a net loss and the pulse peak a net gain. In this case the effectiveness of the pulse shortening mechanism increases as the pulse gets shorter. A steady-state condition is established by the balance between this pulse-shortening mechanism and the pulse broadening processes.

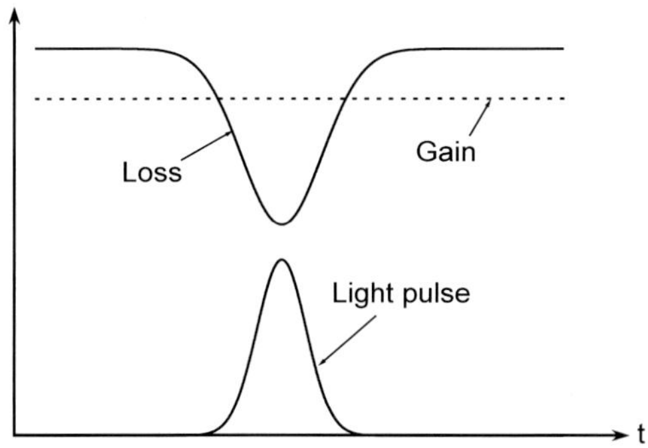

Figure 7. Time-domain description of passive mode locking with a fast saturable absorber.

There are two classes of fast saturable absorbers used for passive mode locking: (i) real fast saturable absorbers and (ii) artificial fast saturable absorbers. The shortest mode-locked pulses have been generated by using artificial saturable absorbers. The most commonly used real saturable absorbers are either solution of fast dye molecules or semiconductors. In the first case cyanine dyes are often used, which are characterized by an upper state relaxation time of a few tens of picoseconds. Mode-locked pulses shorter than a few picosecond cannot be obtained. Semiconductor saturable absorbers have been used in color-center lasers, in coupled-cavity systems, in anti-resonant Fabry–Perot devices and in fiber lasers.

Artificial fast saturable absorbers are based on fast nonresonant optical nonlinearities. The use of such artificial absorbers was first proposed at the beginning of 1970s [24], but their utility was fully recognized only with the emergence of fiber and high-power cw solid-state laser. An extremely important break-through in femtosecond laser technology was made in 1991 with the first demonstration of the self-mode-locked Ti:sapphire laser by Spence et al. [6]. As subsequently explained by Piché [25], this mode-locking technique is based on the optical Kerr effect in the laser crystal. The combination of the Kerr-induced self-focusing of the laser beam, with an intracavity aperture, leads to the generation of an ultrafast artificial saturable absorber. The technique has been called Kerr-lens mode locking (KLM). The optical Kerr effect in the laser crystal gives rise to an intensity dependent change of the refractive index, $\Delta n(\mathbf{r}, t) = n_2 I(\mathbf{r}, t)$, where: n_2 is the nonlinear index coefficient; $I(\mathbf{r},t)$ is the instantaneous laser intensity and $\boldsymbol{r}$ is the beam radial coordinate. This effect is generally due to hyperpolarizability in the medium induced, at high fields, by either a deformation of atomic or molecular electronic orbitals or from a reorientation of elongated molecules (in gas or liquid). For solid media it is due to deformation of the electron cloud of the atom, so that the optical Kerr effect is very fast, with a response time on the order of a rotation period of the outermost electrons of the atom. Due to the "bell" shape of the radial intensity profile of the laser beam, the Kerr effect produces an intensity dependent lensing effect. If we assume that a pulse is already present in the laser cavity, the induced lens changes in time following the temporal intensity profile of the pulse.

This effect determines a change in time of the transverse section of the beam inside the laser cavity. If an aperture of suitable size in placed at a suitable position in the cavity, the pulse tails (which correspond to a larger transverse section of the beam) experience more losses than the peak, which passes almost unchanged, as schematically shown in Figure 8. Pulses get shorter at each round trip in the cavity. This process results in reduced losses for increased intensity, thus leading to the generation of an artificial fast saturable absorber.

Importantly, the self-amplitude-modulation (SAM) induced by the Kerr effect is rather weak, so that it cannot overcome the pulse broadening effect introduced by cavity dispersion in the femtosecond regime. This limits the pulse shortening process well before the limit set by the bandwidth of the gain medium, which, for the commonly used Ti:sapphire (Ti:S) laser would permit pulse durations of a few femtoseconds. When ultrabroadband gain media (with bandwidths as large as 100 THz) are used, cavity dispersion is important in setting the shortest pulse duration achievable in mode locking.

Moreover, the intensity dependent refractive index of the gain medium not only introduces SAM, it also induces a rapid change of the phase of the electric field as a function of time. This process is known as *self-phase modulation* (SPM) and can be described as follows. Consider a Kerr medium and assume that a light pulse is traveling through the medium. After the propagation distance L the accumulated phase is given by:

$$\phi = \omega_0 t - \beta L = \omega_0 t - \omega_0 L(n_0 + n_2 I)/c \tag{7}$$

where n_0 is the unperturbed refractive index of the medium; ω_0 is the pulse central carrier frequency; $\beta = \omega_0 n/c$ is the propagation constant. The instantaneous (angular) carrier frequency of the pulse is given by the temporal derivative of the accumulated phase, and can be calculated, using Equation (7), as

$$\omega = \frac{\partial \phi}{\partial t} = \omega_0 - \frac{\omega_0 n_2 L}{c}\frac{\partial I}{\partial t} \tag{8}$$

From Equation (8), SPM evidently leads to the generation of new spectral components, at lower frequency with respect to ω_0 on the leading edge of the

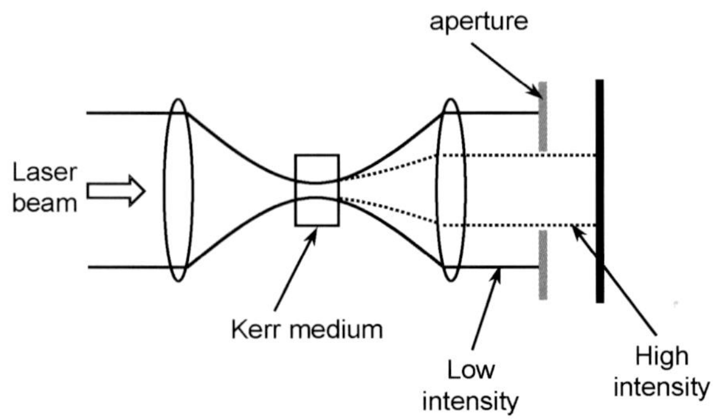

Figure 8. Beam size variation in Kerr-lens mode locking.

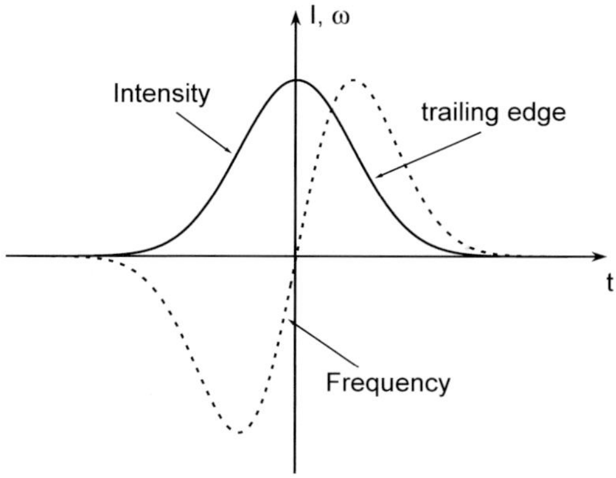

Figure 9. Temporal evolution of the intensity of a laser pulse (solid curve); instantaneous (angular) carrier frequency produced by self-phase modulation (dashed curve).

pulse, and at higher frequency on the pulse trailing edge (we have considered a medium with positive nonlinear index n_2) as shown in Fig. 9. Note that, around the peak of the pulse, the frequency chirp induced by SPM increases almost linearly with time. Neglecting dispersion, the initial pulse electric field, shown in Figure 10(a), does not change its envelope (dashed curve) after propagation in the

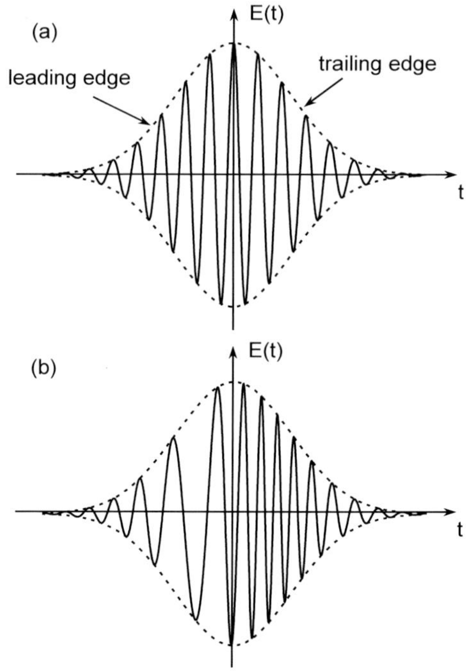

Figure 10. (a) Electric field of a light pulse at the input of a Kerr medium; (b) pulse electric field after propagation in a Kerr medium.

Kerr medium. Figure 10(b) shows the pulse electric field after propagation: on the leading edge of the pulse a red-shift of the instantaneous frequency is evident, while on the trailing edge $\omega(t)$ is blue-shifted. The instantaneous frequency does not change at both the peak and far from the peak of the pulse.

The generation of sub-100 fs pulses in KLM Ti:S lasers is due to a pulse shortening mechanism, resulting from the interplay between SPM induced by the Kerr effect in the laser crystal and a net negative cavity dispersion produced by a suitably designed dispersive delay line introduced in the laser cavity. Therefore, in the following section we briefly review the important concept of group delay dispersion in a dispersive medium, and discuss the essential aspects of the pulse shortening mechanism in a KLM laser for the generation of ultrashort light pulses (down to the sub-10 fs regime).

9.5 Generation of femtosecond pulses: role of cavity dispersion and solitary mode locking

For a given optical system, dispersion determines a frequency dependent propagation time. Spectral components at angular frequency ω are delayed by a *group delay* $T_g(\omega) = \partial\phi/\partial\omega = \phi'(\omega)$, where $\phi(\omega)$ is the pulse phase retardation. Expansion of $T_g(\omega)$ with respect to frequency gives the following expression:

$$T_g(\omega) = \phi'(\omega_0) + \phi''(\omega_0)(\omega - \omega_0) + \tfrac{1}{2}\phi'''(\omega_0)(\omega - \omega_0)^2 + \tfrac{1}{6}\phi''''(\omega_0)(\omega - \omega_0)^3 + \cdots \qquad (9)$$

The first term represents an overall delay of the pulse; the higher order terms of the expansion determine a distortion of the shape of the pulse. The coefficient of the second term, $\phi''(\omega_0)$, is called *group delay dispersion*(GDD), and is a function of reference frequency. $\phi'''(\omega_0)$ and $\phi''''(\omega_0)$ are called third-order and fourth-order dispersion, respectively. Critical values of these dispersion terms above which dispersion causes a significant change of the pulse shape are given by a simple scaling expression: $\phi^{(n)} = \tau_p^n$, where $\phi^{(n)}$ is the nth-order dispersion term and τ_p is the pulse duration. For example, a second order dispersion $\phi^{(2)} = \phi'' = \tau_p^2$ results in a pulse broadening by more than a factor of two. Therefore, dispersion-induced distortions of the pulse shape become increasingly important for decreasing pulse durations.

The optical elements used in a laser cavity usually introduce a positive GDD, which determines a group delay that increases with frequency. This process thus leads to a positive frequency sweep (or *chirp*) of the pulse propagating in the considered optical element. In other words, after the propagation in a dispersive medium, different temporal portions of the pulse have different carrier frequencies. In particular, in the case of positive GDD, the low frequency (red) spectral components of the pulse are located at the leading edge of the pulse, while the high frequency (blue) spectral components are located at the pulse trailing edge. Even a few millimetres of propagation in a transparent medium, such as quartz or sapphire, broadens significantly a pulse with a duration of a few tens of femtoseconds.

The broadening is even accelerated in the presence of SPM, which determines a broadening of the pulse spectrum. To compensate for this positive GDD, a component with a negative GDD has to be inserted into the laser cavity. Different schemes have been employed: the Brewster-angled prism compressor introduced by Fork et al. [26] was the first low-loss source of negative dispersion and has been widely used since its discovery. Pulses as short as 8.5 fs have been demonstrated using only prism compensation [27]. The drawback of this system is that negative GDD is associated with a significant amount of higher-order dispersion, which cannot be lowered or adjusted independently of the desired GDD, thus limiting the bandwidth over which correct dispersion control can be obtained. In 1994, advances in the design of chirped multilayer coatings [28] led to the demonstration of chirped mirrors providing ultrabroadband dispersion control with low losses. Chirped mirrors introduce a frequency-dependent group delay by providing a wavelength-dependent penetration depth of the incident radiation. Dispersion characteristics of chirped mirrors can be tailored to produce the required amounts of GDD as well as higher-order dispersion. Chirped mirrors are now key components in broadband delay lines used for the generation of sub-10 fs pulses directly from laser oscillators.

For the generation of the shortest pulses in a KLM laser, a small but constant intracavity dispersion is required over the full pulse spectrum [29]. The generation of ultrashort pulses is due to a pulse-shortening mechanism based on the interplay between SPM and a net *negative* GDD introduced by prisms or chirped mirrors. This mechanism is called *solitary mode locking* [30]. As previously discussed, the process of SPM results in a frequency chirp increasing linearly (around the pulse peak) with time; conversely, propagation in a medium with a negative GDD gives a frequency chirp that decreases linearly with time. Under appropriate conditions, the two effects cancel each other, thus leading to the formation of an ultrashort pulse. In the framework of the weak pulse-shaping approximation (i.e. when the relative variation, after a cavity round trip, of the amplitude envelope of the pulse electric field is weak) an analitycal theory [29] predicts a steady-state solution for the pulse electric field amplitude of the form $E(t) = E_0\,\mathrm{sech}(t/\tau_0)$, with a minimum τ_0 given by:

$$\tau_0 = \frac{2|D|}{\delta_{\mathrm{SPM}} W_{\mathrm{p}}} \tag{10}$$

and a pulse duration (full-width at half-maximum of the intensity profile) $\tau_{\mathrm{p}} = 1.76\tau_0$. In Equation (10) D (negative) is the net intracavity GDD; W_{p} is the intracavity pulse energy; δ_{SPM} is the SPM coefficient given by $\delta_{\mathrm{SPM}} = 2n_2L/(\lambda_0 w_0^2)$, where: λ_0 is the carrier wavelength of the laser pulse; L is the length of the Kerr medium (typically the gain crystal); w_0 is the $1/e^2$ beam radius. Note that the solitary solution is unstable in a mode-locking laser cavity unless a nonlinear loss mechanism is present, which produces a self-amplitude modulation. This SAM is produce by the Kerr-lens effect.

In the sub-15 fs regime, third-order dispersion (TOD) becomes the main limiting factor in achieving ultrashort pulses. Furthermore, at very short pulsewidths,

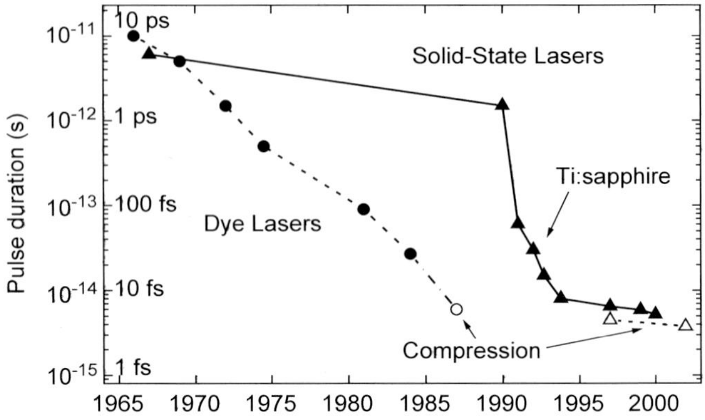

Figure 11. Historical evolution of the ultrafast laser technology.

the weak pulse-shaping approximation is no longer valid. When the changes produced by the individual intracavity elements are large, the pulse may have strongly different durations in different portions of the resonator. In a typical sub-10 fs Ti:S laser, the uncompensated round-trip GDD determined by the gain medium alone introduces a ~100 fs delay between the long and the short wavelength components, so that pulse duration is periodically stretched and recompressed from 100 to sub-10 fs during a round trip in the resonator. In this case the ordering of the intracavity optical elements can be important, particularly in the case of strong SPM.

Since the first discovery of KLM, a dramatic reduction in achievable pulse duration has been obtained, as shown in (Figure 11), which shows the historical evolution of the laser technology for the generation of ultrashort light pulses. Using fused-silica prisms for intracavity dispersion control, pulses in the 10 fs regime were generated in KLM Ti:S lasers, by different research groups [31,32]. With the introduction of chirped mirrors even shorter pulses could be generated. Recently, sub-6 fs pulses have been generated directly from Kerr-lens mode-locked Ti:sapphire lasers [7,8].

9.6 Optical pulse compression: towards the single-cycle regime

Even shorter pulses can be generated by extracavity pulse compression. The general scheme of light pulse compression is the following. The input pulse is first injected into a phase modulator, which broadens the pulse spectrum imposing a frequency chirp (in time). The spectrally broadened and chirped pulse is sent in a dispersive delay line, which re-phases all the new frequency components generated by the phase modulation. Ideally, the dispersive delay line would introduce the opposite chirp on the pulse, thus resulting in the compression of the pulse to its minimum width, $\sim 1/\Delta\omega$, where $\Delta\omega$ is the frequency sweep imposed on the pulse in the first step. This general scheme for compression of optical pulses was independently proposed by Gires and Tournois in 1964 [33] and Giordmaine et al. in 1968 [34]. To compress femtosecond pulses an ultrafast phase modulator has to be used. As first proposed in 1969 by Fisher et al. [35] and by Laubereau [36], it is possible to use the pulse itself to modulate its phase by optical Kerr effect in a suitable medium (SPM). A fundamental requirement for pulse compression is that the Kerr effect is provided by a guiding nonlinear medium. In fact, the spatial intensity profile of the light pulse propagating along a non-guiding medium leads to a spatially non-uniform SPM, so that different frequency chirps are generated across the transverse beam distribution. Spatially uniform spectral broadening can be obtained using a guiding nonlinear medium. In 1981 Nakatsuka et al. [10] performed the first pulse compression experiment using single-mode optical fibers as Kerr medium in the positive dispersion region.

In the last two decades the general scheme of pulse compression described above has been implemented in different ways. Using a single-mode optical fiber as ultrafast phase modulator and a prism-grating compressor, pulses as short as 6 fs at 620 nm were obtained in 1987 from 50 fs pulses generated by a colliding-pulse mode-locking dye laser [11]. In 1997, 13 fs pulses from a cavity-dumped Ti:sapphire laser were compressed to 4.5 fs with the same technique using a compressor consisting of a quartz 45°-prism pair, broadband chirped mirror and a thin-film Gires–Tournois dielectric interferometer compressor [12]. The use of a single-mode optical fiber limits the pulse energy to a few nanojoules. In 1996, using a phase modulator consisting of a hollow fiber filled with noble gases, a powerful pulse compression technique that can handle high-energy pulses was introduced [13]. Implementation of the hollow-fiber technique using 20 fs seed pulses from a Ti:sapphire system and a high-throughput broadband prism chirped-mirror, or chirped-mirror only dispersive delay line has led to the generation of pulses with duration down to 4.5 fs [14] and energy up to 0.55 mJ [15]. In 2003, pulses as short as 3.8 fs were generated by using two gas-filled hollow fiber and a compressor based on a spatial light modulator with liquid crystals [16]. This technique presents the advantages of a guiding element with a large-diameter mode, and a fast nonlinear medium with high damage threshold. Of course, the possibility of taking advantage of the ultrabroadband spectrum, which can be generated by the phase modulation process, is strictly related to the development of dispersive delay lines capable of controlling the frequency-dependent group delay over such a bandwidth.

Alternative methods for phase modulation of short pulses, not based on SPM, have been proposed. One method uses two narrow-band lasers, slightly detuned from the Raman resonance, to generate a spectrum of Raman sidebands, whose Fourier transform is a periodic train of subfemtosecond pulses [37]. A second technique is based on impulsive excitation of molecular motion in Raman-active gases [38].

The potential of stimulated Raman scattering for broadband generation has been demonstrated by several groups. In 2000, using molecular deuterium (D_2), Sokolov et al. [39] demonstrated collinear generation of a Raman spectrum extending over 50000 cm^{-1}. The basic idea of the method is the use of a Raman transition with a sufficiently large coherence so that the generation length and the phase-slip length are of the same order. This coherence is established by driving the molecular transition using two narrow-band lasers, chosen in such a way that the (tunable) frequency difference between the two beams is approximately equal to the fundamental vibrational frequency in deuterium. The induced molecular motion modulates the driving laser frequencies, giving rise to the collinear generation of a very broad spectrum. The generated spectrum has approximately Bessel function sideband amplitudes and in the temporal domain corresponds to a periodic beat of two frequency-modulated signals with center frequencies corresponding to the driving frequencies. Numerical simulation predicts that the group velocity dispersion of the Raman medium itself leads to pulse compression, giving rise to the generation of a train of sub-femtosecond pulses [40].

Another method for molecular phase modulation is based on impulsive excitation of molecular motion in Raman-active gases [38]. In this case a first high-intensity light pulse, with a duration shorter than the molecular vibrational period of a molecular gas, impulsively excites a coherent molecular motion (vibration or rotation) in the gas. A second, relatively weak, delayed pulse propagating in the excited medium is strongly phase-modulated due to the modulation of the gas refractive index induced by the molecular motion. The condition for impulsive excitation is easily met, since the vibrational frequencies of molecular motion in Raman active gases and liquids are typically of the order of several hundreds of cm^{-1}, so that, for exciting pulse durations below 100 fs, the molecular vibrational period turns out to be longer than the laser pulse duration.

9.7 Wavelength tunability by optical parametric amplification

Currently, the most widely used ultrashort laser systems are based on Ti:sapphire and can be tuned only on a small spectral range around 800 nm. For several important applications in nonlinear optics and ultrafast spectroscopy a wider wavelength tunability is strongly required. Parametric processes are now widely used to generate ultrashort pulses tunable in the ultraviolet, visible and near-infrared regions. Parametric devices are based on the use of nonlinear quadratic crystals characterized by a large second-order susceptibility ($\chi^{(2)}$), such as BBO (β-barium borate) or LBO (lithium triborate). In these nonlinear crystals, an high-energy (pump) photon splits into two lower energy (signal and idler) photons, satisfying the energy conservation condition $\hbar\omega_p = \hbar\omega_s + \hbar\omega_i$ (p, s and i stand for pump, signal and idler, respectively). To obtain an efficient process, the interacting beams must also satisfy the momentum conservation (*phase matching*) condition $\hbar k_p = \hbar k_s + \hbar k_i$, where the $\boldsymbol{k}_j$ (j *is* p, s, i) are the wave vectors of the three waves. Phase matching is usually achieved by exploiting the birefringence of a nonlinear crystal. By adjusting the angle between the optical axis of the birefringent crystal and the direction of propagation of the beam, it is possible to change the wavelength for which the phase-matching condition is fulfilled, thus tuning the parametric interaction. Let us consider a negative uniaxial crystal, such as BBO; phase matching can be obtained by propagating the pump beam along the extraordinary (e) axis and signal and idler beams along the ordinary (o) axis, in the so-called type I geometry ($o_{\text{signal}} + o_{\text{idler}} \rightarrow e_{\text{pump}}$). If the three beams are propagating collinearly, the momentum conservation condition becomes the simple scalar equation

$$\omega_p n_e(\omega_p, \theta) = \omega_s n_o(\omega_s) + \omega_i n_o(\omega_i) \quad (11)$$

Equation (11) allows one to obtain the phase matching angle θ, for a given pump wavelength, as a function of signal wavelength. Usually, phase matching can be achieved only over a narrow signal frequency range, thus reducing the gain bandwidth. The situation can be drastically changed by using a *noncollinear* phase matching geometry; in which case the phase matching condition gives rise to the two following scalar equations:

$$\omega_p n_e(\omega_p, \theta) = \omega_s n_o(\omega_s) \cos\alpha + \omega_i n_o(\omega_i) \cos\beta \quad (12)$$

$$\omega_s n_o(\omega_s) \sin\alpha = \omega_i n_o(\omega_i) \sin\beta \quad (13)$$

where α and β are the pump–signal and pump–idler angles, respectively. For a fixed α, Equation (13) allows one to obtain β and Equation (12) to obtain the phase matching angle θ. The angle α offers an additional degree of freedom, which can be used to modify the phase-matching conditions, trying to increase the phase matching bandwidth. This is indeed possible in BBO for a pump wavelength $\lambda_p = 400$ nm: in fact by increasing the angle α the signal wavelength dependence of θ becomes progressively weaker until, for the optimum value $\alpha = 3.7°$, a given crystal orientation ($\theta = 32°$) allows the achievement of phase matching over an ultrabroad bandwidth, extending from 500 to 750 nm. This "magic" angle is by chance also very close to the angle ($\approx 4°$) for which spatial walk-off between pump and signal is compensated, thus enabling high gain. This favourable property of BBO was first recognized by Gale et al. and was initially exploited to build an optical parametric oscillator, pumped by the second harmonic of a Ti:sapphire oscillator, generating visible pulses with duration down to 13 fs [41]. In 1997, the same concept was extended by various groups to optical parametric amplifiers, pumped by the second harmonic of chirped-pulse amplified Ti:sapphire lasers [42–44]. Using this technique, sub-10-fs pulses tunable in the visible have been generated, with energy in the microjoule range.

Few-optical-cycle light pulses tunable in the near-infrared can be generated exploiting nonlinear pulse self-shortening mechanisms in OPAs pumped by Ti:sapphire lasers [45]. High energy infrared tunable femtosecond pulses are of particular interest for applications. Sub-100 fs infrared pulses have been generated using traveling-wave nonlinear frequency converters based on OPAs [45,46]. In 1998 pulses as short as 14.5 fs were generated at 1.5 μm, using an OPA based on type-II phase-matching BBO crystals [47].

9.8 Applications and perspectives: from femtochemistry to attophysics

Many fundamental processes of light-matter interaction take place in nature on extremely short time scales, typically below 1 ps. As an example, electronic motion and dephasing in molecules and solids occurs in the 10 fs time domain, nuclear motion in the 100 fs time scale, and vital functions in living system such as energy relaxation, energy transfer or isomerization occurs within a few ps. Spectroscopy is the natural tool for getting information on the physical mechanisms underlying these phenomena – provided that techniques with suitable time resolution be available. The development of laser sources delivering sub-10 fs light pulses, either directly from an oscillator or by using external compression techniques, allow the traditional fields of time-resolved spectroscopy to be pushed to the extreme. In particular, the generation of powerful light pulses in the 5 fs regime by the hollow fiber compression technique has opened up new frontiers for experimental physics. One of these, the more appealing for future investigation of molecular or solid state physics, is extreme nonlinear optics [48], i.e. the wealth of phenomena taking place when ultrashort pulses are focussed to small spots, thus reaching unprecedented peak intensities so that the electric field of the pulse, rather than the intensity profile, is relevant. The spatial extension of a 5 fs pulse along the propagation direction is limited to a few times the wavelength of the radiation ($\sim$0.5–1 μm in the visible and near-infrared spectral region). Since such pulses are delivered in a nearly diffraction-limited beam they can be focussed to a spot comparable in size to the wavelength, thus leading to a concentration of the light in a volume of a few cubic micrometres. This extreme temporal and spatial confinement allows one to achieve peak intensities higher than $10^{15}\,\mathrm{W\,cm^{-2}}$ with pulse energies in the microjoule range. At these intensity levels the amplitude of the electric field approaches $10^{9}\,\mathrm{V\,cm^{-1}}$, which exceeds the static Coulomb field experienced by outer-shell electrons in atoms, thus leading to optical-field ionization. Using few-optical-cycle pulses the optical-field ionization rate becomes comparable to the laser field oscillation frequency and the electron is set free near the peak of the pulse. The shorter the driving pulse, the stronger is the laser field the electron experiences at the instant of its detachment from the atom. This has opened the way to important applications of intense sub-10 fs pulses in the field of *high-order harmonic generation* in noble gases. The physical processes leading to the generation of extreme-ultraviolet (XUV) or soft-X-ray radiation by high-order harmonic generation, can be understood using the so-called three step model. In the framework of this semiclassical model, an electron exposed to an intense, linearly polarized electromagnetic field is emitted from the atom by tunnel ionization. The freed electron may be driven back towards its parent ion by the external field and, with small probability, it recombines to the ground state, thus emitting a photon with an energy equal to the sum of the ionization potential and the electron kinetic energy gained in the laser field. In the cutoff region (i.e. in the photon energy range close to the highest harmonics) this source was predicted to emit train of attosecond ($1\,\mathrm{as} = 10^{-18}$ s) XUV (or soft X-ray) pulses separated by half of the laser oscillation period. This has been recently experimentally demonstrated: using 40 fs driving pulses Paul et al. [49] have experimentally demonstrated that a group of harmonics (ranging from the 11th to 19th) generated in a jet of argon, are locked in phase and form a train of 250 as pulses, separated by one half cycle of the laser period and extending over about 10 fs. The most promising way to produce *single* attosecond pulses is based on the use of intense few-optical-cycle driving pulses. The cutoff harmonics generated by such pulses were predicted to merge into a continuum, thus generating a single attosecond pulse, as recently experimentally demonstrated [50] with the generation of a 650±150 as pulse, using sub-10 fs driving pulses generated using the hollow fiber compression technique. This important achievement paves the way to the extension of time-resolved spectroscopy into the attosecond domain.

References

1. M. DiDomenico (1964). Small-signal analysis of internal (coupling type) modulation of lasers. *J. Appl. Phys.* **35**, 2870–2876.
2. L.E. Hargrove, R.L. Fork, M.A. Pollack (1964). Locking of He-Ne laser modes induced by synchronous intracavity modulation, *Appl. Phys. Lett.* **5**, 4–5.
3. O.G. Peterson, S.A. Tuccio, B.B. Snavely (1970). cw operation of an organic dye solution laser, *Appl. Phys. Lett.* **17**, 245–247.
4. R.L. Fork, B.I. Green, C.V. Shank (1981). Generation of optical pulses shorter than 0.1 psec by colliding pulse mode locking, *Appl. Phys. Lett.* **38**, 671–672.
5. J.A. Valdmanis, R.L. Fork, J.P. Gordon (1985). Generation of optical pulses as short as 27 femtoseconds directly from a laser balancing self-phase modulation, group-velocity dispersion, saturable absorption, and saturable gain, *Opt. Lett.* **10**, 131–133
6. D.E. Spence, P.N. Kean, W. Sibbett (1991). 60-fsec pulse generation from a self-mode-locked Ti:sapphire laser, *Opt. Lett.* **16**, 42–44.
7. L. Gallmann, D.H. Sutter, N. Matuschek, G. Steinmeyer, U. Keller, C. Iaconis, I.A. Walmsley (2001). Characterization of sub-6-fs optical pulses with spectral phase interferometry for direct electric-field reconstruction, *Opt. Lett.* **24**, 1314–1316.
8. R. Ell, U. Morgner, F.X. Kärtner, J.G. Fujimoto, E.P. Ippen, V. Scheuer, G. Angelow, T. Tschudi, M.J. Lederer, A. Boiko, B. Luther-Davies (1999). Generation of 5-fs pulses and octave-spanning spectra directly from a Ti-sapphire laser, *Opt. Lett.* **26**, 373–375.
9. D. Strickland, G. Mourou (1985). Compression of amplified chirped optical pulses, *Opt. Commun.* **56**, 219–221.
10. H. Nakatsuka, D. Grischkowsky, A.C. Balant (1981). Nonlinear picosecond-pulse propagation through optical fibers with positive group velocity dispersion, *Phys. Rev. Lett.* **47**, 910–913.
11. R.L. Fork, C.H. Brito Cruz, P. Becker, C.V. Shank (1987). Compression of optical pulses to six femtoseconds by using cubic phase compensation, *Opt. Lett.* **12**, 483–485.
12. A. Baltuska, Z. Wei, M.S. Pshenichnikov, D.A. Wiersma (1997). Optical pulse compression to 5 fs at a 1-MHz repetition rate, *Opt. Lett.* **22**, 102–104.
13. M. Nisoli, S. De Silvestri, O. Svelto (1996). Generation of high energy 10-fs pulses by a new pulse compression technique, *Appl. Phys. Lett.* **68**, 2793–2795.
14. M. Nisoli, S. De Silvestri, O. Svelto, R. Szipöcs, K. Ferencz, Ch. Spielmann, S. Sartania, F. Krausz (1997). Compression of high-energy laser pulses below 5 fs, *Opt. Lett.* **22**, 522–524.

15. G. Cerullo, S. De Silvestri, M. Nisoli, S. Sartania, S. Stagira, O. Svelto (2000). Few-optical-cycle laser pulses: from high peak power to frequency tunability, *IEEE J. Selec. Top. Quantum Electron.* **6**, 948–958.
16. B. Schenkel, J. Biegert, U. Keller, C. Vozzi, M. Nisoli, G. Sansone, S. Stagira, S. De Silvestri, O. Svelto (2003). Generation of 3.8 fs pulses from adaptive compression of a cascaded hollow fiber supercontinuum, *Opt. Lett.* **28**, 1987–1989.
17. O. Svelto (1998). *Principles of Lasers*, Plenum Press, New York.
18. W. Koechner (1996). *Solid-State Laser Engineering* (Springer Series in Optical Science). Springer-Verlag, Berlin.
19. N.P. Barnes, J.A. Williams, J.C. Barnes, G.E. Lockard (1988). A self-injection locked, Q-switched, line-narrowed Ti:Al_2O_3 laser, *IEEE J. Quantum Electron.* **24**, 1021–1028.
20. A.J. DeMaria, D.A. Stetser, H. Heynau (1966). Self mode-locking of laser with saturable absorbers, *Appl. Phys. Lett.* **8**, 174–176.
21. C.V. Shank, E.P. Ippen (1974). Subpicosecond kilowatt pulses from a mode-locked cw dye laser, *Appl. Phys. Lett.* **24**, 373–375.
22. R.L. Fork, B.I. Green, C.V. Shank (1981). Generation of optical pulses shorter than 0.1 psec by colliding pulse mode-locking, *Appl. Phys. Lett.* **38**, 671–672.
23. E.P. Ippen (1994). Principles of passive mode locking, *Appl. Phys. B* **58**, 159–170.
24. L. Dahlström (1972). Passive mode-locking and Q-switching of high power lasers by means of the optical Kerr effect, *Opt. Commun.* **5**, 157–162.
25. M. Piché (1991). Beam reshaping and self-mode-locking in nonlinear laser resonators, *Opt. Commun.* **86**, 156–160.
26. R.L. Fork, O.E. Martinez, J.P. Gordon (1984). Negative dispersion using pairs of prisms, *Opt. Lett.* **9**, 150–152.
27. J. Zhou, G. Taft, C. Huang, M.M. Murnane, H.C. Kapteyn, I.P. Christov (1994). Pulse evolution in a broad-bandwidth Ti:sapphire laser, *Opt. Lett.* **19**, 1149–1151.
28. R. Szipöcs, K. Ferencz, C. Spielmann, F. Krausz (1994). Chirped multilayer coatings for broadband dispersion control in femtosecond lasers, *Opt. Lett.* **19**, 201–203.
29. H.A. Haus, J.G. Fujimoto, E.P. Ippen (1992). Analytic theory of additive pulse and Kerr lens mode locking, *IEEE J. Quantum Electron.* **28**, 2086–2096.
30. T. Brabec, C. Spielmann, F. Krausz (1991). Mode locking in solitary lasers, *Opt. Lett.* **16**, 1961–1963.
31. M.T. Asaki, C.-P. Huang, D. Garvey, J. Zhou, H.C. Kapteyn, M.M. Murnane (1993). Generation of 11-fs pulses from a self-mode-locked Ti:sapphire laser, *Opt. Lett.* **18**, 977–979.
32. P.F. Curley, C. Spielmann, T. Brabec, F. Krausz, E. Wintner, A.J. Schmidt (1993). Operation of a femtosecond Ti:sapphire solitary laser in the vicinity of zero group-delay dispersion, *Opt. Lett.* **18**, 54–56.
33. F. Gires, P. Tournois (1964). *C.R. Acad. Sci. (Paris)* **258**, 6112.
34. J.A. Giordmaine, M.A. Duguaym, J.W. Hansen (1968). Compression of optical pulse, *IEEE J. Quantum Electron.* **4**, 252–255.
35. R.A. Fisher, P.L. Kelly, T.K. Gustafson (1969). Subpicosecond pulse generation using the optical Kerr effect, *Appl. Phys. Lett.* **14**, 140–143.
36. A. Laubereau (1969). External frequency modulation and compression of picosecond pulses, *Phys. Lett. A*, **29**, 539–540.
37. S.E. Harris, A.V. Sokolov (1998). Subfemtosecond pulse generation by molecular modulation, *Phys. Rev. Lett.* **81**, 2894–2897.
38. A. Nazarkin, G. Korn, M. Wittmann, T. Elsaesser (1999). Generation of multiple phase-locked Stokes and anti-Stokes components in an impulsively excited Raman medium, *Phys. Rev. Lett.* **83**, 2560–2563.

39. A.V. Sokolov, D.R. Walker, D.D. Yavuz, G.Y. Yin, S.E. Harris (2000). Raman generation by phased and antiphased molecular states, *Phys. Rev. Lett.* **85**, 562–565.
40. A.V. Sokolov, D.D. Yavuz, S.E. Harris (1999). Subfemtosecond pulse generation by rotational molecular modulation, *Opt. Lett.* **24**, 557–559.
41. G.M. Gale, M. Cavallari, T.J. Driscoll, F. Hache (1995). Sub-20-fs tunable pulses in the visible from an 82-MHz optical parametric oscillator, *Opt. Lett.* **20**, 1562–1564.
42. A. Shirakawa, S. Morita, K. Misawa, T. Kobayashi (1997). 100-nm bandwidth noncollinearly phase-matched femtosecond optical parametric amplifier. In: *Tech. Dig. Conference on Lasers and Electro-Optics, Pacific Rim'97* (pp. 18–19, paper TuF2). Optical Society of America, Washington DC.
43. T. Wilhelm, J. Piel, E. Riedle (1997). Sub-20-fs pulses tunable across the visible from a blue-pumped single-pass noncollinear parametric converter, *Opt. Lett.* **22**, 1494–1496.
44. G. Cerullo, M. Nisoli, S. De Silvestri (1997). Generation of 11-fs pulses tunable across the visible by optical parametric amplification, *Appl. Phys. Lett.* **71**, 3616–3618.
45. M. Nisoli, S. De Silvestri, V. Magni, O. Svelto, R. Danielius, A. Piskarskas, G. Valiulis, A. Varanavicius (1994). Highly efficient parametric conversion of femtosecond Ti:Sapphire laser pulses at 1 kHz, *Opt. Lett.* **19**, 1973–1975.
46. K. R. Wilson, V. V. Yakovlev (1997). Ultrafast rainbow: tunable ultrashort pulses from a solid-state kilohertz laser, *J. Opt. Soc. Am. B* **14**, 444–448.
47. M. Nisoli, S. Stagira, S. De Silvestri, O. Svelto, G. Valiulis, A. Varanavicius (1998). Parametric generation of high-energy 14.5-fs light pulses at 1.5 μm, *Opt. Lett.* **23**, 630–632.
48. T. Brabec, F. Krausz (2000). Intense few-cycle laser fields: frontiers of nonlinear optics, *Rev. Modern Phys.* **72**, 545–591.
49. P.M. Paul, E.S. Toma, P. Breger, G. Mullot, F. Augé, Ph. Balcou, H.G. Muller, P. Agostini (2001). Observation of a train of attosecond pulses from high harmonic generation, *Science* **292**, 1689–1692.
50. M. Hentschel, R. Kienberger, C. Spielmann, G.A. Reider, N. Milosevic, T. Brabec, P. Corkum, U. Heinzmann, M. Drescher, F. Krausz (2001). Attosecond metrology, *Nature* **414**, 509–513.

Part II
Spectroscopic and Imaging Techniques (Non-microscopic)

Chapter 10

Autofluorescence spectroscopy of cells and tissues as a tool for biomedical diagnosis

Giovanni Bottiroli and Anna Cleta Croce

Table of contents

Abstract

Several biomolecules involved both in the metabolic processes and in the histological organization of cells and tissues are characterized by fluorescence properties that can be exploited to obtain information on the morpho-functional conditions of the biological substrate, suitable for diagnostic applications.

An overview of the endogenous fluorophores responsible for the autofluorescence is given concerning the photophysical properties and their dependence on the evolution of the biological condition of cells and tissues.

10.1. Introduction

Most of the components of the biological material under excitation at suitable wavelength give rise to a fluorescence emission covering the UV-near-IR spectral range. This emission is called "autofluorescence", or "primary fluorescence" or "light-induced fluorescence" (LIF), to be distinguished from the "induced fluorescence" that is obtained upon labelling the structures under investigation by means of exogenous fluorochromes.

Autofluorescence was the first form of fluorescence to be investigated by microscopy, and was reviewed in detail by Haitinger in 1938 [1]. Autofluorescence is generally much more intense in biological material belonging to the vegetable kingdom than to that of the animals. In fact, plant tissues naturally contain a large variety of fluorochromes that, for their spectral features and quantum efficiency, find wide application as exogenous markers in analytical cytofluorometry. Substances such as quinone, coumarin, cyanine, tetrapyrrole derivatives and many alkaloids are commercially extracted and used as fluorochromes. On this basis, it appears natural that autofluorescence has always been considered a powerful tool in the study of vegetable morphology and physiology [2]. The same cannot be said for autofluorescence in animal tissues. The low spectral specificity and the quite decreased quantum yield meant that, for a long time, autofluorescence was mainly a nuisance for most biological fluorescence microscopists, as it may mimic some specific induced fluorescence and reduce the signal-to-noise ratio of the measurements.

Technological improvements in the fields of excitation sources, light delivery systems and devices for the detection and analysis of fluorescence signals, together with a better knowledge of the photophysical features of the endogenous fluorophores, opened up interesting perspectives for the application of autofluorescence in the characterization of biological animal tissues. Therefore, autofluorescence began to assume importance for two main reasons: because of its widespread occurrence, making it an intrinsic biological parameter, and because of the important role of some autofluorescent materials, making it a specific marker of biological processes.

The works by Chance are the first examples of autofluorescence analysis applied to the dynamic evaluation of the cell metabolism. The reduced pyridine nucleotide

NAD(P)H was localized and quantitatively evaluated in intact cells through its autofluorescence emission, which showed changes in dependence on the aerobiosis/anaerobios conditions and on the glucose supply [3,4].

At present autofluorescence is carefully investigated as an intrinsic parameter of cells and tissues that can be exploited to develop techniques suitable for biomedical diagnosis.

10.2 Autofluorescence in biomedical diagnosis

The overall autofluorescence emission of a tissue is strictly dependent on the chemical nature, the amount, the spatial distribution and the microenvironment of the fluorophores in the biological substrate. Endogenous fluorophores are associated with various biomolecules that can either be responsible for the structural arrangement or involved in the metabolic and functional processes of cells and tissues. To the former case belong proteins, such as collagen, elastin or more generally constitutive proteins; the latter include pyridinic coenzymes (NAD(P)H), flavins, riboflavin, pyridoxine derivatives, and porphyrins, in addition to accumulation products of catabolic processes such as lipofuscins, or more generally, lipopigments.

Changes in the morphological and biochemical properties of cells and tissues, related to physiological state or induced by the occurrence of pathological processes, are expected to modify both the amount and distribution of the endogenous fluorophores, thus affecting the autofluorescence properties.

Differences in both signal amplitude and spectral shape are reported among the leukocyte families that, apart from the cell dimensions, can be related to a different metabolic engagement and/or to a different intrinsic biochemical and physiological conditions [5]. An increase of autofluorescence emission has been reported during the serial passaging of human fibroblasts [6,7]. This occurrence, which can be ascribed to an accumulation of various products of lipid peroxidations (age-pigments, such as lipopigments), was proposed as an index of cell ageing in culture.

The rising of a pathological condition can profoundly affect both the architecture and the metabolism of a tissue. Diabetes mellitus involves an early, non-enzymatic glycosilation of the connective proteins that results in an increase of the autofluorescence signal [8]. Lipid or calcified deposits strongly affect the autofluorescence of normal artery, arising from structural protein fibers of intima, media and adventitia, giving rise to an increase of the long wavelength emission that can be exploited for atherosclerotic plaque diagnosis [9,10].

Autofluorescence analysis for diagnostic purposes is widely applied in oncology, where the rising of a neoplastic lesion can result in alterations of both structural organization and metabolic activity of the tissue. With multilayered epithelial tissues the growing of neoplasia can alter the relative contributions of the histological components in the depth of the tissue that contributes to the overall emission, thus modifying the tissue autofluorescence properties. For colon carcinoma, for example, the tumour invasiveness leads to the lowering or disappearance of submucosa that is replaced by the proliferating neoplastic cells and by their loose connective. This histological alteration results in a strong decrease or loss of the contribution of brightly fluorescing collagen and elastin of the normal submucosa and, therefore, in a reduction of the fluorescence emission amplitude recorded from the surface of the tissue [11–16].

As to metabolic processes, the neoplastic condition is related to alterations in the energetic metabolism that aid the survival and invasiveness of tumour cells, with the prevailing of the anaerobic pathways over the aerobic ones. The phenomenon, already remarked by Warburg in 1956 [17], involves an impairment of the cellular redox state leading to changes in the equilibria among the coenzyme chemical species – mainly nicotinic coenzymes – that result in alterations of both the lineshape and the intensity of the autofluorescence emission [18,19].

The fluorescence characteristics of a biological tissue, as a turbid medium, depend both on the concentration and distribution of the endogenous fluorophore and on the tissue optical properties. The concentration and distribution of non-fluorescent absorbers and scatterers within the tissue will affect the propagation of the light (both excitation and emission), influencing the signal amplitude and the spectral shape of the autofluorescence at the surface of the tissue. The histological and biochemical changes induced by rising of neoplasia can affect the auto-fluorescence emission also through a modification of the tissue optical properties. In fact, rising of neoplasia results in an alteration of histological factors – such as tissue vascularization, ratio between cellular and connectival compartments, and the ratio between cell nucleus and cytoplasm – which strongly affect tissue absorption and scattering. In some cases, such alterations could contribute to enhance the differences in autofluorescence signal measured at tissue surface between normal and diseased conditions, thus improving the diagnostic accuracy of the technique. For example, alterations of the tissue optical properties were considered to explain the differences between the autofluorescence signals of normal and tumour brain tissues, which were larger in the in vivo than in the ex vivo measurements [20]. In other cases, the interplay of tissue optics can make the interpretation of tissue autofluorescence emission rather difficult since the fluorescence spectra measured from bulk tissue differ significantly from those of pure tissue fluorophores. The effects of self-absorption by hemoglobins on the fluorescence spectra from normal and cancerous tissues have been reported [21]. Because removing the effect of tissue optical properties might improve fluorescence-based tissue diagnosis, models have been developed for retrieving an intrinsic fluorescence spectrum from a measured fluorescence spectrum [22].

10.3 Autofluorescence and induced fluorescence

Fluorescence-based diagnostic techniques analyse the emission of both endogenous and exogenous fluorophores. In the former case, alterations of both amount and distribution of naturally occurring fluorophores are exploited to distinguish diseased from healthy tissues. In the latter case, the capability of exogenous fluorophores to localize preferentially in diseased with respect to surrounding healthy tissue allows the lesion to be detected. The exogenous fluorophores most widely used for this purpose are the porphyrin derivatives, namely the hematoporphyrin derivative (HpD) and its aggregate-enriched fraction Photofrin II. These compounds can be excited at different wavelengths from 400 to 620 nm, giving rise to a fluorescence emission with two distinct bands at about 630 and 670 nm.

Porphyrin derivatives were first used in 1961, when Lipson [23] attempted to detect the presence of the tumour lesion in tracheo-bronchial or esophagus tissues through endoscopy; it was already known that porphyrin derivatives preferentially accumulate in tumour tissues. The same derivatives were subsequently used to evidence neoplastic lesions through the differential analysis of the emission spectral profile [24,25]. Due to the presence of aggregates, hematoporphyrin derivatives undergo an intracellular turnover, leading to the appearance of different fluorescent species, the equilibria among which depend on the cellular environment. A relationship between the degree of malignancy and the changes in the spectral shape at different times after administration was observed in the case of colonic neoplasias, which was proved to be due to a different intracellular turnover of the drug [26–28].

At present, the porphyrin-based diagnosis is performed by administering the precursor 5-aminolevulinic acid (ALA) to obtain a selective accumulation of protoporphyrin XI in neoplastic lesions, exploiting the alterations in the heme biosynthetic pathways of tumour cells [29,30]. The greater use of porphyrin derivatives, however, has been for the photodynamic therapy of tumours, due to their ability to activate photodynamic processes that lead to the production of species (singlet oxygen) that are highly toxic for the cells [31].

In diagnostic applications, autofluorescence- and induced-fluorescence-based techniques exhibit respective advantages and disadvantages. Autofluorescence is characterized by a signal amplitude and a spectral selectivity lower than those of induced-fluorescence, because of the relatively low quantum efficiency and of the overlapping of both the excitation and the emission spectra of the several endogenous fluorophores potentially present in the tissue. Conversely, autofluorescence provides real information on the biochemical and physico-chemical nature of the biological substrate since it is directly related to the biomolecules in their natural environment, in the absence of any perturbation that the administration of exogenous substances could induced. On this view autofluorescence has the great advantage of allowing direct, real-time monitoring of changes of the morpho-functional properties of the biological substrate in different physiological, pathological or experimental conditions.

10.4 Endogenous fluorophores

Numerous fluorescent substances of biological origin have been described in the literature. Here, endogenous fluorophores mainly responsible for autofluorescence emission are reported, along with a brief description of their role in the structural arrangement and metabolism of cells and tissues (Table 1).

Table 1. Excitation and emission properties of biological molecules responsible for cell and tissue autofluorescence

Fluorophores	Biomolecules Cell localisation	Excitation Peak position range (nm)	Emission Peak position range (nm)
Aromatic Amino Acid	Proteins		
Residues	A = Phe + Tyr;	240–280	280–350
	B = Trp + Phe + Tyr		
Collagen	Connective tissue	330–340	400–410
Elastin	Extracellular matrix	350, 420	420, 510
Cytokeratins	Epithelia	280, 325	495, 525
Reduced Pyridine	NAD(P)H	330–380	440 (bound)
Nucleotides	(Cofactors in metabolism)		462 (free)
	Mitochondria/cytoplasm		
Flavins,	Riboflavin, FMN, FAD	350–370	480–540
Flavin Nucleotides	(Coenzymes of flavoproteins)	440–450	
	Mitochondria/cytoplasm		
Porphyrins	Prosthetic group of proteins	405	590–690
(Protoporphyrin IX)	Hemoglobin, Myoglobin,	450–630	(630, 690)
	Cytochrome		
(Zinc-protoporphyrin)	Erythroid cells		(595, 635)
Lipofuscins,	Pigments	UV	>540
Lipopigments	(Cell catabolism/cell age)	400–500	
	Cytoplasm		
Vitamins	Vitamin A	370–380	490–510
	Vitamin B6 & Precurors	290–310/	375–395/
		375–395	400–500
Lipids	Arachidonic Acid	320, 350	470–480
	Phospholipids	430–440	520–570
Catecholamines	Neurotransmitters	280–290	320–340
(Serotonin)		305, 360 (dimer),	350, 440 (dimer),
		420 (trimer)	520 (trimer)

10.4.1 Amino acids and proteins

The fluorescence of proteins is due to the presence of the aromatic amino acids tryptophan, tyrosine and phenylalanine [32]. The three amino acids are excited below 280 nm, and they have different excitation and emission spectra, and different quantum efficiencies. Tryptophan exhibits a quantum efficiency (0.21) higher than those of tyrosine (0.20) and phenylalanine (0.04), and an emission maximum at a wavelength (345 nm) longer than those of tyrosine (303 nm) and phenylalanine (282 nm) [33].

According to the kind of aromatic amino acids present, proteins are distinguished as class A (tyrosine and phenylalanine) and class B (tryptophan, tyrosine and phenylalanine) proteins [34]. The fluorescence emission of proteins depends on the amino acid composition, and can be influenced by the structure, the spatial conformation and the microenvironment properties. Actually, tryptophan-containing proteins may show excitation and emission spectra different from those of tryptophan amino acids, depending on the environment and on the presence of other fluorophores in the protein chain [32,34,35].

Proteins show maximum absorption and emission at 240–280 nm and in the 280–350 nm region, respectively. The structural proteins collagen, elastin and cytokeratins differ from this general behaviour. Their amino acid composition and structure are quite different from those of a typical globular protein. These extracellular proteins of the connective tissue exhibit a strong fluorescent emission that is shifted to longer wavelengths (excitation at $\lambda > 300$ nm, emission at $\lambda > 400$ nm).

Collagen is the major extracellular matrix component. At least 11 types of collagen are ranked, according to the composition of the monomeric chains and the degree of polymerization [36]. The relationship between the structural features and the fluorescence properties for each kind of collagen has not been yet fully investigated. Collagens are characterized by the absence of tryptophan, a small amount of tyrosine, and an abundance of phenylalanine. The fluorescence of collagen has an excitation maximum at 330–340 nm and an emission maximum at 400–410 nm – fluorescence properties generally associated with the presence of hydroxylysyl pyridinoline and lysyl pyridinoline, and with the formation of covalent cross-links [37]. The fluorescence of collagen is directly related to the age and degree of maturation [38,39].

Elastin is the major component of the elastic fibers and is found in most connective tissues along with collagen and polysaccharides. The fluorescence of elastin has an excitation maximum at 350 nm and an emission maximum at 420 nm, which seem to be attributed to the presence of a tricarboxylic tri-amino pyridinium ring derivative. These residuals seem to be have important structural implications since they are confined to proteases-resistant regions [40]. Excitation and emission bands at longer wavelengths (420, 510 nm) have also been reported [41].

Collagen and elastin are important in the diagnostic application of autofluorescence since they are highly fluorescent structures and their contribution to the overall autofluorescence depends on the histological organization of the tissue. The changes in the relative amounts of connective components of the arterial walls, in terms of exchange of elastin with collagen and lipids, can be exploited for the diagnosis of atherosclerotic plaques [9,10,42] and for the guidance of laser angioplasty [43]. Collagen and elastin, as the most abundant fluorophores of the submucosa, are very important in the diagnosis of neoplasia of multilayered epithelial tissue. Submucosa is a strongly fluorescent layer due to its richness in both collagen and elastic fibres, which in normal tissue it is located at a depth that is interested by excitation light under the conditions usually employed for diagnostic investigation. Tumour invasiveness induces a mucosal thickening and/or a replacement of the most fluorescent histological component submucosa by lesser fluorescent neoplastic tissue, thus resulting in a lowering of the fluorescence signal in diseased with respect to normal tissue. Examples of the diagnostic value have been reported for different sites, such as colonic mucosa, cervix, bronchi, stomach [14,15,44–47]. The changes in the organization of connective stroma, analysed by means of multispectral imaging autofluorescence microscopy proved also to be useful for the diagnosis of lymph-node diseases, such as Hodgkin's lymphoma or reactive hyperplasia [48].

Cytokeratins are formed by epithelial cells and are obligate heterodimers containing a 1:1 mixture of acidic and basic intermediate filament polypeptides. They are generally found in the epithelia that line the internal body cavities. Each kind of epithelium expresses a characteristic combination of type I and type II keratins. Cytokeratins are characterized by two main excitation bands, peaked at 280 and 325 nm, with two emission bands at 495 and 525 nm. During the active phase of tumour growth cytokeratin levels increase, mainly at the tumour border zone, both for the decomposition of the plasma membranes and for the overexpression by tumour cells. Although cytokeratins contribute appreciably to the epithelium fluorescence signal, the diagnostic importance of their autofluorescence has not yet been adequately considered. Given their important role as tumour markers [49] and that their fluorescence emission can be easily discriminated from those of other endogenous fluorophores – mainly NADH and collagen – [50], cytokeratins could represent a further diagnostic parameter in optical biopsy.

10.4.2 Pyridine nucleotides and flavins as metabolic coenzymes

Pyridinic nucleotides and flavins are the fluorophores mainly responsible for the fluorescence emission rising from the cytoplasm of single cells under exposure to UV light [51]. Their cytoplasmic localization is related to their participation to the most part of metabolic reactions in the cell that lead to energy production, or to anabolic and catabolic functions that, in turn, are energy consuming. In relation to the role of these coenzymes in the energetic metabolism, the term redox fluorometry was defined as the analytical procedure that exploits the intrinsic fluorescence of reduced pyridine nucleotides (NADH, NADPH) and of oxidized flavins to study cellular energy metabolism [52,53].

Aerobic organisms produce energy in the form of ATP by transferring electrons derived from the oxidation of fuel molecules – glucose, fatty acids and, at a lesser extent, proteins – to the ultimate electron acceptor O_2, through a series of biochemical reactions. Electrons are transferred by intermediate carries, which are either pyridine nucleotides or flavins. The production of ATP occurs through three main steps: glycolysis – taking place in the cytoplasm – and the citric acid Kreb's cycle – occurring in mitochondrial matrix – that produce NADH in the reduced form, and the oxidative phosphorylation – taking place in mitochondria inner membrane – leading to the reoxidation of NADH and to the production of ATP. NADH cannot diffuse across the mitochondrial membrane, but an equilibrium between NADH in the cytoplasm and in mitochondria based on "shuttle" systems occurs, that provides for the transportation of the reducing equivalents from the cytosol to mitochondria.

While NADH is mainly involved in reactions leading to energy production, NADPH is used as an electron donor for reductive biosyntheses. NADPH is generated by the pentose phosphate pathway, and is used, for example, for the syntheses of fatty acids, steroid hormones, non-essential amino-acids, deoxyribonucleotides or for the heme degradation to bilirubin. Notably, cells also contain enzymes called transhydrogenases, which transfer an electron from NAD(P)H to $NAD(P)^+$ in a reversible reaction.

The fluorescence properties of NADH are associated with the nicotinamide group in the reduced state, which absorbs at about 340 nm and fluoresces in the 420–490 nm region. No appreciable differences between NADH and NADPH fluorescence have been reported. The fluorescent properties of NAD(P)H change upon binding to enzyme molecules. A blue-shift of the emission peak (from 462 to 440 nm) was reported for the NADH–alcohol dehydrogenase complex [34]. The ratio of quantum efficiency of NAD(P)H in the free and in the bound forms is about 0.33 [18]. The fluorescence decay time of the NAD(P)H bound form is longer than that of the free form: 1.0 ns for NAD(P)H–malate dehydrogenase complex, 0.4 ns for NAD(P)H free form [54].

The excitation/emission properties of the "free groups" flavin mononucleotide (FMN) and flavin-adenin dinucleotide (FAD) in aqueous solution (pH 7) are very close, the absorption maximum being at 445 nm and the emission maximum at 525 nm for both compounds [55]. Quantum efficiencies of 0.25 and 0.05, and decay times of 4.7 ns and 2.3 nm are reported for FMN and for FAD, respectively [55,56]. The fluorescence properties of FAD are strongly affected by the nature of the protein to which the prosthetic group is bound. Early studies assessed that, among flavoproteins engaged in energetic metabolism, only lipoamide dehydrogenase and electron transfer flavoprotein in the mitochondrial matrix contribute significantly to cellular flavoprotein fluorescence. Kunz and Kunz [57] and Kunz [58] established that about 50% of the fluorescence signal of isolated rat liver mitochondria attributable to flavoproteins (FP) comes from the flavin of a-lipoamide dehydrogenase (FP5), a constituent of several different NAD-linked multienzyme complexes; 25% comes from non-NAD-linked electron-transfer flavoprotein (FP3), and the remaining signal is due to ill-defined flavoproteins that can be reduced only by dithionite. Differences in the spectral shapes were also found, the emission of FP5 being more similar to that of free flavin with the main emission in the 510–550 nm region, while that of FP3 showed a maximum at about 480 nm. As pyridine nucleotide is fluorescent in the reduced form (NAD(P)H) and flavin nucleotide are in the oxidized form (FAD), the induction of a cell reduced state – e.g. blocking the respiratory chain by means of potassium cyanide – is expected to produce an increase in the blue (NAD(P)H) fluorescence intensity and, conversely, a decrease in green (FAD) emission. Experimental results, however, may provide evidence that is inconsistent with this assumption [59]. In fact, the changes in the reciprocal redox state of NAD(P)H and flavins in cells undergoing metabolic alterations are "bistable". The redox state can depend upon the antecedent metabolic steady state, so that the same experimental conditions can result in different autofluorescence responses [18]. Nonetheless, in situ analysis of the intrinsic fluorescence properties of the coenzymes proved to be a useful metabolic indicator of the redox state of the cells in the study of cells energetic metabolism. Huang [60] derived an extremely simplified equation describing the redox – fluorescent state of the cells:

$$FADH_2(FP5) + NAD^+ = FAD(FP5) + NADH + H^+$$

The pioneering studies of Chance [3,4] demonstrated that changes in the autofluorescence emission in intact cells in the aerobic – anaerobic transition can be explained by changes in the redox state of NADH coenzyme. Many studies followed, investigating the relationship between cell and tissue fluorescence properties and their energetic metabolism pathways under different experimental conditions [59,61–65]. The cell autofluorescence properties attributable to NAD(P)H were also proved to be related to the functional/transformed status of the cells and to contribute to differentiating the whole fluorescence signal of healthy tissue from that of diseased tissue [66–69]. The blue-shift of the emission spectrum of NAD(P)H in bound with respect to free form allowed changes in the relative presence of the two forms in normal and tumour cells to be seen that were attributed to the decrease in the binding sites for NAD(P)H in the cancerous tissue [18,66,70]. Changes in the contribution of NAD(P)H in the free or in the bound state can be investigated by spectral fitting analysis. When the spectral parameters of pure compounds are defined, this analytical procedure allows the estimation of the contribution of each single fluorophore to the whole emission, similarly to a biochemical analysis (Figure 1).

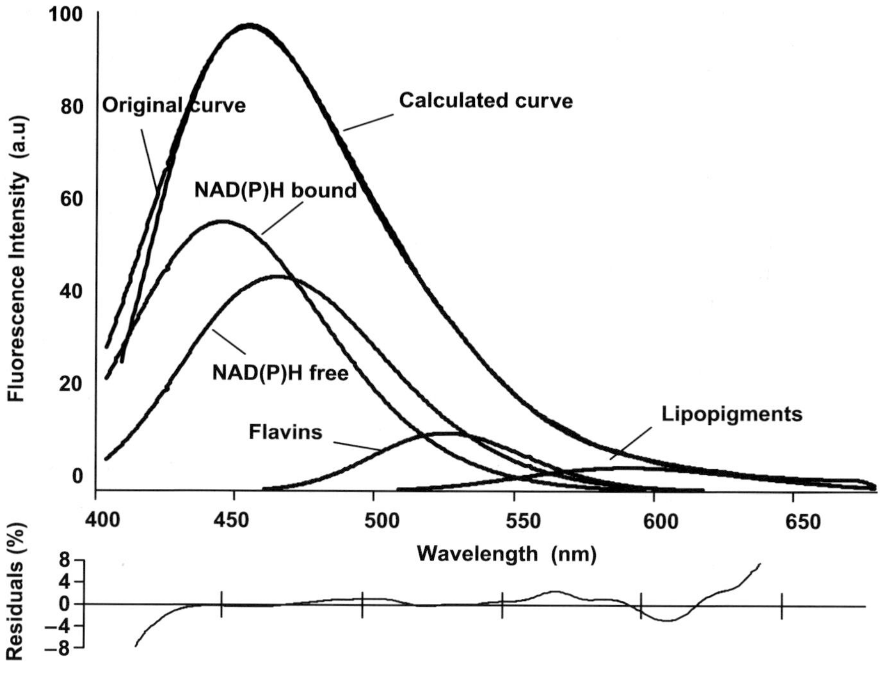

Figure 1. Example of curve fitting of an autofluorescence emission spectrum recorded on 3T3 cultured cells under excitation at 366 nm. The original measured curve and that calculated from the fitting are shown, along with the GMG (half-Gaussian Modified Gaussian) functions representative of the best fitting for the emission of NADH, under free ($\lambda = 463$ nm, FWHM $= 115$ nm) and bound ($\lambda = 444$ nm, FWHM $= 105$ nm) conditions, flavins ($\lambda = 526$ nm, FWHM $= 81$ nm), and lipopigments ($\lambda = 587$ nm, FWHM $= 80$ nm). Coefficient of determination, $r^2 = 0.993$. Results of the residual analysis are shown.

This procedure revealed that an increase in the relative contribution of free NAD(P)H is associated with the transformed condition of the cells. In cultured cell models, normal and transformed fibroblasts, the enhancement of NAD(P)H in the free form was related to an increased anaerobic metabolism and a decrease in the mitochondrial activity [19]. A similar result was obtained in the comparison of the brain tumour tissue glioblastoma and the healthy surrounding tissue, where a prevalence of NAD(P)H in the free form was found in the neoplastic tissue [71].

NAD(P)H autofluorescence analysis is acquiring great relevance for the real time monitoring of the functional-metabolic status of organs, such as heart [72,73] and liver [74–77]. This latter is considered with particular reference to organ transplantation procedure: changes in the relative contribution of fluorophores to the whole emission was assessed during the different transplantation phases, and an inverse relationship between NADH fluorescence signal and APT concentration has been assessed when liver is preserved under aerobic or anaerobic conditions.

10.4.3 Vitamins

Several of the numerous substances generally called vitamins exhibit fluorescence properties. Only a few deserve attention owing to their spectral properties and functional metabolic engagement in tissues under healthy and altered conditions with a view to diagnostic purposes. The FAD precursor vitamin B_2 (riboflavin) was considered in the previous section. Vitamins B_6 and A are considered here.

10.4.3.1 Vitamin B_6 (pyridoxine) and related compounds

Pyridoxine and its metabolites, i.e. pyridoxine-5′ phosphate, pyridoxamine, pyridoxamine-5′-phosphate, pyridoxal-5′-phosphate, which exist in solution as a series of differently charged and tautomeric species, exhibit an appreciable autofluorescence signal. All of these compounds are excited at variable wavelengths in the 290–310 nm spectral region, and fluoresce in the 375–395 nm region, except for pyridoxal-5′-phosphate, which exhibits an excitation maximum ranging from 330 to 390 nm and an emission varying from 400 to 500 nm, mainly depending on the pH of the solution [55]. The contribution of vitamin B_6 to tissue autofluorescence emission was rarely considered, although it participates in numerous reactions concerning metabolism of amino acids, and, perhaps, of lipids, and it is diffused in all the tissues. The difficulty in discriminating the emission of chromophores of vitamin B_6 from those of proteins (collagen or elastin) or of NAD(P)H has been reported by Richards-Kortum [41] in a study investigating the application of excitation emission matrices based method to the spectroscopic diagnosis of colonic dysplasia.

10.4.3.2 Vitamin A (all-trans-retinol)

Vitamin A absorbs in the near-UV region owing to its conjugated polyene structure. Its fluorescence was first reported by Querner in 1932 [78], and subsequently studied by Popper [79] in the human and in the rat. In aqueous media the excitation maximum is at about 325 nm, and the emission maximum is centered at about 490 nm. The emission maximum ranges from 475 to 510 nm, depending on the solvent. The quantum yield of retinol is quite low, ranging from 0.03 to 0.006 [55]. Interaction with proteins, e.g., β-lactoglobulin, considerably affects vitamin A's spectroscopic properties [80]. Although vitamin A autofluorescence observation is limited by a very rapid bleaching, its fluorescence properties were exploited for a serum assay [32,81]. Vitamin A is involved in the accumulation of vitamin A-dependent fluorophores in the degeneration of photoreceptors in the retina [82]. As to in vivo autofuorescence spectroscopy, vitamin A could play an important role in the case of liver, where its presence is related to the role of the organ in the storage and metabolism of retinoids, to be then released in the blood and used by all tissues of the body. Up to now no relationship has been assessed between vitamin A fluorescence and tissue functions that can give useful information for diagnosis of the liver condition.

10.4.4 Lipids

Fatty acids, as such, are generally not fluorescent. Lipid droplets in tissues may appear fluorescent due to other substances dissolved in the lipids, such as retinoids, or their oxidation products. Fluorescence was observed for arachidonic acid, the precursor of prostaglandins, found in the brain, depot fats, and glandular organs, and particularly in the liver. Arachidonic acid is characterized by an excitation in the 300–370 nm range, main peak at about 320 nm, and a broad emission in the 420–550 nm region, emission peak at 470 nm (unpublished observation of the authors). Some lipid derivatives, such as phospholipds and oxidation products, are reported to be fluorescent. Phospholipids can give rise to fluorescence emission in the 520–570 nm range, under excitation at 436 nm, the intensity of fluorescence being a function of the acidity of the phosphate group [32].

10.4.5 Lipopigments (lipofuscins, ceroids and related susbtances)

Lipopigments, namely lipofuscins and ceroids, are a heterogeneous class of macromolecules, mainly composed of the products of peroxidation of fatty acids. They can differ according to the basic structure, the degree of polymerization and oxidation, and presence of non-lipidic material. The products accumulating during ageing are generally called lipofuscins, while the pigments found in relation with pathological processes are called ceroids, although all of them share some physicochemical properties at some moments of their evolution. Lipofuscins in the cells usually appear as cytoplasmic granules, where lipidic chains can be aggregated with proteins, glycoproteins, and can also contain carotenoid derivatives and melanin-related compounds. Lipofuscins, in particular, can derive from the intracellular accumulation of digested biological components, either exogenous or endogenous (lysosomial autodigestion). The composition of ceroids was investigated in the organs of sheep affected by ceroid lipofuscinosis, and found to consist of up to 70%, proteinaceous material, phospholipids and fatty acids, including lysosomal markers, dolichol and ubiquinone. Melanin lipopigments contain pigments derived from the oxidation of tyrosine, and are usually produced in the cells as a defence from photo-oxidation injury [83–87]. The chemical complexity and heterogeneity of these compounds are reflected in the variability of their fluorescence properties. Excitation can be obtained at 340–440 nm, and emission occurs in the yellow-orange region, (450–560 nm) with a great variability in the spectral shape. The rising of fluorescence upon peroxidation of lipids was proved by Tappel [88], who demonstrated that in vitro peroxidation of lysosomes and mitochondria resulted in an increase in amplitude emission, peaked at 460 nm, and in the appearance of a shoulder at about 520 nm. Mitochondria, in particular, gave an emission spectrum quite similar to those of lipofuscins.

At the single cell level the accumulation of lipofuscins is significant in the cell ageing process [7]. It has been experimentally verified that ageing of fibroblasts in culture leads to the accumulation of intracytoplasmic fluorescent granules, attributable to lipofuscins.

A pathological accumulation of lipopigments can occur in the central nervous system. The neuronal ceroid lipofuscinoses comprise a group of progressive neurogenetic diseases that are generally considered to be caused by lysosomal dysfunction, although the exact biochemical defect responsible for the accumulation of the autofluorescent storage material is unknown.

Lipopigments generally do not contribute greatly to tissues fluorescence emission. The presence of lipofuscin-like fluorescent granules in tissues has been described in the colon, where it has been attributed to tissue-dwelling eosinophils in normal mucosa [89], and to accumulation of melanin and saccharide residues containing pigments in colonic melanosis [90]. In relation to neoplasias, accumulation of lipopigments has been described: in the carcinoma of the colon, where, depending on the excitation wavelength, they were hypothesized to contribute to the differences in the autofluorescence spectral shape between the lesion and the non-neoplastic surrounding tissues [14]; in brain tumours, associated with the presence of host infiltrating cells [91]; in the adenomas of the palate, where the pigments contained in plasmacytoid myohepithelial cells helped to explain the pigmentation of salivary gland tumours [92].

10.4.6 Endogenous porphyrins

Naturally occurring porphyrins absorb over a wide range of wavelengths: the main band ($\varepsilon \approx 200{,}000\ M^{-1}\ cm^{-1}$) occurs in the Soret region (about 400 nm), four weaker bands (Q-bands, $\varepsilon \approx 5{,}000$–$15{,}000\ M^{-1}\ cm^{-1}$) occur in the 450–630 nm, which are sensitive to changes in the environment and to the presence of a metal atom. Two main emission bands are generally observed in the 590–700 nm spectral range: peaks at about 630 and 690 and at about 595 and 635 are recorded for protoporphyrin IX and Zinc-protoporphyrin, respectively.

Porphyrins are an ubiquitous class of naturally occuring molecules involved in various biological processes, such as transport of oxygen, catalysis and pigmentation. The tetrapyrrolic ring called protoporphyrin IX is the ultimate precursor during biosynthesis of heme. It is the result of a series of reactions starting from δ-aminolevulinc acid, where key enzymes are δ-aminolevulinate synthase, δ-aminolevulinate dehydrase and ferrochelatase, all of them feedback-inhibited by heme. A tetrapyrrolic ring is part of the structure of vitamin B_{12}. In many species of rodents porphyrins are synthesized and stored in the Harderian gland, which lies within the bony orbit. Its functions remain unclear, but, among other roles, a link in the retinal – pineal – gonadal system has been proposed, and is used as a model in the study of porphyrin biosynthesis [93]. In humans, a small quantity of protoporphyrin IX and zinc protoporphyrin IX is contained in the blood and tissues under healthy conditions. Protoporphyrin IX is excreted only in the bile, while uroporphyrinogen and coproporphyrin are excreted also in urine [94].

A pathological accumulation of porphyrins (porphyrias) can result from specific enzyme defects in the biosynthesis of heme, with the involvement of liver, and with an usually associated photosensitivity.

Spectrofluorometric studies of neoplasias report the spontaneous and occasional occurrence of fluorescence signals in the red region that can be attributed to the emission bands of porphyrin derivatives, for example in lesions of the oral mucosa [95,96], and of the colon [13,97,98]. The presence of endogenous porphyrins in tumour tissues has been attributed to the presence of several species of bacteria in ulcerated and necrotic tissues [95]. It cannot, however, be neglected that changes in the activity of enzymes associated with heme biosynthesis were found in peripheral blood mononucleated cells of patients with lymphoproliferative disorders [99], with epithelial tumours and metastatic spread [100], which may account for the increase in the presence of porphyrins in the tumour tissue [101]. In the latter case it was assessed that porphyrins consisted mainly of coproporphyrin, associated with small amounts of protoporphyrin (10–15%) and traces of uroporphyrin. Alterations in the heme biosynthetic pathways are the basis of obtaining a selective accumulation of protoporphyrin XI upon administration of its precursor 5-aminolevulinic acid (ALA), for photodiagnostic and photodynamic purposes [29]. Finally, we recall the naturally occurring emission band at about 675 nm found in the skin of rats. This has been attributed to pheophorbide *a* and/or to pheophytin *a*, as a degradation product of the chlorophyll present in the animal food [102], which are photodynamically active and can be a possible source of unwanted effects when performing photosensitizing studies in animals models.

10.4.7 Neurotransmitters

Neurotransmitters, or biogenic amines, include adrenaline and nor-adrenaline, derived from dopamine and generally called catecholamines, and serotonin (5-HT), derived from tryptophan. The fluorescence properties of catecholamines are associated with the phenol ring, while those of 5-HT are due to the indole structure. In aqueous solution all these substances exhibit peculiar excitation emission properties in the UV region, similarly to the amino acids from which they are originated, with the excitation maximum at 280–290 nm and the emission at 320–340 nm.

5-HT shows a red-shift of the emission spectrum when the pH is lowered. In 3M HCl solution 5-HT exhibits an emission band centred at about 540 nm, which has been exploited for a 5-HT fluorometric assay on tissue extracts [34]. Tan et al. [103] proved that excitation of 5-HT at longer wavelengths (i.e. 305 nm) by means of a laser source, giving enough power to overcome the low molar absorption coefficient of the compound in this region, resulted in a broad emission band at wavelengths longer than 350 nm, a region suitable for conventional fluorescence microscopy. This fluorescence emission allowed the uptake and depletion of 5-HT in living cells to be monitored by means of fluorescence imaging. Such an approach provides information on 5-HT intracellular distribution, otherwise not attainable with conventional electrochemical techniques [104,105]. 5-HT undergoes oxidation followed by dimerization and trimerization processes leading to a red-shift of the absorption/emission bands at 360/440 nm and 420/520 nm, respectively [106]. This opens up interesting perspectives for an in vivo evaluation of the response of the neurotransmitter level to pharmacological treatments by means of autofluorescence spectroscopy via fiber optic probe [107].

References

1. M. Haitinger (1938). *Fluoreszenzmikroskopie. Ihre Anwendung in der Histologie und Chemie.* Akademische Verlagsgellschaft M.B.H., Leipzig.
2. T.P. O'Brien, M.E. McCully (1981). *The Study of Plant Structure. Principles and Selected Methods.* Termacarphi, Melbourne.
3. B. Chance, V. Legallais (1959). Differential microfluorimeter for the localization of reduced pyridine nucleotide in living cells. *Rev. Sci. Instrum.* **30**, 732–735.
4. B. Chance, B. Thorell (1959). Localization and kinetics of reduced pyridine nucleotide in living cells by microfluorometry. *J. Biol. Chem.* **234**, 3044–3050.
5. M. Monici, R. Pratesi, P.A. Bernabei, R. Caporale, A.C. Croce, G. Bottiroli (1995). Natural fluorescence of white blood cells: spectroscopic and imaging study. *J. Photochem. Photobiol.* **30**, 29–37.
6. J.F. Jongkind, A. Verkerk, W.J. Visser, J.M. Van Dongen (1982). Isolation of autofluorescent 'aged' human fibroblasts by flow sorting. *Exp. Cell. Res.* **138**, 409–417.
7. S.I.S. Rattan, K.D. Keeler, J.H. Buchnam, R. Hollyday (1982). Autofluorescence as an index of ageing in human fibroblasts in culture. *Biosci. Rep.* **2**, 561–567.
8. V.M. Monnier, R.R. Kohn, A. Cerani (1984). Accelerated age-related browning of human collagen in diabetes mellitus. *Proc. Natl. Acad. Sci.* **81**, 583–587.
9. L.I. Laifer, K.M. O'Brien, M.L. Stetz, G.R. Gindi, T.J. Garrand, L.I. Deckelbaum (1989). Biochemical basis for the difference between normal and atherosclerotic arterial fluorescence. *Circulation* **80**, 1893–1901.
10. J.J. Baraga, R.P. Rava, P. Taroni, C. Kittrel, M. Fitzmaurice, M.S. Feld (1990). Laser induced fluorescence spectroscopy of normal and atherosclerotic human aorta using 306-310 nm excitation. *Lasers Surg. Med.* **10**, 245–261.
11. R.M. Cothren, R. Richards-Kortum, M.V. Sivak, Jr., M. Fitzmaurice, R.P. Rava, G.A. Boyce, M. Doxtader, R. Blackman, T.B. Ivanc, G.B. Hayes, M.S. Feld, R.E. Petras (1990). Gastrointestinal tissue diagnosis by laser-induced fluorescence spectroscopy at endoscopy. *Gastrointest. Endosc.* **36**, 105–111.
12. K.T. Schomacker, J.K. Frisoli, C.C. Compton, T.J. Flotte, J.M. Richter, S. Nishlioka, T.F. Deutsch (1992). Ultraviolet laser-induced fluorescence of colonic tissue: basic biology and diagnostic potential. *Lasers Surg. Med.* **12**, 63–74.

13. G. Bottiroli, R. Marchesini, A.C. Croce, M. Dal Fante, C. Cuzzoni, S. Di Palma, P. Spinelli (1993). Autofluorescence of normal and tumour mucosa of human colon. A comprehensive analysis. In: R. Barbour, M.J. Carvlin (Eds), *Physiological Imaging, Spectroscopy, and Early-Detection Diagnostic Methods*, (Volume 1887, p. 205–212). Society Photo-Optical Instrum. Eng., Los Angeles.
14. G. Bottiroli, A.C. Croce, D. Locatelli, R. Marchesini, E. Pignoli, S. Tomatis, C. Cuzzoni, S. Di Palma, M. Dal Fante, P. Spinelli (1995). Natural fluorescence of normal and neoplastic colon: a comprehensive *ex vivo* study. *Lasers Surg. Med.* **16**, 48–60.
15. K. Izuishi, H. Tajiri, T. Fujii, N. Boku, A. Ohtsu, T. Ohnishi, M. Ryu, T. Kinoshita, S. Yoshida (1999). The histological basis of detection of adenoma and cancer in the colon by autofluorescence endoscopic imaging. *Endoscopy* **31**, 511–516.
16. B.W. Chwirot, Z. Michniewicz, M. Kowalska, J. Nussbeutel (1999). Detection of colonic malignant lesions by digital imaging of UV laser-induced autofluorescence. *Photochem. Photobiol.* **69**, 336–340.
17. O. Warburg (1956). On the origin of cancer cells. *Science* **123**, 309-314.
18. J.M. Salmon, E. Kohen, P. Viallet, J.G. Hirscheberg, A.W. Wouters, C. Kohen, B. Thorell (1982). Microspectrofluorometric approach to the study of free/bound NAD(P)H ratio as metabolic indicator in various cell types. *Photochem. Photobiol.* **36**, 585–93.
19. A.C. Croce, A. Spano, D. Locatelli, S. Barni, L. Sciola, G. Bottiroli (1999). Dependence of fibroblast autofluorescence properties on normal and transformed conditions. Role of the metabolic activity. *Photochem. Photobiol.* **69**, 364-374.
20. A.C. Croce, S. Fiorani, D. Locatelli, R. Nano, M. Ceroni, F. Tancioni, E. Giombelli, E. Benericetti, G. Bottiroli (2003). Diagnostic potential of autofluorescence for an assisted intraoperative delineation of glioblastoma resection margins. *Photochem Photobiol.* **77**, 309–318.
21. C.H. Liu, G.C. Tang, W.L. Sha, R.R. Alfano (1990). Effects of self-absorption by haemoglobin on the fluorescence spectra from normal and cancerous tissues. *Lasers Life Sci.* **3**, 167–176.
22. W.C. Lin, S.A. Toms, E.D. Jansen, A. Mahadevan-Jansen (2001). Intraoperative application of optical spectroscopy in the presence of blood. *IEEE J. Selected Top. Quantum Electron.* **7**, 996–1003.
23. R. Lipson, E. Baldes, A. Olsen (1961). The use of a derivative hematoporphyrin in tumour detection. *J. Natl. Cancer Inst.* **26**, 1–8.
24. S. Anderson-Engels, J. Johansson, S. Svanberg, K. Svanberg (1989). Fluorescence diagnosis and photochemical treatment of diseased tissues using lasers. *Anal. Chem.* **61**, 1367A–1373A.
25. C.H. Liu, B.B. Das, W.L. Sha, Glassman, G.C. Tang, K.M. Yoo, H.R. Zhu, D.L. Akins, S.S. Lubicz, J. Cleary, R. Prudente, E. Celmer, A. Caron, R.R. Alfano (1992). Raman, fluorescence, and time-resolved light scattering as optical diagnostic techniques to separate diseased and normal biomedical media. *J. Photochem. Photobiol. B* **16**, 187–209.
26. M. Dal Fante, G. Bottiroli, P. Spinelli (1988). Behaviour of hematoporpyrin derivative in adenomas and adenocarcinomas of the colon: a microfluorimetric study. *Lasers Med. Sci.* **3**, 165–171.
27. G. Bottiroli, A.C. Croce, R. Ramponi (1987). Quantitative analysis of intracellular behaviour of porphyrins. *Photochem. Photobiol.* **46**, 663–667.
28. G. Bottiroli, R. Ramponi (1988). Equilibrium among hematoporphyrin derivative components: influence of the interaction with cellular structures. *Photochem. Photobiol.* **47**, 209–214.

29. J.C. Kennedy, R.H. Pottier, D.C. Pross (1990). Photodynamic therapy with endogenous protoporphyrin IX: basic principles and present clinical experience. *J. Photochem. Photobiol. B* **6**, 143–148.
30. S.L. Marcus, R.S. Sobel, A.L. Golub, R.L. Carroll, S. Lundahl, D.G. Shulman (1996). Photodynamic therapy (PDT) and photodiagnosis (PD) using endogenous photosensitization induced by 5-aminolevulinic acid (ALA): current clinical and development status. *Clin. Laser Med. Surg.* **14**, 59–66.
31. T.J. Dougherty (1989). Photodynamic therapy: status and potential. *Oncology (Huntingt)* **3**, 67–73.
32. S. Udenfriend (1969). Amino acids, amines, and their metabolites. Proteins and peptides. In: *Fluorescence Assay in Biology and Medicine* (Volume II, pp. 195–280). Academic Press, London.
33. F.W.J. Teale, G. Weber (1957). Ultraviolet fluorescence of the aromatic aminoacids. *Biochemistry* **65**, 476–482.
34. S. Udenfriend (1961). Proteins. In: *Fluorescence Assay in Biology and Medicine* (Volume I, pp. 191-227). Academic Press, London.
35. F.W.D. Rost (1995). Autofluorescence in human and animal tissues and cells. In: *Fluorescence Microscopy* (Volume II, pp. 1–15). Cambridge University Press, Cambridge.
36. J.P. Borel, J.C. Monboisse (1993). Les collagènes: pourquoi une telle complexitè structurale? *C.R.S. Biol.* **187**, 124–142.
37. D.R. Eyre, T.J. Koob, K.P. Van Ness (1984). Quantitation of hydroxypyridinum crosslinks in collagen by high-performance liquid chromatography. *Anal. Biochem.* **137**, 380–388.
38. T. Moriguci, D. Fujimoto (1978). Age related changes in the content of the collagen crosslink, pyridinolinde. *J. Biochem.* **94**, 933–935.
39. M. Wolman, D. Fundoianu-Dayan, M. Perlmutter (1985). Blue autofluorescence of human Achilles tendon collagen in relation to aging. *Acta Histochem. Cytochem.* **18**, 439–443.
40. D.P. Thornihill (1975). Separation of a series of chromophores and fluorophores present in elastin. *Biochem J.* **147**, 215–219.
41. R. Richards-Kortum, R.P. Rava, R.E. Petras, M. Fitzmaurice, M. Sivak, M.S.Feld (1991). Spectroscopic diagnosis of colonic dysplasia. *Photochem. Photobiol.* **53**, 777–786.
42. J.M.I. Maarek, L. Marcu, W.J. Snyder, W.S. Grundfest (2000). Time resolved fluorescence spectra of arterial fluorescent compounds: Reconstruction with the Laguerre expansion technique. *Photochem. Photobiol.* **71**, 178–187.
43. A.J. Morguet, R.E. Gabriel, A.B. Buchwald, G.S. Werner, R. Nyga H. Kreuzer (1997). Single-laser approach for fluorescence guidance of excimer laser angioplasty at 308 nm; evaluation *in vitro* and during coronary angioplasty. *Laser Surg. Med.* **20**, 382–393.
44. J. Hung, S. Lam, J.C. LeRiche, B. Placid (1991). Autofluorescence of normal and bronchial tissue. *Lasers Surg. Med.* **11**, 99–105.
45. R. Marchesini, E. Pignoli, S. Tomatis, S. Fumagalli, A.E. Sichirollo, S. Di Palma, M. Dal Fante, P. Spinelli, A.C. Croce, G. Bottiroli (1994). *Ex vivo* optical properties of human colon tissue. *Lasers Surg. Med.* **15**, 351–357.
46. S. Abe, K. Isuishi, H. Taijri, T. Kinoshita, T. Matsuoska (2000). Correlation of *in vivo* autofluorescence endoscopy images with histopathologic findings in stomach cancer. *Endoscopy* **32**, 281–286.
47. I. Pavlova, K. Sokolov, R. Drezek, A. Malpica, M. Follen, R. Richards-Kortum (2003). Microanatomical and biochemical origins of normal and precancerous cervical autofluorescence using laser-scanning confocal microscopy. *Photochem. Photobiol.* **77**, 550–555.

48. L. Rigacci, R. Alterini, P.A. Bernabei, P.R. Ferrini, G. Agati, F. Fusi, M. Monici (2000). Multispectral imaging autofluorescence microscopy for the analysis of lymph-node tissues. *Photochem. Photobiol.* **71**, 737–742.
49. R. Moll, W.W. Franke, D.L. Schiller (1982). The catalog of human cytokeratins. Patterns of expression in normal epithelia, tumours and cultured cells. *Cell* **31**, 11–24.
50. J. Beuthan, O. Minet, G. Müller (1998). Optical biopsy of cytokeratin and NADH in the tumour border zone. *Ann. New York Acad. Sci.* **838**, 150–170.
51. J.E. Aubin (1979). Autofluorescence of viable cultured mammalian cells. *J. Histochem. Cytochem.* **27**, 36–43.
52. R.S. Balaban, L.J. Mandel (1990). Optical methods for the study of metabolism in intact cell. In: J.K. Forskett and S. Grinstein (Eds), *Non-invasive Techniques in Cell Biology* (pp. 213–236). Wiley Liss, New York.
53. B. Chance, P. Cohen, F. Jobsis, B. Schoener (1962). Intracellular oxidation-reduction states *in vivo*. The microfluorometry of pyridine nucleotide gives a continuous measurement of the oxidation state. *Science* **137**, 499-508.
54. J.R. Lakowicz, H. Szmacinski, K. Nowackyk, M.L. Johnson (1992). Fluorescence lifetime imaging of free and protein-bound NADH. *Proc. Natl. Acad. Sci. U.S.A.* **89**, 1271–1275.
55. O.S. Wolfbeiss (1985). The fluorescence of organic natural products. In: S.G. Schulman (Ed), *Molecular Fluorescence Spectroscopy. Methods and Applications; Part I* (pp. 167–370). John Wiley & Sons, New York.
56. G.R. Penzer (1980). Molecular emission spectroscopy (Fluorescence and Phosphorescence). In: S.B. Brown (Ed), *An Introduction to Spectroscopy for Biochemists* (pp. 70–114). Academic Press, London.
57. W.S. Kunz, W. Kunz (1985). Contribution of different enzymes to flavoprotein fluorescence of isolated rat liver mitochondria. *Biochim. Biophys. Acta.* **841**, 237–246.
58. W.S. Kunz (1986). Spectral properties of fluorescent flavoproteins of isolated rat liver mitochondria. *FEBS* **195**, 92–96.
59. B. Thorell (1983). Flow-cytometric monitoring of intracellular flavins simultaneously with NAD(P) levels. *Cytometry* **4**, 61–65.
60. S. Huang, A.A. Heikal, W.W. Webb (2002). Two-photon fluorescence spectroscopy of NAD(P)H and flavoprotein. *Biophys. J.* **82**, 2811–2825.
61. E. Kohen, C. Kohen, J.G. Hirschberg, A. Wouters, B. Thorell (1978). Multisite topographic microfluorometry of intracellular and exogenous fluorochromes. *Photochem. Photobiol.* **27**, 259–268.
62. J.G. Hirschberg, E. Kohen, C. Kohen (1989). High resolution fluorescence and phase microscopy in conjunction with micromanipulation for *in situ* study of metabolism in living cells. In: *Cell Structure and Function by Microspectrofluorometry* (pp. 87–98). Academic Press Inc., New York.
63. M. Dellinger, M. Geze, R. Santus, E. Kohen, C. Kohen, J.G. Hirschberg, M. Monti (1998). Imaging of cells by autofluorescence: a new tool in the probing of biopharmaceutical effects at the intracelllular level. *Biotechnol. Appl. Biochem.* **28**, 25–32.
64. D.W. Piston, S.M. Knobel (1999). Real-time analysis of glucose metabolism by microscopy. *TEM*, **10**, 413–417.
65. G.J. Puppels, J.M.C.C. Coremans, H.A. Briuning (1999). *In vivo* semiquantitative NADH-flurorescence imaging. In: W.T. Moson (Ed), *Fluorescent and Luminescent Probes* (2nd Edn, pp. 457–466). Mason, Life Science Resources Ltd, Cambridge.
66. P. Viallet, E. Kohen, D.O. Schachtschabel, A. Marty, J.M. Salmon, C. Kohen, H.B. Leising, B. Thorell (1978). The effect of atebrine and an acridine analog (BCMA)

on the coenzyme fluorescence spectra of cultured melanoma and Ehrlich ascites (EL2) cells. *Histochemistry* **57**, 189–201.

67. Y.G. Chung, J.A. Schwartz, C.M. Gardner, R.E. Sawaya, S.L. Jacques (1997). Diagnostic potential of laser-induced autofluorescence emission in brain tissue. *J. Korean Med. Sci.* **12**, 135–142.
68. B. Banerjee, B. Miedema, H.R. Chandrasekhar (1998). Emission spectra of colonic tissue and endogenous fluorophores. *Am. J. Med. Sci.* **316**, 220–226.
69. R. Drezek, K. Sokolov, U. Utzinger, I. Boiko, A. Malpica, M. Follen, R. Richards-Kortum (2001). Understanding the contributions of NADH and collagen to cervical tissue fluorescence spectra: modelling, measurements, and implications. *J. Biomed. Opt.* **6**, 385–396.
70. T.G.V. Galeotti, D. Van Rossum, D.H. Mayer, B. Chance (1970). On the fluorescence of NAD(P)H in whole-cell preparation of tumours and normal tissues. *Eur. J. Biochem.* **17**, 485–496.
71. A.C. Croce, S. Fiorani, D. Locatelli, R. Nano, M. Ceroni, F. Tancioni, E. Giombelli, E. Benericetti, G. Bottiroli (2003). Diagnostic potential of autofluorescence for an assisted intraoperative delineation of glioblastoma resection margins. *Photochem. Photobiol.* **77**, 309–318.
72. D. Duboc, P. Abastado, M. Muffat-Joly, P. Perrier, M. Toussaint, C. Marsac, D. Francois, T. Lavergne, J.J. Pocidalo, F. Guerin, et al. (1990). Evidence of mitochondrial impairment during cardiac allograft rejection. *Transplantation* **50**, 751–755.
73. J.F. Ashruf, J.M. Coremans, H.A. Bruining, C. Ince (1995). Increase of cardiac work is associated with decrease of mitochondrial NADH. *Am. J. Physiol.* **269**, H856–862.
74. E.T. Obi-Tabot, L.M. Hanrahan, R. Cachecho, E.R. Beer, S.R. Hopkins, C.K. Chan, J.M. Shapiro, W.W. LaMorte (1993). Changes in hepatocyte NADH fluorescence during prolonged hypoxia. *J. Surg. Res.* **55**, 575–580.
75. B. Vollmar, M. Burkhardt, T. Minor, H. Klauke, M.D. Menger (1997a). High-resolution microscopic determination of hepatic NADH fluorescence for *in vivo* monitoring of tissue oxygenation during hemorrhagic shock and resuscitation. *Microvascular Res.* **54**, 164–173.
76. H. Klauke, T. Minor, B. Vollmar, W. Isselhard, M.D. Menger (1998). Microscopic analysis of NADH fluorescence during aerobic and anaerobic preservation conditions: A non-invasive technique for assessment of hepatic metabolism. *Cryobiology* **36**, 108–114.
77. A.C. Croce, S. Fiorani, I. Freitas, R. Bertone, M.P. Vairetti, A. Ferrigno, G. Bottiroli, D. Neri (2003). Liver metabolic conditions during transplantation: autofluorescence spectroscopy monitoring. *Lasers Surg. Med.*15(suppl.), N183, 51.
78. F. Querner Von (1932). Die paraplasmatischen Einschlüsse der Leberzellen im Fluoreszenzmikroskop und der Leuchtstoff X. *Akad. Anzeiger Akad. Wissenschaft Wein, Math.-Naturw.*, Klasse, No. 18.
79. H. Popper (1944). Distribution of Vitamin A in tissues as visualized by fluorescence microscopy. *Physiol. Rev.* **24**, 205–224.
80. R.D. Fugate, P.S. Song (1980). Spectroscopic characterization of beta-lactoglobulin-retinol complex. *Biochim. Biophys. Acta* **625**, 28–42.
81. J. Kahan (1973). Automated fluorometric assay of serum vitamin A. *Int. J. Vitam. Nutr. Res.* **43**, 127–41.
82. M.L. Katz, G.E. Eldred, W.G. Robinson Jr. (1987). Lipofuscin autofluorescence: evidence for vitamin A involvement in the retina. *Mach. Ageing Dev.* **39**, 81–90.
83. M. Wolman (1980). Lipid pigments (chromolipids): their origin, nature and significance. *Pathobiol. Ann.* **10**, 253–267.

84. A.G.E. Pearse (1985). Pigments and pigment precursors. In: *Histochemistry Theoretical and Applied* (*Vol. II Analytical Technology*), (pp. 874–928). Churchill Livingstone Edinburgh.
85. D.N. Palmer, D.R. Husbands, P.J. Winter, J.W. Blunt, R.D. Jolly (1986). Ceroid lipofuscinosis in sheep. I. Bis(monoacylglycero)phosphate, dolichol, ubiquinone, phospholipids, fatty acids, and fluorescence in liver lipopigment lipids. *J. Biol. Chem.* **261**, 1766–1772.
86. D.N. Palmer, G. Barns D.R. Husbands, R.D. Jolly (1986). Ceroid lipofuscinosis in sheep. II. The major component of the lipopigment in liver, kidney, pancreas, and brain is low molecular weight protein. *J. Biol. Chem.* **261**, 1773–1777.
87. E.A. Porta (2002). Pigments in aging: an overview. *Ann. New York. Acad. Sci.* **959**, 57–65.
88. A.L. Tappel (1973). Lipid peroxidation damage to cell components. *Fed. Proc.* **32**, 1870–1874.
89. D. Barnes, S. Aggrawal, S. Thomsen, M. Fitzmaurice, R. Richards-Kortum (1993). A characterization of the fluorescent properties of circulating human eosinophils. *Photochem. Photobiol.* **58**, 297–303.
90. S.H. Benavides, P.E. Morgnate, A.J. Monserrat, J. Zarate, E.A. Porta (1997). The pigment of melanosis coli: a lectin histochemical study. *Gastrointest. Endosc.* **46**, 131–138.
91. G. Bottiroli, A.C. Croce, D. Locatelli, R. Nano, E. Giombelli, A. Messina, E. Benericetti (1998). Brain-tissue autofluorescence: an aid for intraoperative delineation of tumour resection margins. *Cancer Detect. Prev.* **22**, 330–339.
92. Y. Matsumoto (2001). Lipofuscin pigmentation in pleomorphic adenoma of the palate. *Oral Surg. Oral Med. Oral Pathol. Oral Radiol. Endod.* **92**, 299–302.
93. R.C. Spike, H.S. Johnston, J. McGadey, M.R. Moore, G.G. Thompson, A.P. Payne, (1985). Quantitative studies on the effects of hormones on structure and porphyrin biosynthesis in the Harderian gland of the female golden hamster. 1. The effects of ovariectomy and androgen administration. *J. Anat.* **142**, 59–72.
94. A. Kappas, S. Sassa, R.A. Galbraith, Y. Nordmann (1995). The porphyries. In: C.R. Scriver, A.L. Beaudet, W.S. Sly, D. Valle (Eds), *The metabolic and Molecular Bases of Inherited Disease* (7th edn, Volume II, pp. 2103–2159). McGraw-Hill, New York.
95. D.M. Harris, J. Werkhaven (1987). Endogenous porphyrin fluorescence in tumours. *Lasers Surg. Med.* **7**, 467–472.
96. M. Inaguma, K. Hashimoto (1999). Porphyrin-like fluorescence in oral cancer. *Cancer* **86**, 2201–2211.
97. K.T. Moesta, B. Ebert, T. Handke, D. Nolte, C. Nowak, W.E. Haensch, R.K. Pandey, T.J. Dougherty, H. Rinneberg, P.M. Schlag (2001). Protoporphyrin IX occurs naturally in colorectal cancers and their metastases. *Cancer Res.* **61**, 991–999.
98. B. Mayinger, M. Jordan, P. Horner, C. Gerlach, S. Muehldorfer, B.R. Bittorf, K.E. Matzel, W. Hohenberger, E. Haln, K. Guenter (2003). Endoscopic light-induced autofluorescence spectroscopy for the diagnosis of colorectal cancer and adenoma. *J. Photochem. Photobiol.* **70**, 13–20.
99. N. Schoenfeld, O. Epstein, M. Lahav, R. Mamet, M. Shaklai, A. Atsmon (1988). The heme biosynthetic pathway in lymphocytes of patients with malignant lymphoproliferative disorders. *Cancer Lett.* **43**, 43–48.
100. L. Leibovici, N. Schoenfeld, H.A. Yehoshua, R. Mamet, E. Rakowsky, A. Shindel, A. Atsmon (1988). Activity of porphobilinogen deaminase in peripheral blood mononuclear cells of patients with metastatic cancer. *Cancer* **62**, 2297–2300.
101. G.F. Rubino, L. Rasetti (1966). Porphyrin metabolism in human neoplastic tissues. *Panminerva Med.* **8**, 290–292.

102. G. Weagle, P.E. Paterson, J. Kennedy, R. Pottier (1988). The nature of the chromophore responsible for naturally occurring fluorescence in mouse skin. *J. Photochem. Photobiol.* **2**, 313–320.
103. W. Tan, V. Parpura, G. Haydon, E.S. Yeung, (1995). Neurotransmitters imaging in living cells based on native fluorescence detection. *Anal. Chem.* **67**, 2575–2579.
104. W. Tan, P.G. Haydon, E.S. Yeung (1997). Imaging neurotransmitter uptake and depletion in astrocytes. *Appl. Spectrosc.* **51**, 1139–1143.
105. E.S. Yeung (1999). Following single cell dynamics with native fluorescence microscopy. *Anal. Chem. News & Features* **4**, 522A–529A.
106. http://www.drbio.cornell.edu/Infrastructure/NonlinearMicroscopies_WWW/seroox.htm
107. F. Crespi, A.C. Croce, S. Fiorani, B. Masala, C. Heidbreder, G. Bottiroli (2004). Autofluorescence spectrofluorometry of central nervous system (CNS) neuromediators. *Lasers Surg. Med.* **34**, 39–47.

Chapter 11

Reflectance and transmittance spectroscopy

Enrico Gratton and Sergio Fantini

Table of contents

Abstract

This chapter describes several approaches to the optical study of biological tissue using reflectance and transmittance spectroscopy. This topic has spurred significant research efforts as a result of the relevant physiological and metabolic information provided by the optical data, in conjunction with the safe, non-invasive, and cost-effective optical approach to the study of tissue. We give a brief historical introduction in Section 11.1, followed by a description of the optical absorption and scattering properties of tissue in Section 11.2. Section 11.3 is devoted to continuous-wave (CW) techniques, which can be applied to the study of relatively superficial tissue layers (as is the case for diffuse reflectance imaging and localized measurements using short separations between the illuminating and collecting optodes), as well as deep tissues on the basis of a modified Beer–Lambert law, transport theory, or diffusion theory. The latter model is commonly employed in time-domain and frequency-domain techniques, which are described in Sections 11.4 and 11.5, respectively. Three major application areas of near-infrared tissue studies, namely tissue oximetry, optical mammography, and functional imaging of the brain, are presented in Sections 11.6, 11.7, and 11.8, respectively. Finally, Section 11.9 discusses potential future directions for optical techniques in a number of clinical areas.

11.1 Introduction

The potential of reflectance and transmittance optical techniques for the study of biological tissue has long been recognized. For example, applications in medical diagnostics and monitoring of physiological parameters in vivo were developed as far back as 1929 for optical mammography [1] and 1942 for tissue oximetry [2]. In particular, in the early 1970s, the optical oximetry approach developed into pulse oximetry [3,4], which is routinely used today in hospitals and intensive care units to monitor the oxygen saturation of arterial blood. The demonstration of the applicability of reflectance spectrophotometry to the study of the brain [5] has opened up new opportunities to investigate the cortical architecture in animal models [6] and the brain functional activity in human subjects [7–10]. The introduction of a physical model, namely diffusion theory, to quantitatively describe light propagation in tissue [11], and the development of time-resolved techniques either in the time-domain [11,12] or in the frequency-domain [13–15] to the optical study of tissue have resulted in a further refinement of reflectance and transmittance spectroscopy and imaging of tissue.

The optical study of tissue in vivo is typically performed in reflectance, i.e. with the illumination and collection performed on the same side of the tissue. In many cases, this is dictated by the lack of adequate optical signal transmitted through thick tissues. However, body parts that are relatively thin (for instance, fingers) or present relatively low levels of optical attenuation (for instance, the female breast or the infant's head) lend themselves to transmittance optical studies. In this chapter, we will describe applications based on both reflectance and transmittance geometries, as well as continuous-wave (CW), time-domain (TD), and frequency-domain (FD) optical techniques.

11.2 Optical properties of tissue

The propagation of light in biological tissue can be described in terms of the flow of discrete particles called photons. According to this representation, a light source introduces a given number of photons per unit time at given tissue location and these photons travel inside the tissue following individual paths. The collective propagation of photons along these paths is called *photon migration*. After a photon is introduced inside the tissue by the light source, it can interact with tissue through several mechanisms, including absorption, elastic scattering, and fluorescence. When a photon is absorbed it essentially disappears and transfers its energy to the absorbing center (chromophore). When a photon is elastically scattered by a scattering center in the tissue (for instance a cellular membrane, nucleus, or organelle), its direction of propagation changes while its energy (and wavelength) is essentially not affected. In a fluorescence process, a photon at wavelength λ_x (excitation wavelength) is absorbed and a photon at a longer wavelength λ_m (emission wavelength) is emitted. The fluorescence emission process is not immediate, but takes place with an average delay τ (fluorescence lifetime), typically on a time scale of nanoseconds, with respect to the time of absorption of the excitation photon. These three processes are schematically illustrated in Figure 1. In this chapter, we describe techniques of reflectance and transmittance spectroscopy for the optical study of tissue, which are based on absorption and elastic scattering processes. Fluorescence spectroscopy is described elsewhere in this book.

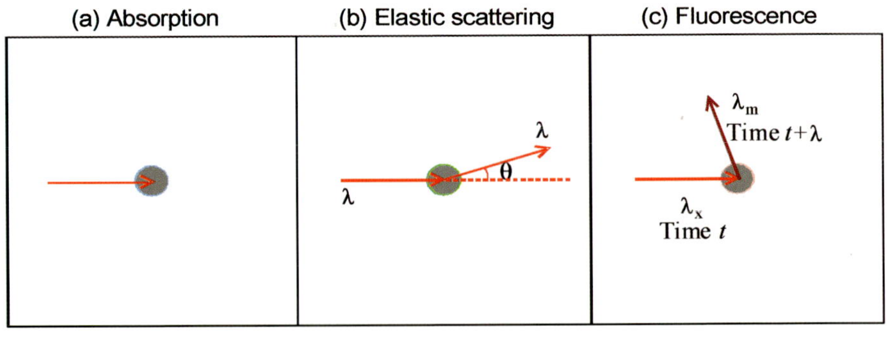

Figure 1. Schematic illustration of (a) absorption, (b) elastic scattering, and (c) fluorescence processes of interaction of photons with chromophores, scattering centers, and fluorophores, respectively, in tissue. In panel (b) λ is the photon wavelength, which is largely unaffected by elastic scattering processes, and θ is the scattering angle. In panel (c), the wavelength of the emission photon λ_m is longer than the wavelength of the incoming (or excitation) photon λ_x, and the emission photon is emitted with an average delay τ with respect to the absorption of the excitation photon.

11.2.1 Absorption

The main absorbers of visible/near-infrared light in bloodperfused tissues are oxy-hemoglobin, deoxy-hemoglobin, and water, with further absorption contributions, which sometimes can be of interest, from myoglobin, lipids, cytochrome oxidase, melanin, bilirubin, etc. The absorption spectra of oxy-hemoglobin (50 μM concentration), deoxy-hemoglobin (50 μM concentration), and water are shown in Figure 2, which is derived from published data of the absorption coefficients of water [16] and hemoglobin [17]. The socalled "medical spectral window" extends from approximately 700 to 900 nm, where the combined absorption of light from hemoglobin and water is minimal (Figure 2). As a result of the relatively smaller absorption, light in this spectral window penetrates deeper in tissues, making it possible to investigate deep tissue sites noninvasively.

The absorption properties of tissue are described by the absorption coefficient (μ_a), which is defined as the inverse of the average photon path length before absorption. From this definition it follows that $1/\mu_a$ is the average distance traveled by a photon before being absorbed. In the near-infrared, typical values of μ_a in tissues range from 0.02 to 0.30 cm^{-1}, so that the mean photon path before absorption thus ranges between 3 and 50 cm.

11.2.2 Scattering

The scattering of photons is due to localized gradients in the refractive index caused by particles that act as scattering centers. The scattering properties are mainly

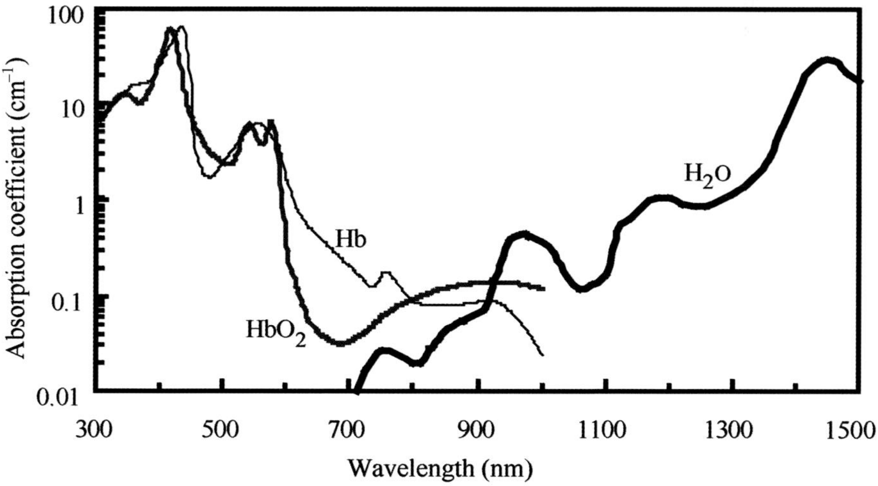

Figure 2. Absorption spectra of three of the most important near-infrared chromophores in tissues, namely oxy-hemoglobin (HbO_2), deoxy-hemoglobin (Hb), and water (H_2O). The absorption coefficient is defined to base *e*. The concentrations of Hb and HbO_2 are both assumed to be 50 μM, a typical value for blood-perfused tissues. Spectra obtained from compiled absorption data for water [16] and hemoglobin [17].

determined by the dimensions of the particles relative to the wavelength of light, and by the difference between the indices of refraction of the particles and the surrounding medium. In biological tissues, the scattering centers include cell nuclei, cell organelles, and cells themselves. In particular, cell organelles such as mitochondria have dimensions comparable to the wavelengths in the medical spectral window (700–900 nm), and their index of refraction is not dramatically different than that of the cytosol. Under these conditions, light scattering is mainly forward directed (i.e. the scattering angle θ shown in Figure 1 (b) is smaller than 90°), as is the case in most biological tissues.

The scattering properties of tissues are described by two parameters: (1) the scattering coefficient (μ_s), defined as the inverse of the average photon path length between successive scattering events; (2) the average cosine of the scattering angle θ, $g = \langle \cos\theta \rangle$. From the definition of μ_s it follows that $1/\mu_s$ is the average distance traveled by a photon between successive scattering events. Even though each scattering event is mainly forward directed, after a large number of collisions a photon loses memory of its original direction of propagation. Under these conditions, the scattering angle θ cannot be measured and we can describe photon scattering in terms of effectively isotropic scattering processes. In tissue spectroscopy, it is customary to define the reduced scattering coefficient ($\mu_s' = (1 - g)\mu_s$) which represents the inverse of the average distance over which the direction of propagation of a photon is randomized. In other words, we can say that $1/\mu_s'$ is the average distance between effectively isotropic scattering events. Of course μ_s' coincides with μ_s in the case of isotropic scattering ($g = 0$). In the optical study of thick tissues, μ_s' is the only measurable scattering parameter. Typical values of μ_s' in biological tissues range from 1 to 50 cm^{-1}, while g is typically 0.8–0.9 [18] (so that μ_s' is about one order of magnitude smaller than μ_s). The average distance traveled by a photon in tissue between effectively isotropic scattering events is typically a few millimetres or less.

11.3 Continuous-wave techniques

Continuous-wave (CW) methods employ light sources (for example arc lamps, light emitting diodes, or laser diodes) with emission properties that are time-independent. While CW methods are typically unable to separate the absorption and scattering contributions to the optical absorbance measured in tissue, they present the advantage of being technically straightforward. For instance, charge coupled device (CCD) cameras can be readily used for reflectance imaging, while photo-multiplier tubes and avalanche photodiodes are typically employed for single-point detection.

11.3.1 Diffuse reflectance imaging

Optical imaging of biological tissue with diffuse illumination and CCD camera detection is a powerful tool to investigate superficial tissue layers with high spatial resolution. A typical experimental setup is illustrated in Figure 3. The general idea

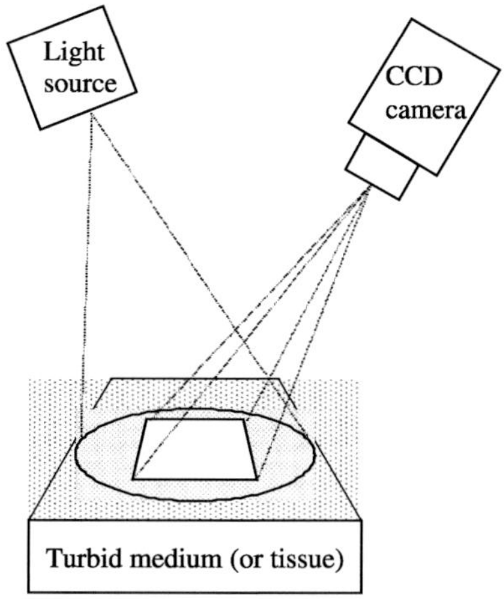

Figure 3. Typical experimental setup for diffuse reflectance measurements. A light source (for instance a laser diode, arc lamp, or incandescent lamp) illuminates a broad area of the sample while a CCD camera images a portion of the illuminated area by collecting diffuse reflectance.

is to uniformly illuminate a broad area of the sample (or tissue) to be investigated, and to collect the diffusively reflected signal over a portion of the illuminated area. Linearly polarized light and a linear analyzer can be used to suppress the detection of specular reflections at the tissue surface. This approach has been used to obtain 50–100 μm resolution in vivo images of the cerebral cortex in animal models [6,19,20] and in humans [21]. The basic approach (see for example Ref. 22) consists of illuminating the cerebral cortex (after removal of the skull and dura) with light of proper wavelength. Spatial uniformity of the cortex illumination is achieved by using multiple light guides. Visible wavelengths (in the 500–600 nm range) are typically used to obtain a high sensitivity to blood vessels and to oxygenation changes, while longer wavelengths (>700 nm) are used to increase the optical penetration depth (in this configuration, the cortex can be investigated up to depths of about 1 mm) and the sensitivity to changes in light scattering. The light reflectance images are analyzed to obtain differential maps that represent the effect of specific cortical activity and provide information on the functional architecture of the cerebral cortex. For example, this method has been used to map the ocular dominance domains in the visual cortex [19,23], and the functional architecture of the somatosensory cortex [24] in animal models. Diffuse reflectance measurements on the skin of human subjects may allow for the discrimination of cutaneous melanoma from other pigmented lesions of human skin [25,26].

The theoretical description of the diffuse reflectance measured from a turbid medium (such as biological tissue) under uniform illumination conditions has been performed according to several different models. The objective of these models is to describe the relationship between the diffuse reflectance (R) and the optical properties of the turbid medium, namely the absorption coefficient (μ_a), the scattering coefficient (μ_s), and the anisotropy scattering factor g. We recall that μ_a is defined as the inverse of the mean distance traveled by the photons in the medium before being absorbed, μ_s is defined as the inverse of the photon mean free path between successive scattering events, and g is defined as the average of the cosine of the scattering angle for photons in the medium. The main theoretical models used to describe the diffuse reflectance are the following.

11.3.1.1 Kubelka–Munk theory

This model is based on a two-flux description of the light propagation in turbid media [27,28], which leads to the following expression for the diffuse reflectance under conditions of perfectly diffuse irradiation of the medium and isotropic light scattering ($g = 0$) [28,29]:

$$R = 1 + \frac{K}{S} - \left(\frac{K^2}{S^2} + 2\frac{K}{S}\right)^{\frac{1}{2}}, \tag{1}$$

where K and S are the Kubelka–Munk absorption and scattering coefficients, respectively, which are related to μ_a and μ_s by the relationship $K/S \approx 2.67\ \mu_a/\mu_s$ [30].

11.3.1.2 Transport theory

Transport theory provides a general description of light propagation in absorbing and scattering media on the basis of the Boltzmann transport equation. This general approach does not provide an analytical expression for the diffuse reflectance. Tabulated values of the diffuse reflectance obtained by solving the Boltzmann transport equation for several values of the ratio μ_a/μ_s and for either isotropic scattering ($g = 0$) or particular cases of anisotropic scattering have been reported [31].

11.3.1.3 Diffusion theory

Even though, strictly speaking, the diffusion approximation to the transport equation is not applicable to the description of the diffuse reflectance obtained under uniform illumination conditions, it has nevertheless been used to obtain an analytical approximation for the diffuse reflectance [32,33]. For collimated, normal irradiation, and a 1.33 refractive index mismatch at the boundary between non-scattering and scattering media, the expression for R derived with diffusion theory is [33]:

$$R = \frac{a'}{1 + 2k(1 - a') + (1 + 2k/3)[3(1 - a')]^{\frac{1}{2}}}, \tag{2}$$

where a' is the optical reduced albedo defined as $\mu_s'/(\mu_a + \mu_s')$, with $\mu_s' = (1 - g)\mu_s$ as the reduced scattering coefficient, and $k = (1 - r_d)/(1 + r_d)$, with r_d as the internal reflection coefficient of the turbid medium.

11.3.1.4 Adding-doubling and Monte Carlo methods

Combining an adding-doubling solution to the transport equation [34] and Monte Carlo methods has resulted in an analytical expression for the diffuse reflectance from a turbid medium where the light scattering is described by the Henyey–Greenstein phase function [35]:

$$R = \mathrm{e}^{\frac{-7.8}{\sqrt{3(1+\mu_s'/\mu_a)}}} \tag{3}$$

11.3.2 Spectral measurements insensitive to scattering changes

As mentioned above, CW methods are usually unable to perform measurements of both absorption and scattering properties of tissue. For instance, the CW reflectance expressions of Equations (1)–(3) all contain the ratio of absorption to scattering coefficients, which prevents the separations of the contributions from these two coefficients to the diffuse reflectance. One way to overcome this limitation of CW tissue spectroscopy is to develop measurement techniques that are relatively insensitive to the scattering properties of tissue. For example, when an illumination optical fiber (or optode) is used to deliver light at one specific location, and a collection optode is used to collect the optical signal at a distance r from the illumination point, there is an optimal optode separation, r, for which the detected optical signal is not affected by changes in the scattering properties of the tissue [36]. The basic idea behind the existence of an optimal source–detector separation that minimizes the sensitivity to the scattering coefficient is that, in the limit of very small separations, the optical signal increases with the scattering coefficient, while the optical signal decreases in the diffusive limit of large separations. Therefore, it is reasonable to expect that, at some intermediate point, the optical signal does not change with the scattering coefficient. Such an optimal source–detector separation, for typical optical properties of biological tissue, was found to be 1.7 mm [36], a value that lends itself to being implemented in a clinical endoscope. The reason that the insensitivity of the optical signal to the scattering coefficient is relevant is the following. The detected optical intensity I in reflectance and transmittance spectroscopy can be described in terms of the Beer–Lambert law:

$$I = I_0 \mathrm{e}^{-\mu_a L} \tag{4}$$

where I_0 is the incident intensity and L is the photon pathlength. In the presence of scattering, L is not the same for all detected photons, so that L in Equation (4) should be replaced by an effective pathlength (which in general differs from the average photon pathlength). If L, or the effective pathlength, is known from calibration measurements or look-up tables and is not affected by the scattering properties of tissues, then one can translate a spectral or temporal variation in I into a corresponding variation in μ_a:

$$\mu_{a2} - \mu_{a1} = \ln\left[\frac{(I/I_0)_1}{(I/I_0)_2}\right] \Big/ L \tag{5}$$

where the subscripts 1 and 2 indicate measurements at different wavelengths or different times. Strictly speaking, the pathlength L also depends on the absorption coefficient itself, but for relatively small values of $\mu_{a2}-\mu_{a1}$ such dependence may be neglected. Being able to apply Equation (5) regardless of the scattering properties of tissue is an important feature that allows, for example, the non-invasive quantitative measurement of drug concentration in tissue [37].

11.3.3 Modified Beer–Lambert law

Another common approach to measuring changes in the tissue absorption using CW methods is based on a generalization of Beer–Lambert law (Equation (4)) in conjunction with the assumption that the scattering properties of tissue do not vary with time. The modified Beer–Lambert equation is the following [12]:

$$I = I_0 e^{-B\mu_a r - G} \tag{6}$$

where B is a pathlength factor that depends on the optical properties of tissue, r is the inter-optode distance (i.e. the geometrical separation between the points of illumination and light collection), and G is an unknown geometry-dependent factor that also accounts for the effect of scattering. If the factor G is constant, then a relatively small temporal variation in the absorption coefficient can be written as:

$$\mu_a(t_2) - \mu_a(t_1) = \frac{1}{Br}\ln\left[\frac{I(t_1)}{I(t_2)}\right] \tag{7}$$

The pathlength factor B is usually estimated from literature values [38–40]. From diffusion theory (see Section 11.3.4), it can be shown that the pathlength factor B is given by the following expressions in terms of the absorption and reduced scattering coefficients [41]:

$$B_{\mathrm{inf}} = \frac{\sqrt{3\mu_s'}}{2\sqrt{\mu_a}} \tag{8}$$

$$B_{\mathrm{seminf}} = \frac{\sqrt{3\mu_s'}}{2\sqrt{\mu_a}} \frac{r\sqrt{3\mu_a\mu_s'}}{\left(r\sqrt{3\mu_a\mu_s'} + 1\right)} \tag{9}$$

where the subscript “inf” refers to an infinite geometry, and the subscript “seminf” refers to the semi-infinite case where the optodes are located at the boundary of a scattering medium that extends indefinitely in one direction.

11.3.4 Diffusion theory

Diffusion theory describes the propagation of photons in optically turbid media. Mathematically, the diffusion equation is a limiting case of the general Boltzmann transport equation [42,43], which describes the propagation of photons in scattering and absorbing media. The key condition that leads to the diffusion approximation, which is often (but not always) fulfilled in the optical study of tissue, is that light propagates in a strongly scattering regime, i.e. $\mu_s' \gg \mu_a$. This condition means that a photon, on the average, undergoes many effectively isotropic scattering events before being absorbed. Under this condition, the CW photon density in the tissue [$U_{CW}(r)$] resulting from a point light source at $r = 0$ obeys the standard diffusion equation [44]:

$$-vD\nabla^2 U_{CW}(r) + v\mu_a U_{CW}(r) = P_0\delta(r), \qquad (10)$$

where v is the speed of light in the tissue, D is the diffusion coefficient defined as $1/[3(\mu_a+\mu_s')]$, P_0 is the source power, and δ is the Dirac delta. The solution to Equation (10) for an infinite medium yields the CW photon density as a function of the distance r from the illumination point [44]:

$$U_{CW}(r) = \frac{P_0}{4\pi vD} \frac{e^{-r\sqrt{3\mu_a\mu_s'}}}{r} \qquad (11)$$

This expression is often used to guide the interpretation of optical data collected in tissue. Data collected in a reflection geometry is often modeled using a semi-infinite model [44], while data in transmission may be modeled using a slab geometry [11], leading to more accurate predictions of the optical signal with respect to the infinite-medium solution of Equation (11).

11.3.5 Determination of tissue optical properties using the spatial non-linearity of diffuse reflectance at short source–detector separations

The solution to the diffusion equation for an infinite medium, which is given by Equation (11), shows that the $\ln(rU_{CW})$ is a linear function of r. It can be shown that, for a semi-infinite geometry, a linear relationship holds between $\ln(r^2U_{CW})$ and r, provided that $r \gg (3\mu_a\mu_s')^{-1/2}$ [44–46]. The non-linearity of the $\ln(r^2U_{CW})$ at relatively short source–detector separations (<1 cm) can be used to measure both the absorption and the scattering properties of tissue from spatially-resolved CW data. By using this approach, Farrel et al. measured the absorption and the reduced scattering coefficients of skin in human subjects in vivo [44].

11.4 Time-domain techniques

Time-domain (TD) studies employ a pulsed light source that emits short light pulses (typically on the order of picoseconds) with a repetition rate of about 1 MHz [11,12]. Mode-locked solid state lasers, or fast laser diodes are used as the light sources. Photomultiplier tubes and microchannel plates in photon counting mode, or streak cameras, are usually employed as the detectors. In the time-domain, one directly measures the time-of-flight distribution of the collected photons, which travel along a set of trajectories that are collectively indicated as a light bundle (see Figure 4(a)).

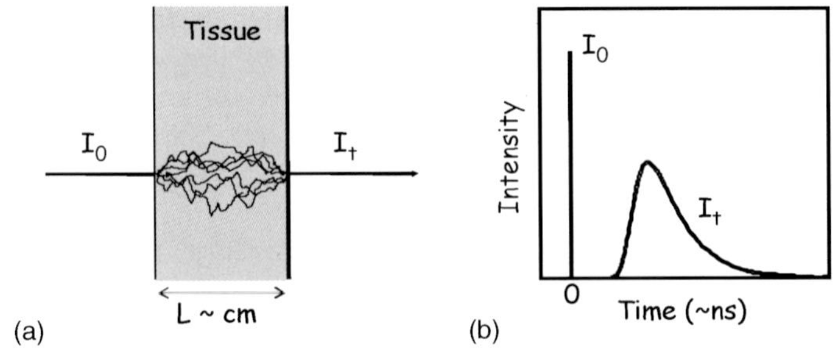

Figure 4. (a) Photons that are incident at one point of the tissue (intensity I_0) and that are detected at another point, in transmission in the case illustrated in the figure, (intensity I_t) travel along a collection of trajectories within the tissue. (b) In the time-domain, I_0 is a short (ps) pulse, while I_t is broadened and delayed on a time scale of nanoseconds as a result of propagation over a tissue thickness L of the order of centimetres.

11.4.1 Diffusion theory in the time domain

In the diffusive regime, which is often established when light propagates over several millimeters or more into tissue, the distribution of photon times-of-flight is given by the solution to the diffusion equation for a point photon source which emits N_0 photons at time $t = 0$. The corresponding source term is written as $N_0\delta(r)\delta(t)$, where δ is the Dirac delta, resulting in Equation (12) for the diffusion equation:

$$\frac{\partial U_{TD}(r,t)}{\partial t} - vD\nabla^2 U_{TD}(r,t) + v\mu_a U_{TD}(r,t) = N_0\delta(r)\delta(t) \qquad (12)$$

where $U_{TD}(r,t)$ is the time-domain photon density at point r and time t, and D is the diffusion coefficient defined in Equation (10). The solution to Equation (12) for an infinite medium is Equation (13) [11]:

$$U_{TD}(r,t) = \frac{N_0}{(4\pi vDt)^{\frac{3}{2}}} e^{\frac{3\mu_s' r^2}{4vt} - \mu_a vt} \qquad (13)$$

The time dependence of U_{TD}, which represents the delay and broadening of the input pulse as it propagates into tissue, is qualitatively shown in Figure 4(b). From Equation (13) one can see that the behavior of U_{TD} at short times is dominated by μ_s' (because of the inverse dependence on t of the first term in the exponent, which contains μ_s'), whereas the behavior at large times is dominated by μ_a (because of the direct dependence on t of the second term in the exponent, which contains μ_a). A fit of Equation (13), or its extension to appropriate geometries such as semi-infinite, slab, cylindrical, or spherical [47], to the experimental data enables one to independently recover μ_a and μ_s' of tissue [11].

11.4.2 Time gating

In spectroscopy applications, time-domain techniques can separately measure the absorption and reduced scattering coefficients of tissue. In imaging applications, where the goal is to measure the spatial distribution of tissue heterogeneities, time-domain methods offer the capability of time-gating to select photons that are associated with a time-of-flight (or, equivalently, with a pathlength) in tissue that is within a given range. Usually, the idea is to select the photons that have traveled the shortest paths, thus narrowing the light bundle illustrated in Figure 4(a) in an attempt to improve the spatial resolution [48]. However, the paucity of weakly scattered photons through tissues that are several centimetres thick poses serious limitations to the use of short time gates. As a result, the increase in spatial resolution afforded by time-gating with respect to CW data does not usually exceed a factor of 2 for a sample thickness of 4 cm and optical properties typical of tissue [49]. An alternative approach to using a fixed-interval time window is to select a variable-interval time window to detect a fixed percentage of the total transmitted photons [50]. Using this variable amplitude of the time window as an imaging parameter, Benaron and Stevenson obtained images of a rat that identified organs such as the heart, liver and spleen [50]. It has also been proposed to use as image parameters the photons collected during various consecutive time windows [51]. This has the advantage that different time windows are affected differently by boundary conditions and geometrical factors, so that one can identify the optimal time window according to the specific requirements of each application.

11.5 Frequency-domain techniques

Steady-state measurements are inadequate to determine the optical parameters of turbid media in the multiple scattering regime. Steady-state measurements only provide the amount of light that has been attenuated after emerging from the tissue. As seen in the previous sections, this amount of light depends on several processes, including absorption, scattering and the index of refraction of the medium. We need additional measurements to determine the optical parameters with measurement at a point. One possibility is to measure the time a photon employs to travel from the light source to the detector, as discussed in Section 11.4. When averaged on many photons, the time course can provide the necessary information to separate the contribution to the light transmitted arising from different processes in the medium. In the simple case of a homogeneous, uniform scattering and absorbing medium we need to measure only the average time it takes the photons to travel from the source to the detector. There are several techniques that have been employed to measure this average time. In general, they fall into two major categories: the time-domain techniques and the frequency-domain techniques.

In this section we discuss the frequency-domain methods [52,13–15]. The major difference between the frequency-domain and the time-domain methods is in the technique used to perform the actual measurements rather than in substantial physical differences. Both for the time-domain and the frequency-domain method we need some sort of modulation of the source intensity, with harmonic content in the frequency-range that is affected by the light transport through the medium. In the frequency-domain method, the light is generally modulated sinusoidally in the frequency range between 50 MHz and 1–2 GHz. This range is necessary to produce appreciable changes in the characteristic modulation of the light as it travels throughout the medium. In the frequency-domain approach, the various harmonics of the light modulation are isolated electronically [53]. Each component is characterized by an amplitude and phase term referred to the amplitude and the phase of the source.

This richer information content with respect to steady state spectroscopy, where the intensity is the only measurable parameter, permits the determination of the optical parameters of the medium.

In principle it is possible to obtain information about the optical parameters of a multiple scattering medium without recurring to time-resolved techniques [44]. For example, the accurate measurements of how the light attenuates as a function of the distance from the light source at the surface of a tissue could be used to determine the optical parameters of the medium. In fact, the light distribution depends both on the scattering and absorption coefficient. Also the angular dependence of the light emitted by the surface depends on these parameters. However, the mathematical relationship of these parameters renders the separation of scattering and absorption problematic and a very high signal-to-noise ratio is needed to separate the contribution of scattering from absorption. For this reason, the most common approach to separately measure scattering and absorption coefficients is based on time-resolved methods.

From the theoretical point of view, there is no basic difference between the frequency-domain and the time-domain. A Fourier transformation will change the expression from one domain into the other. However, there are some important practical differences. In the time domain, the method of the correlated single photon counting is capable of tagging every photon detected with the time of arrival after the laser pulse. However, to process this signal, the detector must separately measure each photon. This is relatively simple to achieve at low photon fluxes but impossible at high fluxes. In a typical experiment, of the order of millions photons per second are collected by the detector fiber and even the faster detectors and counting electronics are unable to resolve them individually. Instead, the frequency-domain method is essentially an analog system. The photons detected are converted in the detector photocurrent. Only the average phase and amplitude modulation is measured. There is no need to separate the contribution of each photon. This technical difference provides much better linearity at high counting rate and much faster measurements for the frequency-domain method.

In the frequency domain technique the light source is modulated sinusoidally at a frequency f in the 50–2000 MHz range depending on the particular implementation [53]. There are two alternative methods for modulating the light intensity. For light sources such as light-emitting diodes (LEDs) or solid-state diode lasers, the current injected into the device is directly modulated [54,55]. Since the diode characteristics (light-emitted versus input current) is highly nonlinear, the light intensity contains the fundamental frequency of modulation and several harmonics.

This is not a problem since the detection system measures each harmonic separately. This is also true for the other type of light sources used in frequency-domain instrumentation, namely mode-locked lasers that produce a very narrow light pulse at high repetition rate. In this case, the source contains a wide range of harmonics that can be filtered and processed individually by the detection electronics [56]. Whatever is the source or method used to modulate the light, the key feature of frequency-domain technology is the modulation of the intensity of the light source. If the intensity is sinusoidally modulated at a frequency f, one can define three parameters: the average intensity (I_{dc} or dc intensity), the amplitude of the intensity oscillations (I_{ac} or ac amplitude), and the phase of the wave (ϕ) with respect to some arbitrary phase, generally the phase of the electronic driving circuit. These parameters, which completely describe and characterize the sinusoidally modulated wave, are the quantities of interest in frequency-domain spectroscopy. On the basis of these definitions, the intensity modulated signal can be written as

$$I = I_{dc} + I_{ac}\cos(\Phi - \omega t) \tag{14}$$

where ω is the angular modulation frequency ($\omega = 2\pi f$ with f modulation frequency). The ratio (I_{ac}/I_{dc}) is called modulation (m). In some applications, phase and modulation are the only parameters of interest.

In a typical experiment, the modulated light source is coupled to an optical fiber which is then put in contact with the tissue of interest. A second optical fiber collects the light at the tissue surface after it has traveled some distance and then is processed by the detection electronics. In most instruments, there are several light sources emitting at different wavelengths and several fibers to collect the light at different points on the tissue surface.

Optical detectors employed in frequency-domain spectroscopy include photomultiplier tubes (PMTs), avalanche photodiodes (APDs), and CCD cameras with image intensifiers. In all cases, the signal at a given frequency is not measured directly. Rather the detector gain is modulated at a frequency that is slightly different from the frequency of the light source. This process, called heterodyning, results in the creation of additional frequencies, in particular at the sum and difference between the frequency of the source and the frequency used to modulate the detector gain. Generally it is the difference frequency that is used. It is possible to show that this difference frequency contains the same phase and amplitude information as the original high frequency, but it appears in a frequency range that is very easy to measure, digitize and process with modern electronic components [56]. Figure 5 shows a typical frequency-domain instrument.

When working in the frequency-domain, it is useful to think of light propagating in a turbid medium in terms of photon density waves. These waves of light intensity modulated at high frequency travel at constant velocity in the medium. Photon density waves display some of the optical phenomena associated with waves such as reflection, refraction and diffraction. The velocity of the propagation of the photon density waves and their attenuation in the turbid medium depend on the medium optical parameters and on the light modulation frequency. In particular, the photon density $U_{FD}(t)$ of the wave at a detector point at the distance r from

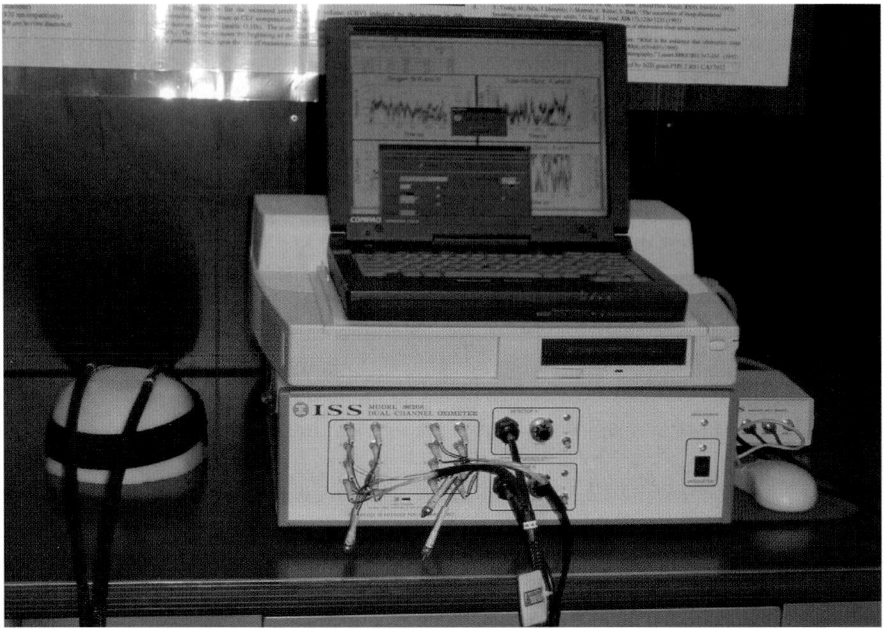

Figure 5. Portable frequency-domain instrument built by ISS Inc, Champaign Il. This instrument has 16 diode laser sources, 2 photomultiplier detectors and operates at 110 MHz modulation frequency. During data acquisition the optical parameters of the medium are continuously displayed.

the source is given by Equation (15), which is valid in the diffusion regime in an infinite and homogeneous medium [57]:

$$U_{\mathrm{FD}}(r,t) = \frac{S}{4\pi v D r}\left[U_{\mathrm{dc}}\mathrm{e}^{-r\sqrt{\mu_a/D}} + U_{\mathrm{ac}}\mathrm{e}^{-kr}\mathrm{e}^{(\Phi_0-\omega t)}\right] \tag{15}$$

where ω is the angular modulation frequency, r is the source–detector separation, v is the speed of light in the medium, D is the diffusion coefficient which is defined as $1/[3(\mu_a+\mu_s')]$ with μ_s' as the reduced scattering coefficient, μ_a is the absorption coefficient, $k = [(v\mu_a - i\omega)/(vD)]^{-\frac{1}{2}}$, ϕ_0 is the phase of the source and S its strength. Equation (15) is the frequency-domain equivalent of Equation (13) in the time domain. From Equation (15), the phase and the amplitude of the wave can be calculated taking the argument (for the phase) and the magnitude (for the amplitude) of Equation (15). In particular, Equation (15) leads to the following expressions:

$$\Phi = r\sqrt{\frac{\mu_a}{2D}}\left(\sqrt{1+x^2}-1\right)^{\frac{1}{2}} \tag{16}$$

$$\ln(rU_{\mathrm{DC}}) = r\left(-\sqrt{\frac{\mu_a}{D}}\right) + \ln\left(\frac{S}{4\pi v D}\right) \tag{17}$$

$$\ln(rU_{AC}) = r\left(-\sqrt{\frac{\mu_a}{2D}}\right)\left(\sqrt{1+x^2}+1\right)^{\frac{1}{2}} + \ln\left(\frac{SA}{4\pi vD}\right) \tag{18}$$

where

$$x = \frac{\omega}{v\mu_a} \tag{19}$$

$$D = \frac{1}{3(\mu_a + \mu_s')} \tag{20}$$

What is important is that from the phase Φ and amplitude U_{AC} of the wave it is possible to extract the reduced scattering and absorption coefficients of the medium. In the next section we discuss how to measure accurately the value of the phase and attenuation of the photon density wave.

11.5.1 Spectroscopy: the multi-distance method

One of the major problems encountered in the practical application of the measurement of the absorption and scattering parameters of turbid media using the front of the photon density wave is that the so-called "source terms", i.e., the phase and the amplitude of the source are not known exactly. In addition, even if they were known exactly, the description of the light transport in the multiple scattering medium is well described by the diffusion approximation only far from the source. We have shown that the extrapolation of the photon density wave phase measured far from the source back to the source can give values of the phase that are unphysical. This apparent contradiction is due to the approximation used to describe the propagation of the photon density wave. However, after traveling for a distance of about 7–8 mm from the source, the propagation of the photon density wave is well described by the diffusion approximation [58]. Since at those larger distances, the photon density wave travels at constant velocity and attenuates like an exponential divided by the distance, if we use the slope of the phase (S_Φ) as a function of distance and the slope of the logarithm of the amplitude (DC or AC) multiplied by the distance (S_{dc} and S_{ac}, respectively), we obtain three straight lines (see Equations (16)–(18)). From each one of a pair of slopes, we can extract the values of the absorption and scattering coefficients in the region in which the photon density wave is propagating according to the following expressions [59]:

$$\text{using DC, Phase: } \mu_a = -\frac{\omega}{2v}\frac{S_{dc}}{S_\Phi}\left(\frac{S_\Phi^2}{S_{dc}^2}+1\right)^{-\frac{1}{2}} \tag{21}$$

$$\text{using Ac, Phase: } \mu_a = \frac{\omega}{2v}\left(\frac{S_\Phi}{S_{ac}} - \frac{S_{ac}}{S_\Phi}\right) \tag{22}$$

$$\text{using DC, AC: } \mu_a = \frac{\omega}{2v}\frac{S_{dc}}{S_{ac}}\left(\frac{S_{ac}^2}{S_{dc}^2} - 1\right)^{-\frac{1}{2}} \tag{23}$$

$$\text{using DC, Phase or using DC, AC: } \mu_s' = \frac{S_{dc}^2}{3\mu_a} - \mu_a \tag{24}$$

$$\text{using AC, Phase: } \mu_s' = \frac{S_{ac}^2 - S_{\Phi}^2}{3\mu_a} - \mu_a \tag{25}$$

There are additional practical advantages of the multi-distance method, one of the most important being the relatively independence of the slope measurement on the curvature of the surface. For practical purposes, we have implemented the measurement of the slope of the phase and amplitude by measuring the amplitude and the phase of the photon density wave at four selected distances from the source. A computer algorithm calculates the average scattering and absorption from those slopes. The recovery of μ_a and μ_s' is very accurate and given the nature of the measurement, i.e., measuring a slope rather than a value, are largely independent of the absorbance of the skin or other local factors [59–61]. Figure 6 shows a typical probe exploiting the multi-distance method. This specific probe was designed to be applied to the forehead. Although the equations we have reported are valid in the infinite medium, similar equations can be written for the semi-infinite medium, which better approximate actual measurement techniques in vivo [45].

The crucial point in the determination of the optical parameters of tissues using the multi-distance method is in how well the actual anatomy realizes the conditions of a semi-infinite medium. We have demonstrated that the effect of the skin absorption is negligible in the multi-distance method, but the effect of large underlying structures, such as bones or fat layers, may strongly modify the modalities of light propagation. We have performed measurements on layered

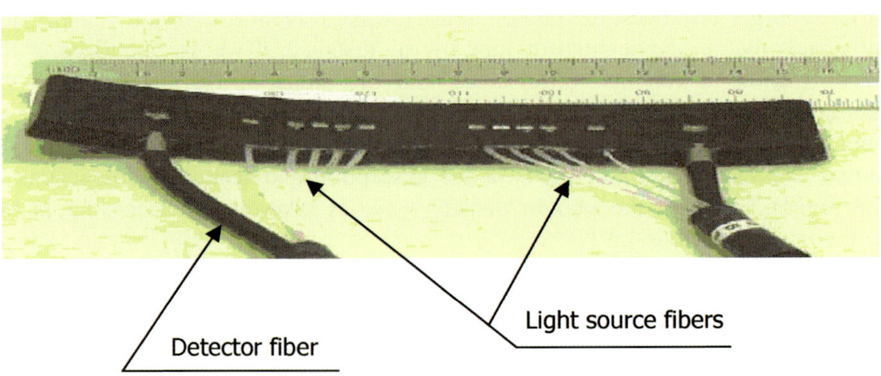

Figure 6. Light from 16 diode lasers at different wavelength are brought to the tissue surface at several distances from the two detector fibers. The detection algorithm automatically calculates the optical parameters using the multi-distance equations.

structures [62,63] to asses how much the multi-distance method is affected by the different layers. The results are quite complex and cannot be easily generalized, since they strongly depend on the optical properties and thickness of the different layers. However, for reasonable parameters of the intervening layers, the multi-distance method may provide one of the most robust approaches. As a general comment, only the full solution of the inverse problem will provide the correct medium parameters for a non-homogeneous medium. In the absence of a fast and reliable solution of the inverse problem, and for measurements of small variations of the optical parameters of one of the layers, the multi-distance method is one of the less affected by systematic errors due to layered structures. In measuring the optical parameters of muscle it appears that the assumptions of a relatively homogeneous mass are well satisfied.

11.5.2 Spectroscopy: multi-frequency

The equations describing the attenuation of photon density waves as they propagate in a multiple scattering medium show that the attenuation and the phase shift, for a given distance r, depend also on the modulation frequency f (or, equivalently, the angular modulation frequency ω). The measurements of phase shift and the modulation ratio as a function of the frequency provide another method to determine the optical parameters of the medium [64,65]. The signal-to-noise ratio analysis and a comparison between the multi-distance and multi-frequency method has been performed [66]. The conclusion is that the two methods are comparable provided that a relatively wide range of frequencies are measured. However, the electronics needed to implement the multi-frequency method is different. The multi-frequency method has the distinct advantage that all the measurements are performed at the same distance. Therefore, artifacts due to local skin heterogeneity are absent. However, the requirement of wide band detectors and electronics reduces the sensitivity of the measurement. In practice, the current implementation of the multi-frequency method utilizes avalanche photodiode detectors that limit the sensitivity [67]. Also the modulation electronics and the frequency analysis are done differently. As a result of the different electronic components, the current instruments have limited penetration. However, they have better capability to accurately recover the optical parameters of the medium.

11.5.3 Imaging: diffuse optical tomography

Once we realized that one can measure the optical properties of tissue, the next step is to build a map of the properties over a large region of the tissue. The basic idea is that the optical parameters are locally different and that each type of tissue has different optical properties. The major complication is that in regions in which there is a superposition of contributions from different tissues it is not possible, with a single frequency-domain measurement, to distinguish the different tissues. In this case, numerous measurements are performed and then an algorithm of inversion is used to estimate the original map of the optical parameters. As computational methods have become faster and more precise, several labs have been working on this possibility. Again, the methods used are classified as CW methods, in which steady-sate light is used and time-resolved methods, either in the frequency or time domain. The tomography issue will be further discussed in the section on optical mammography.

11.6 Near-infrared assessment of oxygenation and pharmacokinetics

One of the unique capabilities that the near-infrared technique provides is the possibility to continuously monitor physiological parameters such as hemoglobin saturation, total hemoglobin content, and blood flow and oxygen consumption. This is due to the non-invasive character of the determination and of the physical principle of the measurement that allows absolute measurements. In this section we discuss the measurements of physiological parameters, their accuracy and under which assumptions we expect that the values measured are accurate.

One of the most important applications of the near-IR technique is the determination of the concentration of the oxy- and deoxy-hemoglobin in tissue. When these two concentrations are know then other parameter such as tissue hemoglobin oxygen saturation and tissue total hemoglobin content can be determined. We have shown in the previous section that it is possible to determine the absolute absorption coefficient at the source wavelength in a uniform homogeneous medium. In tissue, the absorption value is determined by the sum of the contributions of all substances that absorb at each wavelength. To proceed with the measurement of concentrations, we need to account for the relative contributions of the major absorbers. In practice, only a few chromophores are concentrated enough or their extinction coefficient is large enough to substantially contribute to the absorption in the wavelength range from 650 to about 900 nm where most of the measurements are performed. In this region, the major contributions arise from hemoglobin, water and fat. Hemoglobin has two major forms, oxygenated and deoxygenated. There are also other chromophores that can potentially contribute to the absorption, myoglobin, melanin, cytochrome and other substances [68]. However, their contribution in a normal tissue is less than a few percent and we will not discuss their detection in this chapter. Figure 2 shows the spectrum of water, oxy-hemoglobin and deoxy-hemoglobin.

Since in most of the wavelength range considered we have three major contributions to the absorption, we need to determine the absorption coefficient in at least three different bands. However, the contribution of water is less than that of hemoglobin in the wavelength range 700 to 900 nm. In many instruments two wavelengths are used with the purpose to quantify only the hemoglobin and the contribution of water at these wavelengths is estimated and a correction is used. Of course, this method does not provide a measurement of the water content of tissue. A more accurate approach is to determine the absorption at many wavelengths. An interesting approach has been developed in Tromberg's laboratory that uses a combination of frequency-domain and cw technique to effectively expand the number of wavelengths [69]. In this ingenious approach a CW light source is used in addition to the frequency-modulated light source. The absolute absorption coefficient is determined using the frequency-modulated light and a continuous spectrum is measured as provided by the CW white light source. Then an algorithm is used to scale the CW spectrum to match the absolute absorption at these wavelengths that were measured by the frequency-domain method. This methodology allows also the determination of the scattering coefficient at many wavelengths and therefore to measure the wavelength dependence of the scattering coefficient.

The use of near-infrared spectrophotometry (NIRS) to measure hemodynamics and oxygenation in the brain [70,71], the skeletal muscle, muscle flaps [72,73] and individual organs [74] has increased. Measurements on the skeletal muscle have been performed in sports medicine [75], in patients with myopathies [76–78], heart failure [79–81] and peripheral vascular disease (PVD) [82–88]. NIRS has also been used during surgery [89] and in intensive care medicine [90,91]. NIRS is based on the principle of light attenuation and has the unique ability to non-invasively measure at a depth of several centimetres tissue oxygenation and hemodynamics at the level of arterioles, capillaries and venules [92]. Real time measurements (6 Hz), good spatial resolution (5 mm), bedside feasibility, high reproducibility and inexpensiveness could make this method a valuable tool in a clinical setting.

Most of the NIRS-instruments measure at one discrete location and the result is considered to be representative for the entire region of tissue under investigation. However, hemodynamics and oxygenation in the muscle shows a considerable spatial distribution at rest [93–95] and during exercise [96].

11.6.1 Absolute vs. relative measurements

One of the distinctive advantages of frequency-domain near-infrared methodology is the possibility to obtain absolute measurements of concentrations in highly scattering media. The principles of the method are presented in the previous sections. In a uniform and semi-infinite medium, the recovery of the optical parameters can be done using the diffusion approximation of the Boltzmann transport equation. Under this assumption, the method is accurate, in the sense that, with no other prior knowledge, we can obtain the absolute value of the absorption coefficient. If the extinction coefficient is known, we can calculate the concentration of the absorbing substance. To this level, the method is only based on physical principles and it does not require validation. Only when the basic assumption of uniformity and extension is not satisfied do we need to be concerned about the accuracy of the method. Of course, biological tissue is neither homogeneous nor extended. Therefore, in each case we need to re-evaluate our assumption to determine the limit of accuracy of the method. One important consideration is the length scale of the intrinsic heterogeneity of tissues. At the micron level, tissues are composed of cells and sub-cellular components. Those are the elements that give rise to the elementary process of absorption and emission. However, photons travel very fast and visit a relatively large volume of the tissue (in the multiple scattering regime). Their absorption and scattering will depend on some sort of spatial averaging of the intrinsic microscopic heterogeneity. Depending on the local properties of the tissue it is estimated that averaging occurs at least over volumes of several millimetre. Using this type of reasoning we can model the tissue assigning average optical properties corresponding to relatively large structures. For example, when investigating muscles, we divide the tissue into two or more layers, corresponding to the skin, fat layer and muscle. Another important modeling consideration applies to the macro and microvasculature. The large blood vessels appear completely opaque and only the microvasculature contributes to the absorption. Therefore, the accuracy of the determination of the optical parameters depend on how accurate is the modeling rather than on the accuracy of the near-IR instrument per se.

A different kind of approximation is used when only changes in the optical parameters are measured. In this case, the accuracy in the baseline determination is less crucial and the changes in optical parameters can be determined more accurately than the absolute values.

For tissue measurements, the model used should be validated and the accuracy in the determination of the optical parameters should be assessed in every individual situation.

11.6.2 Arterial oxygenation

To separate the contributions of the different vascular compartments, we need some method to distinguish the arterial, venous and tissue blood contribution. The arterial compartment can be distinguished from its characteristic pulsatility. This principle has been largely exploited in the common pulse oxymeter. We describe here a similar approach based on absolute measurement of the oxy- and deoxy-hemoglobin. In fact, one of the possible artifacts of the pulse oxymeter is that it needs a calibration and also an estimate of the differential pathlength factor B defined in Equation (6). Since these parameters can be directly measured by the frequency-domain method on a case by case basis, the measurement of the concentration of oxy- and deoxy-hemoglobin is absolute. However, the total absorption at any given wavelength is due to the contribution of all vascular compartments, not only the arterial blood. The basic idea is to measure only the changes of oxy- and deoxy-hemoglobin associated with the pulse. Since these changes are presumably due only to the addition of arterial blood, the corresponding saturation is related to the saturation of the arterial component. An instrument that implements this principle was constructed [97]. This particular instrument measures the arterial oxygenation in every tissue that has the pulse signal, including the brain. The value of the arterial saturation is compared with the saturation measured by a commercial pulse oxymeter applied to the finger of the same subject.

11.6.3 Venous oxygenation

To distinguish the venous compartment from the tissue and arterial compartment we perform a simple manipulation of the circulation. When a venous occlusion is applied, for example to the calf of a subject, the muscle blood efflux is stopped. As a result there is an accumulation of blood in the muscle. The saturation of this blood is much lower than the arterial saturation and also lower than that of the tissue compartment. Typical accumulation of the blood in a muscle of a subject is shown in Figure 7. The accumulation proceeds through different steps, which also involve the plasticity of the blood vessels. However, when the occlusion is released, the blood rapidly leaves the muscle and the total hemoglobin returns to the baseline level.

We applied venous occlusion to provoke changes in hemodynamic and oxygenation parameters. The venous occlusion method, in comparison to ischemia tests, enables one to measure venous oxygen saturation (SvO_2), hemoglobin flow (HF) and oxygen consumption (VO_2) simultaneously. Moreover, it is less traumatizing to the tissue and can be easily repeated. The calculation of VO_2 by venous occlusion has been validated against an invasive technique [98] with both the venous and arterial occlusion methods providing similar results [99]. The measurements of SvO_2 [100] and BF [101] have been validated against other methods.

For the venous occlusion measurements, a particular distribution of sources and detectors was used to cover a relatively large area of the tissue under study. An area of $18.5 \times 6\ cm^2$ (Figure 8) is covered by the sensor and permits for simultaneous

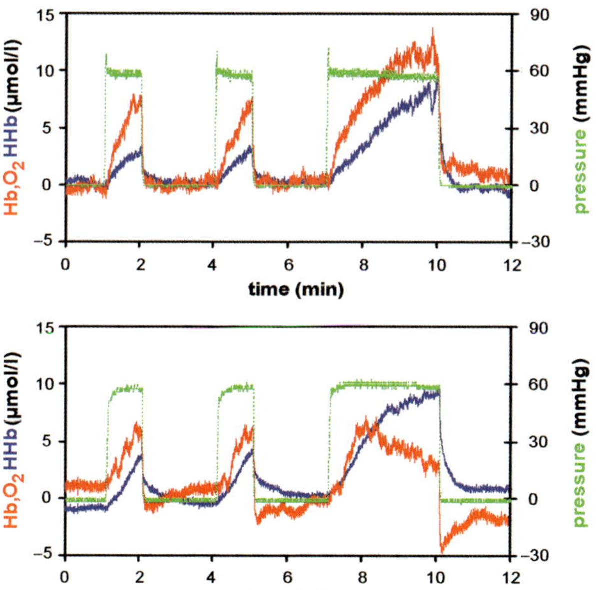

Figure 7. The green line shows the pressure used to producing the venous occlusion. As the occlusion is applied, blood accumulates until a maximum value is achieved. Upon release of the pressure, the oxy- and deoxy-hemoglobin concentration returns to the base value. Note the difference between the top panel (normal subject) and the bottom panel (subject with peripheral vascular disease).

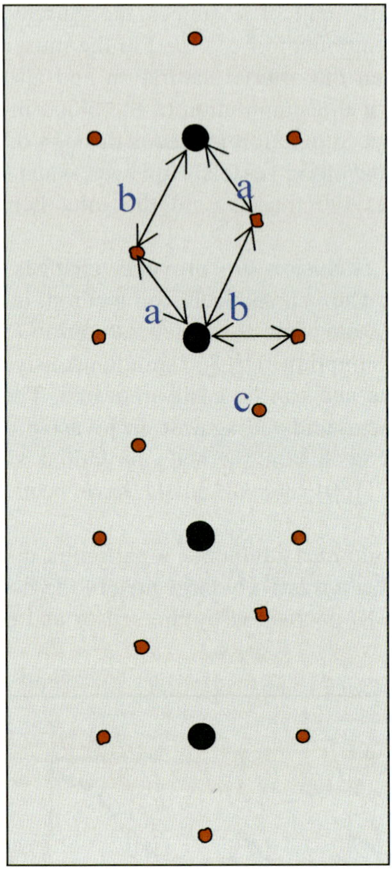

Figure 8. A series of 14 diode laser sources and 4 photomultiplier detectors are used to map a relatively large area (18.5 × 6 cm) of the muscle.

measurements at 22 locations. The four detector fibers are equidistantly located along the centerline of the sensor. The source fibers are arranged at three different distances from the corresponding detector fibers. In the center of the sensor the source–detector distances are 2.4 and 3.5 cm. The inner group of detector and source fibers form three symmetrical parallelograms, which provide a multi-distance approach to determine absolute values for absorption and scattering from which the DPF can be calculated [102]. The distance between a detector and the other corresponding sources is 3.0 cm. Maps over the entire region covered by the sensor were generated at a sample rate of 6 Hz.

The sensor was placed and secured with an elastic band along the lateral side of the calf (lateral gastrocnemius muscle) and a pneumatic cuff was wrapped around the subject's thigh. Venous occlusion was induced (within 2 s) by inflating the pneumatic cuff to a pressure of 60 mmHg. Venous occlusion was maintained for 60 s for the initial two, three or four venous occlusions and for 180 s long for the

Table 1. Mean values ± SD of the hemodynamic and oxygenation parameters

Parameter	Patients	Healthy subjects	P^a
Number of legs	18	16	
$\Delta[O_2Hb]$ (μmol L^{-1})	6.67 ± 3.36	6.85 ± 4.57	ns
$\Delta[HHb]$ (μmol L^{-1})	4.04 ± 1.13	2.36 ± 1.79	0.0002
$\Delta[tHb]$ (μmol L^{-1})	10.71 ± 3.99	9.21 ± 6.33	ns
$\Delta[O_2Hb]$ 180 s (μmol L^{-1})	8.70 ± 5.04	9.80 ± 4.55	ns
$\Delta[HHb]$ 180 s (μmol L^{-1})	9.87 ± 2.82	6.38 ± 3.07	0.0003
$\Delta[tHb]$ 180 s (μmol L^{-1})	18.57 ± 6.27	16.18 ± 7.32	ns
HF (μmol(100 ml)$^{-1}$) min^{-1}	1.34 ± 0.69	1.44 ± 1.17	ns
BF (ml(100 g)$^{-1}$) min^{-1}	0.63 ± 0.29	0.62 ± 0.50	ns
VO_2 (ml(100 g)$^{-1}$) min^{-1}	0.033 ± 0.015	0.022 ± 0.020	0.003
SvO_2 (%)	65.2 ± 13.5	80.8 ± 4.51	0.000004

[a]ns = not significant.

final venous occlusion. At the end of each venous occlusion the pneumatic cuff pressure was quickly released. The pressure curve was recorded with a digital manometer. Consecutive venous occlusions were separated by 2 min rest periods.

Absolute values for absorption and scattering and the DPF were calculated for the three center regions according to the geometry of the sensor (Figure 8) and the frequency-domain equations [41,102]. By combining the attenuation changes with the DPF data [12] quantitative values for relative changes in $\Delta O_2Hb]$ and $\Delta HHb]$ in all 22 locations were obtained. The global mean values for the oxygenation and hemodynamic parameters for both groups, the patients and the healthy subjects, are shown in Table 1. There were significant differences between patients and healthy subjects for $\Delta[HHb]$, VO_2, and SvO_2, while HF and BF were nearly the same. In the patients we found a decrease in $\Delta[O_2Hb]$ during the later part of the 180 s venous occlusion in 44.7% of the measured locations, which is much greater than in healthy subjects (15.3%). Regional mean values for the hemodynamic and oxygenation parameters for the patients are given in Table 2. There were significant differences between the proximal region on the one hand and the two intermediate and distal regions on the other hand. However, the proximal–distal differences were much less pronounced in patients than in healthy subjects (Table 3).

Table 2. Regional mean values ± SD of the hemodynamic and oxygenation parameters

Parameter	Proximal	Intermediate I	Intermediate II	Distal	P^a
$\Delta[O_2Hb]$ (μmol L^{-1})	8.40 ± 5.36	6.05 ± 3.74	5.98 ± 3.07	6.54 ± 3.67	ns
$\Delta[HHb]$ (μmol L^{-1})	4.90 ± 1.81	3.74 ± 1.47	3.75 ± 1.23	3.87 ± 1.62	b
$\Delta[tHb]$ (μmol L^{-1})	13.30 ± 6.41	9.80 ± 4.64	9.73 ± 3.77	10.41 ± 4.63	b
HF in [μmol 100 ml^{-1}min^{-1}]	1.70 ± 1.18	1.21 ± 0.86	1.26 ± 0.65	1.26 ± 0.70	ns
BF in [ml 100 g min^{-1}]	0.78 ± 0.46	0.56 ± 0.37	0.60 ± 0.31	0.59 ± 0.31	ns
VO_2 in [ml 100 g^{-1}min^{-1}]	0.042 ± 0.028	0.030 ± 0.017	0.032 ± 0.014	0.029 ± 0.018	b
SvO_2 (%)	64.2 ± 14.8	66.2 ± 15.8	64.2 ± 17.2	66.0 ± 16.9	ns

[a]ns = not significant; [b]$P < 0.05$.

Table 3. Proximal–distal mean differences for the parameters in patients and healthy subjects

Parameter	Patients	Healthy
$\Delta\Delta O_2Hb]$ ($\mu mol\ L^{-1}$)	1.86 ± 4.18	2.92 ± 2.96^a
$\Delta\Delta HHb]$($\mu mol\ L^{-1}$)	1.03 ± 1.76^b	1.21 ± 0.65^a
$\Delta\Delta tHb]$ ($\mu mol\ L^{-1}$)	2.89 ± 5.20^b	4.13 ± 3.18^a
$\Delta\Delta tHb]$ ($\mu mol\ L^{-1}$)	1.06 ± 6.26	3.76 ± 4.54^a
$\Delta\Delta O_2Hb]$ ($\mu mol\ L^{-1}$)	1.36 ± 6.02	2.30 ± 2.05^a
$\Delta\Delta tHb]$ 180 s ($\mu mol\ L^{-1}$)	2.42 ± 9.38	6.05 ± 3.65^a
HF in [$\mu mol(100\ ml)^{-1} min^{-1}$]	0.44 ± 0.99	0.70 ± 0.60^a
BF in [ml $(100\ g)^{-1}\ min^{-1}$]	0.18 ± 0.44	0.29 ± 0.26^a
VO_2 in [ml $100\ g^{-1}\ min^{-1}$]	0.013 ± 0.022^b	0.011 ± 0.009^a
SvO_2 (%)	-1.8 ± 13.7	-3.9 ± 4.7^b

$^aP<0.005$; $^bP<0.05$; ns = not significant.

NIRS offers the unique capability of simultaneously measuring hemodynamics and oxygenation in small blood vessels, such as arterioles, capillaries and venules, several centimetres deep in the tissue. This feature makes this method complementary to other diagnostic and monitoring methods. By employing a frequency-domain instrument and a special source–detector geometry mapping of these parameters in the muscle tissue becomes feasible.

11.6.4 Tissue oxygenation

As we mention in several parts of this chapter, the near-infrared method measures the contribution of all the hemoglobin compartments (arterial, venous, and capillary). The separation between the compartments is achieved by the different way the arterial, venous and tissue compartment respond to external manipulation. As we have discussed, the arterial compartment is characterized by the pulse, while the venous compartment can be distinguished using venous occlusion. For determining the contribution of the tissue, the signal is first filtered from the pulse and then the average value is taken, integrating on slow changes occurring during several seconds. In general, unless the venous part is specifically evaluated, the measurement of the optical parameters provides an average value of the tissue and the venous part. The values reported in the literature, unless otherwise stated, refer to the average of these two contributions.

11.6.5 Chromophore concentration measurements

One of the early motivations for the development of the methods of diffuse reflectance was for the determination of the concentration of chromophores for photodynamic therapy. There is vast literature concerning the calculation of the light dose delivered at various tissue depths. For large tissue depths, over one centimetre, the methods we have described in this chapter based on illuminating one point of the tissue and collecting the light at a distant points are the most accurate. However, this methodology cannot be used for superficial layers. In this case the general methods of diffuse reflectance are used.

11.7 Optical mammography

The human breast is an ideal sample for near-infrared studies because of its accessibility and relatively low optical absorption, which allows for light transmission through the whole breast. Continuous-wave imaging approaches to optical mammography have a long history, being originally proposed in the 1920s [1]. More recent developments (time-resolved instrumentation and theoretical modeling of light propagation in breast tissue), dating back to the late 1980s to early 1990s, have determined a renewed enthusiasm and spurred a new wave of research efforts in optical mammography. The optical detection of breast cancer in the near-infrared is based on the local structural and functional modifications associated with cancer, and not necessarily on optical signatures associated with cancer cells. These modifications arise from cancer-induced effects such as angiogenesis [103,104], alterations to the blood flow and oxygenation [103] and fibroblast proliferation [104], which in turn affect the optical scattering and absorption properties of breast tissue. However, it is not yet demonstrated that these cancer-induced optical perturbations can lead to an effective optical approach to breast cancer detection on the basis of intrinsic contrast. As a result, it has been proposed to introduce extrinsic dyes as optical contrast agents in optical mammography [105–107].

11.7.1 Continuous-wave optical mammography

Transillumination of the breast for diagnostic purposes was first proposed by Cutler in 1929 [1]. This original approach to the optical study of the human breast was soon abandoned because of the poor clinical performance of the method. In the 1970s and 1980s, technical advances led to two optical techniques for breast imaging called *diaphanography* [108] and *lightscanning* [109], which brought renewed interest for optical mammography, and even led to commercially available optical imagers. These approaches employed a broad beam of visible and near-infrared CW light that illuminated one side of the breast. On the opposite side of the breast, the examiner visually inspected the light transmission pattern and used a video camera for image recording. Despite some encouraging initial results [110], a number of clinical studies conducted in the late 1980s showed that diaphanography and lightscanning are inferior to X-ray mammography both as a screening and as a clinical tool [111,112]. As a result, medical acceptance of diaphanography and lightscanning has been subdued.

Despite the limited success of diaphanography and lightscanning, continuous-wave approaches have been further refined to develop new, more advanced, and more effective instruments for breast imaging. For example, in the early 1990s, Yamashita and Kaneko developed a two-wavelength CW instrument based on a 2D raster scan of two optical fibers, one for illumination and one for light collection, on opposite sides of the breast [113]. The effort of solving the inverse imaging problem, to obtain a spatial reconstruction of the breast optical properties, led the Philips Research Laboratories to develop a CW instrument [114] that performs tomographic imaging by combining multiple optical projections [115]. Absorption and scattering cross-sectional images of the breast have also been reported by Jiang et al., who used a finite element method approach to solving the diffusion equation (direct problem) and a regularized Newton iterative method to determine the spatial distribution of the optical properties (inverse problem) [116].

Recently, Barbour and coworkers have developed a CW tomographic imager especially designed to monitor the temporal evolution of the breast optical properties [117]. This latter approach introduces the novel idea that breast cancer may be detected, and possibly discriminated from benign breast lesions, on the basis of its unique temporal dynamics or response to external perturbations. All of these CW instruments of the latest generation have produced encouraging preliminary data on human subjects.

11.7.2 Time-domain optical mammography

The rich information content of time-domain data and the wide range of possibilities offered for data analysis have prompted the development of several time-domain instruments for optical mammography. Some of these instruments are designed to perform a tomographic reconstruction of the breast optical properties by collecting data either around the pendulous breast [118] or over the planes defined by two glass plates used to slightly compress the breast [107,119]. Other approaches are based on analyzing the temporal distribution of the photons transmitted through the breast by considering different time ranges (time gating) [50,51,120]. The light sources are typically pulsed laser diodes, even if more bulky lasers such as dye lasers and titanium:sapphire lasers have been used to collect spectral data [120].

11.7.3 Frequency-domain optical mammography

A number of frequency-domain approaches for the spectroscopic and imaging study of the breast have recently been developed [67,121–128]. Some of these studies focus on understanding the optical contrast provided by tumors, and the optical properties of breast tissue in general [67,124], or aim at developing new approaches to optical tomography of the breast [123,126]. Frequency-domain optical approaches to breast imaging include a planar compression geometry [121,122,125], a circular array of source and detector optical fibers around the pendulous breast [127], or the use of optically matching fluids where the breast is immersed [126]. Carl Zeiss (Oberkochen, Germany) [121] and Siemens AG, Medical Engineering (Erlangen, Germany) [122] have recently designed and clinically tested frequency-domain prototypes for optical mammography. These prototypes have similar design features. Figure 9 shows a block diagram of the

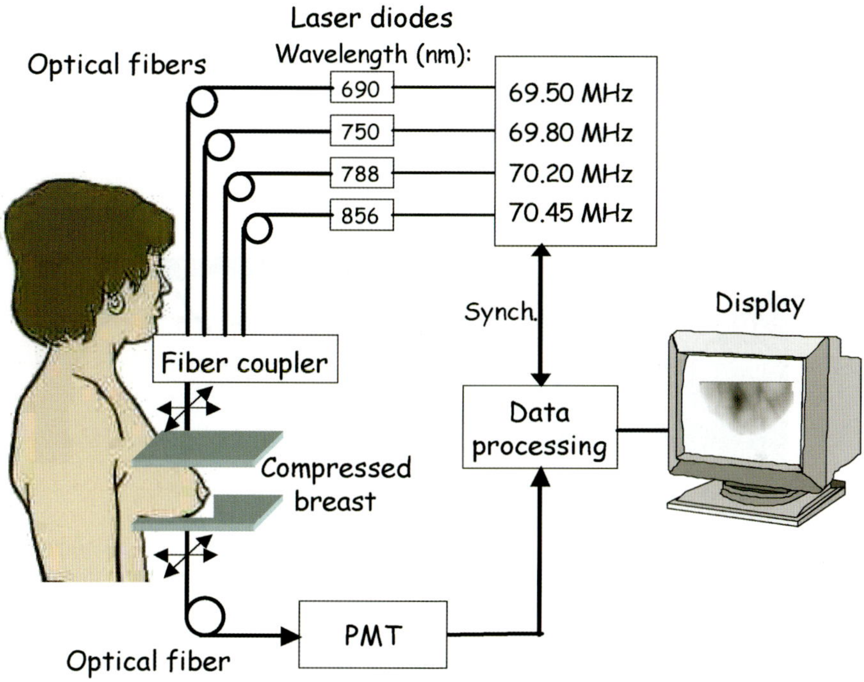

Figure 9. (a) Schematic diagram of the prototype for frequency-domain optical mammography developed by Siemens AG, Medical Engineering [122]. The slightly compressed breast is optically scanned to obtain 2-D projection images at four wavelengths (690, 750, 788, and 856 nm). The optical detector is a photomultiplier tube (PMT).

Siemens prototype. The Zeiss prototype differs in the number of wavelengths (two instead of four) and in the modulation frequency of the laser diode intensity (110 MHz instead of 70 MHz). Figure 10 reports 2D projection images obtained by scanning two collinear optical fibers (one for illumination, one for light collection) placed on the opposite sides of the slightly compressed breast. The total time required to scan the breast is 2–3 min. To generate the mammograms of Figure 10, the frequency-domain optical data have been processed using three algorithms of data analysis, (1) a correction of breast thickness variability and geometrical effects across the image combining amplitude and phase data [129], (2) a second-derivative spatial filter to enhance the display of regions of higher local absorbance [130], and (3) a multi-wavelength data analysis to estimate the oxygenation of detected breast lesions [131]. In particular, the mammograms of Figure 10 report a color-coded representation of hypoxic areas that are superimposed to gray-level representations of regions of negative second-derivative (i.e. local maxima in optical absorbance). As in X-ray mammography, the breast is imaged in two projections, namely craniocaudal (cc) and oblique (ob). Figure 10 reports the craniocaudal (right-hand panel) and oblique (left-hand panel) views of the left breast of a patient with a 3-cm invasive ductal carcinoma. The cancer is readily identified

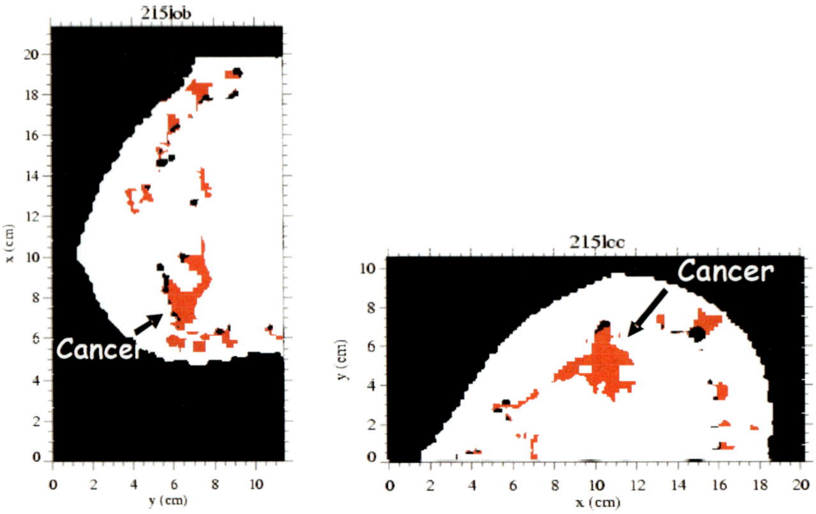

Figure 10. Optical mammograms of the left breast of patient No. 215 in the oblique view (215lob) and craniocaudal view (215lcc). A 3 cm invasive ductal carcinoma is indicated by the arrow, and corresponds to an area of low oxygenation indicated in color.

in the optical mammograms, and corresponds to a low value of oxygenation. A preliminary analysis, based on the criterion that an optical mammogram is positive if it shows a region of abnormal absorbance in both the craniocaudal and oblique views, has lead to a sensitivity (fraction of cancerous breasts successfully detected) of 72% and a specificity (fraction of non-cancerous breasts correctly evaluated as negative) of 52% on a clinical population of 131 patients [132]. This result is consistent with the sensitivity of 73% obtained on 69 patients with the Zeiss prototype [133].

11.8 Near-infrared imaging of the brain

There is a great interest in applying the near-infrared method to detect and characterize aspects of brain circulation. Firstly, circulation problems in the brain are responsible for a number of diseases, including strokes. Secondly, during activation of the brain activity, there is a local change of the concentration of the oxy- and deoxy-hemoglobin due to regional changes in blood flow.

Non-invasive optical studies of the brain can be broadly classified as either "structural" or "functional". Structural information is mainly used to detect tumors or hematomas. Gopinath et al., in 1993 [134], applied a continuous wave near-infrared probe (wavelength of 760 nm) to the frontal, parietal, occipital, parasagittal, and suboccipital regions of the skull to detect intracranial hematomas. Functional information derives from slow and fast optical signals, observed during brain stimulation. Several investigators have performed functional measurements on the motor cortex during motor stimulation [135–138], on the visual cortex during visual stimulation [139–142], and on the frontal region during mental work

[143,144] and on the monitoring of cerebral hemodynamics during sleep [145,146]. All these studies found a significant correlation between the optical signals and cerebral activity.

Brain activity is associated with changes in optical parameters of the tissue, namely the absorption and the reduced scattering coefficients. It is coupled to changes in regional blood flow, blood oxygenation, and metabolism. Hemoglobin and cytochrome-c-oxidase are the only biological compounds in the brain to exhibit variable absorption of near-infrared (NIR) light in response to changes in oxygen variability. NIRS determines changes in cerebral tissue oxygenation and blood volume by measuring fluctuations in concentration of oxy-hemoglobin, deoxy-hemoglobin and reduction–oxidation state of cellular mitochondrial cytochrome aa3 [5,147–149].

The transillumination of the adult brain, using safe illumination power levels, does not appear to be feasible. In a reflection geometry, the average photon penetration depth in a homogeneous tissue having an absorption coefficient μ_a and a reduced scattering coefficient μ_s' is given by $0.5[r/(3\mu_a\mu_s')^{0.5}]^{0.5}$, where r is the distance between the optodes [150]. For a typical tissue $\mu_a = 0.1\ \mathrm{cm}^{-1}$, $\mu_s' = 8\ \mathrm{cm}^{-1}$, $r = 4$ cm, and the average optical penetration is about 0.8 cm. This optical penetration depth, which can be increased by increasing r at the expense of signal-to-noise, indicates that optical methods are limited to probing the outer brain cortex. The optical penetration depth can be significantly affected by the presence of a layered structure such as the one present in the head (skin/scalp-skull-cerebrospinal fluid (CSF)-cerebral tunics-brain). In particular, it has been suggested that the clear layer of CSF may cause a light channeling effect that could reduce (but not prevent) the optical penetration into the brain cortex [151]. While the presence of CSF may have an effect on light propagation in the brain, the light channeling effect in the head is much less significant than the one caused by a transparent layer sandwiched between two scattering materials using a phantom [62]. Additional studies [152] indicated that the presence of CSF layer improves the sensitivity of NIRS signal to absorption changes in the adult brain.

Near-infrared spectroscopy offers the advantage of performing transcranial measurements of changes in cerebral hemodynamics and oxygenation. It is a non-invasive, non-ionizing, portable, bedside method, which can provide real-time measurements of these changes. These characteristics make NIRS the ideal tool to study physiological and pathological processes of the brain, concerning adults and infants, in settings ranging from research and diagnostic laboratories to intensive care units and operating rooms [143,153–155].

Generally, measurement of scattering and absorption coefficients of a particular area of the body can be done non-invasively by placing a light source (emitting light in the near-infrared range–between 700 and 1300 nm) and a detector on two points on the surface of the skin, at a distance of a few cm. Most tissues in the human body are highly scattering.

Although head tissues also absorb light, the absorption coefficient is estimated to be much smaller than the scattering coefficient: the typical value of μ_a in animal tissues is of the order of $0.1\ \mathrm{cm}^{-1}$ [156]. These basic optical properties of head tissues indicate that the Boltzmann's transport equation for photons inside the head can be solved in the diffusion approximation [13,42,157] and that the propagation of photons through the tissue can be described as a diffusion process [27]. This has several consequences of practical importance.

The individual photon trajectories from the source to the detector are highly variable and they make up an extended volume. Modifications in the optical properties occurring within this volume influence the light intensity and the time-of-flight and the size of the volume determines the spatial resolution of the technique (on the order of one cm). Due to the diffusive nature of light propagation in tissues, only photons traveling deeply into the medium are likely to reach the detector. The volume explored by a source–detector pair located on the surface of the medium (head) is a curved spindle with an average depth of about $\frac{1}{2}$ the distance between the source and the detector [52]. Therefore, it is possible to study properties of relatively deep structures (such as the cortex) by placing a source and a detector on the surface of the head. The number of photons reaching the detector (i.e., the attenuation) decreases more than exponentially with the source–detector distance (Equation (17)), whereas the phase delay of the photon density wave is linear with distance (Equation (16)). With adult head tissues, the large attenuation of the light determined by the source–detector distance limits the maximum useful source–detector distance to less than 5–6 cm, and effectively limits the penetration of the technique to less than 1–2 cm from the surface.

11.8.1 CW

As seen in previous sections, in general CW methods do not allow the full determination of the optical parameters of the medium. However, they can be used to detect relative changes. In connection to brain studies, the CW measurements are the most common and they have been extensively used to monitor blood flow, the formation of hematomas and in general to detect brain activity [144]. The reason for using such instrumentation is the low cost, the portability and the possibility to monitor parts of the brain for extended period of times. The limitation is that changes in optical parameters cannot be obtained in absolute terms and that it is difficult to separate changes occurring at superficial layers from effects occurring at the cortex.

In the current clinical practice in the USA, the only commercially available instrument using NIR light for the non-invasive functional study of the brain is the INVOS 3100 systems (Somanetics, Troy MI). Several reports question the validity of the information on brain oxygenation that these instruments provide, as compared with the other near-infrared oximetres used in research [158–160].

11.8.2 Time domain

Time domain studies have also been used to measure the brain optical parameters and to monitor brain oxygenation in infants. One of the first successful monitoring of brain oxygenation was obtained using time-resolved methods by Benaron's group [50]. However, time-resolved methods are in general less sensitive than frequency-domain methods and are also slower. Therefore, most current studies in brain circulation dynamics are performed using frequency-domain instrumentation.

11.8.3 Frequency domain

This section presents relevant examples in two areas of brain circulation research, namely for the assessment of cardiovascular damage in subjects affected by sleep apnea and for the measurements of brain activity following visual stimulation.

Obstructive sleep apnea syndrome (OSAS) has increasingly been recognized as a cause of poor health. In the middle-aged work force, 2% of women and 4% of men meet the minimal diagnostic criteria for the sleep apnea syndrome (apnea/hypopnea score of 5 or higher and daytime hypersomnolence) [161]. In the last two years the national media alone has variously alerted the public on the deleterious effects of OSAS. OSAS is described as a potentially lethal disease because the resulting hypoxia and hypoxemia lead to functional and structural circulation deficits. Altered quality of life, daytime sleepiness, neuropsychological dysfunction and cognitive deficits have been associated with OSAS as well as cardiovascular disease, including systemic and pulmonary hypertension, arrhythmias and ischemic heart disease [162–166]. Since the brain is very sensitive to hypoxia, it has been suggested that the cerebrovascular morbidity is the result of the chronic, cumulative effects of hypoxia and hypoxemia caused by sleep apnea. Cerebrovascular accidents, ranging from transient ischemic attacks to fatal strokes, are closely associated with this syndrome [167,168].

Polysomnography, the standard multi-instrument overnight recordings, can detect sleep apneas and their degree of severity [169,170]. The use of pulse oximetry determines the arterial saturation (SaO_2) and concomitant electroencephalography determines the sleep stage. However, these data do not provide the clinician with the information on cerebral oxygenation and hemodynamics, which are the main parameters one wishes to determine.

Currently, using NIRS it has become possible to continuously and non invasively measure tissue oxygenation and cerebral blood volume with high-time resolution. NIRS measures at the level of arterioles, capillaries and venules [92]. It has been used in sleep research, so far, to identify changes in brain oxygenation and circulation, during sleep states, in healthy newborn infants [171,172] and healthy adult volunteers [145,173]. It has been also employed to detect hemodynamic and metabolic changes in the brain of preterm infants with sleep apnea [174] and adults with OSA [175,176] to assess changes in cerebral oxygenation and blood volume. It is a novel, non-invasive, bedside method that has a promises potential to be a very accurate and comparatively inexpensive diagnostic tool for brain function.

The NIRS parameters, such as oxy-(O_2Hb), deoxy-(HHb), and total hemoglobin (tHb) concentrations, and cerebral tissue hemoglobin oxygen saturation (SO_2) were monitored simultaneously by a frequency-domain, dual channel, two-wavelength (690 and 830 nm) tissue oximeter (OxiplexTS; ISS Inc, Champaign, IL). A near-infrared (NIR) dual sensor probe, having a two-channel configuration, was specially designed for these measurements. The optical source fibers were arranged in pairs, such that each pair contained one fiber connected to a source emitting at each wavelength. The optical signals detected at the tissue surface were guided to the photomultipliers (one per channel) of the oximeter by optical fiber bundles of 3 mm internal diameter each. The output ends of the paired light source fibers were arranged at increasing distances from the input ends of the detector fiber bundles. The four source–detector distances ranged from 1.98 to 4.08 cm (Figure 6).

We applied NIRS during daytime napping and during voluntary breath holding at functional residual capacity (FRC). Different dynamics of O_2Hb, HHb, and tHb concentrations ([O_2Hb], [HHb] and [tHb]) and SO_2 were detected for control and OSA groups during daytime napping and voluntary hypoxia (Figure 11). We recorded reduced cerebral hemodynamic response during breath holding in OSA subjects (Figure 11d) as compared to control non-snorers (Figure 11b). In the OSA subjects, this smaller increase in [tHb] = [O_2Hb] + [HHb] is proportional to the cerebral blood volume and corresponds to the dilatory ability of the cerebral vasculature. A smaller increase, or even a decrease in [O_2Hb] followed by an increase in [HHb] during breath holding, indicates a reduced change in cerebral blood flow that is insufficient to compensate for arterial blood deoxygenation during hypoxia. Hypoxia-induced changes in [O_2Hb], [HHb] (Figure 11c), [tHb], and SO_2 due to sleep apnea were comparable with changes due to breath holding (Figure 11d). These changes were not observed in control subjects (Figure 11a).

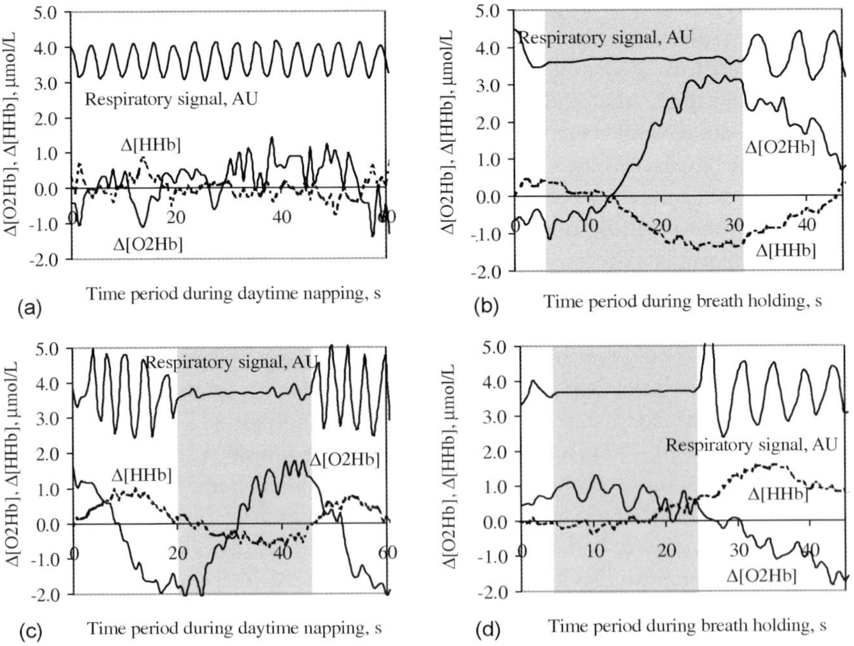

Figure 11. Examples of respiratory signals and changes in cerebral tissue [O2Hb] and [HHb] with respect to baseline levels. Shaded areas correspond to apnea episodes. (a) Control non-snorer (28-years-old) during daytime napping; (b) Control non-snorer (29-years-old) during breath holding; (c) OSA subject (73-years-old) during daytime napping; (d) Same OSA (obstructive sleep apnea) subject during breath holding.

In recent years near-infrared spectroscopy (NIRS) has been proposed as a method to study brain hemodynamics, which is simple and inexpensive compared with such "heavy-duty" methods as functional magnetic resonance imaging (fMRI) and positron emission tomography (PET). Many studies about applying NIRS to monitor functional cerebral hemodynamics have been reviewed in Ref. [70].

For NIRS measurements we use a two-wavelength (758 and 830 nm) frequency-domain (110 MHz modulation frequency) Oximeter (ISS, Champaign, IL), which has sixteen laser diodes (eight per each wavelength) and two photomultiplier tube detectors. At a wavelength of 758 nm, light absorption by the deoxy-hemoglobin (HHb) substantially exceeds absorption by the oxy-hemoglobin (O_2Hb), while at 830 nm the O_2Hb absorption prevails over the HHb absorption. The laser diodes operate in a sequential multiplexing mode with 10 ms "on" time per each diode [138,177]. Light emitted by these laser diodes is guided to the tissue through 10 m long multi-mode silica optical fibers. Two 10 m long glass fiber bundles collect the scattered light and bring it to the detectors. The paired (758 and 830 nm wavelength) source fibers are attached to the probe at 8 positions. Together with two detectors, they provide ten bi-wavelength source–detector channels with a source–detector distance of 3 cm. The probe is centered at the measured C3 position according to the International 10–20 System.

Each exercise run consisted of a 30 s pre-exercise epoch, ten 20 s stimulation epochs separated by ten 20 s control epochs, and a 50 s after-exercise epoch. During stimulation epochs subjects performed light palm squeezing with the right hand following the rhythm presented by a sound system.

Figure 12 shows a map of folding average [HHb] and [O_2Hb] traces. For the light channels situated above the activated area (left part of Figure 12), the characteristic feature was a significant decrease of [HHb] during the stimulation, which was concurrent with a significant increase of the oxy-hemoglobin concentration. Typically rapid changes in [HHb] and [O_2Hb] began 2–3 s after the stimulation onset and continued during the next 7–15 s. The [HHb] then fluctuated near its low level for the rest of the stimulation epoch and the beginning of the resting epoch. During stimulation [O_2Hb] either fluctuated at its high level or exhibited a slight decrease. A rapid recovery toward the baseline level begins 4–6 s after the onset of the rest epoch in both [HHb] and [O_2Hb]. Such behavior of [HHb] and [O_2Hb] in the activated area was qualitatively the same in all six subjects of this study.

No significant decrease in the folding average [HHb] traces concurrent with the significant [O_2Hb] increase was observed during stimulations in the light channels outside the activated area. Usually [HHb] in such channels fluctuated without correlation with the paradigm function, in some cases [O_2Hb], but without significant [HHb] change.

The decrease of [HHb] concurrent with the increase in [O_2Hb] agrees with the results of the previous simultaneous fMRI-NIRS study [178], BOLD signal theory and with the basic knowledge of brain physiology [179]. This indicates that the task-related hemodynamic changes measured by NIRS are not due to artifacts but have intracranial origin.

Brain activity is associated with physiological changes in the optical parameters of the tissue that can be assessed by NIRS. Two major types of signals are reported

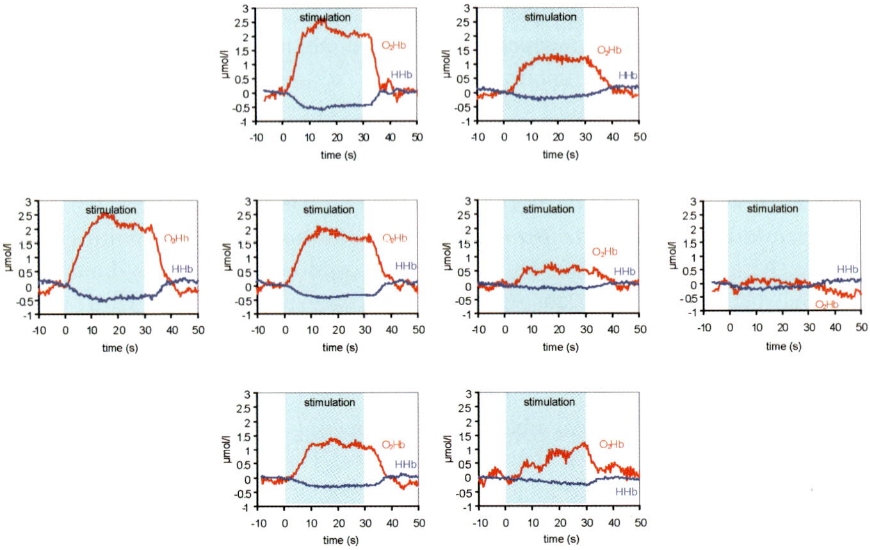

Figure 12. Changes of oxy- and deoxy-hemoglobin concentration during tapping. Stimulation was active during the period marked in blue in the figure.

in the literature: (a) a slow signal in the range of seconds, mainly due to light absorption related in particular to changes in hemoglobin concentration [70,180]; (b) a fast signal in the range of milliseconds which was suggested to be associated with changes in light scattering due to changes in the refractive index at neuronal membranes. This signal has much smaller amplitude than the slow signal and has been described by only a few authors [139].

11.9 Future directions

Near-infrared spectroscopy and imaging are making rapid inroads in clinical and physiological studies. The intrinsic non-invasive character of the technique, the portability and the relatively low cost makes this methodology very attractive in the clinic and for laboratory studies. The information content of the technique is very large. The temporal characteristics are excellent for studies of fast (in the millisecond range) and slow phenomena [137]. As discussed in this chapter, new methods are being proposed that will allow measuring a continuous range of wavelength, thereby vastly increasing the capability to distinguish different chromophores in tissues and to measure the wavelength dependence of the scattering coefficient. The limitation of the method is in the penetration depth in tissue and in the spatial resolution that can be achieved. Penetration and resolution are limited by the physics of light propagation in tissues. In general there is a relationship between penetration and resolution. Superficial tissue layers can be studied with better resolution than deeper layers.

We have mainly discussed mechanisms of contrast provided by naturally occurring substances. There is a rapid developing area in contrast agents for optical tissue imaging, both as absorption and as fluorescence specific contrast agents. In the field of imaging, instruments are becoming more sophisticated and with many source detector pairs. In brain imaging, we are going in the direction of a full-head scanner that can detect brain hemodynamics on the entire brain surface. From the computational point of view, reconstruction algorithms have become faster and more accurate. We can foresee that maps of optical parameters will become common. Perhaps the major progress is in the understanding of the physiological signals. Until recently, it was unclear if diseased tissue will provide any form of optical contrast. Today, due to better in vivo studies, improved measuring modalities and better understanding of the physiology it is clear that many physiological conditions and processes can be distinguished and monitored.

11.9.1 Potential clinical applications

The areas of potential clinical applications are numerous and are growing every day. Near-ir spectroscopy provides a unique tool for the measurements of physiological parameters of tissues. There are two major fields of application of NIRS in the clinic, to obtain functional maps and to monitor changes. Both areas have vast applications in many clinical situations, including neonatal care, brain surgery, organ viability in transplants, monitoring of brain circulation, aging, sport medicine, cancer detection and many others.

Acknowledgements

We thank Dr Francesco Fabbri for useful discussions on the theoretical models for diffuse reflectance imaging, and Erica Heffer for performing the data analysis used to generate Figure 10. We also thank Dr Antonios Michalos, MD for providing the data on the OSA studies, Drs Martin Wolf and Ursula Wolf, MD for the data on the venous occlusion studies and Dr Vlad Toronov for the data on brain activation. We acknowledge support from the National Science Foundation (Award No. BES-93840) and from the National Institutes of Health (Grants P41RR03155, CA57032, and DA14178).

References

1. M. Cutler (1929). Transillumination of the breast. *Surg. Gynecol. Obstet.* **48**, 721–727.
2. G.A. Millikan (1942). The oximeter, an instrument for measuring continuously the oxygen saturation of arterial blood in man. *Rev. Sci. Instrum.* **13**, 434–444.
3. T. Aoyagi, M. Kishi, K. Yamaguchi, S. Watanabe (1974). Improvement of the earpiece oximeter. In: *Abstracts of the Japanese Society of Medical Electronics and Biological Engineering*, Tokyo, Japan (pp. 90–91).
4. Y. Mendelson (1992). Pulse oximetry: theory and applications for noninvasive monitoring. *Clin. Chem.* **38**, 1601–1607.
5. F.F. Jöbsis (1977). Noninvasive, infrared monitoring of cerebral and myocardial oxygen sufficiency and circulatory parameters. *Science* **198**, 1264–1267.

6. A. Grinvald, E. Lieke, R. Frostig, C.D. Gilbert, T.N. Wiesel (1986). Functional architecture of cortex revealed by optical imaging of intrinsic signals. *Nature* **324**, 361–364.
7. B. Chance, Z. Zhuang, C. UnAh, L. Lipton (1993). Cognition-activated low-frequency modulation of light absorption in human brain. *Proc. Natl. Acad. Sci. U.S.A.* **90**, 3770–3774.
8. A. Villringer, J. Planck, C. Hock, L. Schleinkofer, U. Dirnagl (1993). Near infrared spectroscopy (NIRS): a new tool to study hemodynamic changes during activation of brain function in human adults. *Neurosci. Lett.* **154**, 101–104.
9. Y. Hoshi, M. Tamura (1993). Dynamic multichannel near-infrared optical imaging of human brain activity. *J. Appl. Physiol.* **75**, 1842–1846.
10. G. Gratton, P.M. Corballis, E. Cho, M. Fabiani, D.C. Hood (1995). Shades of gray matter: noninvasive optical images of human brain responses during visual stimulation. *Psychophysiology* **32**, 505–509.
11. M.S. Patterson, B. Chance, B.C. Wilson (1989). Time resolved reflectance and transmittance for the non-invasive measurement of optical properties. *Appl. Opt.* **28**, 2331–2336.
12. D.T. Delpy, M. Cope, P. van der Zee, S. Arridge, S. Wray, J. Wyatt (1988). Estimation of optical pathlength through tissue from direct time of flight measurement. *Phys. Med. Biol.* **33**, 1433–1442.
13. E. Gratton, W.W. Mantulin, M.J. vande Ven, J.B. Fishkin, M.B. Maris, B. Chance (1990). The possibility of a near-infrared optical imaging system using frequency-domain methods. *Third International Conf. for Peace through Mind/Brain Science* (pp. 183–189). Hamamatsu City, Japan, August 5–10.
14. B. Chance, M. Maris, J. Sorge, M.Z. Zhang (1990). A phase modulation system for dual wavelength difference spectroscopy of haemoglobin deoxygenation in tissue. *Proc. SPIE* **1204**, 481–491.
15. B.J. Tromberg, L.O. Svaasand, T.T. Tsay, R.C. Haskell, M.W. Berns (1991). Optical property measurement in turbid media using frequency-domain photon migration. *Proc. SPIE* **1525**, 52–58.
16. G.M. Hale, M.R. Querry (1973). Optical constants of water in the 200 nm to 200 μm wavelength region. *Appl. Opt.* **12**, 555–563.
17. S. Prahl (of the Oregon Medical Laser Center, Portland, OR) has tabulated the molar extinction coefficients for oxy-hemoglobin and deoxy-hemoglobin using data from W.B. Gratzer and N. Kollias. These tabulated data are available at http://omlc.ogi.edu/spectra/hemoglobin/summary.html.
18. W.F. Cheong, S.A. Prahl, A.J. Welch (1990). A review of the optical properties of biological tissues. *IEEE J. Quantum Electron.* **26**, 2166–2185.
19. D.Y. Ts'o, R.D. Frostig, E. Lieke, A. Grinvald (1990). Functional organization of primate visual cortex revealed by high resolution optical imaging. *Science* **249**, 417–420.
20. D.M. Rector, R.F. Rogers, J.S. Schwaber, R.M. Harper, J.S. George (2001). Scattered-light imaging in vivo tracks fast and slow processes of neurophysiological activation. *Neouroimage* **14**, 977–994.
21. M.M. Haglund, G.A. Ojeman, D.W. Hochman (1992). Optical imaging of epileptiform and functional activity in human cerebral cortex. *Nature* **358**, 668–671.
22. T. Bonhoeffer, A. Grinvald (1996). Optical imaging based on intrinsic signals. In: A.W. Toga, J.C. Mazziotta (Eds), *Brain Mapping: The Methods*, (pp. 55–97). Academic Press, San Diego, CA.

23. D. Malonek, R.B.H. Tootell, A. Grinvald (1994). Optical imaging reveals the functional architecture of neurons processing shape and motion in owl monkey area MT. *Proc. R. Soc. London B* **258**, 109–119.
24. P.M. Gochin, P. Bedenbaugh, J.J. Gelfand, C.G. Gross, G.L. Gerstein (1992). Intrinsic signal optical imaging in the forepaw area of rat somatosensory cortex. *Proc. Natl. Acad. Sci. U.S.A.* **89**, 8381–8383.
25. R. Marchesini, S. Tomatis, C. Bartoli, A. Bono, C. Clemente, C. Cupeta, I. Del Prato, E. Pignoli, A.E. Sichirollo, N. Cascinelli (1995). In vivo spectrophotometric evaluation of neoplastic and non-neoplastic skin pigmented lesions. III. CCD camera-based reflectance imaging. *Photochem. Photobiol.* **62**, 151–154.
26. B. Farina, C. Bartoli, A. Bono, A. Colombo, M. Lualdi, G. Tragni, R. Marchesini (2000). Multispectral imaging approach in the diagnosis of cutaneous melanoma: potentiality and limits. *Phys. Med. Biol.* **45**, 1243–1254.
27. A. Ishimaru (1978). *Wave Propagation and Scattering in Random Media.* Academic Press, New York.
28. G. Kortüm (1969). *Reflectance Spectroscopy: Principles, Methods, Applications.* Springer-Verlag, New York.
29. P. Kubelka (1948). New contributions to the optics of intensely light-scattering materials. Part I. *J. Opt. Soc. Am.* **38**, 448–457.
30. P.S. Mudgett, L.W. Richards (1971). Multiple Scattering Calculations for Technology. *Appl. Opt.* **10**, 1485–1502.
31. R.G. Giovanelli (1955). Reflection by semi-infinite diffusers. *Opt. Acta* **2**, 153–162.
32. L.O. Svaasand, T. Spott, J.B. Fishkin, T. Pham, B.J. Tromberg, M.W. Berns (1999). Reflectance measurements of layered media with photon-density waves: a potential tool for evaluating deep burns and subcutaneous lesions. *Phys. Med. Biol.* **44**, 801–813.
33. S.T. Flock, M.S. Patterson, B.C. Wilson, D.R. Wyman (1989). Monte Carlo modeling of light propagation in highly scattering tissues -I: Model predictions and comparison with diffusion theory. *IEEE Trans. Biomed. Eng.* **36**, 1162–1168.
34. S.A. Prahl, M.J.C. van Gemert, A.J. Welch (1993). Determining the optical properties of turbid media by using the adding-doubling method. *Appl. Opt.* **32**, 559–568.
35. S.L. Jacques (1996). Reflectance spectroscopy with optical fiber devices and transcutaneous bilirubinometers. In: A.M. Verga Scheggi et al. (Eds), *Biomedical Optical Instrumentation and Laser-Assisted Biotechnology.* (pp. 83–94). (Kluwer Academic Publishers, Netherlands.
36. J.R. Mourant, I.J. Bigio, D.A. Jack, T.M. Johnson, H.D. Miller (1997). Measuring absorption coefficients in small volumes of highly scattering media: source–detector separations for which pathlengths do not depend on scattering properties. *Appl. Opt.* **36**, 5655–5661.
37. J.R. Mourant, T.M. Johnson, G. Los, I.J. Bigio (1999). Non-invasive measurement of chemotherapy drug concentrations in tissue: preliminary demonstrations of *in vivo* measurements. *Phys. Med. Biol.* **44**, 1397–1417.
38. J.S. Wyatt, M. Cope, D.T. Delpy, P. van der Zee, S.R. Arridge, A.D. Edwards, E.O.R. Reynolds (1990). Measurement of optical pathlength for cerebral near infrared spectroscopy in newborn infants. *Dev. Neurosci.* **12**, 140–144.
39. P. Van der Zee, M. Cope, S.R. Arridge, M. Essenpreis, L.A. Potter, A.D. Edwards, J.S. Wyatt, D.C. McCormick, S.C. Roth, E.O.R. Reynolds, D.T. Delpy (1992). Experimentally measured optical pathlengths for the adult head, calf and forearm and the head of the newborn infants as a function of inter optode spacing. *Adv. Exp. Med. Biol.* **316**, 143–153.

40. A. Duncan, J.H. Meek, M. Clemence, C.E. Elwell, L. Tyszczuk, M. Cope, D.T. Delpy (1995). Optical pathlength measurements on adult head, calf and forearm and the head of the newborn infant using phase resolved optical spectroscopy. *Phys. Med. Biol.* **40**, 295–304.
41. S. Fantini, D. Hueber, M.A. Franceschini, E. Gratton, W. Rosenfeld, P.G. Stubblefield, D. Maulik, M.R. Stankovic (1999). Non-invasive optical monitoring of the newborn piglet brain using continuous-wave and frequency-domain methods. *Phys. Med. Biol.* **44**, 1543–1563.
42. K.M. Case, P.F. Zweifel (1967). *Linear Transport Theory*. Addison-Wesley Publishing Co, Reading, MA.
43. J.J. Duderstadt, L.J. Hamilton (1976). *Nuclear Reactor Analysis* (p. 113). Wiley, New York.
44. T.J. Farrel, M.S. Patterson, B. Wilson (1992). A diffusion theory model of spatially resolved, steady-state diffuse reflectance for the noninvasive determination of tissue optical properties *in vivo*. *Med. Phys.* **19**, 879–888.
45. S. Fantini, M.A. Franceschini, E. Gratton (1994). Semi-infinite-geometry boundary problem for light migration in highly scattering media: a frequency-domain study in the diffusion approximation. *J. Opt. Soc. Am. B* **11**, 2128–2138.
46. H. Liu, D.A. Boas, Y. Zhang, A.G. Yodh, B. Chance (1995). A simplified approach to characterize optical properties and blood oxygenation in tissue using continuous near infrared light. *Proc. SPIE* **2389** 496–502.
47. S.R. Arridge, M. Cope, D.T. Delpy (1992). The theoretical basis for the determination of optical pathlengths in tissue: temporal and frequency analysis. *Phys. Med. Biol.* **37**, 1531–1560.
48. K.M. Yoo, B.B. Das, R.R. Alfano (1992). Imaging of a translucent object hidden in a highly scattering medium from the early portion of the diffuse component of a transmitted ultrafast laser pulse. *Opt. Lett.* **17**, 958–960.
49. G. Mitic, J. Kölzer, J. Otto, E. Plies, G. Sölkner, W. Zinth (1994). Time-gated trans-illumination of biological tissues and tissuelike phantoms. *Appl. Opt.* **33**, 6699–6710.
50. D.A. Benaron, D.K. Stevenson (1993). Optical time-of-flight and absorbance imaging of biologic media. *Science* **259**, 1463–1466.
51. D. Grosenick, H. Wabnitz, H.H. Rinneberg, K.T. Moesta, P.M. Schlag (1999). Development of a time-domain optical mammograph and first *in vivo* applications. *Appl. Opt.* **38**, 2927–2943.
52. E. Gratton, W.W. Mantulin, M.J. van de Ven, J.B. Fishkin, M.B. Maris, B Chance, (1993). A novel approach to laser tomography. *Bioimaging* **1**, 40–46.
53. E. Gratton, M. Limkeman (1983). A continuously variable frequency cross-correlation phase fluorometer with picoseconds resolution. *Biophys. J.* **44**, 315–324.
54. M.A. Franceschini, S. Fantini, E. Gratton (1994). LED's in frequency-domain spectroscopy of tissues. In: R.R. Alfano (Ed.), *Advances in Laser and Light Spectroscopy to Diagnose Cancer and Other Diseases SPIE Proc.* **2135**, 300–306.
55. E. Gratton, S. Fantini, M.A. Franceschini, J.B. Fishkin, J. Maier (1994). Near-infrared optical spectroscopy of tissues using an LED frequency domain spectrometer. In: R.R. Alfano, (Ed), *Proc. Advances in Optical Imaging & Photon Migration Topical Mtg. 21* (pp. 278–282) (*Proc. Optical Society of America*).
56. B.M. Feddersen, D.W. Piston, E. Gratton (1989). Digital parallel acquisition in frequency domain fluorometry. *Rev. Sci. Instrum.* **60**, (9), 2929–2936.
57. J. Fishkin, E. Gratton (1993). Propagation of photon density waves in strongly scattering media containing an absorbing "semi-infinite" plane bounded by a straight edge. *J. Opt. Soc. Am. A* **10**(1), 127–140.

58. S. Fantini, M.A. Franceschini, E. Gratton (1997). Effective source term in the diffusion equation for photon transport in turbid media. *Appl. Opt.* **36**, 156–163.
59. S. Fantini, M.A. Franceschini, J.B. Fishkin, B. Barbieri, E. Gratton (1994). Quantitative determination of the absorption spectra of chromophores in strongly scattering media: a light-emitting-diode based technique. *Appl. Opt.* **33**(22), 5204–5213.
60. S. Fantini, M.A. Franceschini, J.S. Maier, S.A. Walker, B. Barbieri, E. Gratton (1995). Frequency-domain multichannel optical detector for noninvasive tissue spectroscopy and oximetry. *Opt. Eng.* **34**, 32–42.
61. S. Fantini, M.A. Franceschini, J.B. Fishkin, E. Gratton (1994). Absolute measurement of absorption and scattering coefficients spectra of a multiply scattering medium. *Proc. SPIE* **2131**, 356–364.
62. M.A. Franceschini, A. Paunescu, S. Fantini, S. Pratesi, J.S. Maier, G.P. Donzelli, E. Gratton (1998). Frequency-domain optical measurements *in vitro* on two- and three-layered tissue-like phantoms and *in vivo* on infant heads. In: J.G. Fujimoto, M. Patterson (Eds), *OSA Trends in Optics and Photonics on Advances in Optical Imaging and Photon Migration* (Volume XXI, pp. 232–236). Optical Society of America, Washington, DC.
63. M.A. Franceschini, S. Fantini, L.A. Paunescu, J.S. Maier, E. Gratton (1998). Influence of a superficial layer in the quantitative spectroscopic study of strongly scattering media. *Appl. Opt.* **37**, 7447–7458.
64. J.B. Fishkin, P.T.C. So, A.E. Cerussi, S. Fantini, M.A. Franceschini, E. Gratton (1995). Frequency-domain method for determining spectral properties in multiply scattering media: methemoglobin absorption spectrum in a tissue-like phantom. *Appl. Opt.* **34**, 1143–1155.
65. J.B. Fishkin, S. Fantini, M.J. van de Ven, E. Gratton (1996). Gigahertz photon density waves in a turbid medium: theory and experiments. *Phys. Rev. E* **53**, 2307–2319.
66. E. Gratton, J.B. Fishkin (1995). Optical spectroscopy of tissue-like phantoms using photon density waves. *Comments Cell. Mol. Biophys.* **8**(6), 309–359.
67. B.J. Tromberg, O. Coquoz, J.B. Fishkin, T. Pham, E.R. Anderson, J. Butler, M. Cahn, J.D. Gross, V. Venugopalan, D. Pham (1997). Non-invasive measurements of breast tissue optical properties using frequency-domain photon migration. *Phil. Trans. R. Soc. London B* **352**, 661–668.
68. S. Wray, M. Cope, D.T. Delpy, J.S. Wyatt, E.O. Reynolds (1998). Characterization of the near infrared absorption spectra of cytochrome aa3 and haemoglobin for the non-invasive monitoring of cerebral oxygenation. *Biochim. Biophys. Acta* **933**(1), 184–192.
69. F. Bevilacqua, A.J. Berger, A.E. Cerussi, D. Jakubowski, B.J. Tromberg (2000). Broadband absorption spectroscopy in turbid media by combined frequency-domain and steady-state methods. *Appl. Opt.* **39**, 6498–6507.
70. A. Villringer, B. Chance (1997). Non-invasive optical spectroscopy and imaging of human brain function. *Trends Neurosci.* **20**, 435–442.
71. T. Wolf, U. Lindauer, U. Reuter, T. Back, A. Villringer, K. Einhaupl, U. Dirnagl (1997). Noninvasive near infrared spectroscopy monitoring of regional cerebral blood oxygenation changes during peri-infarct depolarizations in focal cerebral ischemia in the rat. *J. Cereb. Blood Flow Metab.* **17**(9), 950–4
72. T. Muellner, A. Nikolic, W. Schramm, V. Vecsei (1999). New instrument that uses near-infrared spectroscopy for the monitoring of human muscle oxygenation. *J. Trauma* **46**, 1082–4.
73. M.S. Thorniley, J.S. Sinclair, N.J. Barnett, C.B. Shurey, C.J. Green (1998). The use of near-infrared spectroscopy for assessing flap viability during reconstructive surgery. *Br. J. Plast. Surg.* **51**, 218–26.

74. J. Teller, M. Wolf, H.U. Bucher, M. Keel, S. Fanconi, O. Baenziger (2000). Can near infrared spectroscopy of the liver monitor tissue oxygenation? *Eur. J. Pediatr.* **159**, 549.
75. L. Szmedra, J. Im, S. Nioka, B. Chance, K.W. Rundell (2001). Hemoglobin/myoglobin oxygen desaturation during alpine skiing, *Med. Sci. Sports. Exerc.* **33**, 232–6.
76. V. Quaresima, M. Ferrari (1998). Assessment of quadriceps oxygenation in patients with myopathies by near infrared spectroscopy. *Neurology* **51**, 1238–39.
77. K. Abe, Y. Matsuo, J. Kadekawa, S. Inoue, T Yanagihara (1997). Measurement of tissue oxygen consumption in patients with mitochondrial myopathy by noninvasive tissue oximetry. *Neurology* **49**, 837–41.
78. W. Bank, B. Chance (1994). An oxidative defect in metabolic myopathies: diagnosis by noninvasive tissue oximetry. *Ann. Neurol.* **36** 830–37.
79. R. Belardinelli (1998). Muscle oxygenation kinetics measured by near-infrared spectroscopy during recovery from exercise in chronic heart failure. *G. Ital. Cardiol.* **28**, 866–72.
80. S. Matsui, N. Tamura, T. Hirakawa, S. Kobayashi, N. Takekoshi, E. Murakami (1995). Assessment of working skeletal muscle oxygenation in patients with chronic heart failure. *Am. Heart J.* **129**, 690–5.
81. D.M. Mancini, J.R. Wilson, L. Bolinger, H. Li, K. Kendrick, B. Chance, J.S. Leigh (1994). *In vivo* magnetic resonance spectroscopy measurement of deoxymyoglobin during exercise in patients with heart failure. Demonstration of abnormal muscle metabolism despite adequate oxygenation. *Circulation* **90**, 500–8.
82. T. Komiyama, H. Shigematsu, H. Yasuhara, T. Muto (2000). Near-infrared spectroscopy grades the severity of intermittent claudication in diabetics more accurately than ankle pressure measurement. *Br. J. Surg.* **87**, 459–66.
83. M.C.P. Beekvelt van, W.N.J.M. Colier, R.A. Wevers, B.G.M. Engelen van (2001). Performance of near-infrared spectroscopy in measuring local O_2 consumption and blood flow in skeletal muscle. *J. Appl. Physiol.* **90**, 511–9.
84. C. Casavola, L.A. Paunescu, S. Fantini, M.A. Franceschini, P.M. Lugara, E. Gratton (1999). Application of near-infrared tissue oxymetry to the diagnosis of peripheral vascular disease, *Clin. Hemorheol. Microcirc.* **21**, 389–93.
85. D. Choudhury, B. Michener, P. Fennely, M. Levi (1999). Near-infrared spectroscopy in the early detection of peripheral vascular disease. *J. Vascular Technol.* **23**, 109–13.
86. H.M. Kooijman, M.T. Hopman, W.N. Colier, J.A. van der Vliet, B. Oeseburg (1997). Near infrared spectroscopy for noninvasive assessment of claudication. *J. Surg. Res.* **72**, 1–7.
87. K.K. McCully, L. Landsberg, M. Suarez, M. Hofmann, J.D. Posner (1997). Identification of peripheral vascular disease in elderly subjects using optical spectroscopy. *J. Gerontol. A. Biol. Sci. Med. Sci.* **52**, B159–65.
88. T.R. Cheatle, L.A. Potter, M. Cope, D.T. Delpy, P.D. Coleridge Smith, J.H. Scurr (1991). Near-infrared spectroscopy in peripheral vascular disease. *Br. J. Surg.* **78**, 405–8.
89. T. Kunihara, S. Sasaki, N. Shiiya, T. Murashita, Y. Matsui, K. Yasuda (2001). Near infrared spectrophotometry reflects cerebral metabolism during hypothermic circulatory arrest in adults. *ASAIO J.* **47**, 417–21.
90. C. Terborg, W. Schummer, M. Albrecht, K. Reinhart, C. Weiller, J. Rother (2001). Dysfunction of vasomotor reactivity in severe sepsis and septic shock. *Intensive Care Med.* **27**, 1231–4.
91. J.C. Puyana, B.R. Soller, S. Zhang, S.O. Heard (1999). Continuous measurement of gut pH with near-infrared spectroscopy during hemorrhagic shock. *J. Trauma* **46**, 9–15.

92. H. Liu, B. Chance, A.H. Hielscher, S.L. Jacques, F.K. Tittel (1995). Influence of blood vessels on the measurement of hemoglobin oxygenation as determined by time-resolved reflectance spectroscopy. *Med. Phys.* **22**, 1209–17.
93. V. Quaresima, W.N. Colier, M. van der Sluijs, M. Ferrari (2001). Nonuniform quadriceps O2 consumption revealed by near infrared multipoint measurements. *Biochem. Biophys. Res. Commun.* **285**, 1034–9.
94. S. Nioka, H. Miura, H. Long, A. Perry, D. Moser, B. Chance (1999). Functional muscle imaging in elite and untrained subjects. *SPIE Proc.* **3597**, 282–90.
95. U. Wolf, M. Wolf, J.H. Choi, L.A. Paunescu, L.P. Safonova, A. Michalos, E. Gratton (2003). Mapping of hemodynamics on the human calf muscle with near infrared spectroscopy and the influence of the adipose tissue thickness. *Adv. Exp. Med. Biol.* **510**, 225–30.
96. K.K. Kalliokoski, J. Kemppainen, K. Larmola, T.O. Takala, P. Peltoniemi, A. Oksanen, U. Ruotsalainen, C. Cobelli, J. Knuuti (2000). Muscle blood flow and flow heterogeneity during exercise studied with positron emission tomography in humans. *Eur. J. Appl. Physiol.* **83**, 395–401.
97. M.A. Franceschini, E. Gratton, S. Fantini (1999). Non-invasive optical method to measure tissue and arterial saturation: an application to absolute pulse oximetry of the brain. *Opt. Lett.* **24**, 829–831.
98. S. Homma, T. Fukunaga, A. Kagaya (1996). Influence of adipose tissue thickness on near infrared spectroscopic signals in the measurement of human muscle. *J. Biomed. Opt.* **1**, 418–24.
99. M. Ferrari, T. Binzoni, V. Quaresima (1997). Oxidative metabolism in muscle. *Philos. Trans. R. Soc. London B. Biol. Sci.* **352**, 677–83.
100. C.W. Yoxall, A.M. Weindling (1997). Measurements of venous oxyhaemoglobin saturation in the adult human forearm by near infrared spectroscopy with venous occlusion. *Med. Biol. Eng. Comput.* **35**, 331–6.
101. R.A. De Blasi, M. Ferrari, A. Natali, G. Conti, A. Mega, A. Gasparetto (1994). Nonivasive measurement of forearm blood flow and oxygen consumption be near-infrared spectroscopy. *J. Appl. Physiol.* **76**, 1388–93.
102. D. Hueber, S. Fantini, A.E. Cerussi, B. Barbieri (1999). New optical probe design for absolute (self-calibrating) NIR tissue hemoglobin measurements. *SPIE Proc.* **3597**, 618–31.
103. P. Vaupel, F. Kallinowski, P. Okunieff (1989). Blood flow, oxygen and nutrient supply, and metabolic microenvironment of human tumors: a review. *Cancer Res.* **49**, 6449–6465.
104. S. Thomsen, D. Tatman (1998). Physiological and pathological factors of human breast disease that can influence optical diagnosis. *Ann. New York Acad. Sci.* **838**, 171–193.
105. B. Riefke, K. Licha, W. Semmler (1997). Contrast media for optical mammography. *Radiologe* **37**, 749–755.
106. S. Nioka, S.B. Colak, X. Li, Y. Yang, B. Chance (1998). Breast tumor images of hemodynamic information using a contrast agent with back projection and fft enhancement. In: J.G. Fujimoto, M.S. Patterson (Eds), *OSA Trends in Optics and Photonics Vol. 21, Advances in Optical Imaging and Photon Migration* (pp. 266–270). Optical Society of America, Washington, DC.
107. V. Ntziachristos, A.G. Yodh, M. Schnall, B. Chance (2000). Concurrent MRI and diffuse optical tomography of breast after indocyanine green enhancement. *Proc. Natl. Acad. Sci. U.S.A.* **97**, 2767–2772.
108. C.M. Gros, Y. Quenneville, Y. Hummel (1972). Diaphanologie mammaire. *J. Radiol. Electrol. Med. Nucl.* **53**, 297–306.

109. E. Carlsen (1982). Transillumination light scanning. *Diagn. Imaging* **4**, 28–34.
110. R.J. Bartrum, H.C. Crow (1984). Transillumination lightscanning to diagnose breast cancer: a feasibility study. *AJR* **142**, 409–414.
111. E.A. Sickles (1984). Breast cancer detection with transillumination and mammography, *AJR* **142**, 841–844.
112. A. Alveryd et al. (1990). Lightscanning versus mammography for the detection of breast cancer in screening and clinical practice. *Cancer* **65**, 1671–1677.
113. Y. Yamashita, M. Kaneko (1993). Visible and infrared diaphanoscopy for medical diagnosis. In: G.J. Müller, B. Chance, R.R. Alfano, S.R. Arridge, J. Beuthan, E. Gratton, M. Kaschke, B.R. Masters, S. Svanberg, P. van der Zee (Eds.), *Medical Optical Tomography: Functional Imaging and Monitoring*, (Volume IS11 of SPIE Institute Series), Series, pp. 283–316). SPIE Press, Bellingham, Washington.
114. J.H. Hoogenraad, M.B. van der Mark, S.B. Colak, G.W.'t Hooft, E.S. van der Linden (1998). First results from the philips optical mammoscope. In: D. Benaron, B. Chance, M. Ferrari (Eds), *Photon Propagation in Tissues III* (Proc. SPIE, Volume 3194, pp. 184–190). SPIE, Bellinghom, WA.
115. S.B. Colak, D.G. Papaioannou, G.W.'t Hooft, M.B. van der Mark, H. Schomberg, J.C.J. Paasschens, J.B.M. Melissen, N.A.A.J. Van Asten (1997). Tomographic image reconstruction from optical projections in light-diffusing media. *Appl. Opt.* **36**, 180–213.
116. H. Jiang, N.V. Iftimia, X. Yong, J.A. Eggert, L.L. Fajardo, K.L. Klove (2002). Near-infrared optical imaging of the breast with model-based reconstruction. *Acad. Radiol.* **9**, 186–194.
117. R.L. Barbour, H.L. Graber, Y. Pei, S. Zhong, C.H. Schmitz (2001). Optical tomographic imaging of dynamic features of dense-scattering media. *J. Opt. Soc. Am. A* **18**, 3018–3036.
118. K. Wells, J.C. Hebden, F.E.W. Schmidt, D.T. Delpy (1997). The UCL Multichannel time-resolved system for optical tomography. *Proc. SPIE* **2979**, 599–607.
119. V. Ntziachristos, X.H. Ma, B. Chance (1998). Time-correlated single photon counting imager for simultaneous magnetic resonance and near-infrared mammography. *Rev. Sci. Instrum.* **69**, 4221–423.
120. Cubeddu, A. Pifferi, P. Taroni, A. Torricelli, G. Valentini (1999). Non-invasive absorption and scattering spectroscopy of bulk diffusive media: an application to the optical characterization of human breast. *Appl. Phys. Lett.* **74**, 874–876.
121. M. Kaschke, H. Jess, G. Gaida, J.M. Kaltenbach, W. Wrobel (1994). Transillumination imaging of tissue by phase modulation techniques. In: R.R. Alfano (Ed.), *Advances in Optical Imaging and Photon Migration*, (Proc. Optical Society of America, Volume 21, 88–92).
122. L. Götz, S.H. Heywang-Köbrunner, O. Schütz, H. Siebold (1998). Optische mammographie an präoperativen patientinnen. *Akt. Radiol.*, **8**, 31–33.
123. X.D. Li, T. Durduran, A.G. Yodh, B. Chance, D.N. Pattanayak (1997). Diffraction tomography for biochemical imaging with diffuse-photon density waves. *Opt. Lett.* **22**, 573–575.
124. B.J. Tromberg, N. Shah, R. Lanning, A. Cerussi, J. Espinoza, T. Pham, L. Svaasand, J. Butler (2000). Non-invasive *in vivo* characterization of breast tumors using photon migration spectroscopy. *Neoplasia* **2**, 26–40.
125. M.A. Franceschini, K.T. Moesta, S. Fantini, G. Gaida, E. Gratton, H. Jess, W.W. Mantulin, M. Seeber, P.M. Schlag, M. Kaschke (1997). Frequency-domain techniques enhance optical mammography: initial clinical results. *Proc. Natl. Acad. Sci. U.S.A.* **94**, 6468–6473.

126. S. Zhou, C. Xie, S. Nioka, H. Liu, Y. Zhang, B. Chance (1997). Phased array instrumentation appropriate to high precision detection and localization of breast tumor. In: B. Chance and R.R. Alfano (Eds), *Optical Tomography and Spectroscopy of Tissue: Theory, Instrumentation, Model, and Human Studies II* (Proc. SPIE Volume 2979, 98–106). SPIE, Bellingham, WA.
127. B.W. Pogue, M. Testorf, T. McBride, U. Osterberg, K. Paulsen (1997). Instrumentation and design of a frequency-domain diffuse optical tomography imager for breast cancer detection. *Opt. Expr.* **1**, 391–403.
128. K.T. Moesta, S. Fantini, H. Jess, S. Totkas, M.A. Franceschini, M. Kaschke, P.M. Schlag (1998). Contrast features of breast cancer in frequency-domain laser scanning mammography. *J. Biomed. Opt.* **3**, 129–136.
129. S. Fantini, M.A. Franceschini, G. Gaida, E. Gratton, H. Jess, W.W. Mantulin, K.T. Moesta, P.M. Schlag, M. Kaschke (1996). Frequency-domain optical mammography: edge effect corrections. *Med. Phys.* **23**, 149–157.
130. V.E. Pera, E.L. Heffer, H. Siebold, O. Schütz, S. Heywang-Köbrunner, L. Götz, A. Heinig, S. Fantini. Spatial second-derivative image processing: an application to optical mammography to enhance the detection of breast tumors. *J. Biomed. Opt.* (in press).
131. E.L. Heffer, V.E. Pera, O. Schütz, H. Siebold, S. Heywang-Köbrunner, L. Götz, A. Heinig, S. Fantini. Near-infrared imaging of the human breast: complementing hemoglobin concentration maps with a color-coded display of hypoxic areas. *J. Biomed. Opt.* (submitted).
132. S. Fantini, E.L. Heffer, M.A. Franceschini, L. Götz, A. Heinig, S. Heywang-Köbrunner, O. Schütz, H. Siebold (2000). Optical mammography with intensity-modulated light. In: A.H. Gandjbakhche (Ed) *Proceedings of Inter-Institute Workshop on In vivo Optical Imaging at the NIH* (pp. 111–117). Optical Society of America, Washington, DC.
133. H. Jess, H. Erdl, K.T. Moesta, S. Fantini, M.A. Franceschini, E. Gratton, M. Kaschke (1996). Intensity-modulated breast imaging: technology and clinical pilot study results. In: R.R. Alfano, J.G. Fujimoto (Eds), *OSA Trends in Optics and Photonics on Advances in Optical Imaging and Photon Migration* (Volume 2, pp. 126–129). Optical Society of America, Washington, DC.
134. S.P. Gopinath, C.S. Robertson, C.F. Contant, R.K. Narayan, R.G. Grossman, B. Chance (1995). Early detection of delayed traumatic intracranial hematomas using near-infrared spectroscopy. *J. Neurosurg.* **83**(3), 438–44.
135. G. Gratton, M. Fabiani, M.A. Franceschini, S. Fantini, P. Corballis, E. Gratton (1995). Rapid changes of optical parameters in the human brain during tapping task. *J. Cogn. Neurosci.* **7**(4), 446–456.
136. A. Maki, Y. Yamashita, Y. Ito, E. Watanabe, Y. Mayanagi, H. Koizumi (1995). Spatial and temporal analysis of human motor activity using noninvasive NIR topography. *Med. Phys.* **22**(12), 1997–2005.
137. M.A. Franceschini, V. Toronov, M. Filiaci, E. Gratton, S. Fantini (2000). On-line optical imaging of the human brain with 160-ms temporal resolution. *Opt. Express* **3**, 49–57.
138. V. Toronov, M.A. Franceschini, M. Filiaci, S. Fantini, M. Wolf, A. Michalos, E. Gratton (2000). Near-infrared study of fluctuations in cerebral hemodynamics during rest and motor stimulation: temporal analysis and spatial mapping. *Med. Phys.* **27**(4), 801–815.
139. G. Gratton, A. Sarno, E. Maclin, P.M. Corballis, M. Fabiani (2000). Toward non-invasive 3-D imaging of the time course of cortical activity: Investigation of the depth of the event-related optical signal. *Neuroimage* **11**, 491–504.

140. J.H. Meek, C.E., Elwell, M.J. Khan, J. Romaya, J.S. Wyatt, D.T. Delpy, S. Zeki (1995). Regional changes in cerebral haemodynamics as a result of a visual stimulus measured by near infrared spectroscopy. *Proc. R. Soc. London B. Biol. Sci.* **261**(1362), 351–6.

141. H.R. Heekeren, H. Obrig, R. Wenzel, K. Eberle, J. Ruben, K. Villringer, R. Kurth, A. Villringer (1997). Cerebral haemoglobin oxygenation during sustained visual stimulation – a near-infrared spectroscopy study. *Philos. Trans. R. Soc. London B. Biol. Sci.* **352**(1354), 743–50.

142. R. Wenzel, H. Obrig, J. Ruben, K. Villringer, A. Thiel, J. Bernarding, U. Dirnagl, A. Villringer (1996). Cerebral blood oxygenation changes induced by visual in humans. *J. Biomed. Opt.* **1**(4), 399–404.

143. Y. Hoshi, M. Tamura (1993). Detection of dynamic changes in cerebral oxygenation coupled to neuronal function during mental work in man. *Neurosci. Lett.* **150**(1), 5–8.

144. M. Tamura, Y. Hoshi, F. Okada (1997). Localized near-infrared spectroscopy and functional optical imaging of brain activity. *Philos. Trans. R. Soc. Lond. B. Biol. Sci.* **352**(1354), 737–42.

145. Y. Hoshi, S. Mizukami, M. Tamura (1994). Dynamic features of hemodynamic and metabolic changes in the human brain during all-night sleep as revealed by near-infrared spectroscopy. *Brain Res.* **652**, 257–262.

146. Y. Hoshi, M. Tamura (1997). Fluctuations in the cerebral oxygenation state during the resting period in functional mapping studies of the human brain. *Med. Biol. Eng. Comput.* **35**(4), 328–30.

147. K. Villringer, S. Minoshima, C. Hock, H. Obrig, S. Ziegler, U. Dirnagl, M. Schwaiger, A. Villringer (1997). Assessment of local brain activation. A simultaneous PET and near-infrared spectroscopy study. *Adv. Exp. Med. Biol.* **413** 149–5.

148. M. Wolf, G. Duc, M. Keel, P. Niederer, K. von Siebenthal, H.U. Bucher, Continuous noninvasive measurement of cerebral arterial and venous oxygen saturation at the bedside in mechanically ventilated neonates. *Crit. Care Med.* **25** 1579–82.

149. C. Nolte, M. Kohl, U. Scholz, M. Weih, A. Villringer (1998). Characterization of the pulse signal over the human head by near infrared spectroscopy. *Adv. Exp. Med. Biol.* **454**, 115–23.

150. M.S. Patterson, S. Anderson-Engels, B.C. Wilson, E.K. Osei (1995). Absorption spectroscopy in tissue-simulating materials: a theoretical and experimental study of photon paths. *Appl. Opt.* **34**(1), 22–30.

151. E. Okada, M. Firbank, M. Schweiger, S.R. Arridge, M. Cope, D. Delpy (1997). Theoretical and experimental investigation of near-infrared propagation in a model of the adult head. *Appl. Opt.* **36**(1), 21–31.

152. E. Okada, D.T. Delpy (2000). Effect of scattering of arachnoid trabecule on light propagation in the adult brain. 2000 *Opt. Soc. Am.* 256–8.

153. P.E. Daubeney, D.C. Smith, S.N. Pilkington, R.K. Lamb, J.L. Monro, V.T. Tsang, S.A. Livesey, S.A. Webber (1998). Cerebral oxygenation during paediatric cardiac surgery: identification of vulnerable periods using near infrared spectroscopy. *Eur. J. Cardiothorac. Surg.* **13**(4), 370–7.

154. A.T. Lovell, H. Owen-Reece, C.E. Elwell, M. Smith, J.C. Goldstone (1997). Predicting oscillation in arterial saturation from cardiorespiratory variables. Implications for the measurement of cerebral blood flow with NIRS during anaesthesia. *Adv. Exp. Med. Biol.* **428**, 629–38.

155. P.W. Mc Cormick, M. Stewart, M.G. Goetting, M. Dujovny, G. Lewis, J.I. Ausman (1997). Noninvasive cerebral optical spectroscopy for monitoring cerebral oxygen delivery and hemodynamics. *Crit. Care Med.* **19**(1), 89–97.

156. B.C. Wilson, M.S. Patterson, S.T. Flock (1987). Indirect versus direct techniques for the measurement of optical properties of tissues. *Photochem. Photobiol.* **46**, 601–608.
157. S.J. Madsen, M.S. Patterson, B.C. Wilson, Y.D. Park, J.D. Moulton, S.L. Jacques, M.D. Anderson, Y. Hefetz (1991). Time-resolved diffuse reflectance and transmittance studies in tissue simulating phantoms: A comparison between theory and experiments. In: B. Chance (Ed.), *Time-Resolved Spectroscopy and Imaging of Tissues* (Proc. SPIE, Volume 1431, pp. 42–51. SPIE, Bellingham, WA).
158. E.G. Mc. Keating, J.R. Monjardino, D. Signorini, F.M.J. Souter, P.J. Andrews (1997). A comparison of the INVOS 3100 and the Critikon 2020 near-infrared spectrophotometers as monitors of cerebral oxygenation. *Anaesthesia* **52**(2), 136–40.
159. C.D. Gomersall, P.L. Leung, T. Gin, G.M. Joynt, R.J. Young, W.S. Poon, T.E. Oh (1998). A comparison of Hamamatsu NIRO 500 and the INVOS 3100 near-infrared spectrophotometers. *Anaesth Intensive Care* **26**(5), 548–57.
160. E.M. Nemoto, H. Yonas, A. Kassam (2000). Critical experience with cerebral oximetry in stroke and cardiac arrest. *Crit. Care Med.* **28**(4), 1052–4.
161. T. Young, M. Palta, J. Dempsey, J. Skatrud, S. Weber, S. Badr (1993). The occurrence of sleep-disordered breathing among middle-aged adults. *New England J. Med.*, **328**(17), 1230–5.
162. F. Cirignotta, M. Zucconi, S. Mondini, R. Gerardi, E. Lugaresi (1989). Cerebral anoxic attacks in sleep apnea syndrome. *Sleep* **12**(5), 400–4.
163. H. Palomaki, M. Partinen, S. Juvela, M. Kaste (1989). Snoring as a risk factor for sleep-related brain infarction. *Stroke* **20**(10), 1311–5.
164. H. Palomaki (1991). Snoring and the risk of ischemic brain infarction. *Stroke* **22**(8), 1021–5.
165. H. Palomaki, M. Partinen, T. Erkinjuntti, M. Kaste (1992). Snoring, sleep apnea syndrome, and stroke. *Neurology* **42**(7) Suppl 6:75–81; discussion 82.
166. M.R. Pressman, W.R. Schetman, W.G. Figueroa, B. Van Uitert, H.J. Caplan, D.D. Peterson (1995). Transient ischemic attacks and minor stroke during sleep. Relationship to obstructive sleep apnea syndrome. *Stroke*, **26**(12), 2361–5.
167. O. Marrone, M.R. Bonsignore (1997). Acute and chronic influences of obstructive sleep apnoea on the pulmonary circulation. *Monaldi Arch. Chest Dis.* **52**(3), 263–7.
168. S. Redline, K.P. Strohl (1998). Recognition and consequences of obstructive sleep apnea hypopnea syndrome. *Clin. Chest Med.* **19**(1), 1–19.
169. N.J. Douglas, J.M.A. Thomas (1992). Clinical value of polysomnography. *Lancet* **339**(8789), 347–50.
170. J.M. Di Fry, M.A. Phillipo, K. Curran, R. Goldberg, A.S. Baran (1998). Full polysomnography in the home. *Sleep* **21**(6), 635–42.
171. D.M. Munger, H.U. Bucher, G. Duc (1998). Sleep state changes associated with cerebral blood volume changes in healthy term newborn infants. *Early Hum. Dev.* **52**(1), 27–42.
172. B. Urlesberger, G. Pichler, E. Gradnitzer, F. Reiter, G. Zobel, W. Muller (2000). Changes in cerebral blood volume and cerebral oxygenation during periodic breathing in term infants. *Neuropediatrics* **31**(2), 75–81.
173. S. Shiotsuka, Y. Atsumi, S. Ogata, R. Yamamoto, M. Igawa, K. Takahashi, H. Hirasawa, K. Koyama, A. Maki, Y. Yamashita, H. Koizumi, M. Toru (1998). Cerebral blood volume in the sleep measured by near-infrared spectroscopy. *Psychiatry Clin. Neurosci.* **52**(2), 172–3.
174. O.G. Jenni, M. Wolf, M. Hengartner, K. Siebenthal, M. Keel, H.U. Bucher (1996). Impact of central, obstructive and mixed apnea on cerebral hemodynamics in preterm infants. *Biol. Neonate* **70**(2), 91–100.

175. T. Hayakawa, M. Terashima, Y. Kayukawa, T. Ohta, T. Okada (1996). Changes in cerebral oxygenation and hemodynamics during obstructive sleep apneas. *Chest* **109**(4), 916–21.
176. C. Hausser-Hauw, D. Rakotonanahary, B. Fleury (2000). Obstructive-sleep apnea syndrome: brain oxygenation measured with near-infrared spectroscopy. Preliminary results. *Neurophysiol. Clin.* **30**(2), 113–8.
177. V. Toronov, M. Filiaci, S. Fantini, E. Gratton (1998). Photon density wave correlation spectroscopy detects large-scale fluctuations in turbid media. *Phys. Rev. E.* **58**(2), 2288–2297.
178. A. Kleinschmidt, H. Obrig, M. Requardt, K.D. Merboldt, U. Dirnagl, A. Villringer, J. Frahm (1996). Simultaneous recording of cerebral blood oxygenation changes during human brain activation by magnetic resonance imaging and near-infrared spectroscopy. *J. Cereb. Blood. Flow Metab.* **16**(5), 817–826.
179. V.S. Mattay, D.R. Weinberger (1999). Organization of the human motor system as studied by functional magnetic resonance imaging. *European J. Radiol.* **30**, 105–114.
180. C. Hirth, K. Villringer, A. Thiel, J. Bernarding, W. Muhlnickl, H. Obrig, U. Dirnagl, A. Villringer (1997). Towards brain mapping combining near-infrared spectroscopy and high resolution 3D MRI. *Adv. Exp. Med. Biol.* **413**, 139–47.

Chapter 12

Fluorescence spectroscopy and imaging (non-microscopic)

Part I: Basic principles and techniques

Paola Taroni and Gianluca Valentini

Table of contents

Abstract

Fluorescence techniques are being used more and more both in biology and medicine, and the advent and spread of laser sources has contributed significantly to foster the applications of fluorescence, improving the performance of conventional methods and allowing the introduction of new ones. The basic principles of fluorescence spectroscopy and main fluorescence-related phenomena are outlined. Then, the most wide-spread and novel spectroscopic and imaging techniques (steady-state and time-gated fluorescence spectroscopy, fluorescence quenching, fluorescence resonance energy transfer, multi-photon fluorescence spectroscopy, fluorescence correlation spectroscopy, evanescent wave excited spectroscopy) are described, highlighting their peculiar features and reporting examples of their profitable use. Finally, several applications of fluorescence techniques are described in more detail, covering both biology (e.g. flow cytometry assays for drug discovery/screening, biosensors, DNA-chip reading) and medicine (e.g. detection of tumors and cardiovascular diseases, drug-assisted diagnostics, in vivo molecular imaging), and future trends are outlined.

12.1 Introduction to Part I

Fluorescence spectroscopy is one of the most widely applied spectroscopic techniques in biochemistry and molecular biophysics, and more generally it finds several applications in biology and medicine. These include, for example, the identification and localization of substances as well the detection of pathologies or specific physiological conditions both at a research level (e.g. diagnosis of tumors and atherosclerotic plaques) and as standard clinical routines (e.g. immuno-fluorescence tests for diagnosis of toxoplasmosis, HSV, syphilis, etc.). Such a widespread interest originates from the fact that fluorescence measurements are very sensitive to changes in the structural and dynamic properties of biomolecules and biomolecular complexes, even though they usually provide no detailed structural information. Moreover, they are absolutely non-invasive, which makes them especially appealing for in vivo medical diagnostics.

Either endogenous (i.e. naturally occurring) or exogenous (i.e. externally administered) fluorophores can contribute to the characterization of a specific system or be used for diagnostic purposes. In biological samples, most constituents exhibit fluorescence (e.g. aromatic amino acids and structural proteins). Endogenous substances offer the obvious advantage of not causing undesired or uncontrolled perturbation. However, the simultaneous presence of several contributions, most of which in the same spectral band (blue/green) can limit the sensitivity of methods based on the detection of endogenous emission. Suitable exogenous chromophores with different fluorescence properties can instead be administered to gain specificity or sensitivity to particular physico-chemical properties of the environment (e.g. pH). Exogenous fluorophores, like fluorescein and indocyanine green, are routinely used in ophthalmology [1] and in immuno-fluorescence microscopy or flow cytometry [2]. Other well known and widespread exogenous chromophores are photosensitisers, which combine tumor-localizing properties and strong fluorescence, thus being suitable as therapeutic agents in the photodynamic therapy (PDT) of tumors [3–5] and as valuable markers for diagnosis [5].

Based on the successful application of fluorescence spectroscopy in several fields, fluorescence imaging has developed and spread in use over the years, with the aim of combining the high information content of spectroscopic studies with the abundance and ease of information provided by imaging techniques. Even though the information content of 2D – and sometimes even 3D – imaging is generally somehow reduced as compared with the corresponding spectroscopic point measurements, the possibility to investigate an extended area in real time has made imaging especially appealing for in vivo applications, typically medical diagnostics.

As an introduction, the basic theory and the fundamental phenomena related to fluorescence, as well as the main tools and methodologies of the fluorescence spectroscopist are briefly outlined in the first paragraphs. A short presentation of more traditional fluorescence methods will allow us to highlight typical advantages and limits of fluorescence measurements and show how the peculiar properties of laser sources can be of use in several cases. This information will enable the reader to get a deeper insight into the following applicative sections. The most recent and interesting applications of fluorescence spectroscopy and imaging will be reviewed. This chapter does not intend to be exhaustive, but rather to present the investigative means currently offered by fluorescence, and show how technological advances open unceasingly new ways to fluorescence applications, continuously increasing its potential.

It is probably already clear that very many references can be cited for each topic. Some general references and some more specific ones are quoted in the following. Again, they are intended to provide only a starting point. Special mention, though, should be made of the extensive work of Lakowicz, a milestone in the field, pertaining to both theory [6] and applications, especially but not only in biochemistry (see the 7 volumes of the series *Topics in Fluorescence Spectroscopy*, edited by Lakowicz and published by Plenum Press over the last decade).

12.2 Basic theory

Four fundamental rules should be taken into account to understand the origin of fluorescence emission and its basic features. They can be summarized as follows.

- The Franck–Condon principle: during electronic transitions the nuclei can be regarded as stationary, since electronic reorganization accompanying a transition is much faster than nuclear motion ($\approx 10^{-15}$ s vs. $>10^{-14}$ s). So excitation occurs to vibrationally excited levels of the excited electronic state (Figure 1).
- Emission takes place from thermally equilibrated excited states, that is from the lowest vibrational level of the lowest excited singlet state (Figure 1), because relaxation from the excited vibrational levels (through emission of infrared

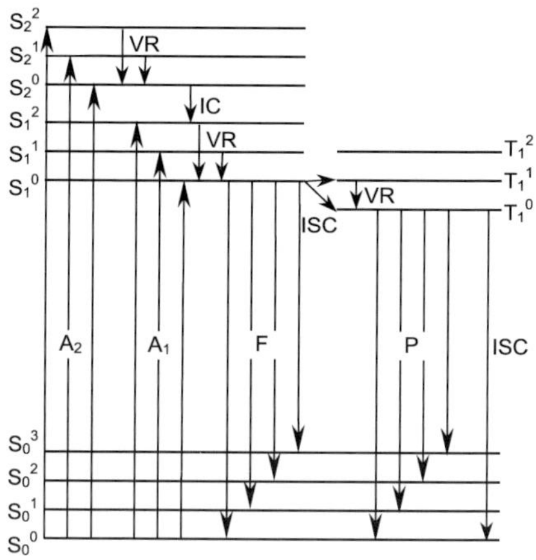

Figure 1. Jablonsky diagram showing excitation and fluorescence emission transitions. A_1 (A_2): absorption to the first (second) excited singlet state; F: fluorescence, IC: internal conversion; VR: vibrational relaxation; ISC: intersystem crossing; P: phosphorescence. ISC and P are reported for completeness, but are not described in the text.

quanta or as kinetic energy lost during collisions) is much faster than emission (10^{-14}–10^{-12} s vs. $\approx 10^{-9}$ s).

In more detail, when the higher vibrational levels of the lower electronic state overlap the lower vibrational levels of the higher electronic state (i.e. the nuclear configurations and energies of the two electronic states are identical) vibrational coupling occurs, and crossover from a higher to a lower excited singlet state is possible. This is a very fast process (10^{-12} s), known as *internal conversion*. If the vibrational levels of the two electronic states are separated, a radiative transition (*fluorescence*) may also occur from the lowest vibrational level of the higher electronic state to any vibrational levels of the lower electronic state. This happens because, even though the lifetime of the fluorescent molecule is of the order of 10^{-9}–10^{-8} s, the actual electronic transition is much faster (10^{-15} s), so that vibrational relaxation cannot take place during the transition. For biomolecules, the frequency of the emitted photon is most often in the ultraviolet (UV) or visible range. Following radiative transition, the molecule undergoes vibrational relaxation to the lowest vibrational level (10^{-14}–10^{-12} s). Which of the two deactivation mechanisms is favored depends on the number of vibrational levels and the difference in energy between the vibrational levels of the two electronic states. The closer the levels are, the more likely internal conversion will occur. This is why molecules with no rigid skeleton, such as aliphatic molecules, rarely exhibit fluorescence, while aromatic molecules, with their rigid ring structures are usually characterized by strong fluorescent emission.

For the same reason, fluorescence occurs almost always only as a deactivation mechanism between the first excited singlet state and the ground state. The higher excited states are generally much closer to each other. Hence, internal conversion is favored and precludes the possibility of fluorescence emission. Consequently, unlike the absorption spectrum, the emission line shape is made of just one band. The presence of more than one band in the spectrum of isolated molecules indicates that more than one species is fluorescing. This does not necessarily mean that different molecules fluoresce. Actually, when the emitting molecules interact with a complex environment or can aggregate, distinct fluorescence bands can be observed, corresponding to different binding sites, states of aggregation, etc.

- The Stokes shift: emission is always of lower energy than absorption (Figure 1). As described above, this is due to nuclear relaxation in the excited state.
- The mirror image rule: fluorescence spectra are mirror images of the lowest energy absorption band. The transition between the lowest vibrational level of the excited electronic state and the lowest vibrational level of the ground state belong to both the absorption and emission spectra. Actually, it is the lowest energy transition of the lowest energy band of the absorption spectrum and the highest energy transition of the emission spectrum. If the two electronic states have the same vibrational structures (i.e. the same vibrational spacings), the longest wavelength absorption band and the emission spectrum will be mirror images.

One characteristic feature of fluorescent emission is the fraction of excited molecules that fluoresce, that is the quantum yield of fluorescence ϕ_f, which can be expressed as:

$$\phi_f = \frac{k_f}{k_f + \sum k_d} = \frac{\tau_f}{\tau_f^o} \tag{1}$$

where k_f is the rate constant for fluorescence (i.e. the probability that the excited molecule will fluoresce); Σk_d is the sum of the rate constants for all radiationless deactivation mechanisms of the first excited singlet state; $\tau_f = (k_f + \Sigma k_d)^{-1}$ is the *lifetime* of the first excited singlet state, i.e. the average time the molecule spends in the excited state; $\tau_f^o = (k_f)^{-1}$ is the radiative lifetime, i.e. the average time the molecule would spend in the excited state if fluorescence were the only deactivation mechanism.

Besides ϕ_f, various other observables can be of interest in the study of fluorescent systems. These include – but do not finish with, as we will see – steady-state intensity, steady-state spectrum, steady-state anisotropy, time-resolved intensity decay (including fluorescence lifetimes) and decay-associated spectra, time-resolved anisotropy decay. Different techniques have been developed for the best assessment of the different observables, which provide distinct pieces of information on the emitting molecule and its microenvironment.

12.3 Fluorescence spectroscopy

12.3.1 Steady-state fluorescence spectroscopy

Fluorescence spectroscopy can be carried out at different levels, from simple measurements of steady-state emission intensity to quite sophisticated time-resolved studies. The information content increases dramatically as the time behavior of fluorescence observables is traced and they are combined in a global analysis of the phenomena of interest. Nevertheless, useful pieces of information can be obtained from steady-state measurements, whose technical requirements, at least in principle, are definitely modest.

The most common observables in steady-state studies include fluorescence intensity (either peak value or integrated over a selected spectral band), emission spectrum (line shape and intensity at different wavelengths), fluorescence anisotropy.

For several fluorophores, k_f (and consequently the emission intensity) is sensitive to local pH, physical factors, such as viscosity, or concentration of specific elements, such as calcium ions. Hence, fluorescence intensity measurements can be used to probe the properties of the local environment of the fluorophore. For this purpose, very often suitable exogenous chromophores are used, characterized by well-known fluorescence features and sensitive to the property of interest.

However, quantitative intensity measurements depend on local changes of fluorophore concentration, excitation intensity and detection efficiency, sample geometry and scattering. Some of these factors (such as those related to source and detection apparatus) can be controlled, while others (such as the strong scattering in biological tissues) cannot easily be accounted for. To eliminate this possible cause of errors, spectrally resolved measurements can be performed, where the line shape, not the absolute intensity value is considered. Fluorescent probes exist that modify their emission line shape in a predictable way, following specific changes in the local environment. In that case, absolute intensity measurements can be replaced by measuring relative intensity at two (or more) wavelengths and ratioing them, so as to eliminate the effects of scattering (which changes slowly with wavelength), geometry, and so on. This type of technique is especially useful when operating in vivo, such as for tumor detection. In such cases accurate absolute intensity measurements are difficult and the "sample" geometry can vary in uncontrolled way from subject to subject and even for repeated measurements on the same subject, thus affecting the results in an unpredictable or barely predictable way.

As an example, a "double ratio" technique was recently applied to the in vivo localization and staging of cervical intraepithelial neoplasia with interesting preliminary results [7]. The "double ratio" is obtained on data collected with excitation at two wavelengths (405 and 435 nm) and detection in two emission bands (550–600 and 660–750 nm), suitably chosen to enhance the difference between diseased and healthy tissue. δ-Aminolevulinic acid (ALA) is administered and favors the formation of Protoporphyrin IX (PpIX) at higher concentrations in lesions. Protoporphyrin IX is strongly fluorescent in the spectral range 660–750 nm under excitation at 405 nm, while the other excitation/emission combination (405 and 550–600 nm, respectively) enhances the endogenous fluorescence, which is usually either uniformly distributed in tissue or weaker in lesions. To estimate the "double ratio", the ratio of the red-to-yellow fluorescence (660–750 to 550–600 nm) at 405 nm excitation is divided by the red-to-yellow ratio at 435 nm excitation. This enhances the detection of lesions, also correcting for different degrees of pigmentation and geometry factors [8].

Fluorophores absorb preferentially light with the electric field parallel to the transition moment of the fluorophore. Both transition moments for absorption and emission have fixed orientations in the fluorophore. Thus, excitation with polarized light causes partially polarized fluorescence emission. Several phenomena can reduce the polarization of fluorescence, e.g. rotational diffusion and transfer of excitation between fluorophores. Consequently, in biophysical studies, an observable that can provide very useful pieces of information is fluorescence anisotropy A, defined as:

$$A = \frac{I_{\parallel} - I_{\perp}}{I_{\parallel} + 2I_{\perp}} \tag{2}$$

where $I_{\parallel}(I_{\perp})$ is the component of the emission intensity, which is parallel (perpendicular) to the electric field of the excitation light. The anisotropy is related to the correlation time τ_c for the diffusion process through the Perrin equation:

$$\frac{A_o}{A} - 1 = \frac{\tau}{\tau_c} \tag{3}$$

where A_o is the limiting anisotropy, that is the anisotropy that would be measured in the absence of rotational diffusion, and depends on the relative orientation of the absorption and emission transition dipoles. Consequently, if the lifetime has already been assessed independently, anisotropy measurements can provide hydrodynamic information on macromolecules and their complexes. Even though some complicating factors are often present (such as the irregular shapes of the macromolecules) [6], in some cases it is even possible to derive an estimate of the molecular weight. Complexation and dissociation phenomena can be investigated by monitoring the fluorescence depolarization. If changes in lifetimes are involved, time-resolved anisotropy measurements are needed.

Steady-state spectrofluorometers are commercially available with different levels of versatility and, consequently, complexity (see, for example, Refs. 9,10), and are routinely used to investigate biomolecules and biological samples. At a very basic level, they allow the measurement of the emission spectrum (fixed excitation and scan over an emission band), as well as of the excitation spectrum (scan over an excitation band and observation at fixed emission wavelength). For a more extensive characterization of the fluorescence properties, bidimensional spectral analysis can be performed, building Excitation–Emission Matrixes (EEMs), from which any excitation or emission spectra can be derived [11]. Calibration of the apparatus with quantum counters, i.e. substances of known ϕ_f and spectral behavior (e.g. Rhodamine b in ethylene glycol, or quinine sulfate in sulfuric acid) allows absolute values to be estimated in excitation/emission spectra.

Most commonly, lamps coupled to monochromators are used as excitation sources, so that light can be provided essentially at any wavelength over a broad band from the ultraviolet to the infrared, as expected from a general-purpose commercial instrument. In specific applications, especially at laboratory level, advantage may be taken of the unique properties of laser beams, especially their monochromaticity and directionality. If several fluorophores were present and just one had to be excited, monochromatic excitation would be required to avoid undesired contributions to the detected emission signal. Even though excitation and emission bands of biomolecules are usually quite broad (i.e. tens of nanometres), a change of very few nanometres in the excitation wavelength can modify dramatically the contributions of different fluorophores to the globally emitted fluorescence. This is the case, for example, for UV-excited fluorescence in human arteries, studied as a possible means for the diagnosis of atherosclerosis. Tuning of the excitation source between 308 and 312 nm moves from an emission line shape dominated by tryptophan fluorescence to a situation in which both tryptophan and structural proteins (elastin and collagen) contribute significantly [12].

Another situation in which laser sources can be useful is the measurement of fluorescence anisotropy. If lamps are used, the excitation wavelengths are usually selected with a monochromator, which causes partial polarization of the transmitted light. Consequently, if an excitation polarizer is used to select either vertical or horizontal excitation polarization, different excitation intensities are obtained, and this needs to be taken into account in the estimate of the fluorescence anisotropy [6]. The use of a laser excitation source, which is inherently monochromatic, and most often linearly polarized, eliminates this problem.

Conversely, low divergence may be of interest when the excitation light has to be coupled into fiber optics with small losses. This is often the case for in vivo measurements on the fluorescence of biological tissues or whenever the sample cannot easily be handled or brought close to the excitation source.

Advantage of laser excitation can also be taken in more sophisticated fluorescence measurements, as with fluorescence resonance energy transfer, which is described in the next section.

12.3.2 Fluorescence resonance energy transfer

Fluorescence resonance energy transfer (FRET) can be exploited as a very powerful means for obtaining dynamic and structural information on macromolecules and molecular complexes. Excitation energy can be transferred from a donor to an acceptor molecule, when they are separated by no more than 70–100 Å. The donor then deactivates with a radiationless transition, while the acceptor can return to its ground state either emitting fluorescence or through a radiationless pathway. The probability and consequently the rate k_{ET} of resonance energy transfer depends markedly on the distance R between the centers of the two molecules:

$$k_{\mathrm{ET}} = \frac{1}{\tau_{\mathrm{d}}}\left(\frac{R_{\mathrm{o}}}{R}\right)^{6} \tag{4}$$

where R_{o} is the distance between the two molecules for which donor fluorescence and energy transfer to the acceptor are equally probable (Förster distance), and τ_{d} is the lifetime of the donor in the absence of the acceptor.

Other factors affect the probability of energy transfer, such as the spectral overlap between donor emission and acceptor absorption, and the relative orientation of the two chromophores (more precisely the relative orientation of their transition dipoles). Thus, FRET provides insight not only about intermolecular distances, but also about the dynamic distribution of distances, and more generally it can give structural information [13,14]. A typical example is the study of protein folding/unfolding, which can be performed at single molecule resolution using FRET [15].

FRET can also be used to investigate self-association and aggregation. For example, dimerization can be detected marking single molecules with either an acceptor (A) or a donor (D). Dimers will be of three types: AA (non-fluorescent), AD (fluorescent), and DD (twice as fluorescent as AD). Taking this into account, the efficiency of dimer formation can be derived from fluorescence measurements, as the fluorescence F relative to the fluorescence F_o in the absence of acceptor is [16]:

$$\frac{F}{F_o} = 1 - Ea \tag{5}$$

where E is the energy transfer efficiency and a is the fraction of molecules that are labeled with the acceptor. If only dimers are formed, the fluorescence intensity is expected to decrease linearly upon increasing the fraction of acceptor-labeled molecules. If self-association led to the formation of trimers or tetramers, a faster fluorescence decrease would be foreseen [6].

One original application of FRET, of increasing interest in recent years, is in the development of molecular beacons. Molecular beacons are oligonucleotides that can recognize the presence of specific nucleic acids in solutions [17–19]. They are usually long molecules, including a quenched fluorophore (see Section 12.3.3), whose fluorescence recovers when hybridization occurs, that is only when the correct nucleotides are recognized, and no mismatch or deletion is present. In more detail, the beacon contains a sequence that is complementary to the target nucleic acid molecule. A fluorescent dye is covalently linked at one end, while a quencher is present at the opposite end. In the initial conformation, the beacon is similar to a loop. Fluorophore and quencher are in close proximity, which minimizes the emitted fluorescence by energy transfer. Upon hybridization, the probe undergoes a conformational change (stretching from a loop to a straight shape) that greatly separates fluorophore and quencher, with a significant increase in the emitted signal. Thus the florescence intensity is a measure of the hybridization efficiency. Further use of molecular beacons is described in Section 12.8.5.

FRET finds most of its applications in fluorescence microscopy, recently also coupled to time-resolved detection (see Section 12.3.5) (this topic is treated in Chapter 13, devoted to optical microscopy).

12.3.3 Photobleaching and fluorescence quenching

In the previous section, attention was drawn to the possibility of reducing (more precisely "quenching") or increasing the emitted fluorescence intensity by energy transfer. Let us now consider more in detail the different phenomena that may lead to a reduction – either reversible or irreversible – in fluorescence intensity.

In principle, the process of excitation and fluorescence emission could be repeated indefinitely in the same molecule. In practice, in some situations and conditions, the fluorescence intensity reduces over time, even if the excitation intensity is kept constant. This phenomenon is called *photobleaching* or *photodestruction*. Frequently, it occurs when the excited state is chemically less stable than the ground state. Molecules that are particularly reactive in the excited state can undergo chemical reactions, leading to loss of fluorescence. These reactions often involve singlet oxygen that can be formed by energy transfer from an excited state of the fluorescent molecule (thus decaying into the first singlet state) to the triplet state of oxygen (forming singlet oxygen). Singlet oxygen is extremely reactive and can easily oxidize the excited fluorophore, thus preventing fluorescence emission. If singlet oxygen is involved, the addition of antioxidants or operation under anoxic conditions can effectively reduce the photobleaching effects.

Photobleaching is often also proportional to (or at least dependent on) the intensity of the excitation light. Thus, weak excitation intensity can significantly reduce photobleaching. This has accounted for especially when collimated laser beams or focussed excitation light is used to increase the detectable signal emitted from weakly fluorescent species.

Fluorescence quenching occurs when the reduction in the emitted intensity is due to (i) intermolecular factors, an external environmental influence, such as the interaction with an external quencher or (ii) intramolecular factors, such as a change of configuration or a substitution.

An external influence can affect either the excited state, causing non-radiative deactivation (*dynamic* or *collisional quenching*), or the ground state, preventing the generation of the excited state through formation of a nonfluorescent complex (*static quenching*) [6].

Both dynamic and static quenching require "contact" between fluorophore and quencher. In collisional quenching, this has to occur during the lifetime of the excited state, meaning that the quencher has to diffuse to the excited fluorophore. Hence, solvent viscosity, reducing diffusion, also reduces quenching. Quenching studies can thus provide information on diffusion rates. They can also be performed to investigate the accessibility and more generally the location of fluorophores, for example in proteins or membranes. The quencher must be chosen so that the membrane is impermeable to it. Then, the fluorescence is unaltered if the fluorophore is located deeply inside the membrane, while it is expected to be increasingly quenched the closer the fluorophore is to the membrane surface. Quenchers with different penetration capacity can also be exploited to better investigate the fluorophore localization [20].

In the simplest case of collisional quenching, when a single reaction causes quenching, the dependence of the reduction in fluorescence intensity F on the quencher concentration $[Q]$ can be described by the Stern–Volmer equation:

$$\frac{F_{\mathrm{o}}}{F} = 1 + K_{\mathrm{D}}[Q] \tag{6}$$

where F_{o} is the fluorescence intensity in the absence of quencher and K_{D} is the Stern–Volmer quenching constant. Interestingly collisional quenching contributes to (non-radiative) depopulation of the excited state. Thus, it is expected to alter the fluorescence lifetime τ of the fluorophore compared with the lifetime τ_o when no quencher is present. More specifically:

$$\frac{\tau}{\tau_o} = \frac{F}{F_o} = \frac{1}{1 + K_D[Q]} \tag{7}$$

When static quenching occurs, the situation can be described analytically by a relationship that closely resembles the Stern–Volmer equation:

$$\frac{F_o}{F} = 1 + K_S[Q] \tag{8}$$

where the quenching constant K_S is the association constant for the formation of the non-fluorescent fluorophore–quencher complex.

As evident from the above equations, it is not possible to discriminate dynamic from static quenching only based on steady state fluorescence intensity measurements. Other elements need to be considered, such as the dependence of dynamic quenching on viscosity or temperature, which increases the diffusion coefficient, thus favoring quenching. Alternatively, time-resolved fluorescence measurements (Section 12.3.5) can be performed, to assess if the fluorescence lifetime depends on the quencher concentration. As described above, dynamic quenching reduces the observed lifetime, introducing an additional mechanism of depopulation of the excited state, while static quenching removes a fraction of fluorescent molecules forming non-fluorescent complexes, but does not affect the lifetime of molecules decaying radiatively.

Also, quenching due to intramolecular factors can effectively be exploited. An application of increasing interest is the production of molecular beacons, where quenching and energy transfer within the same molecule are both exploited to modulate the fluorescence intensity and recognize the presence/absence of a specific species/condition (Section 12.3.2). Another typical application of intermolecular quenching is the study of protein folding at single-molecule level. Each protein is labeled with several fluorophore molecules. In the folded state, the close proximity between fluorophores causes quenching. Unfolding leads to a remarkable increase in fluorescence intensity, as the quenching efficiency reduces with increasing distance between intramolecular fluorophores [21].

Finally, *self-quenching* occurs when quenching of an excited atom or molecule is caused by interaction with another atom or molecule of the same species in the ground state. Self-quenching is often evidenced by a nonlinear relationship between fluorophore concentration and detected fluorescence signal, where instrumental or other causes of nonlinear response have been excluded. Self-quenching occurs primarily in fluorescent dyes with a small Stokes' shift (Section 12.2). A small Stokes' shift means a significant overlap between absorption and emission spectra. So the fluorophores absorb photons at several of the same wavelengths at which they emit, and photons emitted by a given dye molecule may be absorbed by others before they are detected. Self-quenching is obviously favored by high fluorophore concentrations and by any phenomenon that reduces the randomness of molecule distributions (e.g. tendency to aggregation, immobilization on a surface, multiple labelling). In practice, fluorophores with high quantum yield are usually less affected by self-quenching, because they provide a high enough fluorescence signal even at low concentrations.

12.3.4 Multi-photon fluorescence spectroscopy

The peak power readily achieved with femtosecond laser pulses is high enough to induce nonlinear effects in many materials and, in particular, in biological samples. Multi-photon absorption occurs when two or more photons are absorbed simultaneously (within a femtosecond time scale) by the same chromophore and they all contribute to determining the transition from the ground state to the first excited singlet state. Once the molecule is excited, deactivation follows exactly the same rules as considered for single-photon excitation (Introduction).

Such nonlinear effects require excitation with significant power densities. Consequently, the focussed beam of a short pulsed laser source (typically a mode-locked titanium:sapphire laser) is usually exploited. The efficiency of two-photon excitation depends on the square of the excitation intensity, and the intensity itself decreases approximately as the square of the radial distance from the focus. Thus photons are absorbed essentially only in the focal point. Out of focus, the intensity is not high enough for two-photon absorption, and fluorophores present in biological media are usually not effectively excited at long wavelengths. Clearly, the argument holds even better when more than two photons contribute to excitation. This has two positive consequences. First of all, the undesired background fluorescence signal is minimized. In this respect, similar results could also be achieved with a confocal arrangement. However, in the latter case fluorescence photons are generated even out of focus, but they are removed from the detection path, while multi-photon excitation does not elicit fluorescence outside the focal spot. Thus, multi-photon fluorescence spectroscopy can be exploited to study living cells or other specimens with a very limited damage due to photobleaching. As mentioned above, the energy for excitation of fluorescence in biomolecules usually corresponds to UV or blue-green radiation. In biological tissues, these wavelengths have penetration depths in the range of hundreds of microns to a few millimetres. However, light attenuation strongly reduces on going from the UV to the NIR. Consequently, if two infrared photons are used to excite the fluorescence, structures lying deeper in biological media can be investigated.

Since its introduction [22], multi-photon fluorescence excitation has been beneficially exploited for several biological applications, from conventional to more recent and advanced ones, such as the study of neuronal structure and signaling in living animals [23]. Multi-photon excitation has also been coupled to various fluorescence techniques. These obviously include time-resolved fluorescence spectroscopy (Section 12.3.5), since multi-photon excitation is generally achieved with mode-locked laser sources, which are optimal for time-resolved studies. Significant advantages can also be obtained using multi-photon excitation in fluorescence correlation spectroscopy (Section 12.3.6) and evanescent wave induced fluorescence spectroscopy (Section 12.3.7). Notably more information can be obtained by exploiting the same laser beam as used for multi-photon fluorescence excitation also for harmonic generation by multi-photon scattering processes. This can profitably be applied to investigate species with large anharmonic modes, such as collagen [24]. Finally, multi-photon fluorescence spectroscopy is, notably, generally performed by taking advantage of the optical paths of fluorescence microscopes. A detailed description and discussion of the technique can be found in Chapter 16.

12.3.5 Time-resolved fluorescence spectroscopy

As mentioned previously (see Equation (1)), the probability of radiative deactivation of an excited state is strictly related to its lifetime. The study of the time behavior of the emitted fluorescence can thus provide useful insight into the properties of the emitting molecule. In more detail, the fluorescence lifetimes can provide effective means of discrimination among fluorophores. Fluorescent biomolecules generally emit over broad spectral bands. Thus, when more than one fluorophore is present, the emission line shapes often overlap and cannot be separated by steady-state studies, while the time behaviors are usually distinct.

A further interesting feature is the remarkable sensitivity of the decay time to the microenvironment, which allows one to obtain useful information on the localization, binding site, and heterogeneity of the examined fluorophores. This can be especially appealing for the study of biomolecules.

In principle, the time behavior of the emitted fluorescence can be acquired operating in the time domain or dually in the frequency domain. However, the equivalence is true if data are collected at any frequency. In practice, this is definitely not the case, as measurements are usually carried out at a single frequency or a few frequencies at most. Consequently, studies performed in the time domain are expected to provide a higher informative content and they will be described in the following. Detailed treatment of frequency-resolved fluorescence measurements can be found elsewhere [6,25,26].

The abundance of information obtained from time-resolved data as compared with steady state ones comes from considering more aspects of a complex problem. To this purpose, more sophisticated and, until recently, significantly more expensive instrumentation have usually been exploited. This has considerably limited the spread of time-resolved studies in the biomedical community. However, later technological developments have changed the situation. A pulsed laser as the excitation source is required for time-resolved measurements, but now short pulsed diode lasers are available for excitation at several wavelengths from the blue to the red, and a single PC board allows the acquisition of time distributions (see, for example, [27,28]), which previously required the use of several complex and expensive pieces of electronics for time-correlated single-photon counting (TCSPC) (e.g. specifically optimized amplifiers, constant fraction discriminators, time-to-amplitude converters, multi-channel analyzers).

Complex and expensive equipment based on mode-locked laser sources is still needed when high time resolution or excitation at well-defined wavelengths, or tunable over a broad range, are required. Despite rapid and continuous improvements though, current commercial diode lasers provide light pulses with duration of not less than 50–100 ps, and even more at certain wavelengths. Moreover, the spectral width is never below a few nanometres, and only a limited set of central wavelengths is presently available. Similarly, concerning detection, metal package photomultiplier tubes (PMTs) are now available, combining the advantages of low cost, very compact size and low-voltage power supply, but the highest temporal resolution still requires bulky and expensive microchannel plate PMTs.

When several fluorophores are present in a sample, time-gated fluorescence spectroscopy, that is time-resolved measurements (i.e. the acquisition of the fluorescence waveforms) at several wavelengths over a broad spectral range, can provide useful information on the contributions of individual molecular species to the overall emission spectrum. The description of the instrumentation developed at Politecnico di Milano for time-gated fluorescence spectroscopy [29] is reported below to outline the basic principles behind the technique, including TCSPC. Several ultraviolet and visible excitation wavelengths can be obtained with the following sources: a mode-locked argon laser (wavelengths between 351 and 514 nm and pulse duration of 90–120 ps); a mode-locked krypton laser (wavelengths between 413 and 647 nm and pulse duration of 70–90 ps); synchronously pumped dye lasers (Rhodamine 110 and DCM continuously tunable in the range of 530–600 and 600–700 nm, respectively, with pulse duration of few picoseconds). To allow the measurement of lifetimes longer than few nanoseconds, the pile-up of subsequent fluorescence decay waveforms must be avoided. To this purpose, the repetition rate is usually reduced with an external pulse picker (for gas lasers) or a cavity dumper (for dye lasers) from ≅70 MHz to a suitable value, depending on the fluorescence lifetimes to be measured. When the sample is a solution or suspension, it is contained in a quartz cuvette (1 cm pathway) and the fluorescence emitted at 90° (to minimize contributions from direct and scattered laser light) is collected through a scanning monochromator and detected by a double-microchannel plate PMT. For microscopy applications, the laser beam is coupled to the excitation path of a fluorescence microscope attachment for epi-illumination and the monochromator and detector are placed at its exit port.

Fluorescence decay waveforms and spectra are obtained through an electronic chain for TCSPC. This technique is most often the best choice to detect low intensity time-dependent light signals. Its main advantages are the high sensitivity and the good dynamic range. In general, its time resolution is determined by the duration of the excitation pulse and/or by the jitter of the detection apparatus, which can hardly be lower than 30 ps. The time resolution is lower than achievable with a streak camera (i.e. few picoseconds), but TCSPC provides a significantly wider dynamic range, linear time scale, and much smaller noise/background problems.

TCSPC is based on the measurement of the delay between the excitation photon ("start" signal) and the fluorescence photon ("stop" signal). A small percentage of the excitation beam is split off and detected by a fast photodiode. Both the PMT and the photodiode signals are suitably amplified and sent to constant-fraction discriminators (CFDs), which convert the input pulse into an output signal of fixed shape and amplitude, independently of the input amplitude. This improves the accuracy in determining the exact times of excitation and photon emission. CFD outputs are suitably delayed by means of cables and used as the start and stop inputs for a time-to-amplitude converter (TAC). The TAC output is a voltage signal of amplitude proportional to the time delay between start and stop. The use of the PMT signal as the start input allows one to reach higher counting rates and profit by the high repetition rate of the laser pulses. Once a start signal arrives, the TAC waits for a stop signal. If none arrives, after a full time scale, the TAC resets and a new operation can start. The number of excitation signals is much higher ($\sim 10^2$ times) than that of fluorescence signals. If the excitation is used as a start signal, the TAC will spent a lot of time in waiting for a stop signal, which in most cases will not arrive in time. So, the "dead time" is significantly reduced if the TAC operation starts upon arrival of fluorescence photons. The conversion factor (time to voltage) of the TAC can be varied, and its value fixes the measurement time-scale. This allows one to choose the most suitable time-scale for the system to be studied. The TAC output is sent to a multichannel analyzer (operated in the Pulse Height Analyzer mode), which counts the number of TAC outputs whose amplitude falls within the voltage intervals 0–ΔV, ΔV–$2\Delta V$, and so on up to a full scale value. The output of the multichannel analyzer approximates the probability distribution of delays between start and stop events. The smaller ΔV is the better is the approximation. If the photon rate is sufficiently low, this probability distribution corresponds to the fluorescence decay waveform [30]. For a fixed number of channels, the wider is the time-scale, the wider is ΔV, thus implying a reduced accuracy in assessing small time differences. The measured curves are then transferred to a personal computer for multi-exponential fitting with a nonlinear technique, and for the evaluation of the gated spectra. When PC boards for TCSPC are used, integrated electronics replace discrete modules, but the same operations as just described are performed.

A time-gated spectrum is obtained collecting only photons emitted within a selected time-window, characterized by its width W and delay D from the excitation pulse. Fluorescence decays are recorded subsequently at any wavelengths, by scanning the monochromator under computer control. With dedicated software, the gate parameters (W and D) can be selected and the corresponding gated spectrum evaluated by adding, at each wavelength, all the decay counts within the time-gate. A proper choice of width and delay allows one to obtain useful information on the fluorescence features of systems characterized by various lifetimes. In particular, with a narrow gate, shorter than the shortest decay time, and undelayed with respect to the excitation pulse, the contributions of the distinct components are proportional to their initial amplitudes, thus providing information on the relative abundance of the emitting species. In contrast, when the delay is comparable with the lifetime of the longest-living species, its contribution to the spectrum is enhanced with respect to any one else, and sometimes can even be isolated. An example is shown in Figure 2, displaying the fluorescence waveform due to contributions of two species with lifetimes of 0.5 ns (relative amplitude 90%) and 1 ns (relative amplitude 10%). A 100 ps wide, undelayed gate reflects the initial amplitudes of the two emitting species (90% and 10%), while a window open between 5.5 and 6 ns will collect

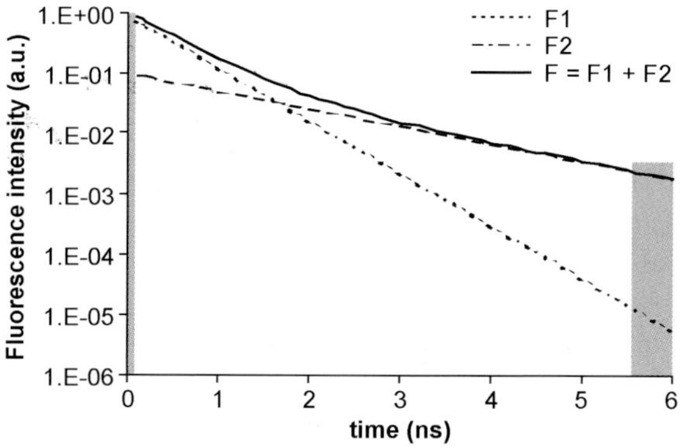

Figure 2. Effect of gate width and delay on detected fluorescence for a waveform due to the contributions of two species: F1 with lifetime of 0.5 ns and relative amplitude of 90%, and F2 with lifetime of 1 ns relative amplitude of 10%.

2–3 orders of magnitude more light from the long-living species, making the other contribution negligible even though its abundance is 9 times higher. Finally, an integrated spectrum is obtained by counting all the TAC outputs at each wavelength. If the time-scale is properly chosen (i.e. essentially no photons are lost, even at long times) it corresponds to a steady-state fluorescence spectrum [29].

The estimate of fluorescence lifetimes is most commonly obtained by interpreting the time-resolved fluorescence waveform at each wavelength λ as a linear combination of a discrete number of exponential functions, each one representing an emitting species with a different lifetime:

$$F(\lambda, t) = \sum_i A_i S_i(\lambda) \exp\left(-\frac{t}{\tau_i}\right) \tag{9}$$

where A_i is the peak amplitude, $S_i(\lambda)$ is the spectral distribution and τ_i the lifetime of the i-th fluorescent component. The best fit of experimental data is achieved by minimizing χ^2 with linear or nonlinear procedures. The latter can provide more accurate estimates, but, if the initial values are too far from the real ones, the process can lead to a secondary minimum, yielding a low χ^2, but not the "correct" lifetimes. To prevent this the fitting process usually implies two subsequent steps: a linear fit is initially performed to provide reasonable guesses of lifetimes and amplitudes to be used subsequently as initial values for a nonlinear least-squares fitting procedure [31].

Fluorescence lifetimes are especially sensitive to the microenvironment. Consequently, a strongly heterogeneous fluorophore environment can cause local changes in lifetimes, which cannot adequately be described by a limited number of discrete lifetimes. A continuous distribution of lifetimes is needed in such cases:

$$F(t) = \int_0^\infty \exp\left(-\frac{t}{\tau_i}\right) \rho(\tau)\, d\tau \tag{10}$$

A stretched exponential function, also known as the Kohlrausch–Williams–Watts function, can be used to represent the continuous distribution of lifetimes [32]:

$$F(t) = F_o \exp\left[-\left(\frac{t}{\tau}\right)^{\frac{1}{h}}\right] \tag{11}$$

where $h \geq 1$ is a heterogeneity parameter. For $h = 1$ Equation (11) degenerates to a mono-exponential decay, corresponding to a homogeneous medium, while the higher the heterogeneity parameter, the wider is the spectrum of lifetimes required to represent the heterogeneity of the sample.

Another interesting possibility for data interpretation relies on the use of maximum entropy methods (MEMs) [33,34], which usually provide a distribution of lifetimes, but can also lead to the estimate of discrete lifetimes. From a mathematical point of view, an inverse problem needs to be solved to interpret a data set in terms of a function. Difficulties arise because the data constitute a limited set and are noisy. Consequently, many different functions are equally consistent with the data set, all providing fits of comparable goodness (i.e. χ^2 value). MEMs extract the most uniform function from the set of feasible ones. Maximizing the entropy imposes no correlation for which there is no evidence in the available data, and correspondingly chooses the function that is minimally structured. To ensure that the recovered distribution agrees with the data, the procedure is subjected to the simultaneous constraint of a decrease in the reduced χ^2. The choice corresponds to the most conservative function that describes the data at a certain χ^2. Different definitions of "entropy" are possible. The Shannon entropy is an example in quite common use. For a random set $\boldsymbol{x}$of N discrete values x_k, each with a probability $p_k = P(x_k)$ and $p = \{p_1, \ldots, p_N\}$, where p is a proper distribution, the Shannon entropy is

$$H(x) = -\sum_k p_k \log(p_k) \tag{12}$$

where 0 log (0) is defined as 0 [35]. In contrast with the conventional multi-exponential analysis, there is no a priori assumption about the shape of the distribution of fluorescence lifetimes. As an example, MEMs can fit the experimental data with the combination of a very large number of lifetimes (e.g., 100–200). Parameter estimation is conceptually straightforward, but the use of MEMs is not yet quite widespread, because they can require considerable computational resources.

As mentioned earlier, fluorescence lifetimes are very sensitive to the fluorophore microenvironment, which make them observables of special interest to investigate heterogeneous environments, local changes in physico-chemical properties, fluorophore localization and binding site, and so forth. Some typical applications of time-resolved fluorescence spectroscopy, such as the discrimination between dynamic and static quenching phenomena, have already been cited (Section 12.3.3). Time-resolved fluorescence spectroscopy can also provide information on dynamic phenomena that are not observed in the stationary state. An example is the energy transfer between molecules of Hematoporphyrin Derivative (HpD, a photosensitizer for PDT [3]) in distinct states of aggregation. When HpD is systemically administered, usually intravenously for PDT, it interacts with the biological substrate and the environment can significantly influence its therapeutic effectiveness. Excited molecules of HpD deactivate either radiatively or by energy transfer, leading to the formation of singlet oxygen, which then oxidizes the tumor tissue. The study of the radiative deactivation as a function of the microenvironment can thus provide useful information on the photosensitizer properties. The drug solution for intravenous injection is similar to a phosphate buffer solution, while a simple model of cell membrane is obtained with cationic detergent (cetyltrimethylammonium bromide, CTAB) solutions above the critical concentration (i.e. when micelles simulating cell membranes are formed). The steady-state fluorescence spectra in these two environments (buffer and 1 mM CTAB) are shown in Figure 3. Upon interpreting the fluorescence waveforms with discrete exponential functions, three lifetimes are detected, of approximately 0.7, 3, and 15 ns. In both conditions, the emission is dominated by the long-living (15 ns) species, but the 615 nm peak observed in buffer corresponds to monomers and oligomers in an unfolded configuration, while the red-shift to 630 nm in CTAB solution is attributed to the hydrophobic environment that causes de-aggregation and unfolding of the end rings of polymeric chains. A more complex, but interesting situation is observed at low CTAB concentrations, in pre-micellar conditions (Figure 4). The steady-state emission is peaked around 660–680 nm (polymeric HpD material interacting with CTAB), with a secondary peak at 630 nm (HpD molecules interacting with CTAB), and only a shoulder at 615 nm (HpD molecules non-interacting with CTAB). With a narrow (0.2 ns) undelayed gate, the contribution at 615 nm is completely absent, while emissions at 615 and at 630 nm are comparable. If all species fluoresce independently of one another, contributions to a narrow undelayed gate should be proportional to the initial amplitudes, thus providing information on relative abundance of the different species. A second gate was also considered, 6 ns wide and delayed by 18 ns with

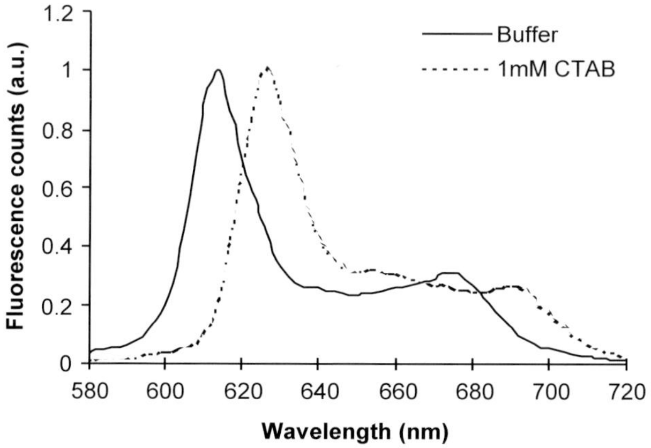

Figure 3. Normalized time-integrated fluorescence spectra of HpD in buffer (solid) and in 1 mM CTAB solution (above critical micelle concentration) (dotted).

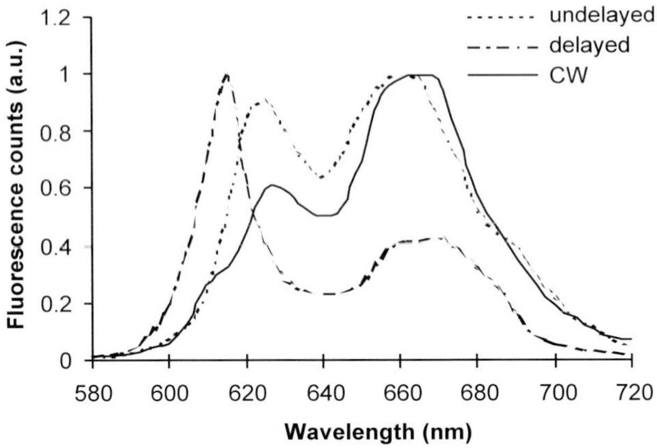

Figure 4. Normalized fluorescence spectra of HpD in 0.005 mM CTAB solution (below critical micelle concentration): time-integrated spectrum (solid), undelayed (0.2 ns width) gated spectrum (dotted), and 18 ns delayed (6 ns width) gated spectrum (dash-dot).

respect to the excitation pulse, to collect essentially only the contribution from the long-lived species. The spectrum measured with the delayed gate peaks at 615 nm, with a non-negligible emission at 660–680 nm. Such emission is characterized by an intermediate lifetime (3 ns), and should not contribute significantly to data collected after 18–24 ns. Thus, the gated spectra cannot be interpreted simply in terms of relative abundance of independent molecular species. The presence of energy transfer from the species emitting at 630 nm to the one at 660 nm can explain the experimental data. Such an hypothesis is confirmed also by the complete absence of the 630 nm (long-living) emission in the delayed spectrum. This observation could not be performed based on steady state measurements, and may be relevant for in vivo applications. In fact, the energy transfer, by altering the relative concentrations of the different molecular species, may alter the interactions of the photosensitizer with the biological substrate and consequently its therapeutic effectiveness.

12.3.6 Fluorescence correlation spectroscopy

If a very small volume (femtolitres) is considered, the probability is that, at any given time, few fluorescent molecules, or even a single one, are present. Over time, molecules will diffuse into and out of the volume, and while inside it they will emit photons. Small molecules, with larger diffusion constants, will enter and leave more frequently than large molecules, but they will stay inside (and emit photons) for shorter times.

When fluorescence correlation spectroscopy (FCS) is performed, fluctuations $\delta F(t)$ of the fluorescence intensity with time around an equilibrium value $\langle F \rangle$ are statistically investigated. For this purpose, the normalized autocorrelation function

$G(t_c)$ is evaluated:

$$G(t_c) = \frac{\langle F(t+\tau_c)F(t)\rangle}{\langle F\rangle^2} = \frac{\langle \delta F(t+\tau_c)\delta F(t)\rangle}{\langle F\rangle^2} + 1 \qquad (13)$$

where τ_c is the correlation time and $\delta F(t) = F(t) - \langle F\rangle$ [36–38]. The autocorrelation function, as its name implies, provides a measure of the self-similarity of the signal over time, and thus estimates the persistence of the information the signal is carrying. Diffusion, local concentrations, triplet formation, quenching mechanisms, chemical reactions, conformational changes occurring during the residence time of molecules in the observation volume, and any other processes leading to statistical fluctuations in the fluorescence signal will induce a characteristic decay of the autocorrelation function. Their presence can thus be recognized by analyzing the time behavior of $G(t)$. Dynamic light scattering also quantifies concentration fluctuations, but the detection sensitivity is considerably improved if fluorescence intensity is chosen as an observable. Fluctuations can be better observed the smaller the number of molecules under investigation. Thus, another advantage of FCS is that it can operate at very low concentrations (nanomolar or even lower), which is difficult with other techniques.

Laser beams can be focussed down to sub-micron sizes in radial direction, but no confinement can be achieved along the beam axis. Thus, to restrict the observation volume, very thin sample holders or "two-dimensional" samples, such as membranes, are needed. This would strongly limit the possibility to perform measurements inside living cells. Thus, to achieve a small enough observation volume, FCS is usually performed in a confocal arrangement, focussing a laser beam down to the resolution limit. The confocal set up is extremely versatile. Most commonly it has been applied to molecules free to move in aqueous solutions, including cellular environments, but it can also be used to investigate molecules fixed on solid surfaces or embedded in solid volumes. Several FCS intracellular applications have been reported [39], but the technique has never become very widespread, because of technical difficulties. The advent of two-photon excitation proved to be a key step. In fact, the excitation volume is inherently smaller (about 0.1 fL) and allows an improved signal quality, especially in turbid samples, like biological ones. In particular, FCS with two-photon excitation has successfully been applied to investigate the intracellular environment, also confirming the typical advantage of two-photon excitation (reduced photobleaching, easier separation of excitation and emission wavelengths, etc.) [40].

12.3.7 Evanescent wave induced fluorescence spectroscopy

If fluorophores are distributed within the sample volume, but only the contribution from surface or close-to-surface molecules is of interest, evanescent waves can provide an interesting means of investigation.

The technique exploits total internal reflection of the excitation light at the sample surface. Thus, an external medium with index of refraction n_e higher than the sample n_s is needed. Usually glass is used for this purpose. A laser beam is skimmed along the internal surface of glass at such an angle that total internal reflection occurs. An evanescent wave propagates outside the glass, i.e. into the sample, but its intensity decays exponentially with depth. The penetration depth depends on both n_s and wavelength, but is typically in the range of 100–150 nm for biological media and visible light. Thus only superficial fluorophores are excited. This makes fluorescence excited by evanescent waves a highly sensitive probe of interfacial environments, which has been widely applied in biochemistry [41] since its introduction a few decades ago [42]. As an example, it can effectively isolate the fluorescence contribution coming from cell membranes, minimizing the background signal due to cytosol, which lies too deep inside to be reached by the evanescent wave.

In particular, the technique has been profitably used to develop fluorescence-based sensors for the rapid detection of small molecules (e.g. antibiotics, pesticides, pharmaceuticals) at very low concentrations levels, which can be of interest in medical diagnostics as well as environmental and food quality control. In this context, a system for the rapid detection of antibiotic residues in milk using disposable chips is under development within the framework of the European Project BIODAM. A recognition layer is immobilized on the sensor surface and competition for antibody binding takes place between the antibiotic residue to be detected in the sample and a known quantity of fluorescently labeled antibiotic added to the sample. The fluorescent signal obtained is high when no antibiotic is present in the sample and low for contaminated samples. From a technical point of view, a laser beam is coupled into a waveguide with a grating. The evanescent wave excites the bound fluorophores and the emitted light is coupled back into the waveguide and out-coupled by the grating. The intensity of the detected signal is proportional to the number of fluorophores on the waveguide surface, thus the fluorescent signal is high for low concentration of analyte.

In general, interfaces provide a heterogeneous environment for molecular species, so that photophysical properties (e.g. fluorescence quantum yield) depend on the precise location of individual fluorophores within this environment. Conventional steady state evanescent wave excited fluorescence spectroscopy can already provide useful information, but additional selectivity is required to obtain a deeper insight into the behavior at the interface. This can be achieved through time-resolution (Section 12.3.5). Variations in the photophysical properties of interfacial species can be detected through changes in fluorescence decay profiles, and the spatial distribution of interfacial species, reflecting the heterogeneity of the environment, can be mapped more closely [43].

A further step is made if evanescent wave induced fluorescence spectroscopy is performed with two-photon [44] or more generally multi-photon excitation. As described previously (Section 12.3.4), the nonlinear nature of multi-photon excitation implies a better confinement of the excited volume. This enhances the selectivity of evanescent wave induced fluorescence spectroscopy, limiting the elicited signal to molecules very close to the surface.

Finally, evanescent wave excitation is very profitably applied in fluorescence microscopy (see Chapter 13), where eliminating out-of-focus fluorescence offers up to a fivefold enhancement of axial optical sectioning compared with confocal and two-photon microscopy [45].

12.4 Fluorescence imaging

As well as the huge field of fluorescence microscopy, addressed in Chapter 13, there are strong motivations both in clinical practice and in molecular biology for moving from fluorescence point measurements to imaging. For what concerns medical diagnosis, the direct visualization of those parts of the body that are under investigation provides immediate information on lesion morphology and extension, is certainly not achievable with point measurements. As mentioned in the Introduction, fluorescence imaging is widely applied in ophthalmology for the diagnosis and staging of retinal pathologies (fluorangiography) [1], and is increasingly used to detect tumors (Sections 12.8.1 and 12.8.2). In molecular biology, fluorescence imaging primarily means the parallel acquisition of thousands of independent measurements at once, thus speeding up the tremendous tasks of DNA sequencing and gene expression profiling. This issue will be considered in Section 12.7.4.

Even though endogenous fluorophores can also be exploited in fluorescence imaging, mainly for tumor detection [46], a definitely better selectivity can be achieved through specific markers showing well-defined fluorescent properties. This is the case for fluorescein or indocyanine green, routinely used in ophthalmology [1], and photosensitizers used for tumor detection [3,5,47]. Markers are also strictly required in all genomic and proteomic procedures.

From a technical point of view, fluorescence imaging can be performed in a straightforward way by illuminating uniformly the area of interest and collecting the fluorescence with a bidimensional detector, such as a CCD video camera (eventually cooled and intensified for the detection of very weak signals). This is the most common option for steady-state measurements (Section 12.4.1), even though a second approach can be considered, based on performing point-measurements and scanning over the area of interest, so as to build a bidimensional matrix of fluorescence data. Especially when sophisticated information has to be recovered, like lifetime or spectral data, or a very high spatial resolution is required, like in DNA microarray reading, this can be an appealing alternative to CCD imaging, even though it usually implies longer measurement times. For fast measurements, fluorescence lifetime imaging can be performed with an intensified CCD video camera whose gain factor is phase modulated (Section 12.4.3) or activated for a very short time synchronously with a pulsed excitation laser (Section 12.4.4). Spatial maps of the fluorescence lifetimes and amplitudes are then reconstructed using suitable image processing algorithms.

12.4.1 Steady-state fluorescence imaging

Steady-state fluorescence imaging is widely used since it requires simple, cheap instrumentation. Distinct fluorophores are discriminated on the basis of differences in their spectra: peak position, shape, intensity (integrated or in a selected band). However, as described previously, visible or near-infrared fluorescence spectra are usually quite broad and spectral overlapping often makes it difficult to distinguish between fluorophores. To improve the specificity of steady-state fluorescence techniques, the operation at multiple excitation and/or emission wavelengths has been suggested (similar to that described for spectroscopy in Section 12.3.1) [48–50]. In this case, images taken at different excitation and/or emission wavelengths are subtracted one from another [51]. A further advantage is obtained if a division instead of a subtraction is performed [52–54]. In fact, in this case not only the selectivity can increase, but the resulting diagnostic parameter is also independent of tissue geometry and spatial profile of the excitation beam.

As an example application of the image ratio technique let us consider the set-up originally developed in Lund, Sweden, for the detection of tumors after sensitization with porphyrins administered either systemically or topically [55–57]. Multi-color imaging is carried out by simultaneously recording three spatially identical, but spectrally different images with the same intensified CCD video camera, using a system developed for endoscopic applications. The images are spectrally separated by means of dichroic mirrors. The first (named A) is in the red region (580–750 nm), the second (B) in the blue region (420–480 nm), and the last (D) in the green-yellow region (480–580 nm). The fluorescence is induced by means of pulsed ultraviolet light (390 nm) from a frequency-doubled Alexandrite laser. The excitation light is delivered by an optical fiber and the fluorescence light is collected through a laryngoscope. The gated image intensifier is synchronized with the laser. A gate width of 500 ns is used to efficiently suppress the room light. Thereby, a normal reflected light image can be recorded simultaneously using a color CCD video camera. Fluorescence images are digitized in a frame grabber and stored on a PC. Calculations are subsequently carried out on the digitized images. The function:

$$F_c = \frac{A - k_1 D}{k_2 B} \tag{14}$$

with A, B, and D as defined above, and k_1 and k_2 being constants with different values for different applications, is used to produce images with optimized contrast between the lesions and the adjacent normal tissue. The PC computes F_c pixel by pixel and feeds the result to a video mixer, where the fluorescence image can be mixed with the normal image from the color CCD camera.

This imaging system has been applied for the detection of neoplastic lesions of the skin, breast, head and neck region and urinary bladder [56,57].

Multi-color imaging is also widely used in confocal scanners devoted to DNA microarray reading (Section 12.7.4).

12.4.2 Time-resolved fluorescence imaging

As extensively shown in Section 12.3.5, the fluorescence lifetime can provide effective means of discrimination among fluorophores. A further interesting characteristic is the sensitivity of the lifetime to the microenvironment and to other important physiological parameters like pH, calcium concentration etc. Conversely, dyes designed to mark specific cellular structures or other targets often show a relatively stable lifetime, which can be considered a distinctive feature of the fluorophore. Therefore, the capability of measuring the lifetime maps may allow one to discriminate different markers to evaluate the distribution of the corresponding targets in the sample under investigation. These and other features make Fluorescence Lifetime Imaging (FLIM) one of the most successful spectroscopic techniques to have been developed in recent years. The practical feasibility of fluorescence lifetime imaging was demonstrated in the early 1990s [58,59] thanks to the availability of fast microchannel plate light amplifiers. These devices allowed scientists to develop instruments to measure the 2D map of the fluorescence lifetime in an extended sample [60–63]. Since then, FLIM has been applied to several fields, including combustion analysis [64], fluorescence microscopy [65,66], tumor detection [67–69], photosynthesis studies [70], capillary electrophoresis [71], DNA sequencing [72] and DNA-chip reading [73].

As shown in Section 12.3.5, assuming the sample under investigation to be a mixture of non-interacting fluorophores, the impulse response function, i.e. the response to a very short excitation pulse, can be described as:

$$F(t) = \sum_i A_i \exp\left(-\frac{t}{\tau_i}\right) \tag{15}$$

where A_i is the amplitude of a generic fluorescence component and τ_i is the lifetime of the corresponding excited state. For each fluorescence component, A_i is the intensity immediately after excitation and is proportional to the concentration of the fluorophore in the mixture.

Indeed, truly multi-exponential systems can hardly be dealt with by fluorescence lifetime imaging because of both the small number (5–10) of time samples and the limited dynamic range and signal-to-noise ratio that characterize FLIM systems. Yet, in many cases the behavior of the fluorescence emission can be assumed to be mono-exponential or bi-exponential. In the former case, one looks for the spatial distribution of the average lifetime. In the latter case, two main tasks can be addressed: (a) a fluorescence component can be enhanced with respect to the other, e.g when an unspecific background emission has to be removed; (b) the relative amplitudes of the two fluorescence components can be estimated, thanks to a discrimination mechanism provided by the lifetime.

The measurements of the fluorescence lifetime maps in a sample can be carried out using two alternative approaches that work either in the frequency or in the time domains. The "frequency" or "modulation" method was developed first since it requires simpler instrumentation; the "time" or "pulsed" method followed shortly afterward. Both are presently capable of picosecond resolution and are successfully applied in many disciplines, in particular in biology and medicine.

12.4.3 Frequency-domain fluorescence imaging

This measurement method relies on excitation light sinusoidally modulated at high frequency, typically in the range 10–100 MHz. As a consequence, the fluorescence emission, which represents the forced response of the system, is modulated at the same frequency. In an optical system sensitive to light intensity the stimulus signal is the superposition of DC and sinusoidal components, because the photon flux is an intrinsically positive quantity (Figure 5). Because of the lifetime exhibited by any fluorescent compound, the modulated emission is delayed in phase by an angle ϕ relative to the excitation. In addition, it is more attenuated than the steady-state (DC) emission, since the sample acts as a low pass filter. Consequently, the modulation depth M, i.e. the amplitude of the sinusoidal component relative to the DC component, is smaller for the emission (F_ω/F_0 with reference to Figure 5) than for the excitation (E_ω/E_0). By measuring the change in modulation depth and the phase delay from input (excitation) to output (fluorescence) at different frequencies one can extract the time constants of the system using a fitting procedure.

However, if the measurements are carried out at a single frequency, as it is usually the case in fluorescence imaging, by looking either at the phase shift or at the change in modulation depth, one gets two different estimations (τ_p and τ_m, respectively) of the average lifetime ($\bar{\tau}$). In the ideal case of mono-exponential decay, τ_p and τ_m are equal and correspond to the actual lifetime, while in the more general case of multi-exponential decay they are weighted averages of the fluorescence decay times τ_i, and they usually differ. In fact, it can be demonstrated that τ_p and τ_m provide an underestimate and an overestimate of $\bar{\tau}$, respectively [6].

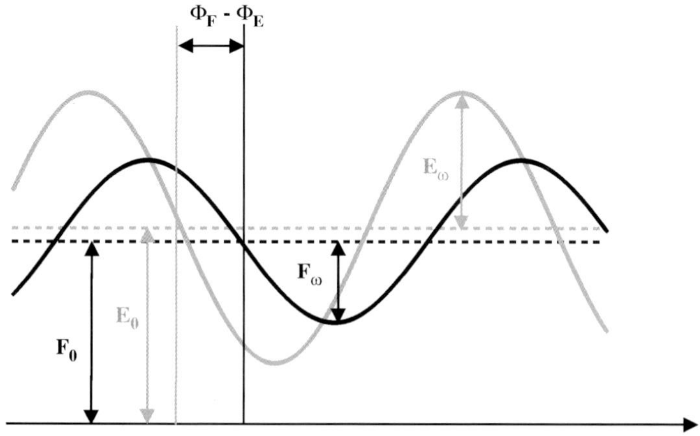

Figure 5. Time profiles of the excitation light (–) and the fluorescence emission (–) in a typical frequency domain fluorometer. E_0 and F_0 indicate the amplitude of the DC components of the excitation and the fluorescence, respectively, while E_ω and F_ω represent the amplitudes of the modulated components. The phase shift between the two curves is also shown ($\phi_\mathrm{F} - \phi_\mathrm{E}$).

While the complete theory of the phase modulation technique can be found elsewhere [6,74,75], the basic principles of fluorescence lifetime measurements are briefly described here.

Let us consider a sample showing a mono-exponential emission:

$$F_{i,j}(t) = A_{i,j}\exp\left(-\frac{t}{\tau_{i,j}}\right) \tag{16}$$

where subscripts i,j refer to image pixel. When the excitation light is sinusoidally modulated at angular frequency ω:

$$E(t) = E_0 + E_\omega \cos(\omega t) \tag{17}$$

the fluorescence $F_{i,j}(t)$ is given by [75]:

$$F_{i,j}(t) = F_{0;i,j} + F_{\omega;i,j}\cos(\omega t + \phi_{i,j}) \tag{18}$$

The signal is made of a DC component ($F_{0;i,j}$) and a sinusoidal component, which oscillates at the same angular frequency as the excitation light, with a phase delay $\phi_{i,j}$.

For a fluorescence sample exhibiting a mono-exponential decay with time constant $\tau_{i,j}$, the phase delay $\phi_{i,j}$ between the emission and the excitation light is given by [75]:

$$\phi_{i,j} = \arctan(\omega\tau_{i,j}) \tag{19}$$

while the modulation depth is reduced by a factor $m_{i,j}$ between excitation and emission:

$$m_{i,j} = \frac{F_{\omega;i,j}/F_{0;i,j}}{E_\omega/E_0} \tag{20}$$

The factor $m_{i,j}$ is also related to the fluorescence lifetime $\tau_{i,j}$ by

$$m_{i,j} = \frac{1}{\sqrt{1 + (\omega\tau_{i,j})^2}} \tag{21}$$

Therefore, for the very simple case of a mono-exponential decay, the measurement of either $\phi_{i,j}$ or $m_{i,j}$ allows one to calculate the lifetime matrix $\tau_{i,j}$ in two alternative ways.

The phase shift and the modulation depth of the fluorescence signal are usually determined using the homodyne technique [76], which transforms high frequency signals into DC signals. Basically, the responsivity *R(t)* of the detector is modulated at the same frequency as the excitation light *E(t)* and the output is integrated to give a DC signal, which is measured as a function of the phase shift between *E(t)* and *R(t)*. Using a MCP light intensifier coupled to a CCD camera, the fluorescence emission is demodulated by the former device and integrated by the latter one. The measurement is a parallel process carried out over all the pixels of the CCD and a lifetime image can be easily calculated. In practice, a sequence of N images of the sample is acquired while the phase difference between *E(t)* and *R(t)* varies from 0 to 2π in equally spaced intervals. Using these images, the demodulation factor $m_{i,j}$ in each pixel as well as the phase shift $\phi_{i,j}$ can be measured. Thereafter, the lifetime map $\tau_{i,j}$ is calculated using either Equation (19) or Equation (21).

The typical instrumentation required to set-up a FLIM system includes a CW laser emitting in the UV or in the visible bands, like an argon laser, an acousto-optical or an electro-optical modulator and a fast image intensifier coupled to a CCD camera. An RF frequency synthesizer, which drives the optical modulator and the light intensifier, and a phase shifter, which provides the variable phase lag between the two devices, complete the system. A crucial point for successful measurements is the phase stability of all the devices, which can be achieved with careful temperature control. The basic apparatus can be coupled to any imaging instrument, like a microscope or an endoscope.

Several applications of fluorescence lifetime have been successfully demonstrated either in biology or medicine. For example, a lifetime fluorescence system has been used to measure the rates of deactivation from the excited state of chlorophyll *a* in leaf samples and single cells. Since transitory changes in the overall fluorescence efficiency of chlorophyll *a* take place in a fraction of a second, an instrument capable of rapidly acquiring lifetime-resolved fluorescence images has been used [70]. FLIM systems for microscopy based on frequency modulation techniques are currently operated in different laboratories [77] and a somewhat unique instrumentation for real time FLIM in endoscopy has been set-up and clinically tested by a research group in Lousanne, Switzerland [78]. The system requires an argon laser for excitation and works at video rate (25 Hz). It shows false color images of the inspected region of the body. Moreover, to allow the physician performing the examination to perceive the morphology of the tissue, the image intensity is proportional to the fluorescence amplitude, while the hue parameter codes the lifetime. The system has been used to explore the bronchi tree in order to detect tumors that are can hardly be discovered by visual inspection.

12.4.4 Time-domain fluorescence imaging

In the time domain approach, the fluorescence lifetime map of a sample is measured using short excitation pulses and a gated detector that acquires only the fluorescence emission falling within a definite time interval. The detector of choice is again a fast light intensifier coupled to a CCD camera. In this case, short voltage pulses drive the photocathode, while the gain is kept constant. In such a way, photoelectrons generated at the photocathode are prevented from entering the MCP for amplification, except during short time intervals (gates) properly delayed with respect to the excitation pulses.

The standard procedure relies on the acquisition of a series of N fluorescence images taken by progressively increasing the delay of the acquisition gate (Figure 6). The experimental details and the algorithms that can be used to measure the lifetime and amplitude maps of one or more fluorescence components (typically up to three) depend on the performances of the image intensifier, which is the main unit of any FLIM system. In fact, while most of the devices currently available are

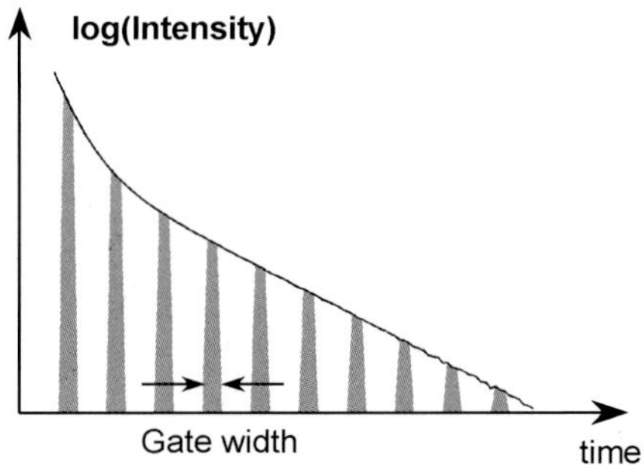

Figure 6. Time domain sampling of a bi-exponential fluorescence emission on a logarithmic scale. The gate, provided by the light intensifier, transforms the continuous intensity curve into a sequence of discrete fluence samples.

capable of a minimum time gate of 3–5 ns, much faster devices, exhibiting a minimum time gate down to 80–100 ps, have been developed only recently. Typical fluorescence decay times of organic compounds fall between few hundreds of picoseconds and several nanoseconds. A fast detector is mandatory for applications that require a precise measurement of subnanosecond lifetimes, while more conventional devices are certainly suitable whenever an estimate of the lifetime in the range of few nanoseconds is required, as it is often the case in clinical diagnosis. Accordingly, different measurement techniques and data processing algorithms tailored for these two families of applications (picoseconds or nanoseconds) have been devised.

For simplicity we will initially consider, once again, a fluorescent sample showing a mono-exponential decay (Equation 16). In the time domain approach the excitation light consists of a sequence of short pulses, having a full-width at half-maximum that depends on the application: around 1 ns for nanosecond measurements and a few picoseconds or less for faster measurements. If the pulse width is well below the decay time τ of the sample, the map of the fluence measured by the CCD when the light intensifier is activated for a gate that starts after a delay d and lasts a time interval w, is given by

$$H_{i,j}(d) = C\int_{d}^{d+w} F_{i,j}(t)\mathrm{d}t = CA_{i,j}\tau_{i,j}\left[\exp\left(-\frac{d}{\tau_{i,j}}\right) - \exp\left(-\frac{d+w}{\tau_{i,j}}\right)\right] \quad (22)$$

where C is a system-dependent constant. Taking the logarithm of Equation (22) gives

$$\ln[H_{i,i}(d)] = \ln(CA_{i,j}\tau_{i,j})\ln\left[1 - \exp\left(-\frac{w}{\tau_{i,j}}\right)\right] - \frac{d}{\tau_{i,j}} \quad (23)$$

This equation shows that, for any pixel (i,j), a linear relationship holds between the logarithm of the fluence $H_{i,j}$ and the delay d, being $\tau_{i,j}$ the inverse of the slope. Then, under the assumption of mono-exponential decay, the lifetime matrix $\tau_{i,j}$ can be easily measured by taking just two fluorescence images after different delays d_1 and d_2 and using the formula:

$$\tau_{i,j} = \frac{d_2 - d_1}{\ln[H_{i,j}(d_1)/H_{i,j}(d_2)]} \tag{24}$$

From Equation (22), a straightforward calculation shows that, once $\tau_{i,j}$ has been found, the map of the relative amplitude $A_{i,j}$ can be obtained from one of the gated images, e.g the one delayed by d_1 (shorter delay), which collected a stronger signal:

$$A_{i,j} = \frac{H_{i,j}(d_1)}{C\tau_{i,j}} \exp\left(\frac{d_1}{\tau_{i,j}}\right)\left[1 - \exp\left(-\frac{w}{\tau_{i,j}}\right)\right]^{-1} \tag{25}$$

Eventually, the matrix $A_{i,j}$ can be normalized to remove the dependence on the instrumental parameter C.

The acquisition of only two images allows one to measure the lifetime map in real time, since image grabbing and processing take a fraction of a second. Nevertheless the instabilities of the laser source and the stochastic gain factor of the image intensifier limit the precision that can be achieved to few percent of the actual lifetime. Thus, whenever possible, even for a mono-exponential decay, it is better to acquire a series of N images ($N = 5$–10). Then, Equation (23) can be easily transformed into a linear system of N equations in the unknown physical quantities A and τ. A least-squares fitting procedure can be used to recover the fluorescence lifetime and amplitude. More robust results are achieved in the multi-image method with respect to the two-image one, thanks to the intrinsic redundancy of data. In fact, the stochastic noise averages out, at least partially, in the extra images of the set, leading to cleaner and more reliable lifetime and amplitude maps.

Fluorescent systems can barely be modeled as truly mono-exponential, due to the ground state heterogeneity of many fluorophores or to the simultaneous presence of several fluorescing molecules in the same specimen, as it often happens in biological samples. In these cases the fluorescence emission after a short excitation pulse can usually be described by Equation (15), which represents a multi-exponential decay. Actually, the fit of a multi-exponential curve is a solved problem that challenges any data acquisition system and analysis algorithm [79]. A precise resolution of a fluorescence signal having more than three exponential components requires many data points, tremendous time stability of the experimental apparatus and a high signal-to-noise ratio, whose actual value depends on the differences in the lifetimes and relative amplitudes of the components to be distinguished. Even though state of the art TCSPC systems (Section 12.3.5) provide measurements with the accuracy required to resolve up to five/six exponential decays, the multi-exponential problem can barely be dealt with using imaging systems. However, in many cases only two main fluorescence components need to be considered. This is quite common in medical diagnosis: a fluorescent dye is used to mark pathological tissues and its emission has to be distinguished from the unspecific background tissue fluorescence. In other applications two markers are used to label targets that are both present in the sample to be investigated. In that case, the emissions of the markers have to be evaluated separately to measure their relative concentration, as in gene expression profiling [80–82]. For these applications, Eq. (15) can be simplified to:

$$F(t) = A_1 \exp\left(-\frac{t}{\tau_1}\right) + A_2 \exp\left(-\frac{t}{\tau_2}\right) \tag{26}$$

In this case three independent parameters have to be determined in each pixel, i.e. the lifetimes τ_1 and τ_2 and the relative amplitude A_1/A_2. A complete solution requires a different approach with respect to the mono-exponential case considered before. In fact, time-consuming iterative algorithms based on a nonlinear least-squares fit must be applied (Section 12.3.5) [83]. For real-time measurements, which are often required in medical diagnosis, the fast processing methods already described for the mono-exponential are used to calculate an effective lifetime τ_{eff} in any pixel. Notably τ_{eff} differs from the average lifetime $\bar{\tau}$. In fact, the effective lifetime τ_{eff} depends on the choice of the acquisition parameters, i.e. the delay and the width of the gates. Nevertheless, in many important cases (Section 12.8.2 for an example) the effective lifetime allows one to easily locate the regions of the fluorescent sample where the concentration of a specific marker is higher.

If a long measurement time can be afforded, many gated images (up to a few hundred) can be acquired with a gate width as short as possible, in an operating regime that resembles a TCSPC measurement (Section 12.3.5). Obviously, this can be done only if the laser source and the experimental apparatus provide enough long-term stability and low jitter. These requirements are met using mode-locked lasers and last generation picosecond image intensifiers. The huge data set contained in the image series can be processed using standard procedures that have already been developed to fit decay curves acquired with TCSPC systems. Several nonlinear algorithms, including Gauss–Newton and Levenberg–Marquardt [31] methods, are already coded and can be effectively applied with minor modifications [84]. Using a workstation based on the latest microprocessors, two maps of lifetimes τ_1 and τ_2 having 128×128 pixels can be calculated from 100 delayed images in a typical processing time of 20 min.

In biological samples there are many situations in which one does not expect a limited number of discrete lifetimes. This is the case, for example, for a fluorophore in a mixture of solvents, such that a range of fluorophore environments exists. Then, the time behavior of the emission is described by a continuous distribution of exponential decays, as already described in Section 12.3.5.

Finally, in many applications the estimate of lifetimes is not the main object of the investigation, but the capability of measuring them offers a powerful means to discriminate amongst different fluorophores. This is the case when the emission is due to $N \geq 2$ markers, whose lifetimes are known and relatively stable. Then, N unknown parameters are represented by the fluorescence amplitudes $A_{m;i,j}$. They can be easily calculated using a fast algebraic algorithm. If interactions amongst fluorophores, like energy transfer or reabsorption, are negligible then the maps of the relative amplitudes $A_{m;i,j}$ lead to the concentration of the fluorophores in any point of the image, which is a valuable information in many practical cases. Under these conditions, the lifetime provides a discrimination mechanism alternative to the spectral one. Some advantages can be pointed out: first of all a single excitation wavelength is required, then one can acquire a set of lifetime images much wider than the number of the unknowns. The redundancy intrinsic in this approach improves the reliability of the results, while the processing time is limited thanks to the use of an algebraic algorithm for data fit. An example of measurements where these concepts have been exploited will be given in Section 12.7.4.

Part II: Biological and medical applications

Laura Marcu

Table of contents

12.5 Introduction to Part II

The use of fluorescence spectroscopy and imaging to characterize biological materials has been widely explored for several decades. The fluorescence signal remitted upon the interaction of excitation light with biological samples contains information regarding the biochemical content, metabolic status, microenvironment, and structure of the investigated sample. This information can be further used for characterization, diagnosis and monitoring of complex biological systems such as tissues and cells. The fluorescence emission of biological samples originates from either endogenous or exogenous fluorescent molecules. Measurements of fluorescence can be conducted (1) in vitro on individual cells or cell culture, (2) ex vivo on tissue specimens removed from humans or animals, and (3) in vivo on cells and tissues in living organisms. While fluorescence-based techniques for in vitro and ex vivo investigations are well-developed and widely applied to investigations of biological systems (e.g. fluorescence microscopy or flow cytometry), the use of fluorescence spectroscopy and imaging for in vivo studies of biological systems and clinical diagnostics is an evolving field.

12.6 Fluorescent probes

12.6.1 Endogenous or intrinsic fluorophores

Several endogenous or naturally occurring fluorophores (Figure 7, [85]) are the origin of biological tissue and cell fluorescence. Among these, five broad groups can be distinguished: aromatic amino acids, structural proteins, enzyme co-factors involved in cellular metabolism, lipopigments and porphyrins.

The main contributors to protein fluorescence are the aromatic amino acids tryptophan, tyrosine, and phenylalanine [86,87]. These fluorophores are characterized by high absorbance at wavelengths below 300 nm and emission in 300–400 nm (tryptophan), 270–350 nm (tyrosine) and 260–300 nm (phenylalanine). The quantum yield of tryptophan (0.14) and tyrosine (0.13) are relatively high when compared to that of phenylalanine (0.02). The amino-acid lifetimes range from 1 to 6 ns depending upon local environment, and this dependency can be exploited for research and diagnostic purposes.

Tissue fluorescence is associated with the structural or extracellular matrix proteins elastin and the collagens. Using chromatographic techniques, the fluorescence of the collagen family of molecules has been attributed to various amino acids and amino acid crosslinks [85,86,87–95]. Several fluorescent crosslinks have been identified including, for elastin, desmosine, isodesmosine and triaminopyridinium derivative, and, for collagen, pyridinoline and pentosidine. Both elastin and collagen have a broad range of absorption spectra from below 300 to above 400 nm. Emission spectra vary with excitation wavelength. For instance, collagen peak emission ranges from 310 nm (excitation 280 nm) to 520 nm (excitation 450 nm). Several components of collagen absorb in the far-UV (aromatic amino acids: phenylalanine, tyrosine), mid-UV (carbonyl components,

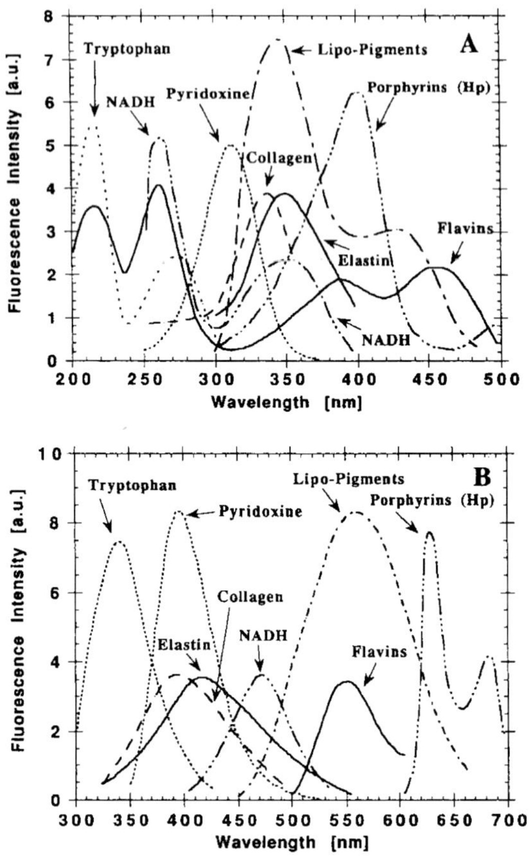

Figure 7. Fluorescence excitation and emission spectra of endogenous tissue fluorophores (from G.A. Wagnieres et al. (1998). *Photochem. Photobiol.*, **68**, 603–632).

aldol condensation crosslinks, keto-amine cross-links) and near-UV (pyridinoline). Consequently, the fluorescence spectra reflect the emission of particular crosslinks that are most efficiently excited within a given UV range [88,93]. Moreover, crosslinks are species-dependent. Differences in fluorescence emission have been identified between, on the one hand, collagen in bovine Achilles tendon, cartilage, bone, dentine and vitreous humor and, on the other hand, collagen in rat tail tendon, skin and cornea [88]. Fluorescence lifetimes also vary with collagen type and/or source [95,96]. Representative time-resolved spectra of the structural proteins elastin and collagen type I from tendon are depicted in Figure 8. Figure 9 depicts the steady-state spectra of several types of mammalian collagens.

The fluorophores involved in cellular metabolism are two enzyme co-factors, the nicotinamide adenine dinucleotide (NADH) and the flavin adenine dinucleotide (FAD) [85–87]. Only the reduced form of NADH and the oxidized form of FAD are fluorescent. Their fluorescence has been described in detail by Lakowicz [87]. NADH absorbs in the 300–390 nm wavelength range and emits in the 400–600 nm

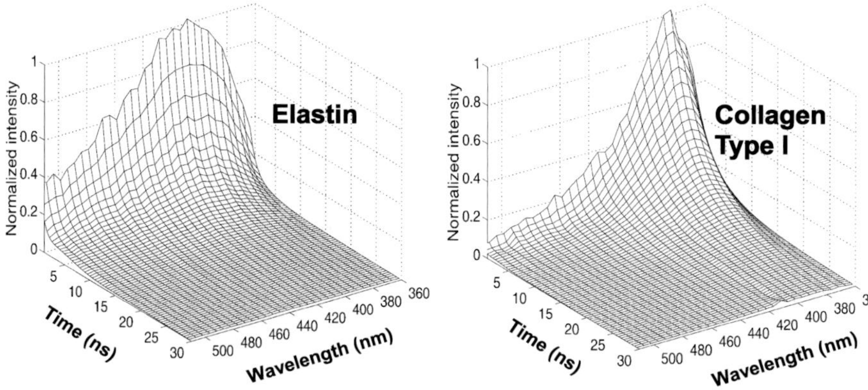

Figure 8. Examples of time-resolved spectra of elastin and collagen type I from bovine Achilles tendon. Adapted from Refs: 179 (L. Marcu et al. 2003)

range (lifetime ~0.4 ns in aqueous buffer). Upon binding to protein, the quantum yield of NADH is reported to increase about four-fold and the lifetime to 1.2 ns. FAD has absorption in the visible range (400–500 nm) and emits around 525 nm. The different spectral and temporal fluorescence behavior of oxidized and reduced forms provides an effective means of determining, non-invasively, the redox state of these physiologically important fluorophores.

Lipopigments in tissue, the end-products of lipid metabolism, have been associated with pathological processes (atherosclerosis, retinal degeneration) and aging. Fluorescent lipopigments include ceroid and lipofuscin. Both absorb light in the UV range and have an intense fluorescence emission in the 450–650 nm range [85,86].

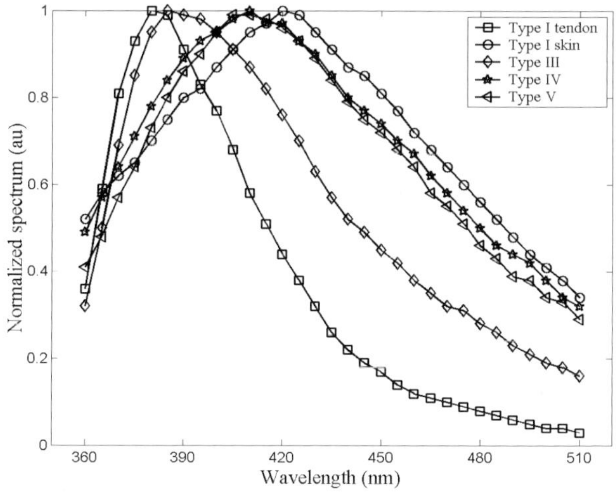

Figure 9. Steady-state fluorescence emission spectra of collagens. Commercial samples of collagen type I from bovine Achilles tendon and calf skin, and placental collagens type III, type IV and type V. Adapted from Ref. 95 (L. Marcu et al. SPIE ...)

Porphyrins are a ubiquitous class of compounds with many important biological representatives, including hemes and chlorophylls. They consist of four pyrrole rings joined by methene bridges. This class of fluorophores absorb visible light, 400–450 nm and have two main emission peaks, around 630 and 690 nm [85,86].

Pathological or physiological changes in tissues and cells result in associated molecular transformations. The various alterations of these tissue fluorophores can be correlated by analyzing the changes of the relative contribution of each fluorescent constituent to the overall fluorescence emission. Owing to the dependence of its properties on the biochemical and histological characteristics of biological cell and tissues, endogenous fluorescence therefore represents an important approach to cell and tissue investigations.

12.6.2 Exogenous or extrinsic fluorophores

Exogenous fluorescent probes are added to investigated samples to provide fluorescence signatures either when biological processes or structures cannot be probed on account of very weak or no intrinsic fluorescence of the biological sample, or when a change of spectral and/or temporal properties of the sample is needed. Numerous exogenous probes for biomedical diagnostics and monitoring have been developed over the past two decades. Comprehensive reviews are presented elsewhere [85–87,97–99]. Several broad classes are distinguished. These include (a) proteins, membranes, organelles and DNA probes, (b) chemical sensing probes, (c) NIR and IR dyes, and (d) special probes such as fluorescent proteins, e.g. Green Fluorescent Protein, (GFP), lanthanides, inorganic nanoparticles, and photosensitizers. Typically, the probes need to be efficiently excited with a specific laser line, while the fluorescence is measured within a narrow emission band. When compared with endogenous fluorophores, exogenous probes have higher quantum efficiency and specificity. They can, however, be subject to instability due to environmental conditions and photobleaching.

Exogenous probes are commonly used in a broad range of cellular spectroscopy and imaging applications, including probing the cellular environment (ionic, pH), cellular interactions, cell sorting, and tracking of drug–cell interaction. The most common exogenous fluorophores for tissue investigations have been developed mainly as PDT photosensitizers (hematoporphyrin derivative, benzoporphyrin derivative, hypericin, δ-aminolevulinic acid) [85,86]. Other probes for in vivo studies include indocyanine green (ICG), GFP, NIR probes, and lanthanide chelates [85,86,98,99]. The pharmacokinetic properties and toxicity of exogenous probes, however, play a crucial role in their applications to in vivo tissue investigations.

12.7 Biological applications

12.7.1 Flow cytometry

Flow cytometry is an optical technique for the measurement of relative size, granularity or internal complexity of single cells within a heterogeneous population of cells [100,101]. The technique involves the analysis of fluorescence and light scatter properties of particles within cells (nuclei, chromosomes) during the single-file passage of those cells through a flow stream and past optical and/or electronic sensors. Using suitable fluorescent probes, it is possible to measure not only structural but also functional properties of cells, such as levels of gene expression and cytoplasmic and mitochondrial membrane potential. Samples are stained with fluorescently-labeled antibodies, with organelles or with molecular probes. These then bind to certain cellular components, though which depends on the specific application.

A typical flow cytometer consists of several basic components, including a light source (typically a laser), a flow chamber and optical system, and photodetectors and processors to convert light signals into analog electrical signals. A schematic representation of the components outlined above is presented in Figure 10. The cytometer identifies the particles by examining the fluorescence emission and scattered light when they pass through the laser beam. Particles are confined to the center of a laminar flow stream, which is intersected by the incident beam in the interaction region. With the aid of specifically designed optics, the light emitted by particles passing through the interaction region is collected in the forward and perpendicular directions relative to the illumination beam. A single-element photodiode detects the intensity of the forward-scatter signal while a photomultiplier-tube monitors the fluorescence at 90°. Fluorescence emission can be measured using both steady-state and time-resolved techniques. Important steps in data acquisition and analysis include, (1) calibration of the instrument, typically using fluorescently labeled polystyrene or latex microbeads, (2) spectral compensation to correct for fluorescent probe spectral overlap, and (3) gating, an analytical process that selects for further analysis those cells with a specific set of characteristics.

Flow cytometry allows simultaneous high-speed quantitative measurements of several different diagnostic parameters in very small volumes. It permits precise characterization of multiple cell subpopulations in complex mixtures, and it can

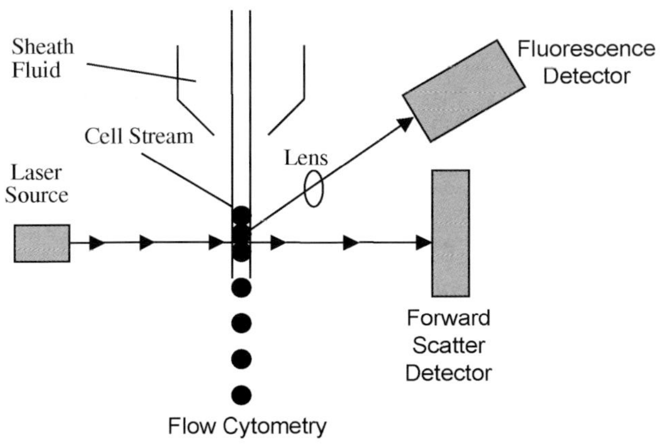

Figure 10. Schematic of flow cytometry process.

sort cells that meet specific criteria. Sorting allows cells with pre-selected characteristics to be diverted from the main flow stream and be collected for further analysis. Cell sorting can be accomplished by two means, electrostatic and mechanical. In electrostatic sorting, the cells of a specific type, after passing the laser interrogation point, are charged and electrostatically deflected to a collection point. This sorting method can be operated at rates of up to 50,000 cells per second. Mechanical sorting employs a mechanical gate that directs a particular type of cell into a desired pathway. This method allows sorting at a maximum rate of only 300 cells per second. However, it is considered to be gentler on cells than the electrostatic method [100,101].

Flow cytometry is widely used in both research (e.g. cell function analysis, multiplexing immunoassays, signal transaction pathways, measurement of gene expression) and in clinical studies (e.g. DNA analysis for tumor ploidy and SPF, HIV monitoring, leukemia and lymphoma immunophenotyping, organ transplant monitoring). Additional information regarding flow cytometry applications can be found at a number of websites [102,103].

12.7.2 Assays for drug discovery/screening

The screening of a drug compound activity against a target (biochemical or cell-based assays) is one of the most important tasks in the relatively new field of "drug discovery". Screening takes place in several stages of the drug discovery process [104–106]. A schema of the latter is presented in Figure 11. The process includes "primary" and "secondary" screening and requires that investigators be able to perform a broad range of assays with widely varying throughput and complexity requirements. "Primary screening" requires the ability to perform assays in a high throughput fashion to allow for the processing of a large number of assays in a single day (typically more than 10,000 data points/day). This 'High Throughput Screening' (HTS) is applied to both biochemical assays (in vitro systems that model the biochemistry of a subset of cellular processes) and cellular screening

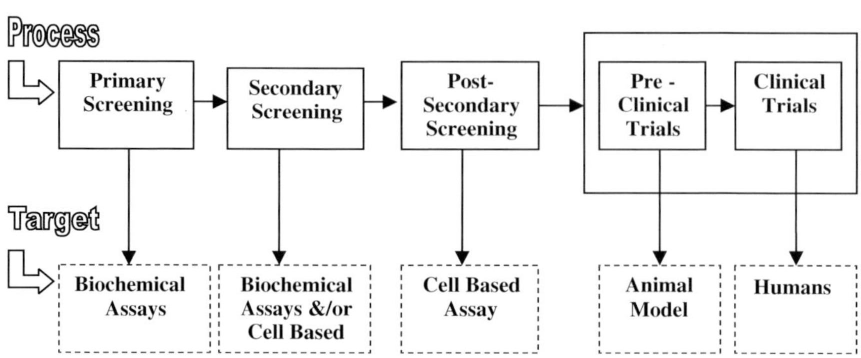

Figure 11. Schematics of the "drug discovery" process.

assays (in vitro live cells). "Secondary screening" differs from "primary screening" with regard to the requirements for lower throughput, the smaller number of components to be tested, and the need for more complex functional assay protocols. "Secondary screening" often involves cellular screening assays.

Fluorescence spectroscopy plays a key role in both primary and secondary screening processes. The screening takes place typically in multiple-well plastic microplates. The spectroscopic instrumentation employed is of various types, being based on steady-state (intensity-based), time-resolved (lifetime-based), FRET (both steady-state and time-resolved), and polarization principles. In addition to endogenous fluorophores (tryptophan, tyrosine, NADH, flavoproteins), other commonly used fluorescence probes in drug discovery are the xanthene dyes, long-wavelength cyanines, metal–ligand complexes, voltage-sensitive dyes, and GFP.

Xanthene dyes and cyanines (a group of dyes that combine relatively long-wavelength absorption with comparatively small molecular size) are typically conjugated with oligonucleotides or proteins to allow targeting and imaging of specific sites in cells. Metal–ligand complexes such as the lanthanide chelates are characterized by a long-lived fluorescence emission (0.5–3 ms), thus allowing it to be distinguished from the short-lived (nanoseconds) autofluorescence of biological samples. Using a time-resolved or time-gated method, both the specificity and the sensitivity of fluorescent measurement from biological samples stained with lanthanide chelates can be enhanced. Voltage-sensitive probes (including JC-1, DiBAC4(3), coumarins) are used to measure the cellular and intracellular membrane potential. GFP and its fluorescent mutant forms also have several unique characteristics that make them suitable for many types of study in both living cultured cells and in organisms. GFP is a very robust protein, resistant to degeneration, and a non-invasive fluorescent marker that can be readily expressed in cells alone or as fusion protein. GFP can function as a cell lineage tracer, a reporter of gene expression, or a measure of protein–protein interaction. A comprehensive review of these fluorescent probes has been presented by Lakowicz [87].

Drug-discovery processes nowadays benefit from a large variety of fluorescence assays able to handle both biochemical (substrate for enzymes, cytochromes activity, molecular binding, reporter genes, protease and pepsin activity) and cell-based (apoptosis, cell proliferation, cytotoxicity, cell migration, cell adhesion) processes [105,106].

12.7.3 Biosensors

A biosensor is an analytical device incorporating two specific biosensing components [107,108]. The first is a biological component (e.g. enzymes, antibodies, lectins, neuroreceptors) that creates a biorecognition event in order to detect chemical or biological species. The second is a physical element that creates a transduction mechanism for converting the physical or chemical response into a detectable signal (e.g. electrical, optical, temperature, mass). Optical biosensors utilize optical techniques to detect and identify biological and chemical species, and fluorescence sensing plays a central role in these techniques. Fluorescence sensing can be achieved in two ways. Firstly, directly, by utilizing a direct change in fluorescence properties brought about either by the analyte binding with the biorecognition element, or by the production of a fluorescent species as a result of a specific reaction. Secondly, indirectly, by utilizing an external probe whose fluorescence property changes in response to the biorecognition of an analyte.

Commonly used fluorescence-based biosensors include: (1) Fluorescence resonance energy transfer (FRET) sensors: these involve an energy transfer which produces a change in the fluorescence emission of the biorecognition element or the fluorescence probe. This technique requires a specific donor–acceptor group within which energy transfer is affected as a result of biorecognition. (2) The molecular beacon-based sensors, which involve an electronic energy transfer between a fluorescent complex and a fluorescent quencher. (3) Fiber-optic sensors: for instance, fluorescence excitation has been used in evanescent wave based sensors. In this case, the fluorescence originates from the analyte specifically binding with a biorecognition element immobilized on the surface of the fiber waveguide [107,108].

Biosensors have the potential not only to reduce the time spent on biological experiments, but also to facilitate a more direct understanding and sensing of molecular interactions. The biosensors could be used in a wide range of applications such as clinical diagnostics, environmental monitoring, drug discovery/development.

12.7.4 DNA-chip reading (P. Taroni and G. Valentini)

The completion of the human DNA sequencing opens up a new era in molecular biology and medicine. Nevertheless, without the identification of the role of the thousands of genes that constitute the human genome, the tremendous amount of data referring to the base sequences will be of limited use. Therefore, the knowledge of the functional correspondence between genes, proteins and diseases is the new challenge for geneticists. Pathological conditions are often the result of an upset in the expressed genes, i.e. genes active in cells. This leads to impairment in the cellular machinery that produces enzymes and proteins, causing physiological disorders. The identification of genes having a specific relevance in certain diseases can be carried out using a technique that is based on the so called DNA-chips or, equivalently, DNA-microarrays [109,110]. Basically, the DNA sequences that represent the coding of many genes are synthesized and arranged on a microscope slide in a regular pattern. Then, the RNA of two samples, e.g. two cell populations, the one to be analyzed and a reference one, is extracted and retro-transcripted to a special form of DNA, called cDNA. The cDNAs of the two samples are labeled with different fluorescence markers and spread onto the chip. Because complementary single-helix DNA strands bind to each other when they come into contact, each gene onto the DNA-chip captures an amount of the two markers that is proportional to how much that specific gene is expressed in the two samples [111,112]. Hence, the map of the relative amplitude of the two markers in the DNA-chip provides the table of the expression of the genes in the sample to be analysed with respect to the reference sample. Using this technique, one can find, for example, which genes are overexpressed or underexpressed in tumor cells, with respect to normal ones.

Other DNA-chips aim at finding in the DNA mutations that are associated with hereditary diseases [113]. In this case, short base sequences, called oligonucleotides and differing from one another in the base sequence, are spotted onto the slide in a regular pattern. Once again, the DNA to be searched for mutation is labeled with a marker and spread onto the slide. Hybridization, i.e. ligation, takes place only in correspondence of the spots, if any, that contain the same mutations present in the DNA under investigation.

The standard technique to read DNA-chips relies on a confocal laser scanner. A microscope slide hosting the DNA-microarray is moved in a raster pattern between two microscope objectives arranged in confocal configuration. Different excitation wavelengths are delivered to the sample, while the fluorescence emission is separated in two or more detection channels with sharp spectral filters. A two channel scanner with excitation and emission wavelengths optimised for a couple of markers emitting in the green and red bands (e.g. cyanine 3–cyanine 5) is used to read cDNA-chips for gene expression profiling.

Even if confocal scanners show high sensitivity and superior spatial resolution, there are reasons for increasing even more the sensitivity of the DNA-microarray reading devices. In fact, in genetic tests the signal is usually very faint since it comes from few molecules. Moreover, interest is in using as little DNA as possible. In fact, it would be beneficial to avoid the time consuming and sometimes unreliable DNA amplification by means of the Polymerase Chain Reaction (PCR). Therefore, any mechanism that can improve the signal-to-noise ratio in fluorescence measurements will meet the favour of geneticists. Among the possible mechanisms, fluorescence lifetime has already been successfully applied to improve DNA sequencing [114–118].

Following the same approach, at Politecnico di Milano (Milan, Italy), fluorescence lifetime imaging has been proposed for DNA-chip reading. For mutation DNA-chips, which have a single marker, the lifetime information was used to remove the background emission coming from the glass substrate and from the chemicals used to process the chips. For expression cDNA-chips, a bi-exponential fit can be applied to discriminate the contribution of two markers in order to calculate their relative concentration. The heart of the fluorescence imaging system is a fast light intensifier optically coupled to a low noise CCD camera. The light intensifier is capable of a minimum gate of 300 ps and has been designed to work at the typical repetition rate of mode-locked lasers (70–80 MHz). The excitation source is a dye laser, synchronously pumped by a second harmonic mode-locked Nd:YLF laser. A cavity dumper reduces the repetition rate to few megahertz. The beam is coupled to a multimode fiber and delivered to the DNA-microarray using a prism positioned for total internal reflection. Excitation light enters the DNA-chip from the back side while fluorescence emission is collected from the front side by means of a high aperture lens. Either dichroic or glass colored filters are used to remove the scattered laser light.

The system was tested on DNA mutation arrays containing 10×10 spots of the same oligonucleotide complementary to a test base sequence simulating the sample to be analyzed. Labeling was performed with cyanine 3 (Cy3, excitation = 550 nm, emission = 570 nm) at 5 concentrations between 0.5 and 10 nM. For each DNA-chip, 44 images, delayed by 250 ps from one another were acquired with an average excitation power of 1 mW at 540 nm. An orange cut-off filter and an interference filter centered at 580 nm were used to reduce the scattered laser light and the background fluorescence. Figure 12 shows the fluorescence image of the microarray processed with the probe at lowest concentration, taken synchronously with the excitation pulses (delay = 0). This image is equivalent to what would be achieved in a steady-state fluorescence measurement. The hybridized spots appear like faint circles arranged in a regular pattern. Despite the narrowband filter, the background fluorescence overcomes the emission of the marker and prevents the recognition of many spots in the array. Nevertheless, using the whole series of time-resolved images Cy3 emission could be discriminated from the background. The lifetime of the background fluorescence was measured in a region of the array without any spot, while the lifetime of the marker was measured in a microarray processed with a probe at high concentration. Mono-exponential emission was assumed for both Cy3 and background fluorescence and the maps of the corresponding amplitudes were calculated. Figure 13 shows the map of the Cy3 amplitude cleaned from the background fluorescence. Comparison of Figures 12 and 13 shows a significant improvement in terms of contrast. Spots that are not clearly discernible in the fluorescence intensity image are well outlined in the amplitude map of the marker. A line profile taken across the first column of oligo spots is also shown in Figures 12 and 13. From these profiles, the increase in the S/N ratio is estimated to be at least 2.5.

The time domain concepts applied to mutation DNA-microarrays could be profitably utilized also for gene expression profiling, provided that the two cDNA

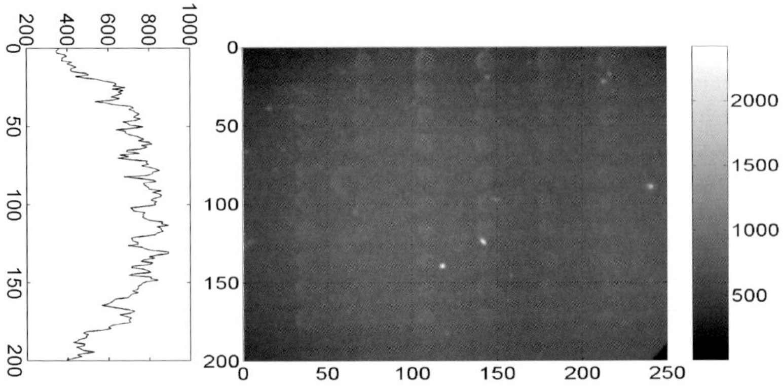

Figure 12. Fluorescence image of a DNA-microarray labelled with Cy3. The image was taken synchronously with the excitation pulses. A line profile across the first column is reported on the left.

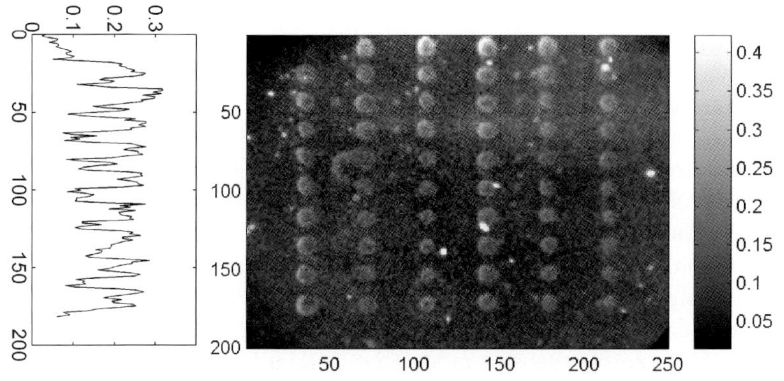

Figure 13. Amplitude of Cy3 fluorescence obtained from a bi-linear fit carried out on a set of time-delayed images of the DNA-microarray considered in Figure 12. A line profile across the first column is reported on the left.

samples are labeled with markers that show similar spectral features, but different lifetimes. Preliminary results indicate that the procedure is feasible even using the markers presently available on the market, which have been designed for spectral discrimination. Yet, it would be preferable to consider markers specifically developed for temporal separation.

Finally, it is worth noting that an intensified CCD camera can map a DNA-microarray in a very short time (few seconds) even if its spatial resolution is limited by that of the intensifier tube. However, high density microarrays could be read with multiple acquisitions. This would allow one to obtain a spatial resolution comparable to the one achievable with the state of the art scanners, while maintaining a higher speed.

The experimental apparatus presently used relies on mode-locked lasers, which are bulky and power consuming. However, solid-state picosecond laser systems emitting in a wide range of wavelengths are available. Using such devices, it is possible to set up a compact instrument to read DNA-microarrays based on time-resolved measurements.

12.8 Medical applications

A current major thrust of fluorescence spectroscopy is in the clinical diagnosis of critical disease in human tissue, for example atherosclerosis and cancer. Reviews in this field have been made by Wagnieres et al. [85]), Bigio and Mourant [119], Andersson-Engels et al. [120], Richards-Kortum and Sevick-Muraca [86], Papazoglou [121], Das et al. [122] and Ramanujam [123]. Both steady-state and time-resolved fluorescence spectroscopy techniques have been employed to obtain information about the pathophysiological and biochemical state of diseased tissues. While steady-state techniques have been extensively investigated and are currently routinely used in research clinics, very few research groups have explored time-resolved techniques [122,124–129]. Fluorescence spectroscopy has been performed in both pre-clinical settings, in various animal models (in vivo) [85,123] and with excised human tissue (ex vivo) [85,86,120–122], and in clinical settings in patients undergoing surgery or endoscopic investigations [123,124,129–131]. Fluorescence from both endogenous fluorophores of tissue (Section 12.6.1) and exogenous probes (Sections 12.8.2 and 12.8.4), such as Hematoporphyrin Derivative, Photofrin, and δ-aminolevulinic acid, have been used in the characterization of malignant transformations [84,85,120,121,123].

A broad range of excitation wavelengths (300–500 nm) have been used for inducing tissue fluorescence. Excitation light sources include continuous wave lasers (He-Cd, argon), intensity-modulated and pulsed lasers (nitrogen, mode-locked argon-pumped dye, excimers), diode lasers, and arc lamps (Hg, Xe). Laser light sources are generally preferred due to their high-efficiency coupling into fiber-optic probes [85,86,120,121].

12.8.1 Oncology – autofluorescence-based techniques

Despite aggressive treatment, including surgical resection, irradiation and chemotherapy, the number of patients diagnosed with cancer has increased dramatically over the past decades [132]. Early detection and diagnosis of malignant transformations in tissue can offer not only survival benefit but also an improved quality of life for cancer patients. Currently, there is no effective diagnostic technique for either early malignant transformations or guidance of surgical resection. Because fluorescence spectroscopy based techniques examine tissue surfaces, rather then tissue volumes, they are particularly suited to such applications. Moreover, optical spectroscopy techniques can be readily adapted to fiber optic probes, endoscopic catheters and biopsy probes, thus allowing remote investigations.

Since 1965, when Lycette and Leslie [133] first reported that fluorescence spectroscopy could be used to discriminate malignant tumors of the esophagus, stomach, breast and thyroid from normal tissue, several research groups have extensively studied ex vivo and in vivo the autofluorescence emission of neoplastic and non-neoplastic tissue in animals and humans. Fluorescence spectroscopy experiments have been conducted on several clinical sites, and include examination of skin, breast, brain, gastrointestinal tract (colon, esophagus), cervix, bronchus, urinary bladder, head and neck (oral cavity, larynx), bile duct, and stomach. Ramanujam (2000) [123] and Wagnieres et al. (1998) [85] have comprehensively reviewed these studies. In addition to steady-state laser-induced fluorescence spectroscopy, a technique currently used in clinical research settings, time-resolved techniques have also been explored by a few research groups [120,122,129,134, 135]). Endogenous fluorophores which play an important role in the fluorescence emission of tumors include amino acids (tryptophan, tyrosine), structural proteins (collagen, elastin), enzyme cofactors (NADH, FAD) and porphyrins [85,123].

A study to assess the ability of fluorescence spectroscopy to detect brain tumor margins intraoperatively in presented below. The degree to which a complete brain resection can be carried out is limited by several factors unique to the central nervous system. Primary brain tumors are infiltrating in nature and the margins of the tumor are frequently indistinct. Current imaging technologies to detect cancer in the central nervous system fail to detect the extent of infiltrating glial tumor [136,137]. Since clinical outcome is closely linked to the extent of surgical resection, which in turn is limited by the inability to visually detect tumor margins, new strategies for intraoperative brain tumor detection are needed.

Recently, steady-state laser-induced endogenous fluorescence has been utilized in surgical oncology for the intraoperative localization of tumor margins [106,108]. By providing an intraoperative optical biopsy of the tissue and delineating the tumor margin, this fluorescence technique may obviate the need for multiple invasive biopsies of the tumor margin. Studies of native brain fluorescence have demonstrated that laser-induced fluorescence (steady-state) techniques can differentiate brain tumors from normal brain tissue both in the human brain and in a rat brain glioma model [138–141]. Moreover, recent studies [134,142] of brain fluorescence have demonstrated that time-resolved data can complement steady-state spectra and provide an additional means of distinguishing between diseased and healthy tissues, thus suggesting the potential use of lifetime fluorescence spectroscopy techniques as intraoperative diagnostic tools for intracranial tumors. Figure 14 shows, for instance, the distinct patterns of the time-resolved fluorescence emission from glioma, meningioma, pituitary tumor and healthy gray matter when excited with a pulsed nitrogen laser (337 nm, 3 ns). The fluorescence emission features of all intracranial tumors were found to be distinct from the surrounding normal brain tissues, white matter and cortex. Endogenous fluorophores likely to contribute to fluorescence emission of investigated tissues include NADH (glioma, pituitary and normal cortex at 450 nm) and collagen (meningioma at 390 nm). Current results demonstrate that both spectral and time-resolved data provide a means of discriminating between diseased and healthy tissues and suggest the potential use of time-resolved fluorescence spectroscopy techniques as intraoperative diagnostic tools for intracranial tumors.

In summary, endogenous fluorescence has been widely explored and shows potential for the detection of both early and advanced neoplasia in screening, diagnostic and monitoring settings. However, the diagnostic potential for guiding surgery has only been evaluated in a limited number of clinical studies. Other optical techniques such as elastic scattering, absorption, and Raman spectroscopy could be used in conjunction with autofluorescence to enhance and complement the diagnostic accuracy of the latter technique [85,123].

12.8.2 Oncology – ALA fluorescence-based techniques (P. Taroni and G. Valentini)

Fluorescence spectroscopy and imaging offer effective opportunities for cancer diagnosis, which have long been carefully investigated [85]. Fluorescence techniques are minimally invasive, relatively inexpensive with respect to other diagnostic

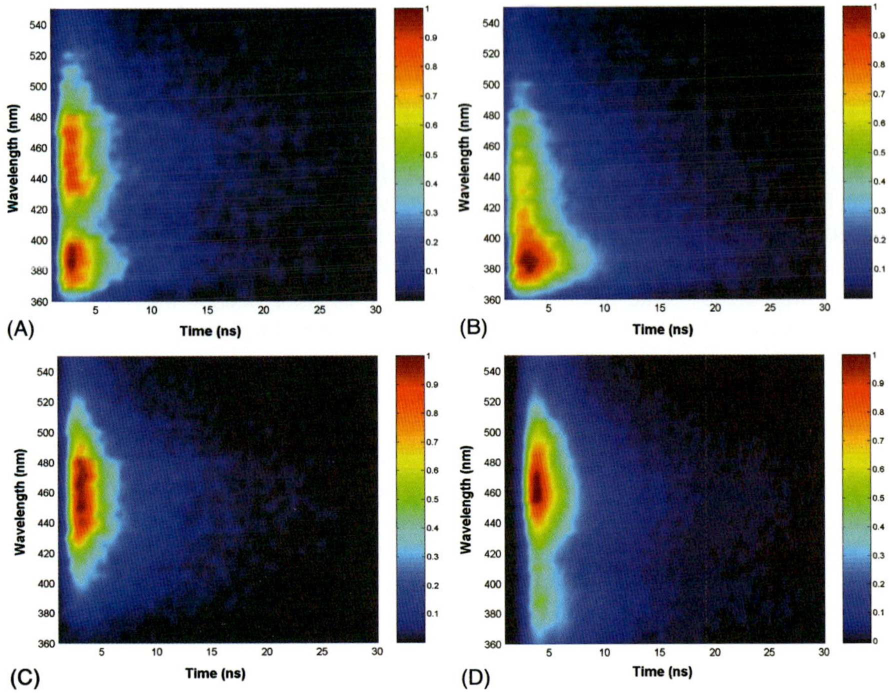

Figure 14. Representative time-resolved fluorescence emission spectra (contour map – distribution of the fluorescence intensity). Color axis represents the fluorescence intensity. (a) Low grade glioma (note: the minimum between peak intensity corresponds to the absorption peak of hemoglobin at ~315 nm), (b) meningioma, (c) pituitary, (d) normal cortex. Note that each type of tissue presents distinct time-resolved fluorescence.

methodologies, and can be easily applied to any part of the human body that can be reached by light, either directly or by means of an endoscope. In diagnostic procedures, fluorescence can in principle provide clues for the detection of several disorders, and, in particular, tumors. Unfortunately, however, a broadband emission is a general characteristic of both cancerous and non-cancerous tissues. Therefore, the mere presence of a fluorescence signal often provides only a limited diagnostic aid. To increase the specificity for tumor detection, the exogenous emission of suitable markers can be considered. To this purpose, great attention has been devoted to photosensitizers originally developed to treat tumors with the photodynamic therapy, i.e. by means of the combination of drugs and activation light [143,144]. Some photosensitizers are also promising for diagnosis since they accumulate in cancerous tissues with a good selectivity, are fluorescent and, last but not least, have already been approved for human administration [145–148]. A very special photosensitizer is δ-aminolevulinic acid (ALA), which is a naturally occurring precursor in the cycle of heme biosynthesis [149–151]. The exogenous administration of this substance bypasses a physiologic regulation mechanism and gives rise to an excess of the intermediate molecule Protorphyrin IX (PpIX), which is photoactive and strongly fluorescent [152,153]. Even though this process takes place also in healthy tissue, in cancerous tissue it is more effective due to enzymatic deficiencies occurring in neoplastic cells. This unbalance in the content of PpIX provides a selectivity criterion that can be profitably exploited to detect tumors by means of fluorescence techniques. In fact, PpIX, when excited around 400 nm, emits a characteristic red fluorescence (peaking around 635 nm) with a lifetime of about 18 ns, i.e. much longer than that of the natural tissue fluorescence (4–6 ns or shorter) [154].

Moreover, an advantage of this particular photosensitization is the possibility to administer the drug topically, thus avoiding skin sensitization, which is usually associated with systemic administration of photosensitizers and requires people to avoid exposure to direct sun light for a long time (i.e. weeks).

Topical administration of ALA is especially suitable for the detection of exposed pathologies, like skin tumors [155–158], even though pigmented lesions, like the very aggressive melanoma, cannot be detected by fluorescence imaging because no appreciable emission can be revealed from these strongly absorbing lesions. This not withstanding, a great interest exists in dermatology for the classification of other lesions, which present a different degree of malignancy [159–161]. In addition, a technique capable of showing the actual extent of the pathologic area would allow surgeons to perform conservative excisions, while reducing the recurrence rate. In the following, we will report on the experience gained at Politecnico di Milano (Milan, Italy), in collaboration with Clinica Multimedica (Milan, Italy), to evaluate the effectiveness of fluorescence lifetime imaging in distinguishing basal cell carcinomas and squamous cell carcinomas from benign lesions [162].

The clinical trial was carried out using the prototype of a clinical diagnostic system that includes the laser source, a nanosecond gated intensified CCD camera, some electronic devices for synchronization and a computer that controls the whole system and processes the images in real time. Details of the experimental set-up can be found elsewhere [163], while the main concepts are summarized below. Excitation light is provided by a dye laser pumped by a sub-nanosecond nitrogen laser. The laser emission peaks around 400 nm with average power of 800 μW. Such a power level does not raise any safety concerns and allows one to perform the diagnostic procedure even on the patient's head. This is very important since most dermatological lesions are on parts of the body exposed to sunlight and in particular close to the eyes. An optical fiber is used to deliver the light to the patient, while the fluorescence signal is collected using high aperture photographic lenses. The system has been designed so that even non-technical personnel can operate it for routine clinical examinations. All the devices are completely controlled via software and some automated procedures help the user to optimize the operating conditions (choice of delays and gain of the light intensifier).

The diagnostic procedure involves the preparation of an ointment made of 2% ALA powder in a lipid emulsion. The ALA ointment is applied onto the lesion with a margin of at least 10 mm in the visibly normal skin and the lesion is covered with an occlusive dressing. The patient is kept at rest for about one hour to allow the metabolism to transform ALA in PpIX. Then, the excess cream is gently removed from the lesion and the fluorescence test is carried out. A preliminary examination is performed using only two gated images. In this case the map of the fluorescence lifetime is calculated in real time using Eq. (24) in Section 12.4.4 and displayed in pseudocolors at 5 frames per second. Then, at least for the cases that present a clinical relevance, a more precise measurement is performed by acquiring 5 images with a 100 ns wide gate delayed by 0, 2, 5, 10 and 20 ns, relative to excitation pulses. The images are processed offline immediately after the acquisition to calculate the maps of the average fluorescence lifetime and of the relative amplitude using the linear fitting algorithms presented in Section 12.4.4. Image processing takes few seconds and can be repeated several times, changing parameters like the set of images included in the linear fit and the color coding to optimize the diagnostic result. Other standard procedures, like patient anamnesis and photographic documentation, complete the examination.

The classification of the lesions from fluorescence images was made according to the following considerations. The malignant character of a lesion is associated with an excess of exogenous fluorescence and a reduction in natural emission [164–166] as compared with healthy tissues [167]. Consequently, the average lifetime in tumors is longer that in healthy tissues since the amplitude of the long living component (PpIX fluorescence) is higher than that of the short living emission (natural fluorescence). The use of fluorescence lifetime as a diagnostic index, instead of intensity or spectral features, gives additional advantages. In fact, the ratio in the denominator of Eq. (24) in Section 12.4.4 provides a normalization that eliminates possible artifacts due to the spatial variation of the excitation light and – most important – gets rid of local differences in skin absorption. These may be due to anomalies in pigmentation, vascularization, or blood perfusion due to an inflammatory status, and might lead to severe artifacts if one looks at the fluorescence intensity. Finally, the diagnostic procedure can be carried out under normal illumination.

Thirty-four patients affected by 48 lesions either malignant (mainly basal cell carcinomas) or benign were included in this trial. For all the patients, the classification of the lesions on the basis of fluorescence lifetime maps was compared to histology, which was routinely performed. A typical fluorescence lifetime image of a malignant lesion is shown in Figure 15(a), while the map of the amplitude is displayed in Figure 15(b). For comparison, the color photo of the lesion is also reported in Figure 15(c). The tumor, which was classified as a basal cell carcinoma, is characterized by a lifetime significantly longer (18 ns) than that of the surrounding healthy tissue (≈10 ns). Notably, the region where the ALA ointment was applied is larger than the lesion itself and can be easily distinguished in Figure 15(a). The fluorescence amplitude is lower in the lesion, as expected. However, the clinical experience demonstrates that, while the fluorescence amplitude presents a strong variability from patient to patient, the lifetime is much more stable and thus reliable for the classification of the lesion. In the present trial, 27 out of 35 malignant lesions were correctly identified, while only 1 out of 13 benignant lesions was misinterpreted. On the basis of these preliminary outcomes, our diagnostic procedure has a sensitivity of 88% and a specificity of 74%.

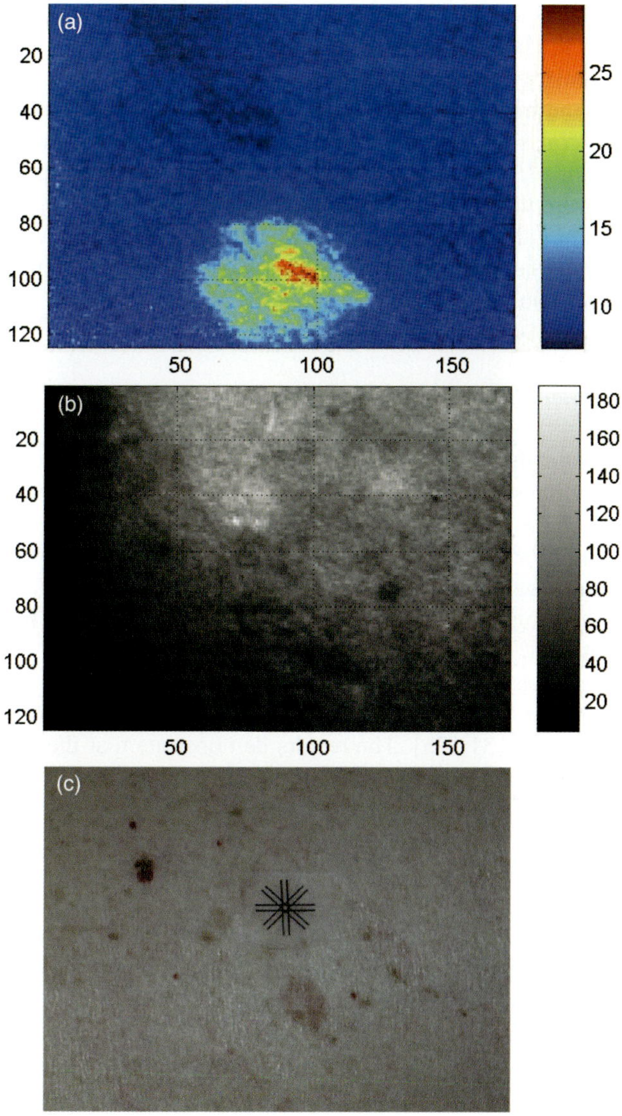

Figure 15. Maps of florescence lifetime (a) and amplitude (b) of the skin of a patient affected by a basal cell carcinoma, after topical application of ALA. The lifetime in the lesion is higher, while the amplitude is lower, as expected for a neoplastic tissue. The corresponding color image is also shown (c).

In conclusion, the topical application of ALA gave interesting results for the detection of malignant skin lesions. Even though in some cases the typical PpIX red fluorescence can be directly observed on the skin lesions of ALA-sensitized patients under CW excitation, this is certainly true only when the marker dose is rather high. When the ALA dose is reduced to a few percent, as required in a minimally invasive diagnostic protocol, the interference of the natural skin fluorescence and the uneven pigmentation prevent reliable tumor identification by means of an intensity-based approach. In such a condition the lifetime technique exhibits its maximum effectiveness, since it provides a comparison between PpIX and endogenous emission, which has been demonstrated to be a preferred indicator for tumor detection [167]. Moreover, the lifetime approach benefits from first order independence from the uneven absorption of the excitation light. Finally, the diagnostic procedure does not require complete darkness. This is very important in terms of applying the technique in the surgical room to demarcate the lesion margins during excision.

12.8.3 Cardiovascular – autofluorescence-based techniques

Atherosclerosis and its manifestations are the leading cause of death in modern societies. Depending on the anatomic location of atherosclerotic plaques, the clinical consequences include myocardial infarction, stroke, limb ischemia and aortic aneurysms [168,169]. Quantitative assessments of atherosclerotic disease during its natural progression and after therapeutic interventions are important in understanding the progression and stabilization of the disease and for the choice of appropriate treatment.

Numerous research groups have investigated the potential of fluorescence spectroscopy for the diagnosis of diseased arterial walls both in vitro [128, 170–174] and in vivo [131,175]. The results demonstrate that there is potential for the application of fluorescence-based techniques for discrimination between normal and advanced lesions [121]. Normal arterial intima (inner layer of the arterial wall) is mainly composed of non-fibrous connective tissue, smooth muscle cells, elastic fibers and collagen, whereas in more advanced lesions the intima undergoes structural disorganization, repair and thickening due several factors, including accumulation of lipids in both intracellular and extracellular space, formation of new fibrous tissue, thrombus formation, and calcification [168,169]. With regard to clinical applications in particular, the studies mentioned above have suggested incorporating spectroscopic evaluation into clinical fiber optic systems that would allow investigation of the changes in intima composition due to disease progression. In this context, initial studies discussed its potential to guide laser angioplasty and to evaluate the likelihood of restenosis, while more recent studies have dealt with the diagnosis of unstable lesions.

Spectroscopic investigations have primarily utilized a steady-state laser-induced fluorescence spectroscopy (LIFS) technique wherein various wavelengths (nm) are used for tissue excitation: 306–310 [173], 308 [176], 325 [172], 337 [174], 458 [177], 476 nm [165].

The choice of excitation wavelength determines which endogenous fluorophores contribute to the emission of arterial wall. The endogenous fluorophores responsible for artery fluorescence are the structural proteins (collagen and elastin) [120,121,128], lipopigments (ceroid and lipofuscin) [128,170], lipid components (e.g. peroxidized lipoproteins) [176], low density lipoproteins (LDL) [176,178], free cholesterol [127,175,179], L-tryptophan [174], β-carotene [170], calcium hydroxypyrophosphate (CaHPO4) [178], and NADH [174]. Except for the 306–310 nm range, small changes in excitation wavelengths do not result in significant changes in the emission spectrum. For instance, at excitation wavelengths above 325 nm the fluorescence of normal arterial wall is largely associated with collagen and elastin emission. The spectra of fibrous (collagenous) lesions, characterized by a narrow band focused in the blue spectral range, closely resembles the emission of Type I collagen [86,174,179]. Whereas, the fluorescence emission of normal arterial wall is broad and has been associated with the fluorescence of elastic fibers mainly localized in the internal elastic lamina. Extended reviews of how the interplay between excitation wavelength and endogenous fluorophores in arterial wall is reflected in the emission spectra of both normal and diseased arterial wall are presented elsewhere [86,121,171,173,175,178,179].

As demonstrated by a few research groups, time-resolved information can improve the specificity of fluorescence measurements in arterial tissue and overcome some limitations of steady-state spectroscopy [173,174,179]. This concept was first introduced by Baraga et al., [173] and further investigated by Andersson-Engels et al. [174]. These studies distinguished between the fluorescence decay characteristics of normal arterial wall and those of fibroatherosclerotic plaque. They also demonstrated that the presence of blood in tissue does not change arterial wall fluorescence decay characteristics, and suggested that time-resolved measurements can enhance diagnosis through the discrimination of biochemical variations in human atherosclerotic lesions. More recently, Marcu et al. [127,179] have also shown that information retrieved from time-resolved spectra of normal and diseased human arterial walls can be used to distinguish between different types of atherosclerotic plaque and, more specifically, to identify lipid-rich lesions. Selected results from this work are given here. Representative time-resolved fluorescence emission spectra (excitation 337 nm, pulse width 3 ns) of human aortic samples are given in Figure 16. Distinct

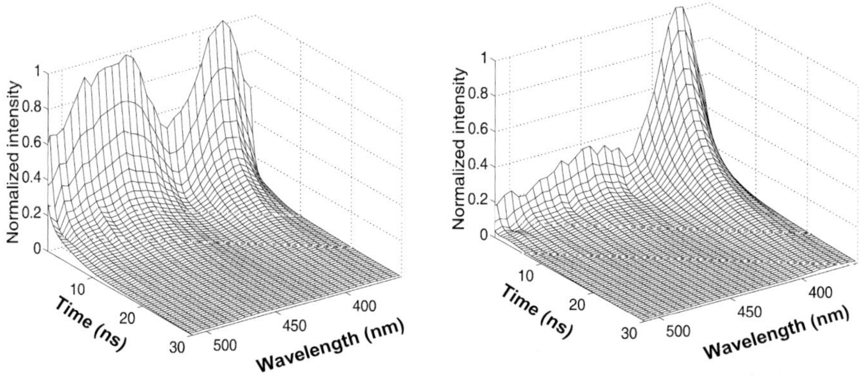

Figure 16. Time-resolved fluorescence spectra of (a) normal aortic wall and (b) collagen-rich (fibrous) aortic plaque. Note: the minimum between peak intensity corresponds to the absorption peak of hemoglobin (~415 nm). Adapted from Ref: 179 (L. Marcu 2003).

emission features were observed for each lesion type in both the spectral- and time-domain. Changes in fluorescence emission characteristics as a function of atherosclerotic plaque formation are shown in Figure 17. These show that analysis of time-resolved fluorescence emission spectra can be used to enhance the

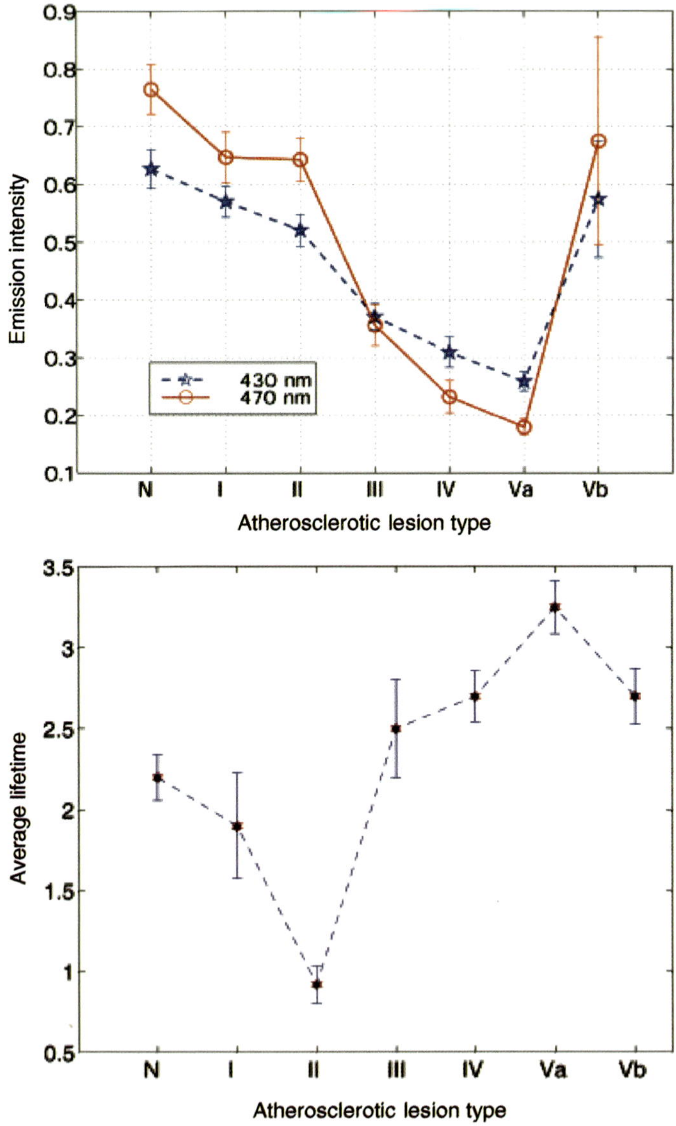

Figure 17. (a) Changes of fluorescence intensity (mean ±SE) as a function of atherosclerotic lesion type for two wavelengths of emission (430 and 470 nm). Results from aorta studies adapted from Ref: 179 (L. Marcu et al. ... 2003). (b) Changes in average lifetime as a function of lesion type for 390 nm emission. Results from coronary artery study adapted from Ref: 127 (L. Marcu et al. 2001)

discrimination between different grades of atherosclerotic lesions, and have demonstrated that a few parameters derived from spectral emission at longer wavelengths (>430 nm) and time-resolved emission at peak emission region (390 nm) can discriminate between all types of arterial samples except for between normal wall and early lesions.

Current research in the field of fluorescence spectroscopy of arterial wall suggests the potential of this technique for the quantitative analysis of atherosclerotic plaque morphological composition, such as the presence of macrophage foam cells, necrotic core, collagen fibers, and lipids. This analysis can be further used to assess plaque instability and the extent of disease progression.

12.8.4 Drug-assisted diagnostics (PDT-related, biodistribution)

The use of optically active exogenous probes has been of research interest in a broad range of diseases, in studies to evaluate the efficacy of drug delivery, and in angiography. The exogenous probes primarily investigated for in vivo diagnosis of tissues involves the use of photodynamic therapy (PDT) agents and other fluorescent agents with absorption bands in the red or NIR region.

PDT has emerged as a promising therapy for diseased tissues (including cancer, cardiovascular, chronic skin, autoimmune, macular degeneration, and antibacterial) by utilizing the activation by light of an exogenous chemical agent, called a photosensitizer. Among widely investigated PDT agents are the porphyrin derivatives, chlorines and bacteriochlorins, benzoporphyrin derivatives, 5-amino-laevulinic acid (ALA), and cationic and dendritic photosensitizers. These drugs have been used intensively as both therapeutic agents in photodynamic therapy and as markers for diseased tissue detection based upon reemitted fluorescence. The first class of PDT agents, widely investigated in the 1980 s, were the hematoporphyrin derivatives (HpD). Although these agents improved the detection of diseased tissues, some major drawbacks were reported. For instance, HpD consists of many different porphyrins with different pharmacokinetic, fluorescence and photodynamic properties, and consequently the degree of correlation between fluorescence and PDT effectiveness varies with tissue type and time of administration [85,86,108]. Moreover, application of these agents has had poor selectivity for early malignant lesions and has resulted in skin photosensitivity.

Some of these disadvantages were overcome by δ-aminolevulinic acid (ALA), a precursor to heme in the heme cycle. Administration of ALA (a non-photoactive substance) results in tissue overexpression of protoporphyrin IX (PpIX), a photoactive and strongly fluorescent substance (absorption at about 400 nm, emission in the 625–725 nm range) with high selectivity to malignant tissue (see also Section 12.8.2). ALA can either be administered orally or applied topically. Also, the pharmacokinetic properties of ALA can be optimized for particular targets. Because both ALA and PpIX are substances naturally present in the body, the toxicity issues are limited [85,86,108].

Indocyanine green (ICG), with an absorption band in the NIR region at ~800 nm and a peak fluorescence emission at 840 nm, is a tricarbocyanine dye tightly bound to serum albumin and partitioned in the vascular space, except in diseases where vascular wall integrity is damaged. Once the dye reaches the affected vascular area, it can be excited using the NIR light and the reemitted light is used to image damaged or diseased tissue areas such as skin burns, subretinal neovascular membranes and leaking vessels, and tumor angiogenesis [85,108].

To improve the efficacy of drug-assisted diagnostic techniques, several important questions need to be addressed, including the mechanism of drug delivery and pharmacokinetics, drug localization and uptake, vascular damage, and immunologic response. Extensive reviews on the mechanism of photodynamic action, localization of photosensitizer and photodamage are presented in [85,86,108].

12.8.5 In vivo molecular imaging (P. Taroni and G. Valentini)

The basic principle of clinical diagnosis is the capability to detect a disease as soon as any change at physiologic or anatomic levels takes place. At the same time, the effectiveness of a therapy is evaluated mainly from symptom regression. Nevertheless, the origin of most diseases can be found at molecular level, and in particular in the impairment of the chemical machinery that manage the production of proteins inside cells. Recent advances in molecular and cell biology, the ability to decode entire genomes and the knowledge of the molecular pathways of many diseases are expected to revolutionize medical practice and biologic research in the 21st century. One of the most promising tools for clinical diagnosis and medical research is the possibility to develop selective markers that address the molecular bases of many diseases. Thus, using suitable imaging systems capable of detecting such markers in living animals and, potentially, in the human body, the onset of a disease can be revealed as early as any change at molecular lever takes place. The combination of high specificity markers with imaging systems operating in vivo represents the new concept of molecular imaging.

While, generally speaking, molecular imaging involves a wide class of imaging methodologies [180], like computer tomography, ultrasounds, magnetic resonance imaging, positron emission tomography, in this chapter only optical imaging is considered.

There are two basic approaches for molecular imaging using optical technologies. The first refers to the detection of photons emitted by chemioluminescent molecules expressed after genetic modification in living animals. The second relies on the more common phenomenon of fluorescence of specific cell targets, even though the good news with respect to traditional labeling is the very precise selectivity of this class of probes, which show molecular specificity. In the following both approaches will be briefly described.

Bioluminescence imaging exploits the emission of visible photons based on energy-dependent reactions catalyzed by luciferase [181]. Luciferase is a family of reporter proteins that can be isolated from a large variety of insects, like fireflies, marine organisms, like some corals, and prokaryotes. Luciferase proteins interact with a substrate (either luciferin or coelenterazin) in an exergonic reaction fueled by ATP, giving an intermediate molecule, which releases photons of visible light

(500–640 nm). The most advanced gene transfer techniques allow biologists to make cells of mammalians, like mice, express the luciferase protein. Transgenic cells of specific tumor lines can be intrinsically labeled by means of the luciferase expression. Then, the transgenic cells are injected in living animals, where they replicate and interact with the host. The systemic administration of the luciferin substrate to the host makes transgenic cells emit light that can be detected using a low noise video camera. The capability to detect, localize and possibly count the marked cells through optical measurements is very useful to study tumor growth, onset of metastasis and cell trafficking. The basic experimental setup developed by Xenogen Corp (IVIS system [182]) to detect the very faint bioluminescence of luciferase consists of a dark imaging chamber provided with a highly sensitive camera cooled down to −120°C for noise reduction. A bioluminescence image of a small animal, typically a mouse, showing a bright spot in correspondence with the location where transgenic cells accumulated deep inside the body, can be taken with an integration time ranging from a minimum of 1 s to a maximum of 10 min. The amount of the collected photons and the shape of the spot can be correlated to the depth of the luminescence site and to the amount of transgenic cells through a suitable model for photon migration in turbid media [183,184].

Using luciferase expressing tumor lines, cancer onset and progression have been monitored [185,186]. Lymphocyte trafficking in vivo has also been studied [187]. A very promising application of molecular imaging is for evaluation of potential antineoplastic therapies. Using bioluminescence, the regression of labeled human cervical carcinoma (HeLa) cells engrafted into immunodeficient mice was assessed [188]. The efficacy of both chemotherapy and immunotherapeutic treatment was evaluated.

Bioluminescence offers the obvious advantage of not requiring external excitation. Moreover, the emitted intensity increases linearly with the concentration over a broad range, thus favouring quantitative measurements, and bioluminescent molecules are scarcely toxic and of small size, which make them easily administered. Conversely, the application of bioluminescence implies gene transfer, making it unfeasible for human use.

Fluorescence can be used to study molecular targets and gene expression in several applications even wider than those allowed by bioluminescence. Two main approaches can be followed. Naturally occurring fluorescent proteins, like the Green Fluorescent Protein (GFP), expressed by some coelenterates, like the jellyfish *Aequorea victoria* can be used. Alternatively, a fluorophore, like a Cyanine dye, is bound to a specific carrier so that the fluorophore activates only when the carrier reaches its target (see also Section 12.3.2). Such a system is called molecular beacon and can be used to track some receptors, like those somatostatin. This is very beneficial for experimental oncology and, potentially, for clinical diagnosis. The experimental apparatus for fluorescence detection either of GFP or molecular beacons relies on a very sensitive video camera and on sharp spectral filters to isolate the target-related emission from other unspecific fluorescence.

The gene of the GFP can be transfected into murine or human cells, which can then be implanted in a living animal. Using a fluorescence imaging system, the marked cells and their replications can be tracked within the body with the aim of studying tumor growth and metastases. Both excitation and detection wavelengths of GFP are below 500 nm, in a spectral range where light penetration in biological tissues is limited due to high attenuation. Moreover, interest is in avoiding undesired background contributions from unspecific endogenous tissue fluorescence, often significant in the blue/green, but generally negligible in the red and near-infrared. Therefore, researchers are looking for mutants of the GFP with excitation and emission bands at a longer wavelengths, like the Red Fluorescent Protein (RFP: 560/580 nm).

Molecular beacons include a fluorophore, which is normally quenched, and do not fluoresce. Upon enzymatic cleavage, which takes place when the beacon reaches its target, the fluorophore separates from the quencher and starts to emit. Many beacons include a near-infrared carbocyanine dye (Cy5.5), which is excited at 670 nm and emits around 700 nm. These wavelengths fall within the spectral window of high transmission through biological tissues and undergo a limited attenuation by naturally occurring absorbers, like hemoglobin.

Even with fluorescence, as much as with bioluminescence, a suitable model for photon migration is required if the aim of fluorescence measurements is not just the qualitative detection of an emission signal (revealing the presence of the target), but rather its accurate localization and quantification [183,184].

Examples of application of GFP-expressing cells include the visualization of pancreatic tumor cell-lines implanted in the pancreas of nude mice [189]. Whole body optical imaging allowed researchers to study primary tumor growth and formation of metastatic lesions that developed in the spleen, bowel, lymph nodes and liver. A similar study was made using GFP-expressing melanoma cells [190].

A molecular beacon based on a cathepsin-B sensitive probe was used at the Center for Molecular Imaging Research (Massachusetts General Hospital, Boston) to compare the invasiveness of two human breast tumor lines implanted in nude mice [191]. Researchers used a probe including multiple fluorochrome residues (Cy5.5) in a configuration that makes them inactive until the probe interacts with cathepsin-B, which is a protease involved in the degradation of the extracellular matrix. After enzymatic cleavage of the probe by cathepsin-B, the fluorochromes are released, resulting in a bright fluorescence signal that can be detected in vivo. Optical imaging of Matrix Metalloproteainase-2 (MMP-2) activity in vivo was demonstrated in a study involving a molecular beacon and two human fibrosarcoma cell lines (MMP-2 positive and negative) implanted in mice [192]. The probe consisted of near-infrared fluorochromes covalently coupled to a poly-L-lysine backbone. Owing to the proximity of the fluorochromes, fluorescence resonant energy transfer (see Section 12.3.2) occurs so that almost no fluorescent signal can be detected in the non-activated state. The cleavage of the backbone by the matrix metalloproteainase activated a strong fluorescence in mice hosting the MMP-2 positive cell line, while mice with the MMP-2 negative cell line showed a significantly lower fluorescence signal. Using similar molecular probes, the activity of metalloproteinase was assessed by in vivo molecular imaging in mice treated with a metalloproteinase inhibitor [193]. A number of other applications of molecular beacons are possible, including tumor detection [194], allele discrimination [195], and hybridization of nucleic acid strands [196].

Techniques of measuring enzyme activity via molecular beacons are expected to have a profound impact on various clinical and experimental studies. A great impact can be foreseen in particular on early detection of tumors and the assessment of drug effectiveness.

12.9 Future trends: synergetic approaches

Although fluorescence-based techniques are very sensitive to biochemical and functional transformations in biological systems, they provide limited structural information. Combining fluorescence techniques with other optical spectroscopy methods (elastic scattering, absorption, Raman) or optical coherence tomography (OCT) can facilitate a complementary and more complete characterization of the optically probed biological systems. Since all optical techniques allow coupling to fiber-optic probes, the information provided by each modality can be obtained at once using the same or parallel pathway for light delivery and collection. Moreover, fluorescence techniques adapted for in vivo investigations can complement other imaging techniques for diagnostics such as MRI, PET, CT and ultrasound. Such synergetic approaches will allow not only a more accurate diagnosis of diseases but also the monitoring of the mechanisms involved in disease progression and the efficacy of therapeutic procedures.

Acknowledgements

This work was supported by the National Institute of Health (Grant # HL 67377) and The Whitaker Foundation.

References

1. J.W. Berkow, R.W. Flower, D.H. Orth, J.S. Kelley (1997). *Fluorescein and Indocyanine Green Angiography* (2nd edn). American Academy of Ophthalmology, San Francisco.
2. J.W. Gratama, J.L. D'Hautcourt, F. Mandy, G. Rothe, D. Barnett, G. Janossy, S. Papa, G. Schmitz, R. Lenkei (1998). Flow cytometric quantitation of immunofluorescence intensity: problems and perspectives. *Cytometry* **33,** 166–78.
3. T.J. Dougherty (1993). Photodynamic therapy. *Photochem. Photobiol.* **58** 895–900.
4. T.J. Dougherty (2002). An update on photodynamic therapy applications. *J. Clin. Laser Med. Surg.* **20,** 3–7.
5. R. Ackroyd, C. Kelty, N. Brown, M. Reed (2001). The history of photodetection and photodynamic therapy. *Photochem. Photobiol.* **74**, 656–669.
6. J.R. Lakowicz (1999). *Principles of Fluorescence Spectroscopy.* Plenum Press, New York.
7. A. Bogaards, M.C. Aalders, C.C. Zeyl, S. de Blok, C. Dannecker, P. Hillemanns, H. Stepp, H.J. Sterenborg (2002). Localization and staging of cervical intraepithelial neoplasia using double ratio fluorescence imaging. *J. Biomed. Opt.* **7**, 215–220.
8. H.J.C.M. Sterenborg, A.E. Saarnak, R, Frank, M. Motamedi (1996). Evaluation of spectral correction techniques for fluorescence measurements on pigmented lesions in vivo. *J. Photochem. Photobiol. B: Biol.* **35**, 159–65.
9. http://www.jyinc.com/

10. http://instruments.perkinelmer.com
11. F. Zuluaga, U. Utzinger, A. Durkin, H. Fuchs, A. Gillenwater, R. Jacob, B. Kemp, J. Fan, R. Richards-Kortum (1999). Fluorescence excitation emission matrices of human tissue: a system for in vivo measurement and method of data analysis. *Appl. Spectrosc.* **53**, 302–311.
12. J.J. Baraga, R.P. Rava, P.Taroni, C. Kittrell, M. Fitzmaurice, M.S. Feld (1990). Laser induced fluorescence spectroscopy of normal and atherosclerotic human aorta using 306-310 nm excitation. *Lasers Surg. Med.* **10**, 245–261.
13. M. Eftink (1990). Fluorescence techniques for studying protein structure. In: C.H. Suelter (Ed), *Protein Structure Determination: Methods of Biochemical Analysis* (Volume 35). John Wiley and Sons, New York.
14. H.C. Cheung (1991). Resonance energy transfer. In: J.R. Lakowicz (Ed), *Topics in Fluorescence Spectroscopy, Vol. 2: Principles*, Plenum Press, New York.
15. A. Deniz, T.A. Laurence, G.S. Beligere, M. Dahan, A.B. Martin, D.S. Chemla, P.E. Dawson, P.G. Schultz, S. Weiss (2000). Single-molecule protein folding: Diffusion fluorescence resonance energy transfer studies of the denaturation of chymotrypsin inhibitor 2. *Proc. Natl. Acad. Sci. U.S.A.* **97**, 5179–5184.
16. W. Veatch, L. Stryer (1977). The dimeric nature of Gramicidin A transmembrane channel: Conductance and fluorescence energy transfer studies of hybrid channels. *J. Mol. Biol.* **113,** 89–102.
17. http://www.molecular-beacons.org
18. S. Tyagi, F.R. Kramer (1996). Molecular beacons: Probes that fluoresce upon hybridization, *Nat. Biotechnol.* **14**, 303–308.
19. S.A.E. Marras, F.R. Kramer, S. Tyagi (2002). Efficiencies of fuorescence resonance energy transfer and contact-mediated quenching in oligonucleotide probes. *Nucleic Acids Res.* **30**, e122.
20. N. Angelini, R. Cubeddu, F. Lenci, A. Losi, A. Pifferi, A. Sgarbossa, P. Taroni, A. Vecli, C. Viappiani (1997). Artificial models of biological photoreceptors: effect of quenchers on the fluorescence properties of hypericin embedded in liposomes. *J. Photochem. Photobiol. B: Biol.* **38**, 245–252.
21. X. Zhuang, T. Ha, H.D. Kim, T. Centner, S. Labeit, S. Chu (2000). Fluorescence quenching: A tool for single-molecule protein-folding study. *Proc. Natl. Acad. Sci. U.S.A.* **97**, 14241–14244.
22. W. Denk, J.H. Strickler, W.W. Webb (1990). Two-photon laser scanning fluorescence microscopy. *Science* **248**, 73-76.
23. F. Helmchen, W. Denk (2002). New developments in multiphoton microscopy. *Curr. Opin. Neurobiol.* **12**, 593–601.
24. R.M. Williams, W.R. Zipfel, W.W. Webb (2001). Multiphoton microscopy in biological research. *Curr. Opin. Chem. Biol.* **5**, 603–608.
25 D.M. Jameson, E. Gratton, R. Hall (1984). The measurement and analysis of heterogeneous emissions by multifrequency phase and modulation fluorometry. *Appl. Spectrose. Rev.* **20**, 55–106.
26. P.C. Schneider, R.M. Clegg (1997). Rapid acquisition, analysis, and display of fluorescence lifetime-resolved images for real-time applications. *Rev. Sci. Instrum.* **68**, 4107–4119.
27. www.Picoquant.de
28. www.Becker&Hickl.de
29. R. Cubeddu, F. Docchio, W.-Q. Liu, R. Ramponi, P. Taroni (1988). A system for time-resolved fluorescence spectroscopy with multiple picosecond gating. *Rev. Sci. Instrum.* **59**, 2254–2258.

30. D.V. O'Connor, D. Phillips (1984). *Time-correlated Single Photon Counting*. Academic Press, London.
31. For example: D.W. Marquardt (1969). An algorithm for least-squares estimation of nonlinear parameters. *J. Soc. Ind. Appl. Math.* **11**, 431–441.
32. K.C. Benny Lee, J. Siegel, S.E.D. Webb, S. Lèvêque-Fort, M.J. Cole, R. Jones, K. Dowling, M.J. Lever, P.M.W. French (2001). Application of the stretched exponential function to fluorescence lifetime imaging. *Biophys. J.* **81**, 1265–1274.
33. T.J. Cornwell, K.F. Evans (1985). A simple maximum entropy deconvolution algorithm. *Astron. Astrophys.* **143**, 77–83.
34. P.J. Steinbach (2002). Inferring lifetime distributions from kinetics by maximizing entropy using a bootstrapped model. J. *Chem. Inf. Comput. Sci.* **42**, 1476–1478.
35. A. Golan, J.M. Perloff (2002). Comparison of maximum entropy and higher-order entropy estimators. *J. Econome.* **107,** 195–211.
36. E.L. Elson, D. Magde (1974). Fluorescence correlation spectroscopy. I. Conceptual basis and theory. *Biopolymers* **13**, 1–27.
37. D. Magde, E.L. Elson, W.W. Webb (1974). Fluorescence correlation spectroscopy. II. An experimental realization. *Biopolymers* **13**, 29–61.
38. W.W. Webb (2001). Fluorescence correlation spectroscopy: genesis, evolution, maturation and prognosis. In: R. Rigler, E.S. Elson (Eds), *Fluorescence Correlation Spectroscopy Theory and Applications* (pp. 305–330). Springer-Verlag, Berlin.
39. P. Schwille (2001). Fluorescence correlation spectroscopy and its potential for intracellular applications. *Cell Biochem. Biophys.* **34**, 383–408.
40. K.M. Berland, P.T.C. So, E. Gratton (1995). Two-photon fluorescence correlation spectroscopy: method and application to the intracellular environment. *Biophys. J.* **68**, 694–701.
41. K. Lok, Y.-L. Cheng, C.R. Robertson (1983). Total internal reflection fluorescence: A technique for examining interactions of macromolecules with solid surfaces. *J. Colloid. Interface Sci.* **91**, 87–103.
42. T. Hirschfield (1965). Total reflection fluorescence. *Can. Spectrosc.* **10**, 128.
43. H. Masuhara, S. Tazuke, N. Tamai, I. Yamazaki (1986). Time-resolved total internal reflection fluorescence spectroscopy for surface photophysics studies. *J. Phys. Chem.* **90**, 5830–5835.
44. I. Gryczynski, Z. Gryczynski, J.R. Lakowicz (1997). Two-photon excitation by the evanescent wave from total internal reflection. *Anal. Biochem.* **247**, 69–76.
45. D. Loerke, B. Preitz, W. Stuhmer, M. Oheim (2000). Super-resolution measurements with evanescent-wave fluorescence excitation using variable beam incidence. *J. Biomed. Opt.* **5,** 23–30.
46. S. Lam, C. MacAulay, J.C. LeRiche, B. Palcic (2000). Detection and localization of early lung cancer by fluorescence bronchoscopy. *Cancer* **89**, 2468–2473.
47. A. Zimmermann, M. Ritsch-Marte, H. Kostron (2001). mTHPC-mediated photodynamic diagnosis of malignant brain tumors. *Photochem. Photobiol.* **74**, 611–616.
48. A. Zangaro, L. Silveira, R. Manoharan, G. Zonios, I. Izkan, R. R. Dasari, M.S. Feld (1996). Rapid multiexcitation fluorescence spectroscopy system for in vivo tissue diagnosis. *Appl. Opt.* **35**, 5211–5219.
49. Y.K.A. Yang, E.J. Celmer, M. Zarawska-Szczepaniak, R.R. Alfano (1997). Fundamental differences of excitation spectrum between malignant and benign breast tissue. *Photochem. Photobiol.* **66**, 518–522.
50. E. Koumantakis, A. Vasileiou, A. Makrigiannakis, E. Unsold, T.G. Papazoglou (1997). Spectral variations of laser-induced tissue emission during in vivo detection of malignancies in the female genital tract. *J. Photochem. Photobiol. B* **40**, 183–186.

51. A.E. Profio, O.J. Balchum, S. Cartens (1986). Digital-background subtraction for fluorescence imaging. *Med. Phys.* **13**, 717–721.
52. R. Baumgartner, E. Unsold (1987). High contrast fluorescence imaging using two-wavelength laser excitation and image processing. *J. Photochem. Photobiol. B* **1**, 130–132.
53. O.J. Balchum, A.E. Profio, N.J. Razum (1987). Ratioing fluorometer probe for localizing carcinoma *in situ* in bronchi of the lung. *Photochem. Photobiol.* **46**, 887–891.
54. M. Sinaasappel, H.J.C.M. Sterenborg (1993). Quantification of the hematoporphyrin derivative by fluorescence measurement using dual-wavelength excitation and dual-wavelength detection. *Appl. Opt.* **32**, 541–548.
55. S. Andersson-Engels, J. Johansson, S. Svanberg (1994). Medical diagnostic system based on simultaneous multispectral fluorescence imaging. *Appl. Opt.* **33**, 8022–8029.
56. K. Svanberg, I. Wang, S. Colleen, I. Idvall, C. Ingvar, R. Rydell, D. Jocham, H. Diddens, S. Bown, G. Gregory, S. Montàn, S. Andersson-Engels, S. Svanberg (1998). Clinical multi-color fluorescence imaging of malignant tumors-initial experience. *Acta Radiol.* **39**, 2–9.
57. S. Andersson-Engels, G. Canti, R. Cubeddu, C. Eker, C. af Klinteberg, A. Pifferi, K. Svanberg, S. Svanberg, P. Taroni, G. Valentini, I. Wang (2000). Preliminary evaluation of two fluorescence imaging methods for the detection and the delineation of basal cell carcinomas of the skin. *Lasers Surg. Med.* **26**, 76–82.
58. T. Ni, L.A. Melton (1991). Fluorescence lifetime imaging: an approach for fuel equivalence ratio imaging. *Appl. Spectrosc.* **45**, 938–943.
59. G. Marriott, R.M. Clegg, T.M. Jovin (1991). Time resolved imaging microscopy. Phosphorescence and delayed fluorescence imaging. *Biophys. J.* **60**, 1374–1387.
60. T.W.J. Gadella, T.M. Jovin, R.M. Clegg (1993). Fluorescence lifetime imaging microscopy (FLIM): Spatial resolution of microstructures on the nanosecond time scale. *Biophys. Chem.* **48**, 221–239.
61. R. Cubeddu, A. Pifferi, P. Taroni, G. Valentini (1995). Tumor detection in mice by measurement of fluorescence decay time matrices. *Opt. Lett.* **20**, 2553–2555.
62. K. Dowling, M.J. Dayel, M.J. Lever, P.M.W. French, J.D. Hares, A.K.L. Dymoke-Bradshaw (1998). Fluorescence lifetime imaging with picosecond resolution for biomedical applications. *Opt. Lett.* **23**, 810–812.
63. G. Wagniéres, J. Mizeret, A. Studzinski, H. van den Bergh (1997). Frequency-domain fluorescence lifetime imaging for endoscopic clinical cancer photodetection: apparatus design and preliminary results. *J. Fluoresc.* **7,** 75–84.
64. T. Ni, L.A. Melton (1996). Two-dimensional gas phase temperature measurements using fluorescence lifetime imaging. *Appl. Spectrosc.* **50,** 1112–1116.
65. A. Periasamy, P. Wodnicki, X.F. Wang, S. Kwon, G. Gordon, B. Herman (1996). Time-resolved fluorescence lifetime imaging microscopy using a picosecond pulsed tunable dye laser system. *Rev. Sci. Instrum.* **67**, 3722–3731.
66. R.M. Clegg, P.C. Schneider, T.M. Jovin (1996). *Biomedical Optical Instrumentation and Laser-Assisted Biotechnology* (pp. 143–156). Kluwer Academic Publisher, Dordrecht.
67. J. Mizeret, G. Wagniéres, T. Stepinac, H. van den Bergh (1997). Endoscopic tissue characterization by frequency-domain flourescence lifetime imaging (FD-FLIM). *Lasers Med. Sci.* **12**, 209–217.
68. R. Cubeddu, G. Canti, A. Pifferi, P. Taroni, G. Valentini (1997). Fluorescence lifetime imaging of experimental tumors in hematoporphyrin derivative-sensitized mice. *Photochem. Photobiol.* **66**, 229–236.
69. R. Cubeddu, A. Pifferi, A. Torricelli, G. Valentini, F. Rinaldi, E. Sorbellini (1999). Fluorescence lifetime imaging: an application to the detection of skin tumors. *IEEE J. Sel. Top. Quantum Electron.* **5**, 923–929.

70. O. Holub, M.J. Seufferheld, C. Gohlke, Govindjee, R.M. Clegg (2000). Fluorescence lifetime imaging (FLI) in real-time (a new technique in photosynthesis research. *Photosynthetica* **38**, 581–599.
71. L.-C. Li, L.B. McGown (1996). On-the-fly frequency-domain fluorescence lifetime detection in capillary electrophoresis. *Anal. Chem.* **68**, 2737–2743.
72. S.J. Lassiter, W. Stryiewski, B.L. Legendre, R. Erdmann, M. Wahl, J. Wurm, R. Peterson, L. Middendorf, S.A. Soper (2000). Time-resolved fluorescence imaging of slab gels for lifetime base-calling in DNA sequencing applications. *Anal. Chem.* **72**, 5373–5382.
73. G. Valentini, C. D'Andrea, D. Comelli, A. Pifferi, P. Taroni, A. Torricelli, R. Cubeddu, C. Battaglia, C. Consolandi, G. Salani (2000). Time-resolved DNA-microarray reading by an intensified CCD for ultimate sensitivity, *Opt. Lett.* **25**, 1648–1650.
74. D.M. Jameson, E. Gratton, R. Hall (1984). The measurement and analysis of heterogeneous emission by multifrequency phase and modulation fluorometry. *Appl. Spectrosc. Rev.* **20**, 55–106.
75. P.C. Schneider, R.M. Clegg (1997). Rapid acquisition, analysis, and display of fluorescence lifetime-resolved images for real-time applications. *Rev. Sci. Instrum.* **68**, 4107–4119.
76. A. Yariv (1988). *Quantum Electronics, 3rd edition.* John Wiley & Sons, New York.
77. T.W.J. Gadella, A. van Hoek, A.J.W.G. Visser (1997). Construction and characterization of a frequency-domain fluorescence lifetime microscopy system. *J. Fluoresc.* **7**, 35–44.
78. J. Mizeret, T. Stepinac, M. Hansroul, A. Studzinski, H. van den Bergh, G. Wagniéres (1999). Instrumentation for real-time fluorescence lifetime imaging in endoscopy. *Rev. Sci. Instrum.* **70**, 4689–4701.
79. A.A. Istratov, O.F. Vyvenko (1999). Exponential analysis in physical phenomena. *Rev. Sci. Instrum.* **70**, 1233–1257.
80. L. Luo, R.C. Salunga, H. Guo, A. Bittner, K.C. Joy, J.E. Galindo, H. Xiao, K.E. Rogers, J.S. Wan, M.R. Jackson, M.G. Erlander (1999). Gene expression profiles of laser-captured adjacent neuronal subtypes. *Nat. Med.* **5**, 117–122.
81. M. Bittner, P. Meltzer, Y. Chen, Y. Jiang, E. Seftor, M. Hendrix, M. Radmacher, R. Simon, Z. Yakhini, A. Ben-Dor, N. Sampas, E. Dougherty, E. Wang, F. Marincola, C. Gooden, J. Lueders, A. Glatfelter, P. Pollock, J. Carpten, E. Gillanders, D. Leja, K. Dietrich, C. Beaudry, M. Berens, D. Alberts, V. Sondak, M. Hayward, J. Trent (2000). Molecular classification of cutaneous malignant melanoma by gene expression profiling. *Nature* **406**, 536–540.
82. P.A. Clarke, R. te Poele, R. Wooster, P. Workman (2001). Gene expression microarray analysis in cancer biology, pharmacology, and drug development: progress and potential. *Biochem. Pharmacol.* **62**, 1311–1336.
83. P.R. Bevington (1969). *Data Reduction and Error Analysis for the Physical Sciences.* McGraw-Hill, New York.
84. W.H. Press, S.A. Teukolsky, W.T. Vetterling, B.P. Flannery (1992). *Numerical Recipes in C: The Art of Scientific Computing.* Cambridge University Press, New York.
85. G.A. Wagnieres, W.M. Star, B.C. Wilson (1998). In vivo fluorescence spectroscopy and imaging for oncological applications. *Photochem. Photobiol.* **68**, 603–639.
86. R. Richards-Kortum, E. Sevick-Muraca (1996). Quantitative optical spectroscopy for tissue diagnostic, *Annu. Rev. Phys. Chem.* **47**, 555–606.
87. J.R. Lakowicz (1999). *Principles of Fluorescence Spectroscopy.* Plenum Press, New York.
88. E.J. Miller, S. Gay (1987). The collagens: An overview and update, *Methods Enzymology* **144**, 3–41.

89. E. Fujimori (1985). Changes induced by ozone and ultraviolet light in type I collagen. Bovine Achilles tendon collagen versus rat tail tendon collagen. *Eur. J. Biochem.* **152**, 299–306.
90. E. Schwartz, F.A. Cruickshank, C.C. Christensen, J.S. Perlish, M. Lebwohl (1993). Collagen alterations in chronically sun-damaged human skin. *Photochem. Photobiol.* **58**, 841–844.
91. R.W. Glanville, D. Breitkreutz, M. Meitinger, P.P. Fietzek (1983). Completion of the amino-acid sequence of the a1 chain from type I calf skin collagen. *Biochem. J.* **215**, 183–189.
92. J. Barnes, L.F. Morton (1976). Presence of Type III collagen in guinea-pig dermal scar. *Biochem. J.* **157**, 263–266.
93. J.M. Menter, G.D. Williamson, K. Carlyle, C.L. Moore, I. Willis (1995). Photochemistry of type I acid-soluble calf skin collagen: dependence on excitation wavelength. *Photochem. Photobiol.* **62**, 402–408.
94. A. Uchiyama et al. (1991). Fluorophores from aging human cartilage. *J. Biochem.* **110**, 714–718.
95. L. Marcu, D. Cohen, J.M. Maarek, W.S. Grundfest (2000). Characterization of type I, II, III, IV, and V collagens by time-resolved laser-induced fluorescence spectroscopy. In *Optical Biopsy III,* SPIE Proc. Volume 3917 pp. 93–101. SPIE, Bellingham, WA.
96. J.M. Maarek, L. Marcu, W.J. Snyder, W.S. Grundfest (2000). Time-resolved fluorescence spectra of arterial fluorescent compounds reconstruction with Laguerre expansion technique. *Photochem. Photobiol.* **71**, 178–187.
97. http://www.molecularprobes.com/
98. W.T. Mason (1999). *Fluorescent and Luminescent Probes for Biological Activity.* Academic Press. Life Science Resources Ltd., Cambridge, UK.
99. R. Kraayenhof, A.W.G. Visser, H.C. Gerritsen (Eds) (2002). Fluorescence Spectroscopy, Imaging and Probes. Springer-Verlag, Berlin, Heidelberg, New York.
100. A.L. Givan (2001). *Flow Cytometry: First Principle* (2nd edn). Wiley-Liss, New York.
101. R. Nunez (2001). *Flow Cytometry for Research Scientists:* Principles and Applications. Horizon Press. Wymondham, Norfolk, UK
102. http://www.isac-net.org
103. http://www.cytometry.org
104. J. Devlin (1997). *High Throughput Screening,* Dekker, New York.
105. http://www.tecan.com
106. http://www.owicki.com
107. F.S. Ligler, C.A. Rawe-Taitt (Eds) (2002). *Optical Biosensors: Present and Future.* Elsevier, Amsterdam.
108. P.N. Prasad (2003). Introduction to Biophotonics. John Wiley & Sons, Inc. Hoboken, New Jersey
109. M.S. Chee, R. Yang, E. Hubbell, A. Berno, X.C. Huang, D. Stern, J. Winkler, D.J. Lockhart, M.S. Morris, S.P.A. Fodor (1996). Accessing genetic information with high-density DNA arrays. *Science* **274**, 610–614.
110. J. Khan, L. Bittner, P.S. Meltzer, J.M. Trent (1999). DNA microarray technology: the anticipated impact on the study of human disease. *Biochim. Biophys. Acta* **1423**, M17–M28.
111. P.A. Clarke, R. te Poele, R. Wooster, P. Workman (2001). Gene expression microarray analysis in cancer biology, pharmacology, and drug development: progress and potential. *Biochem. Pharmacol.* **62**, 1311–1336.
112. L.J. van 't Veer, H. Dai, M.J. van De Vijver, Y.D. He, A.A. Hart, M. Mao, H.L. Peterse, K. van Der Kooy, M.J. Marton, A.T. Witteveen, G.J. Schreiber, R.M. Kerkhoven,

C. Roberts, P.S. Linsley, R. Bernards, S.H. Friend (2002). Gene expression profiling predicts clinical outcome of breast cancer, *Nature* **415**, 530–536.
113. N.P. Gerry, N.E. Witowski, J. Day, R.P. Hammer, G. Barany, F. Barany (1999). Universal DNA microarray method for multiplex detection of low abundance point mutations. *J. Mol. Biol.* **292**, 251–262.
114. H. He, B.K. Nunnally, L.-C. Li, L.B. McGown (1998). On-the-fly fluorescence lifetime detection of dye-labeled DNA primers for multiplex analysis. *Anal. Chem.* **70**, 3413–3418.
115. L.-C. Li, H. He, B.K. Nunnally, L.B. McGown (1997). On-the-fly fluorescence lifetime detection of labeled DNA primers. *J. Chromatogr. B* **695**, 85–92.
116. B.K. Nunnally, H. He, L.-C. Li, S.A. Tucker, L.B. McGown (1997). Characterizaton of visible dyes for four-decay fluorescence detection in DNA sequencing. *Anal. Chem.* **69**, 2392–2397.
117. M. Sauer, J. Arden-Jacob, K.H. Drexhage, F. Göbel, U. Lieberwirth, R. Müller, J. Wolfrum, C. Zander (1998). Time-resolved identification of individual mononucleotide molecules in aqueous solution with pulsed semiconductor lasers. *Bioimaging* **6**, 14–24.
118. U. Lieberwirth, J. Arden-Jacob, K.H. Drexhage, D.P. Herten, R. Muller, M. Neumann, A. Schulz, S. Siebert, G. Sagner, S. Klingel, M. Sauer, J. Wolfrum (1998). Multiplex dye DNA sequencing in capillary gel electrophoresis by diode laser-based time-resolved fluorescence detection. *Anal. Chem.* **70**, 4771–4779.
119. I.J. Bigio, J.R. Mourant (1997). Ultaviolet and visible spectroscopies for tissue diagnostics: fluorescence spectroscopy and elastic-scattering spectroscopy. *Phys. Med. Biol.* **42**, 803–814.
120. S. Andersson-Engels, C. Klinteberg, K. Svanberg, S. Svanberg (1997). In-vivo fluorescence imaging for tissue diagnostics. *Phys. Med. Biol.* **42,** 815–824.
121. T.G. Papazoglou (1995). Malignancies and atherosclerotic plaque diagnosis: is laser-induced fluorescence spectroscopy the ultimate solution? *J. Photochem. Photobiol. B* **28**, 3–11.
122. B.B. Das, F. Liu, R.R. Alfano (1997). Time-resolved fluorescence and photon migration studies in biomedical and model random media. *Rep. Rog. Phys.* **60**, 227–292.
123. N. Ramanujam (2000). Fluorescence spectroscopy of neoplastic and non-neoplastic tissues. *Neoplasia* **2**, 89–117.
124. T. Glanzmann, J.P. Ballini, H. van der Bergh, G. Wagnieres (1999). Time-resolved spectrofluorometer for clinical tissue characterization during endoscopy. *Rev. Sci. Instrum.* **70**, 4067–4077.
125. R. Cubeddu, G. Canti, A. Pifferi, P. Taromi, G. Valenti (1997). Fluorescence lifetime imaging of experimental tumors in hematoporphyrin derivative-sensitized mice. *Photochem. Photobiol.* **66**, 229–236.
126. J.R. Lakowicz, H. Szmacinski, K. Nowaczyk, K.W. Berndt, M. Johnson (1992). Fluorescence lifetime imaging. *Anal Biochem.* **202**, 316–330.
127. L. Marcu, M. Fishbein, J.M. Maarek, W.S. Grundfest (2001). Discrimination of lipid-rich atherosclerotic lesions of human coronary artery by time-resolved laser-induced fluorescence spectroscopy. *Atheroscler. Thromb. Vasc. Biol.* **21**, 1244–1250.
128. J.J. Baraga, P. Taroni, Y.D. Park, K. An, A. Maestri, L.L. Tong, R.P. Rava, C. Kittrell, R.R. Dasari, M.S. Feld (1989). Ultraviolet laser induced fluorescence of human aorta. *Spectrochim. Acta* **45A**, 95–99.
129. M.A. Mycek, K.T. Schomacker, N.S. Nishioka (1998). Colonic polyp differentiation using time-resolved autofluorescence spectroscopy. *Gastrointest. Endosc.* **48**, 390–394.

130. M.F. Mitchell, S.B. Cantor, N. Ramanujam, G. Tortolero-Luna, R. Richards-Kortum (1999). Fluorescence spectroscopy for the diagnosis of squamous intraepithelial lesions of the cervix. *Obstet. Gynecol.* **93**, 462–470.
131. L. Bartorelli, M.B. Leon, Y. Almagor, L. Prevosti, J.A. Swain, C.L. McIntoch, R.F. Neville, M.D. House, R.F. Bonner (1991). In vivo human atherosclerotic plaque recognition by laser-excited fluorescence spectroscopy. *Journal of the American College of Cardiology*, **17**(6), 160B–168B.
132. http://www.nci.nih.gov
133. R.M. Lycette, R.B. Leslie (1965). *Lancet* **2**, 436.
134. L. Marcu, R.C. Thompson, S. Garde, M. Sedrak, K.L. Black, W.H. Yong (2002). Time-resolved fluorescence spectroscopy of primary human brain tumors. In: *Optical Biopsy IV* SPIE Proc. Volume 4613, (pp. 183–187). SPIE, Wellingham, WA.
135. S. Andersson-Engels, J. Johansson, S. Svanberg (1990). The use of time-resolved fluorescence for diagnosis of atherosclerotic plaque and malignant tumours. *Spectrochim. Acta* **40A**, 1203–1210.
136. R.J. Macuiunas (1999). Interactive image-guided surgical technology for glial tumor resection. In: M.S. Berger, C.B. Wilson (Eds), *The Gliomas* (pp. 555–563). W.B. Saunders, Philadelphia.
137. R. Sawaya, M.A. Hammoud, B.E. Ligon, G.N. Fuller (1999). Intraoperative localization of tumor and margin. In: M.S. Berger, C.B. Wilson (Eds), *The Gliomas* (pp. 361–375). W.B. Saunders, Philadelphia.
138. G. Bottiroli, A.C. Croce, D. Locatelli, R. Nano, E. Giombelli, A. Messina, E. Benericetti (1998). Brain tissue autofluorescence: an aid for intraoperative delineation of tumor resection margin. *Cancer Detect Prev.* **22**(4), 330–339.
139. W.C. Lin, S.A. Toms, M. Motamedi, E.D. Jansen, A. Mahadevan-Jansen (2000). Brain tumor demarcation using optical spectroscopy; an in vitro study. *J. Biomed. Opt.* **5**(2), 214–220.
140. Y.G. Chung, J.A. Schwartz, C.M. Gardner, R.E. Sawaya, S.L. Jacques (1997). Diagnostic potential of laser-induced autofluorescence emission in brain tissue. *J. Korean Med Sci* **12**, 135–142.
141. W.S. Poon, B. Ch, K.T. Schomacker, T.F. Deutsch, R.L. Martuza (1992). Laser-induced fluorescence: experimental intraoperative delineation of tumor resection margins. *J. Neurosurg.* **76**, 679–686.
142. L. Marcu, P. Butte, W.H. Yong, R.C. Thompson, K.L. Black, and B. Pikul (2003). Diagnosis of human brain tumors by lifetime fluorescence spectroscopy. *Lasers Surg. Med.* **S15**, 51.
143. T.J. Dougherty (1993). Photodynamic therapy. *Photochem. Photobiol.* **58**, 895-900.
144. R. Ackroyd, C. Kelty, N. Brown, M. Reed (2001). The history of photodetection and photodynamic therapy. *Photochem. Photobiol.* **74**, 656–669.
145. A. Zimmermann, M. Ritsch-Marte, H. Kostron (2001). mTHPC-mediated photodynamic diagnosis of malignant brain tumors. *Photochem. Photobiol.* **74**, 611–616.
146. A.E. Profio (1984). Laser excited fluorescence of hematoporphyrin derivative for diagnosis of cancer. *IEEE J. Quantum Electron.* **QE-20**, 1502–1507.
147. R. Baumgartner, H. Flissinger, D. Jocham, H. Lenz, L. Rupercht, H. Steep, E. Unsold (1987). A fluorescence imaging device for endoscopic detection of early stage cancer – instrumental and experimental studies. *Photochem. Photobiol.* **46**, 759–763.
148. E. Unsold, R. Baumgartner, W. Beyer, D. Jocham, H. Steep (1990). Fluorescence detection and photodynamic treatment of photosensitizied tumors in special consideration of urology. *Lasers Med. Sci.* **5**, 207–212.

149. J.C. Kennedy, R. Pottier, D.C. Pross (1990). Photodynamic therapy with endogenous protoporphyrin IX: basic principles and present clinical experience. *J. Photochem. Photobiol. B* **6**, 143-148.
150. J.C. Kennedy, R.H. Pottier (1992). Endogenous protoporhyrin IX, a clinically useful photosensitizer for photodynamic therapy. *J. Photochem. Photobiol. B* **14**, 275–292.
151. Q. Peng, K. Berg, J. Moan, K. Magne, J.M. Nesland (1997). 5-aminolevulinic acid-based photodynamic therapy: principles and experimental research. *Photochem. Photobiol.* **65**, 235–251.
152. R. Pottier, Y.F.A. Chow, J.-P. LaPlante, T.G. Truscott, J.C. Kennedy, L.A. Beiner (1986). Non-invasive technique for obtaining fluorescence excitation and emission spectra *in vivo*. *Photochem. Photobiol.* **44**, 679–687.
153. J. Bedwell, A.J. MacRobert, D. Phillips, S. Bown (1992). Fluorescence distribution and photodynamic effect of ALA-induced PP IX in the DMH rat colonic tumour model. *Br. J. Cancer* **65,** 818–824.
154. H. Schneckenburger, K. König, K. Kunzi-Rapp, C. Westphal-Frösch, A. Rück (1993). Time-resolved *in-vivo* fluorescence of photosensitizing porphyrins. *J. Photochem. Photobiol. B* **21**, 143–147.
155. Q. Peng, T. Warloe, J. Moan, H. Heyerdahl, H.B. Steen, J.M. Nesland, K.-E. Giercksky (1995). Distribution of 5-aminolevulinic acid-induced porphyrins in noduloulcerative basal cell carcinoma. *Photochem. Photobiol.* **62**, 906–913.
156. K. Svanberg, S. Andersson-Engels, D. Killander, I. Wang, U. Stenram, R. Berg, J. Johansson, S. Svanberg (1994). Photodynamic therapy of non-melanoma malignant tumours of the skin using topical d-aminolevulinic acid sensitization and laser irradiation. *Br. J. Dermatol.* **130**, 741–751.
157. N. van der Veen, H.S. Bruijin, R.J.W. Berg, W. Star (1996). Kinetics and localisation of PpIX fluorescence after topical and systemic ALA application, observed in skin and skin tumours of UVB-treated mice. *Br. J. Cancer* **73**, 925–930.
158. C. af Klinteberg, A.M.K. Enejder, I. Wang, S. Andersson-Engels, S. Svanberg, K. Svanberg (1999). Kinetic fluorescence studies of 5-aminolaevulinic acid-induced protoporphyrin IX accumulation in basal cell carcinomas. *J. Photochem. Photobiol. B* **49**, 120–128.
159. S. Andersson-Engels, G. Canti, R. Cubeddu, C. Eker, C. af Klinteberg, A. Pifferi, K. Svanberg, S. Svanberg, P. Taroni, G. Valentini, I. Wang (2000). Preliminary evaluation of two fluorescence imaging methods for the detection and the delineation of basal cell carcinomas of the skin. *Lasers Surg. Med.* **26,** 76–82.
160. F. Fischer, E.F. Dickson, R.H. Pottier, H. Wieland (2001). An affordable, portable fluorescence imaging device for skin lesion detection using a dual wavelength approach for image contrast enhancement and aminolaevulinic acid-induced protoporphyrin IX. Part I. Design, spectral and spatial characteristics. *Lasers Med. Sci.* **16**, 199–206.
161. F. Fischer, E.F. Dickson, J.C. Kennedy, R.H. Pottier (2001). An affordable, portable fluorescence imaging device for skin lesion detection using a dual wavelength approach for image contrast enhancement and aminolaevulinic acid-induced protoporphyrin IX. Part II. In vivo testing. *Lasers Med. Sci.* **16**, 207–212.
162. R. Cubeddu, A. Pifferi, A. Torricelli, G. Valentini, F. Rinaldi, E. Sorbellini (1999). Fluorescence lifetime imaging: an application to the detection of skin tumors. *IEEE J. Sel. Top. Quantum Electron.* **5**, 923–929.
163. R. Cubeddu, G. Canti, A. Pifferi, P. Taroni, G. Valentini (1997). A real-time system for fluorescence lifetime imaging. In: T. Vo-Dinh, R.A. Lieberman, G.G. Vurek (Eds),

Biomedical Sensing, Imaging, and Tracking Technologies II (Volume 2976, pp. 98–104). SPIE Press, Bellingham, WA.

164. G. Bottiroli, A.C. Croce, D. Locatelli, R. Marchesini, E. Pignoli, S. Tomatis, C. Cuzzoni, S. Di Palma, M. Dal Fante, P. Spinelli (1995). Natural fluorescence of normal and neoplastic human colon: a comprehensive ex vivo study. *Lasers Surg. Med.* **16**, 48–60.
165. K.T. Schomacker, J.K. Frisoli, C.C. Compton, T.J. Flotte, J.M. Richter, N.S. Nishioka, T.F. Deutsch (1992). Ultraviolet laser-induced fluorescence of colonic tissue: basic biology and diagnostic potential. *Lasers Surg. Med.* **12**, 63–78.
166. J. Hung, S. Lam, J.C. LeRiche, B. Palcic (1991). Autofluorescence of normal and malignant bronchial tissue. *Lasers Surg. Med.* **11**, 99–105.
167. C.S. Betz, H. Stepp, P. Janda, S. Arbogast, G. Grevers, R. Baumgartner, A. Leunig (2002). A comparative study of normal inspection, autofluorescence and 5-ALA-induced PPIX fluorescence for oral cancer diagnosis. *Int. J. Cancer* **97**, 245–252.
168. V. Fuster (1996). *Syndromes of Atherosclerosis. Correlations of Clinical Imaging and Pathology.* Armonk, New York.
169. H.C. Stary, A.B. Chandler, R.E. Dinsmore, V. Fuster, S. Glagov, W. Insull, Jr., M.E. Rosenfeld, C.J. Schwartz, W.D. Wagner, R.W. Wissler (1995). A definition of advanced types of atherosclerotic lesions and a histological classification of atherosclerosis. A report from the Committee on Vascular Lesions of the Council on Arteriosclerosis, American Heart Association. *Arterioscler. Thromb. Vasc. Biol.* **15**, 1512–1531.
170. R. Richards-Kortum, R.P. Rava, M. Fitzmaurice, J.R. Kramer, M.S. Feld (1991). 476 nm Excited laser-induced fluorescence spectroscopy of human coronary arteries: Applications in cardiology. *Am. Heart J.* **122**, 1141–1150.
171. L.I. Laifer, K.M. O'Brien, M.L. Stetz, G.R. Gindi, T.J. Garrand, L.I. Deckelbaum (1989). Biochemical basis for the difference between normal and atherosclerotic arterial fluorescence. *Circulation* **80**, 1893–1901.
172. L.I. Deckelbaum, S.P. Desai, C. Kim, J.J. Scott (1987). Discrimination of normal and atherosclerotic aorta by laser-induced fluorescence. *Lasers Surg. Med.* **7**, 330–335.
173. J.J. Baraga, P. Taroni, Y.D. Park, K. An, A. Maestri, L.L. Tong, R.P. Rava, C. Kittrell, R.R. Dasari, M.S. Feld (1989). Ultraviolet laser induced fluorescence of human aorta. *Spectrochim. Acta* **45A**, 95–99.
174. S. Andersson-Engels, J. Johansson, S. Svanberg (1990). The use of time-resolved fluorescence for diagnosis of atherosclerotic plaque and malignant tumours. *Spectrochim. Acta* **40A**, 1203–1210.
175. J. Morguet, R.E. Gabriel, A.B. Buchwald, G.S. Werner, R. Nyga, H. Kreuzer (1997). Single-laser approach for fluorescence guidance of excimer laser angioplasty at 308 nm: Evaluation in vitro and during coronary angioplasty. *Lasers Surg. Med.* **20**, 382–393.
176. A.A. Oraevsky, S.L. Jacques, G.H. Pettit, R.A. Sauerbrey, F.K. Tittel, J.H. Nguy, P.D. Henry (1993). XeCl Laser-Induced fluorescence of atherosclerotic arteries: Spectral similarities between lipid-rich lesions and peroxidized lipoproteins. *Circ. Res.* **72**, 84–90.
177. M. Sartori, K. Weibaecher, G.L. Valderrama, S. Kubodera, R.C. Chin, M.J. Berry, F.K. Tittel, R. Sauerbrey, P.D. Henry (1988). Laser-induced autofluorescence of human arteries. *Circ. Res.* **63**, 1053–1059.
178. J. Morguet, B. Korber, B. Abel, H. Hippler, V. Wiegand, H. Kreuzer (1994). Autofluorescence spectroscopy using a simultaneous plaque ablation and fluorescence excitation. *Lasers Surg. Med.* **14**, 238–248.
179. L. Marcu, W.S. Grundfest, M.C. Fishbein (2003). Time-resolved laser-induced fluorescence spectroscopy for staging atherosclerotic lesions. In: M.A. Mycek,

B. Pogue (Eds), *Handbook of Biomedical Fluorescence* (pp. 397–430). Marcel Dekker, New York.

180. R. Weissleder, U. Mahmood (2001). Molecular imaging, *Radiology*, **219,** 316–333.
181. J.W. Hastings (2001). Bioluminescence. In: N. Sperelakis (Ed), *Cell Physiology Sourcebook: A Molecular Approach* (pp. 1115–1131). Academic Press, New York.
182. www.xenogen.com.
183. V. Ntziachristos, R. Weissleder (2001). Experimental three-dimensional fluorescence reconstruction of diffuse media by use of normalized Born approximation. *Opt. Lett.* **26**, 893–895.
184. J. Wu, L. Perelman, R.R. Dasari, M.S. Feld (1997). Fluorescence tomographic imaging in turbid media using early-arriving photons and Laplace transform, *Proc. Natl. Acad. Sci. U.S.A.* **94**, 8783–8788.
185. M. Vooijs, J. Jonkers, S. Lyons, A. Berns (2002). Noninvasive imaging of spontaneous retinoblastoma pathway-dependent tumors in mice. *Cancer Research* **62**, 1862–1867.
186. A. Wetterwald, G. van der Pluijm, I. Que, B. Sijmons, J. Buijs, M. Karperien, C.W.G.M. Löwik, E. Gautschi, G.N. Thalmann, M.G. Cecchini (2002). Optical imaging of cancer metastasis to bone marrow. *Am. J. Pathol.* **160**, 1143–1153.
187. J. Hardy, M. Edinger, M. Bachmann, R.S. Negrin, C.G. Fathman, C.H. Contag (2001). Bioluminescence imaging of lymphocyte trafficking in vivo. *Exp. Hematol.* **29**, 1353–1360.
188. T.J. Sweeney, V. Mailänder, A.A. Tucker, A.B. Olomu, W. Zhang, Y. Cao, R.S. Negrin, C.H. Contag (1999). Visualizing the kinetics of tumor-cell clearance in living animals. *Proc. Natl. Acad. Sci. U.S.A.* **96**, 12044–12049.
189. M. Bouvet, J. Wang, S.R. Nardin, R. Nassirpour, M. Yang, E. Baranov, P. Jiang, A.R. Moossa, R.M. Hoffman (2001). Real-time optical imaging of primary tumor growth and multiple metastasic events in a pancreatic cancer orthotopic model. *Cancer Res.* **62**, 1534–1540.
190. M. Yang, E. Baranov, P. Jiang, F. Sun, X. Li, L. Li, S. Hasegawa, M. Bouvet, M. Al-Tuvaijri, H. Shimada, A.R. Moossa, S. Penman, R.M. Hoffman (2000). Whole-body optical imaging of green fluorescent protein-expressing tumors and metastases. *Proc. Natl. Acad. Sci. U.S.A.* **97,** 1206–1211.
191. C. Bremer, C. Tung, A. Bogdanov, Jr., R. Weissleder (2002). Imaging of differential protease expression in breast cancer detection of aggressive tumor phenotypes. *Radiology* **222**, 814–818.
192. C. Bremer, S. Bredow, U. Mahmood, R. Weissleder, C. Tung (2001). Optical imaging of Matrix Metalloproteainase-2 activity in tumors: feasibility study in a mouse model. *Radiology* **221**, 523–529.
193. C. Bremer, C.-H. Tung, R. Weissleder (2001). In vivo molecular target assessment of matrix metalloproteinase inhibition. *Nat. Med.* **7**, 743–748.
194. U. Mahmood, C. Tung, A. Bogdanov, R. Weissleder (1999). Near-infrared optical imaging of protease activity for tumor detection. *Radiology* **213**, 866–870.
195. S. Tyagi, D.P. Bratu, F.R. Kramer (1998). Multicolor molecular beacons for allele discrimination. *Nat. Biotechnol.* **16**, 49–53.

Part III
Microscopy Techniques

Chapter 13

Optical microscopy

Herbert Schneckenburger

Table of contents

Abstract

Optical microscopy, including wide-field microscopy, scanning microscopy and micromanipulation, has become an indispensable tool in cell biology and photobiology. The present article is concentrated on wide-field microscopy and covers transillumination and fluorescence microscopy. Basic principles are described as well as advanced techniques, e.g. fluorescence lifetime imaging (FLIM), energy transfer spectroscopy (FRET) and total internal reflection fluorescence microscopy (TIRFM). A few applications are depicted, which may give some insight into a fascinating microcosmos.

13.1 Introduction

The theory of optical microscopy was established in the second half of the nineteenth century by Ernst Abbé (1840–1905), although the first microscopes had been built much earlier. Usually, a microscope is characterized by a highly magnifying objective lens used for imaging a sample with a half-angle σ of the incident cone and a numerical aperture $A = n \times \sin\sigma$, where n is the refractive index of the medium between the sample and the objective lens. The numerical aperture is an important parameter for the quantity of light taken up by a microscope (which is proportional to A^2), for the lateral resolution $\Delta x = \lambda/A$ and for the depth of focus $\Delta z = n \times \lambda/A^2$, with λ being the wavelength of optical radiation.

Epiillumination as well as transillumination are used in optical microscopy. Transillumination is used to visualize a sample (e.g. cell or tissue layer) with high resolution and contrast. Often, contrast enhancing techniques, e.g. phase contrast microscopy or interference contrast microscopy, are applied. Epiillumination is used to visualize non-transparent samples by light that is reflected or diffracted in the reverse direction. Depending on whether direct reflection of incident light is excluded or not, the corresponding technique is named either dark-field microscopy or bright-field microscopy. An epiillumination technique mostly used in life sciences, is fluorescence microscopy. Fluorescence arises upon absorption of light and is related to an electronic transition from an excited state to the ground state of a molecule. Usually, fluorescence is red-shifted as compared with incident light and can be distinguished by using appropriate optical filters. Since the intensity of fluorescence is often, by several orders of magnitude, lower than that of the incident light, fluorescence has to be detected in reverse direction to suppress the excitation light sufficiently well.

Generally, wide-field microscopy and scanning microscopy are distinguished. In the first case a certain part of a sample is illuminated and measured using, e.g., video techniques. In the second case punctual illumination is achieved using a light beam (usually a laser beam) which is scanned over the sample, and the image is created point by point within a memory. This latter technique has several advantages: (1) the optical contrast is enhanced; (2) a highly sensitive photomultiplier (instead of a less sensitive video camera) can be used for image detection, (3) an electronic zoom is created by variation of the scanning amplitude, and (4) the focused laser spot on the sample can be imaged into a pinhole that permits only light from a well-defined layer of the sample to be registered, whereas light from other planes of the sample is excluded. Fluorescence can therefore be recorded layer by layer and combined in a 3-dimensional image. A superposition by out-of-focus images, as occurring in wide-field microscopy, is therefore excluded in confocal laser scanning microscopy (CSLM).

In addition to CLSM, methods of two-photon or multiphoton microscopy have recently been established, where only those molecules of a sample are excited that are precisely in the focus of one or several laser beams, as first described by Denk et al. [1] and later summarized by König [2]. Out-of-focus molecules are not excited and therefore not detected fluorometrically. Using a so-called 4Pi geometry with two opposing lenses of high numerical aperture together with deconvolution techniques, a lateral resolution of 100–150 nm and an axial resolution around 150 nm have been attained [3]. Lateral resolution has been further improved by the so-called stimulated emission depletion microscopy, where a first laser pulse is used to excite a diffraction limited spot of the sample, and a second pulse to deactivate the molecules at the edge of the sample, such that fluorescence arises only from its central part with a diameter down to 33 nm [4].

However, CLSM and multiphoton microscopy need rather complex (and expensive) equipment. In particular, picosecond or femtosecond laser systems are required for multiphoton microscopy. Therefore, wide-field microscopy is still frequently used. This method as well as related new techniques and applications are summarized in the present chapter.

13.2 Transmission microscopy

13.2.1 Principle and theory

The principle of transmission microscopy (Köhler's illumination) is depicted in Figure 1. An image (I) of the object (O) (sample) is formed by the objective lens (Obj). This image can be observed by the eye (E) using the ocular (Oc) (2 lenses) as well as the eye lens. To obtain a homogeneous illumination of the object (O), a homogeneously illuminated field diaphragm (Fd) rather than the inhomogeneous light source (L) is imaged into the sample. The light source itself represents the entrance pupil, which is imaged into the aperture plane (AC) of the condensing lens (C) and further into the aperture plane (AObj) of the objective lens, as depicted in the right part of Figure 1. Such illumination guarantees the most effective and homogeneous imaging. In video microscopy a screen (e.g. film or CCD array) is placed in the image plane (I), thus omitting the ocular (Oc). Objective lenses with 2.5 up to 100 times magnification and oculars with 8 to 12 times magnification are commonly used.

Image formation in a microscope is due to diffraction of incident light by the sample (which acts like an optical grating) and needs several orders of diffraction, at least the 0th and 1st order. Therefore, objects resolved by the microscope must

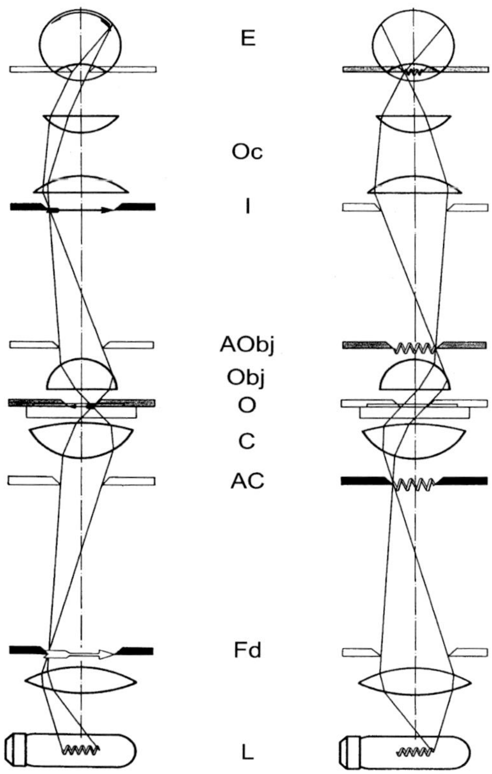

Figure 1. Object ray (left) and pupil ray (right) in a microscope using Köhler's illumination (abbreviations see text); reproduced from Göke [5] with modifications.

have a minimum size such that these diffraction orders can be detected by the objective lens. This results in the equation $\Delta x = \lambda/A$ with Δx being the microscopic (lateral) resolution, λ the wavelength of radiation and A the numerical aperture of the objective lens, if the numerical aperture of the illumination ray is negligible. Otherwise, the numerical apertures of the objective lens and the condensing lens sum up to a maximum of $2A$, thus resulting in

$$\Delta x = \lambda/2A \tag{1}$$

When using a wavelength of 500 nm for irradiation, the microscopic resolution Δx is about 330 nm for a numerical aperture of 0.75 (a typical value for a lens with 40 times magnification) and about 180 nm for a numerical aperture of 1.40 (value of an oil immersion lens with 100 times magnification).

13.2.2 Phase contrast microscopy

Samples in a microscope become visible if the incident light is partly absorbed. However, in thin samples, e.g. cell monolayers, this absorption is often very small, and a phase shift of transmitted light is more easily used for imaging than absorption. Phase contrast microscopy is based on the fact that the phase shift between the 0th order and higher orders of diffraction varies between different parts of the sample. Since the 0th order usually is the most intensive one, this order has to be attenuated (and further shifted in its phase) to get high quality images. This becomes possible when using a ring aperture for the condensing lens which is imaged in the aperture plane of the objective lens (Figure 1, right). Phase contrast objective lenses have a phase-retarding ring at this position which causes a selective attenuation and phase shift of the 0th order, whereas higher orders of diffraction (with different directions of light propagation) are not affected. Therefore, after recombining the manipulated 0th order with the original higher orders of diffraction in the image plane of the microscope, phase contrast microscopy permits high contrast imaging of cells and their substructures, which would not be detected by conventional transmission microscopy. Usually, ring apertures are standardized and used in combination with specific objective lenses.

13.2.3 Interference contrast microscopy

As an alternative to phase contrast microscopy, interference contrast microscopy can be used to detect subcellular structures with high contrast. The principle of this method (Figure 2) is that a linearly polarized light wave is split by a birefringent material (a so-called Wollaston prism) into two perpendicularly polarized waves that transmit to the sample with a slight lateral distance (similar or below the microscopic resolution). These two waves recombine in a second Wollaston prism before reaching a polarizer (analyser) that is placed in front of the detector. This analyser is oriented perpendicularly to the polarization of incident light. Therefore, if no interaction of light with the sample occurs, the two recombining waves result in a wave of the original polarization which cannot pass the analyser, and no detector signal is generated. If, however, the two waves passing the sample are shifted in their phase, the resulting wave (after the second Wollaston prism) is elliptically polarized, and some light can pass the analyser and reach the detector. Therefore, the signal measured by the detector (e.g. CCD camera) reflects the various phase shifts between the two perpendicularly polarized waves all over the sample, resulting in a high contrast image.

In high quality cameras the amplification and offset of the measured signal can be adjusted. If for two adjacent spots of the object, diffraction patterns are detected as depicted in Figure 2(a), a threshold can be introduced, and the signal above the threshold can be amplified (Figure 2(b)). This results in some increase in resolution (about 12%) and a pronounced increase in contrast. Objects of 160 nm diameter can thus be resolved, and objects down to diameters of about 20 nm can be visualized. Figure 3 shows a comparison of conventional transmission, phase contrast and interference contrast microscopy applied to cultivated human glioma cells. Whereas in conventional transmission microscopy the contrast is rather poor, and cells can only be weakly distinguished from the background, cellular

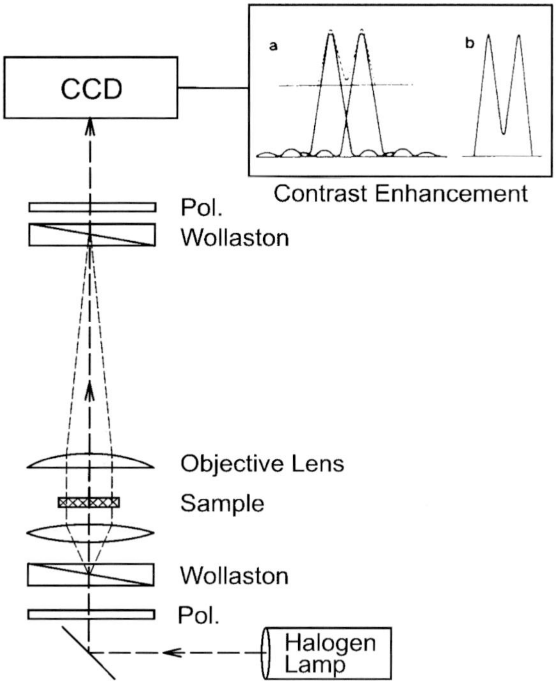

Figure 2. Principle of interference contrast microscopy including contrast enhancement.

substructures are resolved when using phase contrast or interference contrast microscopy. Often the term "differential interference contrast (DIC) microscopy" is used, if the lateral distance between the two perpendicularly polarized light waves is smaller than the microscopic resolution.

13.3 Fluorescence microscopy

13.3.1 Basics

Fluorescence arises from an electronic transition from the lowest excited state to the ground state of a molecule. The energy difference between these states (photon energy) W is correlated with the wavelength λ of radiation according to

$$W = h \times c/\lambda \tag{2}$$

where $h = 6.626 \times 10^{-34}$ J s corresponds to Planck's constant, and $c = 3.00 \times 10^{8}$ m s^{-1} to light velocity in vacuum. Since for many aromatic organic molecules λ is in the range 300–800 nm, conventional microscopic methods can be used. Generally, the rates of radiative (fluorescent) transitions k_{f} and non-radiative transitions k_{nr} originating from the lowest excited molecular level sum up, thus

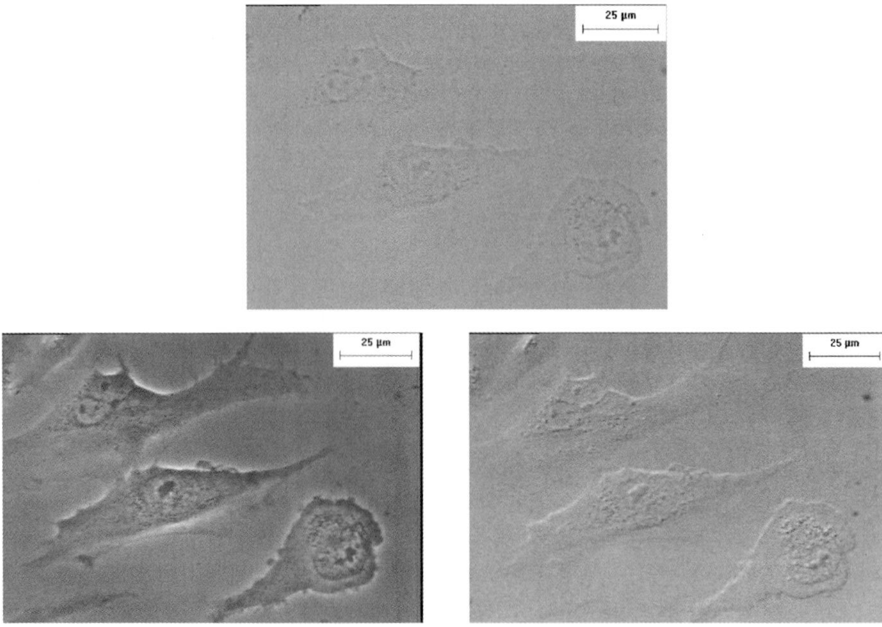

Figure 3. Images of U373MG human glioma cells taken by conventional transmission (top), phase contrast (bottom left) and video-enhanced interference contrast (bottom right) microscopy, using a 40×/0.75 objective lens and a CCD camera (BC2, AVT Horn, Aalen, Germany).

resulting in a total rate $k = k_f + k_{nr}$. The ratio

$$\eta = k_f/k \tag{3}$$

is assigned as the fluorescence quantum yield, whereas the reciprocal

$$\tau = 1/k = 1/(k_f + k_{nr}) \tag{4}$$

corresponds to the fluorescence lifetime of a molecular species. Experimental parameters are often the fluorescence intensity or quantum flux $I(\lambda)$ (measured as W m^{-2} or photons s^{-1}) and the fluorescence lifetime $\tau(\lambda)$, which for many organic molecules is in the picosecond–nanosecond range. For thin samples (e.g. cell monolayers) $I(\lambda)$ is proportional to the concentration c and the fluorescence quantum yield η of relevant fluorophores, whereas $\tau(\lambda)$ may reflect their interaction with adjacent molecules due to different ratios of radiative and non-radiative transitions.

In addition to some endogenous fluorophores, e.g. the amino acids tryptophan and tyrosine, as well as the coenzymes nicotinamide adenine dinucleotide (NADH) and flavin mono- or dinucleotide (FMN/FAD), an increasing number of specific fluorescence markers has become available. These markers include derivatives of fluorescein, rhodamine and cyanines, which may be specific for DNA, membranes, lysosomes or mitochondria. Further fluorophores are used as indicators of signal transduction (e.g. via calcium ions), pHs, reactive oxygen species or membrane potential (for an overview see [6]). The most exciting recent development in the use of fluorescent probes for biological studies has been the introduction of fluorescent proteins. A green fluorescent protein (GFP) is naturally produced by the jellyfish *Aequorea victoria* [7]. After cloning of the GFP gene, various GFP mutants with different excitation and emission properties have been produced. Meanwhile blue, green, yellow or even red fluorescent proteins are distinguished. By fusion of genes coding for a specific cellular protein and a GFP variant, functional fluorescent proteins have been created, permitting site-specific tracking in living cells or even whole organisms [8,9].

A further promising technique is staining with so-called quantum dots, i.e. luminescent nanoparticles of semiconductors such as CdS, ZnS or PbS. These particles had originally been used in material sciences [10] and were only recently introduced as biological labels [11]. In contrast to most organic dyes, quantum dots are characterized by low photobleaching rates and by emission spectra depending on the size of these particles. Therefore, their luminescence properties can be easily varied; however, their photophysical properties should be taken into account. For example, photooxidation of CdSe quantum dots results in some shrinkage of these particles and a concomitant blue-shift of the emission spectrum [12]. Typical luminescence lifetimes of quantum dots are in the range 150 ns–25 μs [13], i.e. considerably longer than those of most organic dyes as well as of cellular autofluorescence. Therefore, it might be rather easy to distinguish their emission from the intrinsic fluorescence of cells or tissues using time-resolving methods (see below).

The fluorescence of most dyes is red-shifted compared with their absorption spectrum (Stokes shift). Therefore, in fluorescence microscopy, a dichroic mirror is often used that permits short-wave excitation light to be deflected onto the sample and long-wave fluorescence to pass the mirror and reach the detector. Dichroic mirrors are usually combined with appropriate bandpass or longpass filters which are optimized for the corresponding dye. When using multi-staining techniques, different detection paths (and sometimes also different excitation sources) have to be used. The lateral resolution in fluorescence microscopy

$$\Delta x = 0.61\lambda/A \tag{5}$$

corresponds to the radius of a diffraction limited spot which is slightly larger than the resolution in transmission microscopy (Equation (1)). When using a wavelength $\lambda = 550$ nm and a numerical aperture $A = 1.40$, a resolution of $\Delta x = 218$ nm is attained.

13.3.2 Video microscopy

In video microscopy a detector target has to be placed at the position of the image plane. Neither an ocular nor a camera objective lens are therefore needed. Often, charge-coupled device (CCD) cameras with typically 800 × 600 picture elements (pixels) are used. Typical pixel sizes are 8 × 8 μm^2. For correct imaging, the pixel size should not be larger than half of the minimum object size. When considering a microscopic resolution of 200 nm in the object plane, this corresponds to 20 μm in the image plane after 100× magnification and implies that microscopic resolution is well retained by the camera system.

Sensitivity (i.e. threshold towards noise) of high-performance CCD cameras with air or Peltier cooling is 1–10 lx, corresponding to about 2×10^{-7} to 2×10^{-6} W cm^{-2} in the green spectral region. Assuming an experiment where a cell monolayer containing a concentration of fluorophores of $c = 10^{-5}$ mol l^{-1} is irradiated with a power density of 50 mW cm^{-2}, one can calculate that the fluorescence quantum yield should be between $\eta = 0.01$ and 0.1 for detection by a CCD camera. While maintaining the camera's sensitivity, a considerable improvement of the signal-to-noise ratio is achieved by integration on the camera chip (typically up to 10 s).

For detection of fluorophores with lower quantum yield or lower concentrations, image intensifying systems can be used. Amplification factors around 10^3 are attained by multichannel plates. In this case, photoelectrons are emitted from a photocathode and amplified within some alveolar structure at a voltage of some hundred volts. Each secondary electron may cause the emission of several (typically 20–200) photons from a phosphor screen to which, again, a high voltage (up to some kV) is applied. Thus, amplification factors of 10^3–10^7 are attained when using one or two multichannel plates. This corresponds to an overall sensitivity down to 10^{-14} W cm^{-2} or about 10^{-20} W pixel^{-1}. Photon counting is therefore possible, if the noise level can be kept sufficiently low, e.g. by cooling with a Peltier element or liquid nitrogen. First reports on single-photon counting imaging were published some 15 years ago [14]. More recently, photon counting imaging has become a valuable technique in single molecule detection (e.g. [15]).

13.3.3 Microspectrofluorometry

Simultaneous measurements of fluorescent dyes and/or intrinsic fluorophores are of increasing interest in cell biology. In addition, the excitation or emission spectrum of a fluorophore may give information about its microenvironment, e.g. pH or interaction with adjacent molecules. Spectral resolution is introduced using, e.g., a set-up as depicted in Figure 4. The wavelength of an excitation lamp (e.g. xenon high-pressure lamp, XBO) is tuned using a monochromator. Excitation light is then deflected by a dichroic mirror and focused into the sample by the microscope objective lens. Due to its longer wavelength, fluorescence radiation passes the dichroic mirror and is again focused in the image plane of the microscope. At this position a diaphragm may be used to select specific objects, e.g. individual cells. A scanning monochromator with a photomultiplier (Figure 4) or alternatively a polychromator with an optical multichannel analyzer (OMA) is used to record the emission spectra. The advantage of the OMA system is that the whole spectrum can be recorded simultaneously, whereas the advantage of the scanning system is its higher sensitivity. The entrance slit of the monochromator is placed either in the exit pupil or in the image plane of the microscope. In the latter case this slit limits the measured object, thus avoiding a separate diaphragm. By transillumination with a lamp emitting a continuous spectrum (e.g. halogen lamp), specific sample areas can be selected for fluorescence measurements. The optical set-up can be modified in many ways, e.g. by replacing the excitation monochromator by interference

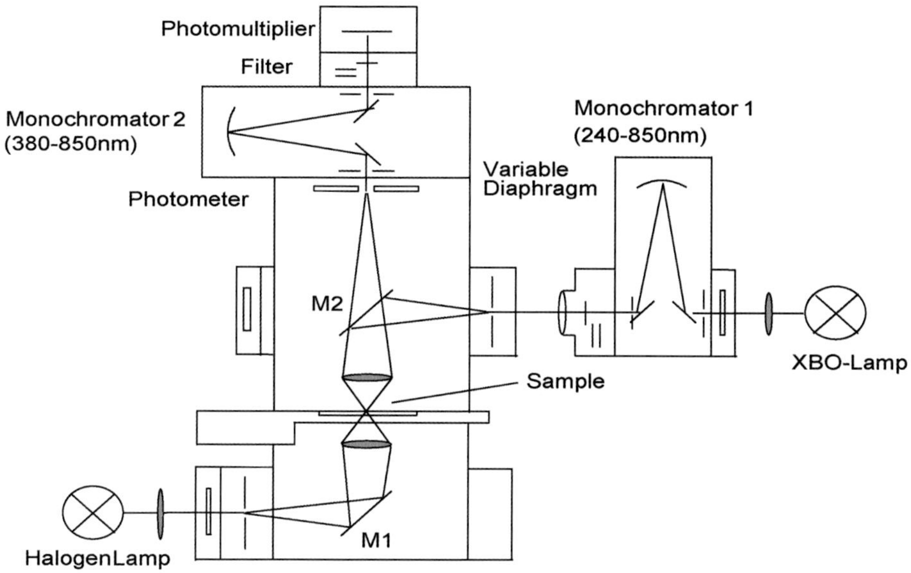

Figure 4. Microscopic set-up for fluorescence excitation and emission spectroscopy with additional transillumination (M_1 = mirror, M_2 = dichroic mirror). Reproduced from [16].

filters or by replacing the excitation lamp by a laser which can be focused to a diffraction limited spot with a radius $r = 0.61\lambda/A$.

Usually, fluorescence spectra are uncorrected, i.e. they are influenced by wavelength-dependent sensitivities of monochromator and detector (emission spectroscopy) or excitation source and monochromator (excitation spectra). Correction is not necessary, if only relative changes of individual emission bands are measured. However, for getting correct spectral intensities $I_F(\lambda)$, a reference source with a well-known spectrum $S(\lambda)$ (e.g. a black body) has to be used. Both signals, $I_F(\lambda)$ and $S(\lambda)$ are "folded" by an instrumental function $G(\lambda)$. When measuring the experimental functions $I_F'(\lambda) = I_F(\lambda) \times G(\lambda)$ and $S'(\lambda) = S(\lambda) \times G(\lambda)$, the true fluorescence spectrum can be calculated using the algorithm

$$I_F(\lambda) = I_F'(\lambda) \times S(\lambda)/S'(\lambda) \tag{6}$$

To obtain spatial and spectral information simultaneously, a spectral imaging system based on Fourier-transform spectroscopy has been applied [17]. Briefly, a Sagnac interferometer is used, where fluorescence light is split into two paths of opposite direction. By moving the beam splitter before recombining the two beams, a phase shift occurs which depends on the wavelength of radiation. Concomitantly, the fluorescence intensity on each pixel is wavelength dependent and is measured as a function of this wavelength. However, several seconds or minutes are needed for acquisition of one "three-dimensional" image $I(x,y,\lambda)$. Multipixel Fourier-transform spectroscopy has been applied for measuring photosensitizers, which are used for photochemotherapy of tumours, within single cells. Monomeric species of protoporphyrin were detected in the plasma membrane, whereas aggregated species accumulated within endosomal and lysosomal compartments [18]. More recently, different conformations (unfolded/folded) of green fluorescent protein (GFP) were found in different cell compartments using multipixel spectral imaging. Thus, pathways of GFP formation, intracellular tansport and changes of conformation could be studied within single cells, as reported by Greenbaum et al. [19].

13.3.4 Time-resolved fluorescence microscopy

When using pulsed light sources and fast detection devices, fluorescence microscopy can be performed with nanosecond or subnanosecond resolution, thus permitting to distinguish between fluorophores of different lifetime. Advantages of time-resolved versus continuous wave (cw) fluorescence measurements are numerous:

- Fluorescence signals of different fluorophores with overlapping emission spectra are resolved.
- Intrinsic fluorescence of cells or tissues can often be suppressed.
- Kinetic reactions, e.g. intermolecular energy transfer, can be evaluated.

Some techniques currently used for fluorescence lifetime measurements in microscopy are given below.

Time-correlated single-photon counting, where single-photon events are detected and correlated with a trigger signal from the light source. Integration over many single photons (typically 10^5–10^6) results in a fluorescence decay curve. Since after each excitation pulse only one photon can be registered, the excitation source should have a high repetition frequency, and simultaneous detection of two or more photons per excitation pulse should be avoided, e.g. by reduction of the counting rate to 5% or less of the excitation rate. An example of a time-correlated single-photon counting measurement of intrinsic cellular fluorescence is given in Figure 5.

Phase fluorometry, where an excitation source is modulated in frequency (using e.g. a Pockels cell), and where a phase shift or a demodulation of the fluorescence signal is measured and correlated with the fluorescence lifetime. If variable modulation frequencies are used, even multiexponential decays can be evaluated, as reported e.g. in Ref. 20.

Fluorescence lifetime imaging (FLIM), where fluorescence images are recorded at variable delay times after excitation by a short laser pulse (Figure 6). In this case, an image intensifier is used that can be gated within nanosecond or subnanosecond time intervals (for details see [16]). Alternatively, fluorescence decay profiles can be measured and evaluated within each pixel of a fluorescence image, as reported in Ref. 21.

Fluorescence decay profiles often show either a monoexponential behaviour according to

$$I(t) = Ae^{-kt} = Ae^{-t/\tau} \tag{7}$$

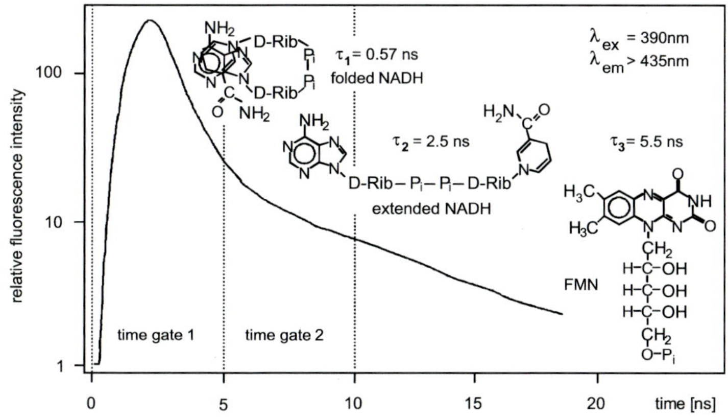

Figure 5. Decay kinetics of intrinsic cellular fluorescence of BKEz-7 endothelial cells from calf aorta after excitation by short pulses (40 ps) of a frequency-doubled laser diode. Excitation wavelength $\lambda_{ex} = 390$ nm; fluorescence measured at $\lambda_{em} \geq 435$ nm. Reproduced from [16].

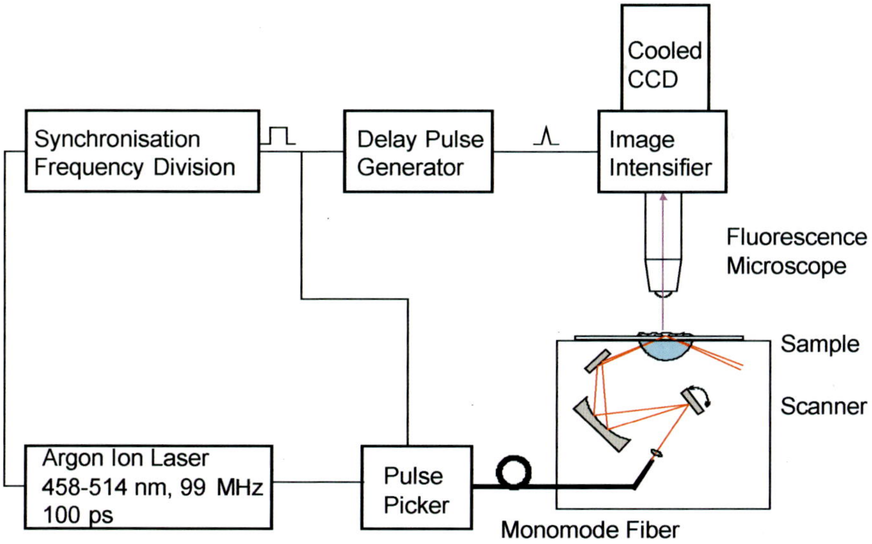

Figure 6. Experimental set-up for fluorescence lifetime imaging (FLIM) using a mode-locked argon-ion laser and an image intensifier with subnanosecond time gates ($\Delta t = 200$–1000 ps; Picostar, LaVision GmbH, Göttingen, Germany). This set-up can also be used for a combination with total internal reflection fluorescence microscopy (see below).

(with k corresponding to the total rate of deactivation of the excited electronic state and τ to the fluorescence lifetime) or a multiexponential behaviour according to

$$I(t) = \sum_i A_i \times e^{-kit} = \sum_i A_i \times e^{-t/\tau i} \tag{8}$$

with A_i corresponding to the amplitude and τ_i to the fluorescence lifetime of an individual component. The relative fluorescence intensity of each component i out of all components j is then given by

$$I_i = \int A_i \times e^{-t/\tau i} dt / \sum_j \int A_j \times e^{-t/\tau j} – dt = A_i \times \tau_i / \sum_j A_j \times \tau_j \tag{9}$$

if the integral is calculated from $t = 0$ to infinity.

An example for a three-exponential decay is given in Figure 5, where intrinsic fluorescence of endothelial cells arises from the coenzymes nicotinamide adenine dinucleotide (NADH), in a folded as well as in an unfolded conformation, and flavin mononucleotide (FMN). Fluorescence lifetimes of 0.57, 2.5 and 5.5 ns were determined from a least-square fitting algorithm after deconvolution. Time gates (as indicated in Figure 5) can be used to calculate fluorescence lifetime images according to

$$\tau_{eff} = \Delta t / \ln(I_1/I_2) \tag{10}$$

where I_1 and I_2 represent the fluorescence intensities measured within two time gates shifted among one another by a delay Δt. τ_{eff} corresponds either to the "effective" fluorescence lifetime of a multiexponential decay or to the real fluorescence lifetime of a monoexponential decay. An example is given in Figure 7, which shows the fluorescence intensity (left) and the effective fluorescence lifetime (right) of BKEz-7 endothelial cells incubated for 30 min with the mitochondrial marker rhodamine 123 (R123) at a concentration of 50 μM. The lifetime image was obtained from two fluorescence images, each recorded within a time gate of 500 ps and a delay Δt of 4 ns between both time gates. The intensity image (left) shows the accumulation of R123 in mitochondria, which is accompanied by a shortening of

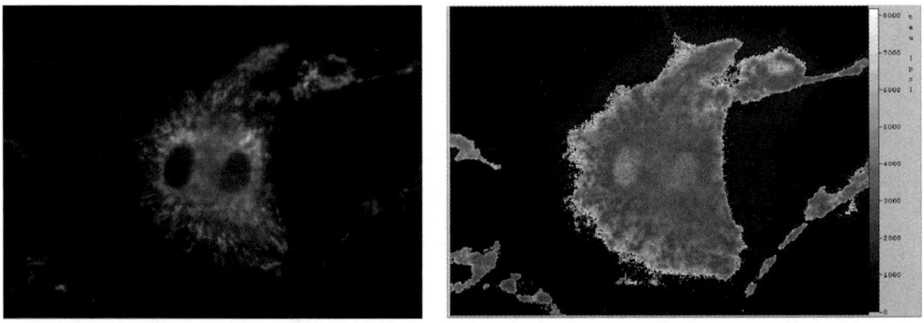

Figure 7. Fluorescence intensity (left) and fluorescence lifetime (right) of BKEz-7 endothelial cells incubated with rhodamine 123 (30 min, 50 μM); Excitation wavelength $\lambda_{ex} = 476$ nm; fluorescence measured at $\lambda_{em} \geq 515$ nm; image size 220×160 μm^2.

the fluorescence lifetime of the dye as demonstrated in the lifetime image (right). According to the Equations (3) and (4) this shortening may be due to fluorescence quenching, i.e. the fluorescence quantum yield η decreases, if the rate of non-radiative transitions k_{nr} increases at high concentration of R123. This possibly originates from an aggregation of the R123 molecules.

Applications of time-gated and frequency modulated imaging techniques include the detection of intracellular calcium (using dyes whose fluorescence lifetimes change upon binding of calcium [23,23]), oxygen [24] or pH [25]. In addition, time-resolving imaging techniques proved useful in studying a class of enzymes (caspases) that play an important role in the initiation and execution of apoptosis. Targeting techniques with mutants of green fluorescent protein (GFP) and the method of non-fluorescent energy transfer (see below) were used [26].

13.3.5 Total internal reflection fluorescence microscopy (TIRFM)

When a beam of light propagating through a medium of refractive index n_1 meets an interface with a second medium of refractive index $n_2 < n_1$, total internal reflection occurs at all angles of incidence Θ greater than a critical angle $\Theta_c = \arcsin n_2/n_1$. Despite being totally reflected, the incident beam establishes an electromagnetic field that penetrates a small distance into the second medium and decays exponentially with perpendicular distance z from the interface according to $I(z) = \mathrm{I}_0 \times \mathrm{e}^{-z/\mathrm{d}(\Theta)}$, where the penetration depth $\mathrm{d}(\Theta)$ for light of wavelength λ is given by

$$d(\Theta) = (\lambda/4\pi)(n_1^2 \times \sin^2\Theta - n_2{}^2)^{-\frac{1}{2}} \tag{11}$$

By variation of Θ, penetration depths can be varied between about 70 and 300 nm, thus giving an axial resolution that is better than in confocal laser scaning microscopy. The method (first reported by Axelrod [27]), therefore, appears to be ideal to examine fluorophores located within the plasma membrane and adjacent parts of the cytoplasm of living cells.

The fluorescence intensity of a dye of concentration $c(z)$ can be calculated, if absorption is integrated over all layers above the interface, resulting in

$$I_{\mathrm{F}}(\Theta) = A \times T(\Theta) \times \int c(z) \times \mathrm{c}^{-z/\mathrm{d}(\Theta)} \mathrm{d}z \tag{12}$$

if all factors that are independent of the angle Θ and the coordinate z are included within the experimental constant A. The solution of this integral is shown in Figure 8 for a homogeneous distribution of fluorophores above the interface (I_c), for a homogeneous distribution in the cytoplasm (I_e) and for a homogeneous distribution in the plasma membrane (I_m), assuming a constant distance of the plasma membrane from the interface. According to Figure 8, it is of great advantage to measure the angular dependence of fluorescence intensity, if information about the cellular location of fluorophores is desired. Details on variable-angle total internal reflection fluorescence microscopy (VA-TIRFM) are given in several articles (e.g. [28–30]).

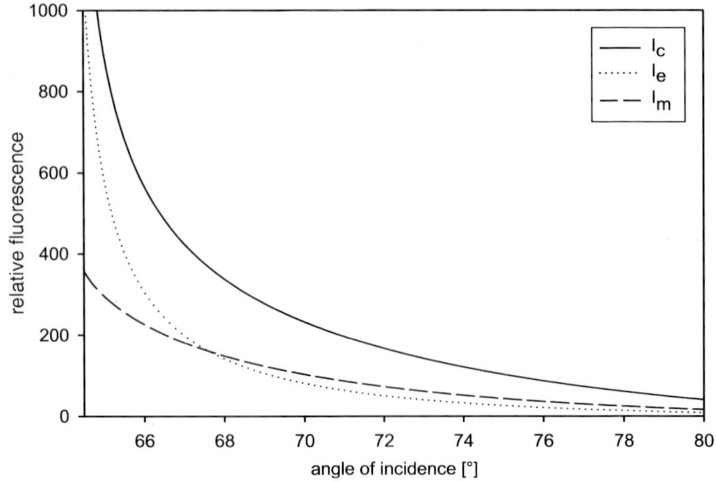

Figure 8. Fluorescence intensity I_F (Θ) for different fluorophore distributions. I_c: continuum with homogeneous distribution of fluorophores; I_e: homogeneous distribution of fluorophores in the cytoplasm; I_m: homogeneous distribution of fluorophores in the plasma membrane. Refractive indices: $n_1 = 1.52$ (glass substrate), $n_2 = 1.37$ (cytoplasm). Assumed distance between substrate and cell membrane: 100 nm. Reproduced from [16].

In general, two different technical solutions for TIR-illumination are known. In one the cell substrate is optically coupled, preferably with immersion oil or glycerol, to a glass or quartz prism. The prism usually has cubic (Figure 9-1a), hemispherical (Figure 9-1b) or hemicylindrical shape. When using a hemicylindrical or a hemispherical prism [31] or when using a multiple laser scanning system [29], the position of the illuminating light spot on the sample is maintained when the angle of incidence is varied (VA-TIRFM). For a prismless TIRF configuration, an objective with a high numerical aperture $A \geq 1.4$ is used for TIR-illumination [32].

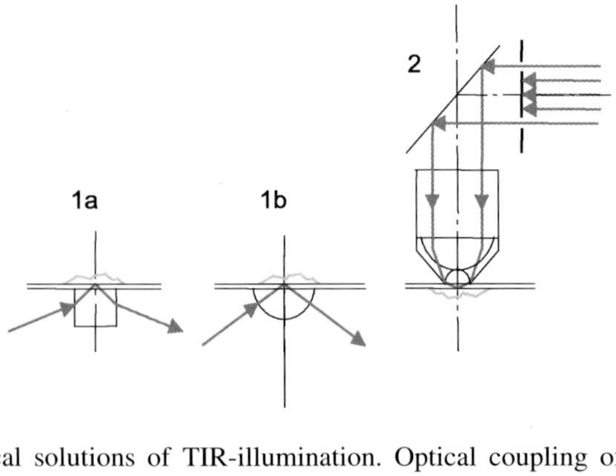

Figure 9. Technical solutions of TIR-illumination. Optical coupling of the cell substrate using a cube shaped (1a) or hemispheric prism (1b); prismless configuration using a high numerical aperture objective lens (2). Reproduced from [16].

Only incident rays traversing a peripheral annulus are allowed to propagate through the objective lens, while central rays are blocked (Figure 9-2). The peripheral rays are incident upon the cell–substrate of the sample at supercritical angles to the cell–substrate interface. This method can be performed with any fluorescence microscope using either a standard mercury arc lamp or a laser. However, variation of penetration depth using different angles of incidence is rather difficult to perform.

So far, total internal reflection fluorescence microscopy (TIRFM) has been applied, e.g., for studies of the topography of cell-substrate contacts [27,33,34], for measuring dynamics [35] or self-association [36] of proteins at membranes, for detection of membrane-proximal ion fluxes [37] or for imaging of endocytosis or exocytosis [38–41]. Previous work of the author's group addressed the investigation of photosensitizers (with tumour-selective properties) in close proximity to the plasma membrane [42,43]. With protoporphyrin IX, a correlation between membrane-associated accumulation of the photosensitizer and cell inactivation was found.

13.3.6 Structured illumination

Usually, in conventional wide-field microscopy the images from the focal plane and out-of-focus parts of a sample are superimposed. On surfaces this problem can be overcome using TIRFM. Optical sectioning (as in confocal laser scanning microscopy), however, becomes possible if structured illumination of a sample is used. For example, a diffraction grating can be imaged in the sample plane, and the intensity resulting from this plane can be described by

$$I = I_0 + I_C \cos\Phi + I_S \sin\Phi \tag{13}$$

where I_0 describes the contribution of a conventional wide-field image, and Φ characterizes a spatial phase due to the grating [44]. If one records three images with the phase angles $\Phi_1 = 0$, $\Phi_2 = 2\pi/3$ and $\Phi_3 = 4\pi/3$ by slightly shifting the grating, one can calculate the relation

$$I_F = [(I_1 - I_2)^2 + (I_1 - I_3)^2 + (I_2 - I_3)^2]^{\frac{1}{2}} \tag{14}$$

where I_0 as well as Φ are eliminated. In other words, an image of the focal plane without intensity modulation is created. Out-of-focus images are not affected by structured illumination and contain only the contribution I_0, which vanishes when applying the Equation (14). Optical sectioning is achieved by moving the sample step by step in a vertical direction by recording three pictures and using Equation (14) in each case. The conventional image I_0 can easily be recovered using the algorithm

$$I_0 = (I_1 + I_2 + I_3)/3 \tag{15}$$

Figure 10 shows structured illumination in the plane of the sample (upper left), and fluorescence images of the membrane marker laurdan [45] applied to U373-MG

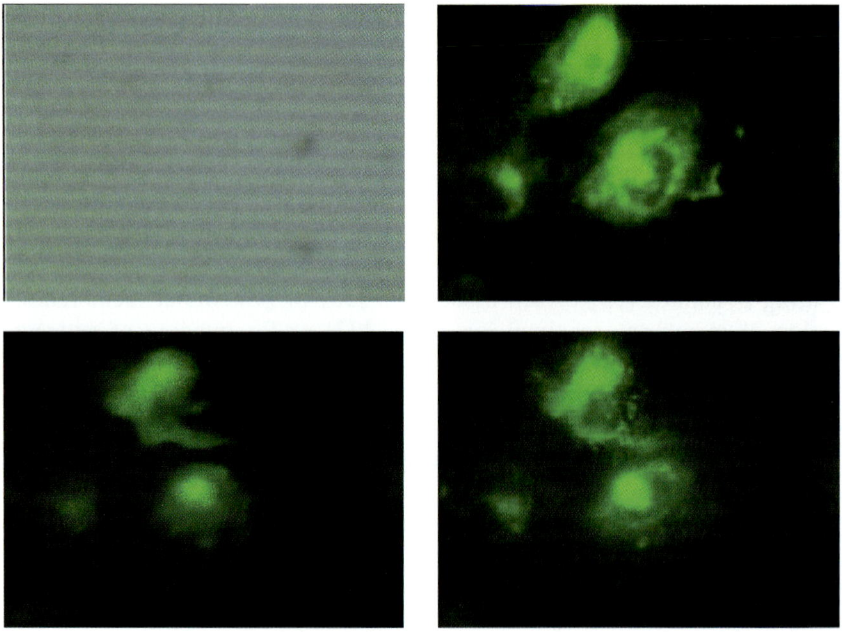

Figure 10. Fluorescence microscopy of U373-MG glial cells using the membrane marker laurdan and structured illumination. Upper left: structured illumination in the sample plane; upper right: fluorescence from the ventral side; lower left: fluorescence from the dorsal side; lower right: fluorescence from whole cells. Image size: 170 × 125 μm. Reproduced from [46].

glial cells are depicted. One image (upper right) shows the ventral side of the cells close to the object slide with fluorescence arising from the plasma membranes and the nuclear membranes, whereas another image (lower left) shows the dorsal side of the cells with fluorescent plasma membranes. Both images result from structured illumination and calculation according to Equation (14). In contrast, calculation according to Equation (15) results in a rather diffuse image of the whole cells (lower left).

The technique of structured illumination has been further improved using (nonlinear) saturated patterned excitation microscopy [47], which provides a further increase in resolution.

13.3.7 Energy transfer spectroscopy (FRET)

13.3.7.1 Basic mechanisms

One of the most interesting mechanisms used in fluorescence spectroscopy and cellular analytics is energy transfer between different molecules in their lowest excited electronic state Three different basic mechanisms, given below, have to be considered.

In the first case a photon emitted by one molecule is reabsorbed by another. In transparent samples this may occur at various distances, if the difference between the energy levels of the excited state and the ground state is the same for both molecules. This reabsorption is therefore of rather little analytical interest, but can cause erroneous measurements.

Energy transfer by direct interaction of optical transition dipoles of a donor and an acceptor molecule is a mechanism of significant analytical importance. This dipole–dipole interaction is proportional to r^{-6} (r being the intermolecular distance) and requires an overlap of the emission spectrum of the donor and the absorption spectrum of the acceptor according to

$$k_{ET} \approx r^{-6} \int \varepsilon_A(\nu) I_D(\nu) \nu^{-4} d\nu \tag{16}$$

with k_{ET} being the rate of energy transfer, $\varepsilon_A(\nu)$ the molar extinction coefficient of the acceptor, $I_D(\nu)$ the flux of emitted photons of the donor and ν the frequency of excitation radiation. Light is absorbed by the donor and emitted by the acceptor, whereas the intermolecular energy transfer is non-radiative. The principle of this so-called Förster mechanism [48] is exemplified in Figure 11 (upper part) for the donor–acceptor pair NADH → rhodamine 123 (R123). The absorption and emission spectra of NADH and R123 are depicted in the lower part of Figure 11. The absorption spectra of NADH (left) and R123 (right) are drawn in full lines, whereas the emission spectra are given by broken lines. A broad overlap of the emission spectrum of NADH and the absorption spectrum of R123 indicates that the condition of resonance is fairly fulfilled. Therefore, the mitochondrial marker R123 [49] is suggested to be suitable to probe selectively mitochondrial NADH [50].

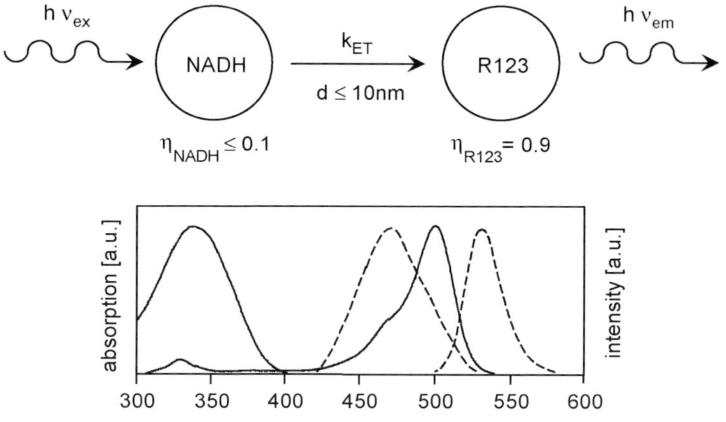

Figure 11. Principle of non-radiative energy transfer from the coenzyme NADH to the mitochondrial marker R123 (top); absorption spectra (full lines) and emission spectra (broken lines) of NADH and R123; spectra of R123 are red-shifted as compared with NADH, with a pronounced overlap of NADH emission and R123 absorption (bottom). Reproduced from [16].

According to Equation (11) fluorescence resonance energy transfer (FRET) is limited to short intermolecular distances of a few nanometres. As a quantitative measure the so-called Förster radius r_0 is used. It is defined as the intermolecular distance where the rate of energy transfer k_{ET} and the sum of all other rates of deactivation of the first excited molecular state S_1 are equal. This can be described by the formula

$$k_{ET} = \tau_0^{-1}(r/r_0)^6 \tag{17}$$

with τ_0 being the lifetime of the excited electronic state in the absence of energy transfer. When using non-radiative energy transfer, a high fluorescence quantum yield η of the acceptor (e.g. $\eta = 0.9$ for R123) is advantageous. Usually, energy transfer via dipole–dipole interaction is restricted to singlet states (with an electron spin $S = 0$) of the donor and the acceptor molecule.

Energy transfer rates can be determined from stationary as well as from time-resolved fluorescence measurements. In the first case, the photon fluxes of donor (I_D) and acceptor (I_A) fluorescence are measured and the ratio

$$I_A/I_D = \tau(\eta_A/\eta_D)k_{ET} \tag{18}$$

is calculated. assuming that the energy transfer rate k_{ET} is proportional to acceptor concentration [51], and that η_A, η_D and τ are known from the literature. Another possibility of calculating k_{ET} is time-resolved fluorescence spectroscopy. If the energy transfer rate k_{ET} is summed up with the rates of radiative (k_F) and non-radiative (k_{nr}) transitions originating from the excited state of the donor molecule, the reciprocal fluorescence lifetime of the donor results in $1/\tau = k_F + k_{nr} + k_{ET}$. In the absence of energy transfer, the reciprocal lifetime of the donor would be $1/\tau_0 = k_F + k_{nr}$. Therefore the energy transfer rate can be calculated according to

$$1/\tau - 1/\tau_0 = k_{ET} \tag{19}$$

In excited triplet states (electron spin $S = 1$) where dipole–dipole interaction becomes negligible, non-radiative energy transfer may arise from an electron exchange mechanism. This process requires a considerable overlap of the electron orbitals of the excited donor molecule and the acceptor molecule in the ground state with spectral characteristics being rather irrelevant. The distance for interaction is rather short (typically no more than 1 nm). Therefore, this process requires direct contact between donor and acceptor molecules, e.g. during diffusion of one of these molecules. Energy transfer rates k_{ET} are often small compared with dipole–dipole interaction, but since the lifetime τ of excited triplet states is usually rather long (μs–ms), the quantum yield of energy transfer $\eta_{ET} = k_{ET}/k = k_{ET} \times \tau$ may still be large. Energy transfer from the excited triplet state of porphyrins, chlorins or related molecules to oxygen molecules, generating highly reactive singlet oxygen, is a well-known process used, e.g., for photodynamic therapy of tumours (for a survey on intracellular reactions see Ref. 52).

13.3.7.2 FRET applications

Energy transfer processes have been used in cellular biology for many years. Most of them are based on resonant dipole–dipole interaction according to the Förster mechanism. In the 1980s microscopic studies on energy transfer provided information on the architecture and intermolecular distances in cell membranes [53,54]. In addition, fluorescence resonant energy transfer (FRET) proved to be useful for measuring the structure and assembly of actin filaments [55] and for the detection of binding sites of enzymes [56].

More recently, measurements of non-radiative energy transfer from the coenzyme NADH to rhodamine 123 (R123; see above) proved to be useful for selective detection of the mitochondrial fraction of NADH, which showed pronounced changes upon inhibition of the respiratory chain [50]. FRET was also used to measure selectively mitochondrial depolarization, which may precede mitochondrial autophagy, apoptosis and necrotic cell death [57]. As already mentioned, FRET is applied increasingly to mutants of green fluorescent protein (GFP). A direct interaction between the two proteins Bcl-2 and Bax, which are involved in the regulation of apoptosis, was proven within individual mitochondria using GFP-Bax and blue fluorescent protein (BFP)-Bcl-2 fusion proteins coexpressed within the same cell [58]. In addition, specific amino acid sequences located between BFP and GFP were cleaved by the enzyme caspase upon induction of apoptosis: non-radiative energy transfer BFP → GFP disappeared, thus allowing activation of specific caspases to be monitored in vitro and in vivo [59]. Since different GFP mutants can be localized on various sites of a protein, conformational changes of proteins (e.g. of calmodulin upon binding of calcium ions) can be measured selectively [60]. This permits visualization of calcium uptake and distribution in single cells. Further applications of FRET have been dedicated to the detection and visualization of GFP-tagged receptors in cells which were focally stimulated by the epidermal growth factor (EGF). Following focal stimulation, a rapid and extensive propagation of receptor phosphorylation over the plasma membrane, resulting in full activation of all receptors, was observed [61]. Similarly, receptor dephosphorylation on the surface of the endoplasmic reticulum was described by Haj et al. [62].

13.4 Perspectives and concluding remarks

Despite an increasing number of applications of confocal or multiphoton laser scanning microscopy, wide-field microscopy is still a valuable tool in biological research with numerous future perspectives, e.g. second harmonic generation [63] or other nonlinear techniques. Further advanced techniques, e.g. fluorescence lifetime imaging (FLIM), energy transfer spectroscopy (FRET) or total internal reflection fluorescence microscopy (TIRFM), have great potential when an increased sensitivity, subcellular resolution or the detection of fast kinetics are required. For example, the microenvironment of specific proteins [64,65] as well as protein–protein interactions [66] could be imaged using FLIM and FRET techniques. In addition, FRET imaging [67,68] is increasingly replacing conventional measurements of non-radiative energy transfer between adjacent molecules. TIRFM measurements, where a thin cellular layer (e.g. plasma membrane) is assessed selectively, are used increasingly to detect single molecules [69] or single ion channels [70] in living cells. Finally, a combination of TIRFM with either laser confocal scanning microscopy [71] or atomic force microscopy [72] appears promising for future applications in cell biology.

Acknowledgements

The author thanks R. Sailer and W.S.L. Strauss for their critical and constructive review of this manuscript, M. Wagner for his contributions on structured illumination microscopy, as well as D. Fritz and D. Wägele for technical assistance.

References

1. W. Denk, J.H. Strickler, W.W. Webb (1990). Two-photon laser scanning microscope. *Science*, **248**, 73–76.
2. K. König (2000). Multiphoton microscopy in life sciences. *J. Microsc.* **200**, 83–86.
3. M. Schrader, S.W. Hell (1996). 4Pi-confocal images with axial superresolution. *J. Microsc.* **183**, 189–193.
4. M. Dyba, S. Hell (2002). Focal spots of size lambda/23 open up far-field fluorescence microscopy at 33 nm axial resolution. *Phys. Rev. Lett.* **88**, 163901.
5. G. Göke (1988). *Moderne Methoden der Lichtmikroskopie* (ISBN 3-440-05765-8). Franckh'sche Verlagshandlung, Stuttgart.
6. J.M. Mullins (1999). Overview of fluorochromes. *Methods Mol. Biol.* **115**, 97–105.
7. C.W. Cody, D.C. Prasher, W.M. Westler, F.G. Prendergast, W.W. Ward (1999). Chemical structure of the hexapeptide chromophore of the Aequorea green-fluorescent protein. *Biochemistry* **32**, 1212–1218.
8. R. Rizzuto, M. Brini, P. Pizzo, M. Murgia, T. Pozzan (1995). Chimeric green fluorescent protein as a tool for visualizing subcellular organelles in living cells. *Curr. Biol.* **5**, 635–642.
9. M. Ikawa, S. Yamada, T. Nakanishi, M. Okabe (1999). Green fluorescent protein as a tool for visualizing subcellular organelles in living cells. *Curr. Top. Dev. Biol.* **44**, 1–20.
10. M.G. Bawendi, M.L. Steigerwald, L.E. Bruns (1990). The quantum mechanics of larger semiconductor clusters ("quantum dots"). *Annu. Rev. Phys. Chem.* **41**, 477–496.
11. M. Bruchez, M. Moronne, P. Gin, S. Weis, A.P. Alivisatos (1998). Semiconductor nanocrystals as fluorescence biological labels. *Science* **281**, 2013–2016.
12. W.G.J.H.M van Sark, P.L.T.M. Frederix, D.J. van den Heuvel, A.A. Bol, J.N.J. van Lingen, C. de Mello Donega, H.C. Gerritsen, A. Meijerink (2002). Time-resolved fluorescence spectroscopy study on the photophysical behaviour of quantum dots. *J. Fluoresc.* **12**, 69–76.
13. J.R. Lakowicz (1999). Novel fluorescence sensing technologies. In: A. Kotyk (Ed), *Fluorescence Microscopy and Fluorescent Probes* (Volume 3, pp. 1–21). Espero Publishing, Prague.
14. W. Mueller-Klieser, S. Walenta, W. Paschen, F. Kallinowski, P. Vaupel (1988). Metabolic imaging of tumours and normal tissues with bioluminescence and photon counting. *J. Natl. Cancer Inst.* **80**, 842–848.
15. Y. Sako, S. Minoguchi, T. Yanagida (2000). Single-molecule imaging of EGFR signalling on the surface of living cells. *Nature Cell Biol.* **2**, 168–172.

16. H. Schneckenburger, R. Steiner, W.S.L. Strauss, K. Stock, R. Sailer (2002). Fluorescence technologies in biomedical diagnostics. In: V.V. Tuchin (Ed), *Handbook of Optical Biomedical Diagnostics* (pp. 825–874). SPIE, Bellingham, WA.
17. Z. Malik, D. Cabib, R.A. Buckwald, A. Talmi, Y. Garini, S.G. Lipson (1996). Fourier transform multipixel spectroscopy for quantitative cytology. *J. Microsc.* **182**, 133–140.
18. Z. Malik, M. Dishi, Y. Garini (1996). Fourier transform multipixel spectroscopy and spectral imaging of protoporphyrin in single melanoma cells. *Photochem. Photobiol.* **63**, 608–614.
19. L. Greenbaum, C. Rothmann, J. Haniana, Z. Malik (2000). Multi-pixel spectral imaging of green fluorescent protein (GFP) in COS-7 cells: Folding kinetics and chromophore formation. In: K. König, H.J. Tanke, H. Schneckenburger (Eds), *Laser Microscopy* Proc. SPIE (Volume 4164, pp. 48–52). SPIE, Bellingham, WA.
20. J.R. Lakowicz, G. Laczko, I. Gryczinski, H. Szmacinski, W. Wiczk (1988). Gigahertz frequency domain fluorometry: resolution of complex decays, picosecond processes and future developments. *J. Photochem. Photobiol. B: Biol.* **2**, 295–311.
21. K. König, U. Tirlapur, C. Peuckert, I. Riemann, A. Bergmann, W. Becker (2000). Time-resolved two-photon imaging of living cells and tissues. *Cell. Mol. Biol.* **46**, 119.
22. J.R. Lakowicz, H. Szmacinski, M.L. Johnson (1992). Calcium imaging using fluorescence lifetimes and long-wavelength probes. *J. Fluoresc.* **2,** 47–61.
23. B. Herman, P. Wodnicki, K. Seongwook, A. Periasamy, G.W. Gordon, N. Mahajan, X.F. Wang (1997). Recent developments in monitoring calcium and protein interactions in cells using fluorescence lifetime microscopy. *J. Fluoresc.* **7,** 85–91.
24. H.C. Gerritsen, R. Sanders, A. Draaijer, C. Ince, Y.K. Levine (1997). Fluorescence imaging of oxygen in living cells. *J. Fluoresc.* **7**, 11–15.
25. H.C. Gerritsen (1996). Confocal fluorescence lifetime imaging. In: J. Slavik (Ed), *Fluorescence Microscopy and Fluorescent Probes* (pp. 35–46). Plenum Press, New York, London.
26. B. Herman, M. Sun, M. Qiu, V. Centonze (2000). Protein interaction of enzymatic activities monitored using FRET. *Cell. Mol. Biol.* **46**, 93.
27. D. Axelrod (1981). Cell-substrate contacts illuminated by total internal reflection fluorescence. *J. Cell Biol.* **89**, 141–145.
28. J.S. Burmeister, A. Truskey, W.M. Reichert (1994). Quantitative analysis of variable-angle total internal reflection fluorescence microscopy (VA-TIRFM) of cell/substrate contacts. *J. Microsc.* **173**, 39–51.
29. B.P. Ölveczky, N. Periasamy, A.S. Verkman (1997). Mapping fluorophore distributions in three dimensions by quantitative multiple-angle total internal reflection fluorescence microscopy. *Biophys. J.* **73**, 2836–2847.
30. K. Stock, R. Sailer, W.S.L. Strauss, M. Lyttek, R. Steiner, H. Schneckenburger (2003). Variable-angle total internal reflection fluorescence microscopy (VA-TIRFM): realization and application of a compact illumination device. *J. Microsc.* in press.
31. M. Oheim, D. Loerke, W. Stühmer, R.H. Chow (1999). Multiple stimulation-dependent processes regulate the size of the releasable pool of vesicles. *Eur. J. Biophys.* **28**, 91–101.
32. A.L. Stout, D. Axelrod, (1989). Evanescent field excitation of fluorescence by epi-illumination. *Appl. Opt.* **28**, 5237–5242
33. G.A. Truskey, J.S. Burmeister, E. Grapa, W.M. Reichert (1992). Total internal reflection fluorescence microscopy (TIRFM) (II) Topographical mapping of relative cell/substratum separation distances. *J. Cell Sci.* **103**, 491–499.
34. J. Hornung, T. Müller, G. Fuhr (1996). Cryopreservation of anchorage-dependent mammalian cells fixed to structured glass and silicon substrates. *Cryobiology* **33**, 260–270.

35. S.E. Sund, D. Axelrod (2000). Actin dynamics at the living cell submembrane imaged by total internal reflection fluorescence photobleaching. *Biophys. J.* **79**, 1655–1669.
36. N.L. Thompson, A.W. Drake, L. Chen, W.V. Broek (1997). Equilibrium, kinetics, diffusion and self-association of proteins at membrane surfaces: Measurement by total internal reflection fluorescence microscopy. *Photochem. Photobiol.* **65,** 39–46.
37. G.M. Omann, D. Axelrod (1996). Membrane-proximal calcium transients in stimulated neutrophils detected by total internal reflection fluorescence. *Biophys. J.* **71,** 2885–2891.
38. W.J. Betz, F. Mao, C.B. Smith (1996). Imaging exocytosis and endocytosis. *Curr. Opin. Neurobiol.* **6**, 365–371.
39. M. Oheim, D. Loerke, W. Stühmer, R.H. Chow (1998). The last few milliseconds in the life of a secretory granule. *Eur. J. Biophys.* **27**, 83–98.
40. A. Rohrbach (2000). Observing secretory granules with a multiangle evanescent wave microscope. *Biophys. J.* **85**, 2641–2754.
41. J.K. Jaiswal, N.W. Andrews, S.M. Simon (2002). Membrane proximal lysosomes are the major vesicles responsible for calcium-dependent exocytosis in nonsecretory cells. *J. Cell Biol.* **159**, 625–635.
42. W.S.L. Strauss, R. Sailer, M.H. Gschwend, H. Emmert, R. Steiner, H. Schneckenburger (1998). Selective examination of plasma associated photosensitizers using total internal reflection fluorescence spectroscopy (TIRFS) – correlation between photobleaching and photodynamic efficacy of protoporphyrin IX. *Photochem. Photobiol.* **67**, 363–369.
43. R. Sailer, W.S.L. Strauss, H. Emmert, K. Stock, R. Steiner, H. Schneckenburger (2000). Plasma membrane associated location of sulfonated meso-tetraphenylporphyrins of different hydrophilicity probed by total internal reflection fluorescence spectroscopy. *Photochem. Photobiol.* **71**, 460–465.
44. M.A.A. Neil, R. Juskaitis, T. Wilson (1997). Method of obtaining optical sectioning by structured light in a conventional microscope. *Opt. Lett.* **22**, 1905–1907.
45. T. Parasassi, G. de Stasio, A. d'Ubaldo, E. Gratton (1990). Phase fluctuation in phospholipid membranes revealed by laurdan fluorescence. *Biophys. J.* **57**, 1179–1186.
46. M. Wagner (2003). *Tiefenauflösende Fluoreszenzmikroskopie mit strukturierter Beleuchtung*, Master's Thesis, Fachhochschule Aalen.
47. R. Heintzmann, T.M. Jovin, C. Cremer (2002). Saturated patterned excitation microscopy – a concept for optical resolution improvement. *J. Opt. Soc. Am. A Opt. Image Sci. Vis.* **19**, 1599–1609.
48. T. Förster (1960). Zwischenmolekularer übergang von elektronenanregungsenergie. *Z. Elektrochem.* **64**, 157–164.
49. L.V. Johnson, M.L. Walsh, L.B. Chen (1980). Localization of mitochondria in living cells with rhodamine 123. *Proc. Natl. Acad. Sci. U.S.A.* **77**, 990–994.
50. H. Schneckenburger, M.H. Gschwend, W.S.L. Strauss, R. Sailer, M. Kron, U. Steeb, R. Steiner (1997). Energy transfer spectroscopy for measuring mitochondrial metabolism in living cells. *Photochem. Photobiol.* **66**, 33–41.
51. H. Port, H. Schneckenburger, H.C. Wolf (1981). Host-guest energy transfer via dipole–dipole interaction in doped fluorene crystals. *Z. Naturforsch. Teil A* **36**, 697–704.
52. A.C.E. Moor (2000). Signalling pathways in cell death and survival after photodynamic therapy. *J. Photochem. Pbotobiol. B: Biol.* **57**, 1–13.
53. P.S. Uster, R.E. Pagano (1986). Resonance energy transfer microscopy: observations of membrane-bound fluorescent probes in model membranes and in living cells. *J. Cell Biol.* **103**, 1221–1234.
54. J. Szöllösi, S. Damjanovich, S.A. Mulhern, L. Tron (1987). Fluorescence energy transfer and membrane potential measurements monitor dynamic properties of cell membranes: a critical review. *Prog. Biophys. Mol. Biol.* **49**, 65–87.

55. D.L. Taylor, J. Reidler, A. Spudich, L. Stryer (1981). Detection of actin assembly by fluorescence energy transfer. *J. Cell Biol.* **89**, 65–87.
56. T.C. Squier, D.J. Bigelow, J.G. deAncos, G. Inesi (1987). Localization of site-specific probes on the Ca-ATPase of sarcoplasmic reticulum using fluorescence energy transfer. *J. Biol. Chem.* **89**, 362–367.
57. J.J. Lemasters, A.L. Nieminen, T. Qian, L.C. Trost, S.P. Elmore, Y. Nishimura, R.A. Crowe, W.E. Cascio, C.A. Brandham, D.A. Brenner, B. Herman (1998). The mitochondrial permeability transition in cell death: a common mechanism in necrosis, apoptosis and autophagy. *Biochim. Biophys. Acta* **1366**, 177–196.
58. N.P. Mahajan, K. Linder, G. Berry, G.W. Gordon, R. Heim, B. Herman (1998). Bcl-2 and Bax interactions in mitochondria probed with green fluorescent protein and fluorescence resonance energy transfer. *Nat. Biotechnol.* **16**, 547–552.
59. N.P. Mahajan, D.C. Harrison-Shostak, J. Michaux, B. Herman (1999). Novel mutant green fluorescent protein protease substrates reveal the activation of specific caspases during apoptosis. *Chem. Biol.* **6**, 401–409.
60. S. Brasselet, E.J.G. Peterman, A. Miyawaki, W.E. Moerner (2000). Single-molecule fluorescence resonant energy transfer in calcium concentration dependent cameleon. *J. Phys. Chem. B* **104,** 3676–3682.
61. P.J. Verveer, F.S. Wouters, A.R. Reynolds, P.I. Bastiaens (2000). Quantitative imaging of lateral ErbB1 receptor signal propagation in the plasma membrane. *Science* **290**, 1567–1570.
62. F.G. Haj, P.J. Verveer, A. Squire, B.G. Neel, P.I. Bastiaens (2002). Imaging sites of receptor dephosphorylation by PTP1B on the surface of the endoplasmic reticulum. *Science* **295**, 1708–1711.
63. P.J. Campagnola, H.A. Clark, W.A. Mohler, A. Lewis, L.M. Loew (2001). Second-harmonic imaging microscopy of living cells. *J. Biomed. Opt.* **6**, 277–286.
64. P.J. Verveer, A. Squire, P.I. Bastiaens (2001). Improved spatial discrimination of protein reaction states in cells by global analysis and deconvolution of fluorescence lifetime imaging microscopy data. *J. Microsc.* **202**, 451–456.
65. K. Suhling, J. Siegel, D. Pholips, P.M. Prench, S. Leveque-Fort, S.E. Webb, D.M. Davis (2002). Imaging the environment of green fluorescent protein. *Biophys. J.* **83**, 3589–3595.
66. M.A. Hink, T. Bisselin, A.J. Visser (2002). Imaging protein-protein interactions in living cells. *Plant Mol. Biol.* **50**, 871–883.
67. A.G. Harpur, F.S. Wouters, P.I. Bastiaens (2001). Imaging FRET between spectrally similar GFP molecules in single cells. *Nat. Biotechnol.* **19**, 167–169.
68. Z. Xia, Y. Liu (2001). Reliable and global measurement of fluorescence resonance energy transfer using fluorescence microscopes. *Biophys. J.* **81**, 2395–2402.
69. Y. Sako, T. Uyemura (2002). Total internal reflection fluorescence microscopy for single-molecule imaging in living cells. *Cell. Struct. Funct.* **27**, 357–365.
70. A. Sonnleitner, L.M. Mannuzzu, S. Terakawa, E.Y. Isacoff (2002). Structural rearrangements in single ion channels detected optically in living cells. *Proc. Natl. Acad. Sci. U.S.A.* **99**, 12759–12764.
71. K. Kawakami, H. Tatsumi, M. Sokabe (2001). Dynamics of integrin clustering at focal contacts of endothelial cells by multimode imaging microscopy. *J. Cell Sci.* **114**, 3125–3135.
72. S. Nishida, Y. Funabashi, A. Ikai (2002). Combination of AFM with an objective-type total internal reflection fluorescence microscope (TIRFM) for nanomanipulation of single cells. *Ultramicroscopy* **91**, 269–274.

Chapter 14

Wide-field autofluorescence microscopy for the imaging of living cells

Franco Fusi, Monica Monici, Giovanni Agati

Table of contents

Abstract

Live cell autofluorescence imaging requires particular attention to select image acquisition techniques. Sensitivity of detection and the viability of the sample are the main considerations. Light microscopy of living versus fixed samples is essentially a trade-off between acquiring images with a high signal-to-noise ratio and damaging the sample under observation [1]. In this chapter we give a brief overview of the main problems that can be encountered with autofluorescence microscopy. Some applications of the technique are reported.

14.1 Introduction

The recent availability of high sensitivity, low noise charge-coupled device (CCD) cameras allows the detection of low-quantum yield autofluorescence signals at a level comparable to that of images obtained with high-quantum yield exogenous markers. Accurate autofluorescence images of cells can be acquired with minimal light exposure and a conventional wide-field microscope [2,3]. Consequently, the possibility of utilizing autofluorescence-based techniques both in research and diagnostics on single living cells or on tissue samples has been reconsidered.

14.2 Fluorescence microscopy

The basic task of a fluorescence microscope is to deliver the excitation light to the specimen and then separate the weaker re-radiating fluorescent light from the much brighter excitation light. This is achieved by using suitable optical filter sets, since the emission spectrum of a fluorophore is shifted toward longer wavelengths with respect to its absorption spectrum. Fluorescence imaging microscopy systems consist of the microscope itself, an illumination source, excitation monochromator (grating or interference filters), a dichroic mirror to separate excitation and fluorescence signals, CCD camera and computer. A spectral analyser is also useful for the identification of optimal excitation and emission wavelengths. The images obtained are digitised and stored for subsequent computation.

Fluorescence microscopy imaging obeys, to a considerable degree, the Lambert–Beer law. The Lambert–Beer law is assumed to be fulfilled when a well-collimated, monochromatic light beam passes through a perpendicular layer of infinitesimal thickness ($\mathrm{d}x$) of a homogeneous diluted solution. More specifically, it is related to the solute itself, which may be considered to be fully transparent, once one has taken into account the contribution of the solvent. In this case the relative attenuation ($-\mathrm{d}I/I_0$) of the beam intensity on crossing the layer $\mathrm{d}x$ is constant:

$$-\mathrm{d}I = \varepsilon c I_0\, \mathrm{d}x \tag{1}$$

where εc represents the "absorption" coefficient of the solute, ε and c being, respectively, the molar absorption coefficient at the excitation wavelength and the molar concentration. I_0 is the incident intensity of the light.

The fraction of light absorbed by a medium is proportional to the thickness of the medium traversed and it is independent (at low intensity) of the light intensity. The intensity of the fluorescence I_{f} emitted from an illuminated fluorochrome is linearly dependent on the absorbed photons:

$$I_{\mathrm{f}} = \Phi_{\mathrm{f}} \mathrm{d}I = \Phi_{\mathrm{f}} \varepsilon c I_0 \, \mathrm{d}x \tag{2}$$

where Φ_{f} is the fluorescence quantum yield and I_0 the incident intensity.

The re-emitted photons are distributed in energy according to the emission spectrum $S_{\mathrm{em}}(\lambda)$, and the probability of emitting a photon at wavelength λ is given by $\Phi_{\mathrm{f}} \cdot S_{\mathrm{em}}(\lambda)$. The emitted light radiates spherically in all directions and only a fraction of it is collected by the objective of the microscope. Moreover, the number of recorded photons is reduced by losses in the optical path and because of the sensitivity of the CCD itself.

Equation (2) can be generalized to calculate the intensity of the light impinging on the single element (pixel, picture element) of the CCD detector. The following equation refers to a single fluorophore j excited with monochromatic light of wavelength λ_{exc} and measured using the filter set i [4]:

$$I_{ij} = t \cdot s \cdot \ell \cdot L(\lambda_{\mathrm{exc}}) \cdot T_{\mathrm{exc}}(\lambda_{\mathrm{exc}}) \cdot \varepsilon_j(\lambda_{\mathrm{exc}}) \cdot c_j \cdot \int_{\lambda} R(\lambda_{\mathrm{em}}) \cdot \Phi_j \cdot S_{\mathrm{em}}(\lambda_{\mathrm{exc}}, \lambda_{\mathrm{em}}) \cdot T_{\mathrm{em}}(\lambda_{\mathrm{em}}) \cdot \mathrm{d}\lambda_{\mathrm{em}} \tag{3}$$

where t is the integration time, l and s represent, respectively, the optical depth of the specimen and the area of the specimen imaged by one pixel of the CCD, ($l{\cdot}s$ is the volume of the sample that is imaged on a single pixel of the CCD). L is the excitation light flux impinging on the sample area s, T_{exc} is the transmission spectrum of the excitation filter set for excitation wavelength λ_{exc}, c_j is the molar concentration and ε_{j} is the absorption coefficient of the fluorophore j. $T_{\mathrm{em}}(\lambda_{\mathrm{em}})R(\lambda_{\mathrm{em}})$ is the fluorescence collection response of the system (defined as the ratio between the pixel counts and the photons emitted from the imaged volume of the sample; $R(\lambda_{\mathrm{em}})$ takes into account the collection response of microscopy optics and detector); $\Phi j S_{\mathrm{em}}(\lambda_{\mathrm{em}})$ is the emitting probability of the excited molecule and $T_{\mathrm{em}}(\lambda_{\mathrm{em}})$ is the transmission spectrum of the filter set i in the emission path.

A biological sample contains many fluorophores that contribute to the total fluorescence. For a filter set i, the total intensity recorded by a single pixel is the sum of the fluorescent emissions from each fluorophore.

$$I_i = \sum_{j=1}^{n} I_{ij} \tag{4}$$

According to Equation (3) and the geometrical theory of optical imaging [5], the intensity of the detected fluorescence signal depends not only on the spectral characteristics and concentration of the fluorophores $\varepsilon \Phi \mathrm{S}_{\mathrm{em}} c_j$, but also on the optical components such as objective, filters and dichroic mirror in the optical path.

If geometrical theory is applied, the excitation light flux $(L(\lambda_{\mathrm{exc}}) \cdot T(\lambda_{\mathrm{exc}}))$ focused on the sample by epiillumination technique depends on the second power of the Numerical Aperture (NA) [6].

Yet fluorescence intensity is proportional to the excitation light flux ($L(\lambda_{exc}) \cdot T(\lambda_{exc})$), until the latter is no longer high enough to quench the fluorophores. The fraction of fluorescence emission that can be collected by the objective lens is also determined by the second power of NA. If M is the total magnification, the detected fluorescence signal is distributed on a surface M^2 times larger than that of the sample, and the intensity impinging on each pixel is proportional to $1/M^2$.

In epifluorescence microscopy, the intensity of the image varies as the fourth power of NA of the objective and varies inversely with the square of M [7]:

$$I \propto \mathrm{NA}_{\mathrm{obj}}^4/M^2 \tag{5}$$

Filters and dichroic mirrors ($T(\lambda_{em})$) reduce the collection factor efficiency too, and other elements present in the microscope's optical path, such as multiple mirrors, lenses and windows, also contribute to light loss. A total collection efficiency ($T(\lambda_{em}) \cdot R(\lambda_{em})$) of 10% represents the upper limit in light collection. Considering an average quantum efficiency of the CCD of about 50% (most CCDs are less efficient in the blue range), the total detection efficiency will be less than 5%.

14.2.1 Detector devices

CCD detectors are ideal for quantitative 2D intensity and position measurements since the signal per image element (pixel) is proportional to the light intensity over a wide range, without introducing any geometrical distortion of the image. At the same time, however, electronic imaging is complicated by the presence of noise, which degrades image quality.

The pixel grey level value acquired by a computer for a single readout of the chip is

$$P = (IQt + N_{\mathrm{T}}t + N_{\mathrm{R}})G \tag{6}$$

where I is the light intensity, Q is the quantum efficiency, t is the integration time, N_{T} is the thermal noise, N_{R} is the readout noise and G is the gain factor to the conversion of accumulated electrons into grey level. N_{R} is independent of the amount of light reaching the camera and is dependent on the readout rate (it is approximately a linear function of the square root of the readout rate). N_{T} is due to the electrons thermally generated within the pixels and collected along with the photoelectrons. These electrons are generated even when the CCD is not illuminated, and are known as dark current noise. By cooling the camera this dark current noise can be drastically reduced (it decreases by about 50% for each 8°C of cooling).

If the response (P) of a camera to a given light level is of the same order of the noise level $(N_{\mathrm{T}}t + N_{\mathrm{R}})G$ the image will not be well resolved. The most important parameter of a camera's response is therefore not the sensitivity itself (proportional to the quantum efficiency Q), but rather the signal-to-noise ratio (S/N) that can be achieved for a given light level reaching the detector

$$\mathrm{S/N} = IQt/(N_{\mathrm{T}}t + N_{\mathrm{R}}) \tag{7}$$

14.3 The wide-field choice

In autofluorescence imaging it is important to consider which microscopy technique will be used: wide-field, laser scanning confocal, or multiphoton.

In the last five years several publications have compared these methods [1,8 and references therein]: wide-field seems to work well with thin (<30 μm) and no-scattering samples, confocal microscopy works well with medium thick (30–80 μm) samples, and multiphoton has proved the best technique for thick samples.

The signal-to-noise ratio is the main parameter to be considered when choosing the best microscope method for the imaging of cell autofluorescence. S/N has two features: one dependent on the sample itself, while the other depends on the noise added to the signal by illumination fluctuations and signal detection [1,8].

A recent quantitative comparison of wide-field and confocal microscopes found that, for thin living cells, a wide-field microscope produced images with better S/N, mainly thanks to a more stable illumination system, while a confocal microscope out-performed the wide-field microscope on thick samples, where the dominant source of noise was out-of-focus light [8]. For very thick samples, multiphoton microscopes have proved to be very effective because of their property of limiting fluorophore excitation and photodamage to a relatively small focal volume [9].

Wide-field images are much less noisy for two reasons. First, illumination is delivered in parallel to all regions of the sample simultaneously, thus eliminating pixel-to-pixel differences in delivered illumination. Second, autofluorescence signal is very low and the cooled CCD camera has detection efficiency 1.5–4 times higher than that of the photomultiplier tubes of the laser-scanning confocal microscope. This means that there is a 4-fold increase in the number of photons detected, and the Poisson shot noise is consequently reduced by a factor of 2 for the same total illumination (which is limited in these living samples because of the possibility of photobleaching and phototoxicity) [10]. Recent literature asserts that it is possible to exceed the fundamental resolution limit of the wide-field fluorescence microscope. The trick is that the limits are valid only to the resolution of the observed optical image, and not to the ability to determine the structure of the sample [11]. These new imaging technologies, which are being used increasingly in live cell imaging, should become routinely available over the next few years [12,13 and the present volume].

14.3.1 Photo-invasivity

Photobleaching and photodamage are two of the most important limitations of fluorescence microscopy in the study of living cells. Fluorescence excitation light causes the formation of free radicals, highly reactive species, that react with cellular constituents and produce oxidative stress. Moreover, the fluorophores undergo molecular degeneration or rearrangement to a non-fluorescent species. In addition, excited fluorophores in a triplet state can generate singlet oxygen, which will react with a wide variety of functional groups on neighbouring biomolecules.

The fact that wide-field requires significantly less excitation light is an important factor for living specimens that may be damaged by excessive exposure. A decrease in excitation light can be achieved either by attenuating light intensity, or by reducing the power of the illumination source. In addition, it is essential to minimize the time of each exposure, since cells may move if the integration time is excessive.

14.3.2 Deconvolution to increase the signal-to-noise ratio

The S/N of images in a live cell experiment is usually limited by the reduction in light intensity and exposure time necessary for a successful experiment. Reduced S/N often compromises resolution, making it difficult to discern cellular structures.

The conventional light microscope collects light from planes above and below the plane of focus. Parts of the cell further from the plane of focus contribute light to the 2D image but this light is more blurred. Unless the cell is essentially flat, these out-of-focus parts of the cell will render the image more hazy. Small adjacent features in the plane-of-focus (X–Y) can be resolved distinctly, but the contrast between them is reduced by out-of-focus fluorescence from the remainder of the cell. Two features that appear to be superimposed in a single two-dimensional microscope image may in fact be significantly separated in the axial direction [14,15].

The effect of out-of-focus contributions may be reduced computationally. The image generated by a microscope is modelled as a convolution of the object (the sample on the microscope) with the microscope's point-spread-function (PSF). The PSF describes how a point in the sample is imaged by microscope optics:

$$\text{Image} = \text{Object} \otimes \text{PSF}$$

where $\otimes$ represents the convolution operator.

The brightness of every point in the image is linearly related through convolution to the fluorescence of each point in the object.

Consequently, the influence of all optics and filters in the light path can be suppressed by computational optical sectioning microscopy, and a deconvolution algorithm can efficiently reverse the loss of contrast, thus compensating for the blurring effect of defocus.

There are two classes of deconvolution algorithms, deblurring algorithms and image restoration algorithms [14–16].

Deblurring algorithms make an estimate of the blurring in a given focal plane and subtract the estimated blur from the plane [14,15].

Deblurring algorithms are fundamentally two-dimensional, because they apply a plane-by-plane operation to each two-dimensional plane of a three-dimensional image stack. In contrast, image restoration algorithms operate simultaneously on every pixel in a three-dimensional image stack and, instead of subtracting blur, they attempt to reassign blurred light to the correct image plane.

The deconvolution procedure produces a large amount of data, and computational time is about 10 times that for confocal methods. However, personal-computer capabilities have increased enormously as regards computational capability and fast memory capacity, so much so that time-consuming iterative calculations can be performed in a few seconds.

Further information on deconvolution image processing techniques can currently be found at <http://microscopy.fsu.edu/primer/digitalimaging/deconvolution/deconvolutionhome.html>.

14.4 Analysis of living samples: problems and considerations

Autofluorescence microscopy is often considered to be a non-destructive tool for living cell imaging. Indeed the possibility of carrying out direct measurements on single viable cells is very attractive. Endogenous fluorophores are molecules with both structural and metabolic functions, and thus, by analysing their emission, we can obtain both morphological and functional information. At the same time, the close relation found between fluorescence pattern and cell morpho-functional state is the basis for cell recognition and the development of new analytical methods. The fact of working on living cells, that is on a system continually interacting with the environment and continually evolving, brings about no few problems.

Often UVA radiation is chosen to excite endogenous fluorophores. UVA exposure, in addition to autofluorescence generation, also induces oxidative stress in cells via type I and type II photo-oxidation processes, resulting in cell damage and cell death [17].

Oxidative stress can be investigated by monitoring the modifications of the autofluorescence pattern and lifetime. Firstly, autofluorescence arises from the cytoplasmic region and appears mostly bound to cytoplasmic organelles, such as mitochondria, while the nucleus is generally non-fluorescent. After UVA exposure, the fluorescence intensity increases and the fluorescence distribution becomes increasingly homogeneous. The nucleus also fluoresces and the nucleoli become the brightest intracellular fluorescent sites [18,19].

The initial decrease is generally explained by a decrease in nicotinamide adenine dinucleotide (NADH) concentration, due to the UVA induced photo-oxidation of the coenzyme. The increase in fluorescence, observed for further UVA exposure, could be explained as an induction of membrane damage and consequent mitochondrial damage, with efflux of NADH in the cytoplasm. The damage of the nuclear membrane could allow the crossing of NADH and its binding to the nucleolus proteins.

In summary, fluorescence excitation radiation utilized in living cell fluorescence microscopy can induce modifications in the cellular redox state and damage the cell structures [20].

Fluorescence relocation combined with a decrease in the average fluorescence lifetime suggests the possibility of an increase in intracellular free NAD(P)H, in comparison with the bound pool. This idea is also supported by the finding of UVA-induced red-shift in the autofluorescence peak [19]. Moreover, a relationship exists between lipofuscin accumulation and oxidative stress. An increase in the rate of lipofuscin accumulation was observed when cells were grown at increasing ambient oxygen concentration [21].

Another problem, not only in autofluorescence techniques but in fluorescence microscopy in general, is photobleaching. The bleaching (chemical change of the fluorophore into a non-fluorescent molecule) often involves reactions with oxygen. Thus, removal of oxygen from the samples reduces photobleaching. Therefore, this procedure is not suitable for analysing living samples. In this case, reduction of photobleaching could eventually be achieved by addition of antioxidants.

Finally, the standardization of the autofluorescence-based techniques needs a deeper knowledge of the correlation between endogenous fluorescence emission and cell biology. The links between autofluorescence and cell cycle, cell age, culture conditions, etc. should be carefully explored. For example, fluorescence is reported to be low in freshly thawed and resided cultures, increasing with time (as cells enter the exponential phase of growth) to a maximum with increased cell numbers. Moreover, the author observed that one line growing in suspension displayed a higher autofluorescence than the same line growing in monolayer [22].

When tissue samples are analysed, care must be taken in interpreting the results of in vitro measurements for tissue diagnostics. The biochemical properties of tissues may change significantly in vitro and in vivo. The ratio NAD+/NADH may change, as well as blood content and oxidation state. These changes can influence UV–visible autofluorescence [23]. Consequently, spectra and fluorescence patterns of tissues in vitro differ from those in vivo. Moreover, in vitro, the fluorescence from NADH appears to decay exponentially with time, while the fluorescence from collagen and FAD remains relatively constant [24].

Bioptical procedures may induce trauma in tissues. Removal of tissues with biopsy forceps could result in local haemorrhage with consequent attenuation of wavelengths due to absorption by oxyhemoglobin [25]. To minimize the problem tissue samples can be obtained from surgical specimens, immediately snap-frozen in liquid nitrogen and stored at $-70°$C, thawed over ice to minimize the structural damage that may occur from rapid defrosting, and moistened to physiologic pH [26].

In conclusion, a proper application of autofluorescence techniques to living sample analysis requires a deeper knowledge of the changes due to the interaction with light excitation and measurement environment, and of course the samples should be maintained under conditions as close as possible to the physiological ones.

14.5 Single living cells

Many authors, utilizing different cell populations, have reported that an autofluorescence pattern reflects intracellular structural organization, and have observed a connection between autofluorescence and the metabolic state of a cell [27 and references reported therein]. These findings accord with the presence, among the principal endogenous fluorophores, of structural proteins and molecules involved in cell metabolism, for example pyridine and flavin nucleotides. Moreover, the discrete cytoplasmic locations of the fluorescence – although to a different extent depending on the cell population – suggest that a high concentration of fluorophores are bound to the organelles involved in their utilization by the cells [22,28].

A brief review on the applications of autoflourescence imaging to living cells and tissue analysis is given below, with particular emphasis on those performed by our group.

14.5.1 Properties of human leukocytes

Research on single living cells principally concerns cells of lymphohemopoietic origin. Studies on the intrinsic fluorescence properties of human leukocytes showed that leukocyte families, i.e. lymphocytes, monocytes, neutrophils and eosinophils, differ in both fluorescence intensity and fluorescence pattern (Figure 1).

Under 365 nm excitation, an emission band centered in the 440–490 nm range, due to the nicotinic coenzymes and derivatives was observed. A shoulder was also detected in the 500–560 nm range, due to flavins and flavin coenzymes. This component of autofluorescence can be better observed with excitation at 436 nm, a wavelength more suitable than 365 nm for exciting flavins. The differences in intensity proved to be related to the concentration of the endogenous fluorophores, possibly reflecting different metabolic rates, rather than to cell dimension calculated in terms of the cytoplasmic to nuclear volume ratio [28].

The intracellular fluorescence pattern shows that emission mostly arises from the cytoplasm, thus confirming that autofluorescence is related to the metabolic processes of cells. Fluorescence from the nucleus is not detectable, while very intense fluorescence originates from the granules of granulocyte cells, due to their lipopigment content [29].

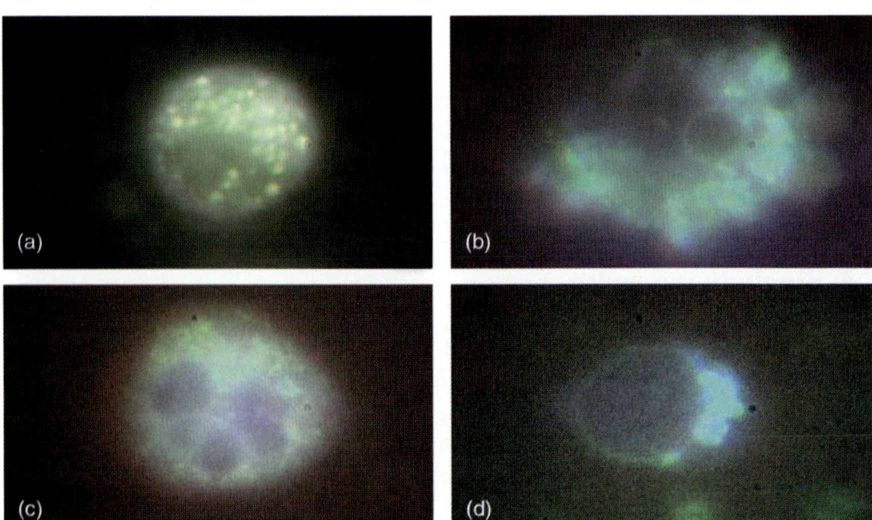

Figure 1. Autofluorescence patterns of human leukocytes: (a) eosinophil granulocyte, (b) monocyte, (c) neutrophyl granulocyte, (d) lymphocyte. Excitation at 365 nm.

14.5.2 Leukaemic cells

Further research has demonstrated the possibility of distinguishing normal cells from leukaemic ones, through their fluorescence emission. Human cell lines of both myeloid and lymphoid origin were considered. Leukaemic cell autofluorescence is generally more intense than that of normal leukocytes, the emission band is broader and the blue peak is red-shifted. The fluorescence pattern of leukaemic cells is characterized by a quite uniform distribution of the fluorescence, and cell structures are barely distinguishable [28] (Figure 2). These data agree with the findings of Croce et al. on normal and transformed fibroblasts [27]. The differences found between autofluorescence of normal and neoplastic cells, both in spectra and fluorescence patterns, can be explained in terms of different contents of pyridine coenzyme forms (oxidized/reduced, free/bound). In neoplastic cells, in contrast to normal cells, the anaerobic component of energy metabolism increases while the aerobic component decreases [30]. The decreased efficiency of the aerobic component gives rise to an increase in the reduced pool of pyridine nucleotides [31]. This may explain the greater fluorescence intensity observed in neoplastic cells. The red-shift and broadness of the peak observed in neoplastic cells may also be explained in terms of prevalence of anaerobic metabolism. This condition involves fewer stages of interaction between coenzymes and enzymes than in aerobic metabolism. As a consequence, the ratio NADH-free/NADH-bound is higher in anaerobic than in aerobic cells. The peak position of the free form (465 nm) is red-shifted in comparison with the peak position of the bound form (444 nm) [32].

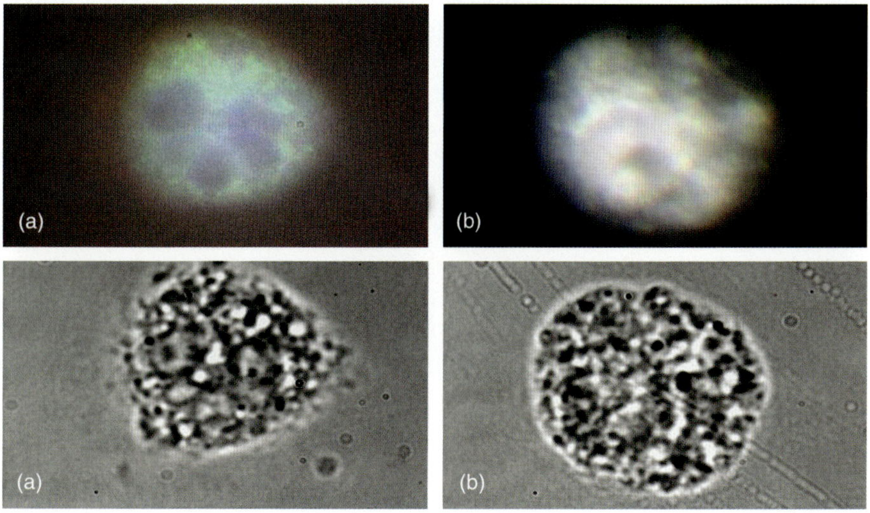

Figure 2. (Top) autofluorescence pattern of (a) human neutrophyl granulocyte and (b) cell from human leukemic cell line U937. Excitation at 365 nm. (Bottom) transmitted light images of the same cells. It can be observed that autofluorescence imaging, with a proper choice of excitation and detection wavelengths, greatly enhances the contrast between the cell images.

14.5.3 Urothelial cells

In urothelial tumor cells autofluorescence excited at 488 nm appeared drastically reduced, in comparison with normal cells. This indicates that the flavin concentration may be significantly lowered, particularly in poorly differentiated types [33]. These findings agree with the results obtained from other authors who showed, by extraction, that riboflavin concentration was reduced in tumor tissues, suggesting a deficient aerobic oxidation system [34].

Data reported in the literature indicate that the decrease in the aerobic metabolism of highly proliferating neoplastic cells could be partly related to the decrease in the mitochondria content found in different neoplastic cells [35]. The different fluorescence distribution observed in the cytoplasm of neoplastic cells (quite homogeneous distribution) compared with that of normal ones (discrete location) could be related to the decrease in the quantity of mitochondria.

14.5.4 Drug interaction

Studies on the interaction between lymphohemopoietic cells and drugs have been performed. Our interest in this subject has mainly been related to the question of antiblastic drug resistance. Some classes of antiblastic drugs have fluorogenic properties. Thus, microspectrofluorometry and multispectral imaging fluorescence microscopy are particularly suitable techniques to perform pharmacodynamic studies on single cells, including uptake, distribution, retention and efflux of drugs. In cells treated with doxorubicin, we could determine the intracellular distribution of the drug, monitor cell autofluorescence changes due to the interaction with the drug, and record drug emission spectra directly from a single cell [36] (Figure 3). Recently, Dellinger et al. proposed a method based on the monitoring of the NAD(P)H fluorescence transients for the probing of biopharmaceutical effects at the intracellular level [37]. Pignon et al., by microspectrofluorometric technique, calculated the retention rate of anthracycline in the nucleus and found that the amount of doxorubicin incorporated into the nucleus was related to the resistant or sensitive character of K562 cells [38].

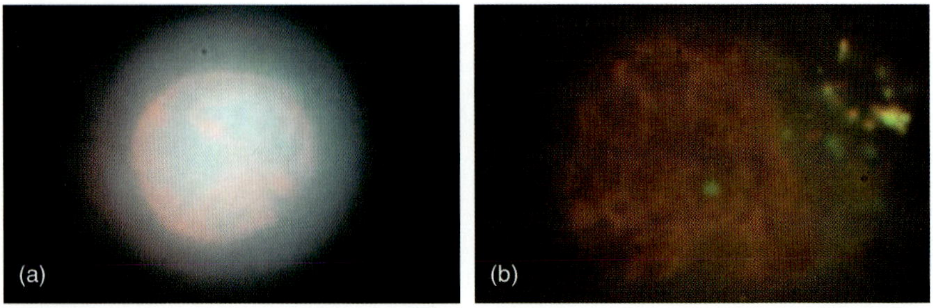

Figure 3. Autofluorescence pattern of cells from human leukemic cell line HL60, treated with doxorubicin. When samples are excitated at 365 nm it is possible to reveal both drug and NAD(P)H fluorescence (a), while excitation at 436 nm is more suitable to monitor the drug fluorescence (b).

On this basis, it should be possible to identify resistant cells, thus providing information of great importance for diagnostic, therapeutic and prognostic evaluations, including the selection of more effective drugs and other substances to induce the reversion of multidrug resistance.

14.5.5 Cell differentiation

The autofluorescence monitoring on living cells is a potentially useful tool for in vitro studying of the cell differentiation processes. Furthermore, different maturation steps can be distinguished on the basis of the cell fluorescence pattern, opening up the possibility for future application of the technique in diagnostics.

The characterization of leukemic cell autofluorescence during differentiation, induced by 12-*O*-tetradecanoylphorbol 13-acetate (TPA) and all-trans retinoic acid (ATRA), has been performed by autofluorescence microspectroscopy and multispectral imaging autofluorescence microscopy.

The results reveal a correlation between cell autofluorescence pattern and cell differentiation degree. When cells differentiate their autofluorescence emission changes, following the morphological and functional rearrangement of cell structures [39]. A decrease in emission intensity and a different distribution of endogenous fluorophores are observed (Figure 4).

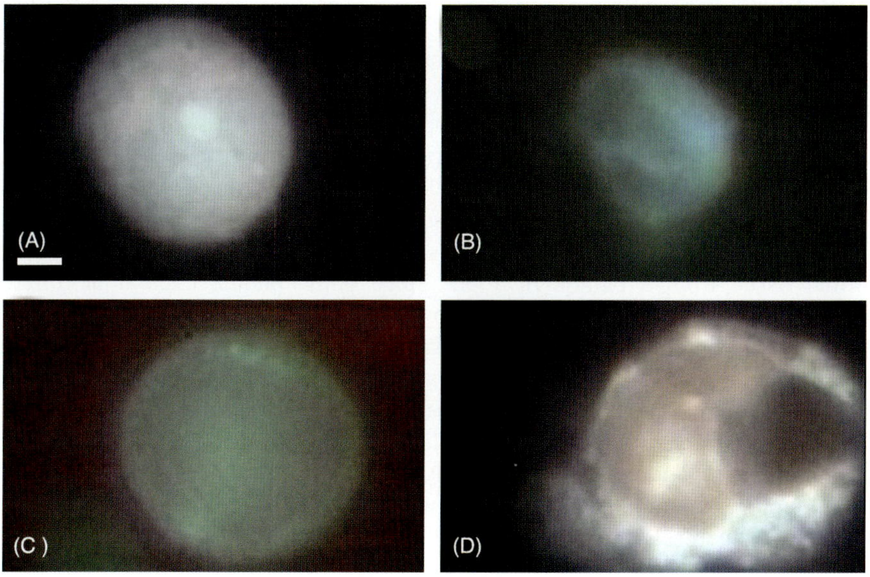

Figure 4. Autofluorescence imaging of HL60 cells before (A) and after (B) treatment with ATRA, a drug inducing differentiation through the granulocytic pathway. Autofluorescence patterns of FLG 29.1 cells before (C) and after (D) the treatment with TPA. The phorbol ester induces in FLG 29.1 cells a process of differentiation through the osteoclastic pathway. Excitation at 365 nm. Scale bar: 5 μm.

The lower autofluorescence intensity observed in differentiated cells is due to (a) the increased importance of the aerobic component of the energetic metabolism and (b) the decrease of autofluorescence arising from the nucleus.

Under differentiating treatment the green component of cell autofluorescence increased, suggesting an increase of the flavin content. These findings are in accordance with the results obtained from other authors showing that flavin concentration was significantly reduced in neoplastic and poorly differentiated cells, suggesting a deficient aerobic oxidation system [33,40].

14.6 Tissue analysis

Although histochemical and immunohistochemical methods are the standard procedures in diagnosis of tissue disorders, useful improvements in evidencing histopathologic manifestations can be obtained with the introduction of tissue autofluorescence analyses. Specific staining methods show up particular tissue structures clearly, but sample preparations can be expensive and time consuming. Furthermore, the labelling implies chemical modification of the target. Autofluorescence diagnosis can be performed in real time because it doesn't require specific treatment of the samples.

14.6.1 Lymph node tissue

Notably, being the active site for the immune system response, lymph nodes are important organs from a diagnostic point of view. Analysis of the proximal lymph nodes is performed for any kind of neoplastic disease to get information on the tumor infiltration. This is of essential help in guiding the surgery for the excision of neoplastic tissues. Therefore, a rapid method for the analysis of lymph nodes has a wide range of interest in clinical oncology.

Using spectrofluorometric techniques alone it is difficult to distinguish a normal tissue from a pathological one since both contain, even if at different concentrations, natural fluorophores with markedly overlapped spectral emissions. With autofluorescence imaging the combination of spectral and spatial resolution provides a suitable characterization of a specific tissue sample, which becomes useful for diagnostic purposes. Therefore, multispectral autofluorescence imaging and spectrofluorometry represent complementary techniques that can provide a more complete quantitative and qualitative analysis of tissue samples.

Multicolor imaging autofluorescence microscopy has been applied to characterize the autofluorescence of lymph node tissues [41]. Figure 5 shows significant examples of autofluorescence images from non-neoplastic and neoplastic lymph node sections respectively. The autofluorescence image of a biopsy section from a hyperplastic lymph node (Figure 5(a)) shows details of a trabecula of connective tissue, which separates two follicles. The typical lymph node organization is maintained. Connective trabeculae are strongly fluorescent, being mainly composed of collagen and elastin. Conversely, a very faint fluorescence characterizes

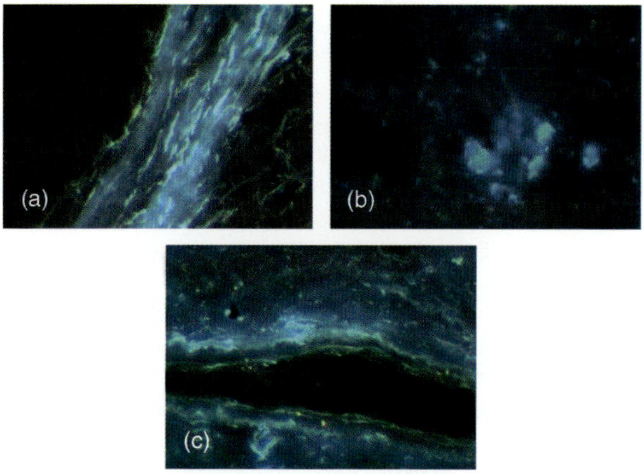

Figure 5. Autofluorescence images of a lymph-node section (10 μ thick). (a) Reactive hyperplasia: a connective trabecula separating two follicles can be observed. (b) Hodgkin's lymphoma, the image shows a group of large and intensely fluorescent cells identified as Reed–Stemberg cells. (c) Detail of a small vessel crossing a connective trabecula. Magnification 10×. Excitation at 365 nm.

the follicles due to the low autofluorescence of lymphocytes, the most highly represented cell population.

The biopsy section from the lymph node of a Hodgkin's lymphoma patient (Figure 5(b)) shows groups of large cells with intense autofluorescence. Comparison with immunohistochemical analysis suggests that these highly fluorescent cells correspond to the Reed–Stenberg's cells typical of Hodgkin's disease. Complete loss of the lymph node structure is evident.

14.7 Perspectives

In recent years, new light microscopy techniques have become available that allow us to image endogenous cellular molecules playing different and important roles in structural and functional processes in biological systems. Consequently, the possibility arises of utilizing autofluorescence-based techniques both in research and diagnostics on single living cells as well as on tissue samples, each time choosing the technique and implementing the instrumental set-up suitable for the different application required.

The resolution attainable by deconvolution algorithms and wide-field microscopy is being enhanced by a constant increase in both the computational capacity of computers and the S/N of CCD cameras.

Autofluorescence imaging of living cells is of great interest as a potential alternative to standard histological preparations for the rapid, low-cost examination of fresh biopsies.

References

1. D.J. Stephens, V.J. Allan (2002). Light microscopy techniques for live cell imaging. *Science* **300**, 82–86.
2. W.A. Carrington, R.M. Lynch, E.D. Moore, G. Isenberg, K.E. Fogarty, F.S. Fay (1995). Superresolution three-dimensional images of fluorescence in cells with minimal light exposure. *Science* **268**, 1483–1486.
3. Y. Hiraoka, J.W. Sedat, D.A. Agard (1987). The use of a charge-coupled device for quantitative optical microscopy of biological structures. *Science* **238**, 36–38.
4. Y. Garini, A. Gil, I. Bar-Am, D. Cabib, N. Katzir (1999). Signal to noise analysis of multiple color fluorescence imaging microscopy. *Cytometry* **35**, 214.
5. M. Born, E. Wolf (1980). *Principles of Optics*. Pergamon Press, New York.
6. J.R. Benford, H.E. Rosenberger, (1978). In: W.G. Driscoll (Ed) *Handbook of Optics: Microscope Objectives and Eyepieces* (Section 6). McGraw-Hill, New York.
7. D.L. Taylor, E.D. Salomon (1989). Basic fluorescence microscopy. *Methods in Cell Biol.* **29**, 207–237.
8. J.R. Swedlow, K. Hu, P.D. Andrews, D.S. Roos, J.M. Murray (2002). Measuring tubulin content in Toxoplasma gondii: A comparison of laser-scanning confocal and wide-field fluorescence microscopy. *Proc. Natl. Acad. Sci. U.S.A.* **99**, 2014–2019.
9. A. Diaspro, C.J.R. Sheppard (2002). Two-photon excitation microscopy: basic principles and architectures. In: A. Diaspro (Ed), *Confocal and Two-photon Microscopy: Foundations, Applications, and Advances.* (pp. 39–74). Wiley-Liss, New York.
10. W.A. Carrington, K.E. Fogarty, L.M. Lifshitz, F.S. Fay, (1989). Three dimensional imaging on confocal and wide field microscopes. In: J. Pawley (Ed), *Handbook of Biological Confocal Microscopy* (pp. 151–161). Plenum, New York.
11. M.G.L. Gustafsson (2000). Surpassing the lateral resolution limit by a factor of two using structured illumination microscopy. *J. Microsc.* **198**, 82–87.
12. A. Egner, S. Jakobs, S.W. Hell (2002). Fast 100-nm resolution three-dimensional microscope reveals structural plasticity of mitochondria in live yeast. *Proc. Natl. Acad. Sci. U.S.A.* **99,** 3370–3375.
13. M.G. Gustafsson (1999). Extended resolution fluorescence microscopy *Curr. Opin. Struct. Biol.* **9**, 627–634.
14. J.R. Swedlow, M. Platani (2002). Live cell imaging using wide-field microscopy and deconvolution *Cell Struct. Funct.* **27**, 335–341.
15. W. Wallace, L.H. Schaefer, J.R. Swedlow (2001). A working-person's guide to deconvolution in light microscopy. *Biotechniques* **31**, 1076–1082.
16. J.G. McNally, W. Fekete, J.A. Conchello (1999). Comparison of 3D microscopy methods by imaging a well-characterized test object, *Proc. SPIE* **2984**, 52–63.
17. R.M. Tyrell, S.M. Keyse (1990). New trends in photobiology. The interaction of UVA radiation with cultured cells. *J. Photochem. Photobiol. B* **4**, 349–361.
18. M. Monici, G. Agati, P. Mazzinghi, F. Fusi, P.A. Bernabei, I. Landini, P. Rossi Ferrini, R. Pratesi (1996). Image analysis of cell natural fluorescence. Diagnostic applications in haematology, *Proc. SPIE* **2928**, 180–185.
19. K. Konig, P.T.C. So, W.W. Matulin, B.J. Tromberg, E. Gratton (1996). Two-photon excited lifetime imaging of autofluorescence in cells during UVA and NIR photostress. *J. Microsc.* **183**, 197–204.
20. K.K. Konig, T. Krasieva, E. Bauer, U. Fiedler, M.W. Berns, B.J. Tromberg, K.O. Greulich (1996). Cell damage by UVA radiation of a mercury microscopy lamp

probed by autofluorescence modifications, cloning assay, and comet assay. *J. Biochem. Optics* **1**, 217–222.

21. G. Gao, U. Johanssopn, I. Rundquist, K. Ollinger (1994). Lipofuscin-induced autofluorescence of living neonatal rat cardiomyocytes in culture. *Mech. Ageing Dev.* **73**, 79–86.
22. J.E. Aubin (1979). Autofluorescence of viable cultured mammalian cells. *J. Histochem. Cytochem.* **27**, 36–43.
23. Bigio, J.R. Mourant (1997). Ultraviolet and visible spectroscopies for tissue diagnostics: fluorescence spectroscopy and elastic-scattering spectroscopy. *Phys. Med. Biol.* **42**, 803–814.
24. K.T. Schomacker, J.K. Frisoli, C.C. Compton, T.J. Flotte, J.M. Richter, N.S. Nishioka, T.F. Deutsch (1992). Ultraviolet laser-induced fluorescence of colonic tissue: basic biology and diagnostic potential. *Lasers Surg. Med.* **12**, 63–78.
25. R. Richards-Kortum, R.P. Rava, R.P. Petras, M. Fitzmaurice, M. Sivak, M.S. Feld (1991). Spectroscopic diagnosis of colonic dysplasia. *Photochem. Photobiol.* **53**, 777–786.
26. B. Banerjee, B. Miedema, H.R. Chandrasekhar (1998). Emission spectra of colonic tissue and endogenous fluorophores. *Am. J. Med. Sci.* **316**, 220–226.
27. A.C. Croce, A. Spano, D. Locatelli, S. Barni, L. Sciola, G. Bottiroli (1999). Dependence of fibroblast autofluorescence properties on normal and transformed conditions. Role of the metabolic activity. *Photochem. Photobiol.* **69**, 364–374.
28. M. Monici, R. Pratesi, P.A. Bernabei, R. Caporale, P. Rossi Ferrini, A.C. Croce, P. Balzarini, G. Bottiroli (1995). Natural fluorescence of white blood cells: spectroscopic and imaging study. *J. Photochem. Photobiol. B* **30**, 29–37.
29. P.F. Weller, S.J. Ackerman, A. Nicholson-Weller, A.M. Dvorak (1989). Cytoplasmic lipid bodies of human neutrophilic leukocytes. *Am. J. Pathol.* **135**, 947.
30. O. Warburg (1956). On the origin of cancer cells. *Science* **123**, 309–311.
31. E.T. Obi-Tabot, L.M. Hanrahan, R. Cachecho, E.R. Beer, S.R. Opkins, J.C.K. Chan, J.M. Shapiro, W.W. La Morte (1993). Changes in hepatocyte NADH fluorescence during prolonged hypoxia. *J. Surg. Res.* **55**, 575–580.
32. J.M. Salmon, E. Kohen, P. Viallet, J.G. Hirschberg, A.W. Wouters, C. Kohen, B. Thorell (1982). Microspectrofluorometric approach to the study of free/bound NAD(P)H ratio as metabolic indicator in various cell types. *Photochem. Photobiol.* **36**, 585–593.
33. M. Anidjar, O. Cussenot, J. Blais, O. Bourdon, S. Avrillier, D. Ettori, J.M. Villette, J. Fiet, P. Teillac, A. De Luc (1996). Laser induced autofluorescence diagnosis of bladder tumors: dependence on the excitation wavelength. *J. Urol.* **155**, 1590–1596.
34. M.A. Pollack, A. Taylor, J. Taylor, R.J. Williams (1942). B Vitamins in Cancerous Tissues I. Riboflavin. *Cancer Res.* **2**, 739–743.
35. P.L. Pedersen (1978). Tumor mitochondria and the bioenergetics of cancer cells. *Prog. Exp. Tumor Res.* **22**, 190–274.
36. M. Monici, F. Fusi, P. Mazzinghi, A. Degli Innocenti o Nocentini, I. Landini, J. Banchelli, B. Bartolozzi, V. Santini, P.A. Bernabei (1999). In: G.J.L. Kaspers, R. Pieters, A.J.P. Veerman (Eds) *Advances in Experimental Medicine and Biology. Drug Resistance in Leukemia and Lymphoma* (Volume III, pp. 457–458). Kluwer Academic/Plenum Publishers, New York.
37. M. Dellinger, M. Geze, R. Santus, E. Kohen, C. Kohen, J.G. Hirschberg, M. Monti (1998). Imaging of cells by autofluorescence: a new tool in the probing of biopharmaceutical effects at the intracellular level. *Biotechnol. Appl. Biochem.* **28**, 25.

38. B. Pignon, H. Morjani, J.P. Vilque, J.M. Millot, G. Simon, B. Lartigue, J.C. Etienne, G. Potron, M. Manfait (1995). In vitro study of THP-doxorubicin retention in human leukemic cells using confocal laser microspectrofluorometry. *Leukemia* **9**, 1361–1367.
39. M. Monici, G. Agati, F. Fusi, R. Pratesi, M. Paglierani, V. Santini, P.A. Bernabei (2003). Dependence of leukemic cell autofluorescence patterns on the degree of differentiation. *Photochem. Photobiol. Sci.* **2**, 981–987.
40. M.A. Pollack, A. Taylor, J. Taylor, R.J. Williams (1942). Vitamins in cancerous tissues. I. Riboflavin, *Cancer Res.* **2**, 739–743.
41. L. Rigacci, R. Alterini, P.A. Bernabei, P. Rossi Ferrini, G. Agati, F. Fusi, M. Monici (2000). Multispectral imaging autofluorescence microscopy for the analysis of lymph-node tissues. *Photochem. Photobiol.* **71**, 737–742.

Chapter 15

Scanning probe microscopy

Cesare Ascoli, Riccardo Gottardi and Donatella Petracchi

Table of contents

Abstract

In this chapter we try to explain the basic physical principles of Scanning Probe Microscopy (SPM) and to discuss the basis of sample preparation. Both these aspects are important to get good images (or even just meaningful results) and to understand the potential of these relatively young experimental methods. We consider in more detail atomic force microscopy (AFM) and scanning near-field optical microscopy (SNOM). Both these techniques are suitable to image biological samples while still having unexplored capabilities for technical applications. An AFM set-up has two possible uses: to image a sample (perturbing it minimally) and to modify it. We give some examples of both these possibilities, which can be alternated on the same sample. However, the resulting panorama is not exhaustive. For what concerns the use of AFM techniques on biological samples it is worth noting that the amount of required knowledge ranges from physics to biochemistry passing through physical chemistry and in particular the physical chemistry of solutions. A strong contribution from molecular biology is also required. Thus, the best thing is to realize that for a fruitful use of AFM really interdisciplinary work is required and an open-minded attitude towards problems that are largely unknown.

15.1 Introduction

Scanning probe microscopy (SPM) identifies a number of microscopic techniques, some of which will be described hereafter, i.e. Scanning Tunneling Microscopy (STM), Atomic Force Microscopy (AFM), Scanning Near-field Optical Microscopy (SNOM or NSOM). All of them are based on the same common principle, namely that of a probe of small dimensions (the meaning of "small" will soon be made clear) that runs over the sample in subsequent scans at a very short distance from the surface; the interaction between the probe and the sample varies during scanning, depending on the sample characteristics (its composition, its surface structure etc.) and this provides the information for sample imaging.

In principle, any kind of possible interaction between the probe and the sample might be used to invent a SPM technique; however, some techniques have proved to be more successful than others. In fact, depending on the sample, one kind of interaction may be more easily observable than another, or may be expected to give more significant information, hence determining the success of one type of SPM over the other in different areas of research.

15.1.1 Scanning tunneling microscope

The first scanning probe microscopy technique was the Scanning Tunneling Microscopy (STM), developed in 1981 by Nobel prize winners Binnig, Rohrer and co-workers [1–3]. As suggested by its name, tunneling is the basis of STM. Thanks to quantum mechanics, particles, such as electrons, may be described by a wave function. Considering this description it becomes clear that, while in classical physics particles can pass a potential barrier only if their energy is higher than the barrier itself, in quantum physics particles have a non-vanishing probability to be found beyond a potential barrier, in a region forbidden by classical physics: this is known as the tunneling effect. In STM [4,5] a sharp tip (the probe) and a conducting sample are separated by an insulator (e.g. a vacuum gap). By tunneling, electrons have some probability to pass from the tip to the conducting sample. When a difference of potential (typically between 1 mV and 1 V) is applied between tip and sample, a tunnel current runs through the insulator (typically between 0.1 nA and 10 nA) for a small enough gap (Figure 1).

Tunneling probability depends exponentially from the tip–sample distance with a very short space constant. This steep dependence makes it possible to achieve atomic resolution in tunneling based microscopy. Typically, decreasing the spacing by 1 Å increases the tunneling current by one order of magnitude [6]. Moreover, since one of the two sides of this conductor – insulator – conductor junction is shaped into a tip, the current flows only through the outmost atoms in the tip. So the tunnel current is shaped into a filament with a width of atomic dimensions (Figure 1) and this allows for a lateral resolution also of atomic dimensions.

By using a piezoelectric drive system the tip scans the sample while the tunneling current is measured. This measurement is then used to image the sample surface.

Two different operating modes can be adopted. In the constant height mode the feedback loop is switched off, while the tip is rapidly scanned at a constant height over the sample (Figure 2(a)). During scanning, tunneling current variations are measured and recorded as a function of the tip position. The recorded currents contain information on the sample topography.

In the constant current mode the feedback moves the tip up and down to keep the tunneling current constant. In this way, in a homogeneous sample, the tip follows the contour of the sample surface at a fixed, unknown distance (Figure 2(b)). A topographic image can be obtained by recording the voltage, which has to be applied to the piezoelectric driver to keep the tunneling current constant. In fact,

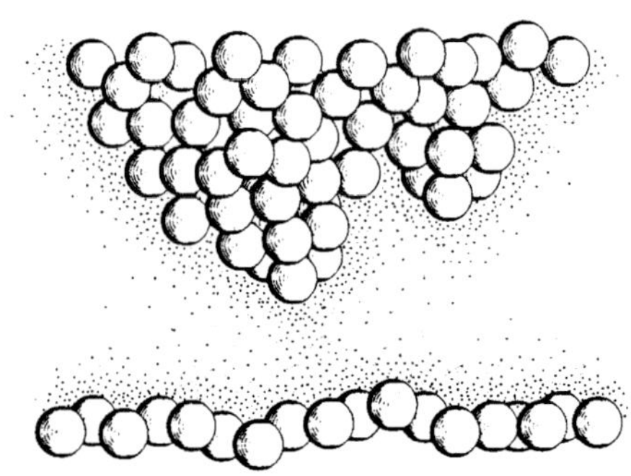

Figure 1. Schematic representation at a microscopic level of a STM tip at a short distance from a conducting surface. Circles represent atoms surrounded by their electronic cloud (the small dots). As the tip approaches the surface, the electronic cloud of the outmost atom of the tip overlaps that of the sample and tunnelling takes place. Redrawn from [4].

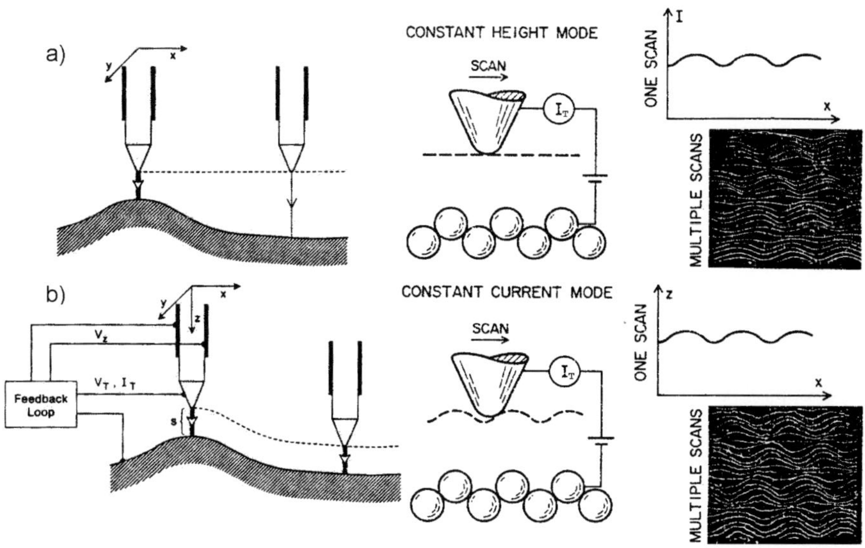

Figure 2. Different operating modes in STM. (a) Constant height mode. The tip is swept on the sample at a constant height; a constant voltage V_T is applied between the tip and the sample to allow the flowing of the current I_T. In the single scan figure the measured current is reported as a function of the x position. Multiple scans give a sort of perspective in the y direction. (b) Constant current mode. A feedback loop controls the piezodrive via the voltage V_z to keep constant s, the tip–sample distance. Deviations of I_T from a preset value act as error signal for the feedback loop. Dotted lines represent the tip track during the scanning of the sample; in the x, z graph z is directly obtained from the applied voltage V_z, by using the known piezodrive characteristics. Redrawn from [4,5].

once the piezoelectric driver characteristic is known, the tip height variations can be easily obtained from the applied voltage. However, it is important to realize that the dependence of the tunneling current on the sample–tip distance is quantitatively different for different materials. Thus the STM "topographic" image of a composite material is actually the image of an isocurrent surface. A similar consideration, however, holds for each scanning probe microscopy, because the concept of "surface" is a macroscopic concept, and it dissolves into many different "surfaces" at the atomic level, depending on the measured interaction.

The constant current mode is particularly fit for surfaces that are not flat on an atomic level, since it allows for the tip to move in the z direction following crests and troughs. Conversely, it has the disadvantage of the response time of the feedback loop, which sets some limitations for the scanning speed.

The constant height mode, because of the faster scan rate, allows for reduced image distortions due to thermal drift. Moreover, it is better fitted for studying dynamic processes on surfaces. It has two main disadvantages, though: it can be applied only to atomically flat surfaces, otherwise the tip might crash into surface protrusions, and it is not always an easy task to extract the topographic information from the tunneling current variations, since the space constant of the tunneling current can change at any point.

Though possessing a high resolution capacity, STM has two major drawbacks: it requires the sample to conduct current, at least up to a certain extent (and this makes it problematic to image most organic materials), and it can better achieve such high resolutions in ultrahigh vacuum (UHV), since the presence of atoms (from impurities or from a medium) interferes with tunneling.

15.1.2 Other SPMs

A SPM technique that has proven very successful has been atomic force microscopy (AFM). More correctly it is referred to as scanning force microscopy (SFM), with a name that better describes its functioning. In fact the probe scans the sample while their interaction (i.e. the force between the two of them) is monitored. The "atomic" in AFM comes from the capacity, under favourable conditions, to achieve atomic resolution.

AFM was first proposed in 1986 [7] as a method for overcoming STM limitations regarding both the sample and the environmental operating conditions. AFM range of applications has grown over the years and today it is used not only to image samples in their native physiological conditions (a particularly important issue in biophysics) but also to explore molecular interaction, for single molecule studies and as a precise nanotool.

The Scanning Near-field Optical Microscopy (SNOM, NSOM or PSTM, Photon Scanning Tunneling Microscopy) uses also a probe proximal to the sample to scan its surface while illuminating it and/or detecting the near-field distribution of the light intensity. In this way it is possible to overcome the half-wavelength limit of the classical optical microscopy (Abbe or Rayleigh criterion), obtaining spatial resolution up to tenths of nanometres. The use of local probes to illuminate and/or to detect light in a near-field configuration was first proposed in 1928 by Synge [8], who also considered the use of piezoelectric actuators [9]. In 1972 Ash and Nichols built the first near-field microscope in the microwaves range [10] while it was only in 1984 that the first SNOM using visible light was developed by two independent groups [11,12].

SNOMs exist now in various configurations, depending on the illumination and detection modes. With this microscopy technique it is possible to study optical properties and to modify samples at a sub-λ level; as a couple of examples, SNOM makes possible to implement evanescent wave optical lithography, and makes it possible to detect the fluorescence of single molecules with spatial resolution.

Among the SPM a very promising approach for biological studies is represented by SICM (Scanning Ion Conductance Microscopy). It is based on the idea, initially proposed by Hansma and co-workers [13], of using as scanning probe a pipette-shaped microelectrode of the type employed by electrophysiologists for intracellular recording. The current flowing in such a probe is limited when the electrode is close to non-conducting (or only partially conducting) object, typically the membrane of a cell. Obviously SICM can only operate in ionic solutions.

After some years of latency this kind of microscopy is now applied successfully to living cells, giving very interesting results. Two points have been of crucial importance for its development: one is the use of a small modulation of the tip–sample distance, and the other is the coupling with classical electrophysiological experiments. The use of a position modulation allows for a better stabilization of the tip–sample distance (particularly critical when working on cellular scale); the coupling with electrophysiological recordings made possible the localization of specific ion channels on the cell surface and their visualization [14,15].

15.2 Atomic force (or scanning force) microscopy

15.2.1 Force detection methods

The first prototype of AFM consisted of a sharp tip (the probe) protruding from an ultra-small cantilever mounted between the sample and an STM tip. In such an experimental setup, as the scanning proceeds (the piezodrive controls the sample displacements), the cantilever is deflected by the sample protrusions and its displacements are monitored via the tunneling current between the metal back of the cantilever and the STM tip (Figure 3). The cantilevers follow Hooke's law,

$$F = -ks$$

where s is the vertical displacement of the tip, k the cantilever spring constant and F is the elastic force. Monitoring the cantilever vertical deflection, it is then possible to infer the AFM tip–sample interacting force. The spring constant is chosen such that the interaction force is not disruptive of the sample, ranging from some μN to hundreds of pN, depending on the sample.

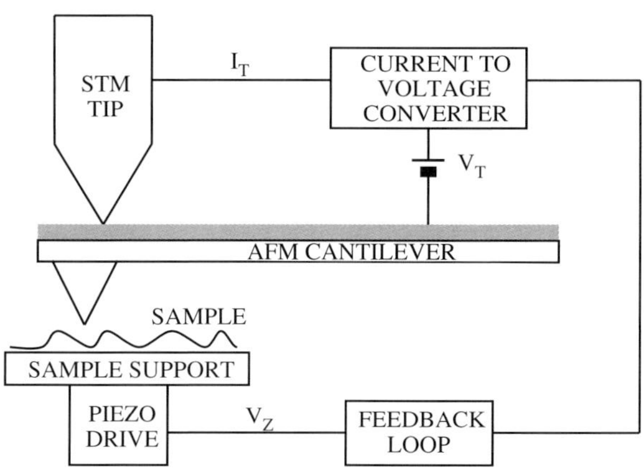

Figure 3. Scheme of the first AFM prototype. A sharp tip is mounted on a small cantilever with a coated back. As the scanning proceeds the AFM cantilever deflections are monitored via the variations of the tunnelling current of the STM mounted on top of the cantilever.

This development of STM allowed imaging of samples at atomic resolutions, no matter whether conducting or not [16]. Moreover it allowed the imaging of liquid covered surfaces, since the tunneling current was not flowing between tip and sample but between tip and the back of the cantilever [17].

However, this architecture allowed only for limited cantilever dynamics, since tunneling current is an extremely critical factor, heavily dependent on the smallest change in the reciprocal positions of STM tip and cantilever. Moreover tunneling current is affected by the presence of contaminants on the cantilever and on the STM tip.

To obtain a more versatile instrument the method of detection employed became that of the optical lever, known as the Poggendorff method.

Conceptually it is very simple: a Gaussian laser beam is focused on the back of the cantilever and the reflected light is collected by a position sensitive photodetector, so the displacement of the laser spot gives a signal proportional to the cantilever deflection and hence to the interacting force (Figure 4). In particular, with a properly mounted four-quadrant photocell, the deflection signal is given by

$$I_D = (I_1 + I_2) - (I_3 + I_4)$$

with $I_{1,2}$ the upper quadrants signal and $I_{3,4}$ the lower quadrants signal in circular numeration, while

$$I_T = (I_1 + I_4) - (I_2 + I_3)$$

gives a measure of the torsion of the cantilever, and hence of the friction the tip encounters while scanning the sample.

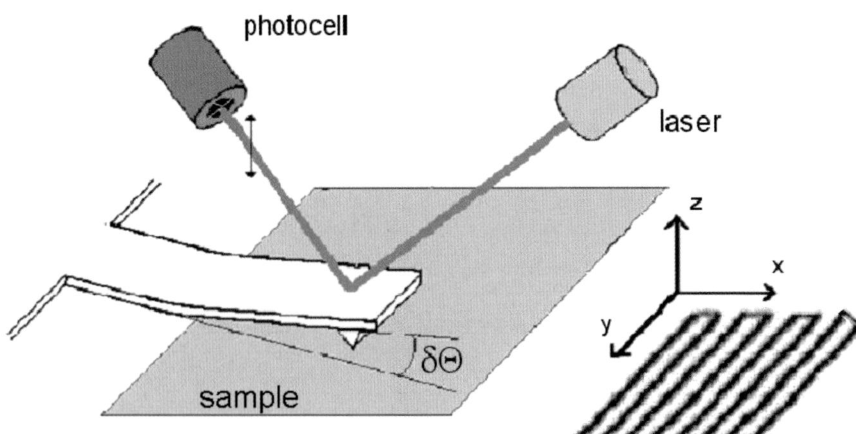

Figure 4. Basic concept of AFM. A laser beam focused on the free end of the cantilever is reflected on a four-quadrant photocell. While the cantilever scans the surface in the x and y directions in a continuous sweep, left-to-right and then right-to-left as in the reproduced track, it is deflected according to the sample profile. The unbalance between the upper and lower quadrants of the position sensitive photodetector is proportional to the deflection angle $\delta\theta$, and hence to the deflection s of the cantilever.

The signal-to-noise ratio of this architecture is also very convenient, in fact it basically depends only on the photodetector shot-noise. The root mean square current shot-noise [18] is

$$i_N = \sqrt{2eI\Delta f}$$

where e is the charge of the electron, I is the generated photocurrent proportional to the laser intensity, and Δf is the detection band.

The laser beam is assumed to be diffraction limited, hence its aperture φ (considering the beam border the point at which the light intensity drops to $1/e$ of the central maximum) [19] is given by

$$\varphi = 2\theta_0 = \frac{2\lambda}{\pi w_0}$$

where θ_0 is the divergence of a Gaussian beam, λ is the laser wavelength and w_0 is the waist radius (Figure 5). Since the laser beam is focused on the cantilever, w_0 is the radius of the laser spot on the back of the cantilever as represented in Figure 4.

A deflection δs of the cantilever tip generates an angular deflection, $\delta\theta$, given by

$$\delta\theta = \frac{3\delta s}{2L}$$

where L is the cantilever length. This is the small angles approximation corrected with a numerical factor of $\frac{3}{2}$ since the cantilever deflection is not linear but it is described by a cubic function (roughly, it does not bend at its junction but at $\frac{2}{3}$ the cantilever length from the tip [20]).

The maximum angular deflection occurs when the reflected beam passes entirely from the lower half to the upper half of the photocell. Considering that an angular deflection of $\delta\theta$ of the cantilever causes a deviation $2\delta\theta$ in the reflected light, one obtains

$$\delta\theta_{max} = \frac{\varphi}{2} = \frac{\lambda}{w_0\pi}$$

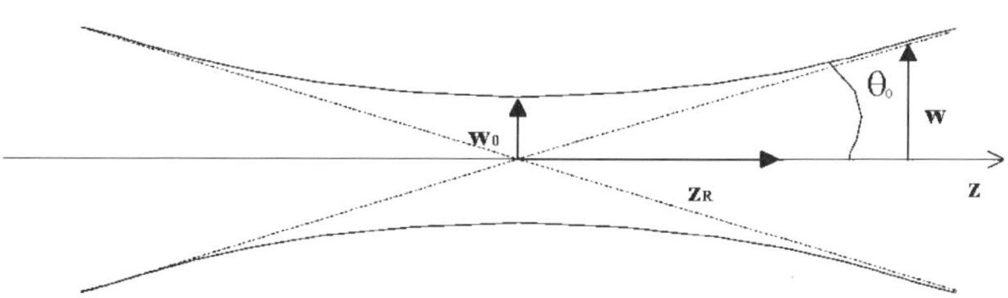

Figure 5. Scheme of a focused Gaussian laser beam. The point of minimal width is the waist, w_0, while z_R is the Rayleigh distance, namely the distance from the waist at which the laser width w is $\sqrt{2}\cdot w_0$; θ_0 is the laser beam divergence.

Hence the maximum measurable vertical displacement of the tip is

$$\delta s_{max} = \tfrac{2}{3} L \delta\theta_{max} = \frac{2\lambda L}{3 w_0 \pi}$$

and it produces a signal $S_{max} = 2I$. A displacement δs causes a signal S given by

$$S = S_{max} \frac{\delta s}{\delta s_{max}} = \frac{3 w_0 \pi I \delta s}{2\lambda L}$$

So, in the end, the signal-to-noise ratio is

$$\mathrm{S/N} = \frac{3 w_0 \pi \delta s}{2\lambda L} \sqrt{\frac{I}{2e\Delta f}}$$

Assuming $\mathrm{S/N} = 1$ and giving to the other variables standard experimental values ($I = 50\,\mu\mathrm{A}$ with an incident light of about 100 μW, $\Delta f = 10\,\mathrm{kHz}$, $\lambda = 670\,\mathrm{nm}$ of a standard red laser diode, $L = 200\,\mu\mathrm{m}$, $w_0 = 20\,\mu\mathrm{m}$) the minimum measurable tip displacement is of the order of $\delta s_{min} \sim 0.1$ Å, which means that AFM can resolve single atoms in height while still scanning at a rather high speed.

Notably, the optical lever is not the only possible method of detection. In fact in other experimental conditions other methods may be more suitable, such as optical interferometry, in which a laser is focused on the cantilever and its reflected light interferes with a reference beam [21,22]. Interference fringe displacements then reveal cantilever deflections. This method has the same signal-to-noise ratio as that of the optical lever [23]. Though entailing a more complicated architecture, optical interferometry is particularly suitable when sample imaging is performed in ultrahigh vacuum or at very low temperatures, since it allows for only the cantilever and sample to be in these hard environmental conditions, connected via an optical fiber to an external photodetector.

15.2.2 Operating modes and cantilever tips

In analogy with STM, in AFM there are two main operating modes: constant height and constant force [16].

As the name suggests, in the constant height mode the tip is swept at a constant height over the sample, generally at high speed, while the cantilever deflections are acquired. It is better to employ this mode only on very flat surfaces, since steep steps cause high interaction forces that may damage the tip or the sample.

In the constant force mode the cantilever deflection is maintained fixed, i.e. the force is kept constant (Hook's law), while the height z of the tip is adjusted accordingly. The signal acquired acts also as the drive of the piezoactuator in the z direction, i.e. the output of the feedback, which stabilizes the cantilever deflection. This mode is typically used at low imaging speeds because of the response time of the feedback loop.

In the constant force mode it is usual to also acquire the so-called error signal. In fact, although the tip deflection is stabilised by the feedback, as the tip encounters a steep height variation the cantilever is transiently deflected and this deflection signal (the error signal) generates the feedback loop response that vertically displaces the cantilever in order to re-establish the previous value of the deflection. The signals acquired, the z signal and the cantilever deflection, give complementary information on the sample (Figure 6(a), 6(b)).

The height signal is a direct measurement of the sample topography, which, however, depends on the applied force. Generally, AFM images are represented as flat 2D images. Such a rendering though does not allow the observer to completely grasp the extent of the information provided, unlike a physical 3D model of the sample. A 2D perspective of a 3D model is a reasonable compromise (Figure 6(c)).

The error signal behaves like the derivative of the topography signal along the scanning line. Its extent is determined by the sampling rate and by the feedback response time. As with all derivatives, the error signal is extremely useful in spotting small features on the sample that might pass unnoticed in the 2D rendering of the topography. Typically error signal maps (usually identified as "deflection maps") are represented as flat 2D images; at first sight they look as if the sample was illuminated in the direction of the tip scan (Figure 6(b)).

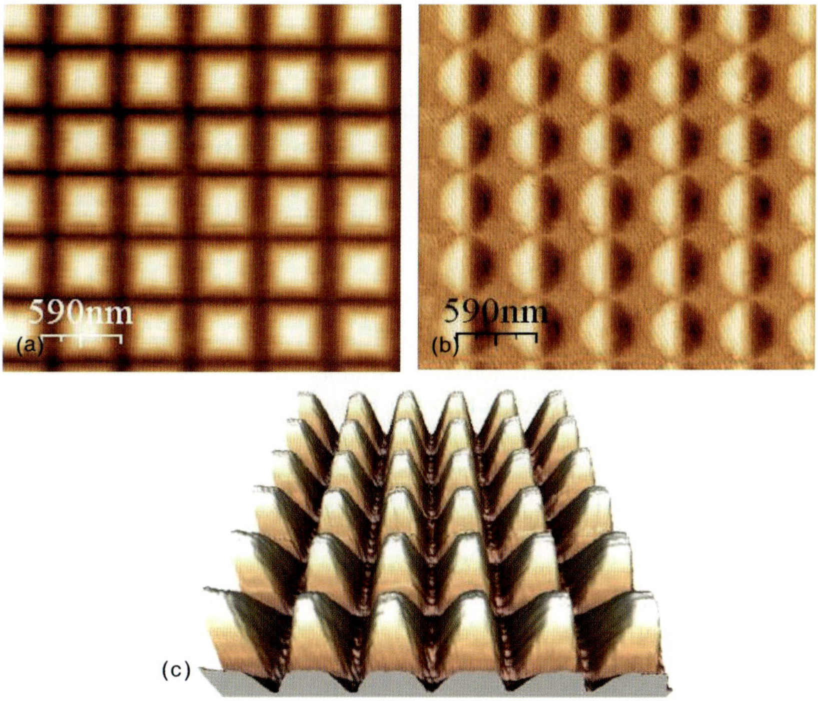

Figure 6. (a) 2D representation of a quantum dot array (height signal). (b) Deflection map acquired in the same scan. The information represented is the cantilever deflections point by point – brighter if the tip moves upwards, darker if it moves downward. Scan direction from right to left. (c) 2D rendering of a 3D representation of the data used in (a). By courtesy of P. Pingue.

During scanning the cantilever is deflected, but lateral forces, which depend mainly on the friction between tip and sample, also twist it. The measurement of the cantilever torsion makes it possible to obtain friction maps, which give qualitatively different information on the sample from that contained in a topographic map. Friction maps have been one of the first steps in the direction of obtaining chemical or so-called spectroscopic information in AFM microscopy [24–26]. An in-depth discussion on tribology at a nanoscopic level has been given by Bhushan et al., (1995) [27].

Other than the conventional imaging modes described above in which the tip is in continuous contact with the sample (in fact they are described generally as "contact mode"), AFM allows also for AC intermittent contact mode imaging. In this mode, generally known as the "tapping mode" [28–32], the cantilever is vibrated perpendicular to the sample at a frequency close to its resonant frequency. When the tip approaches the sample, the vibration, measured via the photodetector, is damped and the damping is used by means of a feedback to keep constant the tip–sample separation. The imaging amplitude in most cases is set to 70–80% of the free oscillating amplitude.

Since the tip touches the sample for a very limited time there is basically no friction, so the friction information is lost, but the risk of damaging tip or sample is greatly reduced compared with contact mode. This confers to AC mode a key role in imaging soft biological samples. Furthermore, beyond the height information obtained from the feedback (Figure 7(A), 7(D)), in AC mode other information is available to the user, in particular the phase signal. In fact the oscillation phase varies during scan compared with that of the drive signal and this phase difference is determined by the nature of the investigated sample: different materials or different features in the same sample have different elastic properties and induce a phase shift. Thus the phase image contains information on the sample composition (Figure 7(B)).

Another way to obtain information, which includes the height topography but goes beyond it, is based on the acquisition of force–distance (FD) curves, which are extensively presented below. Notably, though, acquiring a complete set of force–distance curves provides a lot of information on the sample, ranging from sample elasticity to specific bonds and double layer forces. It is also possible to avoid damaging soft samples because of friction forces; this can be achieved by acquiring an FD curve on a certain position, then the tip, while not in contact, is moved in the x and y directions to another location to acquire the following FD curve [33]. However, much more time is required for the acquisition.

A crucial factor in AFM imaging is the quality of cantilevers and tips. When working in contact the shape of the tip has to be chosen in accordance with the sample features; for instance, tips are required to be long to successfully image samples with deep holes. There are many kinds of commercial cantilevers and different tricks for growing tips on their ends have been developed. Cantilevers are built by microfabrication on silicon wafers, and are made of silicon or silicon nitride. Figure 8(A) shows schematically the shape of a triangular cantilever with a pyramidal tip, while Figure 8(B) is a picture of some chips, each with six

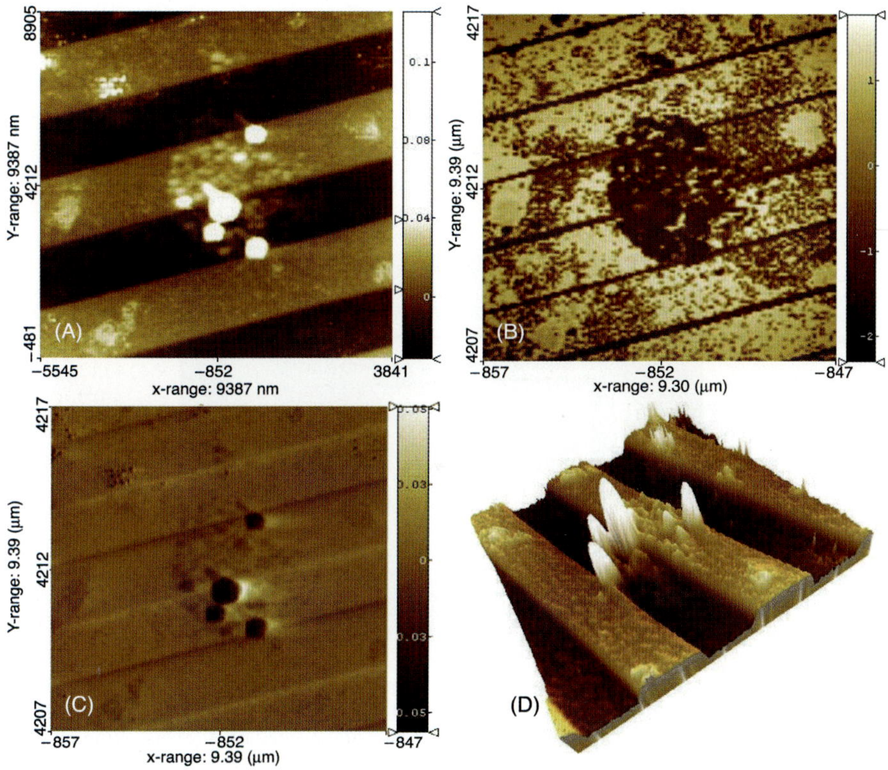

Figure 7. (A) 2D representation of an AFM topography of a standard calibration grating in AC mode (height signal). A good deal of impurities is visible on the grating. (B) Phase image of the same area. It is possible to observe the different phase shifts induced by different impurities and by the grating steps as well as many minor features not clearly identified in the 2D topography. (C) Error signal image of the same area. The difference between the oscillation amplitude and a preset value is the input signal for the feedback loop in the same way that the deflection signal was in contact mode. (D) 3D rendering of the AFM topography. By courtesy of P. Pingue, NEST-INFM and CNR-1B, Pisa.

cantilevers. Rectangular cantilevers are also used. The cantilever's length is of the order of hundreds of micrometres (for instance, in Figure 8(B) the longer is 320 μm, while the shorter one is 85 μm long; the thickness is 0.6 μm). An important parameter is obviously the elastic constant, which ranges from 20 to 0.01 N m^{-1}. Today many different types of cantilevers, also for special purposes, are commercially available; particularly important is the shape of the tip, which is critical for good lateral resolution.

In acquiring force–distance curves (see section 15.2.4 for more details) the cantilever thermal noise may be a limiting factor whose impact can be minimised by an appropriate choice of the cantilever. The criterion to be applied in this case is to choose cantilevers with a high resonance frequency and with relatively low

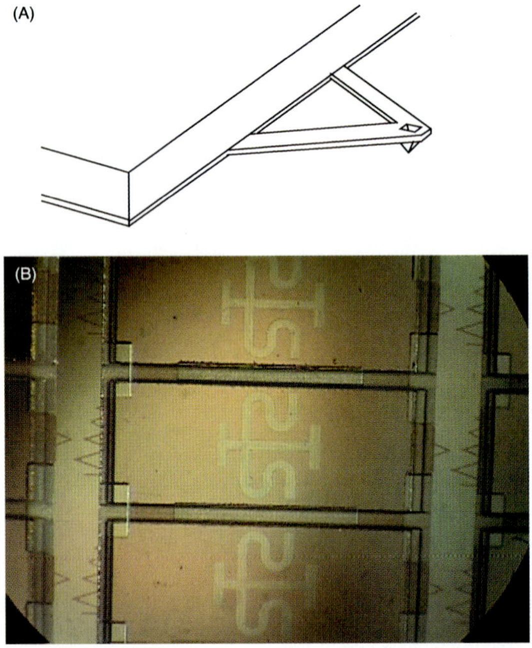

Figure 8. (A) Representation of a chip with a triangular cantilever; the pyramidal tip protrudes from the bottom of the cantilever. (B) A wafer of the former Park Scientific Instruments (PSI), a pioneer in the production of microfabricated cantilevers. Nine chips are visible, each of them with six cantilevers, five triangular plus a rectangular one.

damping. Damping is in fact an aspect of the cantilever–medium interaction, while the other is the amount of thermal noise revealed as "spontaneous" fluctuations of the cantilever deflection. Reducing damping (basically reducing the surface of contact with the surrounding medium) and increasing the resonance frequency are two options to reduce the disturbance due to thermal noise. This last option, i.e. the usefulness of increasing the resonance frequency, becomes clear considering as the thermal noise is spread over the frequency range in which the cantilever can respond to solicitations; for an elastic system this goes from zero to the resonance frequency. A clear paper on this point, which also reports a comparison of thermal noise in AFM and in optical tweezers, is due to Gittes and Schmidt [34]. In fact, thermal noise is often used to characterize nanomechanics sensors [35,36].

However, to choose the appropriate cantilever other factors should be considered. An important one, which can generate a noise far greater than the thermal noise, is due to the interaction of the cantilever itself with the laser light used to detect its deflections. The effects of the laser light on the cantilever are of two different kinds: thermal (a local change of the cantilever temperature, which induces extra fluctuations of the deflection) and directly mechanical (radiation pressure) [37,38]. The noise induced by temperature changes is particularly relevant when working in air or with gold-coated cantilevers that are very useful in specific applications.

AFM techniques are still developing, new sensors and new strategies for scanning are being examined and new possible interactions considered as the basis for future applications of the AFM approach. For example some groups are working on the development of a Magnetic Resonance Force Microscopy [39–43], while the use of ultrasound in AFM is being investigated [44], a new kind of very small cantilever has been introduced [45], and the list could continue.

In general AFM instrumentation can be very sophisticated and expensive, but for specific uses, when very high resolutions or very fast acquisitions are not required, it can be home built using no more than a computer with an analog-to-digital acquisition board and some basic mechanics and electronic aids. Old things, like the pick-up head of old phonographs, can be used as topography detectors [46]. Moreover loudspeakers can be used to scan the sample on a millimetric scanning area, giving an instrument that maintains a very good vertical resolution and a quite good lateral one [47,48].

15.2.3 Is it always possible to obtain a topography with atomic resolution?

Atomic resolution has been achieved in contact mode on a number of inorganic crystals, such as mica, highly oriented pyrolitic graphite (HOPG) and calcite. The strength of AFM is the ability to resolve at the atomic scale surface structures of both conductors and insulators, in air, water and ultrahigh vacuum (UHV), making it a powerful tool in several fields [49–53]. In certain cases AFM has been the only technique capable of accessing materials in the size range from 50 to less than 1 nm. However, most of the reported data show atomic lattices with no well-defined atomic-scale local defects; defects are present only on a larger scale. This is in contradiction with routine STM data that clearly show the presence of defects at the atomic level, so the question arises of what AFM is actually imaging and why it does not achieve a true atomic resolution.

A partial answer can be found if one considers that a typical radius of curvature of the tip is in the range 5–50 nm, while typical lattice constants are some angstroms, so there are 1–2 orders of magnitude between them. An example estimate of the contact radius on a mica surface [54] gives about 1 nm. So the tip contacts simultaneously a number of atoms, and while scanning its vertical displacements reproduce only the lattice period and not the profile of single atoms. This explains why one atom missing at a lattice site is not imaged, since the surrounding atoms maintain the signal periodicity.

However, the tip is not a sphere, thus one could expect to sense, at least partially, the force between the most protruding atom of the tip and a single atom of the sample, similarly to what happens with STM (Figure 1). In AFM this is not straightforward as in STM, since the distance dependence of the tip–sample force is much more gentle than that of the tunneling current (see the following paragraph). Nevertheless, with a low enough net loading force, of the order of 10^{-10} N or less, true atomic resolution was achieved [55], revealing atomic-scale defects.

In addition, true atomic resolution in AC non-contact mode (NCAFM) has been achieved. By oscillating the cantilever tip at a short distance (around the zero force point where long-distance attractive and short-range repulsive forces balance) with an oscillation amplitude so small as to avoid contact, one can measure phase or frequency variations, which give the sample topography [56–59].

When imaging organic samples, the contact mode, though generally more disruptive, can reach a high resolution, and under appropriate conditions (see section 12.2.5) it shows outstanding performance also on delicate organic samples [60–61], resolving single molecules. However, an AC intermittent contact mode can add further information since it gives insights into the nature of the sample via its phase signal; moreover many organic samples can only be imaged in the less disruptive AC mode. The AC mode can achieve high resolution on biological samples comparable to the contact mode [62].

15.2.4 Force–distance curves

Interactions between single atoms obey the Lennard-Jones potential:

$$U(r) = -\frac{A}{r^6} + \frac{B}{r^{12}}$$

where r is the distance between the two atoms and A and B are two parameters characterizing their interaction [63]. The attractive term (the negative one) in particular accounts for the van der Waals interaction potential.

For AFM tip – substrate interaction it is generally impossible to define one basic force law, but a Lennard-Jones-like potential, with repulsive forces at short range and attractive forces at greater distances, is assumed to hold. As sketched in Figures 9 and 10 an interaction force ranging from repulsive to attractive accounts for the basic features of experimental force–distance curves.

When the tip approaches the sample, two forces act on the cantilever: the Lennard-Jones-like tip–sample interaction and the cantilever elastic force.

We call s the cantilever deflection, d the tip – surface distance, and z the vertical position controlled by the piezodrive, which is the position of the cantilever "at rest", i.e. with no interactions taken into account (Figure 9(A)). The height z is equal to the algebraic sum of d and $-s$, provided that the sign of the deflection is assumed to be negative when the cantilever is deflected down. At a given z the cantilever is in equilibrium when the sum of elastic force and Lennard-Jones (L.J). force is zero, i.e. when $F_{\mathrm{L.J.}}(d) = -F_{\mathrm{Elastic}}(s)$. In (Figure 9(B)) both forces are plotted on the same graph, with the elastic force axis inverted. On such a representation z gives the distance between the two vertical axes. As the cantilever approaches the sample, z decreases and the elastic force graph is shifted to the left (Figure 9(B)). The equilibrium point is determined by the intersection of the two graphs.

Let us consider the most general case, when there are three intersections between the two functions.

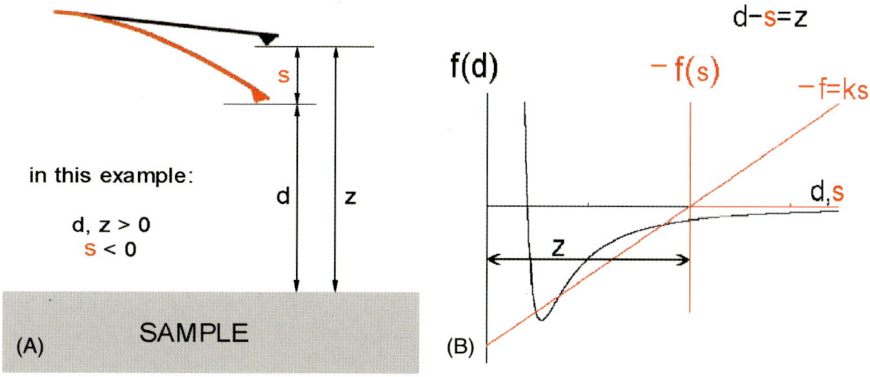

Figure 9. (A) Schematic representation of cantilever and sample relative position: z is the distance (imposed by the piezodrive) between the resting position of the tip and the sample, d is the actual tip–sample distance and s is the deflection of the cantilever; the sign of s is assumed negative when the cantilever is deflected down (attractive force). Very far from the sample (not shown) the cantilever is not deflected, s is zero and d equals z. As tip and sample approach, the Lennard-Jones attractive force becomes significant and the cantilever is deflected down by an amount s. The real tip–sample separation is then d and the sum of d and $-s$ equals z. After a further approach the Lennard-Jones force becomes repulsive and the cantilever is deflected upwards. (B) Lennard-Jones force as function of d (curved line), and elastic force as function of s (straight line) are superimposed with the elastic force axis inverted. The cantilever deflection, due to the balancing of Lennard-Jones and elastic forces, is obtained by the intersection of the two graphs, where $f_{\text{L.J.}}(d) = -f_{\text{Elastic}}(s)$.

The first, to the right, is a point of stable equilibrium. In fact if a perturbation moves the tip to the left, the elastic force wins over the L.J. force and the tip is pushed back; the same occurs on moving the tip to the right. The same holds for the third equilibrium point.

The second equilibrium point, on the other hand, is unstable. A small displacement to the right and the elastic force prevails, pulling to the right till the previous point of stable equilibrium; a small displacement to the left and the tip is pulled by the L.J. force to the left, till the third intersection. Thus s, the measured parameter monitored via the photodetector, depends on z, the piezo-controlled height, and on the equilibrium reached.

Let us briefly consider what happens as the cantilever first approaches the sample and is then retracted (Figure 10(A)).

From right to left and back, we follow a point z representing the piezo-controlled height. Initially, there is only one intersection and, as z approaches, s does not vary significantly. When $z = \gamma$ nothing happens, even if by now there are three intersections and we observe the first significant variations of s as the tip senses an attractive interaction. When $z = \beta$ the right-most stable equilibrium point coincides with the unstable equilibrium point; b is an unstable equilibrium point so the tip jumps to b', which is stable. This phenomenon is called *"jump to contact"* and is clearly reported in Figure 10(B). In b' the tip is in contact with the sample and

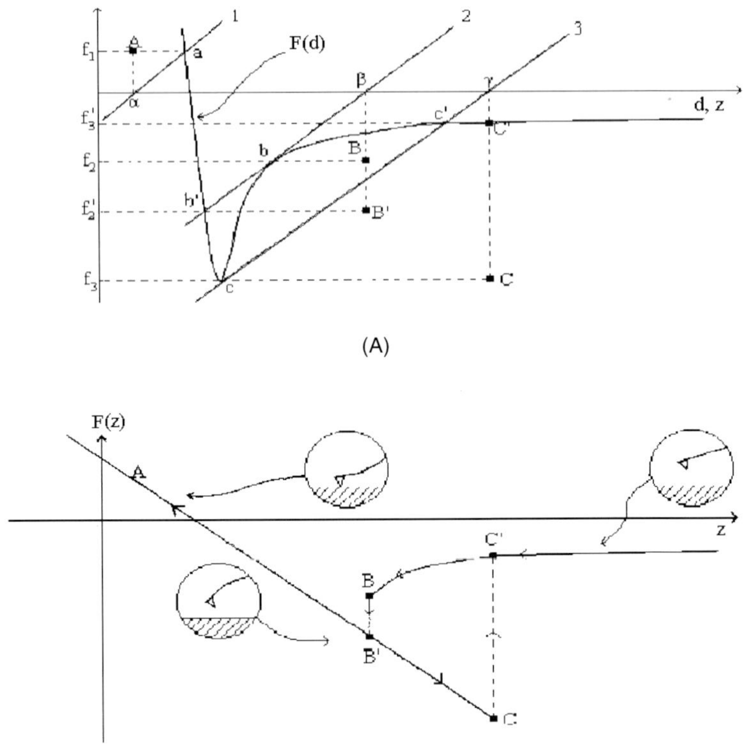

Figure 10. Graphical method to obtain experimental force distance curves from the Lennard-Jones force $F_{\mathrm{L.J.}}$ (A) Lines 1, 2 and 3 represent the elastic force for three different values (α, β, γ) of the height z, which is assigned by the scanning routine and gives the tip–sample distance at which the elastic force is zero. The intersection of these lines with $F_{\mathrm{L.J.}}$ determines the cantilever deflection s, which is measured during the vertical sweep. A single solution is possible for line 1 (intersection a), two solutions for line 2 (intersection points b and b') and two for line 3 (c and c'). Lines between 2 and 3 (not drawn) have three intersections with $F_{\mathrm{L.J.}}$. While the tip approaches the sample the solution moves from c' to b (the corresponding forces can be read on the vertical axis following the dashed lines and range from f_3' to f_2); further approaching tip and sample the solution will jump from b to b'. During the withdrawal a bigger jump will occur between c and c'. (B) Force vs. distance curve during the cantilever approach to and retract from the sample, obtained from the graphs in (A). In the insets a representation of the cantilever deflection is shown for different positions. Redrawn from [66].

a further approach causes a linear increase of s with z since we are in a region where the L.J. force is extremely steep and d is practically constant.

During retraction when $z = \beta$ the stable intersection b$'$ is encountered, so nothing happens and the force vs. distance plot is linear until the intersection $c\,(z = \gamma)$. The point c (Figure 10(A)) is the lowermost intersection and as z increases the only possibility is to jump back to c' and continue from there. Thus, the *"jump-off-contact"* from f_3 to f_3' is observed.

In Figure 10(B) the resulting force vs. distance curve (FD curve) is plotted and since the force is directly proportional to the cantilever deflection, the vertical axis also shows the dependence of s from z.

Force–distance curves give several types of information on the sample, depending on their specific appearance, which in many cases is slightly different from the general one presented above [33,64–66].

When imaging a hard sample the contact line is straight (z and s vary together) and the slope $\Delta s/\Delta z$ is 1. A contact region ($a\ b'c$ region in Figure 10(B)) not perfectly linear indicates a deformation of a soft sample, while the value of the slope is a measure of the ratio between the elastic constant of the sample and that of the cantilever. Thus, by performing a number of FD curves in many nearby sites one may generate an elasticity map of the sample.

Similarly a different tip–sample adhesion (e.g. different extensions in the "jump-off-contact") might be determined by the nature of the materials investigated and may generate an adhesion map of the sample (Figure 11) [33].

Moreover, an erratic behaviour in the FD curve may indicate a change of tip characteristics due to contamination, while a separation between the approaching and retraction linear regions of the FD curve is due to piezodrive hysteresis [53].

The possibility of single molecule force spectroscopy via FD curves is examined below.

15.2.5 Measuring in air, in vacuum, in water

The general form of FD curves is greatly influenced by the medium surrounding the tip. When imaging in air the presence of water vapour always induces the formation

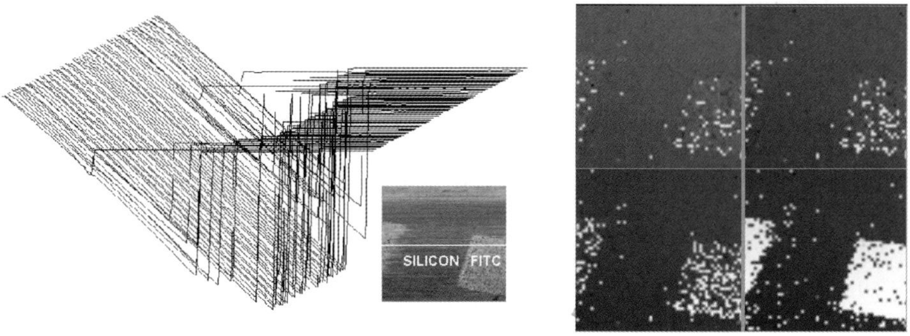

Figure 11. (Left) FD curves made point by point along the scan line evidenced in the inset. Inset: AFM topography of a fluorescein isothiocyanate (FITC) grating on silicon (FITC is placed in 30 × 30 μm squares that are 30 μm apart). (Right) maps obtained by plotting the force measured at given distances (250, 300, 350, 440 nm, respectively) from the position of the maximum preset loading force. This is known as the force-slice representation. In the first one (top left) at some points on FITC the tip has already pulled off the sample; as the distance increases the tip is observed to pull off more and more when on FITC rather than when on silicon. The tip adhesion to silicon is stronger than on FITC. Redrawn from [33].

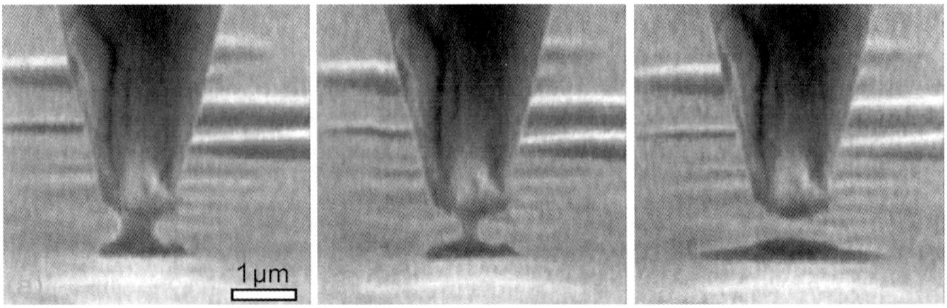

Figure 12. Environmental scanning electron microscopy image of a water meniscus between a tungsten tip and a substrate. As the stylus retracts the meniscus is first stretched and then it breaks. Redrawn from [67].

of a meniscus between tip and sample (Figure 12). The force it exerts is of the order of 30 nN and causes a relevant increase in the "jump-off-contact" distance; in fact the cantilever needs to be deflected much more to apply such a counterforce to break the meniscus and this happens only when the cantilever is farther away from the sample (i.e. z is bigger).

Generally speaking the meniscus contributes to the noise of the system, since it adds strong uncontrolled forces, and represents an obstacle to achieve very high resolution. Conversely, it may play a very important role in nanolithography applications.

One way to avoid the formation of a meniscus is to work in ultrahigh vacuum conditions. An ultrahigh vacuum (UHV) environment presents without doubt the most controlled conditions for AFM imaging since contaminants such as water and hydrocarbons are completely excluded. In this way it is possible to study the fundamental interactions between tip and sample, investigating friction and wear on a nanometre scale, as well as van der Waals forces, surface charges, ferroelectricity and surface magnetism, especially when UHV-AFM is performed in non-contact mode. However there are several experimental problems to be taken into account; first of all, as already mentioned, the difficulties in using the optical lever method of detection. In UHV-AFM interferometry detection proves to be a better choice. Moreover, to achieve UHV conditions, samples to be studied need to be bakeable up to 150°C and to have a vapour pressure below 10^{-11} mbar, which excludes most organic materials and some inorganic ones such as lead and zinc. In the end, UHV-AFM in its different designs appears to be best suited to characterize film growth of insulators and semiconductors [53] and to investigate inorganic crystal surface morphology, resolving monoatomic steps and reaching true atomic resolution [58,68,69]; recent developments have shown that AC non-contact mode UHV AFM can detect even subatomic features (Figure 13) [70,71].

Another way of avoiding a meniscus is to work with the tip and sample immersed in liquid. The difference between the two conditions clearly appears when a FD curve is performed in liquid or in air (Figure 14), showing the extent of the interaction caused by the meniscus.

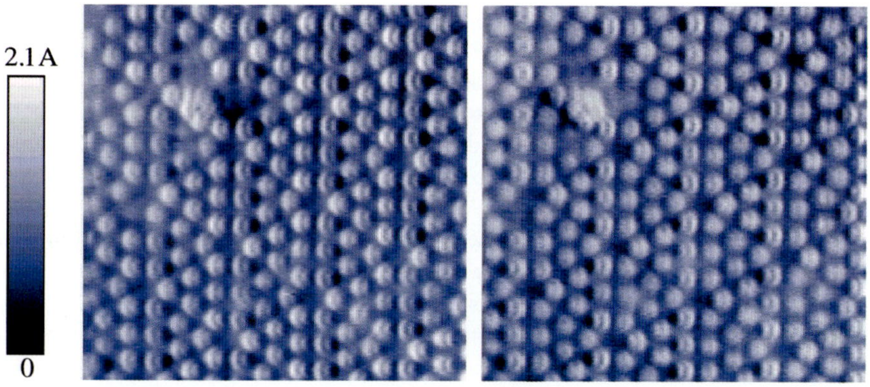

Figure 13. AFM topographic images of Si(111)-(7 × 7) achieved in AC non-contact mode and in UHV. Scanning is from left to right on the left-hand image and from right to left on the right-hand image. Raw data images with a background plane subtraction. It is possible to observe single atoms, lattice defects and sub-atomic features. Redrawn from [70].

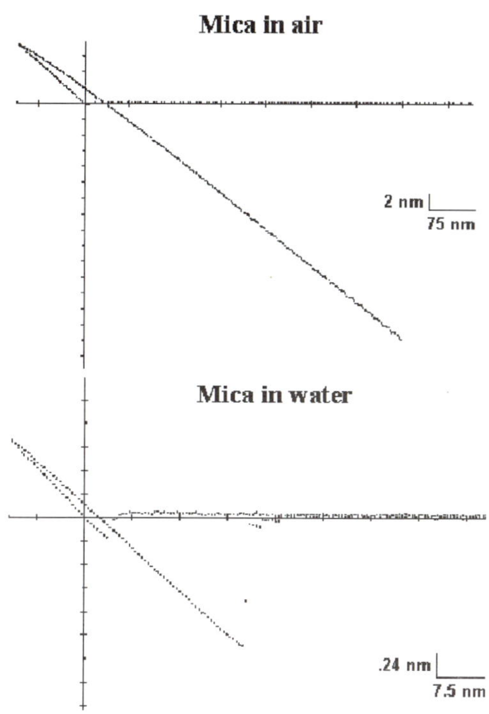

Figure 14. FD curves performed on mica in air (upper curve) and in water (lower curve, different scale). Jump-off-contact forces are ~30 nN in the first case and ~1.5 nN in the second, the difference being determined by the presence of the meniscus when measuring in air. Redrawn from [66].

Certainly the greatest advantage of measuring in liquid is that it allows for a much wider range of materials to be imaged, including biological samples under physiological conditions (buffer solution).

Once some technical details are taken care of in the architecture design (e.g. the laser beam passes a well-defined liquid – air interface and only inert materials are in contact with the liquid), the liquid composition may be adjusted to control tip–sample forces [66,72,73].

To understand this force control mechanism, attention needs to be paid to what happens on a microscopic level to the various components when imaging in water [74]. In fact AFM tips exhibit a net surface charge when in an aqueous solution, due to the dissociation of the tip functional groups. Oppositely charged-ions in the solution respond to this charge by disposing themselves so as to balance it. The final result is a region of counter-ions, some of them bound to the surface (the so-called Stern or Helmholtz layer), most of the others forming an atmosphere of charged particles in rapid thermal motion, the electric double layer. The counter-ions' concentration decreases exponentially, depending on the ions concentration and valence. The same phenomenon occurs on the sample surface and as the tip approaches the sample, the two surface double-layers start to overlap (at a distance of few nm) and an electrostatic interaction arises. As the two come closer, van der Waals forces become important and must be taken into account.

A quantitative description of the interplay of van der Waals and double-layer interactions is provided by the DLVO theory of colloidal stability, after Derjaguin and Landau [75] and Verwey and Overbeek [76].

Generally, van der Waals potential can be considered attractive and independent of electrolyte concentration and pH. Conversely, double-layer interaction is strongly dependant on electrolyte concentration and valence as well as on surface charge. Thus, by adjusting these parameters, while van der Waals forces remain unchanged, the double-layer forces can be modulated, determining whether the net distance-dependent tip–sample non-contact interaction is repulsive, repulsive and attractive, or attractive. Figure 15 gives FD curves at different pH to show the extent of the impart of environmental parameters.

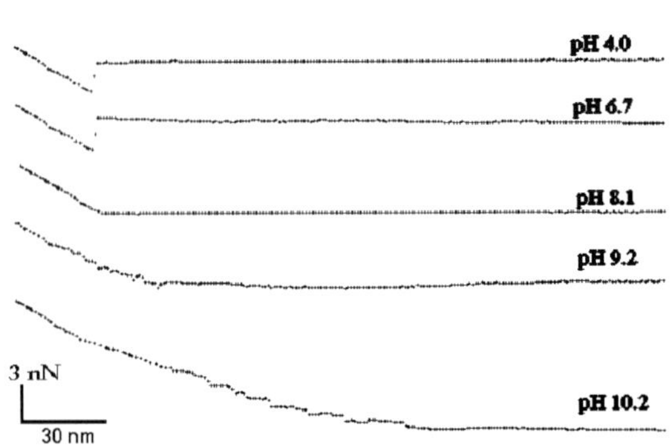

Figure 15. Force–distance curves on stearic acid film in water solutions of different pH (only approach curves are shown). At pH $<$ 8, van der Waals forces dominate, while at pH $>$ 8 the double-layer force becomes predominant. Redrawn from [72].

Figure 16 shows how force–distance curves in solution depend on the electrolyte concentration. The tip–sample contact point is clearly visible (arrow 2 on the first FD curve) as well as the distance at which electrostatic repulsion becomes measurable (arrow 1 on the first FD curve). These two points are the intersection of the FD curve recorded in the indicated solution with the FD curve recorded under no electrostatic repulsion conditions, which is the lower curve in the figure, obtained in an appropriate buffer. The determination of points (1) and (2) for the used buffer also indicates the minimal load force that can be used to be still in the contact region. This indication is useful when imaging soft samples.

15.2.6 Sample preparation, specific adhesion and single molecule studies

One of the main difficulties in AFM imaging is sample immobilization. For many inorganic materials, such as graphite and mica, it is sufficient to glue the sample on a flat support. This is not the case when organic materials are to be imaged: big molecules such as DNA, cells, biological fragments and so forth need to be immobilized onto a support, otherwise the tip would just remove them while scanning.

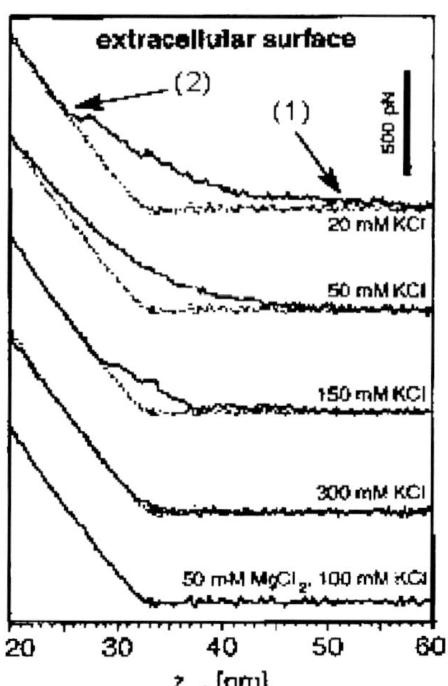

Figure 16. Force–distance curves performed on the extracellular surface of porin OmpF at different electrolyte concentrations and constant pH (7.6). FD curves were recorded during tip–sample approch. Each FD curve is superimposed on the bottom curve, recorded in a buffer containing $MgCl_2$ where no electrostatic repulsion occurs. Arrow (1) indicates the distance at which electrostatic repulsion becomes measurable, arrow (2) shows the tip–sample contact point. Redrawn from [73].

Glass and mica are the most used supports. Depending on the nature of the sample to be studied, glass or mica can be used in their native state or after an appropriate preparation aimed at changing their surface properties [77]. In fact, either the sample is "glued" to the support via a chemical bond or it stays attached to it because of the balance of van der Waals and electrostatic forces between sample and support. Further interactions to be taken into account are hydrophobic and hydrophilic interactions that may prevent, or may cause by themselves, sample adhesion. Moreover all surfaces become covered with hydrocarbons when exposed to ambient air; consequently, the sample support and the AFM probe must be prepared and cleaned from contaminants (e.g. with a detergent or via exposure to UV light) immediately before use.

Glass is a good support in particular for SNOM; its surface, naturally hydrophobic because of hydrocarbon contaminants, may easily become hydrophilic with a glow discharge. Glass surface may also be chemically modified, as in the case of silanized glass, to cause covalent bonding between support and sample.

Coating has been employed to improve the adsorption capability of glass and mica; for instance when the support is coated with poly-L-lysine it offers a positively charged surface on which cells, tissues and plasma membranes, usually negatively charged, adsorb easily. Other procedures require an intermediate step: a gold layer is evaporated onto mica and it is then functionalised with alkanethiols via chemisorption of the sulfur atoms; the sample is then anchored at the head groups at the free end of the alkyl chains [78,79].

A largely used technique though, which is also the simpler from a practical point of view, consists of physisorption of the sample on the support. In fact biological macromolecules are usually charged when in a buffer solution because they expose weak acidic or basic functional groups. The same goes for the support surface; hence the situation is very much similar to the one described above: there are two surfaces (both of them usually flat in this case) that present a surface charge, which is balanced by an electrical double-layer. The DLVO theory [74] may again be applied to determine the buffer conditions necessary to obtain a net attractive force as the result of the van der Waals attraction and double layer repulsion. Thus, once the appropriate electrolyte valence has been determined, sample adhesion may be ensured by adjusting the pH and the electrolyte concentration.

The pH and electrolyte parameters are crucial and may easily determine the non-adsorption of the sample on the support. Moreover a good buffer for sample adsorption is not necessarily good also for sample imaging, since it may set the tip–sample contact point at a force too high or too small to obtain a good resolution; for this reason it is very common to perform sample adsorption in a buffer, then rinse it gently and carry out imaging in a different one. The sample, usually, once firmly attached to the support, stays there even after the buffer is changed.

Once favourable imaging conditions have been set, biomembranes can be imaged in their native state with a lateral resolution down to 0.4–1 nm and a vertical resolution of 0.1–0.2 nm, which means that single molecules in the membrane can be imaged [80]. Figure 17 shows a beautiful image where single molecules are resolved [81].

Figure 17. Topograph of a vescicle of densely packed AqpZ tetramers exposing their extracellular surface. Topograph recorded in buffer solution (17 mM TrisHCl, pH 7.2, 150 mM KCl) in AFM contact mode at an applied force of ~80 pN. Scale bar 100 Å, full gray scale 7 Å, displayed as relieves tilted by 5°. Redrawn from [81].

A beautiful example of imaging conformational changes at molecular resolution is reported in Figure 18 [82]. This image of chaperonine GroEL molecules was obtained using the so-called small cantilevers (not yet commercially available; dimensions of the order of ten micrometres compared to the hundreds micrometres of common cantilevers). Width, thickness and length are reduced so as to maintain the elastic constant in the same range as for the "classical" cantilever. The mass, however, is reduced, so that the resonance frequency is increased and the thermal noise at low frequency is smaller than in the usual cases. Notably, in the right-hand side of the figure, scanning in the y (horizontal in this case) direction was stopped and the same molecules were sampled in subsequent scans, showing the dynamical association and dissociation of GroEL–GroES complexes.

Moreover it is possible to operate the AFM so as to observe the intramolecular forces as has been done with bacteriorhodopsin (BR). BR molecules in their native membrane are assembled into trimers, arranged to form an trigonal bidimensional crystal lattice. When the cantilever tip is gently pressed on one molecule, in a certain percentage of cases, the molecule adsorbs on the tip and is subsequently extracted from the membrane as the tip is retracted. BR, being made of seven transmembrane helices, unfolds in subsequent steps in which the helices unravel in pairs [83]. Figure 19(B) shows a force – extension curve for a single BR molecule. Force – extension curves $f(d)$ are obtained from force – distance curves $f(z)$ by the relation $d=z+s$ explained in Figure 9, and inverting the force axis.

In the case of BR reported in Figure 19 no particular preparation was required to attach the molecules to the stylus, but this is not usually the case. An exciting use of force–distance curves concerns the measurement of the force required to produce unfolding of specific molecular domains or the study of specific bonds between molecular partners (antigen – antibody, receptor and its ligand [84–92]) which requires a specific sample preparation.

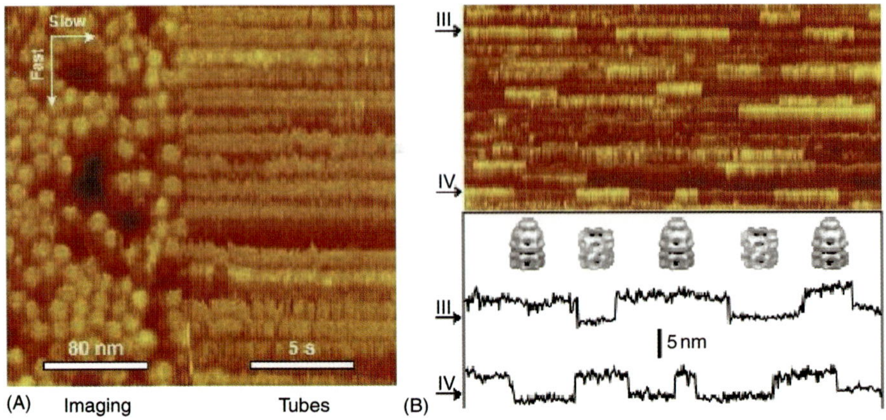

Figure 18. (A) Topography of GroEL proteins adsorbed on mica. The *x* axis in this case is in the vertical direction (indicated by the arrow of the fast scan direction). Half the way through the scanning along the *y* direction (the horizontal one in this case, indicated by the arrow of the slow scan direction) was disabled and thus the tip was continuously scanned on the same line of proteins, generating the protein "tubes" of the right-hand part of the image. Those tubes are the evolution over time of the height of the proteins along that line, monitored with a temporal resolution of 100 ms. (B) GroES (144 nM) and ATP (2.5 nM) are added to the buffer solution and the tubes are imaged. Under those condition the association and dissociation of GroEL–GroES complexes is possible, in fact large repeated variations in height along the length of many tubes are observed. Cross sections of the tubes, indicated with III and IV are reported below, show a stepping of the height between two levels that differ for 3.6±1 nm. This height difference is consistent with that of GroEL and GroEl-GroES complex seen by X-ray crystallography, confirming that the observed phenomenon is the association and dissociation process. Redrawn from [82].

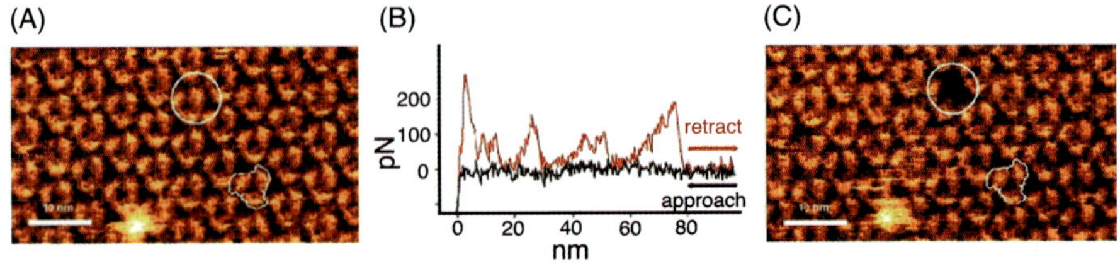

Figure 19. (A) High-resolution AFM images of the cytoplasmic side of purple membranes extracted from *Halobacterium salinarum*. The 2D crystal lattice of bacteriorhodopsin molecules and the organization in trimers are clearly visible. A trimer is contoured, while a circle is traced around a single molecule, the one missing in panel (C). In (B) sudden changes in the force–extension curve (lighter line) show that a molecular bridge was formed and that the protein is stretched. (C) AFM image of the same area (the defect on the lower left-hand side confirms that it is the same area imaged in (A). The circle indicates the single monomer missing. Redrawn from [83].

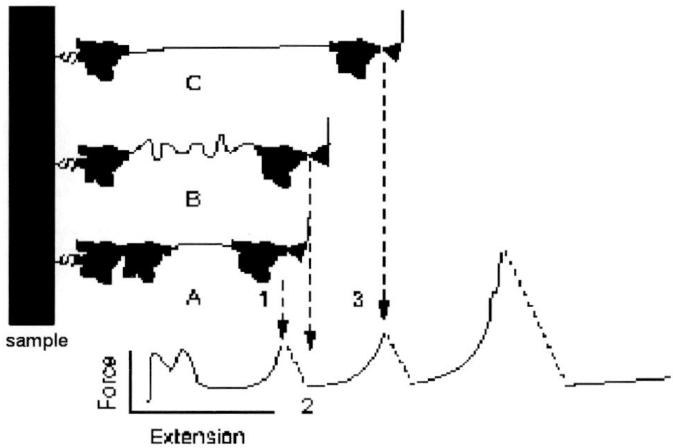

Figure 20. Scheme of force-induced unfolding of titin. The molecular configuration corresponding to the different features of the force–extension curve is shown. In (A) the titin molecule is stretched up to the rupture point of one of its domains (point 1). In (B) as one domain unfolds the cantilever relaxes and the applied force decreases (point 2). As the cantilever retracts the protein is stretched further (C) up to the rupture point of a second domain (point 3), and so forth. Redrawn from [96].

A classical case of repeated unfolding is that of titin [93,94], which is possible thanks to the chemical modification of the COOH terminal by cysteine groups, which allows for the attachment of the molecule to gold particles. Titin unfolding has been studied both by AFM and by optical tweezers [93–95]; Figure 20 shows the scheme of a titin molecule "hooked" to a support and to the AFM stylus [96]. As the stylus is retracted, it stretches the molecule up to the rupture point of one of its domains. Once the domain is unfolded, the cantilever relaxes and pulls again when it finds the resistance of the following domain.

Sometimes molecules unfolded by mechanical stress revert to their original configuration. This is the case of titin and lysozyime [97], shown in Figure 21, which reports a series of repeated force–distance curves, showing the perfect reversibility of this folding/unfolding process. Similar procedures have been applied to investigate the mechanical properties of molecules, together with the strength of their inter- and intra-molecular bonds [98–100].

FD curves can also be used to study specific interactions between molecular partners, by suitable functionalization of sample and tip. Once the stylus has been functionalised, for instance via silanization of the surface and subsequent adhesion of proteins, it can be moved towards the sample so as to induce a chemical bond between the molecular partners present on the support and on the tip. In this way protein – protein interaction forces can be measured. Usually, suitable spacers are also interposed between the support and the molecules to be studied, so that the specific detachment occurs far enough on the FD curve graph from the non-specific adhesion.

Notably, specific adhesion does not occur at any approach of the tip to the sample. When it does not occur the FD curve is very much similar to those in

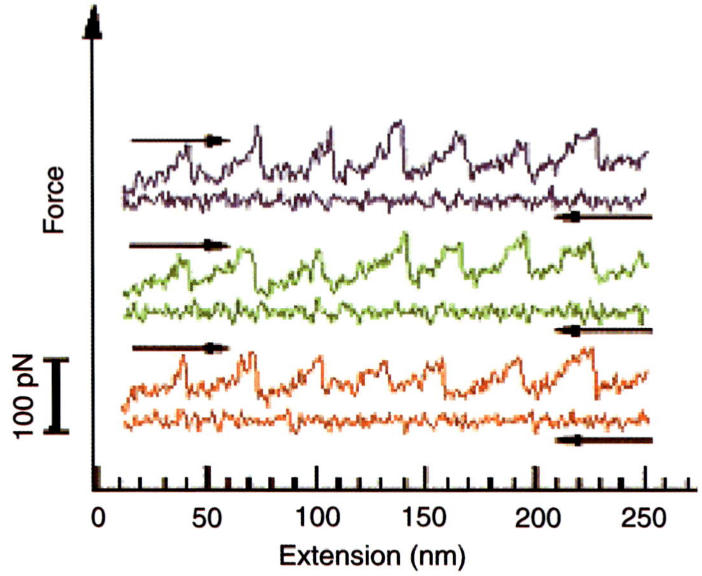

Figure 21. Repeated unfolding–refolding of T4 lysozyme in a polymer. Curves were obtained by stretching and relaxing the same polymer multiple times. Three complete cycles are shown indicating the complete reversibility of the process. The three curves are shifted vertically for clarity so as not to share a common zero force level. Redrawn from [97].

Figure 14, while when it does occur a second jump is visible, following the usual non-specific adhesion peak. This indicates the stretching of a bridge connecting the tip to the support that is longer since it is formed by the two molecules and by the interposed spacers [101]. In fact, inter- and intra-molecular interactions should be studied on a statistical basis, as the rupture of chemical bonds is, intrinsically, a statistical event. Moreover, as a consequence of the statistical nature of the bond breaking, the histogram of the forces at which the bond is broken depends on the velocity at which the stylus is retracted, as clearly evidenced by experimental data reported in Figure 22, which was obtained by a force measuring technique not suitable for imaging [102].

Overall, single molecule force spectroscopy studies may provide a great deal of information regarding the mechanical properties of membrane proteins, protein elasticity [98,103], protein–protein interaction [104,105], and protein structure [106]. These data become particularly important in protein characterisation when unambiguous structural data are not available from other sources, as in the case, for instance, of cadherine [89,90]. This protein is particularly challenging to crystallise; hence, via FD measurements, a model for cadherine–cadherine adhesion was tested and the mechanism of adhesion unveiled.

AFM plays a particular role in studying such a big molecule as DNA, which can be imaged after immobilisation on a support and whose intra-molecular and mechanical properties can be investigated [105,107–109].

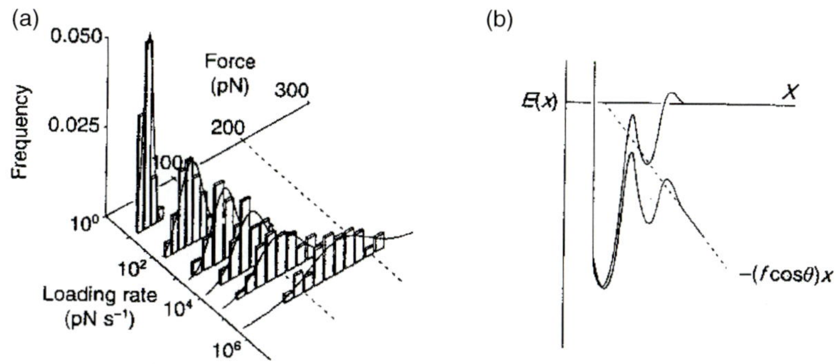

Figure 22. (a) Histogram of biotin–strepavidin bond strength. The histogram peak shifts to the right and the histogram width increases with increased loading rate. (b) Change in the energy profile while the bond is stretched. Redrawn from [102].

15.2.7 AFM as a nanotool

While AFM is used for imaging, the damage that strong load forces could produce on a sample surface must be minimised. However, there is an alternative use of AFM technique that consists in inducing intentionally modifications on the sample. The use of AFM as a nanotool is strictly related to the problem of information storage at the nanometric scale or to nanomanipulation of cellular components, fibrils, and carbon nanotubes.

Generally speaking, there are three main different methods that allow to utilize the AFM in nanolitography.

One method is called local anodic oxidation (LAO). The key to understanding how it works is given in Figure 23. If the tip–sample distance does not vary too much (as in contact mode [110] or in a suitable AC mode [111]) the water meniscus (Figure 12) between tip and sample in a humid atmosphere at room

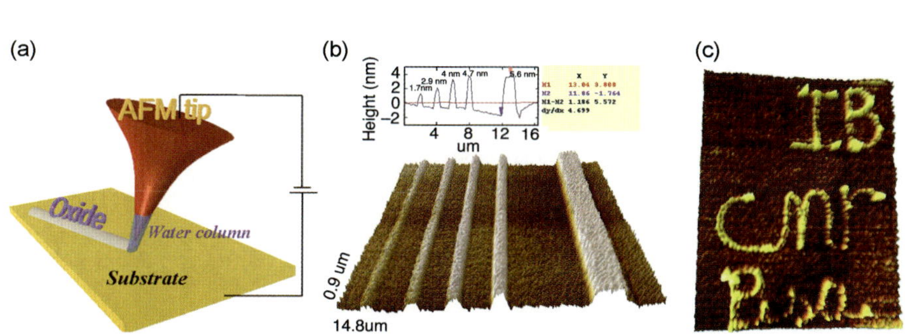

Figure 23. (a) AFM configuration for LAO technique; (b) oxide stripes obtained by LAO on Si at different tip–sample bias; (c) LAO on GaAs in constant-current mode (scan size 5×6 μ).

temperature is quite stable and moves together with the tip during a scan. Employing a conductive tip, and applying a bias between tip and doped semiconductor (Si, GaAs) or metallic (Nb, Al, Ti) samples, an electric current flows and spatially-resolved oxidation of the anodic sample surface occurs. To obtain a uniform oxide layer it is better to stabilize the current flow (surface-conductance variations or sample topography can affect it) instead of the voltage bias by a feedback loop [112]. Figure 23(a)–(c), obtained by a collaboration between CNR-Istituto di Biofisica and NEST-INFM laboratory of Scuola Normale Superiore in Pisa, shows an example of anodic oxidation on Si and GaAs samples. In experiments on a silicon substrate, the oxide layer resulting from different applied bias has been studied spectroscopically; AFM oxidation produces chemically uniform, stoichiometric SiO_2, and its chemical and structural properties do not depend on the applied voltage [113].

A second method is direct writing on the sample, by scratching it with the tip at high applied forces (static ploughing). Diamond coated silicon tips can be used to minimize tip wearing. Grooves down to the 100 nm scale can be so-obtained on semiconductors, metals or polymers [114–117]. Figure 24(a), (b) and (c) gives an example of ploughing on a photoresist; this represents an intermediate step on a sample previously patterned by UV lithography. In fact auto-alignment capability and improved resolution down to a 100 nm scale can easily be obtained in this way, and a modified resist can be successively used as a mask for etching or metal evaporation and lift-off on the substrate [118].

The third method is the so-called "dip pen nanolithography" (DPN) [119]. It works, as its name implies, like an old ink pen: the cantilever tip is plunged into a suitable non-hydrosoluble solution; by capillarity the tip picks up some "ink", and is then employed in contact mode and works just like a pen. This method is very useful to label a part of the sample, or to write patterns with fluorescent chromophores on a nanometric scale [120]. The ink is chosen with a suitable vapour pressure and the amount collected in a single AFM tip "immersion" is sufficient for numerous operations (Figure 25).

Work is in progress in using AFM as a nanotool and it is likely that some interesting applications of AFM technology will appear in the next few years.

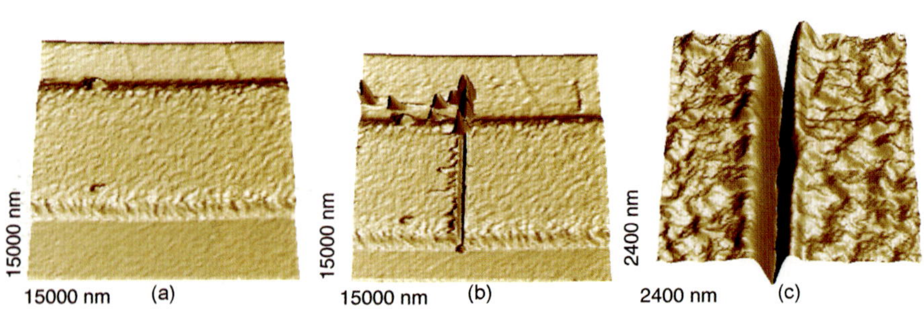

Figure 24. Static ploughing on a photoresist: (a) AFM image of UV-lithography patterned microstructure; (b) groove obtained by direct writing; (c) groove detail.

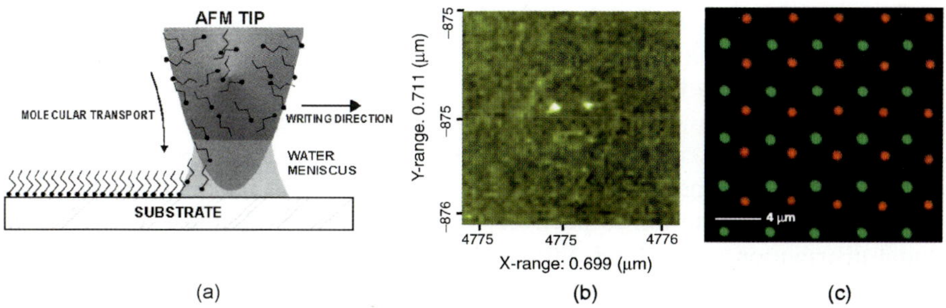

Figure 25. (a) Schematic of DPN process (redrawn from [119]); (b) green fluorescence protein (GFP) "smile" patterned by DPN on GaAs substrate (by courtesy of P. Pingue, NEST-INFM and CNR-IB, Pisa). (c) Combined red-green epifluorescence image of two different fluorophore-labeled sequences (Oregon Green 488-X and Texas Red-X) simultaneously hybridized to a two-sequence array deposited on an SiOx substrate by DPN) (redrawn from [120]).

Recently, IBM in Zurich employed an array of about a thousand AFM tips and a thermo-mechanical writing technique [121] for data storage on a nanometer scale: tiny depressions melted by an AFM tip into a polymer medium represent stored data bits that can then be read by the same tip (Figure 26). The resulting tool, called "millipede", can achieve data densities in the hundreds of Gb in^{-2} range, well beyond the expected limits for magnetic recording (60–70 Gb in^{-2}).

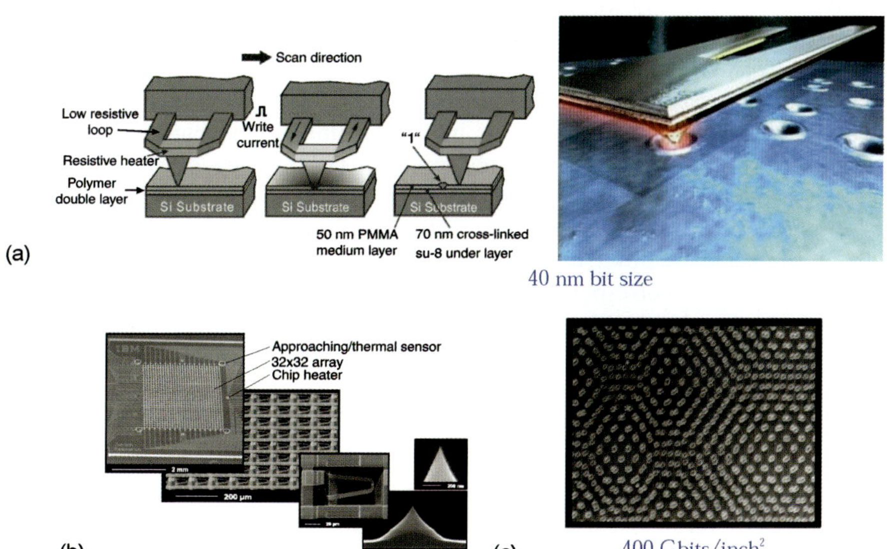

Figure 26. (a) Schematic of thermomechanical writing technique; (b) the "millipede" microfabricated cantilever array; (c) data storage obtained by thermomechanical writing (from http://www.zurich.ibm.com/st/storage).

15.3 Scanning near-field optical microscopy

15.3.1 Basic principles

In classical optical microscopy the resolution limit is assumed to be of the order of the wavelength λ of the illuminating light (actually 0.5λ or 0.62λ according to the Abbe or Rayleigh criterion respectively). This limit in resolution is owing to the spatial distribution of the light diffracted by a narrow slit or by a "point" source (Airy function). In fact, when a plane wave is limited, for instance by an optical system, only a portion of the wave is let through and there is always a spread of the light in directions different from the primitive one. Thus two different objects A and B located at a distance Δx can be detected as distinct in the microscope image until the maximum of the diffraction fringes due to A falls on the first zero of the diffraction pattern due to B; a further decrease of Δx makes it impossible to resolve the two objects.

Quantitatively the Rayleigh criterion tells that two object can be resolved in a microscope image only if their distance is greater than Δx_{min}, given by:

$$\Delta x_{\mathrm{min}} = \frac{1.22\lambda}{2n\sin\alpha} \tag{1}$$

where n is the refractive index of the medium interposed between the sample and the first lens of the optical system and α is half of the angle under which the first lens of the objective is seen from the center of the microscope field. The denominator in Equation (1) is the numerical aperture of the objective.

Rayleigh criterion, however, is not an inviolable theoretical limit. It takes into account only the propagating components of the electromagnetic field, which are commonly called "far field components" and are those involved in traditional optical devices. Nevertheless, light reflected or diffracted by an object contains more information than that carried by its propagating components.

The key point is the diffraction phenomenon [122]. Light diffracted by small objects is initially confined, as happens for the light beyond a screen with small holes. The profile of the light intensity distribution (or that of the electric field) just beyond the screen follows the holes profile; its Fourier transform contains high spatial frequencies, corresponding to the sharp edges of the field profile.

For a screen in the x–y plane with two slits with a separation d and a width L the Fourier transform of the electric field profile at $z = 0$ (just after the screen) is given by:

$$f(k_x)_{z=0} = 4E_0\cos(k_x d)\frac{\sin(k_x L)}{k_x} \tag{2}$$

where the spatial frequency k_x is not limited.

The distribution of the field at a distance z from the screen (or any diffracting object) is obtained by letting each component evolve [122]. However, the amplitude of the wave vector has to be conserved. Thus, due to the high spatial frequencies of a confined field, some of its Fourier components will not propagate very far. In fact, when the projections k_x of the wave vectors in the screen plane is greater than the modulus k_i of the wave vector of the incoming light, k_z will be imaginary because of Pitagora's relation between the amplitude of the wave vector and its components. Physically this means that the intensity of the field will decay in the z direction.

The field at any distance z is therefore obtained using two different expressions corresponding to real or imaginary values of k_z. Each spatial component (whose spatial frequency is given by k_x) is allowed to evolve (to propagate or to decay), and the resulting field is obtained by a composition of all the components according to the two expressions:

$$E(x, z) = \int_{-k_0}^{k_0} \mathrm{d}k_x f(k_x)_{z=0}\, \mathrm{e}^{-jk_z z} \mathrm{e}^{j(\omega t - k_x x - k_y y)} \text{ with } |k_x| < |k_0| \tag{3}$$

$$E(x, z) = \int_{-\infty}^{+\infty} \mathrm{d}k_x f(k_x)_{z=0}\, \mathrm{e}^{-|k_z| z} \mathrm{e}^{j(\omega t - k_x x - k_y y)} \text{ with } |k_x| > |k_0| \tag{4}$$

Figure 27 is obtained by using Equations (2)–(4) and gives the intensity measured on a plane very near to the screen ($z = 10$ nm) and at a distance equal to the wavelength (500 nm). Evidently a loss of details occurs as the field propagates far from the diffracting objects; at a distance equal to λ it becomes almost impossible to detect the existence of two distinct slits.

This example makes it clear that the evanescent part of the field (corresponding to imaginary values of k_z) contains information, on a scale lower than λ, that is lost in the propagating components. Therefore, detecting the evanescent field, or simply

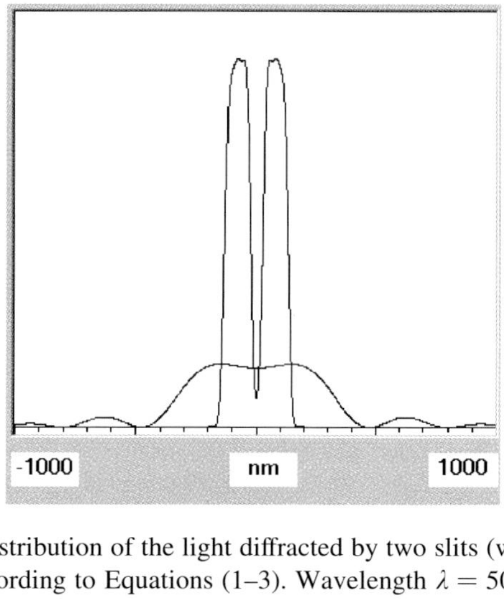

Figure 27. Intensity distribution of the light diffracted by two slits (width 60 nm, separation 20 nm), computed according to Equations (1–3). Wavelength $\lambda = 500$ nm. Vertical scale in arbitrary units. The curve with two separated peaks is obtained at a vertical diplacement $z = 10$ nm, the other one at $z = 500$ nm.

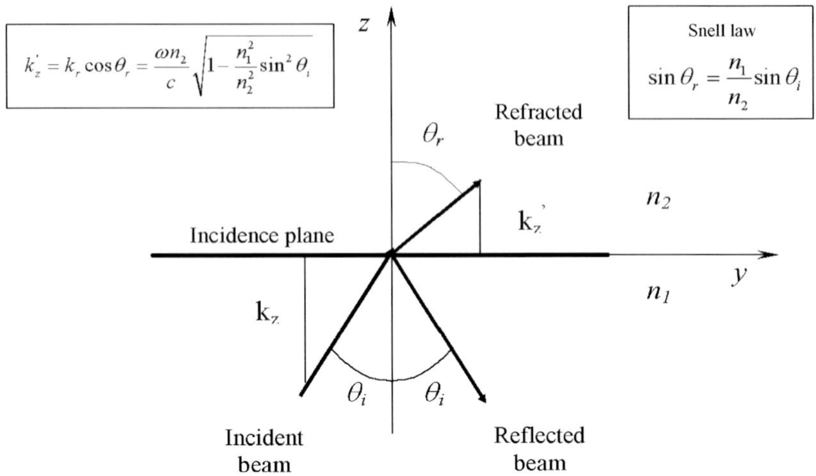

Figure 28. Reflection and refraction at the boundary between two dielectrics. The amplitude of the wave vectors of the incident and refracted beam are given by $k_{\rm i} = \omega n_1/c$ and $k_{\rm r} = \omega n_2/c$. When $n_1 > n_2$ there is a critical incidence angle $\theta_{\rm c}$ such as $\sin\,\theta_{\rm c} = n_2/n_1$. For $\theta_{\rm i} = \theta_{\rm c}$ the refracted beam runs along the boundary plane between the two media. For $\sin\,\theta_{\rm i} > n_2/n_1$ total reflection occurs.

the field near to the sample that diffracts it, can be the basis for an optical microscopy with a resolution overcoming the Rayleigh criterion.

In the case of total reflection the non-propagating part of the field is separated from the propagating one. Total reflection (Figure 28) can occur at the separation between two meida when n_1 is greater than n_2, for instance when light is going from water into air, and depends on the incidence angle. This is why the effect is called fish eye vision, which occurs when one looks up on being under water. As sketched in Figure 29, for the fish the external world is reduced to a disk (with angular aperture $\theta_{\rm c}$).

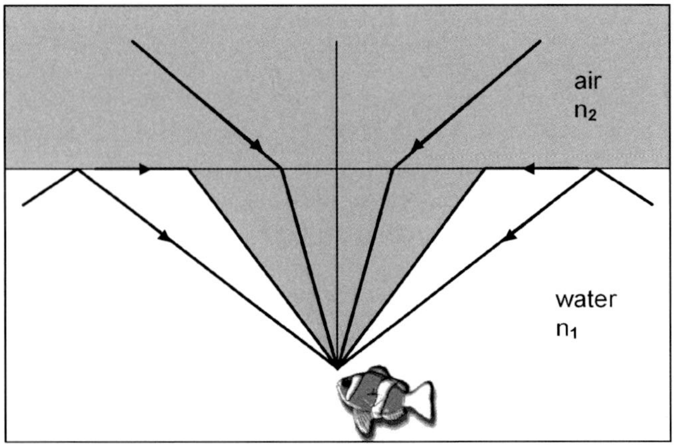

Figure 29. Sketch of fisheye vision.

It is easy to prove (Figure 28) that the component of the wave vector of the refracted beam becomes imaginary when $(n_1/n_2)\sin\theta_i > 1$, namely when total reflection occurs. This means that the refracted field does not propagate, but it exists and decays in the z direction with a space constant d given by:

$$d = \frac{\lambda}{2\pi\sqrt{n_1^2\sin^2\vartheta_i - n_2^2}} \tag{5}$$

For the definition of symbols see Figure 28. For instance, with $n_1 = 1.5$ and $n_2 = 1$ (passage from glass to air) d is about 100 nm when $\lambda = 600\,\text{nm}$. Actually the measurable quantity is not directly the field E but its square, which therefore decays with $d/2$, i.e. with a constant of the order of 50 nm.

However, the evanescent light is not directly visible. The face of a prism where total reflection occurs does not appear illuminated. However, if a third dielectric medium, with $n_3 > n_2$, is moved towards the surface where total reflection occurs, light passes from the first to the third medium, provided that the thickness of the second medium is small enough. This phenomenon, known since Newton's time, is called FTIR (Frustrated Total Internal Reflection) or optical tunnel, because of the analogy with the electronic tunnel.

At the beginning of the 1970s optical tunneling was used to measure nano-movements, as shown in Figure 30(a), where the evanescent field at the surface of a prism is collected by a second prism moved by piezoactuators [123]. The measured decay, reported in Figure 30(b), is not perfectly exponential (a different slope is present close to zero), because of the perturbation of the evanescent field due to the second prism used to collect the light. A computation of the transmission coefficient, made in the mid-1980s [124], confirms the deviation from the exponential decay present in Figure 30.

The detection of the evanescent field in the particular case of total reflection illustrates that, to measure the non-propagating components of the light, it is necessary to go where they are, proximal to the object, and to refract or to scatter them away with a probe [125]. Thus a far-field wave is generated, which acts as a carrier of the near-field information about sub-λ structures on the surface. With the FTIR configuration a very small tip (scattering mode), or a nanometric aperture in a metallic screen (collection mode) is used as a probe.

FTIR is not a unique configuration: the sample can be illuminated with an extended source, detecting the near-field components with a submicrometric aperture in a metallic screen (collection mode) or to illuminate the sample locally (emission mode), and detecting in far field.

A mixture of the collection and emission modes is also possible. Here, the same probe emits and collects the light. When fluorescence of the sample is involved, the two lights can be spectrally separated.

The other important observation is that the measurement itself alters the evanescent field considerably. This is one of the most important issues in SNOM microscopy.

Figure 31 shows a sketch of the principle of near-field microscope. The probe is represented by a hole in a secondary screen that scans the sample. Notably, each

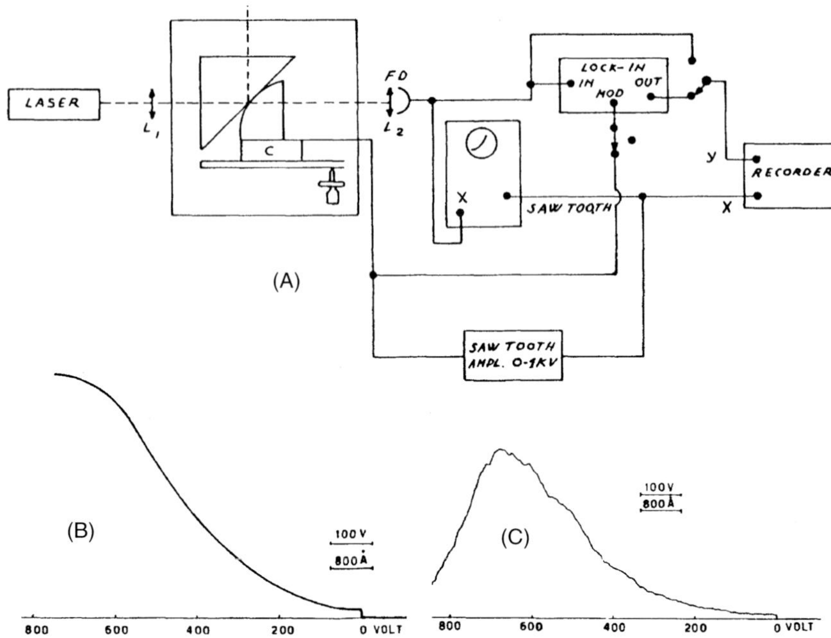

Figure 30. First measurement of distances with sub-Å resolution by using piezoattuators and evanescent field detection (FTIR). (A) Scheme of the apparatus. The saw tooth generator controlled the elongation of ceramic stack C and thus, the approach of the two prisms. (B) Measurement of the transmitted light intensity. In (C) the lock-in amplifier was operating, the position of the piezo drive C was modulated with an amplitude of 2×10^{-2} Å peak to peak and the oscillation amplitude of the optical signal was recorded, reaching a the sensibility of about 10^{-3} Å. Redrawn from [123].

interaction with matter changes the boundary conditions of the Maxwell equations and causes a partial projection in the far field of the evanescent field. The small second hole selects only the near-field components that reach it during its scan trajectory. Only when the holes overlap, can light be transmitted above to the detector.

15.3.2 Probes and operating modes

The most widely used SNOM probes are tapered glass fibers, but micropipettes for electrophysiology or AFM cantilevers have also been used. The smaller the lateral dimensions of the probe, better the resolution in the image, as sketched in Figure 31.

The shape of the probe and the diameter of the hole are critical parameters. The reproducible fabrication of probes with small apertures is still a problem in SNOM microscopy [126–128]. The tapering of optical fibers is obtained by chemical etching. Micropipettes with small apertures are obtained by warming and pulling,

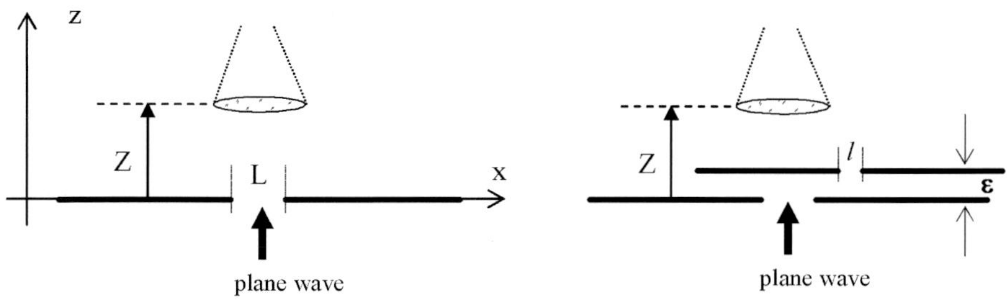

Figure 31. Schematic sketch of classical microscopy and of near-field microscopy. The sample is a screen with a hole, the light detector is far from the sample in both cases. In (A) the detector above the sample will detect the diffracted light; in (B) the distance between the two screens and the width l of the hole in the second screen are small compared to λ. The second screen (the probe) will transmit light only when the two holes overlap.

as usually is done for electrophysiological experiments. In both cases a metal coating of the probe is performed, leaving a small aperture on the tip.

At the beginning of the 1990s common cantilevers for AFM were introduced in SNOM microscopy [129] as probes to detect the optical near-field.

AFM cantilevers in silicon nitride are suitable as near-field probes because they are transparent in a very large range, from 0.2 to 1.5 μm. Moreover their use makes it possible to get, at the same time, topographic information on the sample. Transmission by the tip of the collected light occurs in a semi-angle between 30° and 60° [130]; therefore, large aperture objectives are required to collect the transmitted light. Silicon nitride cantilevers might also act as wave-guides, and this helps in collecting light [131].

The simplest way to prepare metal-coated AFM cantilevers to be used as coupled AFM-SNOM sensors consists of two steps. The tip is metal coated, and then it is repeatedly scanned on a hard alumina surface at such a force that the tip is scratched on its terminal part to expose the silicon nitride substrate. Figure 32(a) shows a cantilever with its gold-coated pyramidal tip; in Figure 32(b), at higher magnification, it is possible to see the hole on the terminal part of the tip. Such holes are in the range 100–300 nm. Recently an improvement was introduced in the fabrication of probes, by growing a metallic tip on the aperture of a conventional fiber probe. Tips were grown by electron beam deposition and then covered with metal. Images with a resolution of 25 nm have been obtained [132].

An interesting book has been dedicated to the fabrication of silicon microprobes for optical near-field applications [133].

The actual dimensions of the aperture in metal-coated probes depend on the penetration δ of the electromagnetic field in the coating metal. Often, aluminium is used, for which δ is 5–6 nm with $\lambda = 500\,\text{nm}$. To have a good screening, at least a coating of about 100 nm is required. The transmission is typically 10^{-4} for an aperture of 80 nm in diameter. The external diameter of a metal-coated probe is of the order of 250 nm, therefore the use of aperture probes is limited to relatively flat samples.

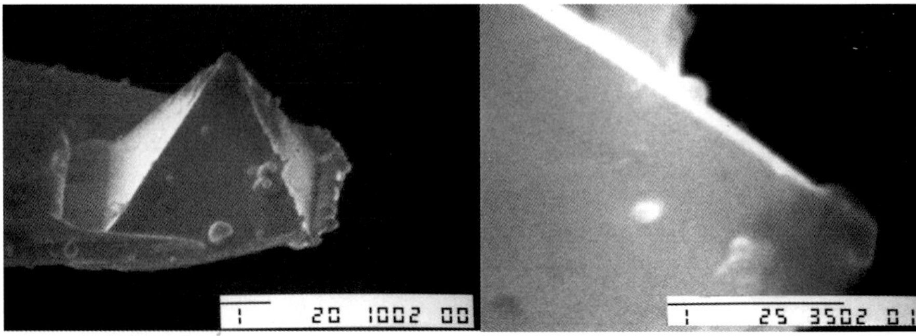

Figure 32. SEM images of a combined AFM/SNOM probe. Left-hand side, the AFM cantilever with its gold coated pyramidal tip; right-hand side, the tip image after AFM scratching; the hole on the terminal part of the tip is visible. Redrawn from [146].

A completely different approach is the so-called "aperture-less probes" [134–136]. Passive probes, like nano-sized gold particles, placed on a tip scanning the sample, scatter light, making possible its detection far from the sample. Recent work on SNOM microscopy is mostly oriented towards aperture-less probes. Single fluorescent molecules (active probes) have been used to detect light or to act as single molecule light source [137,138].

The operating mode in SNOM microscopy is quite different from the other scanning probe microscopies; in fact SNOM has been always associated with some other methods to obtain topographic signals and to control the tip displacement. The image reported in Figure 33 [139], obtained by a microscopic simulation in which sample and probe are schematized as sets of polarizable domains, illustrates how misleading it can be to use the optical signal in controlling the tip position.

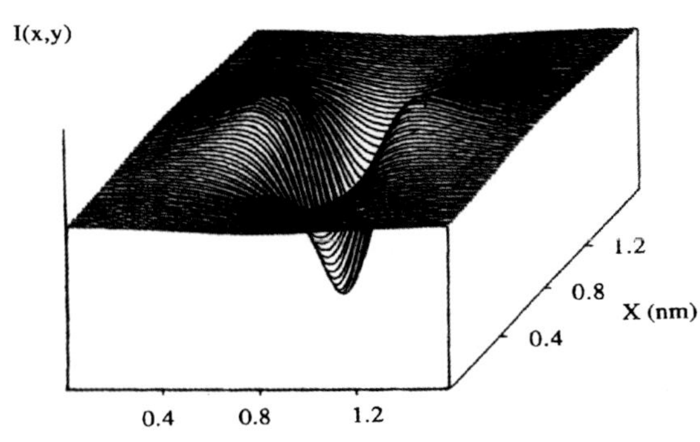

Figure 33. Simulation of near-field imaging of a 1 nm sphere by a collecting probe with terminal radius of 0.25 nm (both are not realistic values). The sphere is illuminated from below under conditions of total reflection. The figure shows the inversion of the contrast in the middle of the figure, where the topographic signal is higher. Redrawn from [139].

Two different methods have been used to control the tip–sample distance. The shearing force between a tapered optical fiber and the surface can be used, by putting the probe in oscillation in the x–y plane (in general using a tuning fork) and recording the oscillation amplitude, which is then used as a feedback signal to stabilize tip–sample distance [140–144]. The other method is associated with the use of AFM cantilevers as optical probes, and allows one to obtain, at the same time, both AFM topographic images and near-field optical images. Nowadays, the combination of SNOM with AFM is quite common. An hybrid AFN/SNOM microscope has been built at IB-CNR in Pisa in collaboration with INOA, Firenze, to study the optical properties of channel waveguides [145,146]. We report here some details of such a microscope (Figure 34), working in the FTIR collection mode.

A different configuration [147,148] is reported in Figure 35. In this last set-up a very compact SNOM head has been developed on the basis of silicon cantilever probes. This head consists of a mirror microscope objective with integrated cantilever holder. It is worth highlighting that the use of a mirror optics is important for spectroscopical investigations from UV to IR.

15.3.3 Sources of artifacts and noise in SNOM imaging

A problem shared by the whole family of near-field optical microscopes concerns the presence of artifacts and the connected issue of image interpretation [149].

There is such a variety of possible sources of artifacts (many of them dependent on the feedback mechanism) that it would be extremely difficult to discuss even a few of them. Conversely, there is a huge literature on this issue that can help the

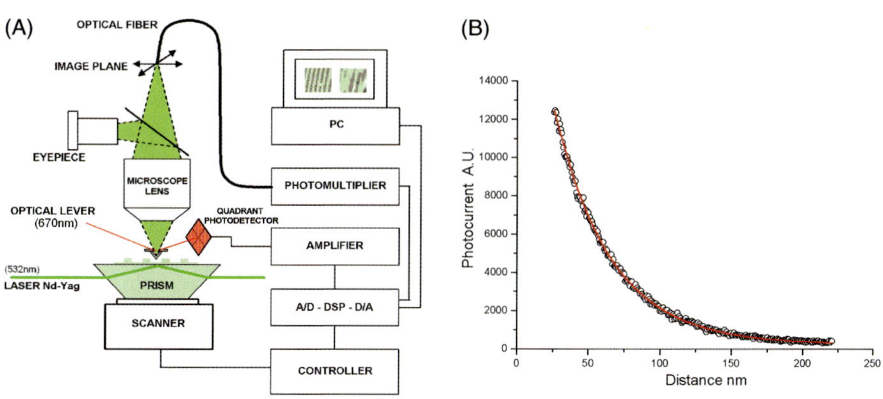

Figure 34. (A) Scheme of the AFM/SNOM built in a collaboration between IB-CNR (Pisa) and INOA (Firenze). The sample is placed on a prism surface and is illuminated by the evanescent field due to total reflection. A modified AFM cantilever is used to measure the evanescent light and to stabilize the tip–sample distance. (B) Optical signal recorded approaching the tip to the sample surface. The solid line is a monoexponential fit with a decay constant of 40 nm. Redrawn from [146].

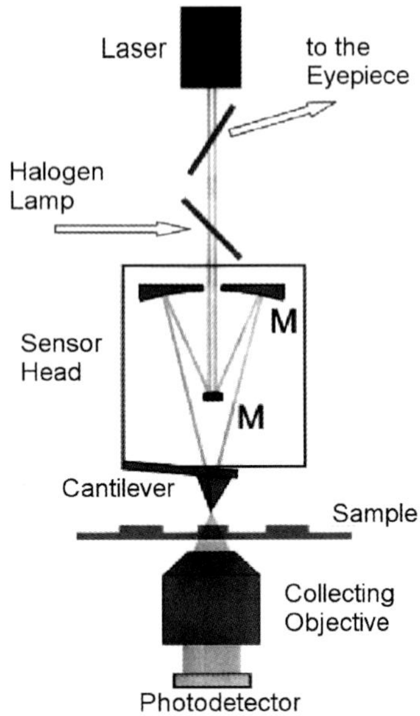

Figure 35. A compact head of an AFM/SNOM built at PTB in Braunschweig, Germany. The microscope optics uses mirrors instead of lenses, so leaving space for the AFM cantilever. Illumination comes through the tip and the photodetector (PD) is placed below. From [147].

researcher interested in the subject [12,150–167]. A widely used method to avoid topographical artifacts is to scan without feedback at a constant tip–sample distance of a few nanometres.

Noise is also due to several causes such as the mesoscopic dimensions of the probe (both hole and coating are really on a scale of just hundreds of atoms) and the fact that forces are acting on it, generated both mechanically and by the interaction with light, whose absorption by the metal coating of tips induces warming and bending of the probe [168]. Moreover light intensity in the near-field region is not uniform, and therefore mechanical forces are exerted on the tip, which can be optically trapped. A different source of disturbance is due to light reflected between the tip walls and the sample, which produces standing waves with interference maxima and minima. This and the perturbation that the probe induces on the evanescent field affect the image. Despite these difficulties beautiful images have been produced by SNOM microscopy.

15.3.4 From imaging of single molecules to the bigger picture

Visualizing and studying single molecules is one of the aims of SNOM. The mere visualization of single molecules by far-field optical microscopy can be achieved using a relatively simple set-up. For instance Figure 36 shows the fluorescence of

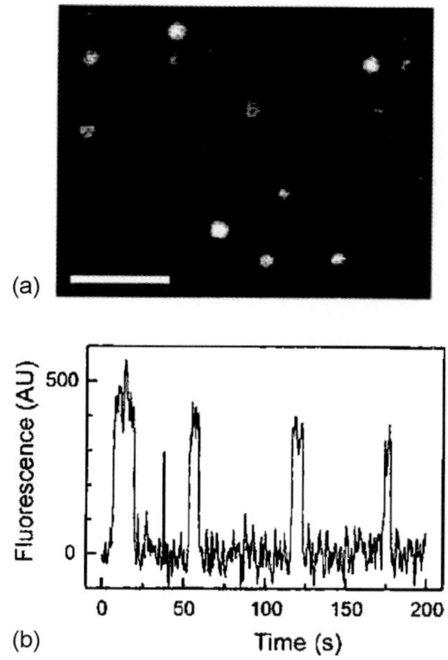

Figure 36. Fluorescence of single molecules of F_1-ATPase engineered in the α subunit in such a way to firmly anchor them on the substrate. The image in (a) was collected by a CCD camera, averaging 16 frames. The bar in the figure corresponds to 2 μm. In (a) each single spot corresponds to a single molecule. Fluorescence occurs only when F_1-ATPase binds with Cy3-ATP present in the solution. Therefore two levels are present in the time recording of the fluorescence of a single spot reported in (b). Redrawn from [169].

single molecules and its dependence on chemical binding [169]. Molecules of F_1-ATPase from *E. coli* were firmly attached on the support; since F_1-ATPase fluorescence depends on its binding with Cy3-ATP, the fluorescence turns on and off, as shown in Figure 36(b). The image reported in Figure 36(a) was obtained by an FTIR microscope: the sample is placed on a support where total reflection occurs and its fluorescence is detected by a CCD camera. Total reflection in this case acts as a very good dark field and the spectral separation between the excitation light and the fluorescence of the molecule is also of help. However, the resolution of the image is that of classical microscopy and the fact that the small bright spots in Figure 36 are single molecules is told by the way in which fluorescence varies with time. A test for single molecule fluorescence is the fact that it appears and disappears in single steps, when a chemical bond is formed or broken, or when photobleaching of the fluorescent molecule occurs.

However, the specificity of SNOM microscopy lies in its ability to obtain sub-wavelength details and, compared with other probe microscopies, SNOM can be easily combined with spectroscopic tools. A good example of this is reported in Figure 37 (see [170] and references therein). Two different molecules are present in the sample (perylene and pyrene) and a phase separation occurs, producing the formation of islands. In Figure 37(a) the islands appear in black and white: the

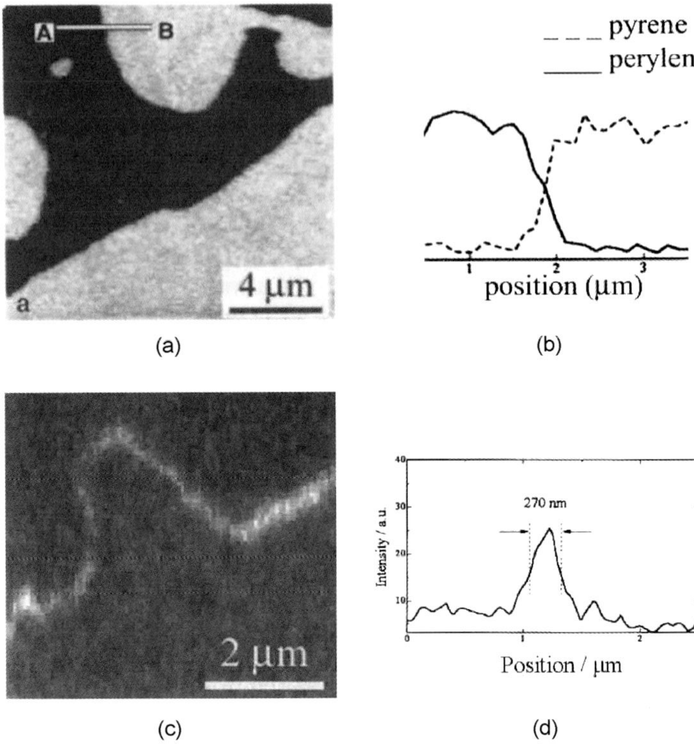

Figure 37. (a) SNOM image of a sample formed of perylene and pyrene molecules; the phase separation of the two components produces islands. The image is obtained by measuring the fluorescence of the pyrene (excited at 325 nm). The fluorescence of perylene (excited at 442 nm) gives a complementary image. In (b) the fluorescence profiles between points A and B is reported. (c) An energy transfer image. Pyrene fluorescence was selectively excited at 325 nm, and the fluorescence of perylene was collected by using a suitable optical filter. Fluorescence occurs only in the narrow region where the two components are mixed. From the fluorescence profile (reported in (d)) the width of the border where the two components are mixed is evaluated to be 270 nm. Redrawn from [170].

image was obtained by a SNOM detecting the fluorescence of perylene or that of pyrene. Perylene fluorescence is excited at 325 nm, that of pyrene is excited at 442 nm. Figure 37(c,d) shows the boundary between two islands. It was obtained by resonant energy transfer, namely by measuring the fluorescence of perylene when that of pyrene was excited. It shows a sharp separation between the two components (about 270 nm).

A further example (Figure 38) shows the specific usefulness of combined AFM/SNOM. The AFM topography of a sample containing DNA (single strand and double strand) is reported together with the SNOM measured fluorescence of labeled DNA (fluorescent dye YOYO-I) [171]. The AFM data reveal that the filament on the left is 0.25 nm high, while those on the right are 0.14 nm high. The width is also different. As the dye labels, preferentially, double-strand

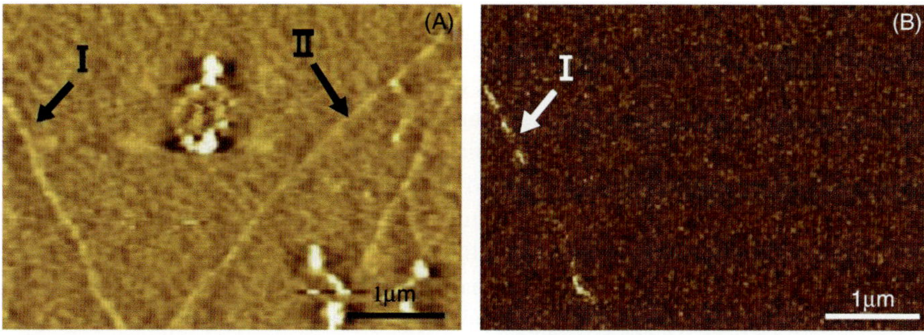

Figure 38. (Left-hand side) AFM image of a sample containing DNA filaments. Different heights (0.25 and 0.14 nm) are evidenced by the AFM data. (Right-hand side) fluorescence image detected by the same multifunctional probe: only the higher filament fluoresces, in agreement with the notion that only double strand DNA bind the fluorescent dye (redrawn from http://www.jaeri.go.jp/english/press/2002/020313/fig01.html).

DNA it is concluded that the two different heights are those of double/single-strand DNA.

As for AFM, SNOM also has different field of application. It has been applied to study quantum wells [172], ferroelectric domain [173], intermolecular coupling in nanometric domains [174], emission spectrum of nanosized particles down to fluorescence resonance energy transfer between a single donor and a single acceptor [175,176]. It is also suitable for studying the co-localization of proteins by dual fluorescence [177]. Moreover, SNOM has been used to study domains in cell plasma membranes [178] and thin film structure and organization [179,180], in particular phase separation as reported above. An important perspective is the use of SNOM as a nanotool, as for instance in nano-patterning of photosensitive polymers or inducing superconductivity at a nanoscale [181,182]. A notable increase in resolution has been achieved in the last few months [183].

15.4 Concluding remarks

The development of microscopy has been always intimately connected with the development of modern science, in particular (but not only) with its ability to study and to try to control life processes. Even from the limited overview reported here it is clear how fast and promising the present development of SPMs is. In particular one of the main functions of SPMs is to modify, in a controlled way, the sample surface. Today, with the required caution and limitations, this is possible, down to the atomic scale. SPMs are the basic key for nanotechnologies.

AFM appears to be particularly suited for studying biological preparation, because of the possibility to work in water and on living cells. However, it is also a powerful tool in nanolithography and in nanomanipulation, and it is easy to predict that these aspects will develop exponentially in the next few years. The interest that companies like IBM show is very evident. SNOM, after some initial difficulties, is now rapidly developing.

The possibility to "go down", to work at a small scale, implies also a new way of thinking. We are used to considering things at a thermodynamic level, and it is now possible to "directly" observe phenomena at an elementary or mesoscopic level. In this respect it appears stimulating to read once more the famous talk that Feynman gave in 1959 at Caltech for the annual meeting of the American Physical Society (it is easy to find it on the net). The idea that "there's plenty of room at the bottom" is now in full operation and the recommendation to build a better microscope for biologists (he was really speaking of a better electron microscope) is now fulfilled.

Acknowledgements

We thank all the people of the AFM group at IPCF-CNR (formerly at IB-CNR PISA) for fruitful discussions and continuous mutual help; in particular, we are indebted to Pasqualantonio Pingue for systematic discussions. We are also grateful to the many students that have worked in the group during their thesis and are now spread around in the world. In particular we want to mention Franco Dinelli, Ciro Cecconi, Brunero Cappella, Lorenzo Lenci, Biagio De Masi, Simona Loi, Claudio Dal Savio; many of the figures reported above are derived from their work.

References

1. G. Binnig, H. Rohrer, C. Gerber, E. Weibel (1982). Tunneling through a controllable vacuum gap. *Appl. Phys. Lett.* **40**, 178–180.
2. G. Binnig, H. Rohrer, C. Gerber, E. Weibel (1982). Surface studies by scanning tunneling microscopy. *Phys. Rev. Lett.* **49**, 57–61.
3. G. Binnig, H. Rohrer, C. Gerber, E. Weibel (1983). 7 × 7 Reconstruction on Si(111) resolved in real space. *Phys. Rev. Lett.* **50**, 120–123.
4. G. Binnig, H. Rohrer (1987). Scanning tunneling microscopy – from birth to adolescence. *Rev. Modern Phys.* **59**, 615–625.
5. P.K. Hansma, J. Tersoff (1987). Scanning tunneling microscopy. *J. Appl. Phys.* **61**, R1–R23.
6. H.J. Guntherodt, R. Wiesendanger (Eds) (1992). *Scanning Tunneling Microscopy I.* Springer Series in Surface Sciences 20. Springer-Verlag, Berlin-Heidelberg.
7. G. Binnig, C.F. Quate, Ch. Gerber (1986). Atomic force microscope. *Phys. Rev. Lett.* **56**, 930–933.
8. E.H. Synge (1928). A suggested method for extending the microscopic resolution into the ultramicroscopic region. *Phil. Mag.* **6**, 356–362.
9. E.H. Synge (1932). An application of piezoelectricity to microscopy. *Phil. Mag.* **13**, 297–300.
10. E.A. Ash, G. Nichols (1972). Super-resolution aperture scanning microscope. *Nature* **237**, 510–512.
11. D.W. Pohl, W. Denk, M. Lanz (1984). Optical stethoscopy: image recording with resolution lambda/20. *Appl. Phys. Lett.* **44**, 651–653.

12. A. Lewis, M. Isaacson, A. Harootunian, A. Murray (1984). Development of a 500 Å spatial resolution light microscope. *Ultramicroscopy* **13**, 227–232.
13. P.K. Hansma, B. Drake, O. Marti, S.A. Gould, C.B. Prater (1989). The scanning ion-conductance microscope. *Science* **243**, 641–643.
14. Y.E. Korchev, M. Milavanovic, C.L. Bashford, D.C. Bennett, E.V. Sviderskaya, I. Vodyanoy (1997). Specialized scanning ion-conductance microscope for imaging of living cells. *J. Microsc.* **188**, 17–23.
15. Y.E. Korchev, Y.A. Negulyaev, C.R.W. Edwards, I. Vodyanoy, M. Lab (2000). Functional localization of single active ion channels on the surface of a living cell. *Nat. Cell Biol.* **2**, 515–619.
16. G. Binnig, Ch. Gerber, E. Stoll, T.R. Albrecht, C.F. Quate (1987). Atomic resolution with atomic force microscope. *Europhys. Lett.* **3**(12), 1281–1286.
17. O. Marti, B. Drake, P.K. Hansma (1987). Atomic force microscopy of liquid-covered surfaces: atomic resolution images. *Appl. Phys. Lett.* **51**(7), 484–486.
18. A. van der Ziel (1955). *Noise*. Chapman and Hall.
19. G.W. Goodman (1996). *Introduction to Fourier Optics* (Chapter 3 and 4). McGraw-Hill International Editions.
20. G. Bernardini (1954). *Fisica Sperimentale* (Volume 1 pp. 396–399). Libreria Eredi Virgilio Veschi, Roma.
21. Ch. Schönenberger, S.F. Alvarado (1989). A differential interferometer for force microscopy. *Rev. Sci. Instrum.* **60**, 3131–3134.
22. D. Rugar, H.J. Mamin, P. Günther (1989). Improved fiber-optic interferometer for atomic force microscopy. *Appl. Phys. Lett.* **55**, 2588–2590.
23. C.A.J. Putman, B.G. de Grooth, N.F. van Hulst, J. Greve (1992). A theoretical comparison between interferometric and optical beam deflection technique for the measurement of cantilever displacement in AFM. *Ultramicroscopy* 42–44, 1509–1513.
24. S. Grafström, M. Neitzert, T. Hagen, J. Ackermann, R. Neumann, O. Probst, M. Wortge (1993). The role of topography and friction for the image contrast in lateral force microscopy. *Nanotechnology* **4**, 143–151.
25. C. Ascoli, F. Dinelli, C. Frediani, D. Petracchi, M. Salerno, M. Labardi, M. Allegrini, F. Fuso (1994). Normal and lateral forces in scanning force microscopy. *J. Vac. Sci. Technol. B* **12**, 1642–1645.
26. F. Dinelli, C.K. Buenviaje, R.M. Overney (2000). Glass transition of thin polymeric films: Speed and load dependence in lateral force microscopy, *J. Chem. Phys.* **113**(5), 2043–2048.
27. B. Bhushan, J.N. Israelachvili, U. Landman (1995). Nanotribology: friction, wear and lubrification at the atomic scale. *Nature* **374**, 607–616.
28. Q. Zhong, D. Inniss, K. Kjoller, V.B. Elings (1993). Fractured polymer/silica fiber surface studied by tapping mode atomic force microscopy. *Surf. Sci. Lett.* **290**, L688–L692.
29. P.K. Hansma, J.P. Cleveland, M. Radmacher, D.A. Walters, P.E. Hillner, M. Bezanilla, M. Fritz, D. Vie, H.G. Hansma, C.B. Prater, J. Massie, L. Fukunaga, J. Gurley, V. Elings (1994). Tapping mode atomic force microscopy in liquids. *Appl. Phys. Lett.* **64**(13), 639–643.
30. C.A.J. Putman, K.O. van der Werf, B.G. de Grooth, N.F. van Hulst, J. Greve (1994). Tapping mode atomic force microscopy in liquid. *Appl. Phys. Lett.* **64**(18), 2454–2456.
31. D. Anselmetti, M. Dreier, R. Lüthi, T. Richmond, E. Meyer, J. Frommer, H.-J. Güntherodt (1994). Biological materials studied with dynamic force microscopy. *J. Vac. Sci. Technol. B.* **12**, 1500–1503.

32. M.A. Lantz, S.J. O'Shea, M.E. Welland (1994). Force microscopy imaging in liquids using ac techniques. *Appl. Phys. Lett.* **65**(4), 409–411.
33. B. Cappella, P. Baschieri, C. Frediani, P. Miccoli, C. Ascoli (1997). Improvements in AFM imaging of the spatial variation of force–distance curves: on line images. *Nanotechnology* **8**, 82–87.
34. F. Gittes, C.F. Schmidt (1998). Thermal noise limitations on micromechanical experiments. *Eur. Biophys. J.* **27**, 75–81.
35. T. Mariani, L. Lenci, D. Petracchi, C. Ascoli (2002). Characterizing tuning forks as nanomechanics sensors through thermal noise spectra measured by a personal computer sound card. *Meas. Sci. Technol.* **13**, 28–32.
36. M.B. Viani, T.E. Schaffer, A. Chand, M. Rief, H.E. Gaub, P.K. Hansma (1999). Small cantilevers for force spectroscopy of single molecules. *J. Appl. Phys.* **86**, 2258–2262.
37. M. Allegrini, C. Ascoli, P. Baschieri, F. Dinelli, C. Frediani, A. Lio, T. Mariani (1992). Laser thermal effects on atomic force microscope cantilevers. *Ultramicroscopy* **42–44**, 371–378.
38. O. Marti, A. Ruf, M. Hipp, H. Bielefeldt, J. Colchero, J. Mlynek (1992). Mechanical and thermal effects of laser irradiation on force microscope cantilevers. *Ultramicroscopy* **42–44**, 345–350.
39. D. Rugar, C.S. Yannoni, J.A. Sidles (1992). Mechanical detection of magnetic resonance. *Nature* **360**, 563–566.
40. J.A. Sidles, J.L. Garbini, K.L. Bruland, D. Rugar, O. Zuger, S. Hoen, C.S. Yannoni (1995). Magnetic resonance force microscopy. *Rev. Modern Phys.* **67,** 249–265.
41. K. Wago, D. Botkin, C.S. Yannoni, D. Rugar (1998). Force-detected electron-spin resonance: Adiabatic inversion, nutation, and spin echo. *Phys. Rev. B* **57**, 1108–1114.
42. C. Ascoli, P. Baschieri, C. Frediani, L. Lenci, M. Martinelli, G. Alzetta, R. Celli, L. Pardi (1996). Micromechanical detection of magnetic resonance by angular momentum absorption. *Appl. Phys. Lett.* **69**, 3920–3922.
43. G. Alzetta, C. Ascoli, P. Baschieri, D. Bertolini, I. Betti, B. De Masi, C. Frediani, L. Lenci, M. Martinelli, G. Scalari (1999). Simultaneous micromechanical and electromagnetic detection of paramagnetic resonance. *J. Magn. Reson.* **141**, 148–158.
44. F. Dinelli, M.R. Castell, N.J. Mason, G.A.D. Briggs, O.V. Kolosov (2000). Mapping surface elastic properties of stiff and compliant materials on the nanoscale using ultrasonic force microscopy (UFM). *Philos. Mag. A* **80**(10), 2299–2323.
45. M.B. Viani, T.E. Schaffer, G.T. Paloczi, L.I. Pietrasanta, B.L. Smith, J.B. Thompson, M. Richter, M. Rief, H.E. Gaub, K.W. Plaxco, A.N. Cleland, H.G. Hansma, P.K. Hansma (1999). Fast imaging and fast force spectroscopy of single biopolymers with a new atomic force microscope designed for small cantilevers. *Rev. Sci. Instrum.* **70**, 4300–4303.
46. T. Mariani, C. Frediani, C. Ascoli (1996). Normal and lateral force images, sub-angstrom height resolution and mid-level lateral resolution with a phonograph cartridge as scanning force sensor. *J. Vac. Sci. Technol. B* **14**, 845–848.
47. T. Mariani, C. Frediani, C. Ascoli (1998). A three-dimensional scanner for probe microscopy on the millimetre scale. *Appl. Phys. A* **66**, S861–S866.
48. T. Mariani, C. Ascoli, P. Baschieri, C. Frediani, A. Musio (2001). Scanning force images through the "Milliscope" – a probe microscope with very wide scan range, *J. Microsc.* **204**, 553–560.
49. G. Binnig (1992). Force microscopy. *Ultramicroscopy* **42–44**, 7–15.
50. J. Frommer (1992). Scanning tunneling microscopy and atomic force microscopy in organic chemistry. *Angew. Chem. Int. Ed. Engl.* **31**, 1298–1328.
51. D. Rugar, P.K. Hansma (1990). Atomic force microscopy. *Phys. Today* **43**(10), 23–30.

52. M. Radmacher, R.W. Tillmann, M. Fritz, H.E. Gaub (1992). From molecules to cells-imaging soft samples with the AFM. *Science* **257**, 1900–1905.
53. D. Sarid (1991). *Scanning Force Microscopy*. Oxford University Press, New York.
54. R.J. Colton, A. Engel, J.E. Frommer, H.E. Gaub, A.A. Gewirth, R. Guckenberger, J. Rabe, W.M. Heckl, B. Parkinson (1998). *Procedures in Scanning Probe Microscopies*. John Wiley & Sons, Switzerland.
55. F. Ohnesorge, G. Binnig (1993). True atomic-resolution by atomic force microscopy through repulsive and attractive forces. *Science* **260**, 1451–1456.
56. A. Schwarz, W. Allers, U.D. Schwarz, R. Wiesendanger (2000). Dynamic-mode scanning force microscopy study of *n*-InAs(110)-(1 × 1) at low temperatures. *Phys. Rev. B* **61**(4), 2837–2845.
57. Y. Sugawara, M. Ohta, H. Ueyama, S. Morita, F. Osaka, S. Ohkouchi, M. Suzuki, S. Mishima (1996). True atomic resolution imaging with noncontact atomic force microscopy. *J. Vac. Sci. Technol. B* **14**(2), 953–956.
58. F.J. Giessibl (1995). Atomic resolution of the silicon (111)-(7 × 7) surface by atomic force microscopy. *Science* **267**, 68–71.
59. R. Lüthi, E. Meyer, M. Bammerlin, A. Baratoff, T. Lehmann, L. Howald, Ch. Gerber, H.J. Güntherodt (1996). Atomic resolution in dynamic force microscopy across steps on Si(111)7 × 7. *Z. Phys. B* **100**, 165.
60. D.J. Müller, G. Büldt, A. Engel (1995). Force-induced conformational change of bacteriorhodopsin. *J. Mol. Biol.* **249,** 239–243.
61. D.J. Müller, F. Schabert, G. Büldt, A. Engel (1995). Imaging purple membranes in aqueous solutions at sub-nanometer resolution by atomic force microscopy. *Biophys. J.* **68**, 1681–1686.
62. C. Möller, M. Allen, V. Elings, A. Engel, D.J. Müller (1999). Tapping-mode atomic force microscopy produces faithful high-resolution images of protein surfaces. *Biophys. J.* **77**, 1150–1158.
63. J.E. Lennard-Jones, B.M. Dent (1926). The forces between atoms and ions. *Proc. Roy. Soc. (London)* **112A**, 230–234.
64. A.L. Weisenhorn, P. Maivald, H.-J. Butt, P.K. Hansma (1992). Measuring adhesion, attraction, and repulsion between surfaces in liquids with an atomic-force microscope. *Phys. Rev. B* **45**(19), 11226–11232.
65. N.A. Burnham, R.J. Colton, H.M. Pollock (1993). Interpretation of force curves in microscopy. *Nanotechnology* **4**, 64–80.
66. B. Cappella, P. Baschieri, C. Frediani, P. Miccoli, C. Ascoli (1997). Force–distance curves by AFM. *IEEE Eng. Med. Biol.* 58–65.
67. M. Schenk, M. Futing, R. Reichelt (1998). Direct visualization of the dynamic behaviour of a water meniscus by scanning electron microscopy. *J. Appl. Phys.* **84**(9), 4880–4884.
68. F.J. Giessibl, Ch. Gerber, G. Binnig (1992). A low-temperature atomic force/scanning tunneling microscope for ultrahigh vacuum. *J. Vac. Sci. Technol. B* **9**, 984–988.
69. F.J. Giessibl, B.M. Trafas (1994). Piezoresistive cantilevers utilized for scanning tunneling and scanning force microscope in ultrahigh vacuum. *Rev. Sci. Instrum.* **65**, 1923–1929.
70. F.J. Giessibl, S. Hembacher, H. Bielefeldt, J. Mannhart (2000). Subatomic features on the silicon (111)-(7 × 7) surface observed by atomic force microscopy. *Science* **289**, 422–425.
71. F.J. Giessibl, S. Hembacher, H. Bielefeldt, J. Mannhart (2001). Response to technical comment: subatomic features in atomic force microscopy images. *Science* **291**, 2509a.

72. B. Cappella, P. Baschieri, M. Ruffa, C. Ascoli, A. Relini, R. Rolandi (1997). Structure and nanomechanical properties of solvent cast stearic acid films in liquid: an atomic force microscopy study. *Langmuir* **15**(6), 2152–2157.
73. D.J. Müller, D. Fotiadis, S. Scheuring, S.A. Müller, A. Engel (1999). Electrostatically balanced subnanometer imaging of biological specimens by atomic force microscope. *Biophys. J.* **76**, 1101–1111.
74. J. Israelachvili (1992). *Intermolecular & Surface Forces, Second Edition.* Academic Press, London.
75. B.V. Derjaguin, L.D. Landau (1941). Theory of the stability of strongly charged lyophobic sols and the adhesion of strongly charged particles in solutions of electrolytes. *Acta Physicochim. URSS* **14**, 633–662.
76. E.J.W. Verwey, J.T.G. Overbeek (1948). *Theory of Stability of Lyophobic Colloids.* Elsevier, Amsterdam.
77. D.J. Müller, A. Engel, M. Amrein (1997). Preparation techniques for observation of native biological systems with atomic force microscopy. *Biosens. Bioelectron.* **8**, 867–877.
78. P. Wagner, M. Hegner, H.-J. Güntherodt, G. Semenza (1995). Formation and in-situ modification of monolayers chemisorbed on template-stripped gold surfaces for operations on nano-scale. *Langmuir* **11**, 3867–3875.
79. P. Wagner, M. Hegner, P. Kernen, F. Zaugg, G. Semenza (1995). N-hydroxysuccinimide ester functionalized self-assembled monolayers for covalent immobilization of biomolecules on gold. *Biophys. J.* **70**, 2052–2066.
80. M. Hegner, A. Engel (2002). Single molecule imaging and manipulation. *Chimia* **56**, 506–514.
81. S. Scheuring, D. Fotiadis, C. Möller, S.A. Müller, A. Engel, D.J. Müller (2001). Single proteins observed by atomic force microscopy. *Single Mol.* **2**, 59–67.
82. M.B. Viani, L.I. Pietrasanta, J.B. Thompson, A. Chand, I.C. Gebeshuber, J.H. Kindt, M. Richter, H.G. Hansma, P.K. Hansma (2000). Probing protein-protein interactions in real time. *Nat. Struct. Biol.* **7**(8), 644–647.
83. F. Oesterhelt, D. Oesterhelt, M. Pfeiffer, A. Engel, H.E. Gaub, D.J. Müller (2000). Unfolding pathways of individual bacteriorhodopsins. *Science* **288**, 143–146.
84. E.L. Florin, V.T. Moy, H.E. Gaub (1994). Adhesion forces between individual ligand-receptor pairs. *Science* **264**, 415–417.
85. D. Leckband, W. Müller, F.J. Schmitt, H. Ringsdorf (1995). Molecular mechanisms determining the strength of receptor-mediated intermembrane adhesion. *Biophys. J.* **69**, 1162–1169.
86. C.A. Helm, W. Knoll, J.N. Israelachivli (1991). Measurement of ligand-receptor interactions. *Proc. Natl. Acad. Sci. U.S.A.* **88**(18), 8169–8173.
87. D. Leckband, F.J. Schmitt, J. Israelachvili, W. Knoll (1994). Direct force measurements of specific and nonspecific protein interactions. *Biochemistry* **33**, 4611–4624.
88. D. Leckband, T.L. Kuhl, H.K. Wang, W. Müller, J. Herron, H. Ringsdorf (2000). Force probe measurements of antibody-antigen interactions. *Methods* **20**, 329–340.
89. S. Sivasankar, W. Brieher, N. Lavrik, B. Gumbiner, D. Leckband (1999). Direct molecular force measurements of multiple adhesive interactions between cadherin ectodomains. *Proc. Natl. Acad. Sci. U.S.A.* **96**(21), 11820–11824.
90. S. Sivasankar, B. Gumbiner, D. Leckband (2001). Molecular mechanism of cadherin binding. *Biophys. J.* **80**, 1758–1768.
91. D. Leckband, T.L. Kuhl, H.K. Wang, W. Müller, J. Herron, H. Ringsdorf (1995). 4-4–20 Anti-fluorescyl IgG Fab' recognition of membrane bound hapten: direct evidence for the role of protein and interfacial structure. *Biochemistry* **34**, 11467–11478.

92. C. Yeung, T. Purves, A.A. Kloss, T.L. Khul, S. Sligar, D. Leckband (1999). Cytochrome c recognition of immobilized, orientational variants of cytochrome b5: direct force and equilibrium binding measurements. *Langmuir* **15**, 6829–6836.
93. M. Carrion-Vasquez, A.F. Oberhauser, S.B. Fowler, P.E. Marszalek, S.E. Broedel, J. Clarke, J.M. Fernandez (1999). Mechanical and chemical unfolding of a single protein: A comparison. *Proc. Natl. Acad. Sci. U.S.A.* **96**(7), 3694–3699.
94. M. Rief, M. Gautel, F. Oesterhelt, J.M. Fernandez, H.E. Gaub (1999). Reversible unfolding of individual titin immunoglobulin domains by AFM. *Science* **276**, 1109–1112.
95. M.S. Kellermayer, S.B. Smith, H.L. Granzier, C. Bustamante (1997). Folding-unfolding transitions in single titin molecules characterized with laser tweezers. *Science* **276**, 1112–1116.
96. D. Leckband (2001). Force as a probe of membrane protein structure and function. *Curr. Opin. Struct. Biol.* **11**, 433–439.
97. G. Yang, C. Cecconi, W.A. Baase, I.R. Vetter, W.A. Breyer, J.A. Haack, B.W. Matthews, F.W. Dahlquist, C. Bustamante (2000). Solid-state synthesis and mechanical unfolding of polymers of T4 lysozyme. *Proc. Natl. Acad. Sci. U.S.A.* **97**, 139–144.
98. D.J. Müller, W. Baumeister, A. Engel (1999). Controlled unzipping of a bacterial surface layer with atomic force microscopy. *Proc. Natl. Acad. Sci. U.S.A.* **96**(23), 13170–13174.
99. A.F. Oberhauser, P.E. Marszalek, H.P. Erickson, J.M. Fernandez (1998). The molecular elasticity of tenascin, an extracellular matrix protein. *Nature* **393**, 181–185.
100. M. Rief, J. Pascual, M. Saraste, H.E. Gaub (1999). Force spectroscopy of spectrin: low unfolding forces in alpha helix bundles. *J. Mol. Biol.* **286**(2), 553–561.
101. P. Hinterdorfer, W. Baumgartner, H.J. Gruber, K. Schilcher, H. Schindler (1996). Detection and localization of individual antibody-antigen recognition events by atomic force microscopy. *Proc. Natl. Acad. Sci. U.S.A.* **93**(8), 3477–3481.
102. R. Merkel, P. Nassoy, A. Leung, K. Ritchie, E. Evans (1999). Energy landscapes of receptor-ligand bonds explored with dynamic force spectroscopy. *Nature* **397**, 50–53.
103. D. Anselmetti, J. Fritz, B. Smith (2000). Single molecule DNA biophysics with atomic force microscopy. *Single Mol.* **1**, 53–58.
104. M.B. Viani, L.I. Pietrasanta, J.B. Thompson, A. Chand, I.C. Gebeshuber, J.H. Kindt, M. Richter, H.G. Hansma, P.K. Hansma (2000). Probing protein-protein interactions in real time. *Nat. Struct. Biol.* **7**(8), 644–647.
105. C. Bustamante, C. Rivetti (1996). Visualizing protein-nucleic acid interactions on a large scale with the scanning force microscope. *Annu. Rev. Biophys. Biomol. Struct.* **25**, 395–429.
106. F. Oesterhelt, M. Rief, H.E. Gaub (1999). Single molecule force spectroscopy by AFM indicates helical structure of poly(ethylene-glycol) in water. *New J. Phys.* **1**, 6.1–6.11.
107. S.B. Smith, Y. Cui, C. Bustamante (1996). Overstretching B-DNA-The elastic response of individual dsDNA and ssDNA molecule. *Science* **271**, 795–799.
108. C. Bustamante, G. Zuccheri, S.H. Leuba, G. Yang, B. Samorì (1997). Visualization and analysis of chromatin by scanning force microscopy. *Methods: Companion Methods Enzymol.* **12**(1), 73–83.
109. G. Zuccheri, A. Scipioni, V. Cavaliere, G. Gargiulo, P. De Santis, B. Samorì (2001). Mapping the intrinsic curvature and flexibility along the DNA chain. *Proc. Natl. Acad. Sci. U.S.A.* **98**(6), 3074–3079.
110. J.A. Dagata, J. Schneir, H.H. Harary, C.J. Evans, M.T. Postek, J. Bennet (1990). Modification of hydrogen-passivated silicon by a scanning tunneling microscope operating in air. *Appl. Phys. Lett.* **56**, 2001–2003.

111. R. Garcia, M. Calleja, P. Pérez-Murano (1998). Local oxidation of silicon surfaces by dynamic force microscopy: Nanofabrication and water bridge formation. *Appl. Phys. Lett.* **72**, 2295–2297.
112. R. Held, T. Heinzel, P. Studerus, K. Ensslin, M. Holland (1997). Semiconductor quantum point contact fabricated by lithography with an atomic force microscope. *Appl. Phys. Lett.* **71**, 2689–2291.
113. M. Lazzarino, S. Heun, B. Ressel, K.C. Prince, P. Pingue, C. Ascoli (2002). Atomic force microscope anodic oxidation studied by spectroscopic microscopy. *Appl. Phys. Lett.* **81**, 2842–2844.
114. T.A. Jung, A. Moser, H.J. Hug, D. Brodbeck, R. Hofer, H.R. Hidber, U.D. Schwarz (1992). The atomic force microscope used as a powerful tool for machining surfaces. *Ultramicroscopy* **42–44**, 1446–1451.
115. L.L. Sohn, R.L. Willet (1995). Fabrication of nanostructures using atomic-force-microscope-based lithography. *Appl. Phys. Lett.* **67**(11), 1552–1554.
116. H. Gobel, P. Von Blanckenhagen (1995). Atomic force microscope as a tool for metal surface modifications. *J. Vac. Sci. Technol. B* **13**(3), 1247–1251.
117. T. Sumonogi, T. Endo, K. Kuwahara, R. Kaneko (1995). Nanoscale layer removal of metal surfaces by scanning probe microscope scratching. *J. Vac. Sci. Technol. B* **13**(3), 1257–1260.
118. P. Pingue, M. Lazzarino, F. Beltram, C. Cecconi, P. Baschieri, C. Frediani, C. Ascoli (1997). Fabrication of hybrid superconductor–semiconductor nanostructures by integrated ultraviolet-atomic force microscope lithography. *J. Vac. Sci. Technol. B* **15**(4), 1398–1401.
119. R.D. Piner, J. Zhu, F. Xu, S. Hong, C.A. Mirkin (1999). "Dip-Pen" nanolithography. *Science* **283**, 661–663.
120. L.M. Demers, D.S. Ginger, S.-J. Park, Z. Li, S.-W. Chung, C.A. Mirkin (2002). Direct patterning of modified oligonucleotides on metals and insulators by dip-pen nanolithography. *Science* **296**, 1836–1838.
121. H.J. Mamin, D. Rugar (1992). Thermomechanical writing with an atomic force microscope tip. *Appl. Phys. Lett.* **61**, 1003–1005.
122. E. Betzig, A. Harootunian, A. Lewis, M. Isaacson (1986). Near-field diffraction by a slit: implications for superresolution microscopy. *Appl. Opt.* **25,** 1890–1900.
123. M. Allegrini, C. Ascoli, A. Gozzini (1971). Measurements of changes in length by an inhomogeneous wave device. *Opt. Commun.* **2**, 435–437.
124. S. Zhu, A.W. Yu, D. Hawley, R. Roy (1986). Frustrated total internal reflection: A demonstration and review. *Am. J. Phys.* **54**, 601–606.
125. R.C. Reddick, R.J. Warmack, T.L. Ferrell (1989). New form of scanning optical microscopy. *Phys. Rev. B* **39**, 767–770.
126. Y. Suh, R. Zenobi (2000). Improved probes for scanning near-field optical microscopy. *Adv. Mater.* **12**, 1139–1142.
127. R. Eckert, J. Freyland, H. Gersen, H. Heinzelmann, G. Schurmann, W. Noell, U. Staufer, N. De Rooij (2001). Near-field optical microscopy based on microfabricated probes. *J. Microsc.* **202**, 7–11.
128. J.T. Krug II, E.J. Sánchez, X.S. Xie (2002). Design of near-field probes with optimal field enhancement by finite difference time domain electromagnetic simulation. *J. Chem. Phys.* **116**, 10895–10901.
129. N.F. Van Hulst, M.H.P. Moers, O.F.J. Noordman, R.G. Tack, F.B. Segerink, B. Bolger (1993). Near-field optical microscope using a silicon-nitride probe. *Appl. Phys. Lett.* **62**, 461–463.

130. J.P. Fillard (1996). Near field optics and nanoscopy. *World Sci. Publish.* (438 pp.) Singapore.
131. X. Bouju, A. Dereux, J.P. Vigneron, C. Girard (1996). Scattering of electromagnetic waves by silicon-nitride tips. *J. Vac. Sci. Technol. B* **14**, 816–819.
132. H.G. Frey, F. Keilmann, A. Kriele, R. Guckenberger (2002). Enhancing the resolution of scanning near-field optical microscopy by a metal tip grown on an aperture probe. *Appl. Phys. Lett.* **81**, 5030–5032.
133. P. Minh, O. Takahito, E. Masayoshi (2002). *Fabrication of Silicon Microprobes for Optical Near-Field Applications* (192 pp). CRC Press, Boca Raton, FL.
134. F. Zenhausern, M.P. O'Boyle, H.K. Wickramasinghe (1994). Apertureless near-field optical microscope. *Appl. Phys. Lett.* **65**, 1623–1625.
135. T.J. Yang, G.A. Lessard, S.R. Quake (1999). An apertureless near-field microscope for fluorescence imaging. *Appl. Phys. Lett.* **76**, 378–380.
136. M. Labardi, S. Patanè, M. Allegrini (2000). Artifact-free near-field optical imaging by apertureless microscopy. *Appl. Phys. Lett.* **77**, 621–623.
137. J. Michaelis, C. Hettich, J. Mlynek, V. Sandoghdar (1999). A single molecule as a probe of optical intensity distribution. *Opt. Lett.* **24**, 581–583.
138. J. Michaelis, C. Hettich, J. Mlynek, V. Sandoghdar (2000). Optical microscopy using single-molecule light source. *Nature* **405**, 325–327.
139. C. Girard, D. Courjon (1990). Model for scanning tunneling optical microscopy: a microscopic self-consistent approach. *Phys. Rev. B* **42**, 9340–9349.
140. E. Betzig, P.L. Finn, J.S. Weiner (1992). Combined shear force and near-field scanning optical microscopy. *Appl. Phys. Lett.* **60**, 2484–2486.
141. J. Barenz, O. Hollricher, O. Marti (1996). An easy to use non optical shear force distance control for near field optical microscopes. *Rev. Sci. Instrum.* **67**, 1912–1916.
142. J-K Leong, C.C. Williams (1996). Shear force microscopy with capacitance detection for near-field scanning optical microscopy. *Appl. Phys. Lett.* **66**, 1432–1435.
143. W.A. Atia, C.C. Davis (1997). A phase-locked shear-force microscope for distance regulation in near-field optical microscopy. *Appl. Phys. Lett.* **70**, 405–407.
144. J. Salvi, P. Chevassus, A. Mouflard, S. Davy, M. Spajer, D. Courjon, K. Hjort, L. Rosengren (1998). Piezoelectric shear force detection: a geometry avoiding critical tip/tuning fork gluing. *Rev. Sci. Instrum.* **69**, 1744–1746.
145. A. Mannoni, F. Quercioli, B. Tiribilli, C. Ascoli, P. Baschieri, C. Frediani (1998). Measuring topography and refractive index of channel waveguides with an Hybrid AFM-SNOM. *J. Lightwave Technol.* **16**, 388–394.
146. C.D. Savio (Sept. 1999). Tesi di laurea in fisica: Misure di caratteristiche ottiche di guide canale con un microscopio ibrido AFM-PSTM.
147. C. Dal Savio, H. Wolff, T. Dziomba, H.A. Fuss, H.U. Danzenbrink (2002). A compact sensor head for simultaneous scanning force and near field optical microscopy. *J. Int. Soc. Precision Eng. Nanotech.* **26**, 199–203.
148. H.U. Danzebrink, D.V. Kazantsev, C. Dal Savio, K. Pierz, B. Guttler (2003). Optical microscope with SNOM option for micro- and nanoanalytical investigations at low temperatures. *Appl. Phys. A* **76**, 889–892.
149. V. Sandoghdar (2001). Trends and developments in scanning near-field optical microscopy. In: M. Allegrini, N. Garcia, O. Marti (Eds), *Proceedings of Int. School Enrico Fermi course CXLIV* (pp. 65–202).
150. G. Kaupp, A. Herrmann, M. Haak (1999). Artifacts in scanning near-field optical microscopy (SNOM) due to deficient tips. *J. Phys. Org. Chem.* **12**, 797–807.

151. P.G. Gucciardi, M. Colocci (2001). Different contrast mechanisms induced by topography artifacts in near-field optical microscopy. *Appl. Phys. Lett.* **79**, 1543–1545.
152. D.W. Pohl, W. Denk, M. Lanz (1984). Optical stethoscopy: image recording with resolution lambda/20, *Appl. Phys. Lett.* **44**, 651–653.
153. E. Betzig, J.K. Trautman, T.D. Harris, J.S. Weiner, R.L. Kostelak (1991). Breaking the diffraction barrier: Optical microscopy on a nanometric scale. *Science* **251**, 1468–1470.
154. P. Hoffmann, B. Dutoit, R.P. Salathé (1995). Comparison of mechanically drawn and protection layer chemically etched optical fiber tips. *Ultramicroscopy* **61**, 165–170.
155. E. Betzig, P.L. Finn, J.S. Weiner (1992). Combined shear force and near-field scanning optical microscopy. *Appl. Phys. Lett.* **60**, 2484–2486.
156. R. Toledo-Crow, P.C. Yang, Y. Chen, M. Vaez-Iravani (1992). Near-field differential scanning optical microscope with atomic force regulation. *Appl. Phys. Lett.* **60**, 2957–2959.
157. B. Hecht, H. Bielefeldt, Y. Inouye, D.W. Pohl, L. Novotny (1997). Facts and artifacts in near-field optical microscopy. *J. Appl. Phys.* **81**, 2492–2498.
158. A. Madrazo, R. Carminati, M. Nieto-Vesperinas, J.-J. Greffet (1998). Polarization effects in the optical interaction between a nanoparticle and a corrugated surface: implications for apertureless near-field microscopy. *J. Opt. Soc. Am. A* **15**, 109–119.
159. P.J. Valle, J.-J. Greffet, R. Carminati (1999). Optical contrast, topographic contrast and artifacts in illumination-mode scanning near-field optical microscopy. *J. Appl. Phys.* **86**, 648–656.
160. J.N. Walford, J.A. Porto, R. Carminati, J.-J. Greffet, P.M. Adam, S. Hudlet, J.-L. Bijeon, A. Stashkevich, P. Royer (2001). The influence of tip modulation in image formation in scanning near-field optical microscopy. *J. Appl. Phys.* **89**, 5159–5169.
161. V. Sandoghdar, S. Wegscheider, G. Krausch, J. Mlynek (1997). Reflection scanning near-field optical microscopy with uncoated fiber tips: How good is the resolution really? *J. Appl. Phys.* **81**, 2499–2503.
162. M. Labardi, S. Patanè, M. Allegrini (2000). Artifact-free near field optical imaging by apertureless microscopy. *Appl. Phys. Lett.* **77**, 621–623.
163. K.D. Weston, J.A. DeAro, S.K. Buratto (1996). Near-field scanning optical microscopy in reflection: a study of far-field collection geometry effects. *Rev. Sci. Instrum.* **67**, 2924–2929.
164. B. Hecht, H. Bielefeldt, D.W. Pohl, L. Novotny, H. Heinzelmann (1998). Influence of detection conditions on near-field optical imaging. *J. Appl. Phys.* **84**, 5873–5882.
165. T. David, C. Chicanne, N. Richard, J.R. Krenn, F. Scheurer, K. Ounadjela, M. Hehn, Y. Lacroute, J.-P. Goudonnet (1999). Application to dielectric, metallic and magnetic samples of a transmission mode scanning near field optical microscope with normal force distance regulation on bent optical fibers, *Rev. Sci. Instrum.* **70**, 4587–4594.
166. R. Brunner, A. Bietsch, O. Hollricher, O. Marti (1997). Distance control in near-field optical microscopy with piezo electrical shear force detection suitable for imaging in liquids. *Rev. Sci. Instrum.* **68**, 1769–1772.
167. M. Labardi, P.G. Gucciardi, M. Allegrini, C. Pelosi (1998). Assesment of NSOM resolution on III-V semiconductor thin films. *Appl. Phys. A: Mater. Sci. Process.* **66**, S397–S402.
168. P.G. Gucciardi, M. Colocci, M. Labardi, M. Allegrini (1999). Thermal expansion effects in near field optical microscopy fiber probes induced by laser absorption. *Appl. Phys. Lett.* **75**, 3408–3410.

169. T. Yanagida, E. Tamiya, H. Muramatsu, P. Degenaar, Y. Ishii, Y. Sako, K. Sako, S. Ohta-Iino, S. Ogawa, G. Marriot, A. Kusumi, H. Tatsumi (2002). Near-field microscopy for biomolecular systems. In: S. Kawata, M. Ohtsu, M. Irie (Eds), *Nano-Optics* (pp. 191–236). Springer-Verlag, Berlin, Heidelberg, New York.
170. M. Fujihira, A. Takahara, O. Karthaus, S. Okazaki, K. Kajikawa (2002). Near-field imaging of molecules and thin films. In: S. Kawata, M. Ohtsu, M. Irie (Eds), *Nano-Optics* (pp. 151–190). Springer-Verlag, Berlin, Heidelberg, New York.
171. J.M. Kim, T. Ohtani, S. Sugiyama, T. Hirose, H. Muramatsu (2001). Simultaneous topographic and fluorescence imaging of single DNA molecules for DNA analysis with a scanning near-field optical/atomic force microscope. *Anal. Chem.* **73**, 5984–5991.
172. R. Cingolani, G. Bastard, M. Labardi, F. Fuso, M. Allegrini, L. Sorba, L. Vanzetti, A. Franciosi (1999). Spatially resolved near field photoluminescence spectroscopy of II-VI quantum well: The role of localized excitons. *J. Appl. Phys.* **86**, 6793–6797.
173. X.K. Orlik, M. Labardi, M. Allegrini (2000). Nanometer scale observation ferroelectric domains using an apertureless near-field optical microscope. *Appl. Phys. Lett.* **77**, 2042–2044.
174. L.F. Lee, A. Adonov, R.D. Schaller, J.M. Frechet, R.J. Saykally (2003). Intermolecular coupling in nanometric domains of light-harvesting dendrimer films studied by photoluminescence near-field scanning optical microscopy (PL SNOM). *J. Am. Chem. Soc.* **125**, 536–540.
175. N.M. Haralampus-Grynaviski, L.E. Lamb, J.D. Simon, J.R. Krogmeier, R.C. Dunn, A. Pawlak, M. Rozanowska, T. Sarna, J.M. Burke (2001). Probing the spatial dependence of the emission spectrum of single human retinal lipofuscin granules using near-field scanning optical microscopy. *Photochem. Photobiol.* **74**, 364–368.
176. T. Ha, T. Enderle, D.F. Ogletree, D.S. Chemla, P.R. Selvin, S. Weiss (1996). Probing the interaction between two single molecules: fluorescence resonance energy transfer between a single donor and a single acceptor. *Proc. Natl. Acad. Sci. U.S.A.* **93**, 6264–6268.
177. T. Enderle, T. Ha, D.F. Ogletree, D.S. Chemla, C. Magowan, S. Weiss (1997). Membrane specific mapping and colocalization of malarial and host skeletal proteins in the Plasmodium falciparum infected erythrocyte by dual-color near-field scanning optical microscopy. *Proc. Natl. Acad. Sci. U.S.A.* **94**, 520–525.
178. J. Hwang, L.A. Gheber, L. Margolis, M. Edidin (1998). Domains in cell plasma membranes investigated by near-field scanning optical microscopy. *Biophys. J.* **74**, 2184–2190.
179. H. Shiku, R.C. Dunn (1999). Domain formation in thin lipid films probed with near-field scanning optical microscopy. *J. Microsc.* **194**, 455–460.
180. H. Shiku, R.C. Dunn (1999). Near-field scanning optical microscopy studies of L-alpha-dipalmitoylphosphatidylcholine monolayers at the air-liquid interface. *J. Microsc.* **194**, 461–466.
181. F. H'dhili, R. Bachelot, A. Rumyantseva, G. Lerondel, P. Royer (2003). Nano-patterning photosensitive polymers using local field enhancement at the end of apertureless SNOM tips. *J. Microsc.* **209**, 214–222.
182. R.S. Decca, H.D. Drew, B. Maiorov, J. Guimpel, E.J. Osquiguil (1999). Inducing superconductivity at a nanoscale: photodoping with a near-field scanning optical microscope. *J. Microsc.* **194**, 407–411.
183. A. Hartschuh, E.J. Sanchez, X. Sunney Xie, L. Novotny (2003). High-resolution near-field Raman microscopy of single-walled carbon nanotubes. *Phys. Rev. Lett.* **90**.

Chapter 16

Confocal and multiphoton microscopy

Alberto Diaspro

Table of contents

Abstract

Confocal and multiphoton excitation microscopy, conceived over twenty and ten years ago, respectively, are increasingly important tools for fluorescence imaging. Notwithstanding this they can be still considered of at an infant stage. Multiphoton excitation (MPE) microscopy takes advantages over both wide field and confocal laser scanning microscopy (CLSM), for the study of the three-dimensional (3D) and dynamic properties of biological systems. The development of mode-locked lasers matched with the increased dissemination of confocal laser scanning microscopes favoured an incredible increase of demands for developing MPE architectures. A simple, fast and comparatively economical method for implementing a two- or multi-photon excitation microscope is still to modify a commercial CLSM, realizing a carefully tailored solution for particular requirements of each application. Three-dimensional fluorescence imaging, photo-induced uncaging of compounds and highly localised photochemistry and micropattering within living cells and tissues are some of the relevant applications in life sciences for CLSM and MPE microscopy methods. Theoretical and experimental bases as well as applications in biology are outlined in this chapter.

16.1 Introduction

There have been various reasons for the continuing growth of interest in optical microscopy during the last few years despite the low resolution with respect to modern scanning probe or electron microscopy [1]. The main reason is that optical microscopy can be still considered unique in its ability to allow the examination of biostructures in a hydrated state in living samples or under experimental conditions that are very close to living or physiological states. This evidence, coupled to the advent of fluorescence labelling, permits the study of the complex and delicate relationships between structure and function in biological systems [2–5].

As recently reminded by Colin Sheppard [6], a microscope is an instrument magnifying objects by means of a specific interaction-more commonly by means of lenses-so as to capture details invisible to the naked eye. Invisible details are not only those related to physical dimensions of the objects. In fact, the so-called contrast mechanisms play a relevant role and increase in number, day by day, due to new inventions around optical microscopy. Inventions in microscopy, stimulated by the needs of scientists, and technology contributed to the evolution of the microscope in its very different modern forms, including different contrast modalities for producing images [1,3,7–9]. Although all far field light microscopes, including conventional, confocal and two-photon microscopes, are limited in the achievable diffraction-limited resolution [10], light microscopy still occupies a unique niche. Its favourable position, especially for applications in medicine and biology, comes from the peculiar ability to image living systems at relatively good spatial resolution. In addition, three-dimensional (3D) optical methods have been widespread since the 1970s [11–19]. During the last 10 years, confocal microscopes have proved to be extremely useful research tools, notably in the life sciences.

Nowadays, optical microscopy has evolved to 3D (*x*–*y*–*z*) and 4D (*x*–*y*–*z*–*t*) analysis allowing researchers to probe even deeper into the intricate mechanisms of living systems [20–25]. Within this scenario, two-photon excitation (TPE) microscopy [26–29], or in general multi-photon excitation (MPE) microscopy [30,31], is probably the most relevant advance in fluorescence optical microscopy since the introduction of confocal imaging in the 1980s [17,21,23,25,32,33]. Notably, its fast and increasing spread has been strongly influenced by the availability of ultrafast pulsed lasers [34,35]; the advances in fluorescence microscopy can be also ascribed to the availability of efficient and specific fluorophores [36–38].

MPE microscopy has a three-dimensional intrinsic ability coupled with five other interesting capabilities. First, MPE greatly reduces photo-interactions and allows imaging of living specimens over long time periods. Second, MPE operates in a high-sensitivity background free acquisition scheme. Third, MPE microscopy can image turbid and thick specimens down to a depth of a few hundreds micrometers. Fourth, MPE allows simultaneous excitation of different fluorescent molecules, reducing colocalization errors. Fifth, TPE can prime photochemical reactions within a subfemtolitre volume inside solutions, cells and tissues. MPE fluorescence microscopy is not only revolutionary in its ability to provide the above mentioned features, together with other practical advantages, but also in its elegance and effectiveness of application of quantum physics [39–41].

Furthermore, this form of non-linear microscopy also favoured the development and application of several investigation techniques starting from two-photon excitation microscopy [42], namely: three-photon excited fluorescence [43,44] second harmonic generation [25,45–48], third-harmonic generation [49,50] fluorescence correlation spectroscopy [51–53] image correlation spectroscopy [54,55], lifetime imaging [30,56–59], single molecule detection schemes [60–66], photodynamic therapies [67] and others [24,68–71].

16.2 Short historical notes

Rough magnifying glasses were used in ancient times, but the evolution of modern microscopes started in the 17th century. Although the first compound microscope was built by Hans and Zacharias Janssen in 1595, Antoni van Leeuwenhoek (1632–1723) managed to make lenses good enough to enable the amazing magnification of about 300× in their very simple microscopes. Thanks to the suggestions of the scientist Robert Hook (1635–1702), around 1670 the instrument maker Christopher Cock built, in London, a very successful compound microscope. Through this instrument Hooke was able to observe cells [71]. Hook's microscope can be regarded as the father of modern instruments. Since Hooke's ornate microscopes [72] and van Leeuwenhoek's single lens magnifiers [73] (Figure 1), the optical microscope has undergone a secure and continuous evolution marked by relevant and revolutionary passages in the last 350 years.

One recent relevant step was the invention of the confocal microscope in its different solutions. In 2000 the Optical Society of America honoured, with the R.W. Wood Prize, Paul Davidovits, M. David Egger, and Marvin Minsky for

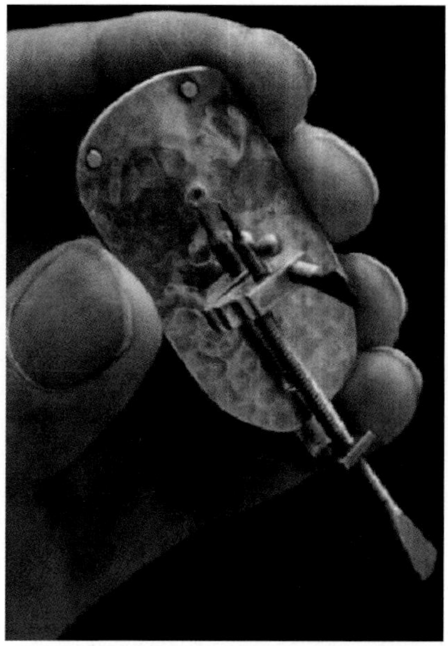

Figure 1. Photo of a replica of the historical van Leuwenhoek's microscope.

"seminal contributions to confocal microscopy". In fact, Minsky in 1957 [74] invented a confocal microscope identical to the concept later developed extensively by Egger and Davidovits at Yale [75,76] and by Sheppard and Wilson in Oxford [16,17,77] and Brakenhoff and colleagues in Amsterdam [15,78]. As reported in the paper "Memoir of an invention" [74], "the circumstances are also remarkable in that Minsky only published his invention as a patent. Yet he not only built a microscope and made it work and it was the kind of prototype of which we would be proud but he showed it to a number of people who went away impressed but nevertheless failed to adopt the concept." It was not until the end of the 1970s, with the advent of affordable computers and lasers, and the development of digital image processing software, that the first single-beam confocal laser scanning microscopes became available in a number of laboratories and were applied to biological and materials specimens. A new revolution followed those times [79]: MPE microscopy. The MPE story, more specifically TPE, dates back to 1931 and its roots are in the theory originally developed by Maria Göppert-Mayer (1931) [80] (Figure 2). The keystone of TPE theory principle is in the prediction that one atom or molecule can simultaneously absorb two-photons in the very same quantum event.

Now, to realise the rarity of the event, one should consider that the adverb "simultaneously" here implies "within a temporal window of 10^{-16}–10^{-15} s": in bright daylight a good one- or two- photon excitable fluorescent molecule absorbs a photon through a one-photon interaction about once a second and a photon pair by two-photon simultaneous interaction every 10 million years [81]. To increase the probability of the event a very high density of photons is needed, i.e. a laser source.

Figure 2. Maria Goppert-Mayer (middle) and friends.

As in the confocal microscope case, the laser is the key for the development and dissemination of the technique. In fact, it was only in the 1960s, after the development of the first laser sources [35,82], that it was possible to find experimental evidence of Maria Göppert Mayer's prediction. Kaiser and Garret (1961)[83] reported two-photon excitation of fluorescence in CaF_2:Eu^{2+} and Singh and Bradley (1964) [84] were able to estimate the three-photon absorption cross-section for naphthalene crystals. These two results consolidated other related experimental achievements obtained by Franken et al. (1961) [85] of second harmonic generation in a quartz crystal using a ruby laser. Later, Rentzepis and colleagues (1970) [86] observed three-photon excited fluorescence from organic dyes, and Hellwarth and Christensen (1974) [87] collected second-harmonic generation signals from ZnSe polycrystals at microscopic level. In 1976, Berns reported about a probable two-photon effect as a result of focusing an intense pulsed laser beam onto chromosomes of living cells [88] and such interactions form the basis of modern nanosurgery [89]. However, the original idea of generating 3D microscopic images by means of such non-linear optical effects was first suggested and attempted in the 1970s by Sheppard, Kompfner, Gannaway and Choudhury at Oxford [45,79,90]. Notably, for many years, the application of two-photon absorption was mainly related to spectroscopic studies [91–94]. The real "TPE boom" took place at the beginning of the 1990s at the W.W. Webb laboratories (Cornell University, Ithaca, NY). However, it was the excellent and effective work done by Winfried Denk and colleagues (1990) [26] that produced the major impact for spreading the technique and that revolutionised fluorescence microscopy imaging. The potentiality of two-photon excited fluorescence imaging in a scanning microscope was rapidly coupled with the availability of ultrafast pulsed lasers. It was the development of mode-locked lasers, providing high peak power femtosecond pulses with a repetition rate around 100 MHz [34,82,95–97], that made possible in practice the fast dissemination of TPE laser scanning microscopy and the flourishing of related techniques in a sort of avalanche effect [24,25,27,28,30,31,68,98].

16.3 Basic principles on confocal and two-photon excitation of fluorescent molecules

16.3.1 Fluorescence and optical sectioning

Fluorescence optical microscopy is very popular for imaging in biology since fluorescence is highly specific either as exogenous labelling and endogenous autofluorescence [2,3,24]. Fluorescent molecules allow both spatial and functional information to be obtained through specific absorption, emission, lifetime, anisotropy, photodecay, diffusion and other contrast mechanisms [25,48]. This means that one can efficiently study, for example, the distribution of proteins, organelles and DNA as well as ion concentration, voltage and temperature within living cells [99–101]. Two-photon excitation of fluorescent molecules is a non-linear process related to the simultaneous absorption of two photons whose total energy equals the energy required for conventional, one-photon, excitation [94,102,103]. In any case the energy required to prime fluorescence is the one sufficient to produce a molecular transition to an excited electronic state. Conventional techniques for fluorescence excitation use ultraviolet (UV) or visible radiation and excitation occurs when the absorbed photons are able to match the energy gap to the excited state. Then the excited fluorescent molecules decay to an intermediate state, giving off a photon of light having an energy lower than the one needed to prime excitation. This means that the energy (E) provided by photons should equal the molecule energy gap (ΔE_g), and considering the relationship between photon energy (E) and radiation wavelength (λ) it follows that

$$\Delta E_g = E = hc/\lambda \tag{1}$$

where $h = 6.6 \times 10^{-34}$ Js is Planck's constant and $c = 3 \times 10^8$ ms^{-1} is the speed of light (considered in a vacuum and at reasonable approximation). Due to energetic aspects, the fluorescence emission is shifted towards a longer wavelength than the one used for excitation. This shift typically ranges from 50 to 200 nm [102,104]. For example, a fluorescent molecule that absorbs one photon at 340 nm, in the ultraviolet region, exhibits fluorescence at 420 nm in the blue region, as sketched in Figure 3.

The possibility of a three-dimensional reconstruction of the distribution of fluorescence within a three-dimensional object like a living cell starting from the acquisition of the two-dimensional distribution of specific intensive properties, for example the fluorescence emitted from chromophores biochemically linked to specific components of the object under study, is one of the most powerful properties of the optical microscope. In fact, this allows complete morphological analysis through volume rendering procedures [101,105] of living biological

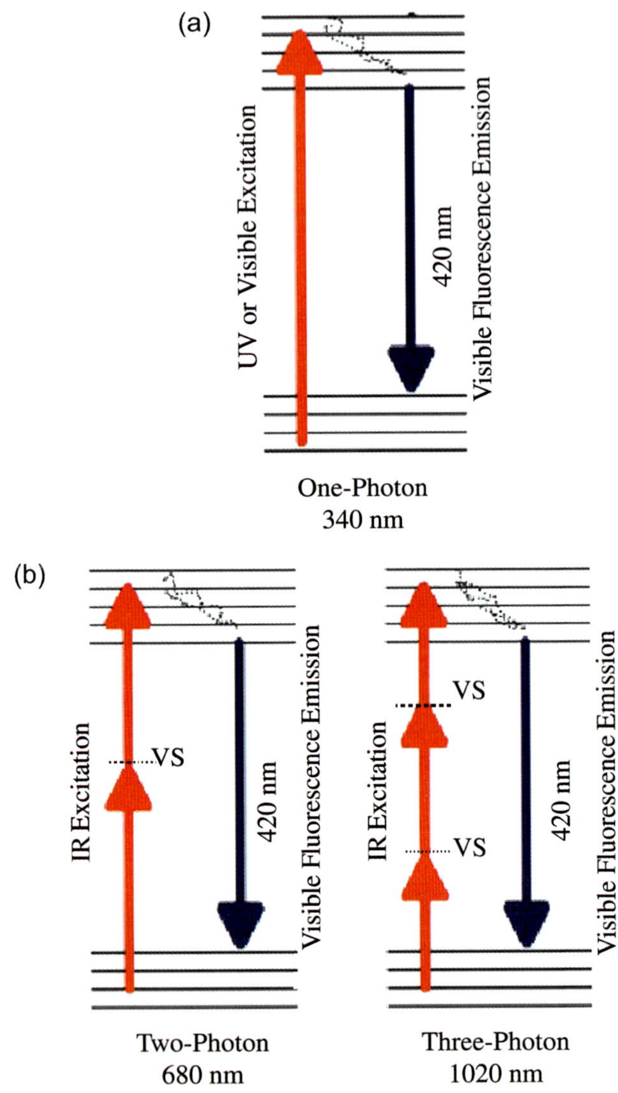

Figure 3. Perrin–Jablonski scheme for conventional (a) and multi-photon (b) excitation of fluorescence. A fluorescent molecule that needs 340 nm photons to be excited can be brought to the excited state within a two- or three-photon scheme at 680 and 1020 respectively. There are no appreciable changes in the emission of fluorescence (all three cases vignette to 420 nm).

specimens where the opportunity of optical slicing allows information to be obtained from different planes of the specimen without being invasive, thus preserving structures and functionality of the different parts [11,12,106].

To image optical slices from a three-dimensional object the so-called optical sectioning technique is used. It is essentially based on a fine z-stepping either of the

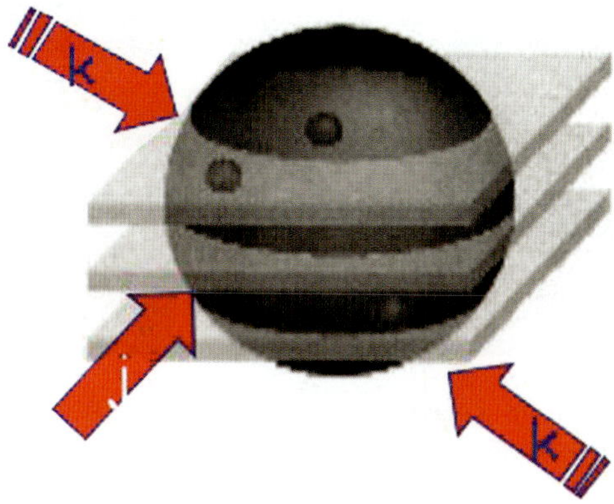

Figure 4. Optical sectioning: current optical plane j affected by information coming from the k-adjacent planes of a thick three-dimensional real sample.

objective or of the sample stage, coupled with the usual x–y image capturing. The synchronous x–y–z scanning allows the collection of a set of two-dimensional images, which are somehow affected by signal cross-talk from other planes from the sample (Figure 4)

The observed image O at a plane j is produced by the true fluorescence distribution I at plane j, distorted by the microscope through S, plus contributions from adjacent k planes and noise N:

$$O_j = I_j S_j + \Sigma_{k \neq j} I_k S_k + N$$

This means that when a set of two-dimensional images is acquired at various focus position and under certain conditions, in principle one can recover the 3D shape of the object, described by the intensive parameter I, by solving the above set of equations. This relationship, usually reported in the Fourier frequency domain [14], links the physical procedure of optical sectioning to some mathematics. In practice, one wants to find the best estimate, according to some criterion, of I through knowledge of the observed images, the distortion or point spread function (PSF) of the image formation system and the additive noise within a restoration scheme classical for space invariant linear systems [14,107,108]. Figure 5 shows an example of digital restoration of microscopic data obtained after solving the appropriate set of equations.

So far, this can be computationally done starting from any data set of optical slices. Recently a WWW service has become available, named Power-Up-Your-Microscope that produces the best estimate of I according to the acquired data set of optical slices. Interested readers can find information and check the service for free through the webpage http://www.powermicroscope.com.

In the next sections it will be evident how part of such image processing can be optically perfomed using some physical stratgems. The first is known as confocal imaging. The second is referred to as multiphoton or two-photon imaging.

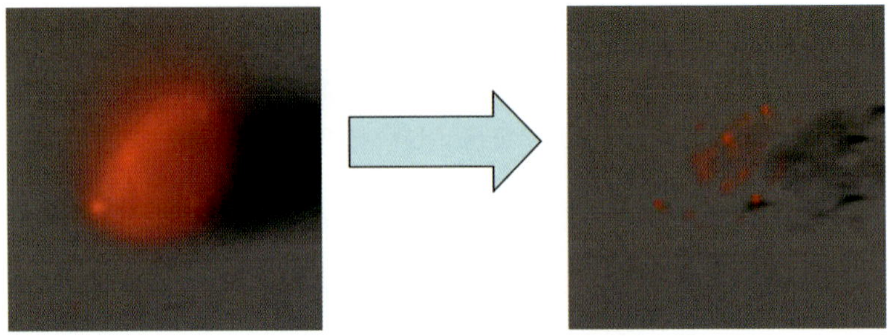

Figure 5. Blurred image from a stack of optical sections (left) as basis for recovery of 3D details visible (right) after computing the best estimate of the intensity distribution within the acquired frame set (www.svi.nl).

16.3.2 Confocal scheme

As reported by Minsky (1961) [74] an ideal microscope would examine each point of the specimen and measure the amount of light scattered, absorbed or emitted by that point, excluding contributions from other parts of the sample from the actual or from adjacent planes. Unfortunately, if we try to obtain images by making many such measurements at the same time then every focal image point will be clouded by aberrant rays of scattered light deflected at points of the specimen that are not the point we are interested looking at, as shown previously. Most of those extra rays would disappear if we could illuminate only one specimen point at a time. There is no way to eliminate every such possible ray, because of multiple scattering, but it is comparatively straightforward to remove all rays not initially aimed at the focal point; using a sort of second microscope (instead of a condenser lens) to image a pinhole (a small aperture in an opaque screen) aperture on a single point of the specimen. This reduces the amount of light in the specimen by orders of magnitude without reducing the focal brightness at all. Even under this condition, some of the initially focused light will be scattered by out-of-focus specimen points onto other points in the image plane, affecting the clarity of the final acquisition, i.e. of the observed image *O*. But we can reject undesired rays, as well, by placing a second pinhole aperture in the image plane that lies beyond the exit side of the objective lens. We end up with an elegant, symmetrical geometry: a pinhole and an objective lens on each side of the specimen. This leads to a situation where one has used two lenses on both excitation and detection sides of the microscope, combining two lenses in a unique effect. The final effect is a better resolving power along the three dimensions. This can be demonstrated by imaging two bright spots positioned close to the optical resolution of the system. Figure 6 shows the resulting images in the distinct cases of a conventional microscope and of a microscope using the above-mentioned stratagem. The price of such a single-point illumination with respect to the classical "one-shot" wide-field illumination-detection case is being able to measure only one point at a time. This is why a confocal microscope must scan the specimen, point by point, which can take a long time because we must add all the time intervals it takes to collect enough light to measure each image point.

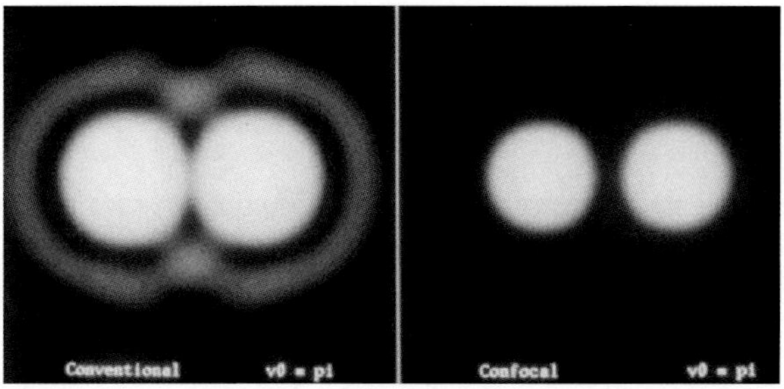

Figure 6. Two point objects at close distance as imaged by conventional (left) and confocal (right) microscope. The superior resolution of the confocal system is apparent [109].

That amount of time was reduced by using a bright light, i.e. with the advent of laser sources. Unfortunately, the final image cannot be viewed directly by eye but has to be built point by point on a computer screen. This is a common fate and drawback of all scanning systems; fast or real time scanning systems are outside the scope of this chapter [1]. Moving the stage with the sample or scanning the beam on the sample was the dilemma. When moving the specimen, the lenses of such a system need to be corrected only for the family of rays that intersect the optical axis at a single focal point. Scanning the beam is far more practical and widely used in most modern architectures. However, this "trick" brings fluorescence microscopy into the domain of three-dimensional fluorescence microscopy as well as computational optical sectioning. In such a classical three-dimensional fluorescence optical microscope the fluorescence process is such that the excitation photons are focused into a diffraction-limited spot scanned on the specimen [17,33]. The three-dimensional ability, i.e. the confocal effect, is obtained by confining both the illuminated focal region and the detected area of the emitted light [6,109]. So far, the light emitted from the specimen is imaged by the microscope objective lens into the image plane. Here a circular aperture (pinhole) is placed in front of a light detector (Figure 7). This pinhole is responsible for rejection of the axial out-of-focus light and of the lateral overlapping diffraction patterns. This produces an improvement of spatial resolution by a factor 1.4 along each direction, resulting in a volume selectivity 2.7 times better than in the wide-field case [15,17,27,110,111]. Besides, it is the physical suppression of the contributions from out of focus layers to image formation that produces the so-called optical sectioning effect. Now, the observed image O at a plane j is produced by the true fluorescence distribution I at plane j, distorted by the microscope through S, plus noise N, if considering an ideal situation where contributions from adjacent k planes can be set to zero: $O_j = I_j S_j + N$.

Image restoration is needed only for correcting from PSF distortions that are less than in the conventional case. Unfortunately, a drawback occurs. In fact, during the excitation process of the fluorescent molecules the whole thickness of the specimen is harmed by every scan, within a hourglass shaped region [13]. This means that even though out-of-focus fluorescence is not detected, it is generated with the negative effect of potential induction of those photobleaching and

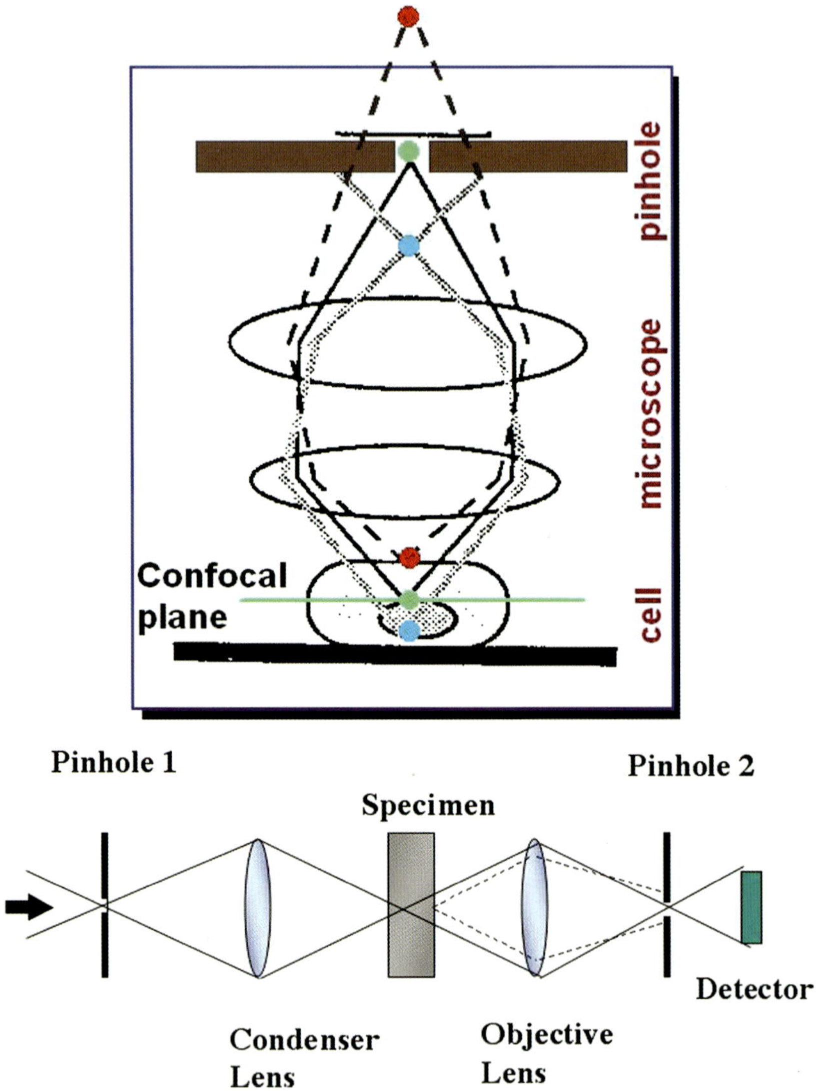

Figure 7. Confocal set-up (above, after Perkin Elmer) and related optical scheme (below). See text for detail.

phototoxicity phenomena previously mentioned. This is particularly serious when there is the need for three-dimensional and temporal imaging coupled to the use of fluorochromes that require excitation in the ultraviolet regime [42,112]. As reported earlier by König and colleagues [113], even using UVA (320–400 nm) photons may modify the biological system activity. DNA breaks, giant cell production and cell death can be induced at radiant exposures of the order of magnitude of $\approx$J/cm^{-2} accumulable during 10 scans with 5 μW laser scanning beam at approximately 340 nm and a 50 μs pixel dwell time. In this context, two-photon excitation of fluorescent molecules provides an immediate practical advantage over confocal microscopy [26,68,109,114–120]. In fact, reduced overall photobleaching and photodamage is generally acknowledged as one of the major advantages of two-photon excitation in laser scanning microscopy of biological specimens [81,121,122]. However, excitation intensity has to be kept low, considering as normal operation mode a regime under 10 mW of average power. When the laser power is increased above 10 mW some nonlinear effects might arise, evidenced through abrupt rising of the signals [123]. Moreover, photo-thermal effects could be induced, especially when focusing on single molecule detection schemes [124].

16.3.3 Multi-photon excitation

We now move from conventional excitation of fluorescence such as the one used in computational optical sectioning and confocal microscopy to a special case of multiphoton excitation, i.e. two-photon excitation. All considerations can be easily extended to the MPE. The physical suppression of contributions from adjacent planes is realized in a completely different way, thus moving again to three-dimensional optical sectioning abilities.

In TPE, two low-energy photons are involved in the interaction with absorbing molecules. The excitation process of a fluorescent molecule can take place only if two low-energy photons can interact simultaneously with the very same fluorophore. As mentioned in the Introduction, the time scale for simultaneity is the time scale of molecular energy fluctuations at photon energy scales, as determined by the Heisenberg uncertainty principle, i.e. 10^{-16}–10^{-15} s [125]. These two photons do not necessarily have to be identical but their wavelengths, λ_1 and λ_2, have to be such that

$$\lambda_{1P} \cong \left(\frac{1}{\lambda_1} + \frac{1}{\lambda_2}\right)^{-1} \tag{2}$$

where λ_{1P} is the wavelength needed to prime fluorescence emission in a conventional one-photon absorption process according to the energy request outlined in Equation (1). This situation, compared to the conventional one-photon excitation process shown in Figure 3(a), is illustrated in Figure 3(b) using a Perrin–Jablonski-like diagram. For practical reasons the experimental choice is usually such that [26,126,127]

$$\lambda_1 = \lambda_2 \approx 2\lambda_{1P} \tag{3}$$

and

$$\Delta E_g = 2hc/\lambda_{1P} \tag{4}$$

Considering this as a non-resonant process and the existence of a virtual intermediate state, one can calculate the resident time, τ_{virt}, in this intermediate state using the time-energy uncertainty consideration for TPE:

$$\Delta E_g * \tau_{virt} \cong \hbar/2 \tag{5}$$

where, $\hbar = h/2\pi$. It follows that

$$\tau_{\text{virt}} \cong 10^{-15} - 10^{-16}\text{s} \tag{6}$$

This is the temporal window available to two photons to coincide in the virtual state.

So far, in a TPE process it is crucial to combine sharp spatial focusing with temporal confinement of the excitation beam.

The process can be extended to n-photons requiring higher photon densities temporally and spatially confined, Figure 3(b). Thus, near-infrared (ca. 680–1100 nm) photons can be used to excite UV and visible electronic transitions producing fluorescence. The typical photon flux densities are of the order of more than 10^{24} photons $\text{cm}^{-2}\,\text{s}^{-1}$, impling intensities around MW-TW cm^{-2} [80,128]. A treatment in terms of quantum theory for two-photon transition has been elegantly proposed by Nakamura [129] using perturbation theory. He clearly described the process by a time-dependent Schrödinger equation, where the Hamiltonian contains electric dipole interactions terms. Using perturbative expansion one finds that the first-order solution is related to one-photon excitation while higher order solutions are related to n-photon ones [94,130].

Now, let us try to discuss TPE on the basis of the following simple assumption: the probability of a molecule undergoing n-photon absorption is proportional to the probability of finding n photons within the volume it occupies at any moment in time [125,131]. The question is: "what is the probability of finding two photons within the interval of time the molecule spends in a virtual state ?" [132].

Here we will refer to the first case earlier discussed by Andrews [131]: what is the probability p_n that n photons are in the same molecular volume? We consider that all the molecules are endowed with a suitable set of energy levels such that all possible n-photon transitions are possible. So far, we consider the relationship between the mean number of photons, M, at any time within a molecular volume and the intensity, I, of the laser beam that is energy per unit area per unit time. Consider a cube of side S through which the photons are delivered within a beam width much larger than S. The mean energy in this cubic box, for a certain wavelength λ, is

$$E_{\text{M}} = Mhc/\lambda \tag{7}$$

Since the cross-sectional area is S^2, and the time needed for each photon to cross the box is S/c, then

$$I = E_{\text{M}}/[S_2(S/c)] = Mhc^2/(\lambda S^3) \tag{8}$$

Recalling that $V = S^3 = V_{\text{m}}/N_{\text{a}}$, where for a molecule the mean volume occupied is the molar volume V_{m} divided by Avogadro's constant, $N_{\text{a}} = 6.022 \times 10^{23}\,\text{mol}^{-1}$, we have

$$M = IV_{\text{m}}\lambda/N_{\text{a}}hc^2 \tag{9}$$

As an example, considering a wavelength of 780 nm delivered at peak intensities of the order of GW cm^{-2} into a reasonable molecular volume of the order of 10^{-4} m^3 mol^{-1}, the resulting M is of the order of 10^{-5}.

Using a Poisson distribution to determine p_n, [125] we get

$$p_n = (M^{-n}/n!)e^{-M} \tag{10}$$

The resulting probability for TPE, $n = 2$, expanding the exponential term in Taylor series for M small and truncating at the first term setting the exponential value to unity, is given by

$$p_2 = M^2/2 \infty kI^2 \qquad k = \text{proportionality factor} \tag{11}$$

Here, the dependence of TPE on I^2 should be evident and has been demonstrated using simple arguments.

Since we have shown that TPE is a process that has a quadratic dependence on the instantaneous intensity of the excitation beam, we can introduce the molecular cross section, as its propensity to absorb in a TPE event photons having a certain energy or wavelength, and refer the fluorescence emission as function of the temporal characteristics of the light, $I(t)$, to it. Calculations can be easily extended to the multiphoton case.

So far, the fluorescence intensity per molecule, I_f (t), can be considered proportional to the molecular cross section $\delta_2(\lambda)$ and to $I(t)$ as:

$$I_f(t) \propto \delta_2 \times I(t)^2 \propto \delta_2 \times P(t)^2\left[\pi\frac{(NA)^2}{hc\lambda}\right]^2 \tag{12}$$

where $P(t)$ is the laser power, and NA is the numerical aperture of the focusing objective lens. The last term of Equation (12) simply takes care of the distribution in time and space of the photons by using paraxial approximation in an ideal optical system [133].

It follows that the time averaged two-photon fluorescence intensity per molecule within an arbitrary time interval T, $\langle I_f(t)\rangle$, can be written as:

$$\langle I_f(t)\rangle = \frac{1}{T}\int_0^T I_f(t)\,dt \propto \delta_2 \times \left(\pi\frac{(NA)^2}{hc\lambda}\right)^2 \frac{1}{T}\int_0^T P(t)^2\,dt \tag{13}$$

in the case of continuous wave (CW) laser excitation.

Now, because the present experimental situation for TPE is related to the use of ultrafast lasers, we consider that for a pulsed laser $T = 1/f_p$, where f_p is the pulse repetition rate [35].

This implies that a CW laser beam, where $P(t) = P_{ave}$, allows transformation of Equation (13) into:

$$\langle I_{f,cw}(t)\rangle \times \propto \delta_2 \times P_{ave}^2 \cdot \left[\pi\frac{(NA)^2}{hc\lambda}\right]^2 \tag{14}$$

Now, for a pulsed laser beam with a pulse width τ_p repetition rate f_p and average power

$$P_{ave} = DP_{peak}(t) \tag{15}$$

where $D = \tau_p f_p$, the approximated $P(t)$ profile can be described as:

$$\begin{aligned} P(t) &= P_{ave}/D \quad \text{for } 0 < t < \tau_p \\ P(t) &= 0 \quad \text{for } \tau_p < t < (1/f_p) \end{aligned} \tag{16}$$

We can write Equation (13) as [134]:

$$\langle I_{f,p}(t)\rangle \propto \delta_2 \frac{P_{ave}^2}{\tau_p^2 f_P^2}\left[\pi\frac{(NA)^2}{hc\lambda}\right]^2 \frac{1}{T}\int_0^{\tau_p} dt = \delta_2 \frac{P_{ave}^2}{\tau_p f_P}\left[\pi\frac{(NA)^2}{hc\lambda}\right]^2 \tag{17}$$

The conclusion here is that CW and pulsed lasers operate at the very same excitation efficiency, i.e. fluorescence intensity per molecule, if the average power of the CW laser is kept higher by a factor of $1/(\sqrt{\tau \cdot f_p})$. This means that 10 W delivered by a CW laser, allowing the same efficiency of conventional excitation performed at approximately 10^{-1}mW, are nearly equivalent to 30 mW for a pulsed laser.

16.3.4 Fluorescent molecules under TPE regime

This leads to the most popular relationship, reported below, that is related to the practical situation of a train of beam pulses focused through an high numerical aperture objective, with a duration τ_p and f_p repetition rate. In this case, the probability, n_a, that a certain fluorophore simultaneously absorbs two photons during a single pulse, in the paraxial approximation, is [26]:

$$n_a \propto \frac{\delta_2 P_{ave}^2}{\tau_p f_p^2}\left(\frac{NA^2}{2\hbar c\lambda}\right)^2 \tag{18}$$

where P_{ave} is the time-averaged power of the beam and λ is the excitation wavelength. Introducing 1 GM Goppert–Mayer = $10^{-58}(m^4 s)$, for a δ_2 of approximately 10 GM per photon [26,135] focusing through an objective of $NA > 1$, an average incident laser power of $\approx$1–50 mW, operating at a wavelength ranging from 680 to 1100 nm with 80–150 fs pulsewidth and 80–100 MHz repetition rate, would saturate the fluorescence output as for one-photon excitation. This suggests that, for optimal fluorescence generation, the desirable repetition time of pulses should be of the order of typical excited-state lifetime, which is a few nanoseconds for commonly used fluorescent molecules. For this reason the typical repetition rate is around 100 MHz. A further condition that makes Equation (18) valid is that the probability for each fluorophore to be excited during a single pulse has to be smaller than one. The reason for this lies in the observation that during the pulse time (10^{-13} s of duration and a typical excited-state lifetime in the 10^{-9} s range) the molecule has insufficient time to relax to the ground state. This can be considered a prerequisite for absorption of another photon pair. Therefore, whenever n_a approaches unity, saturation effects start occurring. The use of Equation (18) allows one to choose optical and laser parameters that maximize excitation efficiency without saturation. It is also evident that the optical parameter for enhancing the process in the focal plane is the lens numerical aperture, *NA*, even if the total fluorescence emitted is independent of this parameter as shown by Xu [135]. This value is confined around 1.3–1.4 as a maximum. Now, one can estimate n_a for a common fluorescent molecule like fluorescein that possesses a two-photon cross-section of 38 GM at 780 nm [134, and next paragraph].

For this purpose, we can use $NA = 1.4$, a repetition rate at 100 MHz and pulse width of 100 fs within a range of P_{ave} assumed as 1, 10, 20 and 50 mW. Substituting the values in Equation (14) we obtain

$$n_a = 38 \times 10^{-58} \frac{P_{ave}^2}{100 \times 10^{-15} \times (100 \times 10^6)^2} \times \left(\frac{(1.4)^2}{2 \times 1.054 \times 10^{-34} \times 3 \times 10^8 \times 780 \times 10^{-9}} \right)^2 \cong 5930 \times P_{ave}^2$$

The final results as function of 1, 10, 20, 50 mw are 5.93×10^{-3}, 5.93×10^{-1}, 1.86, 2.965 respectively. Evidently saturation begins at 10 mW [134,136].

The related rate of photon emission per molecule, at a non-saturation excitation level, in the absence of photobleaching [122,134], is given by n_a multiplied by the repetition rate of the pulses. This means approximately 5×10^7 photons s^{-1} in both cases. Notably, when considering the effective fluorescence emission, one should consider a further factor given by the so-called quantum efficiency of the fluorescent molecules. Below, we report data related to the fluorochrome action cross section that are related to absorption cross section and quantum efficiency. The fluorophore emission spectrum is independent of the excitation mode [135,137]. So far, the quantum efficiency is known from conventional one-photon excitation data [21].

Now, even if the quantum-mechanical selection rules for MPE differ from those for one-photon excitation, several common fluorescent molecules can be used. Unfortunately, knowledge of the one-photon cross section for a specific fluorescent molecule does not allow any quantitative prediction of the two-photon trend. The only "rule of thumb" that can help states that one may expect to have a TPE cross-section peak at double the wavelength needed for one-photon excitation, especially in symmetrical molecules. However, the cross section parameter has been measured for a wide range of dyes [137,138]. Due to the increasing dissemination of TPE microscopy, new "ad hoc" organic molecules, endowed with large engineered two-photon absorption cross sections, have been recently developed [139]. Table 1 summarises the properties of some commonly used fluorescent molecules under two-photon excitation [120,137] [Table 1(A): extrinsic fluorescent molecules; (B) intrinsic fluorescent molecules]. TPE fluorescence from NAD(P)H, flavoproteins [137,140], tryptophan and tyrosine in proteins [141] have been measured. A special mention is due for the green fluorescent protein (GFP), which is an

Table 1. Properties of some commonly used fluorescent molecules under two-photon excitation (TPE). A) TPE cross section of extrinsic (A) and intrinsic (B) fluorophores. Cross section is given in Goppert-Mayer. Quantum yield and fluorescence emission efficiency are also reported, when measured [135]

Fluorophores	λ(nm)	ηδ	δ
(A) Extrinsic Fluorophores			
Bis-MSB	691/700	6.0 ± 1.8	6.3 ± 1.8
Bodipy	920	17 ± 4.9	
Calcium Green 1	800 (780, 820)		
Calcofluor	780/820		
Cascade blue	750	2.1 ± 0.6	
Coumarin 307	776	19 ± 5.5	
CY2	780/800		
CY3	780		
CY5	780/820		
DAPI (free)	700/720	0.16 ± 0.05	
Dansyl	700	1	
Dansyl hydrazine	700	0.72 ± 0.2	
Dil	700	95 ± 28	
Filipin	720		
FITC	740/780/820		
Fluorescein (pH≈11)	780	–	38 ± 9.7
Fura-2 (free)	700	11	
Fura-2 (high Ca)	700	12	
Hoechst	780/820		
Indo-1 (free)	700	4.5 ± 1.3	12 ± 4
Indo-1 (high Ca)	590/700	1.2 ± 0.4	2.1 ± 0.6
Lucifer Yellow	860	0.95 ± 0.3	
Nile Red	810		
Oregon Green Bapta 1	800		
Rhodamine B	840	–	210 ± 55
Rhodamine 123	780–560		
Syto 13	810		
Texas red	780		
Triple probe (Dapi, FITC, and rhodamine)	720/740		
TRITC (rhodamine)	800–840		
(B) Intrinsic fluorophores			
GFP wt	800–850		~6
GFP S65T	~960		~7
BFP	780/820		
CFP	780/840		
YFP	860/900		
EGFP	900–950		
Flavine	~700		~0.8
NADH	~700		~0.02
Phycoerythrin	1064		322 ± 110

important molecular marker [114,142,143]. GFP cross-sections are around 6 GM (800 nm) and 7 GM (960 nm) for wild and S65T type, respectively. As comparison one should consider that the cross-section for NADH, at the excitation maximum of 700 nm, is ca. 0.02 GM [137].

16.3.5 *Optical consequences of TPE*

In terms of optical consequences the two-photon effect has the important consequence of limiting the excitation region to within a sub-femtolitre volume. The 3D confinement of the two-photon excitation volume can be understood based on optical diffraction theory [133]. Using excitation light with wavelength λ, the intensity distribution at the focal region of an objective with numerical aperture $NA = \sin(\alpha)$ is described, in the paraxial regime, by [133,144]:

$$I(u, v) = \left| 2 \int_1^0 J_0(v\rho) e^{-\frac{i}{2} v \rho^2} \rho \mathrm{d}\rho \right|^2 \tag{19}$$

where J_0 is the zeroth order Bessel function, ρ is a radial coordinate in the pupil plane, $u = [8\pi \sin^2(\alpha/2)z]/\lambda$ and $v = [2\pi \sin(\alpha)r]/\lambda$ are dimensionless axial and radial coordinates, respectively, normalized to the wavelength [17]. Now, the intensity of fluorescence distribution within the focal region has a $I(u,v)$ behaviour for the one-photon case and $I^2(u/2,v/2)$ for TPE case as demonstrated above. The arguments of $I^2(u/2,v/2)$ take into proper account that in the latter case one utilizes wavelengths that are approximatively twice the ones used for one-photon excitation. As compared with the one-photon case, the TPE intensity distribution is axially confined [110,116,145]. In fact, considering the integral over v, keeping u constant, its behaviour is constant along z for one photon and has a half-bell shape for TPE. This behaviour, better discussed in Refs. [109–111], explains the three-dimensional discrimination property in TPE.

Now, the most interesting aspect, also predicted by Equations (17) and (11), is that the excitation power falls off as the square of the distance from the lens focal point, within the approximation of a conical illumination geometry. In practice this means that the quadratic relationship between the excitation power and the fluorescence intensity brings about the fact that TPE falls off as the fourth power of distance from the focal point of the objective. This implies that those regions away from the focal volume of the objective lens, directly related to the numerical aperture of the objective itself, do not suffer photobleaching or phototoxicity effects and do not contribute to the signal detected when a TPE scheme is used. Because they are simply not involved in the excitation process, a confocal-like effect is obtained without the necessity of a confocal pinhole. It is immediately evident that also in this case an optical sectioning effect is obtained. In fact, the observed image O at a plane j, produced by the true fluorescence distribution I at plane j, distorted by the microscope through S, plus noise N, holds again the confocal ideal situation where contributions from adjacent k planes can be set to zero as in the confocal situation: $O_j = I_j S_j + N$.

This means that TPE microscopy is intrinsically three-dimensional. The optical sectioning effect is, notably, obtained in a very different way with respect to the confocal solution. No fluorescence has to be removed from the detection pathway. In this case one should be able to collect as much fluorescence as possible. In fact fluorescence can come only and exclusively from the small focal volume traced in Figure 3, which is of the order of a fraction of a femtolitre.

In TPE, over 80% of the total intensity of fluorescence comes from a 700–1000 nm thick region about the focal point for objectives with numerical apertures in the range of 1.2 to 1.4 [15,17,109–111]. This also implies a reduction in background that allows compensation of the reduction in spatial resolution due to the wavelength. The utilisation of infrared wavelengths instead of UV-visible ones also allows deeper penetration than in conventional cases [134,146,147]. In fact, Rayleigh scattering produced by small particles is proportional to the inverse fourth power of the wavelength. Thus, the longer wavelengths used in TPE, or in general in multiphoton excitation, will be scattered less than the ultraviolet–visible wavelengths used for conventional excitation [148]. So far, deeper targets within a thick sample can be reached. Of course, for the fluorescence light, on the way back, scattering can be overcome by acquiring the emitted fluorescence using a large-area detector and collecting not only ballistic photons [127,149,150].

16.4 Confocal and MPE in practice

16.4.1 General aspects

Nowadays, MPE microscopes and architectures, including confocal abilities, are commercially available at very high cost, ranging from 250 k€ to 700 k€. Table 2 gives an overview of market availability.

An MPE microscope can also be constructed from components or, utilising a very efficient compromise, by modifying an existing confocal laser scanning microscope. This last situation allows an effective mix of operational flexibility and of good quality-to-cost ratio. The main elements to realize a MPE architecture including confocal modality are the following: high peak-power laser delivering moderate average power (fs or ps pulsed at relatively high repetition rate) emitting infrared or near-infrared wavelengths (650–1100 nm), CW laser sources for confocal modes, a laser beam scanning system or a confocal laser scanning head, high numerical aperture objectives (>1), a high-throughput microscope pathway, and a high-sensitivity detection system [25,103,115,127,151–160]. Figure 8 shows a general scheme for a MPE microscope also including confocal mode.

In typical MPE or confocal microscopes, images are built by raster scanning the *x–y* mirrors of a galvanometric driven mechanical scanner [33]. This implies that the image formation speed is mainly determined by the mechanical properties of the scanner, i.e. for single line scanning is of the order of ms. Faster beam-scanning schemes can be realised, even if the "eternal triangle of compromise" should be considered for sensitivity, spatial resolution and temporal resolution. In MPE set-ups particular attention should be given to the surfaces of the mirrors and to the way they are mounted on the scanners to get the best reflection efficiency and scanning stability [127,153]. Then, the excitation light should reach the microscope objective passing through the minimum number of optical components and possibly along the shortest path. Typically, high-numerical-aperture objectives, with high infrared transmission, are used to maximise the TPE efficiency

Table 2. Overview of TPE commercial microscopes

Model	Company	Dimension	Pulse width regime	Wavelength range (nm)	Average power (mW)	Laser coupling	Acquisition	Other features
LSM 510 NLO (META)	Zeiss	Compact/ normal	fs	700–900	50	Direct-box/ fiber	Descanned/Non descanned	Spectral capability
MRC 1024 MP	BIO-RAD	Normal/ large	fs	690–1000	Not reported	Direct-box	Descanned/Non descanned	None relevant
Radiance 2000 MP	BIO-RAD	Compact/ normal	fs	690–1000	Not reported	Direct-box	Descanned/ Non descanned	Faster scanning (>750 Hz)
RTS2000 MP	BIO-RAD	Large	fs	690–1000	Not reported	Direct-box	Descanned/ Non descanned	130 frames/s video rate
TCS SP2/AO BS	Leica	Normal/ large	Ps fs available	720–900	Not reported (120 max at the sample)	Fiber/Direct-box	Descanned/ Non descanned	Spectral capability

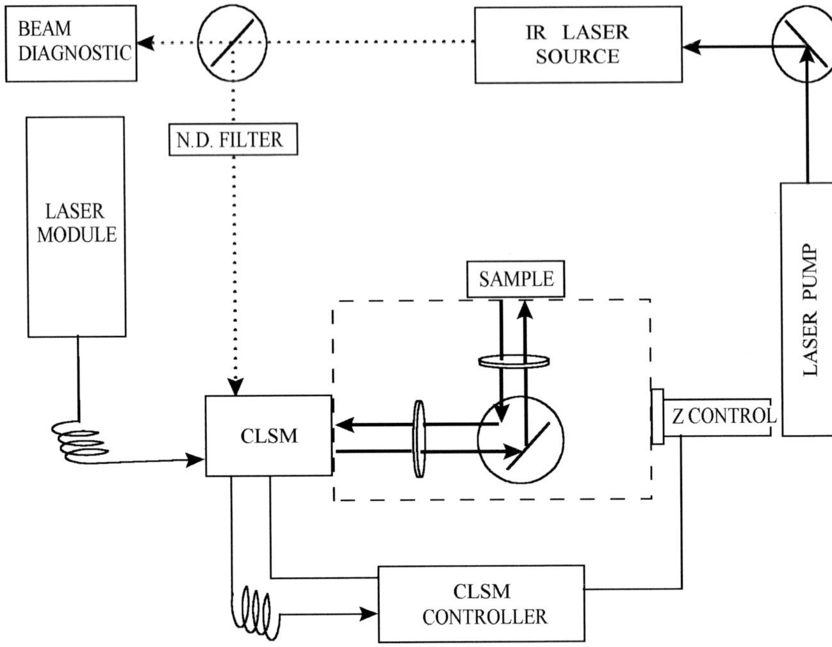

Figure 8. Architecture of a two-photon excitation microscope preserving the conventional confocal ability (Diaspro et al., [188]; www.lambs.it). See text for details.

[127,161]. While the *x–y* scanners provide lateral focal-point scanning, axial scanning can be achieved by means of different positioning devices, the most popular being a belt-driven system using a DC motor and a single objective piezo nano-positioner. Usually, it is possible to switch between confocal and MPE modes retaining *x–y–z* positioning on the sample being imaged. Figures 9(a) and 9(b) show two modern confocal system, both convertible into MPE systems by simple addiction of the proper laser source and some optomechanical adaptation. Acquisition and visualisation are generally completely computer controlled by dedicated software.

Let us consider now two popular approaches that can be used to perform MPE microscopy, namely: descanned and non-descanned mode. The former uses the very same optical pathway and mechanism employed in confocal laser scanning microscopy. The latter mainly optimises the optical pathway by minimising the number of optical elements encountered on the way from the sample to detectors, and increases the detector area. Non-descanned mode is not available on confocal "solo" set-ups. Figure 10 illustrates these two approaches, also including the conventional confocal scheme with a pinhole along with the descanned pathway. MPE non-descanned mode allows very good performances, providing superior signal-to-noise ratios inside strongly scattering samples [115,120,162,163].

In the descanned approach pinholes are removed or set to their maximum aperture and the emission signal is captured using excitation scanning device on the

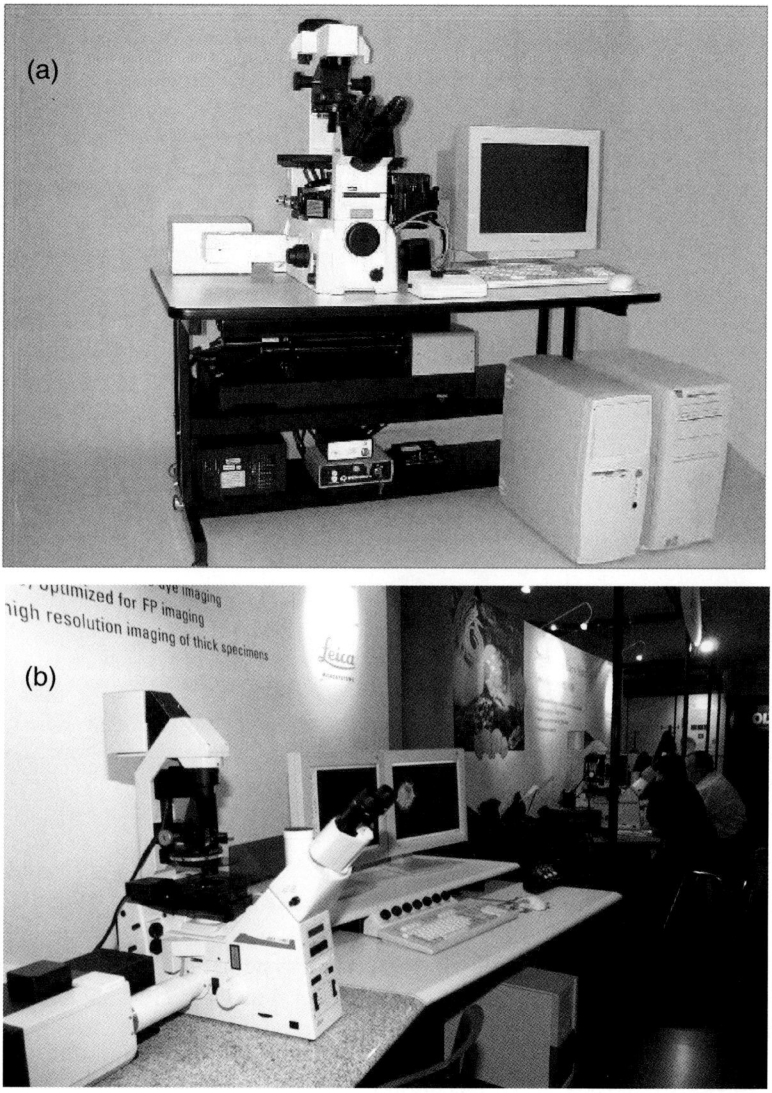

Figure 9. Two new generation confocal systems: a compact architecture by Nikon (a) and a complete system endowed of spectral discrimination by Leica (b). Both systems are adaptable for MPE [126].

back pathway. For this reason it is called the descanned mode. In the latter, the confocal architecture has to be modified to increase the collection efficiency: pinholes are removed and the emitted radiation is collected using dichroic mirrors on the emission path or external detectors without passing through the galvanometric scanning mirrors. A high-sensitivity detection system is another critical issue [120,127,164].

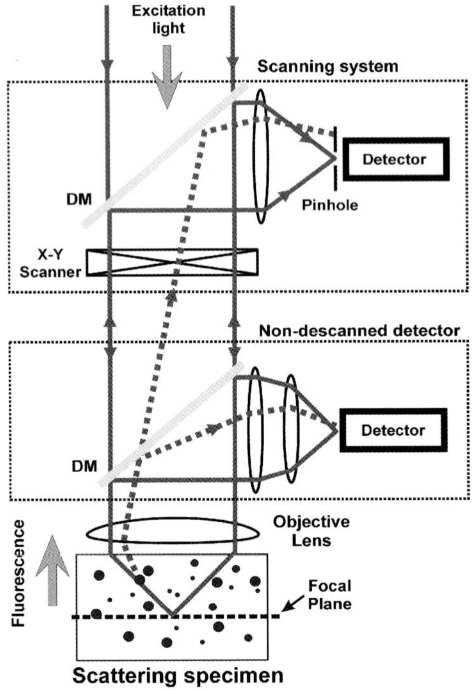

Figure 10. Simplified optical schemes for scanned and non-descanned detection. Confocal pinhole can be used or fully opened. (Courtesy of Mark Cannel, adapted from Soeller and Cannell [149]).

The fluorescence emitted is collected by the objective and transferred to the detection system through a dichroic mirror along the emission path. Due to the high excitation intensity, an additional barrier filter is needed to avoid mixing of the excitation and emission light at the detection system that is differently placed depending on the acquisition scheme being used. Photodetectors that can be used include photomultiplier tubes, avalanche photodiodes, and charge coupled device (CCD) cameras [103,165]. Photomultiplier tubes are the most commonly used. This is due to their low cost, good sensitivity in the blue-green spectral region, high dynamic range, large sensitive area, single-photon counting mode availability [166]. They have a quantum efficiency around 20–40% in the blue-green spectral region that drops down to <1% on moving to the red region. This is a good condition, especially in MPE mode because one wants to reject as much as possible wavelengths above 680 nm which are mainly used for excitation. Another advantage is that the large sensitive area of photomultiplier tubes allows efficient collection of signal in the non-descanned mode within a dynamic range of the order of 10^8. Avalanche photodiodes are excellent in terms of sensitivity, exhibiting quantum efficiencies close to 70–80% in the visible spectral range. Unfortunately, they are high cost and the small active photosensitive area, <1 mm, could introduce drawbacks in the detection scheme and requires special de-scanning optics [167].

CCD cameras are used in video rate multifocal imaging [127,168]. Now, as a further general consideration, to obtain a better spatial resolution in MPE it is also possible to retain the confocal pinhole [46,111,169,170]. Unfortunately, in some practical experimental situations, the low efficiency of MPE fluorescence process may rule out such a solution. However, when pinhole insertion is possible, the major advantage is that the axial resolution can be improved by approximately 40%. However, once the best quality image possible has been obtained then sophisticated mathematical algorithms can be applied to enhance the features of interest to the biological researcher and to improve the quality of data to be used for three-dimensional modelling [14,108,171–173]. Recently, an image restoration web service has been established to get the best quality 3D data set from wide-field, confocal or TPE optically sectioned sample. This tool, called "Power-up your microscope"[174] is available for free at www.powermicroscope.com.

16.4.2 Laser sources

Laser sources, as often happened in optical microscopy, represent an important resource, especially in fluorescence microscopy [34,35]. This section is mainly devoted to laser sources for multiphoton imaging, since CW laser sources are of common utilization today. For non-resonant MPE framework, owing to the comparatively low MPE cross-sections of fluorophores, high photon flux densities are required, $>10^{24}$ photons $cm^{-2}\,s^{-1}$ [30]. Using radiation in the spectral range of 600–1100 nm for MPE, excitation intensities in the MW–GW cm^{-2} are required. This high energy can be obtained by the combined use of focusing lens objectives (see next section) and CW [175,176] or pulsed [26] laser radiation of 50 mW mean power or less [127,177]. TPE microscopes have been realised using CW, femtosecond, and picosecond laser sources [24,25,70]. Since the original successful experiments in TPE microscopy, advances have been made in the technological field of ultrashort pulsed lasers. Nowadays, laser sources suitable for TPE can be described as "turnkey" systems [97,178]. Figure 11 shows the emission range for different laser sources combined with the cross section behaviour of some popular fluorophores. Evidently, the range 700–1050 nm is well addressed by Ti:sapphire lasers. This range of wavelengths is very common because various fluorophores have an excitation range in the conventional one-photon excitation regime within 350–600 nm.

Other laser sources used for TPE are Cr:LiSAF, pulse-compressed Nd:YLF in the femtosecond regime, and mode-locked Nd:YAG and picosecond Ti:sapphire lasers in picosecond regime [34,178]. Moreover, the absorption coefficients of most biological samples, cells and tissues are minimised within this spectral window [120]. Table 3 reports some data about the most commonly used Ti:sapphire laser sources for applications in microscopy and spectroscopy.

Figures 12(a) and 12(b) show two popular lasers sources used for MPE, a classical wide-band Tsunami and a compact new generation Chameleon source (Table 3). These lasers operate in mode-locking mode. This allows the laser to generate a train of very short pulses by modulating the gain or excitation of a

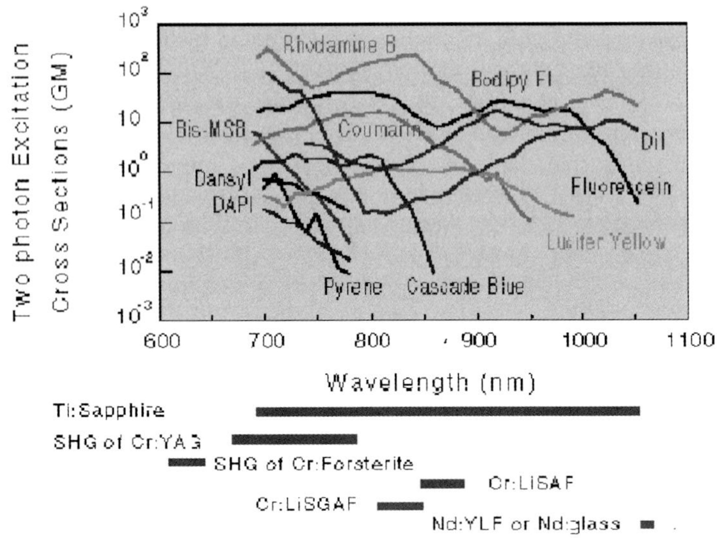

Figure 11. Cross sections of common fluorophores compared with the emission wavelength range available by different commercial laser systems (After Xu et al. [137]).

laser at a frequency with a period equal to the round-trip time of a photon within the laser cavity [35,97]. The resulting pulse width is in the 50 to 150 fs regime. The parameters that are more relevant in the selection of the laser source are average power, pulse width and repetition rate, and wavelength, also according to Equation (18). The most popular features for an infrared pulsed laser are 700 mW–1 W average power, 80–100 MHz repetition rate, and 100–150 fs pulse width. So far, the use of short pulses and small duty cycles are mandatory to allow image acquisition in a reasonable time while using power levels that are biologically tolerable [42,136,179–182]. To minimise pulse width dispersion problems [30] suggests working with pulses around 150–200 nm, which constitutes a very good compromise both for pulse stretching and sample viability. One should always remember that a shorter pulse broadens more than a longer one. Pulse width measurement is a very delicate issue. In fact, because it is not very easy to measure it at the focal volume within the sample, little can be definitely said on it [175,183,184]. Although users do not perform measurement of the pulse width at the sample when they use two-photon microscopy, which would require a specific procedure that, even if not too complex for a researcher in the field, could be irksome for most users, it is a reasonable approximation to assume that at the focal volume a 1.5–2 times temporal pulse broadening occurs using high quality optics [127,184]. As an example, for a measured laser pulse width of about 100 fs, an estimate at the sample is about 150–180 fs under favourable experimental conditions, sample characteristics included. Sample properties are mentioned because for thick samples the role played by thickness, also in terms of pulse width broadening, is not so obvious [115,134,148,185,186].

Table 3. Major commercial laser sources available for TPE

Company/Model	Tuning range	Wavelength (nm)	Pulse width	Average power (mW) for 5 W pump	Dimension	Pump
Spectra Physics–Tsunami	Wide	680–1050	25 fs <100–130 fs up to 100 ps	100–800	Normal/large	Solid state 5–10 W compact
Coherent–Mira	Wide	680–1000	<100 fs 5–10 ps	100–700	Large	Solid state 5–10 W compact
Spectra Physics–Mai Tai	200 nm selectable	720–920 780–980	≈ 100 fs	<1000	Compact	Solid state integrated
Coherent–Chameleon	200 nm selectable	700–760 760–860	≈ 100 fs	<1000	Compact	Solid state integrated

(a) (b)

Figure 12. Ultrafast Titanium-Sapphire laser source: Tsunami-Millennia by Spectra Physics (a). New generation compact ultrafast laser source: Chamaleon by Coherent (b).

As a sort of summary on laser parameters, one should consider keeping the peak power low and the wavelength long.

16.4.3 Lens objectives

Objective lenses influence the performances of any optical microscope, and for a MPE there are system special considerations in view of Equation (18). New technological requisites have to be considered with respect to conventional excitation fluorescence microscope. An adequate transmission in the IR regime has to be coupled with good collection efficiency towards the ultraviolet region. Moreover, the number of components should be minimised without affecting resolution properties in order to reduce pulse widths distortions. Although the collection efficiency of the time-averaged photon flux depends on the numerical aperture of the collecting lens, the total fluorescence generation is independent of the numerical aperture of the focusing lens when imaging thick samples [135]. This is because the increase of intensity, obtained by a sharper focusing (high NA), is counterbalanced by the shrinking of the excitation volume. Thus the total amount of fluorescence summed over the entire space remains constant. The very relevant practical consequence of this is that in TPE measurements on thick samples, assuming no aberrations, the generated fluorescence is insensitive to the size of the focal spot. As a positive consequence, a moderate variation of the laser beam size would not affect the measurements. This is a very efficient condition as, using an appropriate (non-descanned) acquisition scheme, it is possible to collect all the generated fluorescence.

However, the amount of the light delivered or collected depends critically on the cone of light accepted into the objective lens [13]. For point scanning systems, this light cone is related to the numerical aperture of the lens relative to the index of refraction of the immersion medium. The total area of light collected by the lens is then proportional to the square of this ratio. Girkin and

Table 4. Optical lens properties

Objective	Magnification NA	T (%) @ 350 nm	T (%) @ 550 nm	T (%) @ 900 nm
CFI Plan Apochromat	100×Oil IR 1.40	<15	>71	51–70
CFI Plan Fluor	100×Oil 1.30	51–70	>71	>71
CFI Fluor Water dipping	60×1.00	51–70	>71	>71
CFI Plan Fluor	60×Oil 1.25	31–50	>71	51–70
CFI Fluor Water dipping	40×0.80	51–70	>71	>71
CFI Super (S) Fluor	40×Oil 1.30	51–70	>71	>71
CFI Super (S) Fluor	20×0.75	51–70	>71	>71
CFI Fluor Water dipping	10×0.30	51–70	>71	51–70
CFI Super (S) Fluor	10×0.50	51–70	>71	>71

Wokosin [127] reported about the light collection performance of a range of lenses compared to a 1.4 NA 60× oil lens. The 1.4 NA 60× oil and 1.2 NA 60× water lenses have the best collection potential, the 1.3 NA oil lenses can collect 85% compared to the best two lenses, while the 0.75 NA air lenses and 1.0 NA water dipping lens only collect 66% as much light. As usual, a compromise must be reached, as the higher numerical aperture lenses (recently 1.65 by Olympus and 1.45 by Nikon and Zeiss have been available) gain the light collection power at the loss of working distance. Typical working distances for oil immersion lenses with a numerical aperture of 1.4 are around 250 μm (which include the thickness of a cover slip that most commonly is 0.17 mm). This gives a practical working distance of less than 100 μm, making them unsuitable for really deep imaging of intact tissue. As a general rule one selects a lens with as large a numerical aperture as possible giving the working distance required for the preparation under investigation. The 1.4 NA lenses provide a good starting point for TPE lens selection. In fact, the numerical aperture of a lens also determines the ultimate resolution of the optical system [10]. Table 4 reports a collection of transmission data for the Nikon CFI60 series of objective for confocal and MPE imaging [127,161].

König reported an interesting set of data related to pulse broadening due to the microscope optics including lens objectives. Considering a pulse width τ_{in}, the pulse broadening B for a system of lenses can be calculated as $\tau_{\mathrm{out}}/\tau_{\mathrm{in}}$ where τ_{out} is the pulse width at the focal plane [30]: $B = -(\tau_{\mathrm{out}}/\tau_{\mathrm{in}}) = \sqrt{(1 + 7.68(DL/\tau_{in}^2)^2)}$.

The optical dispersion parameter D for glass objectives at 800 nm ranges from 250 to 1600 (fs^2 cm^{-1}) (Table 5). Considering the optical path length factor L, a typical DL value of the whole microscope including oil immersion high NA objective is 5000 fs^2 [30]. Wolleschensky reported a summary of the dispersion parameter for Zeiss microscope objective lenses measured at 800 nm. Ds, in fs^2, are 1714, 1494, 2398, and 1531 (within an error of about 10%) for 40×/0.8 water IR Achroplan, for 63×/0.9 water IR Achroplan, 40×/1.3oil Plan Neofluar, and 20×/0.75 Plan Apochromat, respectively. The pulse broadening factor of a 100 fs pulse was estimated to be between 1.14 and 1.23 [184].

Table 5. Dispersion parameter (D) for different optical materials (see text for details)

D @ 800 nm (fs^2 cm^{-1})	Glass type
+251	CaF^2
+300	Quartz
+389	FK-3
+445	BK7
+1030	SF2
+1600	SF10

16.4.4 Example of a practical TPE microscope realisation

This section is related to the practical realisation of a TPE microscope achieved through minor modifications of a commercial CLSM, in which the ability to operate as a standard CLSM has been preserved [126,187]. This microscope has been established at LAMBS (Laboratory for Advanced Bioimaging, Microscopy, and Spectroscopy, http://www.lambs.it) under the auspices and grants of the National Institute for the Physics of the Matter (INFM, Istituto Nazionale per la Fisica della Materia) in 1999 [126,188]. A scheme of the architecture is sketched in Figure 8. The core of the architecture is a mode-locked Ti:Sapphire infrared pulsed laser (Tsunami 3960 Spectra Physics Inc., Mountain View, CA, USA), pumped by a high-power (5W@532 nm) solid state laser (Millennia V, Spectra Physics Inc., Mountain View, CA, USA). Power and wavelength measurements are performed using an RE201 model ultrafast laser spectrum analyser (Ist-Rees, UK) and an AN2/10A-P model thermopile detector power meter (Ophir, Israel) that constitute the beam diagnostics module of the system. At LAMBS has been realized a compact optical autocorrelator which allows to measure fs laser pulses on the microscope objective plane based on a Michelson interferometer and fluorescence signal [65]. Before entering into the laser scanning head, beam average power is controlled using a neutral density rotating wheel (Melles Griot, USA). For an average power of 20 mW at the entrance of the scanning head, the typical average power before the microscope objective is about 8–12 mW and at the sample is estimated between 2 and 6 mW. We found that at the focal volume a 1.5–1.8 times broadening occurs using a high numerical aperture objective and a reduced amount of optics within the optical path. The scanning and acquisition system for microscopic imaging is based on a commercial single-pinhole scanning head Nikon PCM2000 (Nikon Instruments, Florence, Italy) mounted on the lateral port of a common inverted microscope, Nikon Eclipse TE300. The Nikon PCM2000 has a simple and compact light path that makes it very appropriate for conversion into a two-photon scope [188]. The optical resolution performances of this microscope when operating in conventional confocal mode, and using a 100×/1.3 NA oil immersion objective, have been reported in detail elsewhere and are 178 ± 21 nm laterally and 509 ± 49 nm axially [188]. Under the TPE regime the scanning head operates in the "open pinhole" condition; i.e. a wide-field descanned detection scheme is used [189]. The simple but effective optical path of the PCM2000

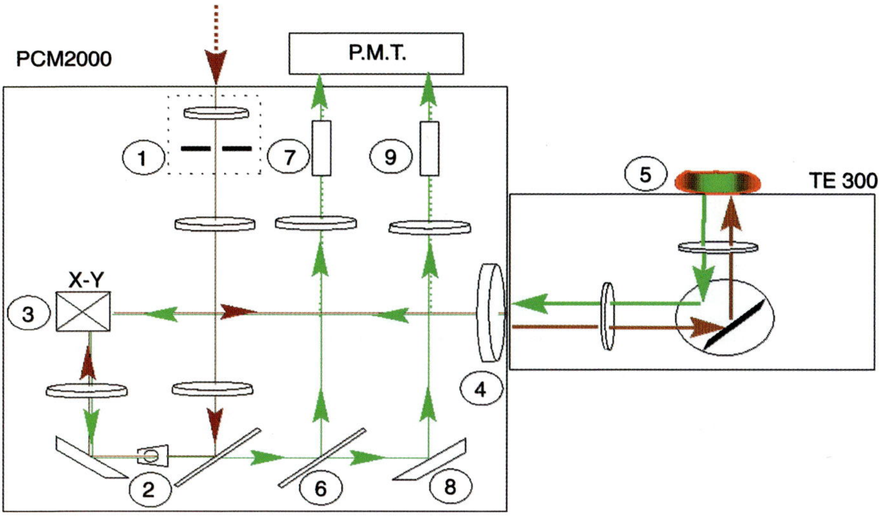

Figure 13. Optical scheme of the confocal scanning head Nikon PCM2000 shown in Figure 9(a). The excitation beam enters into the PCM2000 scanning head through an optical coupler (1) in order to reach the sample on the x-y-z stage (5). The beam passes through the pinhole holder (2) kept in the open position, the galvanometric mirrors (3) and the scanning lens (4). Fluorescence generated from the sample (5) is delivered to the PMT through acquisition channels directed by two selectable mirrors (6,8) via optical fiber (7,9) coupling. One-photon and two-photon mode can be simply accomplished by switching from the single-mode optical fiber (one-photon), coupled to a module containing conventional laser sources (Ar-Ion, He-Ne green), to the two-photon optical coupler (TPOC), allowing Tsunami laser beam delivering (two-photon). Axial scanning for confocal and TPE three-dimensional imaging is actuated by means of two different positioning devices depending on the experimental circumstances and axial accuracy needed, namely: a belt driven system using a DC motor (RFZ-A, Nikon, Japan) and a single objective piezo nano-positioner (PIFOC P-721-17, Physik Instrumente, Germany). The piezoelectric axial positioner allows an axial resolution of 10 nm within a motion range of 1000 nm at 100 nm steps and utilizes a linear variable differential transformer–LVDT–integrated feedback sensor [126].

scanning head is shown in Figure 13. One-photon and two-photon modes can be simply accomplished by switching from the single-mode optical fiber (one-photon), coupled to a module containing conventional laser sources (Ar-Ion, He-Ne green), to the optical path in air, delivering the Tsunami laser beam (two-photon). Switching is realized by the attachment shown in Figure 14. Optical sectioning and switching from confocal to MPE mode are shown in Figures 15 and 16 respectively. A further example of MPE applied to biological imaging is shown in Figure 17.

A high throughput optical fiber delivers the emitted fluorescence from the scanning head to the PCM2000 control unit where photomultiplier tubes (R928, Hamamatsu, Japan [166]) are physically plugged. This solution is particularly useful both for confocal and MPE imaging for three main reasons: (1) electrical

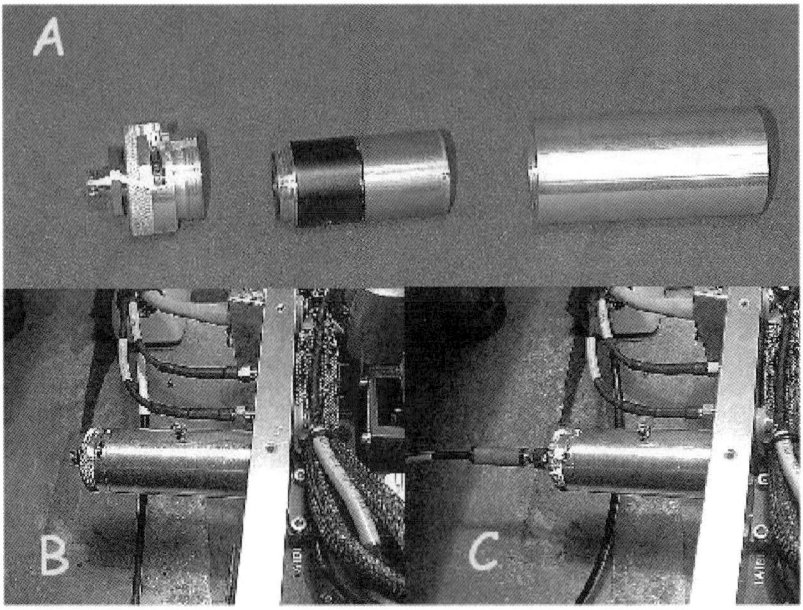

Figure 14. Details of the attachment for switching from confocal to MPE mode, TPOC (two-photon optical coupler). The cost is around 50 euros. (A) Aluminium tube containing a low magnification coupling objective (Edmund Scientific, USA); (B) TPOC plugged at the scanning head input port after removing optical fiber for conventional excitation beam delivery; (C) optical fiber delivering conventional excitation connected through TPOC to the scanning head. By means of a solution adopted in frame (C) it is possible to simply and precisely switch from one- to two-photon excitation mode [126]. (Photograph courtesy of Federico Federici).

noise is reduced; (2) background light noise is reduced; (3) it is possible to directly verify optical conditions, keeping the scanning head without enclosure.

Axial scanning for confocal and MPE three-dimensional imaging is actuated by means of two different positioning devices depending on the experimental circumstances and axial accuracy needed, namely: a belt-driven system using a DC motor (RFZ-A, Nikon, Japan) and a single objective piezo nano-positioner (PIFOC P-721-17, Physik Instrumente, Germany). The piezoelectric axial positioner allows an axial resolution of 10 nm within a motion range of 1000 nm at 100 nm steps and utilizes a linear variable differential transformer – LVDT – integrated feedback sensor.

Acquisition and visualisation are completely computer controlled by a dedicated software, EZ2000 (Coord, The Netherlands; http://www.coord.nl). The main available controls are related to PMTs voltage, pixel dwell time, frame dimensions (1024 × 1024 maximum), field of scan (from 1 to 140 mm using a 100 × objective). To evaluate the performances of the MPE microscope some calibration measurements have to be performed, namely: a fluorescence quadratic behaviour check and point spread function measurements.

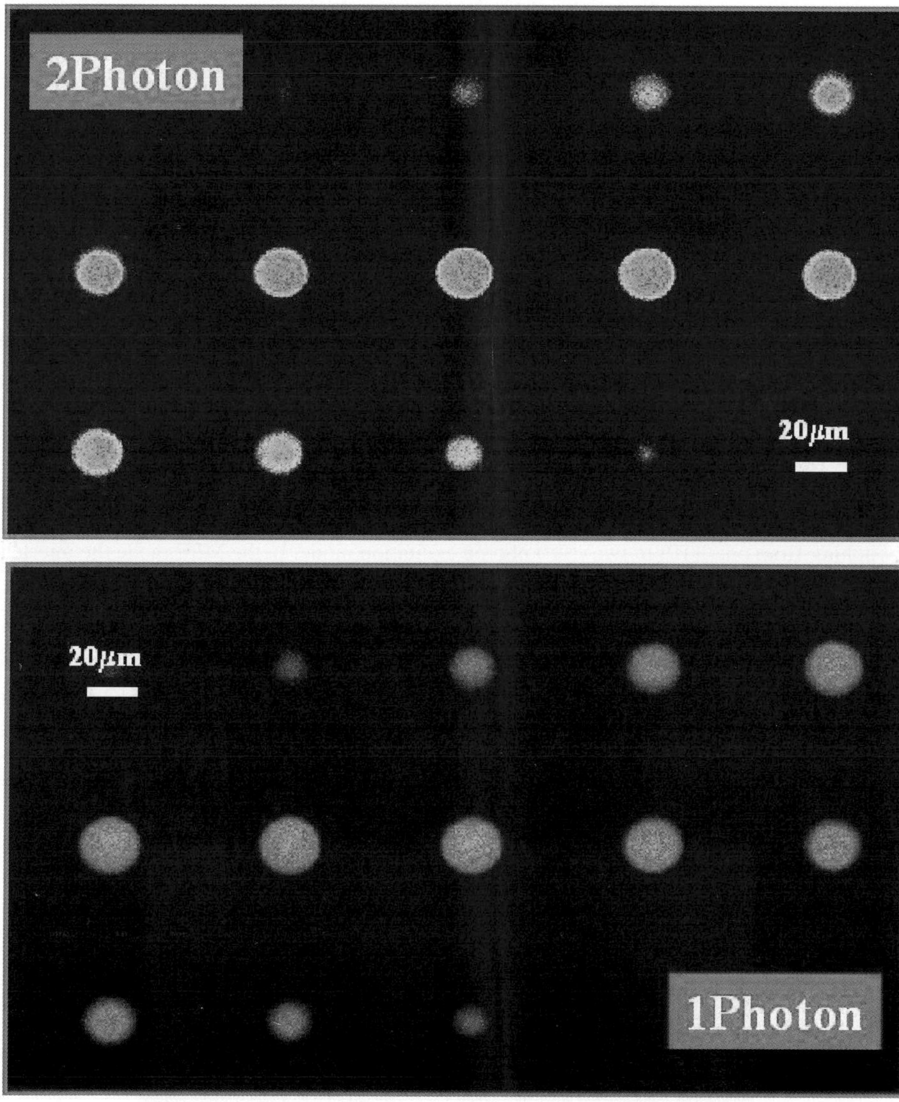

Figure 15. Optical sectioning demonstrated in confocal and TPE mode after switching from one mode to the other using TPOC developed at LAMBS (www.lambs.it).

Point spread function (PSF) measurements are referred to a planachromatic Nikon 100×, 1.4 NA immersion oil objective with enhanced transmission in the infrared region. Blue fluorescent carboxylate modified microspheres 100 nm diameter (F-8797, Molecular Probes, OR, USA) were used. A drop of dilute samples of bead suspensions was spread between two coverslips of nominal thickness 0.17 mm. These microspheres constitute a very good compromise towards the utilisation of subresolution point scatterers and acceptable fluorescence

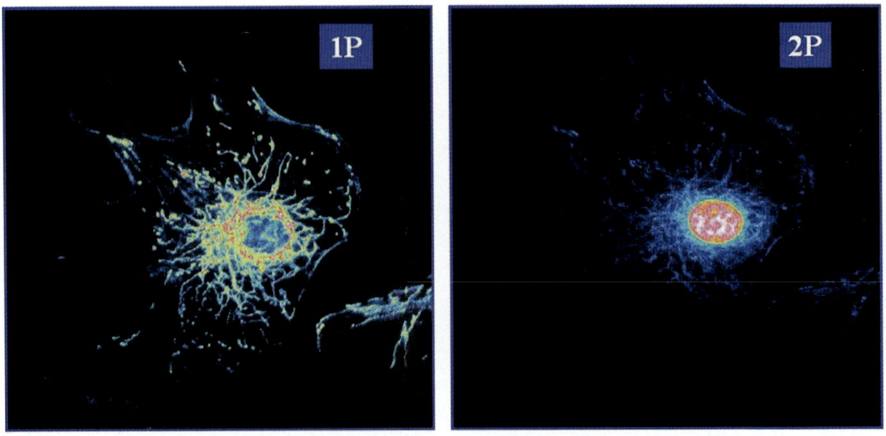

Figure 16. Bovine pulmonary artery endothelial cells (F-147780, Molecular Probes) marked with three different fluorophores for mitochondria, F-actin and DAPI. Image on the left shows mitochondria (red) and F-actin (green) labelled molecules. Imaging to demonstrate switching from onc- (1P) to two- (2P) photon excitation mode. In the MPE mode the internal structure of the nucleus is clearly visibile, i.e. chromatin DNA marked by DAPI, normally excited in the UV. The control of positioning after mode switching is apparent. (Image taken at LAMBS by Federico Federici, National Institute for the Physics of the Matter, INFM).

emission. An object plane field of 18×18 mm was imaged in a 512×512 frame, at a pixel dwell time of 17 ms. Axial scanning has been performed and 21 optical consecutive and parallel slices have been collected at steps of 100 nm. The *x–y* scan step was 35 nm. The scanning head pinhole was set to open position. The measured

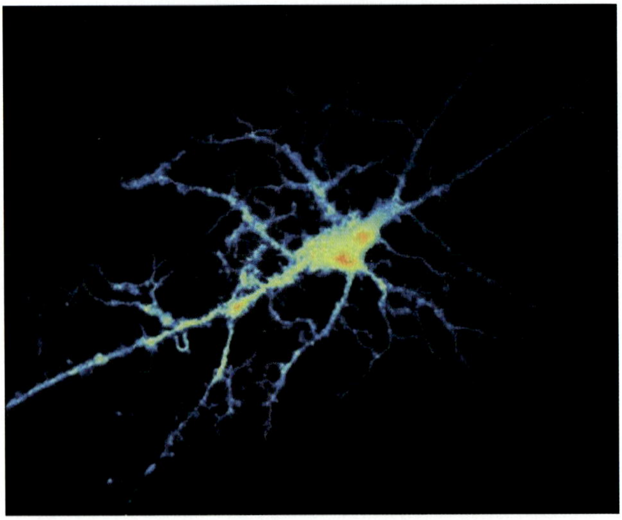

Figure 17. Purkinje cell labelled with Oregon green. Calcium ion concentration is mapped by means of a colour scale from blue (low concentration level) to red (maximum concentration level). (Image taken at LAMBS by Cesare Usai, National Research Council).

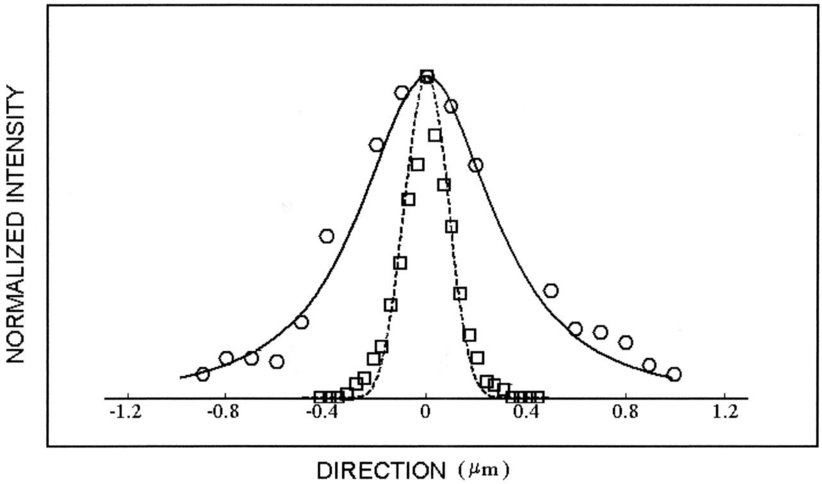

Figure 18. Axial (circles) and lateral (squares) intensity profile of a subresolution fluorescent microsphere to measure the optical resolution properties of an MPE system [126].

full-width at half-maximum (FWHM) lateral and axial resolutions were $210 \times 40\,\text{nm}$ and $700 \pm 50\,\text{nm}$, respectively [126]. Intensity profiles along with the x–y–z directions of experimental data and theoretical expectations are reported in Figure 18. To be sure of operating in MPE, more specifically in the TPE regime

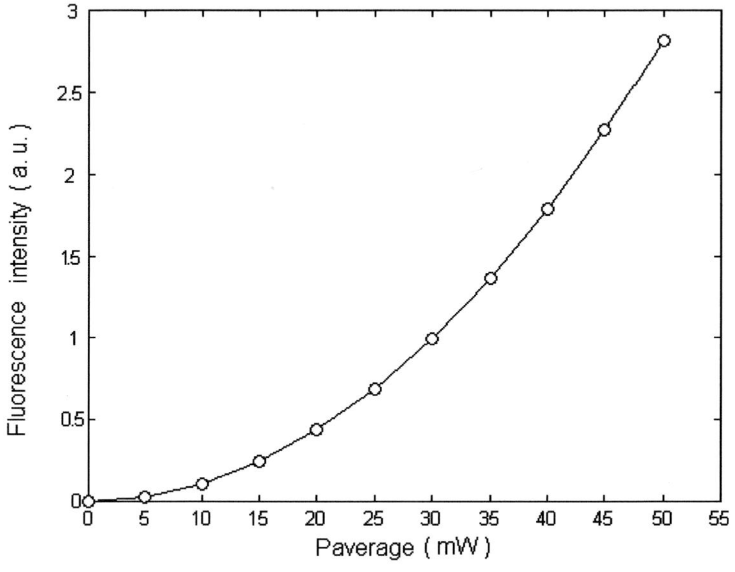

Figure 19. Quadratic behaviour of fluorescence tested with a fluoresceine drop as function of excitation power. The curve shows the fluorescence intensity versus incident power.

Table 6. Comparison between confocal and MPE schemes

	MPE	Confocal
Excitation source	Laser, IR, fs-ps pulsed, 80–100 MHz repetition rate, tunable 680–1050 nm	Laser VIS/UV CW (365, 488, 514, 543, 568, 633, 647 nm)
Excitation/emission separation	wide	Close
Detectors	PMT (typical), CCD, APD	PMT (typical), CCD
Volume selectivity	Intrinsic (fraction of femtolitre)	Pinhole required
Image formation	Beam scanning/AOBS (or rotating disks)	Beam scanning! AOBS (or rotating disks)
Deep imaging	$>500\,\mu m$ (problems related to pulse shape modifications and scattering)	Approx. $200\,\mu m$ (problems related to shorter wavelength scattering)
Spatial resolution	Comparable to confocal, higher signal-to-noise ratio; pinhole increases resolution, good for high fluorescence	Diffraction limited depending on pinhole size
Real time imaging	Possible	Possible
Signal-to-noise ratio	High (especially in non descanned mode)	Good
Fluorophores	All available for conventional excitation plus specifically new designed for MPE	Selected fluorophores depending on laser lines in use
Photobleaching	Only in the focus volume defined through resolution parameters	Within all the double cone of excitation defined by the lens characteristics
Contrast mechanisms	Fluorescence, high order harmonic generation, higher order n-photon excitation, autofluorescence	Fluorescence, reflection, transmission
Commercially available	Yes (but still not mature and too expensive)	Yes (very affordable)

the quadratic behaviour of the fluorescence intensity versus excitation power has to be demonstrated, as plotted in Figure 19. Here the TPE trend was obtained from a solution of fluoresceine. Table 6 summarizes the main differences between a three-dimensional microscope operating in confocal and MPE modes.

16.5 Conclusion

Confocal microscopy is one of the most significant advances in optical microscopy in recent decades, and has become a powerful investigation tool for the molecular, cellular and developmental biologist, materials scientist, biophysicist, and electronic engineer. It is entirely compatible with the range of "classical" light microscopic techniques, and, at least in scanned beam instruments, can be applied to the same specimens on the same optical microscope stage. Its peculiar advantages result in its ability to generate multi-dimensional (*x*–*y*–*z*–*t*) images by non-invasive optical sectioning with a virtual absence of out-of-focus blur, its capacity for multiparametric imaging of multiply labelled samples, and its suitability to investigating at microscopic resolution, large objects thanks to the rejection of scattered light. The advent of confocal microscopy in the mid-1980s favoured the rapid spreading of two- and multi-photon excitation microscopy, and since Denk's report at the beginning of the 1990s there have been dramatic changes in the design of experiments that utilise fluorescent molecules and more specifically in fluorescence three-dimensional optical microscopy. While confocal microscopy is moving to faster architectures in terms of acquisition it is two-photon microscopy that has advanced most in fluorescence optical microscopy. MPE, and more precisely TPE microscopy, constitutes a real progress in science that spread through the scientific community with its intrinsic three-dimensional resolution, the absence of background fluorescence, and the attractive possibility of exciting UV excitable fluorescent molecules, thus increasing sample penetration. In fact, in a TPE scheme two 720 nm photons combine to produce the very same fluorescence as conventionally primed at say 360 nm, and to be utilised in a classical confocal microscope utilising conventional excitation of fluorescent molecules. Excitation of the fluorescent molecules bound to the specific components of the biological systems being studied mainly takes place (80%) in an excitation volume of the order of magnitude of 1 femtolitre. This results in an intrinsic 3D optical sectioning effect. What is invaluable for cell imaging, and in particular for live-cell imaging, is the fact that weak endogenous one-photon absorption and highly localised spatial confinement of the TPE process dramatically reduces phototoxicity stress. The unique characteristics and advantages of MPE can be summarised in the following properties:

(1) Spatially confined fluorescence excitation in the focal plane of the specimen is the key feature of MPE microscopy. It is one of the advantages over confocal microscopy, where fluorescence emission occurs across the entire thickness of the sample being excited by the scanning laser beam. A strong implication is that there is no photon signal from sources out from the geometrical position of the optical focus within the sample. Therefore, the signal-to-noise ratio increases, photo-degradation effects decrease, and optical sectioning is immediately available without the need of pinhole or deconvolution algorithms. Besides, very efficient acquisition schemes can be implemented like the non-descanned one operating at excellent signal-to-noise ratio.

(2) The use of near-IR/IR wavelengths permits examination of thick specimens in depth. This is because, apart from some cases like pigmented samples and portions of the absorption spectral window of water, cells and tissues absorb poorly in the near-IR/IR region. Cellular damage is globally minimised, thus prolonging cell viability and so allowing long-term three-dimensional sessions. Moreover, scattering is reduced and more deep targets can be reached with less problems than with one-photon excitation. The depth of penetration can be up to 0.5 mm. In addition, while in one-photon excitation the emission wavelength is comparatively close to the excitation one (about 50–200 nm longer), in TPE the fluorescence emission occurs at a wavelength substantially shorter and at a larger spectral distance.

Despite the advantages, there are still some practical limitations and open questions that remain to be examined. A severe limitation is given by the high cost of laser sources and by the maintenance costs associated with the limited and unpredictable duration of laser pump diodes. The source costs around 120 K€ and the diodes are 10 k€ each (usually there are two or four diodes for 5 W and 10 W pumps respectively), lasting 2500–5000 h. As other researchers have pointed out, once the technology becomes less expensive and simpler, there will not be a confocal microscope that is not also a two- or multiphoton microscope.

Continuing research in this field is focused on very intriguing problems (the site www.focusonmicroscopy.org offers a complete scenario of the evolution in three-dimensional microscopy over the last five years) like the following: local heating from absorption of IR light by water at high laser power [190] and photo-thermal effects on fluorescent molecules [124], phototoxicity from long wavelength IR excitation and short wavelength fluorescence emission [30,123,147,191], and the development of new fluorochromes better suited for TPE and multiphoton excitation [139].

A major benefit in setting up an MPE microscope is the flexibility in choosing the measurement modality favoured by the simplification of the optical design. In fact, an MPE microscope offers several measurement options without changing any optics or hardware. This means that during the very same experiments one can get real multimodal information from the specimen being studied [48]. Moreover, the usefulness of the MPE scheme is well documented for spectroscopic and life time studies [53,55,58,151,187], for optical data storage and microfabrication [192,193], and for single molecule detection [60,64,120,167,194]. Another very interesting application has been the study of impurities affecting the growth of protein crystals [195] while an important further mention is due also for TPE imaging in plant biology [182] and for measurements in living systems [118,196,197]. The MPE microscope can also be used as an active device – applications are growing in this aspect related to nanosurgery [30], selective uncaging of caged compounds [198], and photodynamic therapy [67,120]. Recently, TPE microscopy, even if in a evanescent-field-induced configuration, has been extended to large area structures of the order of square centimetres [199]. This has application in the realization of biosensing platforms like genomic and proteomic microarrays based upon large planar waveguides. The range of applicability of MPE microscopes is widening rapidly in biomedical, biotechnological and biophysical sciences as well as towards clinical applications [70,71,198].

Acknowledgements

The first Italian TPE architecture realized at LAMBS has been supported by INFM grants. Special thanks go to for Salvatore Cannistraro, Alessandra Gliozzi and Enrico Gratton who believed in the project. This chapter is dedicated to the memory of Mario Arace, Ivan Krekule and Miguel Aguilar. The author would like to thank his wife Teresa and daughter Claudia, and the rabbits, for sharing time with this chapter.

References

1. A. Diaspro (guest editor) (1996). New world microscopy. *IEEE Eng. Med. Biol. Mag.* **15**(1), 29–100.
2. D.J. Arndt-Jovin, R.M. Nicoud, J. Kaufmann, T.M. Jpvin (1985). Fluorescence digital-imaging microscopy in cell biology. *Science*, **230**, 13330–1335.
3. F. Beltrame, B. Bianco, G. Castellaro, A. Diaspro (1985). Fluorescence, absorption, phase-contrast, holographic and acoustical cytometries of living cells. In: A. Chiabrera, H.P. Schwan (Eds), *Interactions between Electromagnetic Fields and Cells* (NATO ASI Series, Volume 97, pp. 483–498). Plenum Press Publishing, New York and London.
4. X.F. Wang, B. Herman (1996). *Fluorescence Imaging Spectroscopy and Microscopy.* Wiley-Liss, New York.
5. B. Herman, H.J. Tanke (1998). *Fluorescence Microscopy*. Springer-Verlag, New York.
6. Sheppard, C.J.R. (2002). The Generalized microscope. In: A. Diaspro (Ed), *Confocal and Two-photon Microscopy: Foundations, Applications and Advances* (pp. 1–18). Wiley-Liss, Inc., New York.
7. F.S. Fay, W. Carrington, K.E. Fogarty (1989). Three-dimensional molecular distribution in single cells analyzed using the digital imaging microscope. *J. Microsc.* **153**, 133–149.
8. P. Benedetti (1998). From the histophotometer to the confocal microscope: the evolution of analytical microscopy. *Eur. J. Histochem.* **42**, 11–17.
9. B. Amos (2000). Lessons from the history of light microscopy. *Nat. Cell Biol.* **2**, E151–E152.
10. E. Abbe (1910). In: O. Lummer, F. Reiche (Eds), *Die Lehre von der Bildentstehung in Mikroskop*. Braunschweig, Germany.
11. M. Weinstein, K.R. Castleman (1971). *Reconstructing 3-D Specimens from 2-D Section Images* (Proc. SPIE Volume 26, pp. 131–138). SPIE Press, Bellingham, WA.
12. D.A. Agard, Y. Hiraoka, P.J. Shaw, J.W. Sedat (1989). Fluorescence microscopy in three-dimensions. *Methods Cell. Biol.* **30**, 353–378.
13. B. Bianco, A. Diaspro (1989). Analysis of the three dimensional cell imaging obtained with optical microscopy techniques based on defocusing, *Cell Biophys.* **15**(3), 189–200.
14. A. Diaspro, M. Sartore, C. Nicolini (1990). Three-dimensional representation of biostructures imaged with an optical microscope: I. digital optical sectioning. *Image Vision Comput.* **8**, 130–141.
15. G.J. Brakenhoff, P. Blom, P. Barends (1979). Confocal scanning light microscopy with high aperture immersion lenses. *J. Microsc.* **117**, 219–232.
16. C.J.R. Sheppard, T. Wilson (1980). Image formation in confocal scanning microscopes, *Optik*, **55**, 331–342.
17. T. Wilson, C.J.R. Sheppard (1984). *Theory and Practice of Scanning Optical Microscopy.* Academic Press, London.
18. K. Carlsson, P.E. Danielsson, R. Lenz, A. Liljeborg, L. Majlof, N. Aslund (1985). Three-dimensional microscopy using a confocal laser scanning microscope. *Opt. Lett.* **10**, 53–55.
19. Shotton (Ed) (1993). *Electronic Light Microscopy. Techniques in Modern Biomedical Microscopy* (Volume 1). Wiley-Liss, New York.
20. P.C. Cheng (Ed) (1994). *Computer Assisted Multidimensional Microscopies.* Springer-Verlag, New York.

21. J.B. Pawley (Ed) (1995). *Handbook of Biological Confocal Microscopy*. Plenum Press, New York.
22. B.R. Masters (1996). *Selected Papers on Confocal Microscopy* (SPIE Milestone Series). SPIE Press, Bellingham, WA.
23. C.J.R. Sheppard, D.M. Shotton (1997) *Confocal Laser Scanning Microscopy*. BIOS, Oxford.
24. A. Periasamy (Ed) (2001). *Methods in Cellular Imaging*, Oxford University Press, New York.
25. A. Diaspro (Ed) (2002). *Confocal and Two-photon Microscopy: Foundations, Applications, and Advances* (pp. 1–576). Wiley-Liss, New York.
26. W. Denk, J.H. Strickler, W.W. Webb (1990). Two-photon laser scanning fluorescence microscopy. *Science* **248**, 73–76.
27. A. Diaspro (guest editor) (1999a). Two-photon microscopy. *Microsc. Res. Technol.* **47**, 163–212.
28. A. Diaspro (guest editor) (1999b). Two-photon excitation microscopy. *IEEE Eng. Med. Biol. Mag.* **18**(5), 16–99.
29. A. Diaspro (1999). Two-photon excitation of fluorescence in three-dimensional microscopy. *Eur. J. Histochem.* **43**, 169–178.
30. K. König (2000). Multiphoton microscopy in life sciences. *J. Microsc.* **200**, 83–104.
31. E. Gratton, N.P. Barry, S. Beretta, A. Celli (2001). Multiphoton fluorescence microscopy. *Methods* 103–110.
32. J.G. White, W.B. Amos, M. Fordham (1987). An evaluation of confocal versus conventional imaging of biological structures by fluorescence light microscopy. *J. Cell Biol.* **105**, 41–48.
33. R.H. Webb (1996). Confocal optical microscopy. *Rep. Prog. in Phys.* **59**, 427–471.
34. E. Gratton, M.J. van de Vende (1995). Laser sources for confocal microscopy. In: J.B. Pawley (Ed), *Handbook of Confocal Microscopy* (pp. 69–97). Plenum, New York.
35. O. Svelto (1998). *Principles of Lasers* (4th edn). Plenum, New York.
36. K. Berland (2002). Basics of fluorescence. In: A. Periasamy (Ed), *Methods in Cellular Imaging* (pp. 5–19). Oxford University Press, New York.
37. P.R. Haughland (Ed) (2002). *Handbook of Fluorescent Probes and Research Chemicals*. Molecular Probes, Eugene, OR.
38. I.S. Harper (2002). Fluorophores and their labeling procedures for monitoring various biological signals. In: A. Periasamy (Ed), *Methods in Cellular Imaging* (pp. 20–39). Oxford University Press, New York.
39. R. Loudon (1983). *The Quantum Theory of Light*. Oxford University Press, London.
40. R.P. Feynman (1985). *QED: The Strange Theory of Light and Matter*. Princeton University Press, Princeton, USA.
41. Y.H. Shih, D.V. Strekalov, T.D. Pittman, M.H. Rubin (1998). Why two-photon but not two photons? *Fortschr. Phys.* **46**, 627–641.
42. W. Denk (1996). Two-photon excitation in functional biological imaging. *J. Biomed. Opt.* **1**, 296–304.
43. S.W. Hell, K. Bahlmann, M. Schrader, A. Soini, H. Malak, I. Gryczynski, J.R. Lakowicz (1996). Three-photon excitation in fluorescence microscopy. *J. Biomed. Opt.* **1**, 71–74.
44. S. Maiti, J.B. Shear, R.M. Williams, W.R. Zipfel, W.W. Webb (1997). Measuring serotonin distribution in live cells with three-photon excitation. *Science* **275**, 530–532.
45. J.N. Gannaway, C.J.R. Sheppard (1978). Second harmonic imaging in the scanning optical microscope. *Opt. Quantum Electron.* **10**, 435–439.

46. R. Gauderon, R.B. C.J.R. Lukins, Sheppard (1999). Effects of a confocal pinhole in two-photon microscopy. *Microsc. Res. Technol.* **47**, 210–214.
47. P. Campagnola, M.D. Wei, A. Lewis, L. Loew (1999). High-resolution nonlinear optical imaging of live cells by second harmonic generation. *Biophys. J.* **77**, 3341–3351.
48. A. Zoumi, A. Yeh, B.J. Tromberg (2002). Imaging cells and extracellular matrix in vivo by using second-harmonic generation and two-photon excited fluorescence. *Proc. Natl. Acad. Sci. U.S.A.* **99**(17), 11014–11019.
49. M. Mueller, J. Squier, K.R. Wilson, G.J. Brakenhoff (1998). 3D microscopy of trasparent objects using third-harmonic generation. *J. Microsc.* **191**, 266–274.
50. J.A. Squier, M. Muller, G.J. Brakenhoff, K.R. Wilson (1998). Third harmonic generation microscopy. *Opt. Express* **3**, 315–324.
51. K.M. Berland, P.T.C. So, E. Gratton (1995). Two-photon fluorescence correlation spectroscopy: method and application to the intracellular environment. *Biophys. J.* **68**, 694–701.
52. P. Schwille, U. Haupts, S. Maiti, W.W. Webb. (1999). Molecular dynamics in living cells observed by fluorescence correlation spectroscopy with one- and two-photon excitation. *Biophys. J.* **77**, 2251–2265.
53. P. Schwille, S. Kummer, A.A. Heikal, W.E. Moerner, W.W. Webb. (2000). Fluorescence correlation spectroscopy reveals fast optical excitation-driven intramolecular dynamics of yellow fluorescent proteins. *Proc. Natl. Acad. Sci. U.S.A.* **97**, 151–156.
54. P.W. Wiseman, J.A. Squier, M.H. Ellisman, K.R. Wilson (2000). Two photon image correlation spectroscopy and image cross-correlation spectroscopy. *J. Microsc.* **200**, 14–25.
55. P.W. Wiseman, F. Capani, J.A. Squier, M.E. Martone (2002). Counting dendritic spines in brain tissue slices by image correlation spectroscopy analysis, *J. Microsc.* **205**, 177–186.
56. K. König, P.T.C. So, W.W. Mantulin, B.J. Tromberg, E. Gratton (1996). Two-photon excited lifetime imaging of autofluorescence in cells during UVA and NIR photostress. *J. Microsc.* **183**, 197–204.
57. T. French, P.T.C. So, D.J. Weaver, Coelho-T. Sampaio, E. Gratton (1997). Two-photon fluorescence lifetime imaging microscopy of macrophage-mediated antigen processing. *J. Microsc.* **185**, 339–353.
58. J. Sytsma, J.M. Vroom, De C.J. Grauw, H.C. Gerritsen (1998). Time-gated fluorescence lifetime imaging and microvolume spectroscopy using two-photon excitation. *J. Microsc.* **191**, 39–51.
59. M. Straub, S.W. Hell (1998). Fluorescence lifetime three-dimensional microscopy with picosecond precision using a multifocal multiphoton microscope. *App. Phys. Lett.* **73**, 1769–1771.
60. J. Mertz, C. Xu, W.W. Webb (1995). Single molecule detection by two-photon excited fluorescence. *Opt. Lett.* **20**, 2532–2534.
61. X.S. Xie, H.P. Lu (1999). Single molecule enzymology. *J. Biol. Chem.* **274**, 15967–15970.
62. M. Sonnleitner, G.J. Schutz, T. Schmidt (1999). Imaging individual molecules by two-photon excitation. *Chem. Phys. Lett.* **300**, 221–226.
63. M. Sonnleitner, G. Schutz, G. Kada, H. Schindler (2000). Imaging single lipid molecules in living cells using two-photon excitation. *Single Mol.* **1**, 182–183.
64. G. Chirico, F. Cannone, S. Beretta, G. Baldini, A. Diaspro (2001). Single molecule studies by means of the two-photon fluorescence distribution. *Microsc. Res. Technol.* **55**, 359–364.

65. F. Cannone, G. Chirico, G. Baldini, A. Diaspro (2003). Measurement of the laser pulse width on the microscope objective plane by modulated autocorrelation method. *J. Microsc.* (OXF), **210**(2), 149–157.
66. G. Chirico, F. Cannone, A. Diaspro (2003b). Single molecule photodynamics by means of one- and two-photon approach. *J. Phys. D: Appl. Phys.* **36**, 1–7.
67. J.D. Bhawalkar, N.D. Kumar, C.F. Zhao, P.N. Prasad (1997). Two-photon photodynamic therapy. *J. Clin. Laser Med. Surg.* **15**, 201–204.
68. A. Diaspro (1998). Two-photon fluorescence excitation. A new potential perspective in flow cytometry. *Minerva Biotechnol.* **11**(2), 87–92.
69. N.S. White, R.J. Errington (2000). Improved laser scanning fluorescence microscopy by multiphoton excitation. *Adv. Imag. Elect. Phys.* **113**, 249–277.
70. B.R. Masters (2002). *Selected Papers on Multiphoton Excitation Microscopy* (SPIE Milestone Series). SPIE Press, Bellingham, WA.
71. A. Periasamy, A. Diaspro (2003). Multiphoton microscopy. *J. Biomed. Opt.* **8**(3), 327–328.
72. R. Hooke (1961). *Micrographia* (facsimile). Dover, New York.
73. B.J. Ford (1991). *The Leeuwenhoek Legacy*. Biopress and Parrand, Bristol and London.
74. M. Minsky (1961). Memoir of inventing the confocal scanning microscope (1988). *Scanning* **10**, 128–138.
75. P. Davidovits, M.D. Egger (1969). Scanning laser microscope. *Nature*, **223**, 831.
76. P.D. Davidovits, M.D. Egger (1971). Scanning laser microscope for biological investigations. *Appl. Opt.* **10**, 1615–1619.
77. C.J.R. Sheppard, A. Choudhury (1977). Image formation in the scanning microscope. *Opt. Acta* **24**, 1051–1073.
78. G.J. Brakenhoff, E.A. van Spronsen, H.T. van der Voort, N. Nanninga (1989). Three-dimensional confocal fluorescence microscopy. *Method. Cell. Biol.* **30**, 379–398.
79. C.J.R. Sheppard, R. Kompfner (1978). Resonant scanning optical microscope, *Appl. Opt.* **17**, 2879–2882.
80. M. Göppert-Mayer (1931). Über Elementarakte mit zwei quantensprüngen. *Ann. Phys.* **9**, 273–295.
81. W. Denk, K. Svoboda (1997). Photon upmanship: why multiphoton imaging is more than a gimmick. *Neuron* **18**, 351–357.
82. F. Wise (1999). Lasers for two-photon microscopy. In: Yuste R., Lanni F., A. Konnerth, (Eds), *Imaging: A Laboratory Manual* (pp. 18.1–9). Cold Spring Harbor Press, Cold Spring Harbor.
83. W. Kaiser, C.G.B. Garrett (1961). Two-photon excitation in CaF_2:Eu^{2+}. *Phys. Rev. Lett.* **7**, 229–231.
84. S. Singh, L.T. Bradley (1964). Three-photon absorption in naphthalene crystals by laser excitation. *Phys. Rev. Lett.* **12**, 162–164.
85. P.A. Franken, A.E. Hill, C.W. Peters, G. Weinreich (1961). Generation of optical harmonics. *Phys. Rev. Lett.* **7**, 118–119.
86. P.M. Rentzepis, C.J. Mitschele, A.C. Saxman (1970). Measurement of ultrashort laser pulses by three-photon fluorescence. *Appl. Phys. Lett.* **17**, 122–124.
87. R. Hellwarth, P. Chistensen (1974). Nonlinear optical microscopic examination of structures in polycrystalline ZnSe. *Opt. Commun.* **12**, 318–322.
88. M.W. Berns (1976). A possible two-photon effect in vitro using a focused laser beam. *Biophys. J.* **16**, 973–977.
89. K. König, I. Riemann, P. Fischer, K.J. Halbhuber (1999). Intracellular nanosurgery with near infrared femtosecond laser pulses. *Cell Mol. Biol.* **45**, 195–201.

90. C.J.R. Sheppard, R. Kompfner, J. Gannaway, D. Walsh (1977). The scanning harmonic optical microscope. *IEEE/OSA Conf. Laser Eng. Appl.* Washington, DC.
91. D.M. Friedrich, W.M. McClain (1980). Two-photon molecular electronic spectroscopy. *Annu. Rev. Phys. Chem.* **31**, 559–577.
92. D.M. Friedrich (1982). Two-photon molecular spectroscopy. *J. Chem. Educ.* **59**, 472–483.
93. R.R. Birge (1986). Two-photon spectroscopy of protein-bound fluorophores. *Acc. Chem. Res.* **19**, 138–146.
94. P.R. Callis (1997). Two-photon-induced fluorescence. *Ann. Rev. Phys. Chem.* **48**, 271–297.
95. D.E. Spence, P.N. Kean, W. Sibbett (1991). 60-fsec pulse generation from a self-mode-locked Ti:sapphire laser. *Opt. Lett.* **16**, 42–45.
96. T.R. Gosnell, A.J. Taylor (Eds) (1991) *Selected papers on Ultrafast Laser Technology* (SPIE Milestone series). SPIE Press, Bellingham, WA.
97. W.G. Fisher, E.A. Watcher, M. Armas, C. Seaton (1997). Titanium:sapphire laser as an excitation source in two-photon spectroscopy. *Appl. Spectrosc.* **51**, 218–226.
98. S.W. Hell (guest editor) (1996). Nonlinear optical microscopy. *Bioimaging* **4**, 121–172.
99. B. Chance (1989). *Cell Structure and Function by Microspectrofluorometry*. Academic Press, New York.
100. R.Y. Tsien (1998). The green fluorescent protein. *Annu. Rev. Biochem.* **67**, 509–544.
101. J.P. Robinson (2001). *Current Protocols in Cytometry*. John Wiley & Sons, New York.
102. J.B. Birks (1970). *Photophysics of Aromatic Molecules*. Wiley Interscience, London.
103. W. Denk, D. Piston, Webb, W.W. (1995). Two-photon molecular excitation in laser scanning microscopy. In: J.B. Pawley (Ed), *Handbook of Confocal Microscopy* (pp. 445–457). Plenum, New York.
104. C.R. Cantor, P.R. Schimmel (1980). *Biophysical Chemistry. Part II: Techniques for the Study of Biological Structure and Function*. Freeman and Co., New York.
105. A. Kriete (Ed) (1992). *Visualization in Biomedical Microscopies.* VCH, Weinheim.
106. D.A. Agard (1984). Optical sectioning microscopy: cellular architecture in three dimensions. *Annu. Rev. Biophys.* **13**, 191–219.
107. M. Bertero, P. Boccacci (1998). *Introduction to Inverse Problems in Imaging*. IOP Publishing, Bristol and Philadelphia.
108. P. Boccacci, M. Bertero (2002). Image restoration methods: basics and algorithms In: A. Diaspro (Ed), *Confocal and Two-photon Microscopy: Foundations, Applications and Advances* (pp. 253–270). Wiley-Liss, Inc, New York.
109. T. Wilson (2002). Confocal microscopy: basic principles and architectures. In: A. Diaspro (Ed), *Confocal and Two-photon Microscopy: Foundations, Applications and Advances* (pp. 19–38). Wiley-Liss, Inc., New York.
110. J. Jonkman, E. Stelzer (2002). Resolution and contrast in confocal and two-photon microscopy. In: A. Diaspro (Ed), *Confocal and Two-photon Microscopy: Foundations, Applications and Advances* (pp. 101–126). Wiley-Liss, Inc., New York.
111. P. Torok, C.J.R. Sheppard, (2002). The role of pinhole size in high aperture two and three-photon microscopy In: A. Diaspro (Ed), *Confocal and Two-photon Microscopy: Foundations, Applications and Advances* (pp. 127–152) Wiley-Liss, Inc., New York.
112. E.H.K. Stelzer, S. Hell, S. Lindek, R. Pick, C. Storz, R. Stricker, G. Ritter, N. Salmon (1994). Non-linear absorption extends confocal fluorescence microscopy into the ultra-violet regime and confines the illumination volume. *Opt. Commun.* **104**, 223–228.
113. K. König, T. Krasieva, E. Bauer, U. Fiedler, M.W. Berns, B.J. Tromberg, K.O. Greulich (1996). Cell damage by UVA radiation of a mercury microscopy lamp probed

by autofluorescence modifications, cloning assay and comet assay. *J. Biomed. Opt.* **1**, 217–22.

114. S.M. Potter (1996). Vital imaging: two-photons are better than one. *Curr. Biol.* **6**, 1596–1598.
115. V.E. Centonze, White, J.G. (1998). Multiphoton excitation provides optical sections from deeper within scattering specimens than confocal imaging. *Biophys. J.* **75**, 2015–2024.
116. M. Gu, C.J.R. Sheppard (1995). Comparison of three-dimensional imaging properties between two-photon and single-photon fluorescence microscopy. *J. Microsc.* **177**, 128–137.
117. D.W. Piston (1999). Imaging living cells and tissues by two-photon excitation microscopy. *Trends Cell Biol.* **9**, 66–69.
118. J.M. Squirrel, D.L. Wokosin, J.G. White, B.D. Barister (1999). Long-term two-photon fluorescence imaging of mammalian embryos without compromising viability. *Nat. Biotechn.* **17**, 763–767.
119. A. Diaspro, M. Robello (2000). Two-photon excitation of fluorescence for three-dimensional optical imaging of biological structures. *J. Photochem. Photobiol. B* **55**, 1–8.
120. P.T.C. So, C.Y. Dong, B.R. Masters, K.M. Berland (2000). Two-photon excitation fluorescence microscopy. *Annu. Rev. Biomed. Eng.* **2**, 399–429.
121. G.J. Brakenhoff, M. Muller, R.I. Ghauharali (1996). Analysis of efficiency of two-photon versus single-photon absorption for fluorescence generation in biological objects. *J. Microsc.* **183**, 140–144.
122. G.H. Patterson, D.W. Piston (2000). Photobleaching in two-photon excitation microscopy. *Biophys. J.* **78**, 2159–2162.
123. A. Hopt, E. Neher (2001). Highly Nonlinear photodamage in two-photon fluorescence microscopy. *Biophys. J.* **80**, 2029–2036.
124. G. Chirico, F. Cannone, G. Baldini, A. Diaspro (2003). Two-photon thermal bleaching of single fluorescent molecules. *Biophys. J.* **84**, 588–598.
125. W.H. Louisell (1973). *Quantum Statistical Properties of Radiation*. Wiley, New York.
126. A. Diaspro (2001). Building a two-photon microscope using a laser scanning confocal architecture, In: A. Periasamy (Ed), *Methods in Cellular Imaging*, pp. 162–179. Oxford University Press, New York.
127. J. Girkin, D. Wokosin (2002). Practical multiphoton microscopy. In: A. Diaspro (Ed), *Confocal and Two-photon Microscopy: Foundations, Applications and Advances* (pp. 207–236). Wiley-Liss, Inc., New York.
128. K. König, H. Liang, M.W. Berns, B.J. Tromberg (1996). Cell damage in near infrared multimode optical traps as a result of multiphoton absorption. *Opt. Lett.* **21**, 1090–1092.
129. O. Nakamura (1999). Fundamentals of two-photon microscopy, *Microsc. Res. Technol.* **47**, 165–171.
130. F.H.M. Faisal (1987). *Theory of Multiphoton Processes*. Plenum Press, New York.
131. D.L. Andrews (1985). A simple statistical treatment of multiphoton absorption. *Am. J. Phys.* **53**, 1001–1002.
132. F.A. Moscatelli (1986). A simple conceptual model for two-photon absorption. *Am. J. Phys.* **54**, 52–54.
133. M. Born, E. Wolf (1980). *Principles of Optics* (6th edn) Cambridge University Press, Cambridge.
134. P.T.C. So, K.H. Kim, C. Buehler, B.R. Masters, L. Hsu, C.Y. Dong (2001). Basic principles of multi-photon excitation microscopy. In: A. Periasamy (Ed), *Methods in Cellular Imaging* (pp. 152–161). Oxford University Press, New York.

135. C. Xu (2002). Cross-sections of fluorescence molecules used in multiphoton microscopy. In: A. Diaspro (Ed), *Confocal and Two-photon Microscopy: Foundations, Applications and Advances* (pp. 75–100) Wiley-Liss, Inc., New York.
136. H.J. Koester, D. Baur, R. Uhl, S.W. Hell (1999). Ca^{2+} fluorescence imaging with pico- and femtosecond two-photon excitation: signal and photodamage. *Biophys. J.* **77**, 2226–2236.
137. C. Xu, J. Guild, W.W. Webb, Denk, W. (1995). Determination of absolute two-photon excitation cross sections by in situ second-order autocorrelation. *Opt. Lett.* **20**, 2372–2374.
138. M. Albota, et al. (1998). Design of organic molecules with large two-photon absorption cross sections. *Science* **281**, 1653–1656.
139. M.A. Albota, C. Xu, W.W. Webb (1998). Two-photon fluorescence excitation crosssections of biomolecular probes from 690 to 960 nm. *Appl. Opt.* **37**, 7352–7356.
140. D.W. B.R. Piston, Masters, W.W. Webb (1995). Three-dimensionally resolved NAD(P)H cellular metabolic redox imaging of the in situ cornea with two-photon excitation laser scanning microscopy. *J. Microsc.* **178**, 20–7.
141. J.R. Lakowicz, I. Gryczynski. (1992). Tryptophan fluorescence intensity and anisotropy decays of human serum albumin resulting from one-photon and two-photon excitation. *Biophys. Chem.* **45**, 1–6.
142. M. Chalfie, Y. Tu, G. Euskirchen, W.W. Ward, D.C. Prasher (1994). Green fluorescent protein as a marker for gene expression. *Science*, **263**, 802–805.
143. M. Chalfie, Kain, S. (Eds) (1998). *Green Fluorescent Protein. Properties, Applications and Protocols*. Wiley-Liss, New York.
144. C.J.R. Sheppard, M. Gu (1990). Image formation in two-photon fluorescence microscopy. *Optik* **86**, 104–106.
145. O. Nakamura (1993). Three-dimensional imaging characteristics of laser scan'fluorescence microscopy: Two-photon excitation vs. single-photon excitation. *Optik* **93**, 39–42.
146. A. Periasamy, C. Noakes, P. Skoglund, R. Keller, A.E. Sutherland (2002). Two-photon excitation fluorescence microscopy imaging in Xenopus and transgenic mouse embryos. In: A. Diaspro (Ed), *Confocal and Two-photon Microscopy: Foundations, Applications and Advances* (pp. 271–284). Wiley-Liss, Inc., New York.
147. K. König, U.K. Tirlapur (2002) Cellular and subcellular perturbations during multiphoton microscopy In: A. Diaspro (Ed), *Confocal and Two-photon Microscopy: Foundations, Applications and Advances* (pp. 191–206). Wiley-Liss, Inc., New York.
148. de Grauw, K., Gerritsen, H. (2002) Aberrations and penetration depth in confocal and two-photon microscopy. In: A. Diaspro (Ed), *Confocal and Two-photon Microscopy: Foundations, Applications and Advances* (pp. 153–170). Wiley-Liss, Inc., New York.
149. Soeller, C. M.B. Cannell (1999). Two-photon microscopy: Imaging in scattering samples and three-dimensionally resolved flash photolysis. *Microsc. Res. Technol.* **47**, 182–195.
150. C. Buehler, K.H. C.Y. Kim, B.R. Dong, Masters, P.T.C. So (1999). Innovations in two-photon deep tissue microscopy. *IEEE Eng. Med. Biol. Mag.* **18**, 23–30.
151. P.T.C. So, K.M. Berland, T. French, C.Y. Dong, Gratton, E. (1996). Two photon fluorescence microscopy: time resolved and intensity imaging. In: X.F. Wang, B. Herman (Eds), *Fluorescence Imaging Spectroscopy and Microscopy* (Chemical Analysis Series, Volume 137, pp. 351–373). J. Wiley & Sons, New York.
152. C. Soeller, M.B. Cannell (1996). Construction of a two-photon microscope and optimisation of illumination pulse duration. *Pfluegers Arch. Eur. J.Physiol.* **432**, 555–561.

153. D.L. Wokosin, J.G. White (1997). Optimization of the design of a multiple-photon excitation laser scanning fluorescence imaging system. In: *Three-dimensional Microscopy: Image, Acquisition and Processing IV* (Proc. SPIE, Volume 2984, pp. 25–29). SPIE Press, Bellingham, WA.
154. S.M. Potter, C.M. Wang, P.A. Garrity, S.E. Fraser (1996). Intravital imaging of green fluorescent protein using two-photon laser-scanning microscopy. *Gene* **173**, 25–31.
155. R. Wolleschensky, T. Feurer, R. Sauerbrey, U. Simon, (1998). Characterization and optimization of a laser scanning microscope in the femtosecond regime. *Appl. Phys. B* **67**, 87–94.
156. W.G. Wier, C.W. Balke, J.A. Michael, J.R. Mauban (2000). A custom confocal and two-photon digital laser scanning microscope. *Am. J. Physiol.* **278**, H2150–H2156.
157. Y.P. Tan, I. Llano, A. Hopt, F. Wurriehausen, Neher, E. (1999). Fast scanning and efficient photodetection in a simple two-photon microscope. *J. Neurosci. Methods* **92**, 123–135.
158. Z.F. Mainen, Malectic-Savic, M., S.H. Shi, Y. Hayashi, R. Malinow, K. Svoboda (1999). Two-photon imaging in living brain slices. *Methods*, **18**, 231–239.
159. A. Majewska, G. Yiu, Yuste, R. (2000). A custom-made two-photon microscope and deconvolution system. *Pflugers Arch. – Eur. J. Physiol.* **441**(2/3), 398–408.
160. V. Iyer, T.M. Hoogland, B.E. Losavio, A.R. McQuiston, P. Saggau (2002). Compact two-photon laser scanning microscope made from minimally modified commercial components. In: A. Periasamy, P.T.C. So (Eds), *Multiphoton Microscopy in the Biomedical Sciences II* (Proc. SPIE, Volume 4620, (pp. 274—280). SPIE, Bellingham, WA.
161. G.S. Benham, S. Schwartz (2002). Suitable microscope objectives for multiphoton digital imaging. In: A. Periasamy, P.T.C. So (Eds), *Multiphoton Microscopy in the Biomedical Sciences II* (Proc. SPIE, Volume 4620, pp. 36–47). SPIE, Bellingham, WA.
162. B.R. Masters, P.T. So, E. Gratton (1997). Multiphoton excitation fluorescence microscopy and spectroscopy of in vivo human skin. *Biophys. J.* **72**, 2405–2412.
163. V. Daria, C.M. Blanca, O. Nakamura, S. Kawata, C. Saloma (1998). Image contrast enhancement for two-photon fluorescence microscopy in a turbid medium. *Appl. Opt.* **37**, 7960–7967.
164. D.L. Wokosin, W.B. Amos, J.G. White (1998). Detection sensitivity enhancements for fluorescence imaging with multiphoton excitation microscopy. *Proc. IEEE Eng. Med. Biol. Soc.* **20**, 1707–1714.
165. D.B. Murphy (2001). *Fundamentals of Light Microscopy and Electronic Imaging*. Wiley-Liss, New York, pp. 1–367.
166. K.K. Hamamatsu Photonics (1999). *Photomultiplier Tubes: Basics and Applications* (2nd edn). Hamamatsu Photonics K.K, Japan.
167. R.A. Farrer, M.J.R. Previte, C.E. Olson, L.A. Peyser, J.T. Fourkas, P.T.C. So (1999). Single molecule detection with a two-photon fluorescence microscope with fast scanning capabilities and polarization sensitivity. *Opt. Lett.* **24**, 1832–1834.
168. K. Fujita, T. Takamatsu, (2001). Real-time in situ calcium imaging with single and two-photon confocal microscopy. In: A. Diaspro (Ed), *Confocal and Two-photon Microscopy: Foundations, Applications and Advances.* Wiley-Liss, Inc., New York.
169. M.B. Cannell, C. Soeller, (1997). High resolution imaging using confocal and two-photon molecular excitation microscopy. *Proc. Royal Microsc. Soc.* **32**, 3–8.
170. A. Periasamy, P. Skoglund, C. Noakes, R. Keller (1999). An evaluation of two-photon excitation versus confocal and digital deconvolution fluorescence microscopy imaging in Xenopus morphogenesis. *Microsc. res. Technol.* **47**, 172–181.

171. D.M. Shotton (1995). Electronic light microscopy – present capabilities and future prospects. *Histochem. Cell Biol.* **104**, 97–137.
172. A. Diaspro, S. Annunziata, M. Robello (2000). Single-pinhole confocal imaging of sub-resolution sparse objects using experimental point spread function and image restoration. *Microsc. Res. Technol.* **51**, 464–468.
173. Carrington, W. (2002). Imaging live cells in 3-d using wide field microscopy with image restoration. In: A. Diaspro (Ed), *Confocal and Two-photon Microscopy: Foundations*, Applications and Advances (pp. 333–346) Wiley-Liss, Inc., New York.
174. A. Diaspro, P. Boccacci, P. Bonetto, M. Scarito, M. Davolio, M. Epifani (2002). Power-up your microscope. www.powermicroscope.com
175. P.E. Hanninen, S.W. Hell (1994). Femtosecond pulse broadening in the focal region of a two-photon fluorescence microscope. *Bioimaging* **2**, 117–121.
176. König K., H. Liang, M.W. Berns, B.J. Tromberg (1995). Cell damage by near-IR microbeams. *Nature* **377**, 20–21.
177. Diaspro and Sheppard (2002). Two-photon excitation microscopy: basic principles and architectures. In: A. Diaspro (Ed), *Confocal and Two-photon Microscopy: Foundations, Applications, and Advances* (pp. 39–74). Wiley-Liss, New York.
178. D.L. Wokosin, V.E. Centonze, J. White, D. Armstrong, G. Robertson, A.I. Ferguson (1996). All-solid-state ultrafast lasers facilitate multiphoton excitation fluorescence imaging. *IEEE J. Select. Top. Quantum Electron.* **2**, 1051–1065.
179. W. Denk, K.R. Delaney, A. Gelperin, D. Kleinfeld, B.W. Strowbridge, D.W. Tank, R. Yuste (1994). Anatomical and functional imaging of neurons using 2-photon laser scanning microscopy. *J. Neurosci. Methods*, **54**, 151–62.
180. König, K., S. Boehme, N. Leclerc, Ahuja, R. (1998). Time-gated autofluorescence microscopy of motile green microalga in an optical trap. *Cell. Mol. Biol.* **44**, 763–770.
181. K. König, A. Gohlert, T. Liehr, I.F. Loncarevic, I. Riemann (2000). Two-photon Multicolor FISH: a versatile technique to detect specific seuqnces within single DNA molecules in cells and tissues. *Single Mol.* **1**, 41–51.
182. U.K. Tirlapur, König, K. (2002). Two-photon near infrared femtosecond laser scanning microscopy in plant biology. In: A. Diaspro (Ed), *Confocal and Two-photon Microscopy: Foundations, Applications and Advances* (pp. 449–468). Wiley-Liss, Inc., New York.
183. J.B. Guild, C. Xu, W.W. Webb (1997). Measurement of group delay dispersion of high numerical aperture objective lenses using two-photon excited fluorescence. *Appl. Opt.* **36**, 397–401.
184. R. Wolleschensky, M. Dickinson, S.E. Fraser (2002). Group velocity dispersion and fiber delivery in multiphoton laser scanning microscopy. In: A. Diaspro (Ed), *Confocal and Two-photon Microscopy: Foundations, Applications and Advances* (pp. 171–190). Wiley-Liss, Inc., New York.
185. M. Gu, X. Gan, A. Kisteman, M.G. Xu (2000). Comparison of penetration depth between two-photon excitation and single-photon excitation in imaging through turbid tissue media. *Appl. Phys. Lett.* **77**(10), 1551–1553.
186. Saloma, C, Saloma-Palmes, C. Kondoh, H. (1998). Site-specific confocal fluorescence imaging of biological microstructures in a turbid medium. *Phys. Med. Biol.* **43**, 1741.
187. A. Diaspro, G. Chirico, F. Cannone, et al. (2001). Two-photon microscopy and spectroscopy based on a compact confocal scanning head. *J. Biomed. Opt.* **6**, 300–310.
188. A. Diaspro, M. Corosu, P. Ramoino, Robello, M. (1999). Adapting a compact confocal microscope system to a two-photon excitation fluorescence imaging architecture. *Microsc. Res. Technol.* **47**, 196–205.

189. A. Diaspro, S. Annunziata, M. Raimondo, M. Robello (1999a). Three-dimensional optical behaviour of a confocal microscope with single illumination and detection pinhole through imaging of subresolution beads. *Microsc. Res. Technol.* **45**(2), 130–131.
190. A. Schonle, S.W. Hell (1998). Heating by absorption in the focus of an objective lens. *Opt. Lett.* **23**, 325–327.
191. R.M. Tyrrell, S.M. Keyse (1990). The interaction of UVA radiation with cultured cells. *J. Photochem. Photobiol. B* **4**, 349–361.
192. B.H. Cumpston, et al. (1999). Two-photon polymerization initiators for three-dimensional optical storage and microfabrication. *Nature* **348**, 51–54.
193. S. Kawata, H.-B. Sun, T. Tanaka. K. Takada (2001). Finer features for functional microdevices. *Nature*, **412**, 697-698.
194. F. Cannone, G. Chirico, A. Diaspro (2003). Two-photon interactions at single fluorescent molecule level. *J. Biomed. Opt.* **8**(3), 391–395.
195. C.L. Caylor, I. Dobrianov, C. Kimmer, R.E. Thorne, W. Zipfel, W.W. Webb (1999). Two-photon fluorescence imaging of impurity distributions in protein crystals. *Phys. Rev. E* **59**, 3831–3834.
196. E.J. Yoder, D. Kleinfeld (2002). Cortical imaging through the intact mouse skull using two-photon excitation laser scanning microscopy. *Microsc. Res. Technol.* **56**(4), 304–305.
197. A. Diaspro, F. Federici, M. Robello (2002). Influence of refractive-index mismatch in high-resolution three-dimensional confocal microscopy. *Appl. Opt.* **41**, 685–690.
198. A. Diaspro, F. Federici, C. Viappiani, S. Krol, M. Pisciotta, G. Chirico, F. Cannone, A. Gliozzi (2003). Two-photon photolysis of 2-nitrobenzaldehyde monitored by fluorescent labeled nanocapsules. *J. Phys. Chem. B*, **107**, 11008–11012.
199. G.L. Duveneck, M.A. Bopp, M. Ehrat, M. Haiml, U. Keller, M.A. Bader, G. Marowsky, S. Soria (2001). Evanescent-field-induced two-photon fluorescence: excitation of macroscopic areas of planar waveguides. *Appl. Phys. B*, **73**, 869–871.

Further reading:

J.D. Axe (1964). Two-photon processes in complex atoms. *Phys. Rev.* **136**, 42–45.

R.R. Birge (1979). A theoretical analysis of the two-photon properties of linear polyenes and the visual chromophores. *J. Chem. Phys.* **70**, 165–169.

W.A. Carrington, R.M. Lynch, E.D.W. Moore, G. Isenberg, K.E. Fogarty, F.S. Fay (1995). Super resolution in three-dimensional images of fluorescence in cells with minimal light exposure. *Science* **268**, 1483–1487.

K. Castleman (2002). Sampling, resolution and digital image processing in spatial and Fourier domain: basic principles. In: A. Diaspro (Ed), *Confocal and Two-photon Microscopy: Foundations, Applications and Advances* (pp. 237–252) Wiley-Liss, Inc, New York.

K.R. Castleman (1996). *Digital Image Processing.* Prentice Hall, Englewood Cliffs, NJ.

R.H. Christie, B.J. Backsai, W.R. Zipfel, et al. (2001). Growth arrest of individual senile plaques in a model of Alzheimer's disease observed by in vivo multiphoton microscopy. *J. Neuroscience* **21**(3), 858–864.

I.J. Cox (1984). Scanning optical fluorescence microscopy. *J. Microsc.*, **133**, 149–153.

I.J. Cox, Sheppard, C.J.R. (1983). Digital image processing of confocal images. *Image Vision Comput.* **1**, 52–56.

A. Diaspro, F. Beltrame, M. Fato, A. Palmeri, P. Ramoino (1997). Studies on the structure of sperm heads of Eledone cirrhosa by means of CLSM linked to bioimage-oriented devices, *Microsc. Res. Technol.* **36**, 159–164.

A. Diaspro, P. Fronte, M. Raimondo, M. Fato, De G. Leo, F. Beltrame, F. Cannone, G. Chirico, P. Ramoino (2002). Functional imaging of living paramecium by means of confocal and two-photon excitation fluorescence microscopy. In: D. Farkas (Ed), *Functional Imaging* (Proc. SPIE, Volume 4622, pp. 47–53). SPIE, Bellingham, WA.

A. Diaspro, D. Silvano, S. Krol, O. Cavalleri, A. Gliozzi (2002) Single living cell encapsulation in nano-organized polyelectrolyte shells. *Langmuir*, **18**, 5047–5050.

C.Y. Dong, B. Yu, L.L. Hsu, P.T.C. So (2002). Characterization of two-photon point spread function in skin imaging applications. In: A Periasamy, P.T.C. So (Eds), *Multiphoton Microscopy in the Biomedical Sciences II* (Proc. SPIE, Volume 4620 pp. 1–8). SPIE, Bellingham, WA.

D. Kleinfeld, W. Denk (2000). Two-photon imaging of neocortical microcirculation. In: R. Yuste, F. Lanni, A. Konnerth (Eds) *Imaging Neurons* (pp. 23.1–23.15). Cold Sping Harbor Laboratory Press, Cold Spring Harbor, New York.

K. König, T.W. Becker, P. Fischer, I. Riemann, K.J. Halbhuber (1999a). Pulse-length dependence of cellular response to intense near-infrared laser pulses in multiphoton microscopes. *Opt. Lett.*, **24**, 113–115.

K. König, U. Simon, K.J. Halbhuber (1996). 3D resolved two-photon fluorescence microscopy of living cells using a modified confocal laser scanning microscope. *Cell Mol. Biol.*, **42**, 1181–1194.

K. König, So, P.T.C., W.W. Mantulin, E. Gratton (1997). Cellular response to near-red femtosecond laser pulses in two-photon microscopes. *Opt. Lett.* **22**, 135–136.

J.R. Lakowicz (1999). *Principles of Fluorescence Microscopy*. Plenum Press, New York.

R.A. Lemons, C.F. Quate (1975). Acoustic microscopy: biomedical applications. *Science*, **188**, 905–911.

Y. Liu, Cheng, D., G.J. Sonek, M.W. Berns, C.F. Chapman, B.J. Tromberg (1995). Evidence of focalized cell heating induced by infrared optical tweezers. *Biophys. J.*, **68**, 2137–2144.

D. Magde, E. Elson, W.W. Webb (1972). Thermodynamic fluctuations in a reacting system: measurement by fluorescence correlation spectroscopy. *Phys. Rev. Lett.* **29**, 705–708.

E.M.M. Manders, J. Stap, G.J. Brakenhoff, van Diel, J.A. Aten (1992). Dynamics of three-dimensional replication patterns during the s-phase analyzed by double labelling of DNA and confocal microscopy. *J. Cell. Sci.* **103**, 857–862.

B.R. Masters, P.T.C. So (1999). Multiphoton excitation microscopy and confocal microscopy imaging of in vivo human skin: a comparison. *Microsc. Microanal.* **5**, 282–289.

L. Moreaux, O. Sandre, J. Mertz (2000). *J. Opt. Soc. Am. B* **17**, 1685–1694.

N.I. Smith, K. Fujita, T. Kaneko, K. Katoh, O. Nakamura, S. Kawata, T. Takamastu (2001). Generation of calcium waves in living cells by pulsed-laser-induced photodisruption. *Appl. Phys. Lett.* **79**, 1208–1210.

D. Ott (2002). Two-photon microscopy reveals tumor development. *Biophot. Int.*, January/February, 46–48.

R. Pike (2002). Superresolution in fluorescence confocal microscopy and in DVD optical storage. In: A. Diaspro (Ed), *Confocal and Two-photon Microscopy: Foundations, Applications and Advances* (pp. 499–524). Wiley-Liss, Inc., New York.

G.T. Rochow, P.A. Tucker (1994). *Introduction to Microscopy by means of Light, Electrons, X Rays, or Acoustics*. Plenum Press, New York and London.

E.J. Sanchez, L. Novotny, G.R. Holtom, X.S. Xie (1997). Room-temperature fluorescence imaging and spectroscopy of single molecules by two photon excitation. *J. Phys. Chem. B*, **101**, 7019–7023.

M. Schrader, S.W. Hell, H.T.M. van der Voort, (1996). Potential of confocal microscope to resolve in the 50-100 nm range. *Appl. Phys. Lett.* **69**, 3644–3646.

P. Schwille (2001). Fluorescence correlation spectroscopy and its potential for intracellular applications. *Cell Biochem. Biophys.* **34**, 383–405.

C.J.R. Sheppard (1977). The use of lenses with annular aperture scanning optical microscopy. *Optik* **48**, 329–334.

C.J.R. Sheppard (1989). Axial resolution of confocal fluorescence microscopy. *J. Microsc.* **154**, 237–241.

P.T.C. So, H. Kim, I.E. Kochevar (1998). Two-photon deep tissue ex vivo imaging of mouse dermal and subcutaneous structures. *Opt. Exp.* **3**, 339–350.

M. Stanley (2001). Improvements in optical filter design. In: A. Periasamy, P.T.C. So (Eds), *Multiphoton Microscopy in the Biomedical Sciences* (Proc. SPIE, p. 4240). SPIE Press, Bellingham, WA.

M. Straub, P. Lodemann, P. Jahn, R. Holroyd, S.W. Hell (2000). Live cell imaging by multifocal multiphoton microscopy. *Eur. J. Cell Biol.* **79**, 726–734.

T. Tanaka, H.B. Sun, S. Kawata (2002). Rapid sub-diffraction-limit laser micro'nanoprocessing in a threshold material system. *Appl. Phys. Lett.* **80**, 312–314.

T. Wilson (1989). Optical sectioning in confocal fluorescent microscope. *J. Microsc.* **154**, 143–156.

T. Wilson (Ed.) (1990). *Confocal Microscopy.* Academic Press, London.

R. Yuste, F. Lanni, A. Konnerth (Eds) (2000). *Imaging neurons: A Laboratory manual.* Cold Spring Harbor Laboratory Press, Cold Spring Harbor, New York.

Part IV
Advanced Imaging Techniques and Novel Ultrasensitive Fluorescence Detection Techniques

Chapter 17

Optical coherence tomography

David D. Sampson and Timothy R. Hillman

Table of contents

Abstract

Optical coherence tomography (OCT) may be described as the extension of low-coherence interferometry to tomographic imaging. It has been primarily directed towards in vivo imaging through thick biological sections, and in particular tissues in the human body, including the eye, gastrointestinal tract, and cardiovascular system. OCT's penetration of highly scattering tissues is limited to a few millimetres, which is lower than ultrasound, magnetic resonance imaging, and X-ray computed tomography, but its resolution for in vivo imaging is higher than these modalities, routinely at around 10 μm, which is sufficient to display clinically relevant morphology, and potentially around 1 μm. After more than a decade of research, OCT is in the early phases of establishing a niche as a medical imaging technology for routine clinical use. In this chapter, we briefly present the early development of OCT in the context of related optical technologies before concentrating primarily on the basic principles of OCT imaging, including the effects of dispersion, noise, and multiple scattering. Finally, we review a selection of the more prominent application areas.

17.1 Introduction

Optical coherence tomography (OCT) combines low-coherence interferometry with lateral point beam scanning to produce two- or three-dimensional images [1]. The low temporal coherence is provided by broadband light and endows the technique with an axial optical sectioning capability – a 'coherence gate'. This capability is similar to that provided by confocal microscopy but the coherence gate does not depend on the aperture of the optical system. OCT has been primarily targeted at in vivo imaging through thick sections of biological systems, particularly in the human body. Largely transparent tissues such as the human eye and developmental biological models such as tadpoles and frogs, as well as turbid, highly-scattering tissues such as the human gastrointestinal tract and cardiovascular system have attracted the most attention.

After more than a decade of research, OCT is in the early phases of establishing a niche as a medical imaging technology for routine clinical use. OCT's penetration of highly scattering tissues is limited to a few millimetres, which is lower than ultrasound, magnetic resonance imaging (MRI) and X-ray computed tomography (CT), but its resolution for in vivo imaging is higher than these modalities, routinely achieving around 10 μm, and potentially around 1 μm. Like ultrasound, OCT's acquisition time is short enough to support tomographic imaging at video rates, which are significantly higher than for CT or MRI, making it more tolerant to subject motion. Ultrasound requires physical contact for good acoustic coupling. OCT however does not require contact so it can be easier to apply, e.g., to the eye. The effective impedance of a medium is also the converse of ultrasound – air spaces represent low impedance for optical wave propagation, but high impedance for ultrasound propagation. Hence, OCT may be used in air-filled hollow organs, whereas ultrasound may not. OCT uses non-ionising radiation at biologically safe levels, allowing long or indefinite exposure times; thus, OCT imaging may be conducted over extended periods if desired. The level of complexity of OCT technology is closer to ultrasound than to CT or MRI and so portable scanners can be realised and costs kept lower to enable a wider range of applications. At present, OCT remains primarily a research tool but investigation into clinical applications and commercial activity are well underway. The clinical applications of most activity and likely comparative advantage are imaging of the eye, primarily the retina, and endoscopic imaging of the gastrointestinal and cardiovascular systems. Most likely, OCT will be in routine clinical use before the end of this decade.

This chapter aims to provide an introduction to OCT technology and to broadly describe its application in the biomedical sphere. There is an emphasis on basic principles that should assist the novice and may be glossed over by the OCT expert. This chapter is written for the 'technical non-specialist' who wishes to examine the relevance of OCT to their own work, or perhaps build an OCT system. To meet this objective in the allotted space, some important areas of OCT have been necessarily omitted; these are mentioned in the concluding remarks. There exists a rapidly expanding body of published material that provides an overview of OCT, generally with a bias towards either the technology or the applications. Works with an emphasis on the technology of OCT include reviews [2–5] and book chapters [6,7]. Works with an applications focus include reviews [8–10], a book chapter [11] and a book [12]. Recent publications include the first edited book devoted solely to OCT [13] and a major review including both technology and applications [14]. There have also been a number of general-interest articles [15–21].

17.1.1 Basic description

Figure 1 shows a schematic diagram of an OCT system in its most basic form. The key components are a Michelson interferometer and a light source with a very broad bandwidth (and correspondingly short coherence time). The coherence time may be thought of as the time by which a light wave can be delayed and remain correlated with itself. It is given approximately by the inverse of the bandwidth and is, therefore, infinite for a monochromatic wave and becomes progressively shorter as the bandwidth increases. The broad bandwidth may be produced by a source with either continuous-wave or short-pulse output.

In the Michelson interferometer, light from the source is split into two paths. In its simplest form, the reference path or arm is terminated by a mirror that is translated to vary the group delay. In the sample path or arm, light is weakly focussed into a sample. A small fraction of the incident light back reflected or back scattered from the sample is captured within the aperture of the sample-path optics. This component combines on the surface of a detector with light reflected from the reference path mirror. For identical polarisations, coherent interference occurs only when the two wavefields are correlated, which requires that the group delays of the two components are matched to within the small range of axial delay

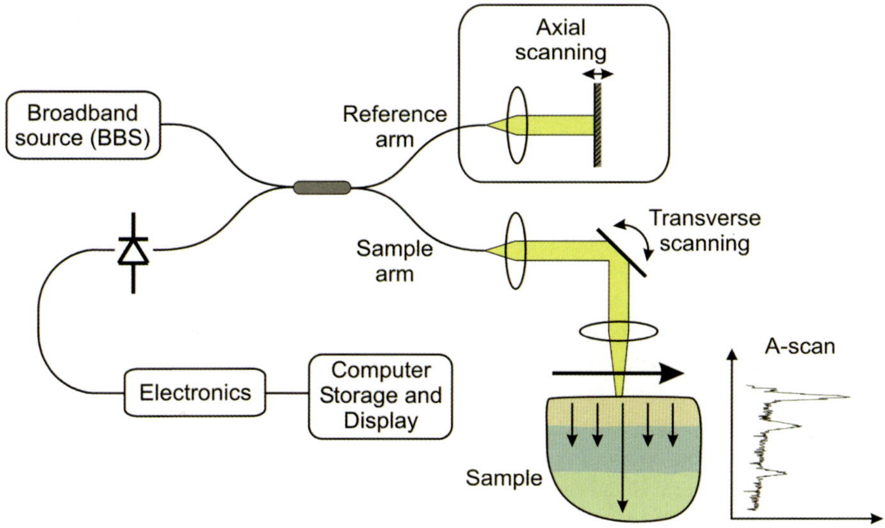

Figure 1. Schematic of an OCT system based on a Michelson interferometer, showing its key components in their simplest forms.

determined by the coherence time of the light. The magnitude of the coherent interference is measured as the reference axial pathlength is scanned by translating the mirror. This corresponds to longitudinal (axial) optical sectioning of the sample, and provides the reflectivity of the sample as a function of group delay. To determine the reflectivity as a function of true physical depth, the group refractive index of the sample medium must be known over the path taken by the wave. So far, the technique as described is usually known as low-coherence interferometry or optical coherence-domain reflectometry.

To construct an image, the beam is then laterally shifted, e.g., by reflection from a tilted mirror in the sample arm, and the axial scan repeated. In this way a two-dimensional slice, i.e., an optical coherence tomograph of the reflectivity profile of a sample, is recorded. A three-dimensional image may be recorded by including a second lateral scan mirror in the sample arm.

Some of the nomenclature used in OCT has been borrowed from ultrasound imaging [22]. An axial line scan is referred to as an A-scan, and when combined with lateral scanning it is referred to as a B-scan or B-mode imaging. Thus, the OCT scan plane is analogous to ultrasound B-mode, and the terms A-scan and B-mode or B-scan are used in the OCT literature. An OCT B-mode image plane is orthogonal to that obtained using confocal microscopy, which we refer to as en face. The OCT scanning engine can, in principle, be configured to record any two-dimensional surface within a sample and is not limited to raster scanning, but en face (usually in the form of optical coherence microscopy) and B-mode are the most common modes. B-mode is more common for two reasons: it provides a depth slice though tissue that is matched to common histological sections; and it is the least technically demanding configuration to build.

17.1.2 Early history and comparison with related optical techniques

OCT may be described as the extension of low-coherence interferometry to tomographic imaging, primarily of biological systems. The term 'optical coherence tomography' was coined in J.G. Fujimoto and co-authors' seminal paper published in *Science* in 1991 [1]. With the benefit of hindsight, the addition of beam scanning to low-coherence interferometry seems an obvious extension, and one which could have been made much earlier, but it was not. The huge consequence of this seemingly modest extension has been the creation of a major new branch of medical imaging, which is now finding clinical application.

Early work on low-coherence interferometry took place more or less contemporaneously with work on time-of-flight reflectometry employing short pulses. The application of short-pulse techniques to biological systems was suggested as early as 1971 [23]. A selection of references to early time-domain ranging in biological media is contained in Ref. 24. Pulse time-of-flight techniques, however, have several disadvantages compared with coherence-domain techniques. The main one, independent of the details of the method employed, is the limit on the time resolution set by the optical pulse width. Until recently, the 100 fs to 10 ps pulse widths of readily available sources have greatly exceeded the few-to-50 fs coherence times of continuous-wave sources, conveying upon time-of-flight techniques much poorer resolution. A popular method of time gating has been the nonlinear cross-correlation of the optical intensity of the probing pulse with a delay-scanned reference pulse. However, the low interaction cross-section of optical nonlinearities has required the use of high optical intensities, and high-intensity probe pulses cause damage in biological media. Linear time-of-flight technologies based on photon counting with picosecond resolution have also been developed [25], but they typically have long acquisition times.

Although not intrinsically a fibre-optic technique, low-coherence interferometry has exploited the development of optical fibre and related photonic technologies since its inception. Low-coherence interferometry was first proposed and demonstrated for micron-resolution optical ranging by three groups in 1987 [26–28]. The motivation for such systems was the characterisation of micro-optical components used in optical fibre communication systems. Youngquist et al. [26] were the first to coin the term 'optical coherence-domain reflectometry' to differentiate continuous ranging over distributed reflections from remote discrete displacement measurements using optical fibre sensors, which was also and still is referred to as low-coherence interferometry [29]. Low-coherence interferometry was applied to the eye as early as 1988 [30,31] and to arteries in 1992 [32]. Early works employed two different configurations. In some works [26,28,32], axial scans were recorded from broadband light coupled into scanning Michelson interferometers in which one arm was terminated by the sample, in an identical manner to OCT. In other works [27,30,31], a coupled pair of interferometers was used in which the sample was one or both paths of the sensing interferometer, which was path matched to a scanning receiving interferometer. This method is equivalent to remote displacement measurement schemes [29] and has the advantage of providing common-path rejection of fluctuations that are present in the optical path connecting the two interferometers, although this advantage has not proven consequential to date.

The first report of OCT imaging [1] contained images of in vitro eye and artery. In vivo measurements using low-coherence reflectometry had already been conducted on the eye [30]. The first in vivo images obtained using OCT were reported in 1993 soon after its initial development [33].

Although OCT may have arisen from low-coherence interferometry, there are several other very closely related coherent microscopies. We shall briefly mention these to provide some perspective on OCT. An alternative description of OCT employing a high-numerical aperture objective in an en face configuration is as a confocal interference microscope employing a low-coherence source. Confocal interference microscopy was first demonstrated in 1979 in transmission [34] and in 1982 in reflection [35], but with highly coherent lasers. Much more recently, confocal microscopy has been demonstrated using a low-coherence source [36], but interference and low-coherence were never brought together prior to the introduction of OCT in 1991. An important distinction must be made between confocal microscopy and OCT, however; OCT has typically employed a low numerical aperture. The resulting weak confocal effect permits axial scanning to be performed without undue attenuation of the beam, enabling axial tomographs in which coherence provides the basis for axial sectioning rather than the confocal effect. Early work on the confocal interference microscope did not demonstrate this mode of operation. The combination of OCT with a strong confocal effect, which was demonstrated in 1994 and termed 'optical coherence microscopy' [37], then introduces the problematic requirement for B-mode operation of synchronous scanning of the confocal and coherence optical sections or 'gates'. Issues surrounding this problem are discussed in Section 17.2.4.

Many coherence-domain techniques are 'full field' in the sense that the entire field of view is illuminated simultaneously, as in conventional microscopy, not point scanned as in confocal microscopy or OCT. Interference microscopy that provides visualisation of interference patterns superimposed on conventional images is as old as microscopy itself [38]. Such interferograms recorded in transmission or reflection from thin sections provide some quantitative information on surface profile, sample thickness, or internal structures. The adaptation of interference microscopy to automated surface profiling, with nanometer accuracy over ranges of tens to hundreds of microns, took place in the 1980s, motivated by the semiconductor industry's need to control the quality of integrated circuits [39,40]. Such techniques have been variously termed white-light interferometry, coherence-probe or coherence-scanning microscopy, correlation microscopy and interference microscopy [41]. Coherence-probe techniques are based on use of a spatially-extended white light source to fully illuminate the surface to be characterised and the imaging of the light reflected from the surface combined with light from the reference path using an area detector [42]. Such interferometric systems may be thought of as an array of OCT systems operating in parallel. The sample is axially scanned with sub-micron step size and each pixel in the image records an axial interferogram. The surface profile is determined from the peak of the envelope and the phase of the central fringe. The extension of parallel detection techniques from surface profilometry to sub-surface optical sectioning has been a much more recent development that has yet to fulfil its promise. Recent developments are described in Ref. 43.

The above discussion places into context the initial development of OCT and its relationship with related optical technologies. Before continuing it is worth summarising the key advantages and disadvantages of OCT to be explored in subsequent sections. The advantages are summed up by the following points.

- The strength of the optical sectioning is conveyed by heterodyne interferometric detection. Because of this OCT can image to greater depths than confocal microscopy in highly-scattering tissues.
- The axial resolution of the optical section is not dependent on the numerical aperture of the optical system, which bestows an advantage when the numerical aperture is limited, e.g., in the human eye or in endoscopic imaging.
- Point scanning avoids crosstalk from neighbouring lateral sites in the sample.
- In common with confocal microscopy, high-speed scanners make real-time operation at video rates feasible.
- Images in depth, which match the orientation of conventional histological sections in many fields of medicine, are readily produced, often making the interpretation of OCT images more readily interpretable by clinicians than en face images.
- The point-scanning feature can be implemented in fibre optics, which makes endoscopic and catheter-based imaging possible.

OCT possesses one important disadvantage compared with other optical techniques:

- OCT images are subject to the corrupting effects of speckle, i.e., the coherent interference of multiple lightwaves, which can limit image fidelity and resolution, as well as depth.

17.2 Theory of optical coherence tomography

Here we present a comprehensive theoretical description of the most common form of optical coherence tomography. We commence with a description of a static Michelson interferometer illuminated with broadband light as a scalar, time-invariant linear system. This is a convenient formalism to elucidate the complex filtering properties of the interferometer and the sample, which include dispersion, a prominent feature in OCT imaging. The formalism is a suitable basis for so-called spectral-domain OCT, but we do not describe it here. The axial resolution is considered in terms of the coherence time and the conventions used in OCT are contrasted with those used elsewhere. The transverse resolution is described and the basis of optical coherence microscopy (OCM) is explained. The constraint of time invariance of the linear system is then removed and we consider generation and detection of the OCT signal under various forms of scanning. Finally, we provide a noise analysis which includes the effects of balanced detection and use it to highlight the trade-offs in system design.

17.2.1 Michelson interferometer with distributed reflectance: linear systems description

We consider a Michelson interferometer, shown schematically in Figure 2, illuminated by polychromatic light described by a real-valued, time-varying electric field $e^{(r)}(t)$. The interferometer is modelled as a one-dimensional linear shift-invariant system [44] in which the sample arm and reference arm are described by complex impulse responses $h(t)$ and corresponding transfer functions $H(\nu)$, subscripted by 'S' or 'R', respectively, where ν represents frequency. These functions are not normalised in order that they may describe attenuation or gain, and are related by the inverse Fourier-transform equation $h(t) = \int_{-\infty}^{+\infty} H(\nu)\exp(-j2\pi\nu t)\mathrm{d}\nu$. We follow Goodman [45] in representing $e^{(r)}(t)$ by its analytic continuation $e(t)$, defined by

$$e(t) \equiv e^{(r)}(t) + je^{(i)}(t) \tag{1}$$

where $e^{(i)}(t)$ is the Hilbert transform of $e^{(r)}(t)$ and $e^{(r)}(t) = \mathrm{Re}\{e(t)\}$. This formalism conveys the advantage that the power spectrum of $e(t)$ is zero for negative frequencies and it is determined by

$$e(t) = 2\int_0^{\infty} E(\nu)\exp(-j2\pi\nu t)\mathrm{d}\nu \tag{2}$$

where $E(\nu)$ is the two-sided, complex spectral density of the real signal, $e^{(r)}(t)$.

The instantaneous intensity $I(t)$ is defined by $I(t) \equiv |e(t)|^2$. The finite response time of the detection process is taken into account by the impulse response $h_\mathrm{D}(t)$ so that the detected intensity is $I^\mathrm{D}(t) = |e(t)|^2 \otimes h_\mathrm{D}(t)$, where $\otimes$ denotes convolution. In the frequency domain, we have $I^\nu \equiv \Im\{I^\mathrm{D}(t)\} = [E(\nu) \otimes E^*(-\nu)]H_\mathrm{D}(\nu)$, where $*$ represents the complex conjugate, $\Im$ is the Fourier transform, defined for a

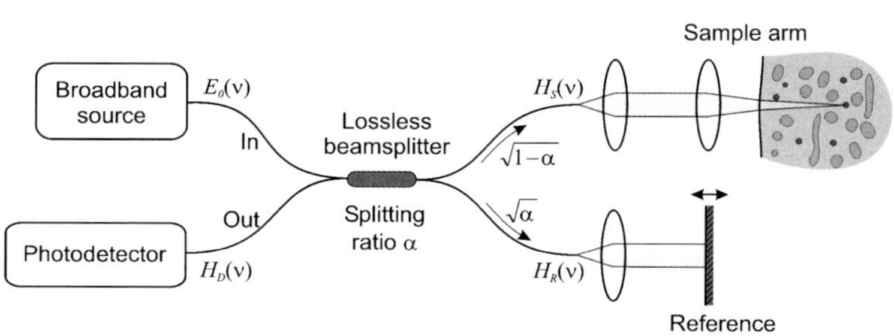

Figure 2. Schematic of a Michelson interferometer denoting some quantities used in the analysis.

function $g(t)$ as $\Im\{g(t)\} \equiv \int_{-\infty}^{\infty} g(t)\exp(j2\pi ft)\mathrm{d}t$, and the convolution is understood to be with respect to v. The operation within the square brackets is equivalent to an autocorrelation. The expression embodies the mixing process which results in the generation and baseband filtering of difference frequencies, total extinction of the sum frequencies. Hence, whereas v in the arguments of $E(v)$ takes on values pertaining to optical frequencies, v in the argument of $H_\mathrm{D}(v)$ takes on baseband values [46].

To start with, we treat the interferometer in the most general terms possible without imposing any specific form on its transfer functions, and determine the general filtering properties of the interferometer on the incident optical spectrum. In the sample arm, we assume ideal plane wave propagation for both incident and backscattered waves. Modifications to account for the confocal effects of focussing lenses and the structure of reflectors will be considered in Section 17.2.4.

We assume a lossless beamsplitter with intensity splitting ratio α. A lossless beamsplitter is required to justify the assumption that the outputs of the interferometer are in exact antiphase. In practice, deviations from exact antiphase are rarely observed, but can be induced by loss [47]. Such deviations are most important when the phase of the interference is to be determined or used to evaluate the envelope. Conservation of energy in a lossless device imposes a relative phase shift of π for reflection versus transmission in a bulk-optic beamsplitter, or $\pi/2$ in a fibre-optic coupler [48]. We consider polarised light at the input of the interferometer described by the analytic signal $e_0(t)$. We further assume that the interferometer and sample contain no birefringence and, hence, there is no alteration of the state of polarisation of the input light, justified on the grounds of simplicity. The reflected electric fields at the ports of the interferometer labelled 'Out' and 'In', respectively, in Figure 2 are given by

$$e_\mathrm{Out}(t) = \sqrt{\alpha(1-\alpha)}h_\mathrm{S}(t) \otimes e_0(t)\exp(j\pi/2) + \sqrt{\alpha(1-\alpha)}h_\mathrm{R}(t) \otimes e_0(t)\exp(j\pi/2) \quad (3)$$

and

$$e_\mathrm{In}(t) = \alpha h_\mathrm{S}(t) \otimes e_0(t)\exp(j\pi) + (1-\alpha)h_\mathrm{R}(t) \otimes e_0(t) \quad (4)$$

In both cases, the path difference between the interfering beams could be represented by a round-trip optical delay τ in the sample arm term, i.e., the term $h_\mathrm{S}(t) \otimes e_0(t)$ might be replaced with $h_\mathrm{S}(t) \otimes e_0(t-\tau)$. Such an approach is particularly suitable for the case of a single plane reflecting sample. However, it has not been used here since we wish to describe a distributed reflectance. Furthermore, the use of a simple delay τ cannot account for the fact that the interferometer response is dependent on the sample spatial axial coordinate and that the delay itself depends on the frequency components of the source. Hence, the effect of the time delay is absorbed into the sample arm impulse response $h_\mathrm{S}(t)$; this will be analysed in Section 17.2.3. The detected intensity at the port labelled 'Out' is then given by

$$I_\mathrm{Out}^\mathrm{D}(t) = \alpha(1-\alpha)\left\{ \begin{array}{l} |h_\mathrm{S}(t) \otimes e_0(t)|^2 + |h_\mathrm{R}(t) \otimes e_0(t)|^2 \\ +2\mathrm{Re}\left\{[h_\mathrm{S}(t) \otimes e_0(t)][h_\mathrm{R}^*(t) \otimes e_0^*(t)]\right\} \end{array} \right\} \otimes h_\mathrm{D}(t) \quad (5)$$

It is instructive to consider the frequency-domain transfer function of the interferometer. The transfer function takes into account phase and group delay, as well as dispersion, but assumes there is no birefringence in the interferometer or the sample. The Fourier transform of Equation (5) is given by

$$I_{\mathrm{Out}}^{v} = \alpha(1-\alpha)\left\{\begin{array}{l} [H_{\mathrm{S}}(v)E_0(v)] \otimes [H_{\mathrm{S}}^{*}(-v)E_0^{*}(-v)] \\ +[H_{\mathrm{R}}(v)E_0(v)] \otimes [H_{\mathrm{R}}^{*}(-v)E_0^{*}(-v)] \\ +[H_{\mathrm{S}}(v)E_0(v)] \otimes [H_{\mathrm{R}}^{*}(-v)E_0^{*}(-v)] \\ +[H_{\mathrm{S}}^{*}(-v)E_0^{*}(-v)] \otimes [H_{\mathrm{R}}(v)E_0(v)] \end{array}\right\} H_{\mathrm{D}}(v) \qquad (6)$$

We now invoke the assumption that the coherence time of the source is much less than the averaging time of the detector; so that we may set $h_{\mathrm{D}}(t) = 1$ and $H_{\mathrm{D}}(v) = \delta(v)$, where δ is the Dirac delta function. The coherence time is considered in detail in Section 17.2.2. Since the axial resolution of OCT depends on employing a broadband source with typically sub-picosecond coherence time, and real-time imaging requires detector response times in the sub-microsecond range, this turns out to be a surprisingly good assumption. Setting $h_{\mathrm{D}}(t) = 1$ would seem to contradict our intuitive understanding that an OCT system requires a finite detector bandwidth because it is a dynamic system based on scanning. In fact, scanning implies that an OCT system taken as a whole is not time invariant. Hence, we must interpret our current theoretical formalism as describing the OCT system in a static mode. We will consider how then to incorporate scanning in Section 17.2.5. Substituting $H_{\mathrm{D}}(v) = \delta(v)$ into Equation (6), we obtain

$$\begin{aligned} I_{\mathrm{Out}}^{v} &= \alpha(1-\alpha)\left\{\begin{array}{l} \int_{-\infty}^{+\infty} H_{\mathrm{S}}(w)E_0(w)H_{\mathrm{S}}^{*}(w-v)E_0^{*}(w-v)\mathrm{d}w \\ + \int_{-\infty}^{+\infty} H_{\mathrm{R}}(w)E_0(w)H_{\mathrm{R}}^{*}(w-v)E_0^{*}(w-v)\mathrm{d}w \\ + \int_{-\infty}^{+\infty} H_{\mathrm{S}}(w)E_0(w)H_{\mathrm{R}}^{*}(w-v)E_0^{*}(w-v)\mathrm{d}w \\ + \int_{-\infty}^{+\infty} H_{\mathrm{S}}^{*}(w-v)E_0^{*}(w-v)H_{\mathrm{R}}(w)E_0(w)\mathrm{d}w \end{array}\right\}\delta(v) \\ &= \alpha(1-\alpha)\left\{\begin{array}{l} \int_{-\infty}^{+\infty} |H_{\mathrm{S}}(w)|^2 G_0(w)\mathrm{d}w + \int_{-\infty}^{+\infty} |H_{\mathrm{R}}(w)|^2 G_0(w)\mathrm{d}w \\ + 2\mathrm{Re}\left\{\int_{-\infty}^{+\infty} H_{\mathrm{S}}(w)H_{\mathrm{R}}^{*}(w)G_0(w)\mathrm{d}w\right\} \end{array}\right\}\delta(v) \end{aligned} \qquad (7)$$

where w is a variable of integration equivalent to the optical frequency, and at the second step we have made the substitution $v = 0$ in accordance with the delta function. The power spectral intensity of the incident light $G_0(v)$ is related to the field spectral density by $G_0(v) = |E_0(v)|^2$. The equivalent equation for the output port labelled 'In' is given by

$$I_{\mathrm{In}}^{\nu} = \left\{ \begin{array}{l} \alpha^2 \int\limits_{-\infty}^{+\infty} |H_{\mathrm{S}}(w)|^2 G_0(w)\mathrm{d}w + (1-\alpha)^2 \int\limits_{-\infty}^{+\infty} |H_{\mathrm{R}}(w)|^2 G_0(w)\mathrm{d}w \\ -2\alpha(1-\alpha)\mathrm{Re}\left\{ \int\limits_{-\infty}^{+\infty} H_{\mathrm{S}}(w)H_{\mathrm{R}}^*(w)G_0(w)\mathrm{d}w \right\} \end{array} \right\} \delta(\nu) \qquad (8)$$

Equations (7) and (8) describe in the most general terms the linear filtering function performed by the interferometer on the input light. The first two terms are only sensitive to both the amplitude response of the respective transfer functions, but the cross term is sensitive to both amplitude and phase responses.

We can write the last term in Equation (7) in the form of an inverse Fourier transform to be evaluated at the origin, giving

$$I_{\mathrm{Out}}^{\nu} = \alpha(1-\alpha) \left\{ \begin{array}{l} \int\limits_{-\infty}^{+\infty} |H_{\mathrm{S}}(w)|^2 G_0(w)\mathrm{d}w + \int\limits_{-\infty}^{+\infty} |H_{\mathrm{R}}(w)|^2 G_0(w)\mathrm{d}w \\ + 2\mathrm{Re}\left\{ \int\limits_{-\infty}^{+\infty} H_{\mathrm{S}}(w)H_{\mathrm{R}}^*(w)G_0(w)\exp(-j2\pi w t')\mathrm{d}w \Big|_{t'=0} \right\} \end{array} \right\} \delta(\nu) \qquad (9)$$

By inverse Fourier transforming Equation (9), we may write the detected intensity in an alternative form given by

$$I_{\mathrm{Out}}^{\mathrm{D}} = \alpha(1-\alpha) \left\{ \begin{array}{l} \lim\limits_{T\to\infty} \frac{1}{T} \int\limits_{-T/2}^{+T/2} |h_{\mathrm{S}}(t) \otimes e_0(t)|^2 \mathrm{d}t + \lim\limits_{T\to\infty} \frac{1}{T} \int\limits_{-T/2}^{+T/2} |h_{\mathrm{R}}(t) \otimes e_0(t)|^2 \mathrm{d}t \\ +2\,\mathrm{Re}\left\{ \Gamma(t') \otimes h_{\mathrm{S}}(t') \otimes h_{\mathrm{R}}^*(-t')|_{t'=0} \right\} \end{array} \right\} \qquad (10)$$

where we have dropped the time dependence of $I_{\mathrm{Out}}^{\mathrm{D}}$ contained in Equation (5) since Equation (10) is based on the assumption of time invariance. In writing Equation (10) we have invoked the Wiener–Khinchin Theorem (as we have previously done implicitly in the progression from Equation (5) to Equation (7)), which makes the connection between the autocorrelation function of $e_0(t)$, $\Gamma(t')$, and the spectral density $G_0(\nu)$, given by

$$\Gamma(t') \equiv \lim_{T\to\infty} \frac{1}{T} \int\limits_{-T/2}^{+T/2} e_0(t+t')e_0^*(t)\mathrm{d}t = \int\limits_{-\infty}^{+\infty} G_0(\nu)\exp(-j2\pi\nu t')\mathrm{d}\nu \qquad (11)$$

The inclusion of the exponential factor in the last integrand of Equation (9) is not entirely artificial. If the optical delay τ had been explicitly incorporated into Equations (3) and (4) instead of being absorbed into $h_{\mathrm{S}}(t)$, the convolution in the interferometric term would not necessarily have to be evaluated at the point $t'=0$; instead, it would be evaluated at $t'=\tau$, and become simply $\Gamma(\tau)\otimes h_{\mathrm{S}}(\tau)\otimes h_{\mathrm{R}}^*(\tau)$. $\Gamma(\tau)$ is itself an analytic signal referred to as the self-coherence function. If we define the sample and reference intensities $I_{\mathrm{S,R}} \equiv \lim_{T\to\infty}(1/T)\int_{-T/2}^{+T/2}|h_{\mathrm{S,R}}(t)\otimes e_0(t)|^2\mathrm{d}t = \int_{-\infty}^{+\infty}|H_{\mathrm{S,R}}(\nu)|^2 G_0(\nu)\mathrm{d}\nu$, then we may further simplify Equation (10) to give

$$I_{\mathrm{Out}}^{\mathrm{D}} = \alpha(1-\alpha)\left[I_{\mathrm{S}} + I_{\mathrm{R}} + 2\mathrm{Re}\left\{ \Gamma(t') \otimes h_{\mathrm{S}}(t') \otimes h_{\mathrm{R}}^{*}(-t')\Big|_{t'=0} \right\} \right] \tag{12}$$

and the equivalent expression for the input port is

$$I_{\mathrm{In}}^{\mathrm{D}} = \alpha^2 I_{\mathrm{S}} + (1-\alpha)^2 I_{\mathrm{R}} - 2\alpha(1-\alpha)\mathrm{Re}\left\{ \Gamma(t') \otimes h_{\mathrm{S}}(t') \otimes h_{\mathrm{R}}^{*}(-t')\Big|_{t'=0} \right\} \tag{13}$$

17.2.1.1 Reduction to a Michelson interferometer with discrete reflectors

We pause to consider the conventional Michelson interferometer, i.e., we consider the sample and reference arms to be lossless, non-dispersive, and terminated in reflectors. The interferometer is characterised by round-trip, frequency-independent group delays τ_{S} and τ_{R}, respectively, and frequency-independent intensity reflectivities R_{S} and R_{R}, respectively. Thus, we have $h_{\mathrm{S}}(t) = \sqrt{R_{\mathrm{S}}}\delta(t-\tau_{\mathrm{S}})$ and $h_{\mathrm{R}}(t) = \sqrt{R_{\mathrm{R}}}\delta(t-\tau_{\mathrm{R}})$ and, equivalently, $H_{\mathrm{S}}(v) = \sqrt{R_{\mathrm{S}}}\exp(j2\pi v\tau_{\mathrm{S}})$ and $H_{\mathrm{R}}(v) = \sqrt{R_{\mathrm{R}}}\exp(j2\pi v\tau_{\mathrm{R}})$, for all v. It is frequency independence that allows us to use the same symbol for the group delay in $h(t)$ and the phase delay in $H(v)$. We assume that $\alpha = \frac{1}{2}$ and define $I_0 \equiv \int_{-\infty}^{+\infty} G_0(v)\,\mathrm{d}v$ as the incident optical intensity. Equation (7) then becomes

$$I_{\mathrm{Out}}^{v}(\tau) = \left\{ \frac{I_0}{4}(R_{\mathrm{S}} + R_{\mathrm{R}}) + \frac{\sqrt{R_{\mathrm{S}}R_{\mathrm{R}}}}{2} \int_{-\infty}^{+\infty} G_0(w)\cos(2\pi w\tau)\mathrm{d}w \right\} \delta(v) \tag{14}$$

where $\tau = \tau_{\mathrm{S}} - \tau_{\mathrm{R}}$ and we have recognised explicitly the dependence of I^{v} on τ. The intrinsic filtering of the source input spectrum can be seen in the integrand in Equation (14), i.e., the interferometer functions as an optical transversal filter with a period dependent on the differential delay τ. The detected intensity is given by

$$\begin{aligned} I_{\mathrm{Out}}^{\mathrm{D}}(\tau) &= \frac{I_0}{4}(R_{\mathrm{S}} + R_{\mathrm{R}}) + \frac{\sqrt{R_{\mathrm{S}}R_{\mathrm{R}}}}{2}\mathrm{Re}\left\{ \int_{-\infty}^{+\infty} G_0(w)\exp(-j2\pi w\tau)\mathrm{d}w \right\} \\ &= \frac{I_0}{4}(R_{\mathrm{S}} + R_{\mathrm{R}}) + \frac{\sqrt{R_{\mathrm{S}}R_{\mathrm{R}}}}{2}\mathrm{Re}\left\{ \Gamma(\tau) \right\} \end{aligned} \tag{15}$$

For the input port we obtain

$$I_{\mathrm{In}}^{\mathrm{D}}(\tau) = \frac{I_0}{4}(R_{\mathrm{S}} + R_{\mathrm{R}}) - \frac{\sqrt{R_{\mathrm{S}}R_{\mathrm{R}}}}{2}\mathrm{Re}\left\{ \Gamma(\tau) \right\}$$

We now consider the form of the interferogram and for convenience set $R_{\mathrm{S}} = R_{\mathrm{R}} = 1$. It is conventional to define the complex degree of coherence $\gamma(\tau)$ by $\gamma(\tau) \equiv \Gamma(\tau)/\Gamma(0)$, with magnitude between 0 and 1, so that the outputs of the interferometer may be expressed as $I_{\mathrm{Out,In}}^{\mathrm{D}}(\tau) = I_0\{1 \pm \mathrm{Re}[\gamma(\tau)]\}/2$. Further, following Goodman [45], the complex degree of coherence can be expressed as $\gamma(\tau) = |\gamma(\tau)|\exp\{-j[2\pi\bar{v}\tau - \alpha'(\tau)]\}$, where $\alpha'(\tau) = \arg\{\gamma(\tau)\} + 2\pi\bar{v}\tau$ and $\bar{v}$ is the mean frequency given by $\bar{v} = \int_{-\infty}^{+\infty} vG_0(v)\mathrm{d}v / \int_{-\infty}^{+\infty} G_0(v)\mathrm{d}v$. The oscillatory term at the mean frequency in $I_{\mathrm{Out,In}}^{\mathrm{D}}(\tau)$ arises from the Fourier-transform relation between the complex degree of coherence and the source power spectral density. For all symmetric functions $G_0(\nu)$, the argument of the inverse Fourier transform, is given entirely by $2\pi\bar{\nu}\tau$, i.e., $\alpha'(\tau) = 0$. The outputs of the interferometer may then be expressed in the familiar form $I^{\mathrm{D}}_{\mathrm{Out,In}}(\tau) = I_0\{1 \pm |\gamma(\tau)|\cos(2\pi\bar{\nu}\tau)\}/2$, which as a function of τ comprises a carrier of frequency $\bar{\nu}$ modulated by a symmetric envelope determined by the coherence function.

It is worth briefly noting the conditions under which the oscillatory term is chirped and the envelope is not symmetric. In the absence of dispersion, the envelope is always symmetric since $G_0(\nu)$ is a real function with symmetric inverse Fourier transform. The carrier is chirped when $G_0(\nu)$ is asymmetric, however, because away from $\tau = 0$, $\alpha'(\tau)$ is non-zero. In the presence of dispersion, both carrier chirping and asymmetric envelopes are possible, as considered in more detail in Section 17.2.3.

17.2.2 Coherence length and axial resolution

Corresponding to the complex degree of coherence $\gamma(\tau)$, we define a normalised power spectral density as $g_0(\nu) \equiv G_0(\nu)/\int_{-\infty}^{\infty} G_0(\nu)\,\mathrm{d}\nu$. Consider the specific example of a Gaussian power spectral density given by

$$g_0(\nu) = \frac{2}{\Delta\nu}\sqrt{\frac{\ln 2}{\pi}}\exp\left[-\left(2\sqrt{\ln 2}\,\frac{\nu-\bar{\nu}}{\Delta\nu}\right)^2\right] \tag{16}$$

where $\Delta\nu$ is the full-width at half-maximum (FWHM) of $g_0(\nu)$. A Gaussian spectrum has generally been preferred in OCT because it corresponds to a Gaussian coherence function, which contains no sidelobes and decays more rapidly than other bell-shaped curves. The corresponding complex degree of coherence is given by

$$\gamma(\tau) = \exp\left[-\left(\frac{\pi\Delta\nu\tau}{2\sqrt{\ln 2}}\right)^2\right]\exp(-j\,2\pi\bar{\nu}\tau) \tag{17}$$

Given the potential for confusion in the definition of the resolution of an OCT signal based on the width of this function, we shall spell out both the conventional definitions and those used in OCT. The width of the coherence function for an arbitrary power spectrum is characterised by the coherence time $\tau_{\mathrm{c}} \equiv \int_{-\infty}^{+\infty} |\gamma(\tau)|^2\mathrm{d}\tau = \int_{-\infty}^{+\infty} [g_0(\nu)]^2\mathrm{d}\nu$. Using this definition, the coherence time of the Gaussian coherence function is given by $\tau_{\mathrm{c}} = \sqrt{2\ln 2/\pi}/\Delta\nu$. Sometimes the coherence time is defined in terms of the full-width at half-maximum, in which case $\tau_{\mathrm{FWHM}} = 4\ln 2/\pi\Delta\nu$. It is convenient to define an associated optical (group) length in a medium, the coherence length, given by $\tilde{l}_{\mathrm{c}} \equiv c\tau_{\mathrm{c}}$ where c is the speed of light in vacuo. To avoid confusion, we will henceforth denote optical lengths with a tilde to distinguish them from physical lengths. It is also convenient to express the coherence length in terms of analogously defined wavelength-domain quantities $\bar{\lambda}$ and $\Delta\lambda$, where $\bar{\lambda} = c/\bar{\nu}$ and $\Delta\lambda = \bar{\lambda}^2\Delta\nu/c$, yielding for a Gaussian power spectral density $\tilde{l}_{\mathrm{c}} = \sqrt{2\ln 2/\pi}(c/\Delta\nu) = \sqrt{2\ln 2/\pi}(\bar{\lambda}^2/\Delta\lambda)$.

There are two changes to these definitions required to conform to the definition of axial resolution used in OCT [49]. Firstly, because of the prevalence of the Gaussian coherence function, it has become conventional to use the full-width at half-maximum as the measure of width. Secondly, it has become conventional to express the resolution in terms of the one-way optical pathlength difference $\tilde{l}_g$, which is related to the group delay difference τ_g by $\tau_g = 2\tilde{l}_g/c$, where the factor of two arises because of the double pass [50]. The group delay difference describes the Gaussian envelope of the complex degree of coherence, as will be further discussed shortly. The half-width at half-maximum coherence length of a Gaussian coherence function is given by $\tilde{l}_c^{\mathrm{HWHM}} = (2\ln 2/\pi)\bar{\lambda}^2/\Delta\lambda$. Because of the double pass, this quantity exactly corresponds to the FWHM of the OCT axial response, i.e.,

$$\tilde{l}_{\mathrm{FWHM}} = \frac{2\ln 2}{\pi}\frac{c}{\Delta\nu} = \frac{2\ln 2}{\pi}\frac{\bar{\lambda}^2}{\Delta\lambda} \tag{18}$$

Since the coherence length relates to the envelope of $\gamma(\tau)$, the corresponding physical one-way pathlength in a medium l_{FWHM} depends on the group index, n_g, according to $\tilde{l}_{\mathrm{FWHM}} = n_g l_{\mathrm{FWHM}}$, hence the resolution is always better in a medium than in air. Figure 3 shows a plot of the complex degree of coherence and the corresponding power spectral density, and indicates the terms defined thus far.

17.2.3 Phase delay, group delay and dispersion

To assess the impact of phase delay, group delay and dispersion on the OCT signal, we consider the transfer functions separated into amplitude and phase responses as $H_S(\nu) = |H_S(\nu)|\exp(j\phi_S)$ and $H_R(\nu) = |H_R(\nu)|\exp(j\phi_R)$, where the phases ϕ_S and ϕ_R are in general functions of frequency. The OCT signal depends on the phase difference, $\phi = \phi_S - \phi_R$, as embodied in Equation (9), which may be written in terms of a Taylor series expansion as

$$\phi(\nu) = \phi(\bar{\nu}) + (\nu - \bar{\nu})\left.\frac{d\phi}{d\nu}\right|_{\bar{\nu}} + \frac{1}{2}(\nu - \bar{\nu})^2\left.\frac{d^2\phi}{d\nu^2}\right|_{\bar{\nu}} + \frac{1}{6}(\nu - \bar{\nu})^3\left.\frac{d^3\phi}{d\nu^3}\right|_{\bar{\nu}} + \cdots \tag{19}$$

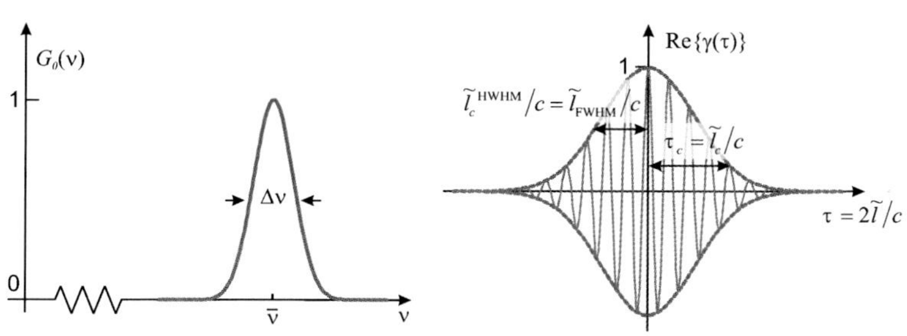

Figure 3. Power spectral density and complex degree of coherence, indicating the relationship between the coherence length and OCT resolution.

where the subscripts signify that the derivatives are to be evaluated at the mean frequency $\bar{\nu}$. We will see that the third-order term in the expansion is needed to describe the impact of dispersion on 1300 nm OCT systems. The first two terms in the outermost curly brackets of Equation (7) are independent of delay and, hence, it is only necessary to consider the cross term, which generates the signal of interest in OCT. As an aside, we note the need to modify the definition of mean frequency used in Equation (19) to account for amplitude filtering:

$$\bar{\nu} \equiv \frac{\int_{-\infty}^{+\infty} \nu |H_S(\nu)||H_R(\nu)|G_0(\nu)\mathrm{d}\nu}{\int_{-\infty}^{+\infty} |H_S(\nu)||H_R(\nu)|G_0(\nu)\mathrm{d}\nu} \tag{20}$$

Considering only the first four terms of the expansion, we may write the last term inside the outermost curly brackets of Equation (7) as

$$2\mathrm{Re}\left\{\int_{-\infty}^{+\infty} H_S(\nu)H_R^*(\nu)G_0(\nu)\mathrm{d}\nu\right\} =$$

$$2\mathrm{Re}\left\{\begin{array}{l}\exp(j2\pi\bar{\nu}\tau_p) \\ \int_{-\infty}^{+\infty}\left[\begin{array}{l}|H_S(\nu)||H_R^*(\nu)|G_0(\nu)\times \\ \exp\left(j\left\{(\nu-\bar{\nu})2\pi\tau_g + (\nu-\bar{\nu})^2 2\pi L D_\nu + \frac{1}{3}(\nu-\bar{\nu})^3(2\pi)^2 L D_\nu^{(1)}\right\}\right)\end{array}\right]\mathrm{d}(\nu-\bar{\nu})\end{array}\right\} \tag{21}$$

where L is the one-way physical pathlength corresponding to the unmatched dispersion and we have defined the phase delay $\tau_p \equiv \phi(\bar{\nu})/2\pi\bar{\nu}$, group delay

$$\tau_g \equiv \frac{1}{2\pi}\frac{\mathrm{d}\phi}{\mathrm{d}\nu}\bigg|_{\bar{\nu}},$$

group-delay dispersion (GDD)

$$D_\nu \equiv \frac{1}{2\pi}\frac{1}{2L}\frac{\mathrm{d}^2\phi}{\mathrm{d}\nu^2}\bigg|_{\bar{\nu}},$$

and second-order GDD

$$D_\nu^{(1)} \equiv \frac{1}{(2\pi)^2 2L}\frac{\mathrm{d}^3\phi}{\mathrm{d}\nu^3}\bigg|_{\bar{\nu}}.$$

The GDD and second-order GDD are related to the first-order dispersion parameter $D_\lambda \equiv (1/2L)(\mathrm{d}\tau_g/\mathrm{d}\lambda)|_{\bar{\lambda}}$ and second-order dispersion parameter $S_\lambda \equiv (1/2L)\times(\mathrm{d}^2\tau_g/\mathrm{d}\lambda^2)|_{\bar{\lambda}}$ commonly used in fibre-optic communications by $D_\nu = -(c/\bar{\nu}^2)D_\lambda$ and $D_\nu^{(1)} = (c^2/2\pi\bar{\nu}^4)S_\lambda + (c/\pi\bar{\nu}^3)D_\lambda$, respectively.

In the absence of dispersion, Equation (21) as a function of delay represents an envelope-modulated carrier. The carrier frequency is determined by the mean optical frequency and the phase delay difference between the sample and the reference arms. The phase-delay difference is related to the one-way optical pathlength difference $\tilde{l}$ by $\tau_p = 2\tilde{l}/c$, and $\tilde{l}$ to the refractive index by $\tilde{l} = \int_{path} n(z) dz$, or for a homogeneous medium simply by $\tilde{l} = nl$. The envelope responds to changes in the group delay $\tau_g(\bar{\nu}) = 2\tilde{l}_g/c$, and $\tilde{l}_g$ is similarly related to the group index by $\tilde{l}_g = \int_{path} n_g(z) dz$. The group and phase indexes are related to each other by

$$n_g = n + \bar{\nu} \left.\frac{dn}{d\nu}\right|_{\bar{\nu}}$$

When the delay is scanned by a plane mirror axially translated in air, the resulting changes in group and phase delays are equal. But such equality is not generally the case when a frequency-domain scanning delay line is employed (considered in Section 17.3).

In a medium, the phase and group delays of a given physical path are not equal since the refractive index is wavelength dependent. Additionally, the wavelength dependence of the refractive index can be nonlinear, resulting in dispersion. If this dependence is not matched over the paths traversed in the sample and reference arms of the interferometer, the result is dispersive spreading of the envelope and concomitant reduction in its peak value, as well as chirping of the carrier. To consider the impact of dispersion on the OCT signal, we again consider the specific case of a Gaussian coherence function [51]. Setting for simplicity $|H_S(\nu)| = |H_R(\nu)| = 1$, substituting Equation (16) into Equation (21), and performing the integration gives for the detected intensity I^D corresponding to the interference term

$$I^D = 2\mathrm{Re}\left\{\exp(j2\pi\bar{\nu}\tau_p)\Im\left\{\exp[-(A + jB)u^2 + jCu^3]\right\}\right\} \quad (22)$$

where $u = \nu - \bar{\nu}$, $\Im$ is the Fourier transform from the u-domain to the τ_g-domain, $A = 4\ln 2/(\Delta\nu)^2$, $B = 2\pi L D_\nu$, and $C = 4\pi^2 L D_\nu^{(1)}/3$. In the absence of second-order group delay dispersion, i.e., if $C = 0$, Equation (22) can be solved exactly and the broadened envelope remains Gaussian [52]. In this case, the axial FWHM is broadened to

$$\tilde{l}_D = \tilde{l}_{FWHM}\sqrt{1 + \left(\frac{\pi}{2\ln 2} D_\nu (\Delta\nu)^2 L\right)^2} = \tilde{l}_{FWHM}\sqrt{1 + \left(\frac{\pi}{2\ln 2}\frac{c}{\bar{\lambda}^2} D_\lambda (\Delta\lambda)^2 L\right)^2} \quad (23)$$

This expression demonstrates the equivalence of normal dispersion ($D_\lambda < 0$) and anomalous dispersion ($D_\lambda > 0$) in broadening of the Gaussian response function. For non-Gaussian functions, narrowing as well as broadening can take place. Narrowing is inevitably accompanied by the introduction of large sidelobes.

When $C \neq 0$, the integral in Equation (22) cannot be solved exactly, but the approximate width of the function can still be evaluated based on the assumption that it remains nearly Gaussian [52] and is given by

$$\tilde{l}_\mathrm{D} = \tilde{l}_\mathrm{FWHM}\sqrt{1+\left(\frac{\pi}{2\ln 2}D_\nu(\Delta\nu)^2L\right)^2+\left(\frac{\pi^2}{4\sqrt{2}(\ln 2)^{3/2}}D_\nu^{(1)}(\Delta\nu)^3L\right)^2}$$
$$= \tilde{l}_\mathrm{FWHM}\sqrt{1+\left(\frac{\pi}{2\ln 2}\frac{c}{\bar{\lambda}^2}D_\lambda(\Delta\lambda)^2L\right)^2+\left(\frac{\pi}{8\sqrt{2}(\ln 2)^{3/2}}\frac{c}{\bar{\lambda}^3}(\bar{\lambda}S_\lambda+2D_\lambda)(\Delta\lambda)^3L\right)^2} \tag{24}$$

The broadening factor in Equation (24) contains an implicit dependence on $\tilde{l}_\mathrm{FWHM}$, which can be made explicit by substituting Equation (18) into Equation (24) to obtain

$$\tilde{l}_\mathrm{D} = \tilde{l}_\mathrm{FWHM}\sqrt{1+\left(\frac{2\ln 2}{\pi}c\bar{\lambda}^2\frac{1}{\tilde{l}_\mathrm{FWHM}^2}D_\lambda L\right)^2+\left[\frac{(\ln 2)^{3/2}}{\sqrt{2}\pi^2}c\bar{\lambda}^3\frac{1}{\tilde{l}_\mathrm{FWHM}^3}(\bar{\lambda}S_\lambda+2D_\lambda)L\right]^2} \tag{25}$$

The sources of unmatched dispersion in OCT imaging are components in the instrument itself [53] or the sample. Dispersion due to the instrument occurs most commonly through the use of fibre components. Typical values of D_λ and S_λ for non-dispersion modified single-mode fibre are $D_{800} = -100\,\mathrm{ps\,nm^{-1}\,km^{-1}}$, $S_{800} = 0.5\,\mathrm{ps\,nm^{-2}\,km^{-1}}$, $D_{1300} = 0\,\mathrm{ps\,nm^{-1}\,km^{-1}}$ and $S_{1300} = 0.09\,\mathrm{ps\,nm^{-2}\,km^{-1}}$ [54], where the units of $\mathrm{km^{-1}}$ refer to the physical length.

Figure 4 shows the broadening factor due to unequal fibre lengths calculated from Equations (23) and (24) as a function of the mismatch length L, scaled for convenience by a suitable value L_0, for 10 and 1 µm axial resolutions. Thus, the amount of excess single-mode fibre required to double the axial resolution when $\tilde{l}_\mathrm{FWHM} = 10\,\mu\mathrm{m}$ is 20 mm at 800 nm, and 540 mm at 1300 nm; when $\tilde{l}_\mathrm{FWHM} = 1\,\mu\mathrm{m}$, the values are 0.20 and 0.54 mm, respectively. Equation (25) shows that the broadening rapidly becomes a linear function of mismatched length.

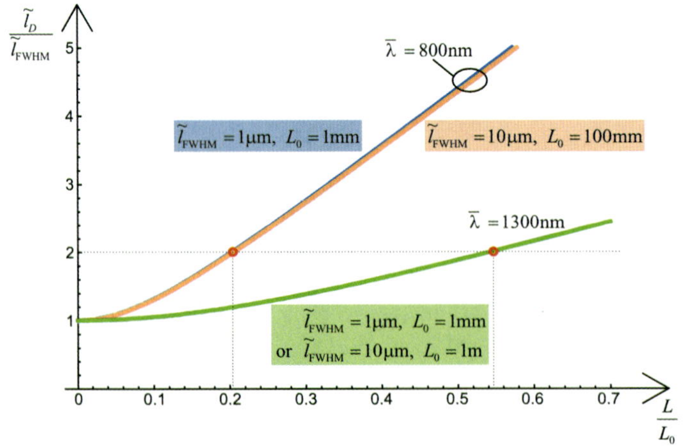

Figure 4. Dispersion broadening factor versus scaled optical fibre length mismatch. Circles mark factor-of-two broadening for the four cases considered.

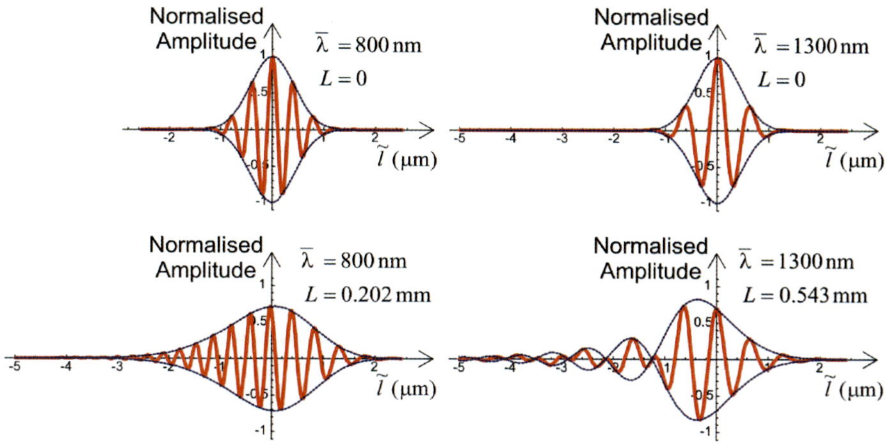

Figure 5. Interferograms at 800 and 1300 nm for a 1 μm resolution and no dispersion (upper) and for sufficient dispersion to cause a factor-of-two broadening of the envelopes (lower) (In the sample, we assume $n = n_g$).

Dispersion balancing at 1300 nm is roughly an order of magnitude less sensitive to fibre length matching than at 800 nm for 10 μm resolution, but this advantage is eroded at 1 μm resolution because of the stronger dependence of the broadening factor upon the second-order GDD.

Figure 5 shows plots of interferograms based on a Gaussian coherence function versus one-way physical pathlength at 1 μm axial resolution for 800 and 1300 nm wavelengths and for the cases of no dispersion and broadening sufficient to double the axial response according to Equation (24), obtained by numerically solving Equation (22). For the 800 nm case, the asymmetry of the broadened interferogram envelope shows that the effect of the non-zero $D_\nu^{(1)}$ is not negligible at 1 μm resolution. However, the $D_\nu^{(1)}$ term in Equation (25) depends on $1/\tilde{l}^3_{\text{FWHM}}$, whereas the D_ν term depends on $1/\tilde{l}^2_{\text{FWHM}}$, so that the asymmetry would be considerably reduced at 10 μm resolution.

Within the assumption that the phase difference $\phi(\nu)$ is adequately represented by the four-term Taylor series expansion in Equation (19), then the broadening factor in Equations (24) and (25) gives the exact increase in the root-mean-square width of the squared interferogram envelope, no matter what its functional form. However, if the broadened envelope deviates significantly from Gaussian in shape, then there will be some error in using it to assess the FWHM. This deviation is evident in Figure 5. For the 800 and 1300 nm cases, respectively, the FWHM broadens by 2.6% and 32% less than the factor of two expected for a Gaussian. These percentages are 0.03% and 32% for broadening of a 10 μm resolution. Thus, Figure 4 is a useful guide to determine the dispersion matching requirements in an OCT system.

There are additional consequences of dispersion evident in Figure 5 – reduction of the envelope amplitude and chirping of the carrier. Assuming the unbroadened and broadened axial response are described by Gaussian functions, conservation of energy implies that the product of the square of the peak amplitude and width of a signal is a constant; hence, if the width doubles the peak is reduced by $1/\sqrt{2}$.

Figure 5 bears this out, to the extent that the lower curves are indeed Gaussian. Chirping of the carrier is evident in Figure 5, particularly in the lower left curve. The impact of this chirp is generally small, depending on the processing of the signal (discussed in Section 17.2.5).

Unbalanced dispersion caused by material dispersion in the system components does not vary during an axial scan and, hence, it may be readily eliminated by compensation. Thus, while balancing dispersion in the instrument is a practical issue in system implementation in order to maximise the resolution, it is not a fundamental limitation. The frequency-domain delay line (Section 17.3) is particularly advantageous for dispersion balancing because it can independently alter dispersion and delay. Dispersion can also be compensated post detection if the full phase and amplitude information is available [55–57].

Dispersion in the sample itself is a more fundamental issue because, since it is not known a priori and depends on the total pathlength in the sample, it cannot be balanced at all positions during an axial scan. At 10 μm resolution, sample dispersion is only really a significant problem in OCT imaging over long lengths in the eye [58], but at 1 μm resolution it becomes an important issue [59]. In Ref. 59, such resolution required adjustment of the dispersion balance for each sample depth. Post-processing-based compensation of OCT images with known, scan-independent dispersion imbalance has been demonstrated [56], as has compensation for the depth-dependent dispersion of 144 μm of glass under very broadband illumination [60]. However, these compensation techniques all require significant processing to generate the compensated scans and images, a difficulty for real-time operation. Alternatively, if the reference arm is made to have the same scanlength-dependent dispersion as the sample, then dispersion compensation may be performed in real time with each scan. This has been demonstrated using the frequency-domain optical delay line [61] and is described in Section 17.3. As OCT resolution improves towards 1 μm, real-time dispersion compensation is expected to become much more important.

17.2.4 Lateral resolution and confocal optical sectioning

Whilst the coherence length of the source is the main determinant of the OCT axial resolution (Equation (17)), the transverse resolution is entirely dependent on the sample arm optics. When a collimated Gaussian beam with ($1/e$ field) radial beam width W'_0 is incident upon a lens with focal length f, assuming no truncation of the beam by the lens, the width of the focused beam waist will be given by $W_0 = (1/\pi)(\bar{\lambda}f/W'_0)$ [62] (independent of the sample refractive index in the paraxial approximation), corresponding to a FWHM transverse resolution of (cf. [63]):

$$l_{\mathrm{T}} = \left(\frac{2\sqrt{\ln 2}}{\pi}\right)\left(\frac{\bar{\lambda}f}{W'_0}\right) \tag{26}$$

and a depth of focus equal to twice the Rayleigh range:

$$2z_{\mathrm{R}} = \frac{2\pi n W_0^2}{\bar{\lambda}} \approx \frac{2n\bar{\lambda}}{\pi(\mathrm{NA})^2} \tag{27}$$

where NA represents the numerical aperture of the beam given approximately by $W_0^{'}/f$. We observe from Equation (15) that the interferometric component of the OCT signal is dependent on the square root of the reflected power and, hence, on the magnitude of the Gaussian beam field. Combining the width of a propagating Gaussian beam with the OCT axial point-spread function, we can calculate the envelope of the three-dimensional OCT point-spread function when the optical path matchpoint is located at the beam waist. A contour map of this function is shown in Figure 6, and the shape of the FWHM sample volume is indicated. Ref. 64 takes further the notion of OCT resolution, providing a definition of the full interferometric three-dimensional point-spread function of an OCT scanner, even allowing for low-angle multiple scattering.

We now examine the interferometric component of the detector response as a function of the sample reflectivity function. We shall neglect the difference between group and phase delay in the sample and reference arms of the interferometer, and express the general delay difference as $\tau = 2\tilde{l}/c$. We define sample- and reference-arm optical pathlengths $\tilde{l}_S$ and $\tilde{l}_R$, with their origins suitably located such that the optical pathlength difference is $\tilde{l} = \tilde{l}_S - \tilde{l}_R$. With a single sample reflector, this equation, along with Equations (17) and (18), may be substituted into the interferometric component of Equation (15) to yield:

$$I^D(\tilde{l}) = \frac{\Gamma(0)\sqrt{R_S R_R}}{2} \exp\left[-\left(\frac{2\sqrt{\ln 2}\tilde{l}}{\tilde{l}_{FWHM}}\right)^2\right] \mathrm{Re}\left\{\exp(-j\bar{k}(2\tilde{l}))\right\} \tag{28}$$

where $\bar{k} = 2\pi/\bar{\lambda}$ is the mean wavenumber. Now, the sample reflectivity R_S must be modified to take account of the illumination/collection properties of the optical

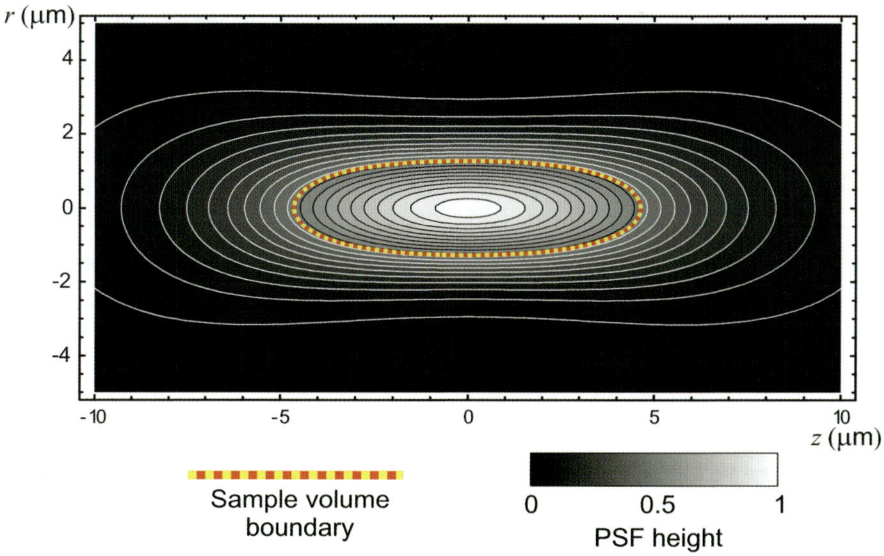

Figure 6. Contour map of the OCT point-spread function in the transverse (r) and axial (z) directions, with the FWHM sample volume indicated. (Parameters used: NA = 0.26, $n = 1.3$, $\bar{\lambda} = 1300$ nm, $\Delta\lambda = 54$ nm.)

system, which generates a confocal effect. It should, in principle, be corrected by a confocal axial response function $h_F[(l_S - l_F)/n]$, where l_F is the axial location of the focal plane [3,65]. (The dependence of this function on sample phase refractive index assumes the paraxial approximation.) Accounting for the axially-distributed square-root reflectivity $\sqrt{R_S}(l_S)$, and ignoring constant factors, Equation (28) becomes

$$I^D(\tilde{l}_R) \propto \int_0^\infty \sqrt{R_S}(l_S)\sqrt{h_F\left(\frac{l_S - l_F}{n}\right)} \exp\left[-\left(\frac{2\sqrt{\ln 2}\tilde{l}}{\tilde{l}_{FWHM}}\right)^2\right] \cos(2\bar{k}\tilde{l})\, \mathrm{d}\tilde{l}_S \tag{29}$$

Correction of the reflectivity is often ignored since the width of the confocal axial response function is generally much greater than the coherence length of the source. Indeed, herein lies the difference between the distinct modalities of OCT and optical coherence microscopy (OCM). In OCT, a low-NA objective lens is used to ensure a long depth of focus in the axial direction, so that the confocal axial response remains approximately constant across the length of the scan. In OCM, a high-NA objective is utilised to provide a very narrow axial confocal response (and consequently a limited field of view). Figure 7 elucidates this point.

We now examine the form of the confocal axial response function $h_F[(l_S - l_F/n)]$ for the specific case of a fibre-optic implementation. Intuitively, the fibre-optic system provides confocal optical sectioning since the single-mode fibre tip acts as a confocal 'pinhole', for which the transmitted and collected light must occupy the same spatial mode. Additionally, only the light scattered from the sample into this mode will generate an interference signal with the reference beam. Ref. 66 treats this problem rigorously assuming use of collimating lens/objective (confocal) lens pair, and determines that the fibre-optic confocal microscope behaves as a coherent microscope, and the axial response to a transverse plane reflector obeys Equation (30):

$$h_F\left(\frac{l_S - l_F}{n}\right) = \left|\frac{A[1 - \exp(-A + ju)]}{(A - ju)[1 - \exp(-A)]}\right|^2 \tag{30}$$

where $A = [(2\pi a_0 r_0)/\lambda d]^2$ is a dimensionless 'confocal parameter', in which a_0 is the radius of the objective lens, r_0 is the fibre spot size and d the focal distance of

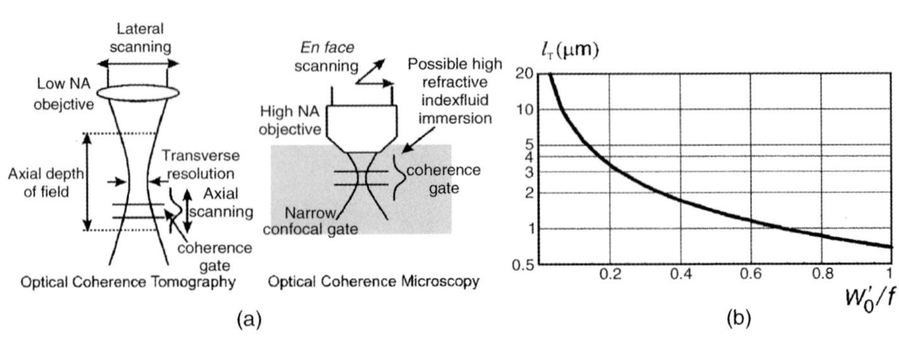

Figure 7. (a) Sample optics for OCT and OCM. Adapted from [65]. (b) FWHM transverse resolution versus Gaussian beam width scaled by focal length.

the collimating lens. The parameter A is comparable to the normalised pinhole radius in a hard-aperture confocal microscope. The axial optical coordinate is $u = (8\pi/\lambda)[(l_S - l_F)/n]\sin^2(\alpha/2)$, where $\sin\alpha = a_0/\sqrt{a_0^2 + f^2}$ is the numerical aperture of the confocal lens.

Equation (29) shows that an axial scan essentially takes the form of a convolution between the confocal response-modified sample reflectivity and the axial coherence function. The multiplicative effect of both gates in the OCM case can be appreciated by considering a plane reflector being passed through the stationary, co-located confocal and coherence gates ($n_g l_F = \tilde{l}_R$). Then expressing the received signal in terms of the axial optical location of the reflector $\tilde{l}_P$, with unit reflectivity, gives

$$I^D(\tilde{l}_P) \propto \sqrt{h_F\left(\frac{\tilde{l}_P - \tilde{l}_R}{n_g n}\right)} \exp\left[-\left(\frac{2\sqrt{\ln 2}(\tilde{l}_P - \tilde{l}_R)}{\tilde{l}_{FWHM}}\right)^2\right] \cos\left[2\bar{k}(\tilde{l}_P - \tilde{l}_R)\right] \quad (31)$$

which is the product of a confocal and a coherence function. Figure 8 demonstrates the enhanced optical sectioning effect of the inclusion of a confocal gate. The OCM response combines both the narrow peak of the confocal response with the rapid attenuation of OCT far from the optical path matchpoint.

In general, it is necessary to use small numerical apertures for deep imaging into samples, a trade-off between transverse resolution and working distance. Numerical apertures of 0.05 to 0.1 are typical, giving Rayleigh ranges at 1300 nm of 170 and

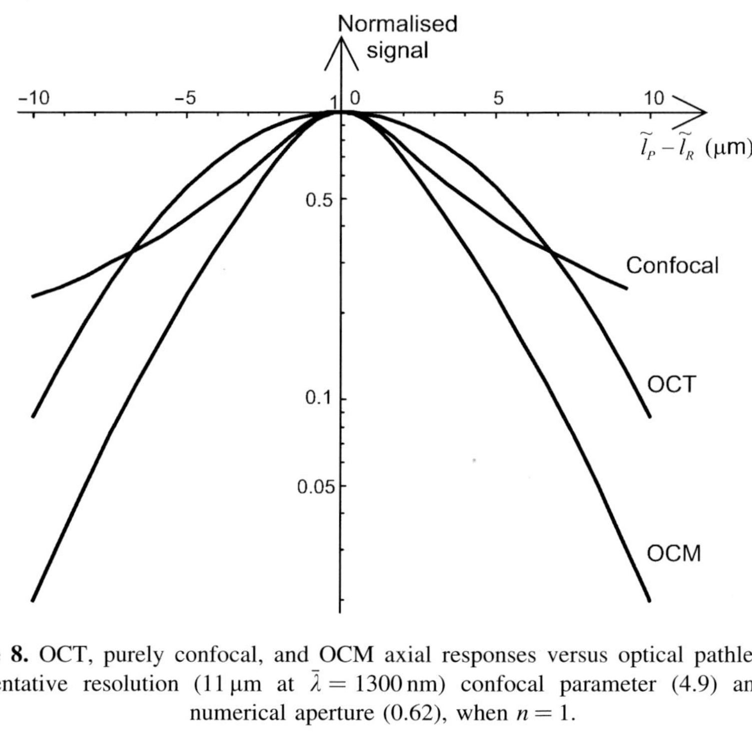

Figure 8. OCT, purely confocal, and OCM axial responses versus optical pathlength for representative resolution (11 μm at $\bar{\lambda} = 1300$ nm) confocal parameter (4.9) and beam numerical aperture (0.62), when $n = 1$.

41 μm, respectively. However, as indicated by Figure 8, the OCM modality actually produces a narrower response function than either OCT or confocal microscopy. This is due to enhanced out-of-focus rejection, since to contribute to the signal, light must be both coherently detected and occupy the receiver spatial mode, conditions that are satisfied by the ballistic light from the focal plane, but by very little of the multiply scattered light. Effective maximal imaging depths may be calculated by comparing the relative amounts of each that contribute to the image [37].

A fundamental difference between confocal microscopy and OCT should be noted here. In a confocal microscope, the image information is contained in a DC intensity term: the measured signal is equivalent to a convolution between the low-pass confocal point-spread function and the sample power reflectivity. In OCT, axial image information is contained only within the heterodyne band-pass cross term and, hence, demodulation schemes can be used to separate it from the DC intensity terms to retrieve the sample reflectivity distribution [34].

OCM is faced with the technical problems associated with coupling together both a confocal and a coherence gate [65]. In the en face imaging approach, this problem is minimized since no axial scanning is necessary. However, in B-mode operation, difficulties arise due to the dependence upon sample refractive index of the rate at which the two gates traverse the medium. Specifically, a reference arm scan of distance Δz results in coherence gate shift of physical distance $\Delta z/n_g$, whereas an objective translation of Δz yields a confocal gate shift of physical distance $n\Delta z$, assuming that the paraxial approximation holds. (This approximation will not hold for a high-NA objective unless index matching fluid is used between the objective and the sample.) As such, sophisticated focus-tracking techniques are necessary to perform axial scanning [67,68]. These must be dynamic and, in principle, not deterministic in nature if they are to be effective in navigating samples with randomly-varying refractive index. The scheme in Ref. 67 is particularly appealing. It is based on mounting the reference mirror and the sample on the same linear scanning stage. Shifting the stage by Δz shifts the coherence gate by $2\Delta z/n_g$ and the confocal gate by $n\Delta z$. If $n_g = n = \sqrt{2}$, which fortuitously is a reasonable approximation for a wide range of biological samples, the shifts in gate positions are equal.

17.2.5 Time-varying electrical signal

We now consider the correspondence between the time-varying electrical signal arising from pathlength scanning and the linear systems description given above. Discussions of this topic are also found in Refs. 7 and 69. Since OCT systems commonly operate by scanning the reference arm group delay, we must relax our previous assumption of time invariance and consider a detector impulse response that is no longer simply equal to a constant. We now consider a time-varying linear transfer function corresponding to the changes induced by scanning.

We assume the interferometer reference path is scanned with corresponding time-dependent group delay $\tau_g(t)$ and phase delay $\tau_p(t)$. We set the dispersion to zero for simplicity and also assume that the reflectors in the sample are well separated so that no distortion of the envelope occurs through coherent interference (i.e. no speckle). The intensity on the surface of the detector can then be described by (cf. Equation (12))

$$I^{D}_{Out}(t) = \alpha(1-\alpha)\Big\{I_S(t) + I_R(t) + 2\sqrt{I_S(t)I_R(t)}\left|\gamma(\tau_g)\right|\cos\left[2\pi\bar{\nu}\tau_p(t) + \phi(t)\right]\Big\} \otimes h_D(t) \qquad (32)$$

where ϕ is a phase to account for deliberately imparted phase modulation as well as undesired phase fluctuations caused, e.g., by changes in $\bar{\nu}$, or in pathlengths as a result of mechanical instability. The self-mixing terms $I_S(t)$ and $I_R(t)$ have a low-pass frequency response that is independent of the optical frequency and bandwidth. If $I_R \gg I_S$, as should be the case, then $I_S(t)$ can be neglected. However, $I_R(t)$ can not necessarily be neglected because its amplitude modulation (caused by non-ideal scanning) could be important. The cross-correlation term has a band-pass response with centre frequency determined by the rate of change of phase delay, i.e., $\bar{f} = \bar{\nu}(\mathrm{d}\tau_p/\mathrm{d}t) + (1/2\pi)(\mathrm{d}\phi/\mathrm{dt})$. If we neglect for the moment the $\mathrm{d}\phi/\mathrm{d}t$ term and consider the source of pathlength scanning to be linear over a physical length l in time Δt, with velocity $\mathrm{v} = l/\Delta t$ and an associated phase velocity $\mathrm{v_p}$, then since $\tau_p = 2l/\mathrm{v_p}$, the resulting carrier (fringe) frequency is $\bar{f} = \bar{\nu}(2\mathrm{v}/\mathrm{v_p})$. If the pathlength scanning is achieved by translating a retroreflector in air, then the carrier frequency is given by the corresponding Doppler shift, i.e., $\bar{f} = (\bar{\nu}/c)2\mathrm{v} = (2\mathrm{v}/\bar{\lambda})$. Thus, an OCT system using a source with a 1 μm mean wavelength and a retroreflector scanning at 10 mm s^{-1} will generate a carrier of frequency 20 kHz. We note that real-time operation requires much faster scanning – equivalent to linear velocities of the order of 1 m s^{-1}, and a corresponding carrier frequency of the order of 2 MHz. Note that the carrier frequency will also be altered by any relative motion of the scatterers in the sample.

The spectral bandwidth (FWHM) Δf of the detected electrical signal is determined by the width of the envelope, i.e., of $|\gamma(\tau_g)|$, and the rate at which it is scanned, $\mathrm{d}\tau_g/\mathrm{d}t$. Thus, the bandwidth is proportional to the rate of change of group delay scaled by the inverse of the envelope width, $\Delta f = \Delta\nu(\mathrm{d}\tau_g/\mathrm{d}t)$. Again, for a translating retroreflector with linear velocity v, the signal bandwidth is given by $\Delta f = \Delta\nu(2\mathrm{v}/c) = 2\mathrm{v}(\Delta\lambda/\bar{\lambda}^2)$, and we also have $\bar{f}/\Delta f = \bar{\nu}/\Delta\nu = \bar{\lambda}/\Delta\lambda$. Using the same parameters as above and assuming a Gaussian coherence function with 10 μm resolution gives, for velocities of 10 mm s^{-1} and 1 m s^{-1}, bandwidths of 880 Hz and 88 kHz, respectively. Figure 9 is a schematic diagram illustrating the frequency spectrum of the OCT signal, as well as its various noise components to be considered in the next section.

The detected OCT signal is a replica of the correlation-domain signal downshifted by the factor $2\mathrm{v}/\mathrm{v_p}$ with an envelope scaled by $2\mathrm{v}/\mathrm{v_g}$. Only phase-correlated fields in the square-law mixing process contribute to the signal. For incoherent sources, this implies reference and sample arm fields at the same optical frequency. As a result, it is possible to define an isomorphic mapping of optical frequency to electrical frequency [69]. The mixing of reference and sample arm

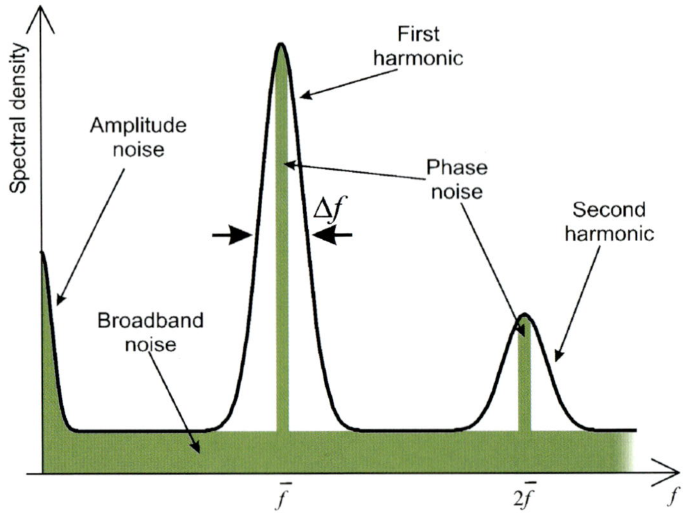

Figure 9. OCT signal electrical spectral density, including noise. Amplitude noise arises from fluctuations in the reference signal I_R induced by unstable scanning, for example. Broadband noise includes shot noise, electrical noise, and source-induced and cross-beat intensity noise. Second and higher harmonics depend on the phase modulation.

fields with uncorrelated phases results in noise. For incoherent sources, fields at different optical frequencies are uncorrelated and so all optical frequency differences map to a given electrical frequency, which is a form of intensity noise (discussed in Section 17.2.6).

The above discussion has not explicitly considered the filtering of the detected signal as described by the transfer function of the electronics. We can immediately appreciate that electronic filtering has the propensity to distort the detected signal. As a result, there is a trade-off between resolution and noise suppression, which is considered in Section 17.2.6.

17.2.5.1 Generation of the carrier frequency

There are two scenarios in OCT imaging when the scanning engine does not generate a carrier and, thus, additional phase modulation is required. The first is when axial scanning is performed using a particular configuration of the frequency-domain optical delay line that results in a constant phase delay as the group delay is scanned (Section 17.3). The carrier frequency can then be conveniently set by the additional source of phase modulation to avoid aliasing and to avoid overlap with the spectrum of any amplitude modulation [70]. The second is imaging en face since there is no axial scanning. Such phase modulation is generally applied in the form of a linear ramp or a sinusoid. If a linear ramp of period $T = 1/\bar{f}$ is used such that $\bar{\nu}\tau_p = mt/T$, for m an integer and $0 < t < T$, then, neglecting the instantaneous flyback, the phase modulation produces a pure carrier. This method of signal processing was first introduced for extraction of phase shifts in optical fibre sensors [71]. Application of a linear phase ramp enables envelope detection to proceed in the same manner as when using a scanning mirror, with the exception that $\bar{f}$ may be chosen for convenience and Δf is not related to it. However, if the magnitude of the phase ramp is not exactly a multiple of 2π or, more commonly, the modulation signal is a filtered version of the ramp, harmonics are generated. In particular, if the phase modulation imparted is a sinusoid, then the resulting cross-correlation term is described by the well-known Bessel spectrum [72], given in our case by

$$I^{\mathrm{D}}(t) = 2\alpha(1-\alpha)\sqrt{I_{\mathrm{S}}(t)I_{\mathrm{R}}(t)}|\gamma(\tau_{\mathrm{g}})| \times \left\{ \begin{array}{l} \cos\phi(t)\left[J_0(\phi_{\mathrm{M}}) + 2\sum\limits_{m=1}^{\infty} J_{2m}(\phi_{\mathrm{M}})\cos 2\pi(2m)\bar{f}t \right] \\ +\sin\phi(t)\left[2\sum\limits_{m=0}^{\infty} J_{2m+1}(\phi_{\mathrm{M}})\sin 2\pi(2m+1)\bar{f}t \right] \end{array} \right\} \otimes h_{\mathrm{D}}(t) \quad (33)$$

where ϕ_{M} is the amplitude of the imparted phase modulation, and $\phi(t)$ represents all other sources of phase fluctuation. In en face systems, the random phase $\phi(t)$ shifts the bias point on the axial OCT response, resulting in randomly-varying frequency components, including complete fading of the fundamental when modulating about a bright or dark fringe. This is known as 'signal fading' and is also present in optical fibre sensors [72]. It is not present in B-mode OCT as long as the rate at which the axial interferogram envelope is generated greatly exceeds the rate of phase fluctuation, in which case the higher harmonics merely represent a loss of signal and a reduction in signal-to-noise ratio. Signal fading can be overcome by selecting the phase modulation amplitude to equalise the electrical powers in the first and second harmonics. These powers are proportional to $\sin^2\phi(t)[J_1(\phi_{\mathrm{M}})]^2$ and $\cos^2\phi(t)[J_2(\phi_{\mathrm{M}})]^2$, respectively, and if the modulation amplitude is chosen as $\phi_{\mathrm{M}} = 0.84\pi = 2.6$, then $J_1 = J_2$, and the sum of the powers is independent of the phase fluctuation $\phi(t)$. Podoleanu et al. have developed novel alternatives to external phase modulation for carrier generation in en face OCT imaging [73,74].

17.2.5.2 Relationship to the image system parameters

Before considering precisely how to extract the envelope and phase of the signal, we consider the relationship of the OCT signal's carrier frequency and bandwidth to the imaging system parameters. We consider only the B-mode operation of the imaging system but other modes of operation are similarly described with suitable redefinition of terms. We assume an image frame rate F, a sample transverse resolution l_{T}, and range L_{T}. For N image pixels per resolution element, there are $NL_{\mathrm{T}}/l_{\mathrm{T}}$ pixels in the transverse dimension, i.e., $NL_{\mathrm{T}}/l_{\mathrm{T}}$ transverse lines per frame and an A-scan rate of $NL_{\mathrm{T}}F/l_{\mathrm{T}}$. Assuming a linear group-delay scan over optical length range $\tilde{L}_{\mathrm{g}}$, the signal bandwidth is given by

$$\Delta f = 2\Delta\nu \frac{NL_{\mathrm{T}}F}{l_{\mathrm{T}}} \frac{\tilde{L}_{\mathrm{g}}}{\mathrm{v}_{\mathrm{g}}} \quad (34)$$

As an example, consider a representative real-time system operating at 8 frames per second with 256×256 pixels over a 1.5 mm transverse by 1.25 mm axial range in air, a 25 μm transverse resolution and 10 μm axial resolution, a 1300 nm mean wavelength and a Gaussian spectrum. These parameters correspond to 4.3 and 2.0 pixels per resolution element in the transverse and axial directions, respectively, a 2048 lines per second axial scan rate, 2.6 m s^{-1} axial scan velocity in air, and a 232 kHz bandwidth. If the axial scan was performed by linearly translating a mirror, then the mean frequency would be governed by $\bar{f}/\Delta f = \bar{\nu}/\Delta \nu$, giving a nominal 4 MHz carrier frequency.

17.2.5.3 Electronic demodulation to extract envelope and phase

Ideally, we would like to fully determine the impulse response $h_S(t)$ corresponding to a given voxel of the sample, which requires full determination of the amplitude and phase of the interference term corresponding to that location. Often, knowledge of $|h_S(t)|$ is sufficient, requiring the determination of the envelope of the signal described by Equation (32). Several methods are able to be used to extract the envelope and phase of the OCT signal. Many OCT measurements have been based on incoherent demodulation, often termed envelope detection, i.e., a quantity ideally proportional to $\sqrt{I_S(t)I_R(t)}|\gamma(\tau)|$ is measured as the group delay is scanned and the interferometric phase is not measured. However, the phase provides information on sample scatterer velocity, spectral filtering and birefringence. Compared with the envelope, measurement of the phase requires significantly higher scanning system stability. For a 10 μm coherence length and 1 μm mean wavelength, the ratio of the envelope width to the fringe width, given by $4\bar{l}_c/\bar{\lambda}$, is a factor of 40. Such high resolution demands proportionately greater stability of the delay scanning system, and for translating mirror delay lines usually requires a reference interferometer to independently monitor the mirror position [75]. It also potentially increases the digital sampling rate required by the same factor.

We briefly consider first methods, that record both envelope and phase, and then the most common method that records only the envelope, and finally an optical-domain method that facilitates signal processing.

- *Direct digitisation*

Direct digitisation of the detected signal $I^D(t)$ enables the envelope and phase information to be extracted from its analytic continuation $\mathrm{I}^D(t)$ [76,77], i.e.,

$$\mathrm{I}^{\mathrm{D}}(t) = I^{\mathrm{D}}(t) + \frac{j}{\pi}\mathrm{PV}\left\{\int_{-\infty}^{\infty} \frac{I^{\mathrm{D}}(t)}{t' - t}\,\mathrm{d}t'\right\} \tag{35}$$

where PV{} denotes Cauchy principal value, under the assumption that the envelope of $I^D(t)$ varies slowly compared with the phase. The digitisation rate must satisfy Nyquist's theorem and, thus, be greater than or equal to twice the highest frequency component in the signal. We can truncate the power spectrum and approximate the highest frequency as $\bar{f} + \Delta f$. When the mean frequency and bandwidth are related by $\bar{f}/\Delta f = \bar{\nu}/\Delta \nu$, a sampling rate of at least $2(\bar{f} + \Delta f)$ is required, giving 4.2 Msamples per second for the example given in the previous section. If the scanning engine provides control over $\bar{f}$, the minimum value it could be set to is $\bar{f} = \Delta f$, requiring a digitisation rate of at least $4\Delta f$, but this would clearly violate the assumption of a slowly-varying envelope, and so could not be achieved in practice. A further consequence of the reduced digitisation rate would be a reduction in the resolution of the measured phase.

Direct digitisation imposes a very significant computational overhead which makes real-time operation very difficult to attain, although, recently, impressive performance has been achieved [78]. Hyle Park et al. developed real-time software capable of displaying a 2048 × 256 pixel envelope image at one frame per second, as well as images at lower resolution of flow and birefringence derived from the phase. Their two-channel system employed a carrier generated by a frequency-domain delay line of $\bar{f} = 625$ kHz and had a bandwidth (calculated from Equation (34)) of $\Delta f = 200$ kHz, and a per-channel sample rate of 2.5 Msamples per second, which corresponds to $3(\bar{f} + \Delta f)$ [78].

Such demanding performance suggests digital signal processing implementations may be attractive alternatives; although work is underway, none has yet been reported. At the Msample-per-second rates required for real-time imaging, the quantisation of the sampling system has generally been linear with 12 bits of dynamic range (72 dB), which may be less than the full dynamic range of the OCT A-scan signal. The impact of this restriction has yet to become clear, but for many applications it should not be important as long as the weakest image-bearing signals are detected.

- *Coherent (synchronous) demodulation*

An alternative to direct digitisation is synchronous demodulation based on mixing, as shown in Figure 10. Here, a reference input to the mixer must be available from the pathlength scanner and used in combination with the low-pass filter to down-shift the signal to baseband [75]. This reduces the required digitisation rate to $2\Delta f$. If the reference is phase-synchronised to the scanner, then the envelope is the amplitude of the output signal. If synchronisation cannot be achieved, e.g., when the reference signal is derived indirectly from a scanning translation stage, then two mixers are required to extract the in-phase I and quadrature Q signals from which the desired envelope can be obtained from $\sqrt{(I^2 + Q^2)}$ and the desired phase (modulo π) from $\arctan(I/Q)$. Such schemes can be implemented with a laboratory lock-in amplifier, but these instruments are generally restricted to low frequencies (less than 100 kHz) and require very stable reference signals. An alternative is to derive the reference from the signal itself using a phase-locked loop, which has the advantage of directly outputting the carrier frequency in real time [79].

- *Incoherent envelope detection*

Incoherent envelope detection involves down-shifting the signal to baseband by the process of rectification or squaring followed by low-pass filtering (Figure 10). For ideal rectification, the signal envelope is undistorted and, therefore, the positive frequency bandwidth is $\Delta f/2$. For ideal squaring, the signal envelope is narrowed and distorted and the bandwidth is increased. If the detected signal is described by a Gaussian envelope, then its response is narrowed by a factor of $\sqrt{2}$ and the bandwidth is increased by the same factor. Commonly, envelope detection and

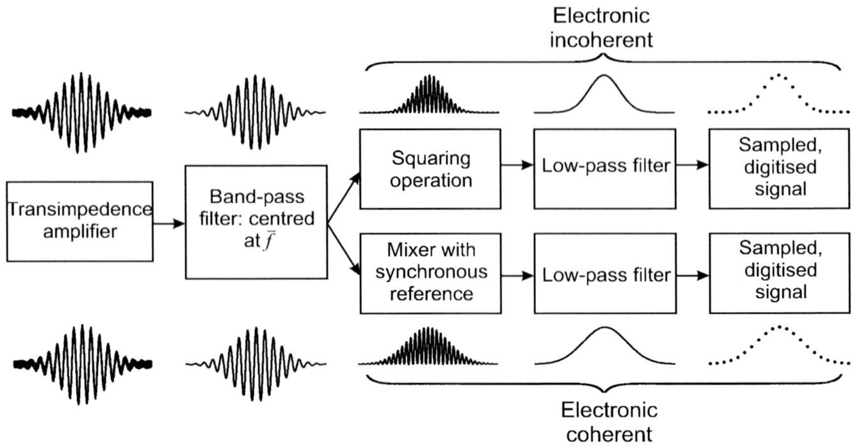

Figure 10. Electronic demodulation of the OCT signal. The upper branch shows electronic incoherent demodulation via squaring and the lower branch shows electronic coherent demodulation via mixing with a synchronous reference signal.

dynamic range compression have been achieved simultaneously using a logarithmic demodulator. Performing a logarithmic transformation on the signal further distorts it and broadens the response. Thus, the requirements of the low-pass filter vary according to the details of the envelope detection, as does the required sample rate. A constraint on the demodulator is the requirement that the mean frequency $\bar{f}$ be sufficiently well separated from the highest demodulated frequency component to enable effective removal of the 'carrier' by the low-pass filter. In practice, this places a lower bound on the number of fringes per envelope.

17.2.5.4 Optical-domain method to extract envelope and phase

It is possible to utilise polarisation to generate simultaneously signals at the interferometer output in phase quadrature. For example, if a reference beam of circular polarisation combines with a sample beam of linear polarisation oriented at 45° to the axis, then in-phase and quadrature signals are generated directly. These signals may be digitised and subsequently combined to form the analytic signal $\mathrm{I}^{\mathrm{D}}(t) = I + jQ$ instead of requiring generation of the Hilbert transform [77]. This method trades off the greater complexity of polarising optics with simpler post-processing. Notably, it has the specific advantage of not depending on the phase factor changing much more rapidly than the envelope, which allows one to dispense with the need to generate a carrier by high-frequency phase modulation. At the same time, however, measurement of the sample-related phase requires its slew rate to significantly exceed that caused by random fluctuations.

17.2.6 Noise sources, sensitivity and dynamic range

We now need to consider explicitly the conversion of optical intensity into photocurrent to analyse the noise processes in OCT. We define a responsivity ρ that represents the conversion efficiency from optical power into photocurrent. The responsivity and the quantum efficiency η at frequency ν are related by $\rho = \eta e / h\nu$. Since we are dealing with broad optical bandwidths, the dependence of the quantum efficiency and responsivity on wavelength may need to be taken into account. The signal-to-noise ratio, SNR, is conventionally defined in terms of the electrical power, or equivalently in terms of the photocurrent, as $\mathrm{SNR} \equiv \overline{i^2}/\sigma^2$, where i is the photocurrent averaged over the detector response time and $\sigma^2 = \overline{i^2} - \bar{i}^2$ is the variance of the photocurrent. In general, the SNR is a time-dependent quantity that will depend on the scan delay as well as on whether it is the envelope or full signal that is to be measured. We consider the maximum SNR of the OCT envelope arising from a single isolated reflector. We assume the envelope has been derived through coherent demodulation. The case of incoherent demodulation has been treated in Ref. 69. We first consider the SNR at the output port and then consider how this is modified for detection at the input port. The peak of the signal envelope at the output, assuming no attenuation by the coherent demodulator, is given by (cf. Equation (32)) $\overline{i^2} = 2[\alpha(1-\alpha)]^2 \rho^2 I_S I_R |\gamma(0)|^2$. Since $|\gamma(0)|^2 = 1$ by definition, we henceforth omit this factor. The sources of noise to be taken into account are shot noise, phase noise, intensity noise, and thermal/excess noise of the receiver. The sources of noise are assumed to be independent and so their respective variances are summed to give the total noise variance. We now consider each source of noise in turn.

17.2.6.1 Shot noise

Shot noise is additive white Gaussian noise that arises from the stochastic nature of the photon arrival times. The detected noise power is proportional to the average optical power incident on the photodetector. Along with the signal, the noise in the signal pass-band is down-shifted to baseband and low-pass filtered. The noise power, σ^2_{Sh}, is given by

$$\sigma^2_{\mathrm{Sh}} = 2\rho e\, \alpha(1-\alpha)(I_S + I_R)B \tag{36}$$

where e is the electronic charge and B is the noise-equivalent low-pass bandwidth of the receiver. This bandwidth is potentially affected by the band-pass filter as well as by the low-pass filter (Figure 10). The shot noise-limited SNR is then given by

$$\mathrm{SNR}_{\mathrm{Sh}} = \frac{\rho\alpha(1-\alpha) I_S I_R}{e(I_S + I_R)B} \tag{37}$$

Assuming at the receiver the intensity from the reference arm dominates over that from the sample arm, which is the usual situation for weakly reflecting samples, and an ideal 50/50 coupler, we obtain [49]

$$\mathrm{SNR}_{\mathrm{Sh}} = \frac{\rho I_S}{4eB} = \frac{1}{4}\frac{\eta}{h\nu}\frac{I_S}{B} \tag{38}$$

As noted in Ref. 8, Equation (38) corresponds to the $I_S/2h\nu$ photons per second returning from the sample arm generating $\eta(I_S/2h\nu)$ electrons per second, giving a total of $\eta(I_S/2h\nu)T$ detected electrons in response time $T \approx 1/2B$. We note that our definition of I_S differs from Ref. 8 by a factor of $\alpha = 1/2$. Thus, if shot noise dominates over all other noise sources, the SNR is independent of the power in the reference arm.

17.2.6.2 Phase noise

Phase drifts and fluctuations in the interferometer have already been mentioned. Such fluctuations are caused by changes in ambient conditions, including thermal expansion, mechanical vibration, acoustic noise, as well as by instability in the mean wavelength. Although such phase noise generally has a $1/f$ spectral characteristic, it directly modulates the carrier $\bar{f}$ through $d\phi/dt$, as shown schematically in Figure 9, and, hence, cannot be removed by filtering. It is rarely explicitly mentioned in the literature and is invariably dealt with by designing the system to minimise its effects. Long optical pathlengths in each arm make the interferometer more susceptible to the effects of vibration; this susceptibility can be improved by employing common-path interferometer configurations. Because its noise variance is not known a priori, and because it can, in principle, be eliminated, we do not consider it further here.

17.2.6.3 Intensity noise

As mentioned in Section 17.2.5, whenever the detected optical field components are uncorrelated, i.e., incoherent, their mixing or beating, arising from square-law detection, generates noise [45]. The resulting instantaneous intensity is stochastic with large fluctuations. For band-limited detection, these fluctuations are smoothed by averaging over the response time, but residual noise remains, referred to variously as excess intensity noise, beat noise, or phase-induced intensity noise. Since the photocurrent is proportional to the intensity, the noise power is proportional to the product of the intensities from which it arises. The spectral density of this noise is proportional to the cross-correlation of the respective optical spectral densities attenuated by the low-pass response of the receiver [46]. The effects of intensity noise generated by incoherent light are significant in diverse fields, including fibre-optic gyros [80] and fibre-optic communications, in which incoherent light commonly occurs in the form of amplified spontaneous emission and the resulting noise is often referred to as 'spontaneous-spontaneous beat noise' [81]. In OCT, such intensity noise arises from the common use of incoherent light sources such as light-emitting and superluminescent diodes, and semiconductor optical amplifiers. When using such sources, additional noise arises from the beating of incoherent fields generated by residual, unwanted reflections in the OCT system with delay differences that exceed the coherence time of the light [82–84].

For coherent communication systems dual balanced detection can be used to eliminate common-mode (local oscillator) intensity noise [85]. The two photocurrents from the outputs of the device which combines the signal with the local oscillator are subtracted. In OCT, the equivalent process requires simultaneous detection at both interferometer ports. Balanced detection rejects the excess intensity noise inherent in the source because it is correlated and in phase at the input and output ports of the interferometer and, thus, it is cancelled by subtraction. Conversely, the OCT signals are in antiphase and, hence, subtraction doubles the signal photocurrent (quadruples the signal power). Cancellation does not take place, however, for the additional noise generated by incoherent beating between fields derived from sample and reference arm reflections, for the same reason that the signal is not cancelled. Let us briefly contrast these effects with the effect of balanced detection on the shot noise. Since the shot noise processes at the two detectors are uncorrelated, balanced detection doubles the noise power but quadruples the signal power, increasing the shot noise-limited SNR by a factor of two.

Coherent pulsed sources have also been used in OCT systems. Theoretically, the intensity noise of light from these sources is extremely small [86]. In practice, the intensity noise in such sources is often much higher than this limit and dominated by technical noise, but this is also cancelled by balanced detection. Finally, intensity noise generated by instabilities in the scanning system can also be present, and this is typically low-pass in character. The sources of intensity noise are included in the schematic representation of the detected OCT signal and noise shown in Figure 9.

To evaluate the intensity noise generated by incoherent sources, we divide the intensities from the reference and sample arms into coherent and incoherent components such that $I_R = I_{Rc} + I_{Ri}$ and $I_S = I_{Sc} + I_{Si}$. There are four corresponding field terms and, thus, sixteen beat terms made up of four self-beat (source intensity noise) and twelve cross-beat terms, two of which represent the OCT signal and the rest arise from residual reflections. The resulting excess intensity noise power at a single photodetector may be written [45] as

$$\sigma_{Ex}^2 = [\rho\alpha(1-\alpha)]^2 \begin{bmatrix} (I_{Rc})^2 + (I_{Sc})^2 + (I_{Ri})^2 + (I_{Si})^2 \\ +2I_{Rc}I_{Ri} + 2I_{Sc}I_{Si} + 2I_{Ri}I_{Sc} \\ +2I_{Rc}I_{Si} + 2I_{Ri}I_{Si} \end{bmatrix} \frac{1+P^2}{2} 2\tau_c B \tag{39}$$

where P is the degree of polarisation, which is assumed to be the same for light from the sample and reference arms, and we neglect any optical filtering in the sample and reference arms. In a well-designed system, we expect a strong reference signal and low residual reflections such that $I_{Rc} \gg I_{Ri}, I_S$, which leaves $(I_{Rc})^2$ as the dominant term so that

$$\sigma_{Ex}^2 \approx [\rho\alpha(1-\alpha)]^2 (I_{Rc})^2 \frac{1+P^2}{2} 2\tau_c B \tag{40}$$

Thus, the SNR for single-ended detection may be written

$$\mathrm{SNR}_{Ex} = \frac{2I_{Sc}}{I_{Rc}} \frac{1}{(1+P^2)\tau_c B} \tag{41}$$

We see from Equation (41) that increasing the power of the source will not improve the SNR. Using polarised light reduces the SNR by a factor of two relative to unpolarised light. Decreasing the coherence time improves the SNR because, prior to low-pass filtering, the same noise power is spread over a greater electrical bandwidth. Decreasing the receiver bandwidth increases the SNR because less noise power is contained in-band.

Balanced detection attenuates all the self-beat terms by the common-mode rejection ratio CMR but doubles the cross-beat terms. The reduction of the self-beat noise causes the cross-beat noise to become significant, for which the dominant term is $I_{\mathrm{Rc}}I_{\mathrm{Si}}$. Thus, the noise power, σ^2_{Bal}, is given by

$$\sigma^2_{\mathrm{Bal}} \approx [\rho\alpha(1-\alpha)]^2\left[\frac{(I_{\mathrm{Rc}})^2}{\mathrm{CMR}} + 8I_{\mathrm{Rc}}I_{\mathrm{Si}}\right]\frac{1+P^2}{2}2\tau_c B \tag{42}$$

and the associated SNR is given by

$$\mathrm{SNR}_{\mathrm{Bal}} = \frac{I_{\mathrm{Sc}}}{\left(\frac{I_{\mathrm{Rc}}}{8\mathrm{CMR}} + I_{\mathrm{Si}}\right)}\frac{1}{(1+P^2)\tau_c B} \tag{43}$$

17.2.6.4 Electrical and thermal noise

The electrical noise contributed by the receiver typically exceeds the limit set by the thermal noise in the transimpedance resistor by around 5–10 dB. We assume that the electrical noise is well described as additive white Gaussian noise. We can express the electrical noise power σ^2_{El} in terms of the thermal noise using

$$\sigma^2_{\mathrm{El}} = M\frac{4kTB}{R_{\mathrm{f}}} = \mathrm{NEP}\times B \tag{44}$$

where M is a multiplier to account for the excess over the expected thermal noise, k is Boltzmann's constant, T is absolute temperature, R_{f} is the effective transimpedance of the receiver, and NEP is the noise equivalent power per Hertz of the receiver. The associated SNR is, thus, given by

$$\mathrm{SNR}_{\mathrm{El}} = \frac{[\alpha(1-\alpha)]^2\rho^2 I_{\mathrm{S}}I_{\mathrm{R}}R_{\mathrm{f}}}{2MkTB} \tag{45}$$

17.2.6.5 Dynamic range versus sensitivity

The terms dynamic range, signal-to-noise ratio, and sensitivity are used somewhat interchangeably in the OCT literature. The sensitivity of an OCT system is the minimum reflectivity R_{min} that can be detected, which we assume to correspond to an SNR of one. If we assume $I_{\mathrm{S}} = I_0 R_{\mathrm{min}}$, then a unity shot noise-limited SNR corresponds to a sensitivity given by

$$R_{\mathrm{min}} = \frac{eB}{\rho\alpha(1-\alpha)I_0} \tag{46}$$

We shall see below that shot noise-limited sensitivity is not readily attainable. The dynamic range is defined as the range over which a signal can be detected. The maximum detectable signal cannot exceed the value obtained when the sample is a correctly aligned plane mirror.

17.2.6.6 Total signal-to-noise ratio and operating noise regimes

For single-ended detection, the total SNR, including all noise sources, is

$$\mathrm{SNR_{Single}} = \frac{\overline{i^2}}{\sigma^2_{\mathrm{Sh}} + \sigma^2_{\mathrm{Ex}} + \sigma^2_{\mathrm{El}}} \tag{47}$$

After dual balanced detection, we have

$$\mathrm{SNR_{Dual}} = \frac{4\overline{i^2}}{2\sigma^2_{\mathrm{Sh}} + \sigma^2_{\mathrm{Bal}} + 2\sigma^2_{\mathrm{El}}} \tag{48}$$

Note that, as with shot noise, the thermal noise component in Equation (48) has twice the power relative to single-ended detection as the noise processes are uncorrelated between the two photodetectors.

To find the sensitivity, we set the SNR in Equations (47) and (48) to unity and solve for R_{Sc}. Figure 11 shows the inverse of the sensitivity plotted as a function of the reflectivity of the reference arm R_{Rc}. Figure 11(a) is for the case of a single detector and Figure 11(b) is for the case of balanced detection with 25 dB common-mode rejection. The plots illustrate, for typical parameters given in the figure caption, the relative importance of the various noise processes. Figure 11(a) shows that, without balanced detection, there is a trade-off between excess intensity noise and electrical noise which results in an optimum value of power from the reference arm. The reference arm power will typically need to be attenuated to achieve the maximum sensitivity. For the parameters used to generate the plot, the optimum R_{Rc} is -30 dB. For lower values, the electrical noise dominates, and for higher values, excess intensity noise dominates. Figure 11(b) shows that, with balanced detection, the sensitivity is higher and more closely approaches the limits set by shot and cross-beat noise, and the effect of the reference power level on the

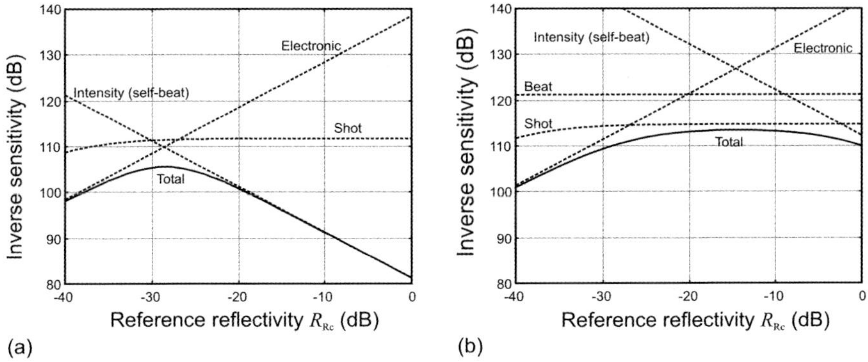

Figure 11. (a) Inverse sensitivity versus reference reflectivity, R_{Rc}, for a single-ended detection system, for the parameters: $I_0 = 10$ dBm, $R_{\mathrm{Si}} = -40$ dB, $\bar{\lambda} = 1300$ nm, $\Delta\lambda = 50$ nm, a polarised Gaussian-spectrum light source, $R_{\mathrm{f}} = 10\,\mathrm{k}\Omega$, and passband with $\Delta f = 100$ kHz. Dashed curves show the contribution of each of the noise components to the overall sensitivity. (b) Inverse sensitivity versus R_{Rc} for a balanced-detection system with the same parameters as (a), and common-mode rejection of 25 dB.

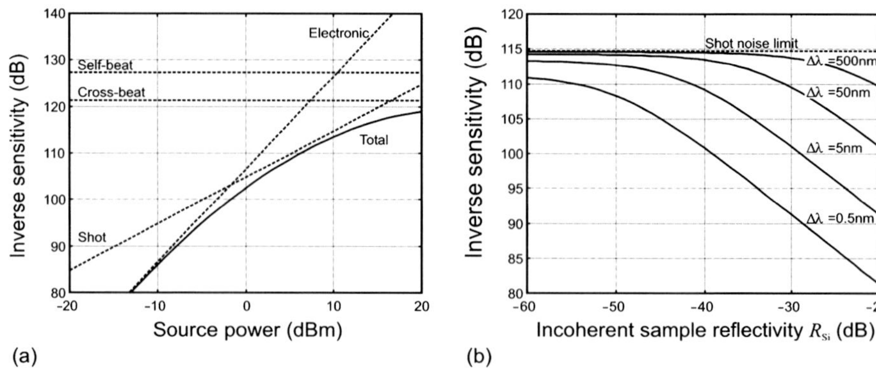

Figure 12. (a) Inverse sensitivity versus source optical power for the same parameters as Figure 11(b), with $R_{Rc} = -15$ dB. (b) Maximum sensitivity for a balanced-detection system with a 50:50 splitting ratio, calculated versus R_{Si} for several FWHM source bandwidths.

sensitivity is considerably reduced, with a much higher optimum R_{Rc} (−15 dB). Figure 12(a) is a plot of inverse sensitivity versus source optical power for the same parameters as Figure 11(b) with $R_{Rc} = -15$ dB. At $I_0 = -10$ dBm, electrical noise dominates, at 5 dBm shot noise dominates, and at 20 dBm cross-beat intensity noise is the most dominant. This plot shows that even with common-mode rejection of the I_{Rc}^2 (self-beat) excess intensity noise, the sensitivity is ultimately limited by the cross-beat components, particularly the $I_{Rc}I_{Si}$ component, which points to the importance of eliminating stray reflections in the sample arm optics. Figure 12(b) shows the optimum sensitivity achievable with balanced detection and a 50:50 splitting ratio for several FWHM source bandwidths. The curves show the feasibility of achieving shot-noise-limited detection, as well as, for a given source power, the ease of approaching this limit with a wider optical bandwidth.

17.2.6.7 Signal-to-noise ratio at the input port and of other interferometer designs
At the input port, as the splitting ratio is varied, the shot and excess intensity noise vary according to changes in the detected reference arm power. The signal-to-noise ratios can be found from the expressions already presented by making the following substitutions: in Equation (37) for shot noise-limited, $\alpha(1-\alpha) \rightarrow \alpha^2$; and in Equation (41) for intensity noise-limited, $1 \rightarrow \alpha^2/(1-\alpha)^2$. There is no change to the electrical noise-limited SNR since the signal power is the same at both ports.

The availability of optical circulators has opened up a range of possibilities for optical architectures that can more effectively utilise the optical power available. These architectures and their performance have been surveyed by Rollins et al. [83] for two-path interferometers lacking a common-mode path, and by Beddows et al. [87], who include common-mode two-path interferometers. Their collective results show that use of a power-efficient configuration, plus a carefully selected coupler splitting ratio, can give extra improvement in sensitivity or, alternatively, allow the use of a lower-power source to achieve a given sensitivity.

17.2.6.8 Sensitivity-resolution trade-off
Hee has highlighted the trade-offs in a time-domain OCT system between optical power, imaging speed, signal-to-noise ratio, and axial resolution [69]. We follow his arguments here. Assuming shot noise-limited performance and a noise-equivalent bandwidth B equal to the low-pass filter bandwidth $\Delta f/2$, using Equation (18) and $\Delta f = 2\mathrm{v}\Delta\lambda/\bar{\lambda}^2$, we can recast Equation (38) in terms of the optical power from the sample, axial resolution, and linear scanning velocity, as

$$\frac{\mathrm{SNR}_{\mathrm{Sh}}\mathrm{v}}{\tilde{l}_{\mathrm{FWHM}} I_{\mathrm{S}}} = \frac{\eta}{h\nu}\frac{\pi}{8\ln 2} \tag{49}$$

The right-hand side of this expression is a constant and, hence, it neatly captures the inherent trade-offs in altering the key parameters of the OCT system. Currently, the most desirable goal would be to reduce the axial resolution from the readily achievable 10 to 1 μm. Equation (49) shows that to achieve this without compromising the sensitivity would require a concomitant increase in source power by an order of magnitude or a reduction in scanning speed by an order of magnitude.

In fact, it is not possible to simultaneously achieve the shot noise-limited SNR and the resolution given in Equation (49). Attainment of the shot noise limit requires a receiver with a low-pass response matched to the envelope signal. Assuming synchronous coherent demodulation and a Gaussian coherence function, this implies the use of a low-pass filter with a half-Gaussian response matched to the signal, i.e., with positive-frequency bandwidth $\Delta f/2$. Because the filter impulse response is then also Gaussian, its convolution with the OCT axial response results in a system response time increased by a factor of $\sqrt{2}$. The degree of broadening caused by demodulation also depends to some extent on the method employed. Swanson et al. [49] recommended a pass-band of $2\Delta f$ prior to incoherent envelope demodulation (presumably rectification), implying an equivalent base bandwidth of Δf and broadening by a factor of $\sqrt{5}/2 = 1.12$. When incoherent envelope detection is implemented using a squaring operation, the signal bandwidth is increased by a factor of $\sqrt{2}$ and, therefore, a low-pass filter with bandwidth $\Delta f/\sqrt{2}$ should be used, resulting in a broadening of the response time by a factor of $\sqrt{6}/2 = 1.22$.

Important caveats apply on the utility of Equation (49) and the relevance of the shot noise limit. Firstly, the limiting factor on the sensitivity is likely to be the excess intensity noise (single detector) or cross-beat noise (balanced system). Secondly, such limits take no account of the often dominant deleterious effects of multiple scattering and speckle, to be considered in Sections 17.4.2 and 17.4.3, respectively.

17.3 Frequency-domain delay scanning

A wide range of delay scanning technologies have been developed for interferometry, pulse shaping and characterisation, optical signal processing, and optical communications. Many of the technologies relevant to OCT are surveyed in Ref. 88. In this section, we describe the delay line technology that has emerged as the best currently available for OCT – the so-called Fourier-domain rapid-scanning optical delay line, which we will refer to as the frequency-domain optical delay line (FD-ODL).

FD-ODL, originally developed for applications in pico- and femto-second pulse shaping and dispersion control [89], was first adapted to OCT by Tearney et al. [90]. The delay line is based on the Fourier correspondence of a linear phase ramp in the frequency domain with a time shift. The phase ramp is imparted by an angle-scanned mirror in which tilt angle is nearly linearly mapped to delay. In the form suitable for OCT, it was first described by Kwong et al. [91], who outlined its salient features and coined the term 'rapid scanning optical delay' or 'RSOD'. After its first report in OCT [90], it was further refined by Rollins et al. [92], who reported high-speed operation based on a resonant galvanometer mirror, by Silva et al. [93], who reported long delays (exceeding 25 mm), and finally by Zhao et al. [76], who incorporated electro-optic phase modulation. The delay line conveys many advantages, including high-speed [92] (10 s of meters per second), high linearity (at the few-percent level or better), long range [93] (typically millimeters), phase control [90], and variable dispersion [92]. We review below the operation of the FD-ODL in OCT systems, establish its operating ranges and consider its features, including long-range scanning and axial depth-dependent compensation of dispersion.

17.3.1 Principles of operation

Figure 13 shows a schematic diagram of the FD-ODL and a basic OCT system. Fibre coupling to the FD-ODL is assumed as this sets the most stringent conditions upon the optical beam properties. Light from the fibre is collimated, dispersed by the grating, and focused by a lens located at a distance from the grating approximately equal to its focal length. The grating and fibre are configured such that the mean wavelength of the light is diffracted along the optical axis of the lens, or parallel to it. A tiltable mirror with its pivot located in the other focal plane of the lens and its tilt axis in the focal plane reflects the beam. As can be seen from Figure 13, after spectral recombination, the reflected beam returns to the fibre collimated but offset from the input by a distance that depends on the angle of reflection. To overcome the associated scan angle-dependent coupling loss, a mirror is used to create a double pass which, for perfect optical elements, removes this problem and, as a beneficial side-effect, doubles the delay.

Despite its apparent simplicity, the FD-ODL contains a number of subtle features. Without compromising the description of its key features, we may use the paraxial approximation to describe the FD-ODL, and further assume that a wavelength component of the light directed to and from the grating from the direction of the lens is incident upon and exits the grating at the same angle to the grating normal, which is a good approximation. For a single pass, the phase difference ϕ between the field components at wavenumber k and mean wavenumber $\bar{k}$ is given by [61,94]

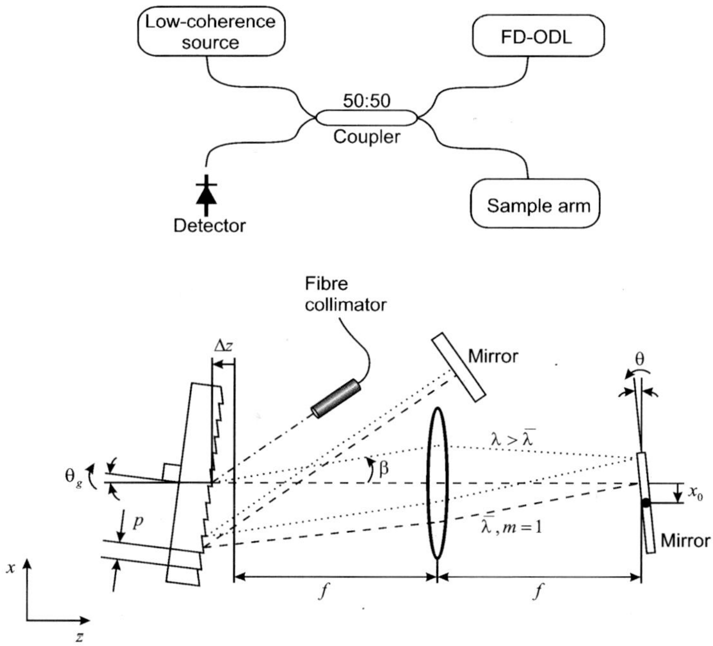

Figure 13. Schematics of (upper) scanning Michelson interferometer employing the FD-ODL and (lower) the FD-ODL. m indicates diffraction order. All angles are shown for positive values.

$$\phi(k) = 2x_0\theta\bar{k} + \left(2x_0\theta + \frac{4\pi m f\theta}{p\bar{k}\cos\theta_g}\right)(k-\bar{k}) - \left(\frac{8\pi^2m^2\Delta z}{p^2\bar{k}^3\cos^2\theta_g} + \frac{8\pi^2m^2f\theta\sin\theta_g}{p^2\bar{k}^3\cos^3\theta_g}\right)\frac{(k-\bar{k})^2}{2!} \tag{50}$$

where the symbols and signs are defined in Figure 13. This expression includes the terms needed to calculate the group delay dispersion (GDD) introduced by the delay line and its slope. In Figure 14, the phase difference is plotted against wavenumber for representative experimental parameters and offsets of the pivot point of the galvanometer mirror, x_0. Neglecting the scan-angle-independent phase difference between the interferometer arms, the resulting phase and group delays, for a single pass of the grating, are given by

$$\tau_p = \frac{1}{c}\frac{\phi(\bar{k})}{\bar{k}} = \frac{2\theta x_0}{c} \tag{51}$$

and

$$\tau_g = \frac{1}{c}\phi'(\bar{k}) = \frac{2\theta x_0}{c} + \frac{4\pi m\theta f}{cp\bar{k}} \tag{52}$$

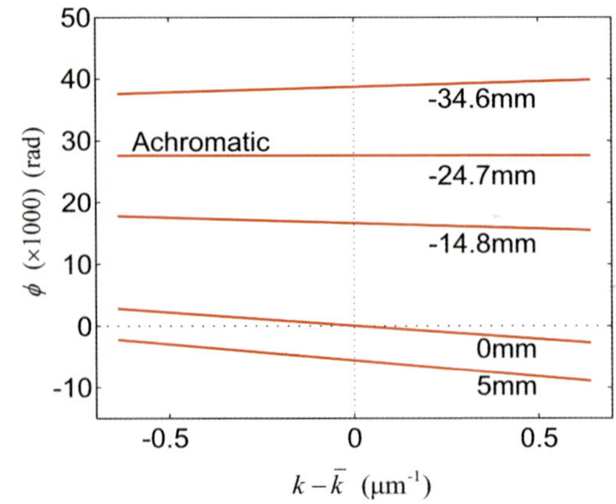

Figure 14. Phase difference versus wavenumber for a range of offsets, x_0, and $\theta = 5°$. Normal operating regime for OCT is $x_0 \geqslant 0$.

In general, then, phase and group delay are not equal, which in an OCT interferogram may be described as unequal apparent motions of the fringes and envelope as the mirror angle is scanned. In fact, when the pivot offset x_0 is set to zero, no phase modulation results and an envelope which is apparently fringe free is observed. In Figure 14, this situation is conveyed by the zero intercept and the finite slope. This is not to say that an OCT signal so-generated is phase-independent, however, because it remains sensitive to fluctuations in the phase difference between the reference and sample paths. Examples of such measured OCT signals are shown in Figure 15. When the tilt-mirror offset is set to a positive value, as defined in Figure 13, phase modulation is re-established, confirmed by the finite value of ϕ at $k - \bar{k} = 0$ in Figure 14. The number of fringes appearing under the envelope asymptotically approaches that due to a conventional translating mirror as the phase delay approaches the group delay, i.e., as the second term on the right-hand side of Equation (52) becomes negligible.

As described in Section 17.2.5.3, phase modulation is commonly employed in OCT signal processing. The decoupling of the carrier frequency from the scan rate provided by the FD-ODL, which is not possible with a translating mirror, conveys great flexibility in choice of the carrier to suit the signal processing. Either the FD-ODL is set to zero pivot offset and the phase modulation is provided externally to the delay line [76], or the offset is chosen to produce a convenient carrier [78].

When the offset is negative, it is possible to choose a value for which the two terms in Equation (52) cancel. At this value, the envelope is stationary during scanning and the fringes move. The phase shift resulting from the mirror tilt is then achromatic over the bandwidth of the FD-ODL [95] (Figure 14).

The origin of the group delay in the FD-ODL lies in the transverse offsets of the reflected wavelength components at the grating [91,94]. This is substantially unaltered as the pivot offset is changed, as seen by the slight change in slope versus

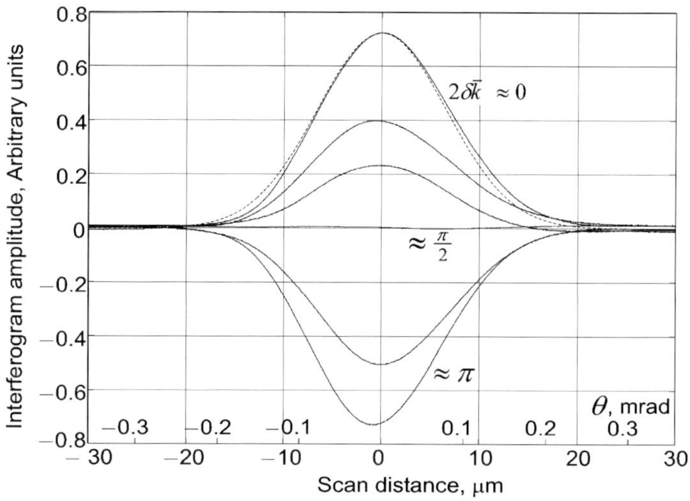

Figure 15. Measured interferograms (solid curves) for $x_0 = 0$ at various phase differences. Dashed curve is a theoretical plot for zero phase difference. Reproduced from [94] with permission.

positive offset in Figure 14. The origin of the phase delay is in the on-axis displacement of the tilted mirror from the focal plane [94].

17.3.2 Scan range limits

For most OCT applications, tissue turbidity limits scan ranges to a few millimetres. However, there are some measurements, including the length of the human eye, and the size and shape of large hollow organs, that require ranges in the tens of millimetres. The basic limit on the scan range of the FD-ODL is set by the aperture of the optics. The range corresponding to the intersections of the mean wavelength $\bar{\lambda}$ with the lens edges is given by $A\bar{\lambda}/p$, where A is the aperture [93]. This limit assumes that the double-pass mirror does not cause vignetting at zero tilt angle, which would occur if the mirror were placed in the plane of the system, and can be avoided by using a polarisation-based double pass [93] or an out-of-plane geometry [76]. In practice, this limit cannot be reached because of vignetting, and because of lens aberrations which cause wavelength-dependent beam walk-off at the fibre. We have modelled this behaviour using a ray tracing package (Zemax™) and have shown that the most critical requirements are the use of an achromatic doublet lens, and the use of a zero-offset design. Figure 16 shows a comparison of the results of modelling and experiment for a delay line with theoretical maximum delay length of 26.6 mm. The offset pivot displays significantly smaller range, because the beam focussed onto the galvanometer mirror is not reflected from the focal plane of the lens except at zero tilt angle and, thus, it is not properly re-collimated by the lens. This causes the return beam to the fibre to have a scan-dependent longitudinal focus error (i.e. blurring),

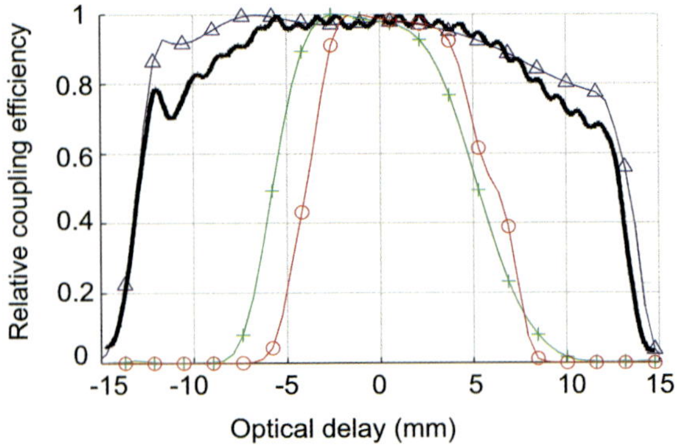

Figure 16. Relative coupling efficiency versus delay for the configurations: (Δ) on-pivot; (+) off-pivot; (○) on-pivot with a singlet lens. The heavy solid line is a measured response for the on-pivot configuration.

which reduces coupling efficiency. This effect combined with chromatic aberration makes the on-pivot performance using a singlet lens comparable.

17.3.3 Bandwidth limits

To examine the limits on the bandwidth supported by the FD-ODL, a ray-tracing model has been created with the following components: 6.1 mm focal-length aspheric fibre collimator, 400 l mm^{-1} grating and 60 mm achromat lens. The diffraction grating is assumed to have constant efficiency with wavelength and the source spectrum is centred at 1315 nm and has uniform intensity over 500 nm. The model assumes zero pivot offset and that diffracted rays follow a co-planar return path. Figure 17 shows the variation of mean wavelength, FWHM bandwidth and relative coupling efficiency versus optical delay (in mm). The FWHM bandwidth is reduced to 400 nm at zero delay but still exceeds 200 nm at 3 mm delay. The model predicts a shift in mean wavelength over the scan range; relative to long wavelengths, short wavelengths are more strongly attenuated for positive delays and less strongly for negative delays. The model also predicts strong changes in spectral shape with delay (not shown); for negative delays spectra are symmetric and single peaked, whereas, for long positive delays they are double peaked. The relative coupling efficiency (over all wavelength components) versus scan distance is reasonably flat; 41% at 3 mm delay. Somewhat improved performance could be achieved by replacing the lens with a parabolic reflector.

17.3.4 Dispersion behaviour

Dispersion additional to material dispersion in the FD-ODL arises when the grating is not in the focal plane of the lens and when the grating is tilted relative to the optical axis. These two cases differ with respect to their dependence on the scan angle. Scan-independent dispersion caused by grating translation (the Δz term in

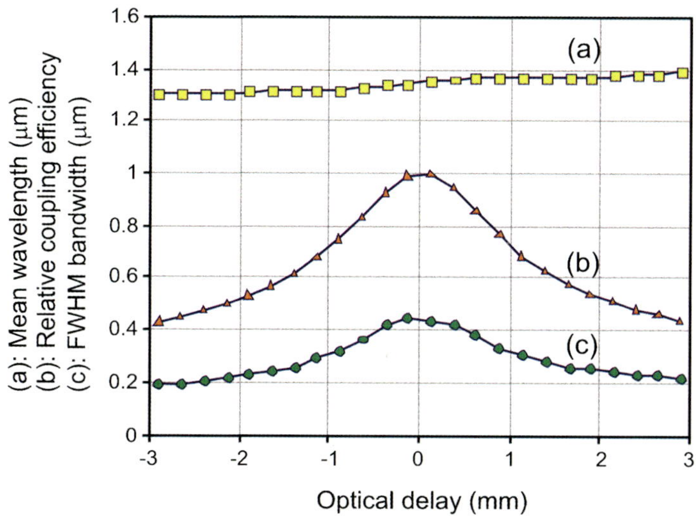

Figure 17. Simulated results for (a) mean wavelength, (b) relative coupling efficiency, and (c) FWHM bandwidth versus delay of an FD-ODL illuminated with a 500 nm bandwidth source centered at 1315 nm.

Equation (50)) is very useful for balancing dispersion in the two arms of the interferometer. The effects of dispersion were described in Section 17.2.3. However, there are limitations: grating translation to compensate the GDD cannot independently compensate higher-order dispersion, and the compensation depends on the sample pathlength [61]. Scan angle-dependent dispersion can be used to compensate for sample dispersion during a scan, and this is demonstrated in Figure 18, in which the GDD caused by 4 mm of lithium tantalate is compensated in real time by correct choice of grating tilt angle [61]. As described in Section 17.2.3, compensation becomes a significant issue as bandwidth increases towards micron-order coherence lengths.

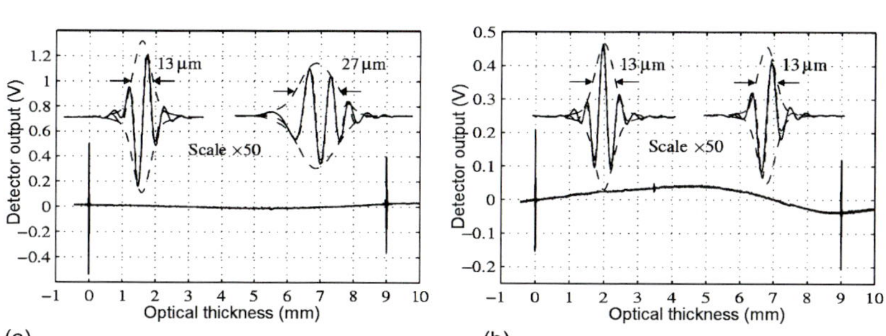

Figure 18. (a) Single OCT A-scan through 4 mm of $LiTaO_3$ with grating set normal. (b) Same as (a) but grating set at an angle of 7.5°. The expanded scale response in (a) shows the broadening at the end of the scan caused by dispersion, and its dynamic compensation using the FD-ODL is shown in (b). Reproduced from [61] with permission.

17.4 OCT image formation in turbid media

In most samples of interest, scattering dominates strongly over absorption [96]. Hence, we now consider the effects of scattering on image formation and neglect the effects of absorption. In Section 17.2, the system response was couched in terms of the spatially-resolved reflectance of the sample. This reflectance would ideally be determined only by the local variations in scattering, in which case a diffraction-limited image of the spatially-resolved reflectance of the sample could be formed. For this to be the case, light must propagate in the sample without scattering except for a single backscatter. Such a process occurs with a small but finite probability. Much more likely, light undergoes many scattering events and this multiply scattered light degrades the fidelity with which the local reflectance is determined.

The extent of multiple scattering is related to the distribution of the propagation delay of the scattered light. Figure 19 indicates schematically the arrival times for light propagating through a turbid medium. In transmission, the small fraction of light which propagates without scattering is termed the ballistic light [97]. In reflection, ballistic light must undergo a single backscatter to be detected, and so the term 'singly backscattered' has commonly been used. More commonly, some light undergoes a finite and variably small number of low-angle forward scattering events that result in a continuous distribution of early arrival times – this is the so-called 'snake' light, often termed 'low-angle multiple scattering' in the OCT literature. Finally, the largest proportion of the light undergoes many scattering events, the so-called diffuse light, which causes it to be decorrelated with spatial location in the medium, rendering it impossible to form an image in the conventional sense. The relative fractions of the three types of light depend on the properties of the medium. To determine ball-park figures, consider transmission in a homogeneous medium in which absorption can be neglected. We define the optical depth $\mathrm{OD} \equiv \mu_t z$, where μ_t is the total extinction coefficient (equal in this case to the scattering coefficient μ_s), and z is the thickness of the medium. Assuming $\mathrm{OD} = 3$, a Henyey-Greenstein scattering function with anisotropy

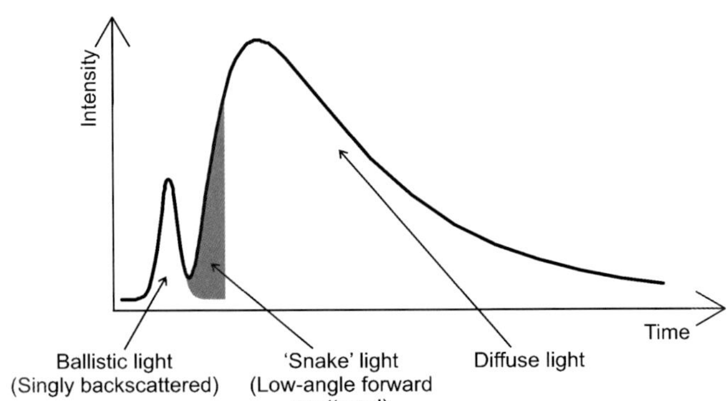

Figure 19. Schematic of arrival times of light detected after propagation through a highly scattering medium. Labels in parentheses indicate the terms commonly used in OCT.

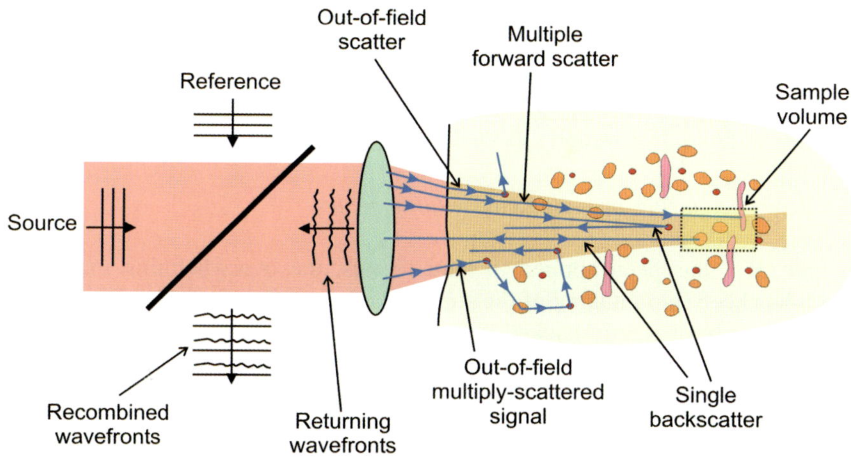

Figure 20. Schematic of the scattering processes and resulting wavefront distortion that affect OCT images.

parameter $g = \langle \cos\theta \rangle$ of 0.85, and that the trajectory of 'snake' light does not deviate from the optical axis by more than an angle of $\theta = 10°$, gives 5, 20, and 75%, respectively for the proportions of ballistic, snake, and diffuse light.

The resolution of an OCT system is limited by the factors set out in Section 17.2. Whether these resolution limits can be attained, though, depends on the effects of the sample on the ballistic light that contributes the diffraction-limited image signal in OCT and on the relative proportion of multiple scattering and its origins. Figure 20 attempts to portray schematically the main degrading influences:

(i) Absorption and scattering of light out of the field of view of the system causes extinction of the propagating wave.
(ii) Multiple low-angle forward scatter (within the field of view of the system) causes wavefront distortion of the incident and backscattered beam, resulting in beam spreading and a loss of transverse spatial coherence.
(iii) Multiple single back scattering from inside the sample volume is coherently detected and causes noise-like modulation of the signal (which has been termed signal-bearing speckle [98]).
(iv) Multiple scattering from outside the sample volume causes a signal to be generated that does not contain any information concerning the sample volume (termed signal-degrading speckle [98], or when averaged, termed diffuse reflectance).

These effects are the primary ones that are present in all measurements to a greater or lesser degree. There are a range of other degrading effects, however, that are worth mentioning:

- Uncompensated dispersion of the incident and backscattered waves.
- Polarisation state rotation caused by the birefringence of the medium.

- Filtering of the optical power spectrum caused by the wavelength dependence of the scattering and absorption by the sample.

Before considering the main effects in more detail, we first consider the sources of contrast in biological samples as well as how these are connected to tissue state.

17.4.1 Light scattering in biological cellular media

Although OCT has been applied to hard tissues, the overwhelming majority of research has been into imaging of the superficial layers of soft tissues. These layers consist primarily of tightly-packed groups of cells held in place by a network of elastic fibres. Scattering is considered to arise primarily from the spatially-varying refractive-index continuum of such structures, as indicated schematically in Figure 21. Determination of the in vivo refractive index distribution and/or the associated scattering coefficients with sufficient resolution to attribute precise properties to sub-cellular constituents remains to be achieved, but significant progress has been made. The majority of studies directed towards the understanding of light scattering from tissues have been conducted in vitro on cell suspensions [99–104]. Two paths have been followed–determination of the angular variation of single scattering [99,102,104], and determination of the phase contrast produced by cells [100,101,103]. These studies have broadly confirmed the expectations from Mie theory. At low forward angles, light in tissues is scattered primarily by large

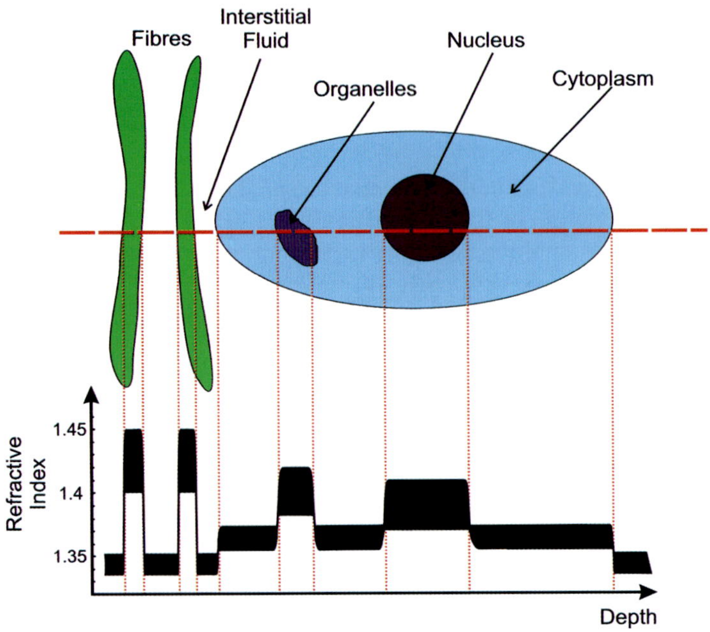

Figure 21. Schematic of the microscopic structure and refractive index of soft tissue. Expected index ranges are shown for each constituent. Dimension of typical human cells in soft tissue is 10–20 μm. Adapted from [101].

spheroid objects, i.e., by the cells themselves. At slightly larger angles, perhaps up to a few tens of degrees, scattering is dominated by cell nuclei, and at very large angles, including backscattering, scattering is dominated by small structures within cells. The ratio of forward to backscattering is typically in excess of 10^3. There has been some quantification of the contributions from mitochondria [102,104], but not of the other membranous structures in cells. Membranes, made up of phospholipid bilayers 5 to 10 nm thick, form the outer layer of cells and cell nuclei, and are also present in other structures. Mitochondria contain folded membranes, and other cytoplasmic structures (e.g., the endoplasmic reticulum and the Golgi apparatus) contain layers of membranes.

The conditions under which such cellular and sub-cellular structures are rendered observable are not completely clear. For example, purely confocal reflectance images taken in vitro have failed to clearly display cell membranes [103], but confocal images of skin taken in vivo have revealed such membranes in some but not all layers of the epidermis [105]. Cell membranes and nuclei are the dominant scatterers in OCT images of largely transparent tadpoles (see Section 17.5). A key issue is the strong dependence of the scattering coefficient on the refractive-index difference between structures and the surrounding medium. Modelling of a single cell suspended in a medium has predicted a reduction in scattering by a factor of nearly four when the index differences were reduced by 0.02 to match the cytoplasm to the surrounding medium [104]. Such effects have been studied since the 1950s [106]. Since the index of the intracellular medium is not precisely known, experiments conducted in vitro on cells suspended in a medium may not reproduce in vivo conditions [103]. In particular, low-angle forward scattering is expected to be less prominent in vivo because of the much closer index matching of cell membranes to the surrounding medium, but this has not yet been confirmed. A consequence of the lack of precise values for the in vivo refractive indices of cells, nuclei, cytoplasm and organelles, is the use of a wide variety of values in models and the conduct of research on the effects of varying the refractive index [104].

Scattering is correlated with tissue state through the influence of pathology, size, shape, and organisation of cells on effective scatterer size distribution and refractive index. Epithelial dysplasia, the abnormality associated with cancers and pre-cancers, is characterised by nuclei which are enlarged, pleomorphic (irregular in shape and size distribution), and occupy more of the cell volume [107]. Also, because they divide more rapidly than normal cells and often have extra chromosomes, the protein concentration of such nuclei is often much higher, resulting in a higher refractive index [104]. Changes in scattering over time have been observed in vivo from nuclei during cell mitosis with OCT [108], in vitro on application of a contrast agent with confocal microscopy [104], and in vivo, attributed to dysplasia [109], with elastic light scattering spectroscopy. There is much less evidence of such correlations other than from cell nuclei. Changes in the angular distribution of scattering, attributed to mitochondrial swelling during apoptosis (programmed cell death), have been observed in vitro [110]. At the level of large aggregations of cells, the refractive index is sensitive to variations in hydration [111], calcification [112], and tumour malignancy [113], but there has not yet been sufficient research performed to establish the utility of these findings.

Although some landmark progress has been made [108,109], the correlation of scattering with the state of tissue is an immature field of research. In vivo subcellular resolution histology remains a 'holy grail' for OCT and other light scattering techniques. OCT, at least, has made significant practical progress in this direction [108] but it is apparent that it will be a difficult goal to attain.

17.4.2 Extinction and multiple scattering

Since ballistic light is expected to carry the majority of the image information, it is useful to consider a simple model to predict its signature in an OCT A-scan. The extinction-single backscatter model incorporates factor (i), extinction by absorption or wide-angle scattering, in our list of four main influences given. In this section, we will also consider the modifying effects of factors (ii), low-angle forward multiple scattering, and (iv), wide-angle multiple scattering, on the OCT image. In the next section, we will consider the effects of the other factor, as well as the statistical properties of multiple scattering, and strategies for the amelioration of the effects of multiple scattering. In the context of OCT imaging, since the effects of multiple scattering all result from coherent interference, the term multiple scattering may be used interchangeably with speckle.

The extinction-single backscatter model considers the optical power incident on the sample, αI_{S0}, to be extinguished during propagation to the backscattering location l_S, characterised by the position-dependent total extinction coefficient $\mu_t(l)$, so that the optical power $\alpha I_{S0}\exp\{-\int_0^{l_S}\mu_t(l)\mathrm{d}l\}$ is incident on the sample volume. A fraction of this power is backscattered, characterised by the backscattering coefficient $\mu_b(l_S)$ ($m^{-1}\,sr^{-1}$). In a model of the sample as a distribution of discrete scatterers, this coefficient is related to the single-particle backscattering cross-section, σ_b (m^2), by $\mu_b = N\sigma_b$, where N is the number density and $\sigma_b(l_S) \equiv 4\pi z^2 I_b(0)/I_{Si}(l_S)$, where $I_b(0)$ is the optical power density at the collection aperture in the absence of the medium and $I_{Si}(l_S)$ is the optical power density incident on the particle [114]. σ_b, also known as the radar cross-section, is the area required to intercept the correct amount of incident power to produce the actual power density at the collection aperture when the backscattering is extended isotropically. An assumption inherent in this approach is that the scatterers in the sample volume may be represented by a single backscattering coefficient, independent of the resulting coherent interference. The breakdown of this assumption is examined in the next section. The backscattered power is further extinguished on the return path and, finally, we must consider the aperture function of the sample-arm optics. Thus, the term I_S in the expression for the signal power $\overline{i^2} = 2[\alpha(1-\alpha)]^2\rho^2 I_S I_R|\gamma(0)|^2$ is replaced by [13,115]

$$I_S = I_{S0}\int_L A\left(\frac{l_S - l_F}{n}\right)\mu_b(l_S)\exp\left\{-2\int_0^{l_S}\mu_t(l)\mathrm{d}l\right\}\mathrm{d}l_S \tag{53}$$

where $A[(l_S - l_F)/n]$ describes the aperture function of the sample optics, is the product of the confocal response function (Equation (30)) and the solid angle over which the collection optics operate, and the region L contains all axial locations within a coherence length of the optical path matchpoint Equation (53) assumes we have measured the OCT signal at the peak of the envelope ($|\gamma(0)|^2 = 1$), and due to integration over the region L, includes only contributions from within the sample volume.

Hence, in the single-scattering regime, an OCT A-scan in a homogeneous, turbid medium shows a sharp peak at the air–sample interface followed by an exponential decay. In any medium, the OCT A-scan signal is dominated by the μ_t term because of the long pathlengths involved but the local contrast is provided primarily by variations in μ_b. Different structures in the sample contribute differently to μ_t and μ_b and, hence, the two quantities are not necessarily strongly correlated [116]. In principle, μ_t can be determined from the exponential decay if A is known, under the assumption that ballistic light dominates over multiply scattered light. Figure 22 shows the measured average signal strength, on both linear and log scales, versus pathlength for a mirror placed at successive discrete depths in an Intralipid™ solution [117]. The ballistic light is indicated by the sharp peak. The multiple scattering tail is clearly evident from the outset and begins to approach the magnitude of the ballistic signal at a depth of around nine mean free paths, where the mean free path $l_{MFP} \equiv 1/\mu_t$. The ballistic component of scattered light obeys the Lambert–Beer law – exp $(-\mu_t l_S)$ dependence in a homogeneous sample to depths of 15 to 20 mean free paths in transillumination [118]. Eventually, departures from the Lambert–Beer law arise from the wavelength dependence of the scattering parameters. An alternative and rather elegant visualisation of the ballistic light is contained in Ref. 119.

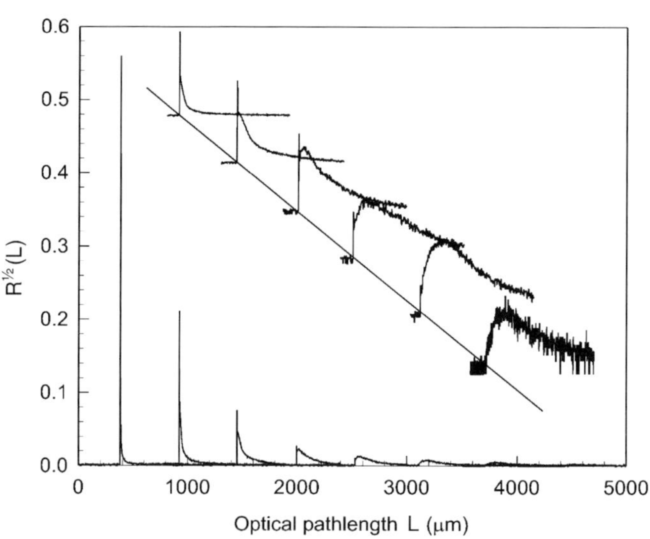

Figure 22. Square-root reflectivity versus optical pathlength from a mirror at a series of locations in Intralipid solution ($\mu_s = 50\,cm^{-1}$, $\mu_a = 0.1\,cm^{-1}$, $g = 0.84$, $n = 1.33$). Reproduced from [117] with permission.

Multiply scattered light can lead to wide variations in the apparent value of μ_t and a dependence of the measured value on the experimental setup [116,117,120,121]. In early work, Schmitt et al. [116] showed that the measured extinction coefficient depended strongly on beam diameter in the sample and to a slightly lesser degree on wavelength. In their experiment, increasing the beam diameter in the sample by shifting the focal point of the objective to a greater depth, with all else equal, increased the measured extinction of rat artery. The main effect of shifting the focal point was to increase the transverse cross-section of the medium probed by the beam. Schmitt et al. supposed that this in turn increased the size range of structures contributing to low-angle forward scattering which led to a greater degradation of the transverse spatial coherence of the beam. As a result, mixing with the reference beam was less efficient and so produced a higher extinction. Pan et al. [117] showed that the OCT signal was significantly enhanced above what was to be expected based on separately performed narrow-beam, angle-gated backscattering measurements. In their experiment, the higher numerical aperture of the OCT system compared with the angle-gated measurement led to the capture of a greater proportion of all types of multiply scattered light, thus increasing the overall signal and producing a lower extinction than expected. These two works highlight the difficulties in using OCT to determine an absolute, transferable, and setup-independent value for the average extinction coefficient of a highly scattering medium. Measurement of changes in μ_t would appear to be more feasible. Measurement of such changes has been used in a range of applications, including to characterise Brownian motion in liquids [121], to detect apoptosis in cells [122], to detect morphological changes in the bladder [123], and to detect blood glucose concentration [124].

The effects of multiple scattering were further examined by Yadlowsky et al. [120], commencing with the observation that the signal strength in the OCT image of a fingernail viewed through the 300 μm thick cuticle was much less than expected based on the signal strength for direct viewing of the nail, and the more or less constant signal strength of the underlying dermis when viewed through either region. This same effect can be observed in more recently published images [78]. Yadlowsky et al. conducted a series of experiments in which A-scan signals were detected from small isotropic scatterers overlaid by suspensions of varying concentrations and particle sizes. They concluded that multiple low-angle forward scattering increased the strength of the probe field, and the probe field increased with particle size, consistent with the higher anisotropy of scattering. The effect on the image, though, depended on the size of the structures in the sample volume. They claimed higher effective backscattering would be observed only from structures smaller than the transverse correlation length of the probe field (a measure of the transverse spatial coherence), because in effect these structures did not see the wavefront distortion caused by low-angle multiple forward scattering. Thus, to reconcile their conclusions with the fingernail observation, we must conclude that the nail is dominated by scatterers that are large compared with the correlation length and the dermis must contain a relatively larger concentration of smaller scattering structures than the nail. Their second contribution was to demonstrate the presence of an isotropic non-localised haze at depths below high densities of isotropic scatterers, which is an emphatic demonstration of factor (iv) in our list, i.e., multiple scattering from outside the sample volume.

Quantitation of the contributions of multiple scattering to OCT images remains a difficult problem. For homogeneous suspensions of varying turbidity, observations of the distinct change in slope of the measured signal versus depth, quantifying the transition from the ballistic to the diffuse scattering regime, have been made [125–127]. A clear example of this is shown in Figure 23, which plots the average OCT signal obtained from transillumination of a suspension of 1 μm latex microspheres. There is surprisingly little unequivocal data on the impact of multiple scattering on image quality. Figure 24 is reproduced from Izatt et al. [125] and shows en face images of a standard target imaged through the latex-sphere suspension corresponding to the first emergence of the multiple scattering. What is striking is the strong degradation in image quality caused by shifting the reference delay over a small delay range from the ballistic peak to the diffuse peak. However, translating these results to a detailed understanding of reflectance imaging in heterogeneous tissues is not straightforward. Schmitt and Knüttel [128] have produced a simulation which partially takes into account the effects of multiple scattering. Their simulated images show the improvement obtained through

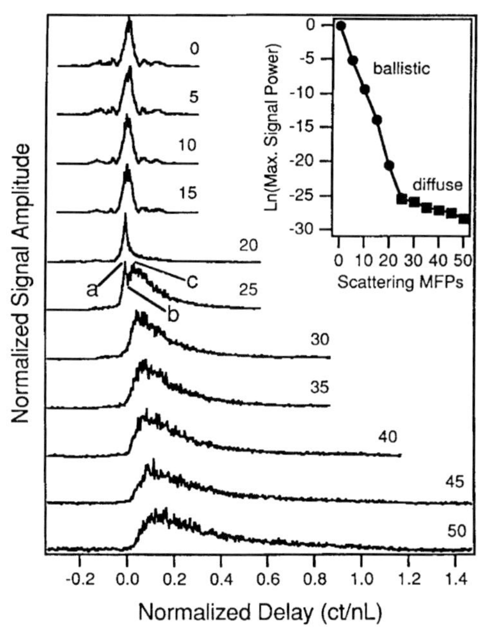

Figure 23. Coherence-gated scattering profiles in transillumination through different concentrations of 1 μm diameter microsphere solutions equivalent to from 0 to 50 mean free paths. Letters a, b, and c correspond to images in Figure 24. The apparent narrowing of the signal at 20 mfps is caused by a reduction of the coherence length as a consequence of increasing the power of the mode-locked titanium:sapphire laser source. Inset: Logarithmic attenuation (relative to water) of the ballistic (circles) and diffuse (squares) peak powers obtained from the scattering profiles. Reproduced from [125] with permission.

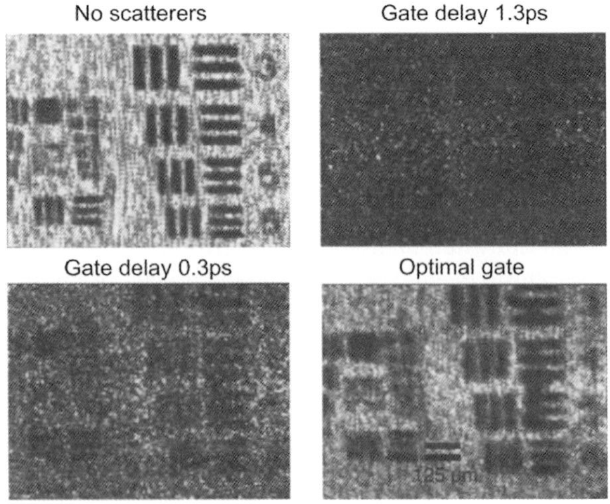

Figure 24. En face coherence-gated transillumination images of an opaque resolution target embedded in pure water and a solution of microspheres in water equivalent to 27 scattering mean free paths estimated from Mie theory. Upper left: Pure water. Remaining images correspond to labelled points in Figure 23: 'a' ballistic peak – lower right; 'b' 0.3 ps – lower left; 'c' diffuse peak at 1.3 ps – upper right. Reproduced from [125] with permission.

decreasing the coherence length of the source, and the lack of improvement obtained by increasing the numerical aperture of the objective when multiple scattering dominates.

Based on the above discussion, it seems reasonable to expect that most published OCT images of highly-scattering samples are affected by multiple scattering, i.e., speckle noise, including those obtained by time averaging. Single-scan data shows much more dramatically the effect of speckle than the averaged data presented above. Figure 25 is an example of a single A-scan and its time-averaged equivalent taken in an Intralipid solution. Time averaging of multiple images improves the image quality when uncorrelated motion of the scatterers, e.g., due to Brownian motion and convection in liquids, causes the speckle pattern to change over time, as is the case for the sample used to generate Figure 25.

However, for most in vivo applications such long-term averaging is not feasible, and so the speckle noise must be tolerable, or other means must be undertaken to mitigate its effects. We consider speckle in more detail in the next section.

17.4.3 Speckle

Despite its ubiquitous presence and many passing references to it in published papers, speckle in OCT systems has not been extensively studied [75,98,119,121,128–132]. It has been studied extensively, though, in closely related optical systems, as well as in other imaging modalities, most notably ultrasound, astronomy and synthetic aperture radar. The study of laser speckle for biological applications has been extensively reviewed by Tuchin [133].

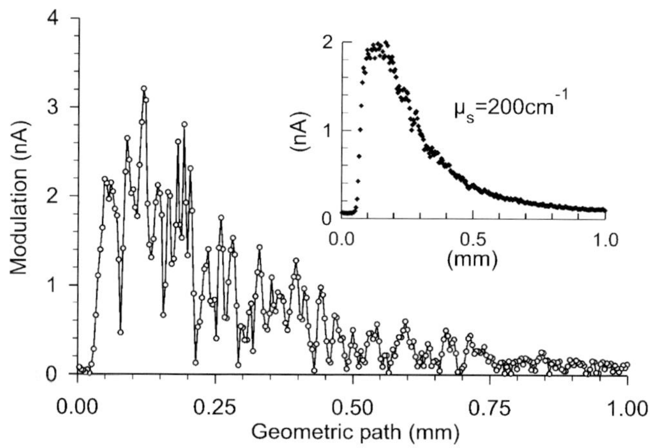

Figure 25. OCT A-scan and averaged A-scan through Intralipid solution showing the noise inherent in a single scan and its improvement through averaging 10 scans. Reproduced from [126] with permission.

17.4.3.1 Coherent imaging and multiple scattering

Of the three classes of multiple scattering described in the introduction to Section 17.4, only one bears any relation to the sample volume and, thus, is fundamentally unavoidable. The probability of multiple uncorrelated, or partially-correlated, single-backscattering events from within the sample volume is high. Hence, even in the absence of multiple scattering from outside the sample volume, an image would be corrupted by the coherent interference of the multiple fields. Speckle from such multiple singly-backscattered light would only be absent if the backscattered and reference wavefronts were exactly matched, which would require as a sample a plane reflector in the focal plane of the sample objective. To show the prevalence of multiple single backscattering, consider a suspension of 0.3 μm diameter polystyrene spheres of index 1.57 in water with a representative scattering coefficient of $\mu_s = 50\,\text{cm}^{-1}$. At a wavelength of 1 μm, the mean number of scatterers in a 10 μm cube is around 200. This should be taken as a lower limit for biological tissue since, as discussed, a large fraction of the total backscattering cross-section of tissues is thought to come from small sub-wavelength size structures with much lower refractive index contrast than in this example [100,101,104].

Recalling our linear systems framework, then, the sample could be represented with an impulse response $h_S(t)$ describing a set of discrete reflectors $h_S(t) = \sum_i \sqrt{R_i}\delta(t-\tau_i)$, on the understanding that such reflectors are no longer well separated, as we have assumed previously. The discussion in Section 17.4.1 has highlighted, though, that biological samples are probably more accurately described by a refractive-index continuum, $n(\vec{r})$, where $\vec{r}$ represents a three-dimensional spatial location vector, and a local backscattering coefficient $\mu_b(\vec{r})$, which is a function of the refractive index gradient $\nabla n(\vec{r})$. However, the one-dimensional discrete model is useful for exploring the basic issues. Figure 26 shows

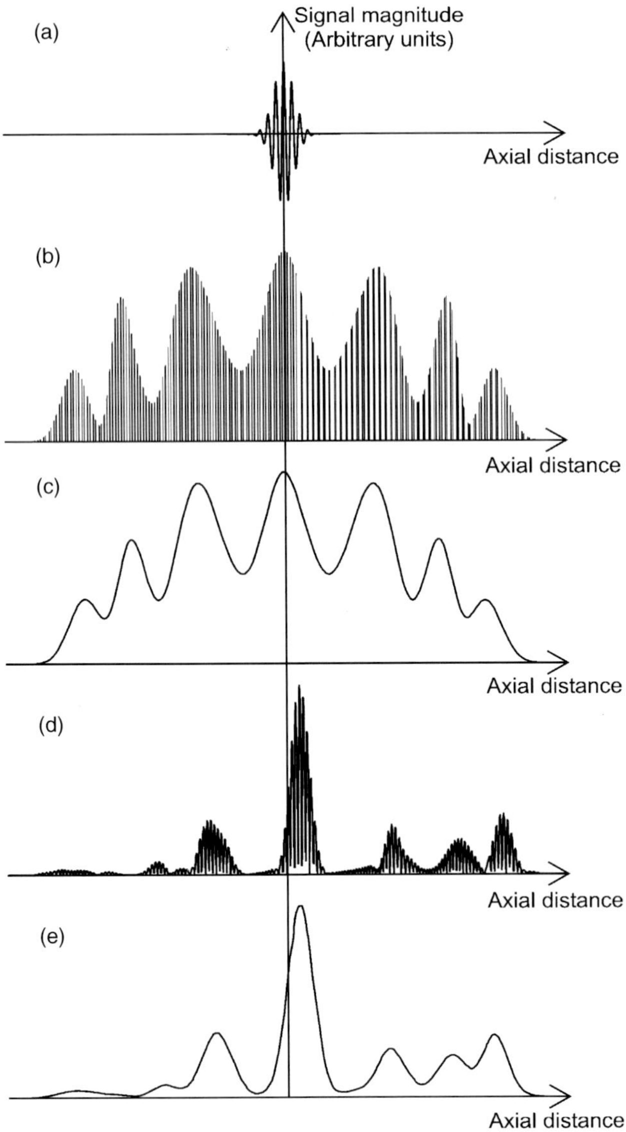

Figure 26. Simulation of the distortion of an OCT A-scan caused by the coherent detection process. (a) Axial PSF with Gaussian envelope. (b) Ideal backscatter profile generated by a dense random distribution of point scatterers with variable cross sections. (c) Incoherent signal formed by convolving the backscatter profile with the envelope of the OCT PSF in (a). (d) Coherent signal formed by convolving the backscatter profile with the OCT PSF in (a). (e) Low-pass filtered version of the coherent signal. Comparison of (c) and (e) with (b) highlights the distortion introduced by coherent interference of close-spaced scatterers.

an example of a random axial distribution of discrete particles that demonstrates the principle that the images of all non-planar scattering objects are distorted by the coherent detection process. Coherent interference between scatterers results in an image (speckle pattern) that does not faithfully reproduce the backscattering cross-section and is sensitive to the phase differences between the scattered fields. In principle, the inverse problem of determining the structure of the sample from the measured coherent image could be solved, but in practice it has not been shown to be feasible.

To explore the distortion introduced by coherent imaging systems further, we write Equation (29) for the OCM/OCT optical transfer function in the simplified and analytic form:

$$\mathrm{I}^{\mathrm{D}}(\tilde{l}_{\mathrm{R}}) = K\sqrt{R_{\mathrm{LS}}}(\tilde{l}_{\mathrm{R}})\sqrt{h_{\mathrm{LS}}(\tilde{l}_{\mathrm{R}})} \otimes \gamma_{\mathrm{LS}}(\tilde{l}_{\mathrm{R}}) \tag{54}$$

where K is a constant, $\sqrt{R_{\mathrm{LS}}}(\tilde{l}_{\mathrm{R}}) = \sqrt{R_{\mathrm{S}}}(\tilde{l}_{\mathrm{R}}/n_{\mathrm{g}})$ is the optical pathlength-resolved square-root reflectance,

$$h_{\mathrm{LS}}(\tilde{l}_{\mathrm{R}}) = h_{\mathrm{F}}\left(\frac{\tilde{l}_{\mathrm{R}} - n_{\mathrm{g}}l_{\mathrm{F}}}{nn_{\mathrm{g}}}\right)$$

is the confocal axial point-spread function, and

$$\gamma_{\mathrm{LS}}(\tilde{l}_{\mathrm{R}}) = \exp\left[-\left(\frac{2\sqrt{\ln 2}\tilde{l}_{\mathrm{R}}}{\tilde{l}_{\mathrm{FWHM}}}\right)^2\right]\exp[-j\bar{k}(2\tilde{l}_{\mathrm{R}})]$$

is the modified coherence function of the light. Equation (54) highlights the fact that convolution with the coherence function is the root cause of the problem. This problem may be further appreciated by transferring it to the spatial-frequency domain by Fourier transforming both sides of Equation (54) with respect to spatial frequency k_l to give

$$\Im\{\mathrm{I}^{\mathrm{D}}\}(k_l) = K\left(\Im\left\{\sqrt{R_{\mathrm{LS}}}\right\}(k_l) \otimes \Im\left\{\sqrt{h_{\mathrm{LS}}}\right\}(k_l)\right)\Im\{\gamma_{\mathrm{LS}}\}(k_l) \tag{55}$$

Schmitt et al. [98] have proposed that the length scales of useful variations in tissue are set by the scales of the continuous structures, with the largest being a blood vessel and the smallest a protein macromolecule. If we assume appropriate lengths of 100 µm and 0.01 µm, respectively, a coherence length of 10 µm, mean wavelength of 1300 nm, and a numerical aperture of 0.5, then we can plot representative values for the magnitudes of $\Im\{\sqrt{R_{\mathrm{LS}}}\}(k_l) \otimes \Im\{\sqrt{h_{\mathrm{LS}}}\}(k_l)$ and $\Im\{\gamma_{\mathrm{LS}}\}(k_l)$, as shown in Figure 27. This figure highlights the inadequacy of the band-pass nature of the OCT transfer function in responding to the full range of spatial frequencies in the image, $\Im\{\sqrt{R_{\mathrm{LS}}}\}(k_l)$. It suggests that improved performance would be obtained by operating with a shorter coherence length. Note that the confocal response h_{LS} is low-pass in the spatial-frequency domain but effectively up-converted by convolution with the reflectance function, rather than the coherence function.

The other multiple scattering mechanisms of low-angle forward scattering and wide-angle scattering further modify the speckle pattern shown in Figure 26. The

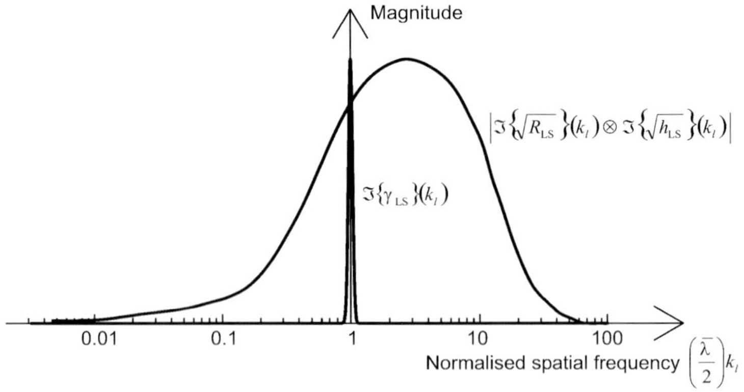

Figure 27. Estimated spatial frequency content of biological tissue compared with the spatial frequency response of OCT/OCM assuming $n_g = 1$, for a 10 μm coherence length, 1300 nm mean wavelength, and 0.5 numerical aperture.

relative importance of and distinctions between the three regimes of multiple scattering have not been thoroughly investigated. Schmitt et al. [98] point out, though, that scattering in or near the sample volume tends to emphasise low spatial frequencies in the recorded speckle pattern, whereas, wide-angle scattering tends to emphasise high spatial frequencies.

17.4.3.2 Statistics of OCT speckle

We firstly briefly review the three common cases of speckle statistics in optics and then consider the relevant form for OCT. Speckle arising from the superposition of a large number of monochromatic, polarised waves of random, independent, identically-distributed amplitude and uniformly-distributed phase is described by a random phasor sum and the intensity I is described by a negative exponential probability density function $p(I)$ given, for non-negative I, by [45]

$$p(I) = \frac{1}{\langle I \rangle} \exp\left(-\frac{I}{\langle I \rangle}\right) \tag{56}$$

where the angle brackets denote ensemble average. Equation (56) assumes that the speckle is 'well developed', which means that processes such as spatial averaging by a finite-area detector or temporal averaging are absent. Speckle is often characterised by the ratio of the standard deviation to the mean intensity, $\sigma_I/\langle I \rangle$, termed the contrast ratio, which in this case is unity. If the waves are unpolarised then, due to the independence of the two orthogonally-polarised components, the resultant distribution of well-developed speckle is the convolution of the right-hand side of Equation (56) with itself, giving [45]

$$p(I) = \left(\frac{2}{\langle I \rangle}\right)^2 I \exp\left(-2\frac{I}{\langle I \rangle}\right) \tag{57}$$

The contrast ratio for unpolarised speckle takes the value $1/\sqrt{2}$. Thus, unpolarised speckle is intrinsically less corrupting than polarised speckle. The third case of interest corresponds to randomly polarised waves in the presence of a strong signal, which modifies the distribution as follows:

$$p(I) = \frac{1}{\langle I \rangle} \exp\left(-\frac{I + S}{\langle I \rangle}\right) I_0\left(\frac{2\sqrt{IS}}{\langle I \rangle}\right) \tag{58}$$

where S is the intensity of the strong signal, I_0 is a modified Bessel function of the first kind, zero order, and $\langle I \rangle$ is the mean intensity in the absence of the strong signal. This distribution very rapidly approaches a Gaussian as S exceeds $\langle I \rangle$.

For OCT, a strong and well-determined constant signal is provided by the reference, denoted E_R. If the ith scattering signal is denoted by E_{Si}, then in simple terms the output of the interferometer is given by

$$I = |E_R|^2 + \left|\sum_i E_{Si}\right|^2 + 2\mathrm{Re}\left\{\sum_i E_R^* E_{Si}\right\} \tag{59}$$

The inclusion of all terms on the right-hand side of Equation (59) corresponds to the constant strong signal case which leads to $\sqrt{I}$ being Rician-distributed and, in the limit that the reference is strong and the signal weak, it is Gaussian distributed. The situation in OCT is somewhat different, however, because of the phase modulation that is usually present. The modulation of the field in the reference arm means that the first two terms in Equation (59) are at baseband, but the noise in the third (mixing) term is centered on the passband frequency. If we assume the passband is sufficiently far above the baseband to enable separation by filtering, we can neglect the first two terms. Furthermore, we can assume for convenience that the reference phase is stable and set it to zero, giving for the mixing term at the passband frequency

$$I_{INT} = 2\sqrt{I_R}\mathrm{Re}\left\{\sum_i E_{Si}\right\} \tag{60}$$

This expression is valid for detection of the full interferogram and would result in a Gaussian probability density function for I_{INT}. The factor $\sqrt{I_R}$ on the right-hand side of Equation (60) just scales the result and represents the heterodyne advantage. The envelope of the real part of a function, however, is not just a mapping of the negative values onto the positive values – it is essentially a determination of the amplitude, hence, we obtain

$$I_{ENV} = 2\sqrt{I_R}\left|\sum_i E_{Si}\right| \tag{61}$$

The second term in Equation (61) is the magnitude (amplitude) of a random phasor sum, which is described by a Rayleigh distribution given by

$$p(I_{ENV}) = \frac{I_{ENV}}{\sigma_I^2} \exp\left(-\frac{I_{ENV}^2}{2\sigma_I^2}\right) \tag{62}$$

which is in agreement with Fercher et al. [14]. The Rayleigh distribution has a contrast ratio of $\sqrt{(4/\pi - 1)} = 0.52$, which is about half that for non-interferometric fully-developed speckle described by Equation (56). The OCT signal power is proportional to I_{ENV}^2 and Schmitt et al. [98] have shown that Equation (57) provides a good fit to experimental data when I is substituted with I_{ENV}^2. The OCT probability density function would be further modified by taking the logarithm of the envelope, as is commonly done in demodulating and presenting OCT images.

All the density functions presented above are based on the assumption that there are sufficient component phasors to justify the use of the central limit theorem. It is conceivable that the central limit theorem would not hold for small sampling volumes when the effective number of scatterers is small, and near the surface of a sample when the contributions from low-angle and wide-angle multiple scattering are small. In this regime, the speckle pattern may contain information about the sample.

17.4.3.3 Methods of speckle reduction

Since speckle is ubiquitous in OCT imaging, one might consider its utilisation, as has been the case in closely related optical systems [133], rather than its avoidance. There is one example of this reported by Schmitt [130], who demonstrated the use of two-dimensional cross-correlation of static speckle to track the local motion in a sample undergoing compression. The motivation was the use of the elastic or shear modulus of a tissue as a contrast mechanism for detecting lesions in soft tissue. More generally, however, the motivation for the study of speckle has been the removal of its distorting effects. Speckle reduction techniques may be broadly classified into those involving modifications to the experimental apparatus and those involving post processing of the recorded images. Here we consider modifications to the apparatus and in the next section we consider post-processing techniques.

It has been claimed (although not shown) that employing unpolarised light would reduce the speckle contrast by up to a factor of $\sqrt{2}$ [98]. Employing unpolarised light in OCT requires engineering the OCT system to be free from polarisation-dependent loss, birefringence and polarisation mode dispersion in order not to introduce distortions into the axial response function. An alternative approach is to average the OCT envelope signal to convert $\sqrt{R}(l)|\cos\phi(l)|$, where ϕ is a phase representing the distorting effects of speckle, into $\left\langle\sqrt{R}(l)|\cos\phi(l)|\right\rangle \propto \sqrt{R}(l)$, where the angle brackets denote the averaging procedure. This averaging is achieved by probing a sample-induced ensemble of values of ϕ without perturbing $\sqrt{R}(l)$. For an average over M such measurements, the addition on an intensity (envelope) basis which is implied leads to an improvement, characterised by a decrease in the contrast ratio by $\sqrt{M}$ for fully-developed speckle [67]. Averages may be formed over space, optical frequency, or time, but convey a potential loss of resolution, either temporal or spatial, which implies a trade-off. Many reported results are simply averages over multiple sequential measurements and rely on the fluctuations inherent in the sample to probe a sufficient range of phases to improve the overall image quality without introducing motion artefacts. Temporal averaging has been studied to some degree by Pan et al. [134] and by Rollins et al. [92] for various in vivo samples.

A spatial-diversity OCT system was developed by Pagnoni et al. [135] in which eight frames are simultaneously recorded, each with 5 μm lateral offset. The averaging of these frames substantially improves the image quality and, in addition,

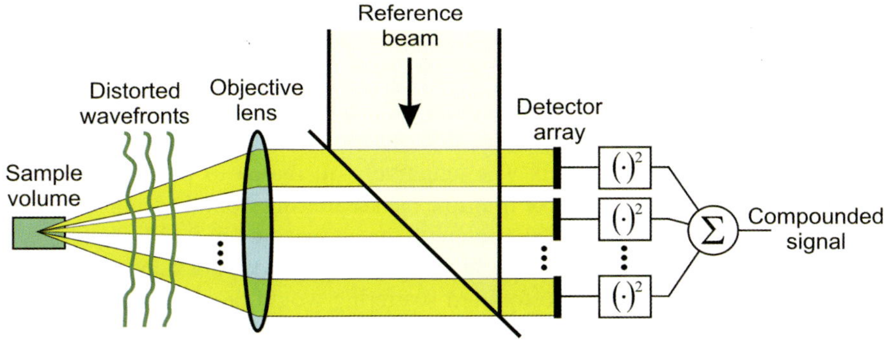

Figure 28. Principle of the angular compounding method. Each detector element receives light backscattered at a different angle. The detected signals are squared prior to summation. Adapted from [129].

conveys a three-dimensional impression, at the expense of transverse resolution. Images of skin produced by this system can be seen in Section 17.5.2. Another spatial diversity scheme was demonstrated by Bashkansky and Reintjes [131]. Schmitt et al. [67,129] introduced an alternative form of spatial diversity into an OCT system by employing a segmented detector and summing incoherently the resulting interference signals, as indicated schematically in Figure 28. They investigated equal weighting in the summations as well as an adaptive weighting scheme [136]. Their free-space system performs averaging of the backscattered signals from different angular apertures. Although only a quadrant photodetector ($M = 4$) was used, implying a maximum improvement in the contrast ratio of $\sqrt{M} = 2$, the visual impact of the consequent reduction in speckle from images of human skin taken in vivo is substantial. Figure 29 shows the significant improvement obtained with this system. The accompanying theoretical loss of resolution, $M/2 = 2$, is not apparent in the images in Ref. 129. The explanation lies in the effect of multiple scattering in a highly scattering sample such as skin. The diffraction-limited

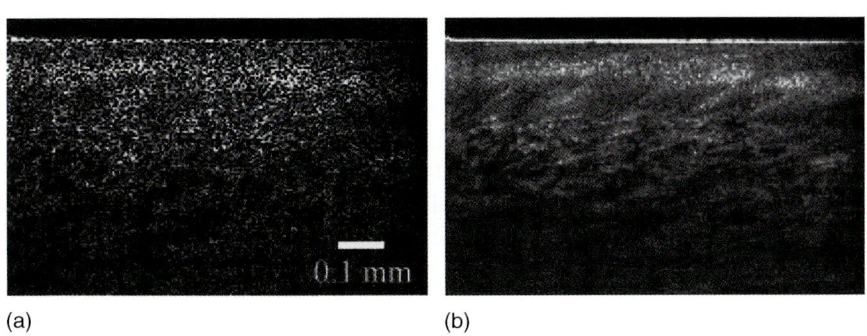

Figure 29. OCT images before and after angle compounding. The uncompounded image in (a) is formed from one element of the quadrant detector. To form the compounded image in (b), the square root of the sum of the squares of the interference signals from all four elements was calculated for each pixel. Reproduced from [98] with permission.

transverse resolution is doubled by the larger detector aperture, but the image is blurred by multiple scattering, which is characterised by the transverse coherence length. If multiple scattering rather than diffraction limits the transverse resolution, then spatial diversity will improve the image quality without reducing resolution. A second key point demonstrated in Ref. 129 was the lack of correlation between the signals from each detector. This was tentatively attributed to low-angle forward scattering rather than the effects of multiple backscattering from the sample volume. This attribution appears in contradiction with the observation in Ref. 98 by the same authors that the character of speckle in images of skin does not depend on depth, since the influence of low-angle forward scattering would be expected to increase with depth. This is important because the presence of correlation would suggest that the speckle contains valuable information about the sample volume, whereas, its absence suggests that the speckle is simply an unwanted corruption of the desired information. Correlation would only be anticipated, though, when the effective number of scatterers in the sample volume is very small [137]. As pointed out in Ref. 129, further work is needed to confirm the origins of the observed loss of correlation.

The shot noise-limited signal-to-noise ratio scales as $1/\sqrt{M}$ in spatial diversity schemes, implying a higher source power requirement to reach the shot noise limit. Such schemes are awkward to implement in a fibre-optic configuration, and such a configuration has not been demonstrated to date. The sample rate for real-time operation scales with M, the number of detectors, which is also awkward. Spatial diversity could be implemented alternatively by scanning a reduced-size reference beam across the surface of a single detector of the same area, but such scanning is not straightforward at the speeds required for real-time operation [98]. In this context, it is important to realise that it is diversity in the sample path rather than diversity in the reference path that leads to effective averaging.

Frequency diversity has been investigated only cursorily to date [98]. Frequency diversity implies division of the received optical signal into bands with the consequent very undesirable reduction in axial resolution. The reduction in resolution could be restored by the use of a strong confocal response, with its own set of difficulties. A second major difficulty is the relatively weak decorrelation with wavelength of well-developed speckles originating from the sample volume. However, decorrelation may be strengthened by the effects of low-angle forward scattering in the overlying medium. Given the need for a broad bandwidth to sufficiently decorrelate the measured signals, it is reasonable to ask whether the broad bandwidth is more effectively used in a coherent detection mode. A detailed study on how OCT speckle contrast is altered by coherence length remains to be performed. It is apparent, though, from Figure 27 that the spatial frequency coverage should improve inversely with the coherence length.

17.4.4 Post-processing methods for image enhancement

It is worth broadening the discussion here from mitigation of the effects of speckle to image enhancement in general. The main types of processing that have been or could be used fall into three categories: basic image enhancement techniques applied to the OCT envelope image; deconvolution using a priori knowledge of the system response, applied to the envelope or potentially to the full interferometric signal; and other phase-domain techniques. Deconvolution has mostly been applied in the conventional manner for resolution enhancement, but invariably it also impacts on the speckle noise in images. In the following discussion, we firstly consider approaches based on the OCT envelope signal and then consider phase-domain techniques that require access to the interferometric phase information. Research into OCT image processing is in its early stages and a clear idea of the costs and benefits must await more comprehensive studies.

17.4.4.1 Methods applied to the envelope signal

Basic image enhancement includes signal averaging, as already discussed, and low-pass spatial filtering, which involves a loss of resolution. A more sophisticated form of filtering has been applied to endoscopic images of coronary arteries [138]. The rotating kernel transformation technique involves the local averaging of image pixel (envelope) intensities over a line of fixed length and carefully chosen angle. Its intended use was the automatic identification of boundaries in images to help speed up the process of diagnosis.

Xiang et al. [139] noted that wavelet filtering techniques had been successfully employed in speckle noise reduction for medical ultrasound and synthetic-aperture radar images. They applied a wavelet filter with nonlinear thresholding of the wavelet coefficients to the OCT envelope signal output from their OCT system with four-fold spatial diversity [67]. They characterised the images using two measures: a signal-to-noise ratio defined as the ratio of the mean to standard deviation of the pixel magnitudes in a defined area of the image; and a contrast-to-noise ratio for a region of interest referenced to a featureless region of the image, defined as the ratio of the difference in means of the two regions to their RMS standard deviation. They noted average improvements in the two measures by factors of 10 and 1.5, respectively, as well as the preservation of sharp boundaries.

Image deconvolution involves the restoration of the representation of an object by reversing known distortion. It is a more sophisticated approach than the image enhancement techniques described above, which are essentially designed to improve the presentation of an image [140]. The deconvolution procedure, even in the absence of noise, does not necessarily lead to a unique, realisable solution. It is critically sensitive to noise and often involves long computation times. Several simple attempts [141,142] have been made to improve resolution by one-dimensional deconvolution of the OCT envelope employing inverse Fourier transformation. In each case, the methods were applied to one-dimensional images of stacked microscope slides. The methods demonstrated about a two-fold improvement in resolution. In one case, the expected sharp degradation of the signal-to-noise ratio was also demonstrated [142]. A similar one-dimensional technique was applied to two-dimensional images of fresh onion by Kulkarni et al. [143]. They quantified the resolution improvement as a factor of three and noted a 32 dB increase in the noise floor of an image originally with −102 dB sensitivity [143]. Interestingly, they also noted that, to obtain their results, the measured auto- and cross-correlation envelope responses were required to be measured to sub-micron accuracy, which necessitated use of an auxiliary interferometer to calibrate the reference delay to the required accuracy. These early results confirmed the findings in many other fields [144] that approaches that are more robust to noise were required.

Kulkarni et al. [145] went on to show the much more robust performance of one-dimensional constrained iterative deconvolution on two-dimensional images of fresh onion. The resolution improvement was not compromised but the reduction in dynamic range was reduced to 2 dB. The differences in the blurred and unblurred images shown in the paper are emphatic. Another one-dimensional iterative technique has been applied to remove the side lobes in the envelope point-spread function deliberately introduced by dispersion [146]. The idea behind the work was to use dispersion to narrow the main lobe of the coherence function, with the side effect of producing unwanted side-lobes. The side-lobes were then to be removed by iterative deconvolution. From the results shown for a glass slide and fresh onion, the overall benefit of the procedure appears to be marginal.

The pointwise subtractive iterative technique known as CLEAN was first applied in one-dimensional form on two-dimensional OCT images by Zhou and Schmitt [147], and followed up in two-dimensional form by Schmitt [64]. CLEAN was originally developed in radio-astronomy to remove the side-lobes introduced by the ringing response of sparse-array imaging systems from strong reflectors in order to reveal close-by weak reflectors [148]. Schmitt's contribution [64] is noteworthy in a number of respects. It was the first application of a two-dimensional deconvolution and the first to employ a point-spread function that broadened with depth to take account of the progressive loss of transverse spatial coherence with depth. Schmitt recognized the problem inherent in the CLEAN algorithm – it was designed to clean up point images of targets that are separated by more than the width of the main lobe of the point-spread function. When applied to multiple unresolved targets it produces corrugation artefacts. To account for this, he employed a modified algorithm that had already been proposed to deal with this problem in astronomical images. Importantly, he accounted for the problem in a second way by applying CLEAN to images in which speckle was already reduced by four-fold spatial-diversity detection [67]. CLEAN was shown to be effective in improving the resolution of images of latex spheres in gelatin, at the expense of signal-to-noise ratio. Figure 30 shows in vivo images of skin taken from Ref. 64. It was noted that some structures not discernable in the original image were revealed by the CLEAN process. In addition the corrugations caused by multiple unresolved scatterers were not completely eliminated and had produced some artefacts. More recently, Piao et al. have applied with some success a less sophisticated one-dimensional CLEAN to reduce the side-lobes caused by a structured source spectrum [149].

CLEAN has some intrinsic ability to interpolate gaps in the spatial frequency response of an optical system, which suggests it could help ameliorate the effects of destructive interference. However, its shortcomings are also well recognized. Given the purpose for which CLEAN was developed, it would be somewhat fortuitous if it were proven to be the most effective means of dealing with speckle noise in

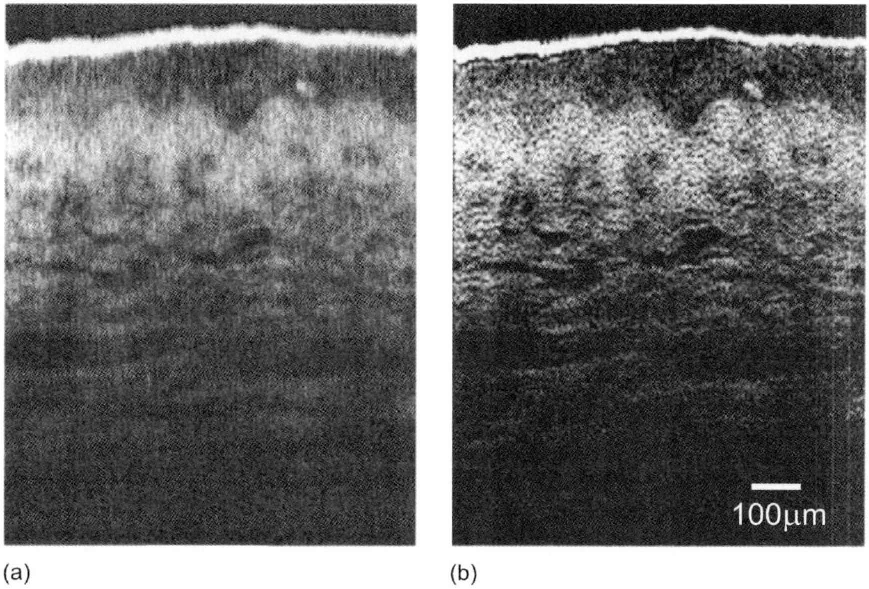

Figure 30. (a) OCT image of the dorsal side of the skin of the index finger between the phalangeal joints. Speckle noise has already been reduced by incoherent addition of signals from four elements of a quadrant detector. (b) Image after deconvolution using a modified CLEAN algorithm. Both images are displayed on the same grey scale after square-root compression. Reproduced from [64] with permission.

OCT images. However, the images produced in Ref. 64 are the most convincing demonstration of improvement considered so far.

Before moving on to methods that operate in the complex domain, it is worth pointing out a basic problem with deconvolution approaches based on the OCT envelope signal. If we denote the object by o, the image by i and the smearing point-spread function by s then, assuming linearity, we may write a simple equation upon which the deconvolution schemes described so far are based:

$$i_{\text{incoherent}} = s_{\text{incoherent}} \otimes o \tag{63}$$

However, this equation does not strictly represent the envelope detection process. What is actually measured may be more accurately represented by

$$\begin{aligned} \langle i_{\text{coherent}} \rangle &= \langle s_{\text{coherent}} \otimes o \rangle \\ &= H\{|s_{\text{coherent}} \otimes o|\} \end{aligned} \tag{64}$$

where $\langle\,\rangle$ indicates the image envelope, and H indicates a spatial low-pass filter. The equivalence of these two equations is only evident for the special case of well-separated scatterers, i.e., scatterers separated by at least several coherence lengths. Samples such as well-spaced spheres and fresh onion largely satisfy this condition, and this perhaps explains the success in these cases. Highly-scattering samples with heterogeneous scatterer size distributions such as skin certainly do not meet this criterion and it is expected, therefore, that, other things being equal, methods that exploit the fully-fringed interferogram should enjoy greater success. These methods are considered in the next section.

17.4.4.2 Methods applied to the full interferometric signal

Access to the full amplitude and phase information contained in the interferometric signal provides enhanced opportunities for post-processing to improve the quality of images. Scope exists to reduce the speckle, to be considered below, as well as to alter the shape of the response function, and compensate for dispersion.

The only application of phase-domain techniques specifically to reduce speckle noise is the so-called zero adjustment procedure (ZAP) [75]. ZAP was originally developed to reduce speckle in ultrasound images [150]. Its goal is to 'fill in' the dark speckles in an image, which it does by detecting the spikes of the instantaneous spatial frequency associated with dark speckles. It then uses rotation in the complex plane of the zeroes of a Z-transform of a short A-scan segment to 'fill in' the speckle. Thus, it is a one-dimensional technique applied to two-dimensional images. A major problem in performing ZAP is determining by how much a zero should be rotated, which corresponds to making an estimate of the separation of scatterers that cause the speckle. Thus, the ZAP procedure assumes speckle originates from the sample volume only and deals only with destructive interference. Images of skin in vivo in Ref. 75 were obtained by processing all four images from a four-fold spatial diversity system and then averaging the results. The ZAP-processed images do not show significant improvement over the unprocessed image; on the contrary they exhibit some blurring of boundaries.

We conclude the discussion on methods of speckle reduction and the deconvolution of OCT images with some general observations. The most robust method of speckle reduction is likely to be the use of more optical bandwidth to achieve stronger coherence gating. Despite significant progress in micron-resolution OCT [13], the overall feasibility of this approach in soft tissue imaging has not yet been demonstrated. The most successful speckle-reduction technique demonstrated to date is the spatial diversity scheme based on angle reported by Schmitt [129], which has formed the basis for several subsequent investigations. It would be very interesting to see an implementation of this principle in optical fibre. The success of deconvolution techniques is more difficult to ascertain for several reasons. Firstly, many investigations have been restricted to samples in which widely separated scatterers dominate, which is not representative of the vast majority of biological samples of interest. It is expected that envelope deconvolution would break down when multiple unresolved scatterers are present, although this has not yet been explicitly shown. Furthermore, these investigations have been limited to one-dimensional deconvolution and have not exploited the transverse dimension nor accounted for such features as the variation of the point-spread function with depth. Secondly, the point of departure for much of Schmitt's seminal investigations has been an image in which speckle reduction by spatial diversity has already been applied. Hence, it is difficult to gauge the effectiveness of the post-processing methods as the only form of speckle reduction. It may be that spatial diversity or post processing may be separately used to achieve similar ends, but this is not yet known. An important consideration in the utility of post processing is the time required to perform the process. Some techniques, particularly the iterative ones, are computationally expensive, and could only ever be used in applications that do not require real-time performance. In this respect, hardware implementations have an advantage. Finally, deconvolution involving the coherent point-spread function has hardly been examined [64]. This is not an easy route to take because of the much greater constraints on the measurement system and the unusual symmetry of the point-spread function, but it is theoretically more rigorous and ultimately may lead to a more effective route to solving the inverse problem when unresolved multiple scattering is present, as pointed out by Schmitt [64].

17.5 Applications

OCT imaging has been performed on many samples and considered for many applications – too many to be usefully surveyed here. In this section, we provide a brief overview of a selection of significant in vivo applications of OCT. The survey is meant to be illustrative and by no means exhaustive. It will give a sense of the state of the art and capabilities of OCT technology. More detailed information is provided in Ref. 13.

17.5.1 Posterior human eye

The human eye was the first object imaged in vivo by OCT [33,151] and more effort has been expended on OCT imaging of the eye than on any other object. A search of current journal contents will reveal that the field is now so vast and specialised that it would merit a review in its own right. A short overview is provided in Ref. 152. OCT imaging of the eye has been conducted in the 800 nm band because the eye more strongly absorbs in the 1300 nm band. The optical properties of eye tissues are discussed in Ref. 153. OCT performs similarly to the scanning laser ophthalmoscope (SLO) in the anterior eye, where the confocal sectioning is not limited by aberrations or the aperture of the eye. However, it has a striking advantage in imaging the posterior eye because its axial resolution of 10 μm greatly exceeds the 300–400 μm achieved by the SLO [154]. The axial resolution of the SLO in the posterior eye is limited by the numerical aperture of the eye and its aberrations. The resolution can be improved by adaptive optics, but this significantly complicates the instrumentation [155]. For OCT systems, the achievable transverse resolution is at best 10 μm [156], but is usually between 15 and 40 μm. An important distinction between OCT and the SLO is their respective axial and en face orientations. However, en face OCT imaging of the eye has been demonstrated [157]. OCT also has advantages over ultrasound, which requires contact with the eye, implying the need for anaesthesia and patient discomfort. Typical clinical ultrasound instruments achieve a depth resolution of around 150 μm. High-frequency transducers operating at 100 MHz have reduced this resolution to around 20 μm but at the expense of a penetration depth reduced to 4 mm [158].

Various features peculiar to OCT imaging of the eye are worth noting. Safety standards limit the optical power incident on the eye to the sub-milliwatt regime [156]. Eye tissues such as the cornea and lens are nominally transparent and so very weakly backscatter – typically −50 to −100 dB of the incident power. Thus, images of these structures are well described by an extinction-single backscatter model. The retinal layers present a different picture. Hammer et al. [159] have used a single backscatter approach to estimate the scattering coefficient of the retina from OCT data and obtained a value of $\mu_s = 12\,\text{mm}^{-1}$, giving a scattering mean free path of 83 μm. The retina is about 200 μm thick at the fovea but it can attain thicknesses of 600 μm, e.g., in the presence of age-related macular degeneration [160]. More recent high-resolution imaging has shown that the layers of the retina vary considerably in their backscatter coefficient, and this is correlated with the orientation of the structures within them [152,156]. Thus, in retinal imaging the impact of highly reflecting layers may be more significant than the effects of multiple scattering.

The relative transparency of structures in the eye results in images with a wide dynamic range that are not well portrayed with an 8-bit grey scale. As a result, a false-colour scale, in which the logarithm of the signal strength is mapped to a rainbow order of colours, has become commonly used [12]. Other notable features of the eye include its dispersion, having an average value about 10% greater than that of water [51], and the birefringence of the cornea and the nerve fibre layer of the retina. The eye is capable of comparatively high-velocity motion, which can be involuntary or caused by poor fixation. The resulting motion artefacts must be accounted for by sufficient acquisition speed or by postprocessing. The assumption that the macular surface should be smooth has been used to eliminate high-spatial-frequency ripples caused by motion [158]. An alternative approach to eliminating motion artefacts, in at least the axial direction, is to employ a surface reflection from the cornea as the reference signal [30]. This approach results in a reduced signal-to-noise ratio, and causes difficulty in matching the wavefronts of the two interfering waves [161]. It has not been shown to be superior to the simpler Michelson interferometer. Figure 31 shows an early OCT image of the anterior chamber of the normal eye [12]. Backscatter from transparent structures such as the cornea and lens is visible, as well as from nominally opaque structures such as the sclera and iris. Figure 32 shows an early OCT image of a normal fovea and optic disc, demonstrating differentiation of some layers of the retina [12], and corresponding fundus photograph. Spectacular improvements have been demonstrated with an ultrahigh-resolution OCT system [156] which has achieved 3 μm resolution at the retina and 2 μm resolution at the cornea.

Clinical assessment with OCT will need to demonstrate new capabilities compared with current mainstays such as fluorescein angiography and visual field testing [152]. OCT is somewhat complementary to fluorescein angiography since it enables examination of non-vascular structures. OCT will also need to overcome the potential barrier presented by its histological section when ophthalmologists are conventionally presented with en face optical images of the fundus [157]. This requires laying a substantial foundation of clinical experience, which is now well

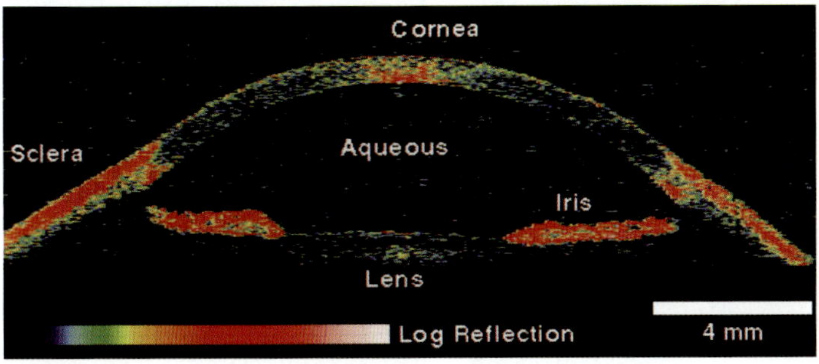

Figure 31. OCT image of a normal anterior eye chamber presented in false colour to increase the dynamic range of scattering that can be visualised. Reproduced from [12] with permission.

underway. The indications are that OCT will eventually become the tool of choice for diagnosis of diseases of the retina.

OCT imaging has enabled a new understanding of several retinal conditions, including macular oedema arising from retinal diseases such as diabetic retinopathy, inflammatory disease or macular degeneration, as well as of the role of vitreous adhesions in these conditions [152,162]. A major clinical thrust is the use of OCT to detect the reduction in nerve fibre layer thickness accompanying axon loss which is thought to be brought on by glaucoma. This reduction is thought to precede visual field defects and identifiable cupping of the optic-nerve head and so OCT is expected to enable earlier detection of the onset of glaucoma [163,164].

Many of the clinical studies of the retina have been conducted with commercial instruments available from Humphrey Instruments [165]. These instruments have been designed for imaging of the posterior eye and were first released in 1996. The latest version is specified to perform up to 400 A-scans per second with an axial resolution of 10 μm or better in tissue at a wavelength of 820 nm and a transverse pixel size of 20 μm. An example of a retinal nerve fibre thickness analysis from this system is shown in Figure 33.

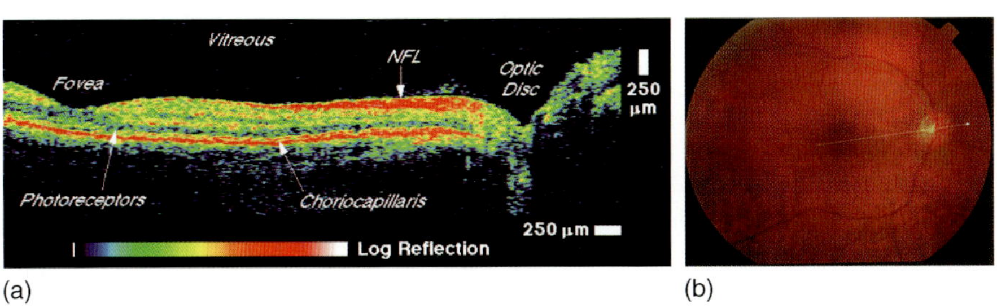

Figure 32. (a) OCT image of a normal fovea and optic disc taken along the papillomacular axis. (b) Corresponding fundus photograph with the location of the OCT transverse scan marked. Reproduced from [12] with permission (Copyright © 1996, American Medical Associaton).

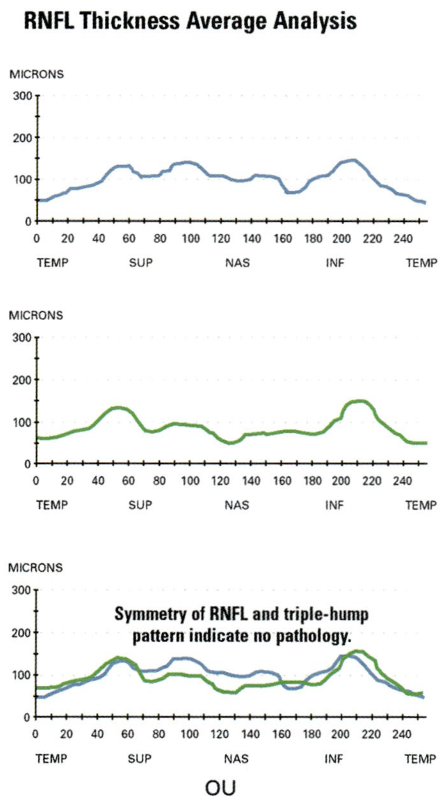

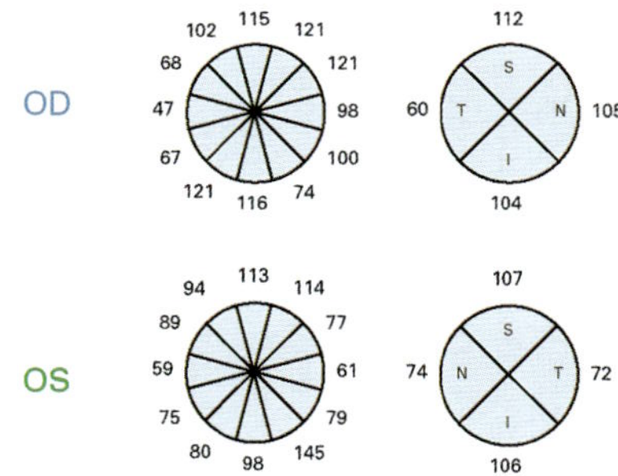

PERSONAL	
Name	Cupping Asymmetric
BirthDate	XXXX
SCAN	
ScanType	Fast RNFL Thickness (3.4)
ScanDate	03/02/2002
ScanLength	10.87

	OD (N=3)	OS (N=3)	Diff(OD-OS)
Smax/Imac	0.965	0.89	0.075
Smax/Tavg	2.283	1.903	0.381
Imax/Tavg	2.367	2.139	0.228
Smax/Navg	1.292	1.851	-0.559
Max-Min	98	101	-3
Smax	137	137	0
Imax	142	154	-12
Savg	112	107	5
Iavg	104	106	-2
Avg. Thickness	95.583	89.833	5.75

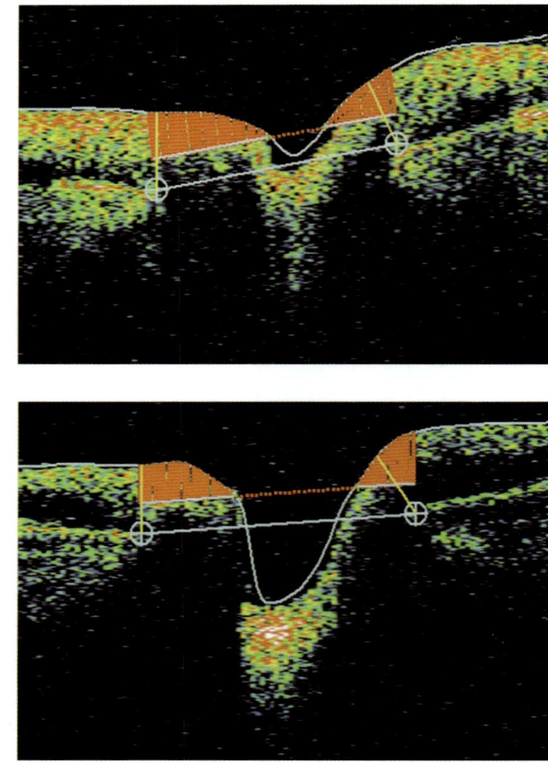

Figure 33. Example images and output from a commercial OCT instrument, including analysis of retinal nerve fibre layer thickness. Courtesy of Humphrey Instruments [165].

One final notable point is that since OCT measures optical length, conversion into physical length or thickness requires the assumption of an appropriate refractive index, the significance of which has not been discussed in the literature. In Ref. 156, a value of 1.36 was assumed for retinal layers, and in Ref. 30 an average value for the full axial length of the eye of 1.401 is specified. Expected values given in Ref. 166 for the cornea, aqueous humour, lens, and vitreous body range from 1.336 to 1.408.

17.5.2 Human skin

Skin is a highly scattering, layered media and, thus, OCT images are limited in depth and resolution by multiple scattering. Pan and Farkas have estimated the scattering coefficient to be in the range of 30–40 cm^{-1} at 1300 nm and three times larger than this at 800 nm [134]. Skin has attracted a good deal of attention because of its accessibility and the variation in its morphology. Many researchers have published images of their own fingers and fingernails!

Figure 34 shows a schematic diagram of skin morphology [167]. It consists of a surface layer of dead keratinocytes, the stratum corneum, which is generally flat and around 10 μm in thick, except on the palms of the hand and soles of the feet, where its thickness can exceed 100 μm. The next layer, the epidermis, is typically 50 to 100 μm thick and largely made up of closely packed cells. The boundary between the basal layer of the epidermis and the upper layer of the dermis, the papillary dermis, consists of a series of undulations similar in topography to an egg carton, although the shapes of the protrusions vary and can be more complex. The protrusions of the papillary dermis into the epidermis, each containing blood capillary loops, are known as the dermal papillae and in section the resulting ridges protruding into the dermis are known as rete ridges. The undulations can be as thick as the epidermis itself or almost entirely absent depending on the location on the body and the age of the subject. The papillary dermis, typically 200 μm thick, is mainly composed of connective tissue and the blood capillary loops. The connective tissue consists mainly of thin collagen fibrils and elastic fibres. The papillary dermis also contains some specialised cells at a much lower density than the epidermis. The size of the fibres and blood vessels increases with depth in the lower dermal layer, the reticular dermis, which is typically 2–3 mm thick. OCT images do not generally extend below the reticular dermis into the subcutaneous fat.

Perhaps the best images of skin have been produced by Knüttel's group at ISIS Optronics [168] using their commercially available instrument based on multiple 1300 nm OCT systems operating in parallel [135]. Figure 35 shows an OCT image of the palm of the hand. In common with m vost other OCT images, no cellular structures can be resolved. The gross architecture of the skin is, however, clearly evident. The stratum corneum is clearly differentiated from the epidermis, two sweat ducts with their characteristic spiral shape are clearly visible, the architecture of the dermal papillae is visible, and horizontal striations representative of the fibre structures of the upper dermis can be seen. In OCT images of skin, the epidermis appears brighter than the dermis. The stratum corneum can appear brighter or darker

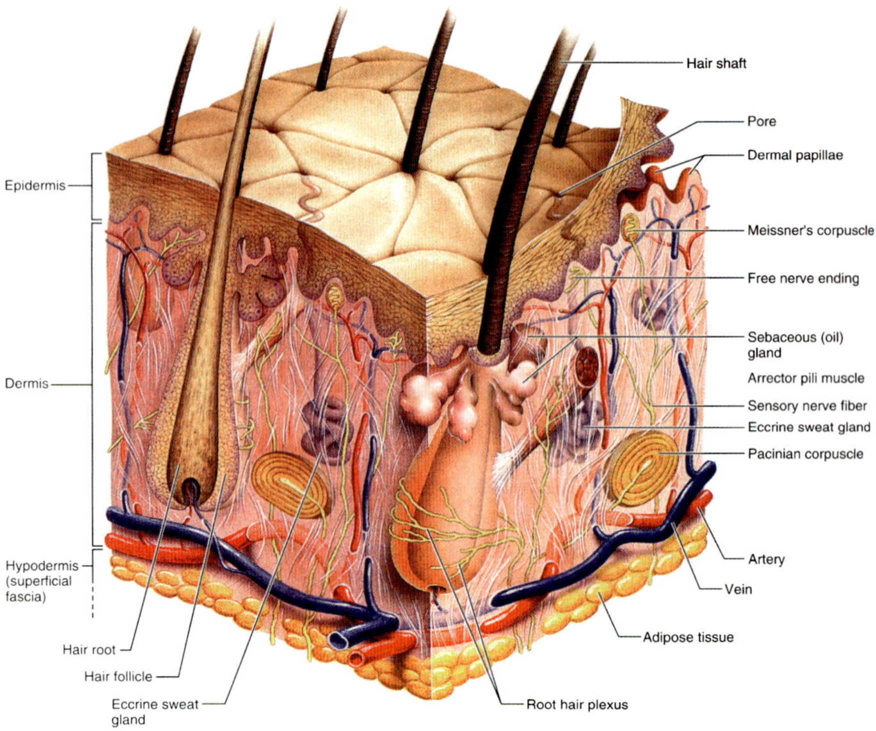

Figure 34. Three-dimensional schematic of the structure of human skin. The epidermal and dermal layers have been pulled apart at the right-hand corner to reveal the dermal papillae. Reproduced from Figure 5.4, page 152 of [167]. Copyright © 2001 by Benjamin Cummings. Reprinted by persmission of Pearson Education, Inc.

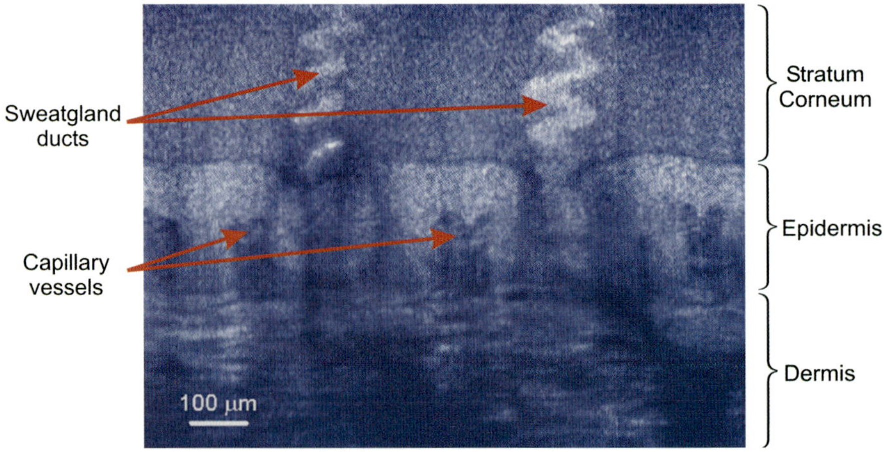

Figure 35. OCT image of the palm of the hand obtained in vivo, showing sweatgland ducts in the stratum corneum, and blood capillary loops in the dermal papillae. Courtesy of ISIS Optronics, GmbH [168].

than the epidermis depending on its thickness, and on the presence or absence of matching liquids. Blood vessels generally appear dark, which is thought to be due to absorption by haemoglobin. Pan and Farkas have estimated the absorption coefficient of haemoglobin to be $4\,\text{cm}^{-1}$ at 1300 nm [134]. It has also proven possible, although difficult, to image blood flow in capillaries using the Doppler shift from moving scatterers in blood [76], as shown in Figure 36. Clearly, given the scale of the image, only a small part of the blood plexus has been detected. Better appreciation of the skin's architecture is obtained from three-dimensional visualisation, which has been attempted by various groups [134,157,168], with some impressive early results, including the visualisation of some cylindrical structures in the dermis attributed to blood vessels [168].

There are several notable issues related to OCT imaging of skin, some of which are also applicable to other highly scattering samples. Firstly, gross structures can lead to the generation of large artefacts such as shadows in OCT images. In particular, non-contact probes lead to images that can be highly distorted by the wavefront distortions introduced by surface topography. This is generally avoided by utilising a glass plate to flatten the skin. In addition, an index matching liquid such as glycerol is used to reduce the surface reflection. Shadows also appear beneath structures such as sweat ducts, as can be seen in Figure 35. Secondly, the gross morphology in OCT images of skin appears relatively insensitive to the acquisition time, which has often been on the order of 30 s or more, or involved averaging of frames over equivalent acquisition times. Thirdly, most OCT imaging of skin has been conducted at 1300 nm. Pan and Farkas have conducted a co-registered comparison of 800 and 1300 nm systems and demonstrated qualitatively the greater depths possible at 1300 nm. Quantifying the difference is not straightforward, however, because it is not obvious at what depth the multiple scattering washes out the image. Fourthly, comparison of OCT images with histopathology is complicated by a number of factors. Since OCT sections constitute a slice of width

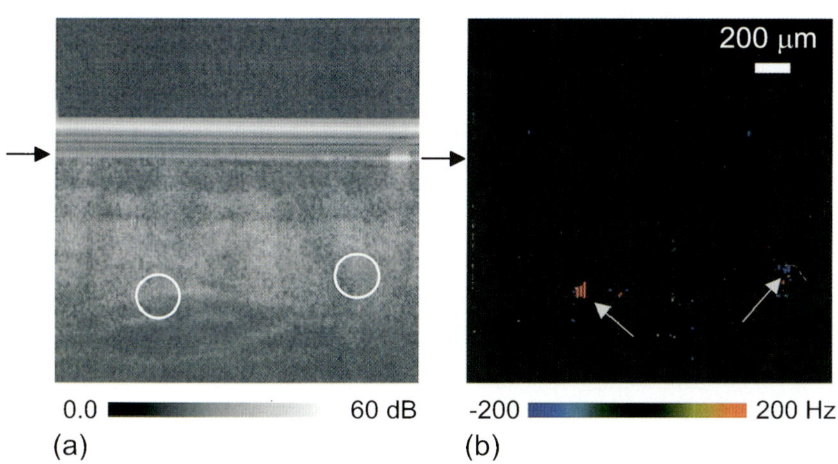

Figure 36. OCT images of structure (a) and blood flow (b) in human skin (palm) obtained in vivo. Reproduced from [76] with permission.

of around 10 μm or so, locating and producing the correct slice of the excised sample for histological imaging is extremely difficult and no procedure for doing this has yet emerged. As a result, there are very few OCT/histology pairs that show any evidence of correlation. The correlation is further complicated by the fact that OCT measures axial optical length and by the variable shrinkage of excised samples, as well as other artefacts such as folding and curling that can be induced during processing. Thus, validation against the gold standard of histological images has been qualitative at best.

Efforts to utilise OCT in clinical applications present a mixed picture. Various skin pathologies have been studied by two groups in particular [169, 170]. Without the ability to visualise cells and subcellular structures, indications of pathology have mostly relied on gross architectural changes or abnormal changes in scattering. A huge variety of skin lesions have been studied. Figure 37 shows an example of an OCT image of a benign naevus. The bright regions in the dermis are likely to correspond to intradermal melanocytes, since melanin provides strong contrast in purely confocal images of skin [171]. There are several important potential applications, e.g., the delineation of tumour borders to aid in planning surgical excision, and the differentiation of benign naevus from malignant melanoma. However, convincing evidence of the feasibility of these applications remains to be presented. One interesting example of an application of Doppler OCT to skin has been the monitoring of blood flow in port wine stain birth marks before and after laser treatment [172]. Another interesting example is the possibility of detection of burn-damaged skin through polarisation-sensitive OCT [173]. A third interesting example has been the monitoring of the average refractive index of skin using a modified form of OCT [174]. The average refractive index is altered by the application of a range of cosmetically and pharmaceutically active compounds. In general, determination of refractive index may offer a new avenue for diagnostic applications of OCT [175].

The level of clinical activity surrounding the application of OCT to skin is much lower than the activity relating to the human eye. As a result, it is not yet clear whether compelling clinical applications for OCT imaging of skin will emerge.

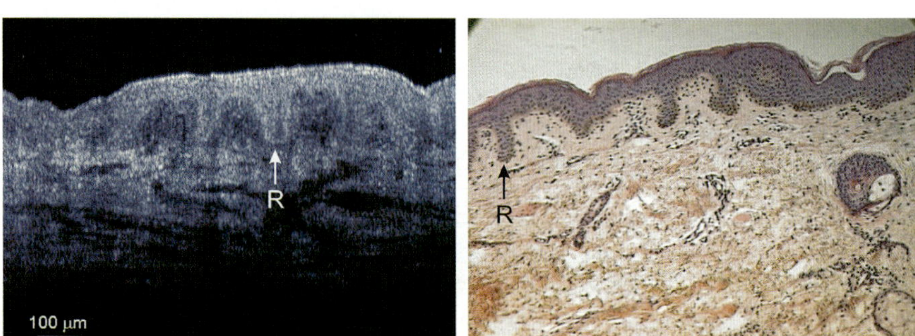

Figure 37. OCT image and nearby histology of a naevus located on the lower arm. The structure marked 'R' is observed in both images. Enlarged Rete ridges can be observed along the dermal/epidermal junction. Courtesy of ISIS Optronics, GmbH [168].

However, the extent of the potential applications surveyed here suggests that, in time, clinical activity will increase.

17.5.3 Endoscopic applications

The term 'endoscopic OCT' has been used as a catch-all to describe OCT imaging in which the sample optics are contained within an endoscope or similar surgical instrument, in a catheter, or in a needle. The first reports of endoscopic OCT imaging in vivo were published in 1997 [176,177]. In Ref. 176, a rabbit trachea and oesophagus were scanned in vivo using a probe rotating inside a catheter to provide a signal similar to a scanning radar [178]. This probe design, shown schematically in Figure 38, has become widely used. In Ref. 177, various human mucosal tissues were imaged in vivo, including the oesophagus, larynx, stomach, urinary bladder, and uterine cervix. In this case, the OCT system employed a forward-directed

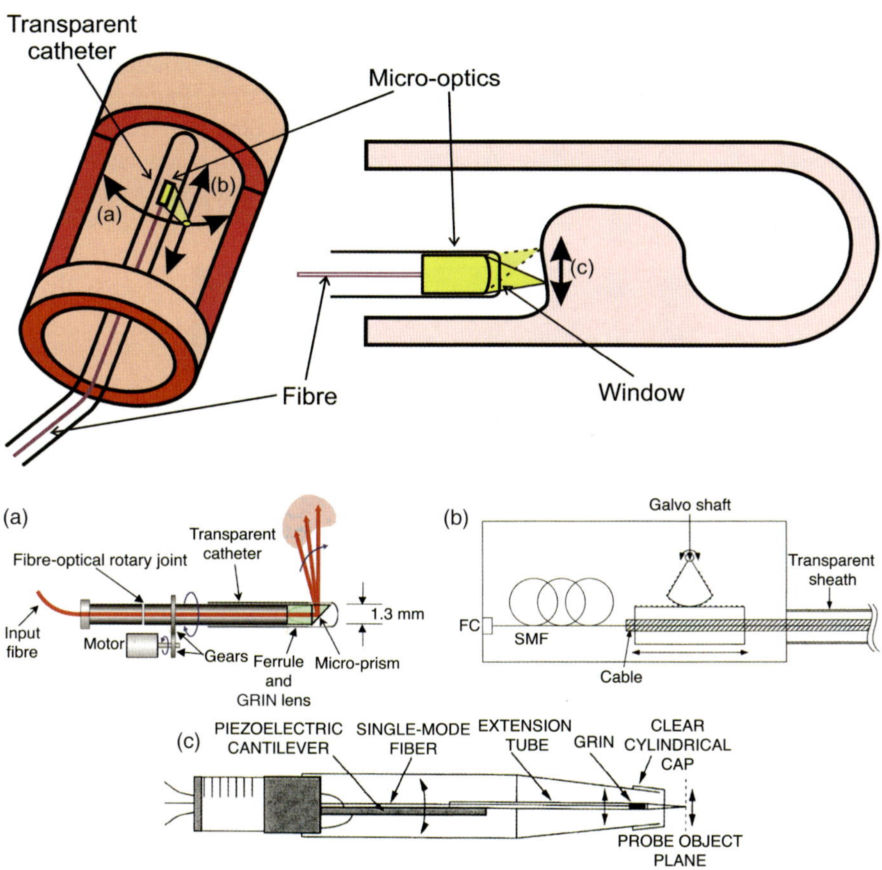

Figure 38. Scan geometries of catheter/endoscopic OCT probes and examples of their implementation for (a) radial scanning (after [178]); (b) side scanning (reproduced from [180] with permission); (c) forward scanning (reproduced from [179] with permission).

scanning system based on a galvanometer mirror, shown schematically in Figure 38, and was located in the biopsy channel of a standard gastro-endoscope. Several other designs for forward-imaging probes have also been reported [179]. It has become apparent that penetration depth and image contrast are greatly enhanced when the probe is in contact with the tissue being imaged. This requirement cannot generally be met for the full scan in the radial-scanning probe. Motivated by the need for continuous contact, a side-scanning design has been developed [180], shown schematically in Figure 38. More recently, a radial OCT scanner has been implemented in needle format, in which the whole needle rotates with the probe [181].

Numerous applications have been surveyed using endoscopic OCT [13,182]. Despite the limitation of the penetration depth to a few millimetres or less, the high resolution of OCT makes it attractive in a number of areas. Promising applications include early detection and ongoing monitoring of neoplasias and tumours, and guidance of surgery and intervention, including guidance of conventional excisional biopsies. Human tissues that have been surveyed in vivo include the gastrointestinal tract, reproductive tract (via both colposcopy and laparoscopy), urinary tract, respiratory tract, and coronary arteries. In coronary arteries, the detection and characterisation of atherosclerotic plaque and monitoring of the integrity of devices such as stents have been the main objectives. Figure 39 shows a cross-sectional OCT image of the human oesophagus obtained using a radial scanner [183] in which much of the expected morphology is in evidence, including clearly delineated mucosa, submucosa, and muscularis propria. Much of the

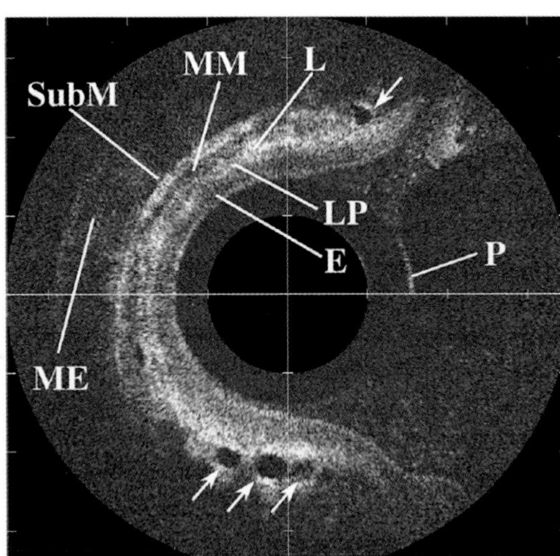

Figure 39. OCT image of the human oesophagus in vivo using a radial scanner. 'P' outer surface of catheter, 'E' squamous epithelium, 'LP' lamina propria, 'MM' muscularis mucosa, 'subM' submucosa, 'ME' muscularis externa, 'L' lymph nodules, arrows denote blood vessels or glands. The catheter is in contact with the tissue over a region centered in the direction of top-left. Tick marks denote 1 mm intervals. Reproduced from [183] with permission.

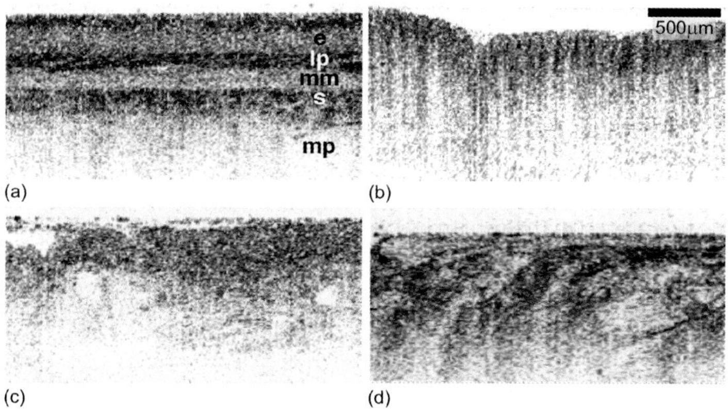

Figure 40. OCT images of the upper gastrointestinal tract obtained in vivo using a linear scanner. (a) Normal oesophageal wall showing layered structure. 'e' epithelium, 'lp' lamina propria, 'mm' muscularis mucosa, 's' submucosa, 'mp' muscularis propria. (b) Normal gastric mucosa. Vertical bands correspond to crypt and pit architecture. (c) Barret's oesophagus showing absence of layered structure, irregular surface and presence of epithelial glands. (d) Oesophageal adenocarcinoma, showing morphologic disorganisation. Reproduced from [185] (Copyright 2002, with permission from Association of University Radiologists and Elsevier).

human gastrointestinal tract has been surveyed using OCT imaging – oesophagus, small intestine, and colon, and a comprehensive discussion of this is contained in Ref. 184. One of the applications receiving most attention is the detection of oesophageal metaplasia, a common benign precursor of adenocarcinoma of the oesophagus and known as Barrett's oesophagus. Barrett's oesophagus is characterised by disruption of the uniformity of the mucosal layers [185], as evidenced in Figure 40, which demonstrates OCT images of normal oesophageal tissue, Barrett's oesophagus, and adenocarcinoma, obtained using a linear scanner.

Atherosclerotic plaques are located in the arterial wall and make up a pool of relatively compliant lipid enclosed in a fibrous cap. Some plaques are prone to rupture, resulting in acute myocardial infarction (heart attack), which is often fatal. OCT provides higher resolution images than either of the currently available imaging modalities, magnetic resonance and intravascular ultrasound, and provides the opportunity to identify the vulnerability of plaques to rupture. Figure 41 shows an intracoronary OCT image obtained with a radial scanner operating at 1300 nm compared with a corresponding image obtained using intravascular ultrasound. The clear superiority of the OCT images is evident. The superiority of the technology and the importance of the application has led to a recent focus on this area for commercialisation by Lightlab Imaging [186]. Figure 42 shows an OCT image of a complex plaque acquired in vivo. There remain several important limitations in the present state of intracoronary OCT imaging [185]. Firstly, high-resolution imaging of arterial substructure has not been possible in blood-filled arteries. Blood has been displaced for imaging by a saline flush; typically 8–10 ml. Although it has been reported that flushing has been well tolerated by patients, it has not always

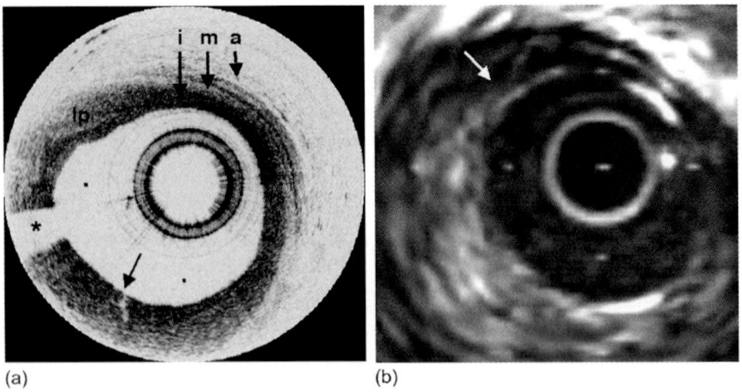

Figure 41. (a) OCT image of a coronary artery obtained in vivo using a radial scanner. 'i' intima, 'm' media, 'a' adventitia, 'lp' lipid-rich region, * obscuration caused by metal guidewire used to position OCT catheter, 'arrow' a fissure present in the fibrous plaque sector. (b) Corresponding intravascular ultrasound image. Arrow denotes location of guidewire. Reproduced from [185], Copyright 2002, with permission from Association of University Radiologists and Elsevier.

been successful in removing blood sufficiently to generate a high-resolution image. Furthermore, the window available for imaging is about 2 s in duration, which limits the area which can be screened [185]. A second issue is the rotation speed of the probe, typically 4–8 Hz, which is insufficient to completely avoid motion artefacts generated at the heart rate [187].

17.5.4 Biology

17.5.4.1 Developmental biology of animals

Boppart and colleagues from J.G. Fujimoto's group have conducted extensive studies on OCT imaging of primarily *Xenopus laevis* (African clawed frog), but

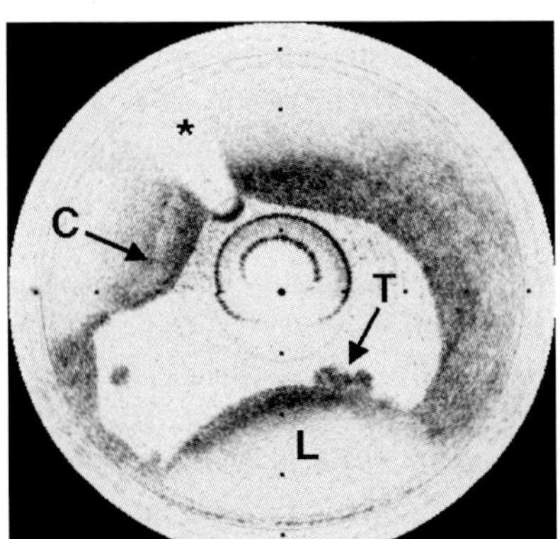

Figure 42. OCT image of a complex plaque obtained in vivo using a radial scanner. * obscuration caused by metal guidewire used to position OCT catheter, 'C' calcific nodule, 'L' lipid-rich region, 'T' thrombus. Tick marks denote 500 μm intervals. Reproduced from [185], Copyright 2002, with permission from Association of University Radiologists and Elsevier.

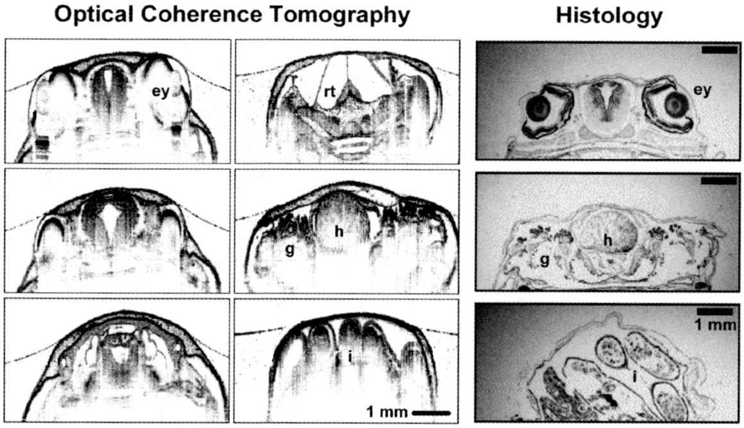

Figure 43. OCT images obtained in vitro of the developing biology of *Rana pipiens* (Stage 49 - 12 day) and corresponding histology. Structures including the eye (ey), respiratory tract (rt), gills (g), heart (h), and gastrointestinal tract (i) are observed. Reproduced from [189], Copyright 1996, with permission from Elsevier.

have also studied *Rana pipiens* (Leopard frog) tadpoles and *Brachydanio rerio* (Zebra fish) embryos and eggs. Their work, all conducted at a wavelength of 1300 nm, is extensively reviewed in Refs. 11 and 188. OCT provides the opportunity to study morphology, both statically and dynamically, over large fields of view, typically 3 × 3 mm, and through millimetre thicknesses of overlying tissue. Long observation periods are possible without harm to the sample because of the relatively low light levels used, the absence of exogenous substances, and the absence of physical contact with the imaging apparatus. There have also been some noteworthy inroads in obtaining sufficient resolution with OCT to visualise sub-cellular processes. Figure 43 shows OCT images acquired in vitro and the

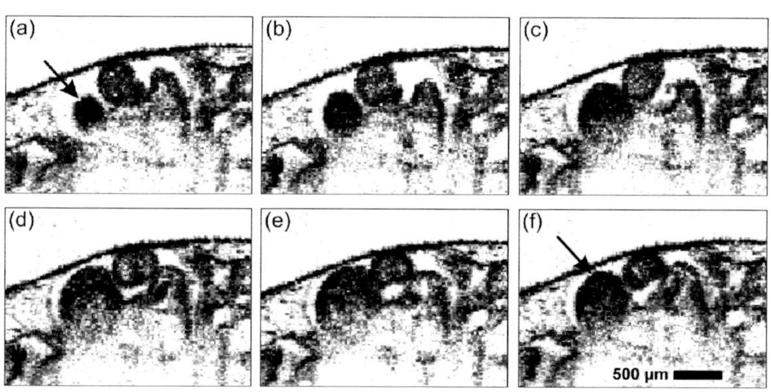

Figure 44. OCT images obtained in vivo of a complete heartbeat cycle of *Xenopus laevis*. Sequence runs from (a) to (f). (a): Beginning of diastole with emptied ventricle (arrow). (f): Beginning of systolc with filled ventricle (arrow). Bar represents 500 μm. Reproduced from [188] (after [190]) with permission. Copyright 1997, National Academy of Sciences, USA.

corresponding histology of a developing *Rana pipiens* tadpole [189]. One notable feature is the excellent correlation of the images with histology. Many structures in such models are much more transparent than human tissues, but some structures, such as the iris and sclera in the eye, can still be sufficiently scattering to cause shadowing. However, shadowing can often be overcome by re-orientation of the sample. Tissue structures in such models can be differentiated by their varying degrees of scattering; cartilage is highly backscattering and fluid-filled cavities are weakly backscattering.

The study of morphology requires only static samples, whether they are imaged in vitro or in vivo; hence, Boppart et al. have quoted image acquisition times in the range 10–30 s to enable the highest signal-to-noise ratio possible and, thus, optimise the image contrast. However, it is also possible to perform functional OCT imaging in real time and this has also been powerfully demonstrated by Boppart et al. and by Rollins et al. [92]. Rollins et al. demonstrated real-time imaging at up to 32 frames

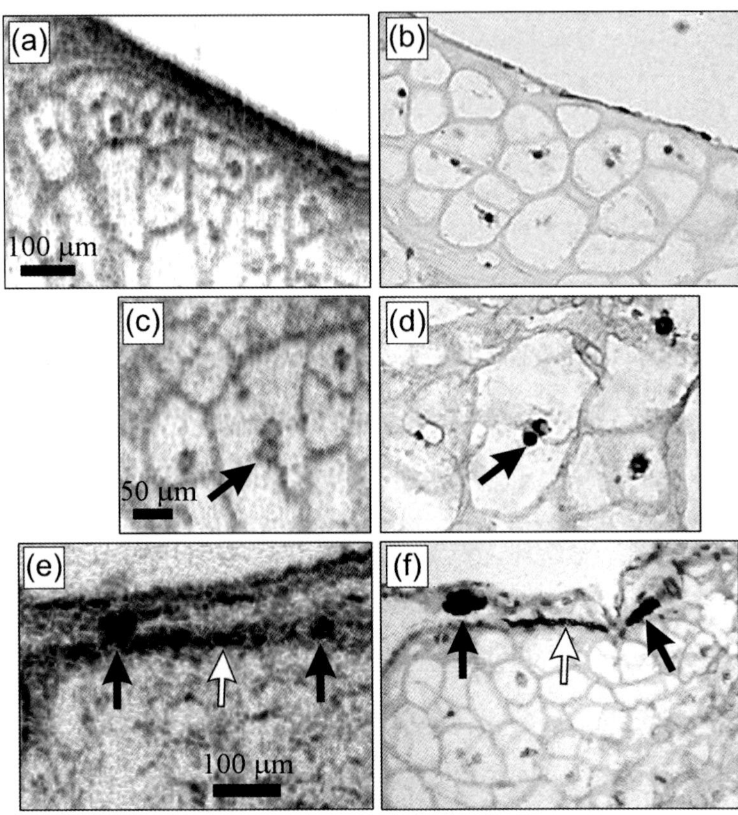

Figure 45. Ultrahigh-resolution OCT image obtained in vitro and associated histology of *Xenopus laevis* mesenchymal cells (a)–(d) and neural crest melanocytes (e),(f). Arrows in (c) and (d) show two cell nuclei and membrane synthesis of a cell undergoing mitosis. (e) black arrows indicate melanocytes and white arrow indicates a melanin layer. Reproduced from [108] with permission from *Nature*.

per second of *Xenopus laevis* heart, murine (mouse) eye, and human skin. The murine eye was reportedly imaged simply by holding a live unanaesthetised animal in the hand [92]! Figure 44 shows a *Xenopus laevis* heartbeat sequence; each image was acquired in 250 ms [190]. An excellent real-time movie of a similar sample can be viewed in Ref. 92. Such functional imaging can be used to study the effect of pharmacological agents on physiology, e.g., as reported in Ref. 13.

Cellular level resolution has been achieved in the *Xenopus laevis* model [58,108], representing a major technical achievement. Image acquisition times were not given, but it can be assumed that that long acquisition times were necessary at such high resolution. Figure 45 shows *Xenopus laevis* mesenchymal cells and neural crest melanocytes imaged with high-power pulses at 1300 nm from a Cr^{4+}-forsterite laser with axial and transverse resolutions of 5 and 9 μm, respectively, as well as their associated histology. It has been noted that correlation with histology at the cellular level is complicated by the changes in cellular morphology that take place between OCT imaging and subsequent euthanising of the sample, but the image shows strong correlations. This system was used to demonstrate the feasibility of studying cell mitosis as well as cell migration. Figure 46 shows an even more compelling image of *Xenopus laevis* at axial and transverse resolutions of 1 and 5 μm, respectively. This image was recorded at 800 nm using high-power pulses from a titanium:sapphire laser. The high transverse resolution required a high-numerical aperture objective lens with Rayleigh range of 25 μm. Thus, OCT images were acquired with the focus at different depths within the sample, as well as compensating at each depth for sample dispersion. The image in Figure 46 is, thus, synthesised from a large set of images. Despite the technical difficulties, the

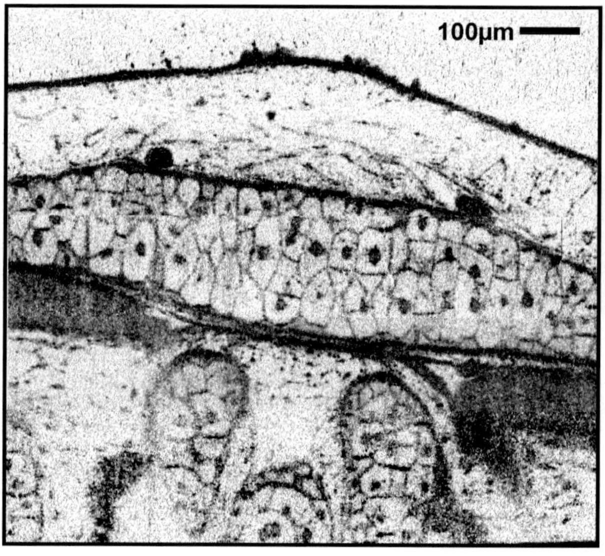

Figure 46. Ultrahigh-resolution composite OCT image of *Xenopus laevis* tadpole obtained in vivo. Reproduced from Figure 13.15 in [11] (after [59]) with permission. Copyright 2003, Springer-Verlag.

image demonstrates the achievement of resolutions approaching those achieved in confocal or multiphoton microscopy, but at up to 1 mm into the comparatively highly scattering sample.

17.5.4.2 Plants

Non-destructive three-dimensional microscopic imaging of intact plants at penetration depths beyond 200 μm remains a challenge to confocal and multiphoton microscopy. To address this challenge, an en face OCM system has been developed [191,192], based on a superluminescent diode source at 850 nm, achieving axial and transverse resolutions of 15 and 5 μm, respectively. Efforts have been focussed on

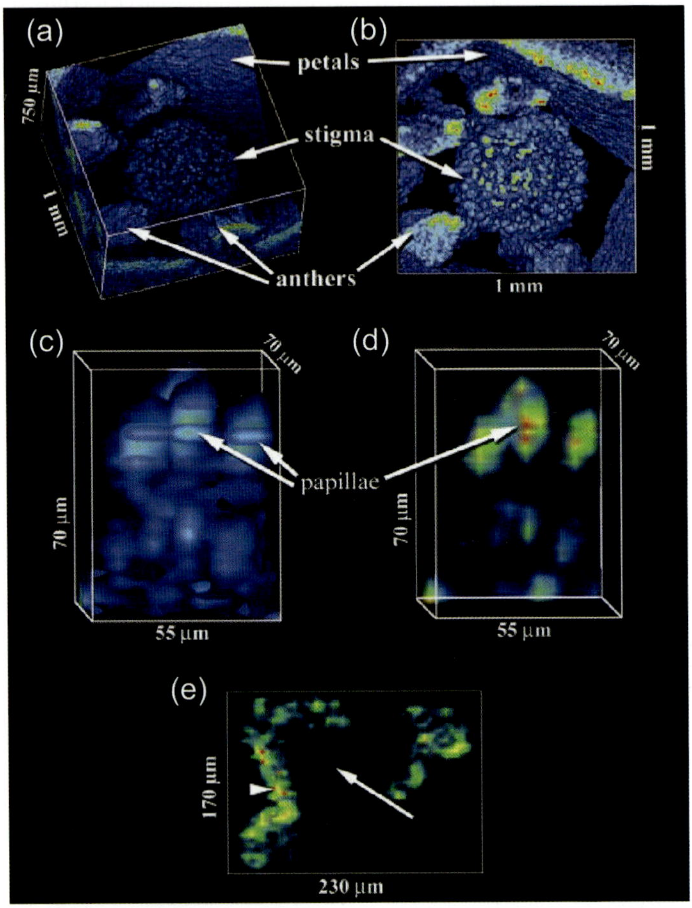

Figure 47. Three-dimensional reconstructions from OCM images of an open *Arabidopsis* flower. (a) Image showing petals, stigma and anthers. (b) Same image as (a) but seen from above and with cropping at top and bottom. (c,d) Stigmatic papillae rendered with two different algorithms. (e) En face 6 μm section through the anther near its junction with the filament. Arrowhead: Region of high OCM signal. Arrow: Low signal may represent gaps between anther locules or top of the filament. Reproduced from [192] with permission from Blackwell Publishing.

three-dimensional imaging of plant morphology. As expected, individual cells have not been observed. Figure 47 shows a three-dimensional reconstruction of an open *Arabidopsis* flower, graphically illustrating the potential of the system. One interesting observation made in Ref. 192 is the absence of observed signal from plant cell walls. It is not immediately apparent why cell walls do not contribute to the OCM signal in the case of plants, but as seen here contribute so strongly in the case of animals.

17.6 Conclusion

In meeting the objectives stated in the introduction, several important aspects of OCT have been omitted and it is worth noting at least some of them. Optical spectral-domain techniques offer a number of potential advantages for OCT, including higher speed and sensitivity spectroscopy of samples, the removal of the need for a scanning reference delay, the recording of the amplitude and phase information contained in the interference signal, and the opportunity to exploit multiple-element detectors. Doppler OCT, i.e., the determination of scatterer flow velocity from the full interferogram signal, has been mentioned on several occasions but has not been discussed in detail. Polarisation-sensitive OCT can map the sample birefringence, and has shown early promise in assessing tissue alteration through heating and burning, as well as assessment of the cornea and retina. Further information on these techniques is found in Refs. 13 and 14.

Recently, there has been an increase in research activities surrounding ultrahigh-resolution OCT, with a number of groups having now reported resolutions in the micron regime [59,193,194]. Real-time time-domain OCT imaging at such resolutions will be governed by the optical power/resolution/signal-to-noise ratio trade-off discussed in Section 17.2.6, as well as the need to overcome the degrading influence of scan-dependent sample dispersion and multiple scattering. Ultrahigh-resolution OCT offers tantalising prospects but many issues remain to be explored before its performance limits are clearly established.

The combination of OCT with other imaging modalities or with other probes is in its early stages. There is significant activity in the combined effects of different modalities, including OCT and sound [195], and OCT and nonlinear optical processes [196].

On the clinical front, in vivo confocal microscopy of skin and internal organs is paralleling developments in OCT. Multiphoton and other nonlinear microscopies are also rapidly developing. In the light of these technologies, it is pertinent to question what comparative advantages OCT offers. OCT out-performs confocal microscopy in applications in which the numerical aperture of the beam is intrinsically restricted, e.g., in the human retina, but in vivo confocal imaging of skin has demonstrated cellular resolution at the level of the basal layer of the epidermis, whereas OCT has not. OCT would appear to be unparalleled in its assessment at modest resolutions of structures buried deep in tissues, such as coronary plaque. The routine clinical application of OCT imaging in most cases still awaits the amassing of a body of evidence that would confirm its comparative advantage in clinical practice. This process is slowed somewhat by limited access to OCT technology due to the technical difficulty in developing it and the modest level of commercial activity. The weight of evidence, though, suggests that it is simply a matter of time before OCT imaging becomes an established clinical modality.

Acknowledgements

We wish to acknowledge the contributions of all of our colleagues at the Optical + Biomedical Engineering Laboratory, past and present, who have worked on aspects of OCT; especially Steven Adie, Sergey Alexandrov, Julian Armstrong (especially for his contributions to the section on noise), Philippe Lauper, Matthew Leigh, Simon Moore, Nicholas Price, Dilusha Silva, Elwyn Smith, Ian Walton, and Andrei Zvyagin.

References

1. D. Huang, E.A. Swanson, C.P. Lin, J.S. Schuman, W.G. Stinson, W. Chang, M.R. Hee, T. Flotte, K. Gregory, C.A. Puliafito, J.G. Fujimoto (1991). Optical coherence tomography, *Science* **254**, 1178–1781.
2. A.F. Fercher (1996). Optical coherence tomography. *J. Biomed. Opt.* **1**, 157–173.
3. J.A. Izatt, M.D. Kulkarni, H.-W. Wang, K. Kobyashi, M.V. Sivak Jr. (1996). Optical coherence tomography and microscopy in gastrointestinal tissues. *IEEE J. Selected. Top. Quantum Electron.* **2**, 1017–1028.
4. B.R. Masters (1999). Early development of optical low-coherence reflectometry and some recent biomedical applications. *J. Biomed. Opt.* **4**, 236–247.
5. J.M. Schmitt (1999). Optical coherence tomography (OCT): A review *IEEE J. Sel. Topics Quantum Electron.* **5**, 1205–1215.
6. V.V. Tuchin (2000). *Tissue Optics* (pp. 187–242). SPIE Press, Bellingham, WA.
7. A.M. Rollins, S. Yazdanfar, S. Radhakrishnan, V. Westphal, M.V. Sivak, J.A. Izatt (2002). Real-time imaging of microstructure and blood flows using optical coherence tomography, In: V.V. Tuchin (Ed), *Handbook of Optical Biomedical Diagnostics* (pp. 939–986). SPIE Press, Bellingham, WA.
8. M.E. Brezinski, J.G. Fujimoto (1999). Optical coherence tomography: High-resolution imaging in nontransparent tissue, *IEEE J. Selected Top. Quantum Electron.* **5**, 1185–1192.
9. J. Van Dam, J.G. Fujimoto (2000). Imaging beyond the endoscope, *Gastrointestinal Endoscopy* **51**, 512–516.
10. J.G. Fujimoto, C. Pitris, S.A. Boppart, M.E. Brezinski (2000). Optical coherence tomography: An emerging technology for biomedical imaging and optical biopsy. *Neoplasia* **2**, 9–25.
11. S.A. Boppart (2003). Optical coherence tomography. In: P. Török, F.-J. Kao (Eds), *Optical Imaging and Microscopy* (pp. 309–337). Springer-Verlag, Berlin.
12. C.A. Puliafito, M.R. Hee, J.S. Schuman, J.G. Fujimoto (1996). *Optical Coherence Tomography of Ocular Diseases*, Slack, Inc., Thorofare New Jersey.
13. B.E. Bouma, G. Tearney (Eds) (2001). *Handbook of Optical Coherence Tomography*. Marcel Dekker, Inc., New York.
14. A.F. Fercher, W. Drexler, C.K. Hitzenberger, T. Lasser (2003). Optical coherence tomography – principles and applications. *Rep. Prog. Phys.* **66**, 239–303.

15. J.A. Izatt, M.R. Hee, D. Huang, E.A. Swanson, C.P. Lin, J.S. Schumann, C.A. Puliafito, J.G. Fujimoto (Oct. 1993). Micron-resolution biomedical imaging with optical coherence tomography. *Opt. Photon. News* 14–19.
16. J.A. Izatt, M.D. Kulkarni, K. Kobayashi, M.V. Sivak, J.K. Barton, A.J. Welch (May 1997). Optical coherence tomography for biodiagnostics *Opt. Photon. News* 41–47.
17. S.M. Reiss (Oct. 1999). Optical coherence tomography: An imaging modality for the new millennium, *Biophoton. Int.* 36–45.
18. J.G. Fujimoto, W. Drexler, U. Morgner, F. Kärtner, E. Ippen (Jan. 2000). Optical coherence tomography: High resolution imaging using echoes of light. *Opt. Photon. News* 24–31.
19. S.M. Reiss (Dec. 2001). OCT update: Pushing a promising technology to market. *Biophoton. Int.* 40–45.
20. Several articles in *Opt. Photon. News* (Apr. 2002)
21. S. Boppart (Sept. 2002). Looking below the surface. *OE Magazine* 28–31.
22. F.W. Kremkau (1998). *Diagnostic Ultrasound: Principles and Instruments* (5th edn). W.B. Saunders, Philadelphia, PA.
23. M.A. Duguay, A.T. Mattick (1971). Ultrahigh-speed photography of picosecond light pulses and echoes. *Appl. Opt.* **10**, 2162–2170.
24. J.G. Fujimoto (Ed) (1993). Special issue on time resolved imaging and diagnostics in medicine. *Opt. Photon. News* (Oct. 1993)
25. D.V. O'Connor, D. Phillips (1984). *Time-correlated Single-Photon Counting*. Academic Press, New York.
26. R.C. Youngquist, S. Carr, D.E.N. Davies (1987). Optical coherence-domain reflectometry. *Opt. Lett.* **12**, 158–160.
27. K. Takada, I. Yokohama, K. Chida, J. Noda (1987). New measurement system for fault location in optical waveguide devices based on an interferometric technique. *Appl. Opt.* **26**, 1603–1606.
28. B.L. Danielson, C.D. Whittenberg (1987). Guided-wave reflectometry with micrometer resolution. *Appl. Opt.* **26**, 2836–2842.
29. Y.J. Rao, D.A. Jackson (1996). Recent progress in fibre optic low-coherence interferometry. *Meas. Sci. Technol.* **7**, 981–999.
30. A.F. Fercher, K. Mengedoht, W. Werner (1988). Eye length measurement by interferometry with partially coherent light. *Opt. Lett.* **13**, 186–188.
31. C.K. Hitzenberger (1992). Measurement of corneal thickness by low-coherence interferometry. *Appl. Opt.* **31**, 6637–6642.
32. X. Clivaz, F. Marquis-Weible, R.P. Salathé, R.P. Novàk, H.H. Gilgen (1992). High-resolution reflectometry in biological tissues. *Opt. Lett.* **17**, 4–6.
33. E.A. Swanson, J.A. Izatt, M.R. Hee, D. Huang, C.P. Lin, J.S. Schuman, C.A. Puliafito, J.G. Fujimoto (1993). *In vivo* retinal imaging by optical coherence tomography. *Opt. Lett.* **18**, 1864–1866.
34. G.J. Brakenhoff (1979). Imaging modes in confocal scanning light microscopy (CSLM). *J. Micros.* **117**, 233–242.
35. D.K. Hamilton, C.J.R. Sheppard (1982). A confocal interference microscope. *Opt. Acta* **29**, 1573–1577.
36. P. Corcuff, C. Bertrand, J.-L. Lévèque (1993). Morphology of human epidermis in vivo by real-time confocal microscopy. *Arch. Dermatol. Res.* **285**, 351–355.
37. J.A. Izatt, M.R. Hee, G.M. Owen, E.A. Swanson, J.G. Fujimoto (1994). Optical coherence microscopy in scattering media. *Opt. Lett.* **19**, 590–592.
38. W. Krug, J. Rienitz, G. Schulz (1964). *Contributions to Interference Microscopy.* Hilger & Watts, London.

39. M. Davidson, K. Kaufman, I. Mazor, F. Kohen (1987). An application of interference microscopy to integrated circuit inspection and metrology. *Proc. SPIE* **775**, 233–241.
40. P. de Groot, L. Deck (1995). Surface profiling of white light interferograms in the spatial frequency domain. *J. Mod. Opt.* **42**, 389–401.
41. K.G. Larkin (1996). Efficient nonlinear algorithm for envelope detection in white light interferometry. *J. Opt. Soc. Am. A* **13**, 832–843.
42. T.R. Corle, G.S. Kino (1996). *Confocal Scanning Optical Microscopy and Related Imaging Systems.* Academic Press, USA.
43. H. Saint Jalmes, M. Lebec, E. Beaurepaire, A. Dubois, A.C. Boccara (2001). Full-field optical coherence microscopy. In: B.E. Bouma, G. Tearney (Eds), *Handbook of Optical Coherence Tomography* (pp. 299–333). Marcel Dekker, Inc., New York.
44. B.E.A. Saleh, M.C. Teich (1991). *Fundamentals of Photonics*, Appendix B, (p.928). John Wiley & Sons, New York.
45. J.W. Goodman (1985). *Statistical Optics.* John Wiley & Sons, New York.
46. M. Nazarathy, W.V. Sorin, D.M. Baney, S.A. Newton (1989). Spectral analysis of optical mixing measurements. *J. Lightwave Technol.* **7**, 1083–1096.
47. I. Sanz, M.A. Muriel (1992). Measurement technique for characterization of 2×2 couplers. *Electron. Lett.* **28**, 1303–1305.
48. P. Green Jr. (1993). *Fiber Optic Networks* (p. 75). Prentice-Hall, Englewood Cliffs, NJ.
49. E.A. Swanson, D. Huang, M.R. Hee, J.G. Fujimoto, C.P. Lin, C.A. Puliafito (1992). High-speed optical coherence domain reflectometry. *Opt. Lett.* **17**, 151–153.
50. U. Wiedmann, P. Gallion, G-H. Duan (1998). A generalised approach to optical low-coherence reflectometry including spectral filtering effects. *J. Lightwave Technol.* **16**, 1343–1347.
51. C.K. Hitzenberger, A. Baumgartner, W. Drexler, A.F. Fercher (1999). Dispersion effects in partial coherence interferometry: implications for intraocular ranging. *J. Biomed. Opt.* **4**, 144–151.
52. G.P. Agrawal (2001). *Nonlinear Fiber Optics* (3rd edn, p. 80). Academic Press, San Diego.
53. B. Bouma, G.S. Tearney, S.A. Boppart, M.R. Hee, M.E. Brezinski, J.G. Fujimoto (1995). High resolution optical coherence tomographic imaging using a mode-locked $Ti:Al_2O_3$ laser source. *Opt. Lett.* **20**, 1486–1488.
54. E.-G. Neumann (1988). *Single-mode Fibers: Fundamentals.* Springer-Verlag, Berlin.
55. E.D.J. Smith, N. Wada, W. Chujo, D.D. Sampson (2001). High resolution OCDR using 1.55 μm supercontinuum source and quadrature spectral detection. *Electron. Lett.* **37**, 1305–1307.
56. J.F. de Boer, C.E. Saxer, J.S. Nelson (2001). Stable carrier generation and phase-resolved digital data processing in optical coherence tomography. *Appl. Opt.* **40**, 5787–5790.
57. D.L. Marks, A.L. Oldenbourg, J.J. Reynolds, S.A. Boppart (2003). Digital algorithm for dispersion correction in optical coherence tomography for homogeneous and stratified media. *Appl. Opt.* **42**, 204–217.
58. C.K. Hitzenberger, A. Baumgartner, W. Drexler, A.F. Fercher (1999). Dispersion effects in partial coherence interferometry: implications for intraocular ranging. *J. Biomed. Opt.* **4**, 144–151.
59. W. Drexler, U. Morgner, F.X. Kärtner, C. Pitris, S.A. Boppart, X.D. Li, E.P. Ippen, J.G. Fujimoto (1999). *In vivo* ultrahigh-resolution optical coherence tomography. *Opt. Lett.* **24**, 1221–1223.
60. A.F. Fercher, C.K. Hitzenberger, M. Sticker, R. Zawadzki, B. Karamata, T. Lasser (2001). Numerical dispersion compensation for partial coherence interferometry and optical coherence tomography. *Opt. Express* **9**, 610–615.

61. E.D.J. Smith, A.V. Zvyagin, D.D. Sampson (2002). Real-time dispersion compensation in scanning interferometry. *Opt. Lett.* **27**, 1998–2000.
62. B.E.A. Saleh, M.C. Teich (1991). *Fundamentals of Photonics* (pp. 80–107). John Wiley & Sons, New York.
63 J.G. Fujimoto (2001). Optical coherence tomography: Introduction. In: B.E. Bouma, G. Tearney (Eds), *Handbook of Optical Coherence Tomography* (pp. 1–40). Marcel Dekker, Inc., New York.
64. J.M. Schmitt (1998). Restoration of optical coherence tomography images of living tissue using the CLEAN algorithm. *J. Biomed. Opt.* **3**, 66–75.
65. H.-W. Wang, J.A. Izatt, M.D. Kulkarni (2001). Optical coherence microscopy. In: B.E. Bouma, G. Tearney (Eds), *Handbook of Optical Coherence Tomography* (pp. 275–298). Marcel Dekker, Inc., New York.
66. M. Gu, C.J.R. Sheppard, X. Gan (1991). Image formation in a fiber-optical confocal scanning microscope. *J. Opt. Soc. Am. A* **8**, 1755–1761.
67. J.M. Schmitt, S.L. Lee, K.M. Yung (1997). An optical coherence microscope with enhanced resolving power in thick tissue. *Opt. Commun.* **142**, 203–207.
68. F. Lexer, C.K. Hitzenberger, W. Drexler, S. Molebny, H. Sattmann, M. Sticker, A.F. Fercher (1999). Dynamic coherent focus OCT with depth-independent transversal resolution. *J. Mod. Opt.* **46**, 541–553.
69. M.R. Hee (2001). Optical coherence tomography: Theory. In: B.E. Bouma, G. Tearney (Eds), *Handbook of Optical Coherence Tomography* (pp. 41–66). Marcel Dekker, Inc., New York.
70. Z. Chen, T.E. Milner, J.S. Nelson (1997). Optical Doppler tomographic imaging of fluid flow velocity in highly scattering media. *Opt. Lett.* **22**, 64–66.
71. D.A. Jackson, A.D. Kersey, M. Corke, J.D.C. Jones (1982). Pseudoheterodyne detection scheme for optical interferometers. *Electron. Lett.* **18**, 1081–1082.
72. A. Dandridge (1991). Fiber optic sensors based on the Mach-Zehnder and Michelson interferometers, In: E. Udd (Ed), *Fiber optic sensors: An Introduction for Engineers and Scientists* (pp. 271–323). John Wiley & Sons, New York.
73. A.Gh. Podoleanu, G.M. Dobre, D.J. Webb, D.A. Jackson (1996). Coherence imaging by use of a Newton rings sampling function. *Opt. Lett.* **21**, 1789–1791.
74. A.Gh. Podoleanu, G.M. Dobre, D.A. Jackson (1998). *En-face* coherence imaging using galvanometer scanner modulation. *Opt. Lett.* **23**, 147–149.
75. K.M. Yung, S.L. Lee, J.M. Schmitt (1999). Phase-domain processing of optical coherence tomography images. *J. Biomed. Opt.* **4**, 125–136.
76. Y. Zhao, Z. Chen, C. Saxer, S. Xiang, J.F. de Boer, J.S. Nelson (2000). Phase-resolved optical coherence tomography and optical Doppler tomography for imaging blood flow in human skin with fast scanning speed and high velocity sensitivity. *Opt. Lett.* **25**, 114–116.
77. Z. Ding, Y. Zhao, H. Ren, J. Stuart Nelson, Z. Chen (2002). Real-time phase-resolved optical coherence tomography and optical Doppler tomography. *Opt. Express* **10**, 236–245.
78. B. Hyle Park, M.C. Pierce, B. Cense, J.F. de Boer (2003). Real-time multi-functional optical coherence tomography, *Opt. Express* **11**, 782–793.
79. A.V. Zvyagin, J.B. FitzGerald, K.K.M.B. Silva, D.D. Sampson (2000). Real-time detection technique for Doppler optical coherence tomography. *Opt. Lett.* **25**, 1645–1647.
80. H.C. Lefevre (1993). *The Fibre-optic Gyroscope.* Artech House, Boston.
81. E. Desurvire (1994). *Erbium-doped Fiber Amplifiers*, John Wiley & Sons, New York.
82. K. Takada (1998). Noise in optical low-coherence reflectometry. *IEEE J. Quantum Electron.* **34**, 1098–1108.

83. A.M. Rollins, J.A. Izatt (1999). Optimal interferometer designs for optical coherence tomography. *Opt. Lett.* **24**, 1484–1486.
84. G.J. Pendock, D.D. Sampson (1997). Noise in coherence-multiplexed optical fiber systems. *Appl. Opt.* **36**, 9536–9540.
85. G.P. Agrawal (1992). *Fiber-optic Communication Systems*, (p. 255), John Wiley & Sons.
86. D. von der Linde (1986). Characterization of the noise in continuously operating mode-locked lasers. *Appl. Phys. B* **39**, 201–217.
87. R. Beddows, S.W. James, R.P. Tatam (2002). Improved performance interferometer designs for optical coherence tomography, In E. Udd, R.O. Claus (Eds), 15th *International Conference on Optical Fiber Sensors* (pp. 527–530).
88. A.M. Rollins, J.A. Izatt (2001). Reference optical delay scanning. In: B.E. Bouma, G. Tearney (Eds), *Handbook of optical coherence tomography* (pp. 99–123). Marcel Dekker, Inc., New York.
89. C. Froehly, B. Colombeau, M. Vampouille (1983). Shaping and analysis of picosecond light pulses. In: E. Wolf (Ed), *Progress In Optics XX*, (pp. 63–153) North-Holland Publishing Company Amsterdam.
90. G.J. Tearney, B.E. Bouma, J.G. Fujimoto (1997). High-speed phase- and group-delay scanning with a grating-based phase control delay line. *Opt. Lett.* **22**, 1811–1813.
91. K.F. Kwong, D. Yankelevich, K.C. Chu, J.P. Heritage, A. Dienes (1993). 400-Hz mechanical scanning optical delay line. *Opt. Lett.* **18**, 558–560.
92. A.M. Rollins, M.D. Kulkarni, S. Yazdanafar, R. Ung-arunyawee, J.A. Izatt (1998). *In vivo* video rate optical coherence tomography. *Opt. Express* **3**, 219–229.
93. K.K.M.B.D. Silva, A.V. Zvyagin, D.D. Sampson (1999). Extended range, rapid scanning optical delay line for biomedical interferometric imaging. *Electron. Lett.* **35**, 1404–1406.
94. A.V. Zvyagin, E.D.J. Smith, D.D. Sampson (2003). Delay and dispersion characteristics of a frequency-domain optical delay line for scanning interferometry. *J. Opt. Soc. Am. A* **20**, 333–341.
95. A.V. Zvyagin, D.D. Sampson (2001). Achromatic optical phase shifter/modulator. *Opt. Lett.* **26**, 187–189.
96. V.V. Tuchin (2000). *Tissue Optics*, (pp. 1–108), SPIE Press, Washington.
97. L. Wang, P.P. Ho, C. Liu, G. Zhang, R.R. Alfano (1991). Ballistic 2-D imaging through scattering walls using an ultrafast optical Kerr gate. *Science* **253**, 769–771.
98. J.M. Schmitt, S.H. Xiang, K.M. Yung (1999). Speckle in optical coherence tomography. *J. Biomed. Opt.* **4**, 95–105.
99. R.A. Meyer (1979). Light scattering from biological cells: dependence of backscatter radiation on membrane thickness and refractive index. *Appl. Opt.* **18**, 585–588.
100. J. Beuthuan, O. Minet, J. Helfmann, M. Herrig, G. Müller (1996). The spatial variation of the refractive index in biological cells. *Phys. Med. Biol.* **41**, 369–382.
101. J.M. Schmitt, G. Kumar (1998). Optical scattering properties of soft tissue: a discrete particle model. *Appl. Opt.* **37**, 2788–2797.
102. J.R. Mourant, J.P. Freyer, A.H. Hielscher, A.A. Eick, D. Shen, T.M. Johnson (1998). Mechanisms of light scattering from biological cells relevant to noninvasive optical-tissue diagnostics. *Appl. Opt.* **37**, 3586–3593.
103. C. Smithpeter, A. Dunn, R. Drezek, T. Collier, R. Richards-Kortum (1998). Near real time confocal microscopy of cultured amelanotic cells: sources of signal, contrast agents and limits of contrast. *J. Biomed. Opt.* **3**, 429–436.
104. R. Drezek, A. Dunn, R. Richards-Kortum (1999). Light-scattering from cells: finite-difference time-domain simulations and goniometric measurements. *Appl. Opt.* **38**, 3651–3661.

105. M. Rajadhyaksha, S. Gonzalez, J.M. Zavislan, R.R. Anderson, R.H. Webb (1999). *In vivo* confocal scanning laser microscopy of human skin II: Advances in instrumentation and comparison with histology. *J. Invest. Dermatol.* **113**, 293–303.
106. R. Barer (1957). Refractometry and interferometry of living cells. *J. Opt. Soc. Am.* **47**, 545–556.
107. R.S. Cotran, S.L. Robbins, V. Kumar (1994). *Robbins' Pathological Basis of Disease*. W.B. Saunders, Philadelphia, PA.
108. S.A. Boppart, B.E. Bouma, C. Pitris, J.F. Southern, M.E. Brezinski, J.G. Fujimoto (1998). *In vivo* cellular optical coherence tomography imaging. *Nat. Med.* **4**, 861–865.
109. V. Backman, M.B. Wallace, L.T. Perelman, J.T. Arendt, R. Gurjar, M.G. Müller, Q. Zhang, G. Zonios, E. Kline, T. McGillican, S. Shapshay, T. Valdez, K. Badizadegan, J.M. Crawford, M. Fitzmaurice, S. Kabani, H.S. Levin, M. Seiler, R.R. Dasari, I. Itzkan, J. Van Dam, M.S. Feld (2000). Detection of preinvasive cancer cells. *Nature*, **406**, 35–36.
110. N. Boustany, S.C. Kuo, N.V. Thakor (2001). Optical scatter imaging: subcellular morphometry *in situ* with Fourier filtering. *Opt. Lett.* **26**, 1063–1065.
111. A. Knüttel, S. Bonev, C. Kugler (2001). In vivo evaluation of locally mapped refractive indices with OCT. In: V.V. Tuchin et al. (Eds), *Progress in Biomedical Optics and Imaging* (Proc. SPIE Volume 4251, pp. 136–143). SPIE, Bellingham, WA.
112. V.V. Tuchin (1995). Optical and thermal properties of biological tissue, In: G. Müller, A. Roggan (Eds), *Laser-induced interstitial thermotherapy*, SPIE Press, Bellingham, WA.
113. B.B. Das, F. Liu, R.R. Alfano (1997). Time-resolved fluorescence and photon migration studies in biomedical and random media. *Rep. Prog. Phys.* **60**, 227–292.
114. C.F. Bohren, D.R. Huffman (1983). *Absorption and Scattering of Light by Small Particles*, John Wiley & Sons, Inc., New York.
115. J.M. Schmitt, A. Knüttel, R.F. Bonner (1993). Measurement of optical properties of biological tissues by low-coherence reflectometry. *Appl. Opt.* **32**, 6032–6042.
116. J.M. Schmitt, A. Knüttel, M. Yadlowsky, M.A. Eckhaus (1994). Optical-coherence tomography of a dense tissue: statistics of attenuation and backscattering. *Phys. Med. Biol.* **39**, 1705–1720.
117. Y. Pan, R. Birngruber, J. Rosperich, R. Engelhardt (1995). Low-coherence optical tomography in turbid tissue: theoretical analysis. *Appl. Opt.* **34**, 6564–6574.
118. G. Popescu, A. Dogariu (2000). Ballistic attenuation of low-coherence optical fields. *Appl. Opt.* **39**, 4469–4472.
119. G. Häusler, J.M. Herrmann, R. Kummer, M.W. Lindner (1996). Observation of light propagation in volume scatterers with 10^{11}-fold slow motion. *Opt. Lett.* **21**, 1087–1089.
120. M.J. Yadlowsky, J.M. Schmitt, R.F. Bonner (1995). Multiple scattering in optical coherence microscopy. *Appl. Opt.* **34**, 5699–5707.
121. D.A. Boas, K.K. Bizheva, A.M. Siegel (1998). Using dynamic low-coherence interferometry to image Brownian motion within highly scattering media. *Opt. Lett.* **23**, 319–321.
122. F.J. van der Meer, D.J. Faber, M.C. Aalders, J. Perrée, T.G. Van Leeuwen (2001). Detection of apoptosis by optical coherence tomography (OCT). In: V.V. Tuchin, J.A. Izatt, J.G. Fujimoto (Eds), *Coherence Domain Optical Methods in Biomedical Science and Clinical Applications V* (Proc. SPIE Volume 2451, pp. 165–169). SPIE, Bellingham, WA.
123. T.-Q. Xie, M.L. Zeidel, Y. T. Pan (2002). Detection of tumorigenesis in urinary bladder cancer with optical coherence tomography. *Opt. Express* **10**, 1431–1443.

124. K.V. Larin, M.S. Eledrisi, M. Motamedi, R.O. Esenaliev (2002). Noninvasive blood glucose monitoring with optical coherence tomography. *Diabetes Care* **25**, 2263–2267.
125. J.A. Izatt, M.R. Hee, D. Huang, J.G. Fujimoto, E.A. Swanson, C.P. Lin, J.S. Schuman, C.A. Puliafito (1993). Optical coherence tomography for medical diagnostics. In: G.J. Müller (Ed), *Medical Optical Tomography: Functional Imaging and Monitoring* (Volume IS11, pp. 450–472). SPIE, Bellingham, WA.
126. Y. Pan, R. Birngruber, R. Engelhardt (1997). Contrast limits of coherence-gated imaging in scattering media. *Appl. Opt.* **36**, 2979–2983.
127. D.J. Smithies, T. Lindmo, Z. Chen, J.S. Nelson, T.E. Milner (1998). Signal attenuation and localization in optical coherence tomography studied by Monte Carlo simulation. *Phys. Med. Biol.* **43**, 3025–3044.
128. J.M. Schmitt, A. Knüttel (1997). Model of optical coherence tomography of heterogeneous tissue. *J. Opt. Soc. Am. A* **14**, 1231–1242.
129. J.M. Schmitt (1997). Array detection for speckle reduction in optical coherence tomography. *Phys. Med. Biol.* **42**, 1427–1439.
130. J.M. Schmitt (1998). OCT elastography: imaging microscopic deformation and strain of tissue. *Opt. Express* **3**, 199–211.
131. M. Bashkansky, J. Reintjes (2000). Statistics and reduction of speckle in optical coherence tomography. *Opt. Lett.* **25**, 545–547.
132. S.J. Kirkpatrick, D.D. Duncan (2002). Optical assessment of tissue mechanics. In: V.V. Tuchin (Ed), *Handbook of Optical Biomedical Diagnostics* (pp. 1037–1084). SPIE Press, Bellingham, USA, WA.
133. V.V. Tuchin (Ed) (2002). *Handbook of Optical Biomedical Diagnostics*, SPIE Press, Bellingham, WA.
134. Y. Pan, D.L. Farkas (1998). Noninvasive imaging of living human skin with dual-wavelength optical coherence tomography in two and three dimensions. *J. Biomed. Opt.* **3**, 446–455.
135. A. Pagnoni, A. Knüttel, P. Welker, M. Rist, T. Stoudemayer, L. Kolbe, I. Sadiq, A.M. Kligman (1999). Optical coherence tomography in dermatology. *Skin Res. Technol.* **5**, 83–87.
136. K.M. Yung, S.H. Xiang, J.M. Schmitt (1998). Adaptive beamforming for optical coherence tomography of biological tissue, in *Proc. 20th Annual Conf. of the IEEE Eng. In Med. Biol. Soc.* **20**, (pp. 891–894).
137. E. Jakeman (1984). Speckle statistics with a small number of scatterers. *Opt. Eng.* **23**, 453–461.
138. J. Rogowska, M.E. Brezinski. (2000) Evaluation of the adaptive speckle suppression filter for coronary optical coherence tompography imaging. *IEEE Trans. Med. Imaging* **19**, 1261–1266.
139. S.H. Xiang, L. Zhou, J.M. Schmitt (1998). Speckle noise reduction in optical coherence tomography, *Proc. SPIE* **3196** 79–88.
140. R.H.T. Bates, M.J. McDonnell (1986). *Image Restoration and Reconstruction.* Clarendon Press, Oxford.
141. M. Bashkansky, M.D. Duncan, J. Reintjes (1998). Signal processing for improving field cross-correlation function in optical coherence tomography. *Appl. Opt.* **37**, 8137–8138.
142. R.K. Wang (1999). Resolution improved optical coherence-gated tomography for imaging through biological tissues. *J. Mod. Opt.* **46**, 1905–1912.
143. M.D. Kulkarni, W. Tobocman, J.A. Izatt (1996). High resolution optical coherence tomography using deconvolution. In: R.R. Alfano, J.G. Fujimoto (Eds), *Advances in Optical Imaging and Photon Migration* (OSA TOPS Volume 2, pp. 227–230). OSA, Washington.

144. P.A. Jansson (1984). *Deconvolution with Applications in Spectroscopy*. Academic Press, New York.
145. M.D. Kulkarni, C.W. Thomas, J.A. Izatt (1997). Image enhancement in optical coherence tomography using deconvolution. *Electron. Lett.* **33**, 1365–1367.
146. I. Hsu, C. Sun, C. Lu, C.C. Yang, C. Chiang, C. Lin (2003). Resolution improvement with dispersion manipulation and a retrieval algorithm in optical coherence tomography. *Appl. Opt.* **42**, 227–234.
147. L. Zhou, J.M. Schmitt (1997). Deconvolution and enhancement of optical coherence tomograms, *Proc. SPIE* **2981**, 46–57.
148. A.R. Thompson, J.M. Moran, G.W. Swenson, Jr. (1986). *Interferometry and Synthesis in Radio Astronomy*. Wiley, New York.
149. D. Piao, Q. Zhu, N.K. Dutta, S. Yan, L.L. Otis (2001). Cancellation of coherent artefacts in optical coherence tomography. *Appl. Opt.* **40**, 5124–5131.
150. A.J. Healey, S. Leeman (1993). Speckle reduction methods in ultrasound pulse-echo imaging. *Acoustic Sens. Imaging* **369**, 68–76.
151. A.F. Fercher, C.K. Hitzenberger, W. Drexler, G. Kamp, H. Sattmann (1993). In vivo optical coherence tomography. *Am. J. Ophthalmol.* **116**, 113–114.
152. H. Rashed, J. Izatt, C. Toth (Apr. 2002). Optical coherence tomography of the retina. *Opt. Photon. News*, 48–51.
153. V.V. Tuchin (2000). *Tissue Optics* (pp. 109–133). SPIE Press, Bellingham, WA.
154. J.F. Bille, A.W. Dreher, G. Zinser (1990). Scanning laser tomography of the living human eye. In: B.R. Masters (Ed), *Noninvasive Diagnostic Techniques in Ophthalmology*, (chapter 28), Springer-Verlag, Berlin.
155. A. Roorda, D.R. Williams (February 1997). New directions in imaging the retina. *Opt. Photon. News* 23–29.
156. W. Drexler, U. Morgner, R.K. Ghanta, F.X. Kärtner, J.S. Schuman, J.G. Fujimoto (2001). Ultrahigh resolution ophthalmic optical coherence tomography. *Nat. Med.* **7**, 502–507.
157 A. Gh. Podoleanu, J.A. Rogers, D.A. Jacson, S. Dunne (2000). Three dimensional OCT images from retina and skin. *Opt. Express* **7**, 292–298.
158. M.R. Hee, J.A. Izatt, E.A. Swanson, D. Huang, J.S. Schuman, C.P. Lin, C.A. Puliafito, J.G. Fujimoto (Jan./Feb. 1995). Optical coherence tomography for ophthalmic imaging. *IEEE Eng. Med. Biol.* 67–76.
159. M. Hammer, D. Schweitzer, E. Thamm, A. Kolb (2000). Optical properties of ocular fundus tissues determined by optical coherence tomography. *Opt. Commun.* **186**, 149–153.
160. A.E. Elsner, A. Remsky, J.P. Walker, G.L. Wing, P.A. Raskauskas, D.C. Fletcher, L.M. Kelly, C. Kiesel (Jul. 2000). Imaging in the aging eye. *Opt. Photon. News* 20–25.
161. A. Baumgartner, C.K. Hitzenberger, H. Sattmann, W. Drexler, A.F. Fercher (1998). Signal and resolution enhancements in dual beam optical coherence tomography of the human eye. *J. Biomed. Opt.* **3**, 45–54.
162. M.J. Rivellese, C.A. Puliafito (2001). Optical coherence tomography in the diagnosis and management of posterior segment disorders. In: B.E. Bouma, G. Tearney (Eds), *Handbook of Optical Coherence Tomography* (pp. 471–485). Marcel Dekker Inc., New York.
163. P. Carpineto, M. Ciancaglini, E. Zuppardi, G. Falconio, E. Doronzo, L. Mastropasqua (2003). Reliability of nerve fiber layer thickness measurements using optical coherence tomography in normal and glaucomatous eyes. *Ophthalmol.* **110**, 190–195.
164. V. Guedes, J.S. Schuman, E. Hertzmark, G. Wollstein, A. Correnti, R. Mancini, D. Lederer, S. Voskanian, L. Velazquez, H.M. Pakter, T. Pedut-Kloizman, J.G. Fujimoto (2003). Optical coherence tomography measurement of macular and nerve fiber layer thickness in normal and glaucomatous human eyes, *Ophthalmol.* **110** 177–189.

165. http://www.humphrey.com/index.html, as at 15 June 2003.
166. M. Niemz (2002). *Laser-Tissue Interactions*, (2nd edn), (p. 165), Springer-Verlag, Berlin.
167. E.N. Marieb (2001). *Human Anatomy and Physiology*, (5th edn), Benjamin Cummings, San Fransisco.
168. http://www.isis-optronics.de, as at 15 June 2003.
169. N.D. Gladkova, G.A. Petrova, N.K. Nikulin, S.G. Radenska-Lopovok, L.B. Smopova, Yu. P. Chumakov, V.A. Nasonova, V.M. Gelikonov, G.V. Gelikonov, R.V. Kuranov, A.M. Sergeev, F.I. Feldchtein (2000). *In vivo* optical coherence tomography imaging of human skin: norm and pathology. *Skin Res. Technol.* **6**, 6–16.
170. J. Welzel (2001). Optical coherence tomography in dermatology: A review. *Skin Res. Technol.* **7**, 1–9.
171. M. Rajadhyaksha, M. Grossman, D. Esterowitz, R.H. Webb, R.R. Anderson (1995). *In vivo* confocal scanning laser microscopy of human skin: Melanin provides strong contrast. *J. Invest. Dermatol* **104**, 946–952.
172. Y. Zhao, Z. Chen, C. Saxer, Q. Shen, S. Xiang, J.F. de Boer, J.S. Nelson (2000). Doppler standard deviation imaging for clinical monitoring of in vivo human skin blood flow. *Opt. Lett.* **25** 1358–1360.
173. J.F. de Boer, S.M. Srinivas, A. Malekafzali, Z. Chen, J.S. Nelson (1998). Imaging thermally damaged tissue by polarization sensitive optical coherence tomography. *Opt. Express* **3**, 212–218.
174. A. Knüttel, M. Boehlau-Godau (2000). Spatially confined and temporally resolved refractive index and scattering evaluation in human skin performed with optical coherence tomography. *J. Biomed. Opt.* **5**, 83–92.
175. S.A. Alexandrov, A.V. Zvyagin, K.K.M.B.D. Silva, D.D. Sampson (2003). Bifocal optical coherence refractometry of turbid media. *Opt. Lett.* **28**, 117–119.
176. G.J. Tearney, M.E. Brezinski, B.E. Bouma, S.A. Boppart, C. Pitris, J.F. Southern, J.G. Fujimoto (1997). *In vivo* endoscopic optical biopsy with optical coherence tomography. *Science* **276**, 2037–2039.
177. A.M. Sergeev, V.M. Gelikonov, G.V. Gelikonov, F.I. Feldchtein, R.V. Kuranov, N.D. Gladkova (1997). *In vivo* endoscopic OCT imaging of precancer and cancer states of human mucosal. *Opt. Express* **1**, 432–440.
178. G.J. Tearney, S.A. Boppart, B.E. Bouma, M.E. Brezinski, N.J. Weissman, J.F. Southern, J.G. Fujimoto (1996). Scanning single-mode fiber optic catheter-endoscope for optical coherence tomography *Opt. Lett.* **21**, 543–545.
179. S.A. Boppart, B.E. Bouma, C. Pitris, G.J. Tearney, J.G. Fujimoto, M.E. Brezinski (1997). Forward-imaging instruments for optical coherence tomography. *Opt. Lett.* **22**, 1618–1620.
180. B.E. Bouma, G.J. Tearney (1999). Power-efficient nonreciprocal interferometer and linear-scanning fiber-optic catheter for optical coherence tomography. *Opt. Lett.* **24**, 531–533.
181. X. Li, C. Chudoba, T. Ko, C. Pitris, J.G. Fujimoto (2000). Imaging needle for optical coherence tomography. *Opt. Lett.* **25**, 1520–1522.
182. F.I. Feldchtein G.V. Gelikonov, V.M. Gelikonov, R.V. Kuranov, A.M. Sergeev, N.D. Gladkova, A.V. Shakov, N.M. Shakova, L.B. Snopova, A.B. Terent'eva, E.V. Zagainova, Yu. P. Chumakov, I.A. Kuznetzova (1998). Endoscopic applications of optical coherence tomography. *Opt. Express* **3**, 257–270.
183. A.M. Rollins, R. Ung-arunyawee, A. Chack, R.C.K. Wong, K. Kobayashi, M.V. Sivak Jr., J.A. Izatt (1999). Real-time *in vivo* imaging of human gastrointestinal ultrastructure

by use of endoscopic optical coherence tomography with a novel efficient interferometer design. *Opt. Lett.* **24**, 1358–1360.
184. N.S. Nishioka, S. Brand, B.E. Bouma, G.J. Tearney, C.C. Compton (2001). Gastrointestinal applications of optical coherence tomography. In: B.E. Bouma, G. Tearney (Eds), *Handbook of Optical Coherence Tomography* (pp. 673–691). Marcel Dekker, Inc., New York.
185. B.E. Bouma, G.J. Tearney (2002). Clinical imaging with optical coherence tomography. *Acad. Radiol.* **9**, 942–953.
186. http://www.octimaging.com/index.html, as at 15 June 2003.
187. G.J. Tearney, B.E. Bouma, M.S. Shishkov, K. Schlendorf, S. Houser, D.-H. Kang, H.T. Aretz, T.J. Brady, I.-K. Jang (2001). Optical coherence tomography in cardiology. In: B.E. Bouma, G. Tearney (Eds), *Handbook of Optical Coherence Tomography* (pp. 693–703). Marcel Dekker, Inc., New York.
188. S.A. Boppart, J.G. Fujimoto, M.E. Brezinski (2001). Optical coherence tomography and developmental biology. In: B.E. Bouma, G. Tearney (Eds), *Handbook of Optical Coherence Tomography* (pp. 505–538). Marcel Dekker, Inc., New York.
189. S.A. Boppart, M.E. Brezinski, B.E. Bouma, G.J. Tearney, J.G. Fujimoto (1996). Investigation of developing embryonic morphology using optical coherence tomography. *Develop. Biol.* **177**, 54–63.
190. S.A. Boppart, G.J. Tearney, B.E. Bouma, J.F. Southern, M.E. Brezinski, J.G. Fujimoto (1997). Noninvasive assessment of the developing *Xenopus* cardiovascular system using optical coherence tomography. *Proc. Natl. Acad. Sci. USA* **94**, 4256–4261.
191. J.W. Hettinger, M. de la Peña Mattozzi, W. Meyers, M.E. Williams, A. Reeves, R.L. Parsons, R.C. Haskell, D.C. Petersen, R. Wang, J.I. Medford (2000). Optical coherence microscopy: a technology for rapid in vivo non-destructive visualization of plants and plant cells. *Plant Physiol.* **123**, 3–15.
192. A. Reeves, R.L. Parsons, J.W. Hettinger, J.I. Medford (2002). In vivo three-dimensional imaging of plants with optical coherence microscopy. *J. Microsc.* **208**, 177–189.
193. B. Povazay, K. Bizheva, A. Unterhuber, B. Hermann, H. Sattmann, A.F. Fercher, W. Drexler, A. Apolonski, W.J. Wadsworth, J.C. Knight, P.St.J. Russell, M. Vetterlein, E. Scherzer (2002). Submicrometer axial resolution optical coherence tomography. *Opt. Lett.* **27**, 1800–1802.
194. Y. Wang, Y. Zhao, J.S. Nelson, Z. Chen, R.S. Windeler (2003). Ultrahigh-resolution optical coherence tomography by broadband continuum generation from a photonic crystal fiber. *Opt. Lett.* **28**, 182–184.
195. J.O. Schenk, M.E. Brezinski (2002). Ultrasound induced improvement in optical coherence tomography (OCT) resolution. *Proc. Natl. Acad. Sci. USA* **99**, 9761–9764.
196. K.D. Rao, M.A. Choma, S. Yazdanfar, A.M. Rollins, J.A. Izatt (2003). Molecular contrast in optical coherence tomography by use of a pump-probe technique. *Opt. Lett.* **28**, 340–342.

Chapter 18

Laser optoacoustic imaging

Steven L. Jacques

Table of contents

Abstract

Optoacoustic imaging is an imaging modality based on pressure waves generated in light-absorbing objects by pulsed lasers. An array of acoustic detectors record the time-resolved arrival of pressure waves from a region of interest. Backprojection of these time-resolved recordings yields a reconstruction of the original energy deposition. Such imaging is suitable for imaging blood vessels, hemorrhages, and pigmented lesions such as melanoma.

18.1 Introduction

Optoacoustic imaging or *photoacoustic imaging*is an imaging modality that uses pressure waves induced by a pulsed laser (Figure 1). A pulsed laser source illuminates a tissue and generates heat in an absorbing object within a tissue. The thermoelastic expansion of the heated object creates a pressure wave that propagates to the tissue surface where an array of detectors records the time-resolved arrival of pressure. A computer algorithm backprojects these time-resolved recordings into the tissue to map the source of the pressure waves, which is a map of the distribution of heat deposition.

Optoacoustic imaging is a special case of *thermoacoustic imaging*. In general, the pulsed energy source of thermoacoustic imaging can be a pulsed laser, a pulse of radiofrequency(RF) or microwave frequency electromagnetic energy, or any energy source that heats tissue. The principles of such imaging are the same regardless of the energy source. With pulsed RF or microwave energy, the heating of tissues varies with

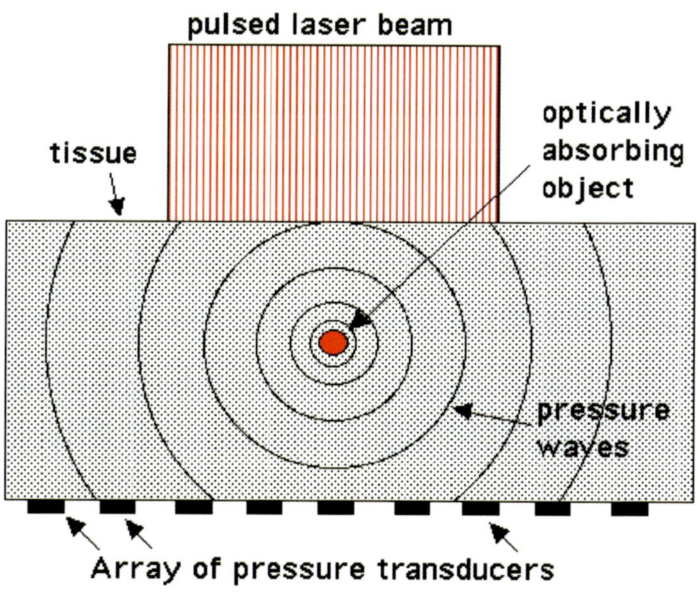

Figure 1. Optoacoustic imaging involves the generation of pressure waves by absorption of pulsed laser light by an optically absorbing object within a medium such as tissue.

the amounts of free and bound water, salt concentration and protein and lipid content. With pulsed laser energy, there is selective heating of blood vessels whose hemoglobin strongly absorbs light, melanin pigmentation as in hair follicles, or water.

A key distinguishing characteristic of thermoacoustic imaging is that the pressure waves are generated in the absorbing objects within the tissue, so the objects become a source of pressure wave signal. This characteristic contrasts with ultrasound imaging in which a pressure wave launched at the tissue surface propagates into the tissue and is reflected or perturbed by the object. Contrast in ultrasound imaging depends on an object's mechanical impedance mismatch relative to the surrounding tissue which perturbs transmission or causes reflectance of the pressure waves. Contrast in thermoacoustic imaging depends on the selective absorption of pulsed energy by the object.

Optoacoustic imaging is also very different from diffuse light tomography in which photons propagate into a tissue, are perturbed by an absorbing object such as a blood vessel, and the perturbation propagates back to the surface for detection. An analogy (Figure 2) might be the task of finding a ninja dressed in black who is hiding in the woods at night. One approach is to use a flashlight and look for the how the ninja perturbs or reflects the light from the flashlight. A second approach is to have the ninja hold a candle. It is easier to find a ninja holding a candle. By generating pressure waves in the absorbing object, the object becomes a source of signal, like the ninja holding a candle.

Optoacoustic imaging can be implemented using different wavelengths of light. The absorption spectrum of the object can be determined from the wavelength dependence of the mapping of heat deposition due to the laser. For example, imaging with two wavelengths that are absorbed differently by oxy-hemoglobin and deoxy-hemoglobin can allow mapping of the blood content and mixed arteriovenous blood oxygen saturation.

After delivering photons into the tissue, the photons must diffuse through the tissue as they undergo multiple photon scattering events. The photons diffuse down to a perturbing object such as a blood vessel that absorbs some of the light, then the

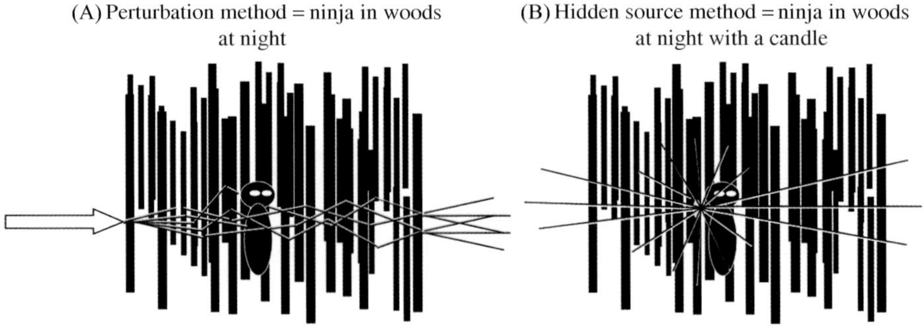

Figure 2. Two types of imaging use either perturbation methods or hidden source methods. Perturbation methods are like trying to find a black ninja in the woods at night with a flashlight. Hidden source methods are like trying to find a ninja holding a candle in the woods at night. Optoacoustic imaging is a hidden source method. The pressure waves generated in an object are analogous to the candle held by the ninja.

perturbation diffuses back to the surface for detection. Hence, optical imaging with diffuse light has poor spatial resolution because it is based on a diffusion process. But the optoacoustic signal generated by the object's absorption of diffuse photons yields a sharply defined pressure wave that mimics the shape of the object. This pressure wave propagates to the tissue surface with only slight acoustic scattering. Hence, the acoustic detectors on the tissue surface can detect a relatively sharp image of the object. The spatial resolution of optoacoustic imaging is equal to the laser pulse duration t_{laser} times the speed of sound c_s, $c_s t_{\text{laser}}$. A Q-switched laser has a 10 ns pulse duration and yields a spatial resolution of about 15 μm. Such resolution can be achieved in the first mm of superficial tissue.

There is a frequency-dependent viscoelastic attenuation of pressure waves by tissue that acts as a low-pass filter, which removes high spatial frequencies from the pressure wave. Such attenuation unsharpens the image. Therefore, when imaging at depths of a cm or more in soft tissues, the highest acoustic frequency that is transmitted is about 3 MHz and spatial resolution is about 1 mm, not 15 μm.

In summary, optoacoustic imaging combines the spectral capabilities of optical spectroscopy with the spatial resolution of ultrasound.

18.1.1 History

The photoacoustic effect was discovered by Alexander Bell in 1880 [1] who heard a "pure musical tone" emanating from an enclosed gas that absorbed a modulated beam of light. The effect has been widely used ever since. The interested reader is referred to a review by Gusev and Karabutov [2]. Recent work in thermoacoustic imaging is being pursued by many groups around the world, for example as reported at the annual conference on Biomedical Optoacoustics [3]. An early example was reported by Kruger [4] who demonstrated photoacoustic imaging using a pulsed laser and an absorbing object within a milk-like solution as an experimental model. Kruger's work eventually has led to using microwave energy sources for imaging breast cancer [5] and examples of his work can be found at the website http://optosonics.com. Another early example was Oraevsky et al. [6] who used photoacoustic imaging of the superficial layers where a pulsed laser enters a tissue to determine the optical properties of the tissue. Oraevsky's work has led to using laser sources to image breast cancer based on hemoglobin absorption. Such quantitative use of photoacoustic imaging is also being developed for transcranial monitoring of the oxygenation of the brain [7].

18.2 Basic principles and theoretical background

18.2.1 Velocity potential and pressure

Consider a small mass m (kg) that is traveling at a velocity v (m s^{-1}) and strikes a wall (Figure 3). The mass elastically bounces off the wall and now travels at a velocity $-v$. Hence, the change in velocity Δv is negative and equals $-2v$. If the process of bouncing off the wall takes a time period Δt(s) then the force F(N or kg m s^{-2}) exerted

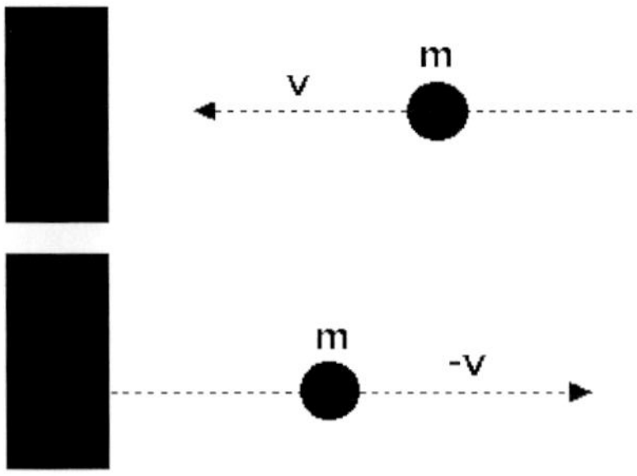

Figure 3. A mass (m) travelling with velocity (v) strikes a wall and is elastically reflected backward with a new velocity ($-v$). The positive force on the wall is proportional to the negative change in velocity ($-\Delta v$).

on the wall equals the change in momentum per unit time or $-m\Delta v/\Delta t$. The exerted force is positive and proportional to the negative of the change in velocity. Similar to the above example, a positive pressure P ($N\,m^{-2}$ or $J\,m^{-3}$ or Pa) generated by thermoelastic expansion in a material with density ρ (kg m^{-3}) is proportional to the negative of change in a parameter called the velocity potential ϕ ($m^2\,s^{-1}$):

$$P = -\rho \frac{\partial \phi}{\partial t} \tag{1}$$

Conversely,

$$-\phi(t) = \int_0^t \frac{P(t)}{\rho} \mathrm{d}t \tag{2}$$

The negative velocity potential $-\phi$ is also proportional to the density of energy deposition W (J m^{-3}) that is located a distance $r = c_s t$ away from the point of observation, where t is the time after the initial energy deposition. The calculation in analytic form is

$$-\phi(t) = \frac{\beta}{4\pi\rho C_p} \int_V \frac{W(r)}{r} \delta(r - c_s t) \mathrm{d}V \tag{3}$$

where β is the thermal expansivity ($^\circ C^{-1}$) which describes strain ($\Delta L/L$, where L is length) per degree, ρ is density (kg m^{-3}), and C_p is specific heat (J kg^{-1}). The $\delta(r-c_s t)$ is a Dirac delta function that is non-zero when r equals $c_s t$. The integral of all energy deposition W in an incremental shell at the surface of a sphere with radius $c_s t$ contributes to the velocity potential observed at the center of the sphere.

As a special example, consider a uniform energy deposition W at time zero throughout an infinite medium with no boundaries. At time t, a detector detects pressure that is proportional to the negative change in velocity potential per unit time ($-\mathrm{d}\phi/\mathrm{d}t$). The incremental $-\mathrm{d}\phi$ is due to W deposited in a spherical shell centered on the detector with a shell volume of $4\pi r^2 \mathrm{d}r$ where $r = c_s t$ and $\mathrm{d}r = c_s \mathrm{d}t$:

$$-\mathrm{d}\phi = \frac{\beta}{4\pi\rho C_p} \frac{W}{r} 4\pi r^2 \mathrm{d}r = \frac{\beta}{\rho C_p} W r \mathrm{d}r = \frac{\beta c_s^2}{\rho C_p} W \mathrm{d}t \tag{4}$$

Therefore, the pressure is

$$P = -\rho \frac{\partial \phi}{\partial t} = \frac{\beta c_s^2}{C_p} W = \Gamma W \tag{5}$$

The result shows how the pressure P is proportional to the energy deposition W. The parameter $\Gamma = \beta c_s^2 / C_p$ is called the Grüneisen coefficient (dimensionless) and equals about 0.12 for aqueous media at room temperature. The units of pressure P and energy deposition W are the same, $1\text{Pa} = 1\,\text{J}\,\text{m}^{-3}$. An energy deposition W equal to $1\ \text{J}\,\text{cm}^{-3}$ or $10^6\ \text{J}\,\text{m}^{-3}$ will raise the temperature of water by 0.239°C, and generate a pressure of $0.12 \times 10^6\ \text{J}\,\text{m}^{-3}$ or Pa, which equals 1.2 bar.

18.2.2 Computation of ϕ and P

The analytic expression Equation (3) can be restated in a form that allows approximate evaluation by a discrete computation. Such computations allow the prediction of time-resolved pressure generated at any point of observation for any arbitrary shape of object. The velocity potential ϕ ($\text{m}^2\,\text{s}^{-1}$) is computed by summing all the energy deposition W ($\text{J}\,\text{m}^{-3}$) over a volume V (m^3) that has occurred at a distance $r \pm c_s \Delta t/2$ (m) from the detector:

$$-\phi(k) = \frac{\beta}{4\pi\rho C_p} \frac{1}{\Delta t} \sum_j \frac{W(j)V(j)}{r(j)} \tag{6}$$

where Δt equals the size of the time bins, and the time of pressure arrival at the detector is $t = r(j)/c_s$. The index k refers to the time bin of the velocity potential, $k = \text{round}(t/\Delta t)$. The index j refers to the jth volume element of the entire volume being considered. Each jth volume element has a unique value of W, V, and r. The time-resolved pressure P is approximately calculated as

$$P(k) = \rho \frac{-\phi(k+1) - [-\phi(k)]}{\Delta t} \tag{7}$$

If detected $-\phi$ increases with time, i.e., $-\phi(k+1) > -\phi(k)$, then pressure $P(i)$ is positive.

To illustrate the use of Equations (6) and (7), consider the velocity potential and pressure observed by a detector located 1 cm from a sphere of uniform energy deposition $W = 10^6\text{J}\,\text{m}^{-3}$, within an infinite aqueous medium (Figure 4). The values of β, ρ, and C_p are approximately $2.29 \times 10^{-4}\,°\text{C}^{-1}$, $1000\,\text{kg}\,\text{m}^{-3}$, and $4184\,\text{J}\,\text{kg}^{-1}$. The speed of sound in the medium is about $1480\ \text{m}\,\text{s}^{-1}$. A collection shell of radius r and volume $4\pi r^2 \text{d}r$ is centered on the detector and intersects the energy deposition W that will arrive at the detector at time r/c_s after the energy deposition occurs. Figure 4 illustrates the collection shells that intersect the sphere at different times. The solid lines indicate shells that intersect the sphere of energy deposition W, and the dashed lines indicate shells that do not intersect any energy deposition and consequently yield zero contribution to the negative velocity potential $-\phi$. The $-\phi$ at first increases with time because the shells of collection intersect increasingly larger regions of the sphere of deposition. As the shells pass the center of the sphere, they begin to intersect increasingly smaller regions and $-\phi$

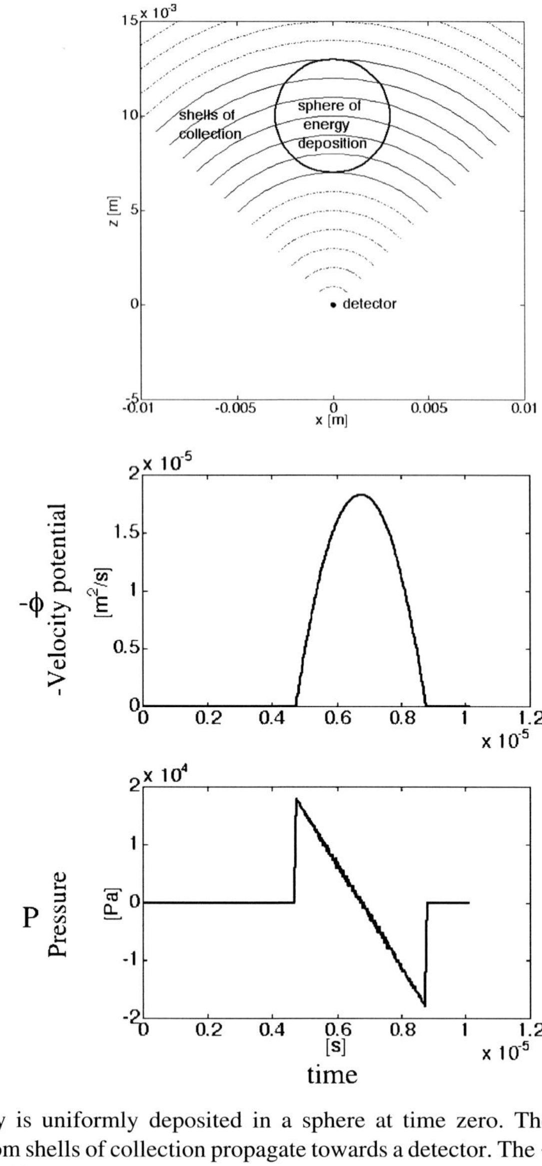

Figure 4. Energy is uniformly deposited in a sphere at time zero. The negative velocity potential ($-\phi$) from shells of collection propagate towards a detector. The $-\phi$ first increases as the collection shell intersects increasingly larger portions of the sphere, reaches a peak when the collection shell reaches the center of the sphere, then decreases as the shell intersects smaller portions of the sphere. The pressure is proportional to the time derivative $-\mathrm{d}\phi/\mathrm{d}t$, so the pressure is initially positive as $-\phi$ is increasing, then negative as $-\phi$ is decreasing.

decreases. Therefore, the pressure P is initially positive, as $-\phi$ increases, then negative as $-\phi$ decreases.

In summary, although one measures pressure P, the negative velocity potential $-\phi$ is the parameter that is proportional to the energy deposition W and is useful for image reconstruction.

18.2.3 Reconstruction of the image of W from pressure recordings

Let an array of detectors on the tissue surface yield a set of time-resolved recordings of arriving pressure, $P(t)$. How shall these time-resolved data be backprojected into the tissue to specify the source of the pressure waves, in other words the spatial distribution of the heat deposition from the pulsed laser?

Figure 5 illustrates the basic concept of backprojection and the limits to resolution. A circle of detectors are distributed around a central region that is to be imaged. An absorbing object, for example a small blood vessel, is shown in the central region. The collection cone of one of the detectors is shown, illustrating how a detector may have a directionality to its receiving function. In this example, a simple cone of uniform collection is denoted. During the measurement, a portion of the absorbing object generates a thermoelastic expansion. This portion of the object is denoted as the jth volume element with a volume $V(j)$ and an energy of deposition $W(j)$ located a distance $r(j)$ from the detector. The factor $W(j)V(j)/r(j)$ contributes to the velocity potential arriving at the detector at time $t = r/c_s$, where r is the distance between the detector and this portion of the object. Equation (6) describes the $-\phi(k)$ that arrives at time $t(k)$ at the detector, and Equation (7) describes the pressure that the detector will record. During backprojection, however, this $-\phi(k)$ contribution is backprojected into the larger backprojection volume $V_b(k,i)$ rather than into the true source volume $V(j)$. This $V_b(k,i)$ depends on

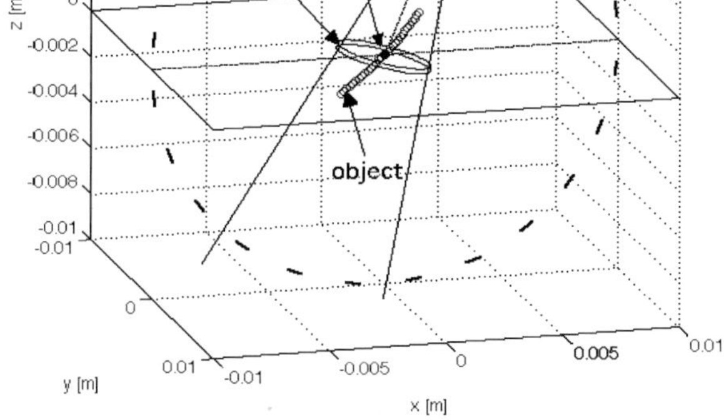

Figure 5. An array of detectors surrounds a region of interest that contains an absorbing object such as a blood vessel. The collection cone for one of the detectors is shown. The portion of the object whose energy deposition $W(j)$ contributes to the pressure wave that emanates from the volume $V(j)$ and propagates a distance $r = c_s t$ to the detector. During image reconstruction, this pressure signal is backprojected into the disk-shaped volume $V_b(k)$ where k is a time index.

the antenna or receiving function of the detector, and in this example equals a disk with an area equal to the cross-section of the collection cone at a distance $c_s t$ from the detector and thickness $c_s \Delta t$. Hence, the measurement of the product $W(j)V(j)$ is backprojected as a predicted product $W'(k,i)V_b(k,i)$, in other words the backprojection assigns measurement of $W(j)V(j)$ into a larger volume $V_b(k,i)$ with a lower energy density $W'(k,i)$. Under special conditions, such as an isotropic acoustic detector within a homogeneous medium with uniform energy deposition, the shells of collected $W(j)V(j)$ can match the shells of backprojected $W'(k,i)V_b(k,i)$ and there is perfect reconstruction of the original W. However, experimental systems will usually involve a mismatch between $V(j)$ and $V_b(k,i)$.

Let us summarize the backprojection calculations. Assume that the detectors have yielded a set of time-resolved pressure recordings $P(k,i)$ where $k = \text{round}(t/\Delta t)$, which is an integer serving as the time index for time $t(k)$, and Δt is the time step of the data acquisition. The integer i indicates the ith detector in the array of N_i detectors. The volume of interest is divided into N_j subvolumes with individual volumes $V(j)$.

Step 1: Use Equation (2) to convert $P(k,i)$ into the time-resolved negative velocity potential, $-\phi(k,i)$, by integration:

$$-\phi(k,i) = \frac{\Delta t}{\rho} \sum_{kk=1}^{k} P(kk,i) \tag{8}$$

Step 2: Rearrange Equation (6) to solve for $W'(j)$, which denotes the *predicted* energy deposition in the jth volume based on the measured $-\phi$:

$$W'(j) = \frac{4\pi\rho C_p \Delta t}{\beta} \sum_{i=1}^{N_i} \sum_{k=1}^{N_k} \frac{-\phi(k,i)c_s t(k)}{V(k,i)} [c_s t(k) - r(j,i)\Delta t] \tag{9}$$

where $(c_s t(k) - r(j,i) \leq \Delta t)$ is a Boolean operator that is 1 if true and 0 if false. Only $-\phi(k,i)$ acquired at time $t(k)$ satisfying this Boolean condition will contribute to the $W'(j)$. The contribution $-\phi(k,i)c_s t(k)$ is normalized by the backprojection volume $V_b(k,i)$ to yield an energy density that contributes to $W'(j)$. Usually, a given kth contribution from an ith detector, $-\phi(k,i)c_s t(k)/V_b(k,i)$, will contribute to $W'(j)$ in several jth volumes because the backprojection volume $V_b(k,i)$ is larger than the mapping voxel $V(j)$. One must pay attention to the details of how the backprojection volumes align with the mapping voxels to properly conserve energy, an issue not fully discussed here. The total energy of deposition Q(J) is calculated based on the $W'(j)$ calculated by Equation (9) using the measured $-\phi(k,i)$:

$$Q = \sum_{j=1}^{N_j} W'(j)V(j) \tag{10}$$

$W'(j)$ constitutes the image reconstruction of the original $W(j)$. The following steps, 3 and 4, are optional but usually will improve the accuracy of the image reconstruction.

Step 3: A mask can eliminate many of the unwanted ghost images generated in W' by step 2. If one is imaging a region with one or more discrete objects of optical absorption embedded in a background medium that is either non-absorbing or else is a low-level uniform background absorption, then one can create a mask, $M(j)$ congruent with $W'(j)$, for which each jth voxel equals zero if any of the detectors do not assign any $W'(j)$ above that expected for a uniform background value $W_{\text{background}}$. Otherwise, $M(j)$ equals 1. Essentially, each detector has a veto power to insist that no positive value should be assigned to $W'(j)$-$W_{\text{background}}$. The mask is generated by evaluating the $W'(j)$ for each ith detector, denoted as $W'(j,i)$, evaluating the Boolean value ($W'(j,i)$ > background) for each detector, and taking the product of Boolean values for all detectors:

$$M(j) = \prod_{i=1}^{N_i}\left[W'(j,i) > W_{\text{background}}\right] \tag{11}$$

The mask M then multiplies, voxel by voxel, the incremental image $W' - W_{\text{background}}$ to yield a new W':

$$W'(j) = M(j)K\left[W'(j) - W_{\text{background}}(j)\right] + W_{\text{background}}(j) \tag{12}$$

where

$$K = \frac{\sum\limits_{j=1}^{N_j} M(j)\left[W'(j) - W_{\text{background}}(j)\right]}{Q - \sum\limits_{j=1}^{N_j} W_{\text{background}}(j)} \tag{13}$$

Multiplication by the factor K maintains conservation of energy. The ghost images of W' from step 2 are somewhat eliminated in the new W' image.

Step 4: An iterative algorithm can further improve the image reconstruction. Paltauf et al. [8] reported a scheme based on the van Cittert algorithm. The original set of measurements using all detectors can be denoted as ϕ. Let Equation (8) be summarized by a forward calculation denoted by the function A, where $\phi = A(W')$. Equations (9–13) can be summarized by an inverse calculation denoted by the function B, where $W' = B(\phi)$. The iterative algorithm is summarized:

$$\begin{aligned} W'_1 &= B(\phi) \\ \phi_1 &= A(W'_{(1)}) \\ &\text{for } i = 2{:}N \\ \Delta W' &= B(\phi - \phi_{i-1}) \\ W'_i &= W'_{i-1} + \Delta W' \\ \phi_i &= A(W'_i) \\ &\text{end} \end{aligned} \tag{14}$$

The algorithm lets the residual discrepancy, $\phi - \phi_{i-1}$, between measurement and predicted measurement after $i-1$ iterations be backprojected as $B(\phi - \phi_{i-1})$ to yield a corrective $\Delta W'$ which is added to the current image W'_{i-1} to yield a new image W'_i. Iteration continues until no further improvement in the image occurs.

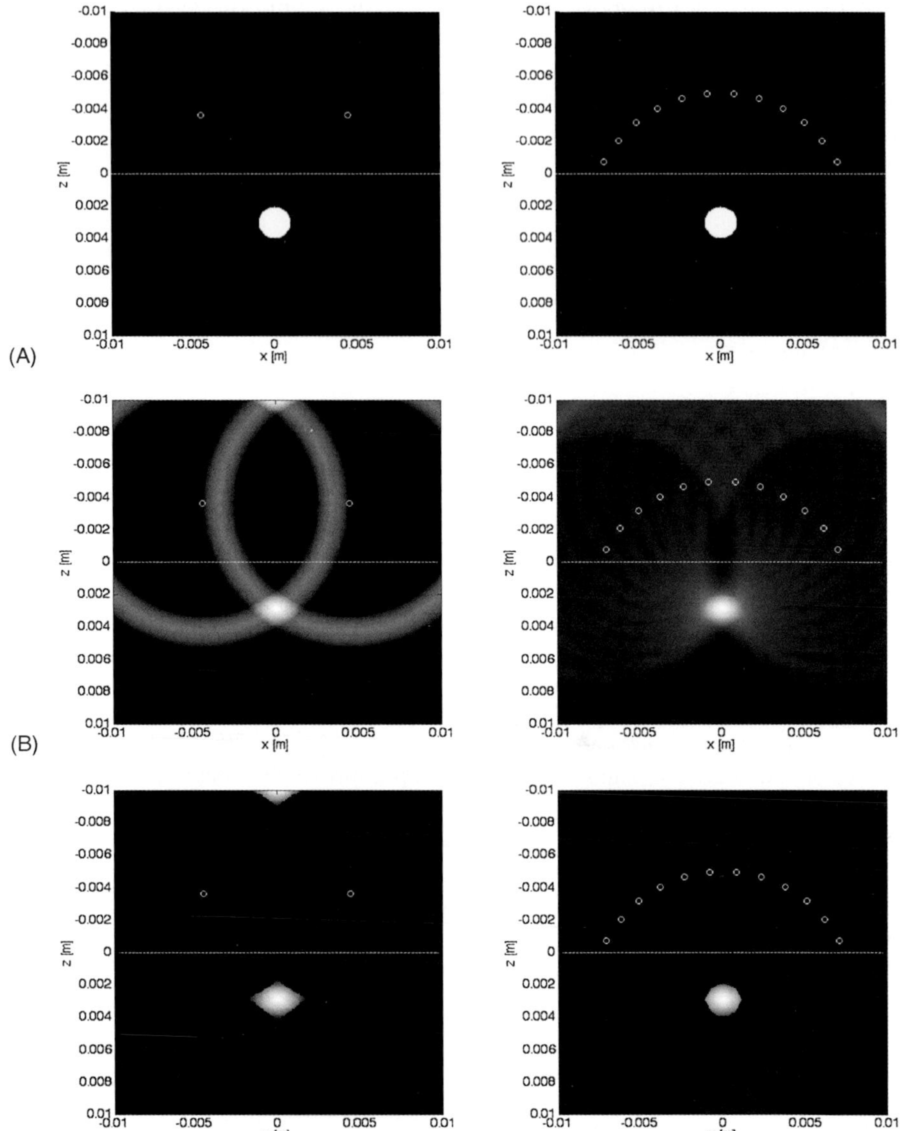

Figure 6. Image reconstruction for a sphere of uniform energy deposition, using either 2 (left-hand side) or 12 isotropic acoustic detectors. (A) Geometry of the sphere and detectors. (B) Backprojection using step 2, which generates many image ghosts. (C) Backprojection using steps 2 and 3 in which a mask eliminates most of the image ghosts.

There are important details for implementing this algorithm, such as not allowing any negative predicted $W_i^{\prime}$ and ensuring conservation of energy.

Figure 6 illustrates the backprojection of simulated $-\phi$ recordings for 2 detectors and for 12 detectors with isotropic collection arranged in an arc around a central spherical object with uniform energy deposition W. Figure 6(A) shows the geometry of the detectors and the spherical object. Figure 6(B) shows the how backprojection of the volumes $V_b(k,i)$ are spherical shells of thickness $c_s\Delta t$, and their summation yields the backprojected image W'. The non-zero contributions to W' from each detector overlap to specify the location of the sphere of energy deposition. However, most of the spherical shell of backprojected W' from each detector is a ghost image. Figure 6(C) shows the effect of multiplying W' by a mask M, eliminating most of the ghost image.

The image reconstruction scheme described here is just one of several possible approaches. For example, Liu [9] describes another reconstruction algorithm for thermoacoustic imaging.

18.3. Experimental apparatus

18.3.1 Pulsed laser

The basic experimental approach is to use a pulsed laser to generate pressure in absorbing objects such as a blood vessel or melanin pigment within a tissue. The pressure generated will be maximized if the laser pulse duration, t_{laser} (s), is sufficiently short that pressure cannot propagate out of the absorbing object, with width or diameter d (m), during the laser pulse. Hence, pressure can build up before dissipating by propagation. This condition is called the stress confinement condition and is summarized as

$$t_{laser} \ll \frac{d}{c_s}$$

Dingus and Scammon [10] pointed out that if a stress relaxation time τ is defined as $c_s t_{laser}/d$, then the maximum pressure generated in the absorbing object will decrease proportionally to a factor A that is unity for short laser pulses and drops toward zero for long laser pulses:

$$A = \frac{1 - e^{-\tau}}{\tau} \quad \text{where } \tau = \frac{c_s t_{laser}}{d}$$

A drops to 0.90 when τ equals 0.2 and drops to 1/2 when τ is 1.6. A Q-switched laser will typically provide pulses with 10 ns duration, and A is 0.90 for a 74 μm diameter object and 0.5 for a 9.2 μm diameter object.

The best possible spatial resolution of optoacoustic images is $c_s\Delta t$ where Δt equals the laser pulse duration t_{laser}. For the 10 ns laser and aqueous medium where $c_s = 1480\ \mathrm{m\,s^{-1}}$, the best possible spatial resolution is 14.8 μm. The A value for an object whose size matches the best possible spatial resolution is 0.63.

18.3.2 Piezoelectric transducers

The most common approach to measurement of pressure is to use piezoelectric transducers such as PVDF film or lithium niobate crystals, which generate electric charge on their surfaces when strained by a pressure wave. When such piezoelectric devices are monitored by a high impedance amplifier, the charge accumulates on the capacitance of the device to yield a voltage that is sensed by the amplifier. Voltage is proportional to charge that is proportional to strain that is proportional to pressure. The voltage per pressure (V Pa^{-1} or V bar^{-1}) depends on the area and thickness and composition of the piezoelectric device. Oraevsky et al. [6] reported a 10^{-7} V Pa^{-1} or 10 mV bar^{-1} calibration for a lithium niobate detector. One of our students, John Viator [12], built a PVDF device with a detection area of 1 mm^2 and thickness of 25 μm which had a 5.5 mV bar^{-1} calibration.

18.3.3 Optical transducers

Optical techniques can be applied to pressure measurements. One approach is to sense the displacement of the medium–air surface as a pressure wave generated within the medium arrives at the surface. Jacques et al. [12] reported such a sensor based on a common-path interferometer using a He-Ne laser, which demonstrated a calibration in the range of 30–100 mV bar^{-1} and a noise level of 10–30 mbar (1 mV). Beard and Mills [13] implemented a detector based on a Fabry–Perot interferometer. Another approach is to sense the change in refractive index in a medium at the surface interface between the tissue and an applied glass plate, based on the change in reflectance at this interface. Paltauf and Schmidt-Kloiber [14] demonstrated such an optical sensing of acoustic waves, which is discussed in the next section.

18.4. Current results

18.4.1 Phantom experiment

Paltauf et al. [14] reported a demonstration of optoacoustic imaging using an optical pressure transducer, as shown in Figure 7. A Q-switched pulsed laser delivered light from above a chamber containing a clear phantom tissue with absorbing spheres made of acrylamide gel. The phantom consisted of a water layer with overlying clear mineral oil and the gel spheres floated on top of the water – oil interface. The spheres had strong optical absorption, i.e., an absorption coefficient $\mu_a = 60\,cm^{-1}$such that the 1/e penetration depth was 170 μm. A glass prism contacted the bottom of the phantom and a He-Ne laser beam illuminated a 120 × 250 μm spot on the bottom of the phantom. Reflectance from the prism–water interface was directed to a photodiode detector, and this reflectance varied as pressure waves arrived at the water–prism interface and affected the refractive index mismatch. The probe beam was translated to each of nine positions in a 3 × 3 array on the bottom of the phantom, and a time-resolved pressure recording

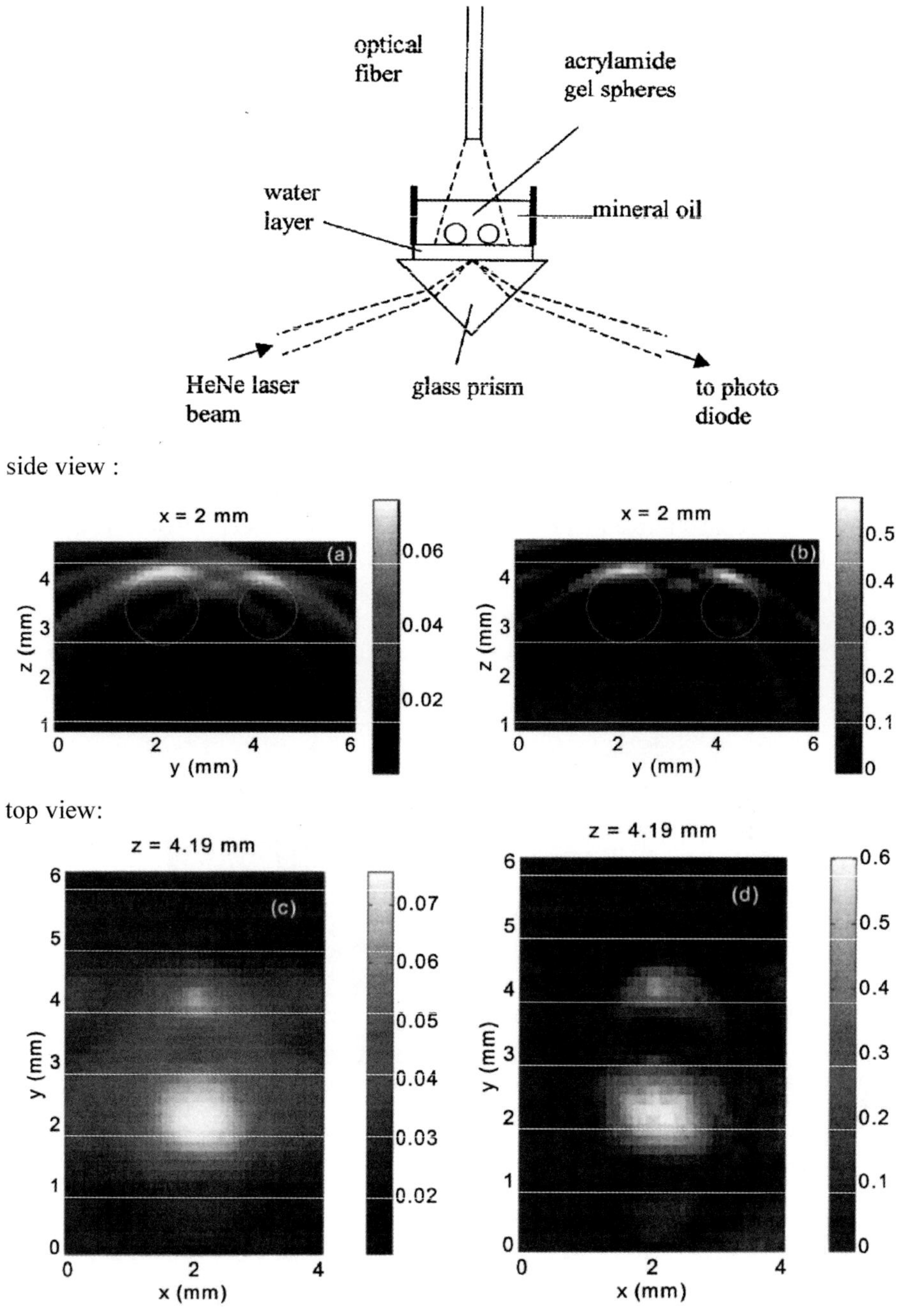

Figure 7. Experimental image reconstruction of two absorbing spheres within a mineral oil medium floating on a layer of water. The measurements used an optical detector in which the reflection of a He-Ne laser off the prism–water interface varied as pressure waves generated in the spheres by pulsed laser irradiation from above arrived at the interface. Time-resolved measurements of pressure were taken at an array of nine detector positions. Top [(a) and (b)] and side views [(c) and (d)] of the reconstructed images are shown for two cases: (1) the first estimate based on step 2 only (see Equation 8), and (2) an improved estimate based on 11 iterations of step 4 (see Equation (14)). The gray level scales are in J cm^{-3}. Circles in (a) and (b) indicate the positions of the absorbing spheres. Reprinted with permission from Paltauf et al. [14].

was made at each position. The pressure recordings $P(k,i)$ were analyzed according to steps 1, 2, and 4 outlined in Section 18.2.3. Figures 7(a) and 7(c) show side and top views of the image reconstruction after applying steps 1 and 2 only. Figures 7(b) and 7(d) show side and top views of the image reconstruction after applying 11 interactions of step 4. The images show the upper surface of the absorbing spheres where the strongest absorption occurred, which are present as arcs of strong energy deposition W. The iterative algorithm improved the images by partially removing the ghost images.

18.4.2 In vivo measurements of portwine stain lesions of skin

Viator et al. [15] reported optoacoustic measurements of the depth profile of portwine stain lesions in human skin. Portwine stain lesions are vascular abnormalities characterized as enlarged venous blood vessels. Figure 8(A) shows the construction of a PVDF piezoelectric transducer consisting of a circular film of PVDF attached to a small cylindrical coaxial cable. The back of the PVDF film contacted the central coaxial conductor, and the front of the conductor

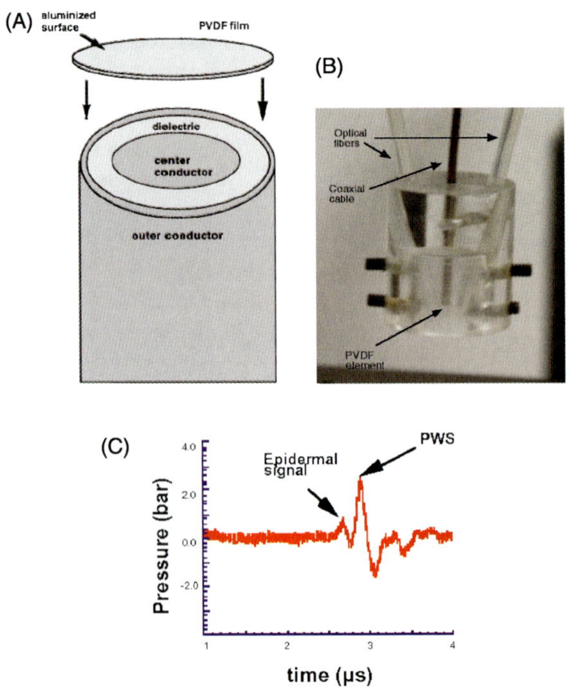

Figure 8. Experimental image of the depth profile of a port wine stain lesion using an optoacoustic probe. (A) The probe was constructed with PVDF film attached to the end of a small coaxial cable. (B) A photo of the probe. (C) Time-resolved trace of pressure showing the signals from the pigmented epidermis and the underlying portwine stain lesions. Reprinted with permission from Viator et al. [15].

was electrically connected to the outer coaxial conductor by conductive paint. The device was coupled to the skin by a small chamber filled with water (Figure 8(B)). Two optical fibers delivered pulsed laser light from a Q-switched 2nd-harmonic Nd:YAG laser (532 nm wavelength, 8 mJ pulse^{-1} per fiber, 4 ns pulse duration, 4 mm diameter illumination spot size). A typical time-resolved recording of pressure is shown in Figure 8(C). The depth profile of the portwine stain skin site shows the overlying pigmented epidermis with its absorbing melanin content, and the deeper vascular structure of the portwine stain lesion.

18.5 Conclusions

Optoacoustic imaging offers a means of imaging tissue on the basis of optical absorption as a contrast mechanism. Different wavelengths can be used so that spectroscopically weighted images can be generated. Although the light is multiply scattered as it diffuses down to an absorbing object such as a blood vessel, the pressure wave generated by thermoelastic expansion due to energy deposition in the object can have sharp edges and faithfully record the shape of the object. Such pressure waves can propagate back to the tissue surface for detection without significant acoustic scattering. Viscoelastic attenuation will attenuate the higher acoustic frequencies of the pressure wave when attempting to image deeper depths of a cm or more. However, imaging superficial tissue layers can be accomplished with spatial resolution of the order of 10s of μm.

Optoacoustic imaging is still in its early stages of development. The availability of increasingly small, reliable, turn-key and inexpensive pulsed lasers is driving the development of this imaging modality. The microfabrication of detector arrays is a current challenge.

Acknowledgements

I wish to thank Guenther Paltauf who worked as a Visiting Professor and John Viator who worked as a graduate student in my laboratory. This work is currently supported by an NIH Bioengineering Partnership Grant (RO1-ER000224).

References

1. A.G. Bell (1880) *Am. J. Sci.* **20**, 305.
2. V.E. Gusev, A.A. Karabutov (1993). *Laser Optoacoustics* (transl. K. Hendzel). American Institute of Physics, New York.
3. A.A. Oraevsky (2000). Biomedical Optoacoustics. *Proceedings of SPIE* (Volume 3916). SPIE Bellingham, WA.
4. R.A. Kruger (1994). Photoacoustic ultrasound. *Med. Phys.* **21**(1), 127–131.

5. R.A. Kruger, K.D. Miller, H.E. Reynolds, W.L. Kiser, Jr, D.R. Reinecke, G.A. Kruger (2000). Breast cancer in vivo: contrast enhancement with thermoacoustic CT at 434 MHz-feasibility study. *Radiology* **216**(1), 279–283.
6. A.A. Oraevsky, S.L. Jacques, F.K. Tittel (1997). Measurement of tissue optical properties by time-resolved detection of laser-induced transient stress. *Appl. Opt.* **36**, 402–415.
7. R.O. Esenaliev, I.V. Larina, K.V. Larin, D.J. Deyo, M. Motamedi, D.S. Prough (2002). Optoacoustic technique for noninvasive monitoring of blood oxygenation: a feasibility study. *Appl. Opt.* **41**(22), 4722–4731.
8. G. Paltauf, J.A. Viator, S.A. Prahl, S.L. Jacques (2002). Iterative reconstruction algorithm for optoacoustic imaging, *J. Acoust. Soc. Am.* **112**(4), 1536–1544.
9. P. Liu (1997). The P-transform and photoacoustic image reconstruction. *Phys. Med. Biol.* **43**, 667–674.
10. R.S. Dingus, R.J. Scammon (1991). Grueneisen-stress induced ablation of biological tissue. *Proc. of SPIE* (Volume 1427, pp. 45–54). SPIE, Bellingham, WA.
11. J.A. Viator (2002). *Photoacoustic Imaging*, PhD dissertation, Oregon Health & Science University, Portland, Oregon.
12. S.L. Jacques, P.E. Andersen, S.G. Hanson, L.R. Lindvold (1998). Non-contact detection of laser-induced acoustic waves from buried absorbing objects using a dual-beam common-path interferometer. *Proc. of SPIE* (Volume 3254, pp. 307–318). SPIE, Bellingham, WA.
13. P.C. Beard, T.N. Mills (1997). Extrinsic optical-fiber ultrasound sensor using a thin polymer film as a low-finesse Fabry–Perot interferometer. *Appl. Opt.* **35**, 663–675.
14. G. Paltauf, H. Schmidt-Kloiber (1997). Measurement of laser-induced acoustic waves with a calibrated optical transducer. *J. Appl. Phys.* **82**, 1525–1531.
15. J.A. Viator, G. Au, G. Paltauf, S.L. Jacques, H. Ren, Z-P Chen, J.S. Nelson (2002). Clinical testing of a phototacoustic probe for port wine stain depth determination. *Lasers Surg. Med.* **30**, 141–148

Chapter 19

Polarized light imaging of tissues

Steven L. Jacques and Jessica C. Ramella-Roman

Table of contents

Abstract

Polarized light imaging offers images whose contrast is based on the scattering of light by the ultrastructure of a tissue, i.e., the nuclei, mitochondria, membranes (endoplasmic reticulum, other organelles), and fibers such as collagen or actin-myosin. Polarized light imaging is implemented in a complete form using 16 images involving four types of polarized light for both illumination and detection, which are summarized using the 16-element Mueller matrix to characterize each pixel of an image. Alternatively, a simplified form of imaging acquires only two images, using linearly polarized light, that are algebraically combined to create an image that is based only on photons scattered from the superficial tissue layers and rejects multiply scattered photons from deeper tissue layers. Hence, the simple polarization imaging is useful in surveying superficial tissues such as skin to find the margins of skin pathology.

19.1 Introduction

What is polarized light? Why should one consider using polarized light to image tissues? In this introduction, the motivation for using polarized light in imaging is outlined and the nature of polarized light is described.

19.1.1 Polarized light scattering as a contrast mechanism

Optical scattering of polarized light offers a mechanism of image contrast. Previous experience with scattered light has considered the multiply scattered light from a bulk thickness of tissue, i.e., the light reflected from an in vivo tissue site. The randomization of photon trajectories, coherence, and polarization properties due to multiple scattering has yielded the impression that light scattering is a relatively uninteresting contrast mechanism. But photons that have scattered only once or twice retain significant information about the tissue structures that scatter light. Significant information remains in the trajectories of scattered photons and in their wavelength-dependent polarization and coherence properties. Such singly or doubly scattered photons offer a contrast mechanism that has not been fully utilized or investigated. Imaging modalities that depend on photons that are only singly or doubly scattered include polarized light imaging, optical coherence tomography, and reflectance-mode confocal microscopy.

The reason that sunsets appear red and the overhead sky appears blue is understood to depend on the angular and wavelength dependence of photon scattering by very small particles in the atmostphere. Scattering as a contrast mechanism can provide a signature or fingerprint that characterizes the ultrastructure of a tissue or cell. The size, density and shape of the nucleus will influence the wavelength-dependent and angle-dependent scatter of photons, allowing a means to detect early dysplasia or to map local spread of cancer. The onset of apoptosis (programmed cell death) is marked by mitochondrial changes which can be detected as changes in photon scattering. The size distribution of collagen fiber bundles affects photon scattering. These structures are in the 1 to 10 μm diameter range, which is usually larger than near-ultraviolet, visible and near-infrared wavelengths (0.3–1.3 μm). Photon scattering by such large structures is often called Mie scattering and scattering from a size distribution of large structures behaves roughly as λ^{-b} for $b \approx 0.5$–1.5 and where λ is the wavelength of light. Membranes of the endoplasmic reticulum and other organelles (~100 nm), and the ~70 nm molecular periodicity of collagen fibrils are small compared with typical wavelengths and contribute to the so-called Rayleigh-limit scattering that spectrally behaves as λ^{-4}. These wavelength dependencies of scattering for different size particles have correlated angular dependencies of scattering. Photon scattering offers a signature or fingerprint that characterizes the ultrastructure within each pixel of an image. A great advantage of scattering as a contrast mechanism is that contrast is obtained without introducing exogenous reagents into the body.

Polarized light imaging adds to the information contained in singly and doubly scattered photons by tracking the polarization state of the photons. This chapter discusses this contrast mechanism.

19.1.2 Polarized light scattering as a gating mechanism

Optical scattering of polarized light offers a mechanism of gating or selecting the photons accepted for creation of an image. When polarized light illuminates a tissue, the photons scattered back out of the tissue by one or two single scattering events retain their polarization state to some degree. However, photons which undergo multiple scatterings before escape are randomized with respect to their polarization state, and the ensemble of escaping multiply scattered photons behaves as unpolarized light. Imaging with polarized light offers an opportunity to select for just the photons that have scattered only once or twice from the superficial tissue layers and to reject the photons that have penetrated deeply and scattered multiple times. This gating mechanism is discussed in this chapter.

19.1.3 Historical note

The use of polarized light in imaging has long been recognized. Anderson [1] described the practice in dermatology of viewing skin illuminated with linearly polarized light using polarized glasses. The illumination light is oriented in one direction and the doctor observes the skin through polarized glasses with linear polarization filters that can be aligned either parallel or perpendicular to the orientation of the illumination light. When observing with "parallel" glasses, the doctor sees the glare off the skin surface and surface details are enhanced. When observing with "perpendicular" glasses, the surface glare is blocked and the doctor sees a portion of the multiply scattered light backscattered from the deeper skin tissue layers. Jacques et al. [2,3] have modified this scheme to enhance what might be called the "subsurface glare" while rejecting the surface glare, and this technique is described here as POL imaging. Backman et al. [4] have used the wavelength dependence of scattered polarized light to measure the nuclear size distribution of mucosal tissue and cells in an effort to map the onset of neoplasia.

The broader approach towards polarized light imaging uses both linearly and circularly polarized light to fully characterize how a tissue modifies the polarization of incident light when the tissue scatters light. While polarized light has been used in various other fields of optics, and has been commonly used in microscopy of thin tissue sections, the behavior of polarized light when backscattered from a bulk thickness in vivo tissue is still not fully described. This method is being investigated by several laboratories around the world and is still a relatively young field.

The interested reader is referred to a recent special issue of the *Journal of Biomedical Optics* on "Tissue Polarimetry" [5] which collects a set of articles on imaging and measurements with polarized light. These articles provide in depth bibliographies on the field and are a good starting point for further study.

19.2 Basic principles and theoretical background

19.2.1 What is polarized light?

Light consists of photons that are a localized electromagnetic wave that is observed to propagate like a particle in a straight-line trajectory when measured experimentally. The electric field E of an electromagnetic wave can be described as the vector sum of two electrical field components, called $E_{\parallel}$ and $E_{\perp}$, that are perpendicular to each other. The relative magnitude and phase of these $E_{\parallel}$ and $E_{\perp}$ electric fields specify the polarization status of the photon, as described in Figure 1 in which $E_{\parallel}$ is aligned with the x axis and $E_{\perp}$ is aligned with the y axis. The x,y and z axes of Figure 1 are consistent with the right-hand rule specifying that propagation is along the positive z axis. How the total electric field E is divided into $E_{\parallel}$ and $E_{\perp}$ depends on the frame of reference chosen to define x and y. In other words, the description of the polarization state as $E_{\parallel}$ and $E_{\perp}$ depends on the frame of reference, but the total E is the real field and is uniquely defined regardless of the frame of reference.

While the magnitude and phase of $E_{\parallel}$ and $E_{\perp}$ can take on any values yielding a continuum of possible states, there are six types of polarized light that are commonly used to experimentally characterize the polarization state of light (Figure 2). Let $E_{\parallel}$ be aligned parallel to the x-axis in the horizontal surface (x–z plane) of our experimental table, and $E_{\perp}$ be aligned perpendicular to the surface of our experimental table. Then the six types of polarized light are:

1. H: The vertical wave ($E_{\perp}$) component has zero magnitude, and the total wave is a horizontal linearly polarized photon (called H) ($\phi = 0°$).
2. V: The horizontal wave component ($E_{\parallel}$) has zero magnitude, and the total wave is a vertical linearly polarized photon (called V) ($\phi = 90°$).

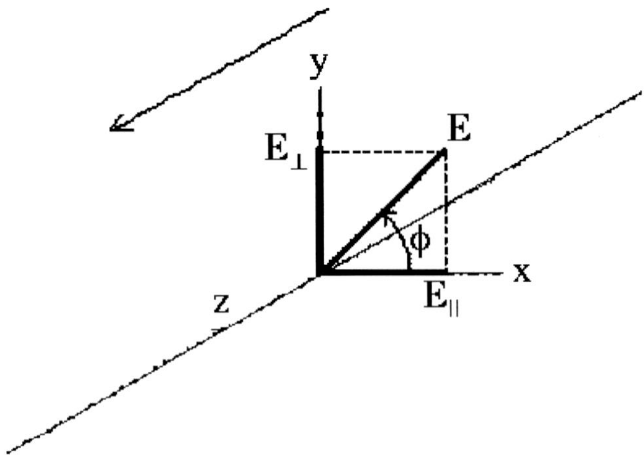

Figure 1. $E_{||}$ and $E_\perp$ components of the total electric field E are shown aligned with the x and y axes, respectively. The angle ϕ indicates the orientation of the E field at a moment in time, viewed as the wave approaches the observer. The wave is propagating along the z-axis towards the lower left, towards the observer.

3. P^+: The two components $E_{||}$ and $E_\perp$ are aligned in phase and equal in magnitude, and the sum of the two waves is a $+45°$ linearly polarized photon (called P^+) ($\phi = 45°$).
4. P^-: The two waves are 180° out of phase but equal in magnitude, and the sum of the two waves is a photon that is a $-45°$ linearly polarized photon (called P) ($\phi = -45°$).
5. R: The two wave components are equal in magnitude but $E_\perp$ leads $E_{||}$ by 90° in phase, and the sum of the two waves is a right circularly polarized photon (called R) (ϕ rotates counterclockwise as the photon approaches the observer).
6. L: The two wave components are equal in magnitude but $E_\perp$ lags $E_{||}$ by 90° in phase, and the sum of the two waves is a left circularly polarized photon (called L) (ϕ rotates clockwise as the photon approaches the observer).

Later in this chapter, experimental measurements of the intensity associated with these types of polarized light will be used to characterize the light delivered to and scattered from tissues.

A simple example of polarized light is the common experience of polarized sunglasses that use a linear polarization filter to reduce glare (Figure 3). When sunlight reflects off a smooth surface such as body of water or a road, photons that are linearly polarized parallel to the surface will strongly reflect while photons linearly polarized perpendicular to the surface will less strongly reflect. If one views the glare off a horizontal smooth road surface through polarized sunglasses that incorporate a vertically oriented linear polarization filter, the sunglasses will block the dominant glare due to photons vibrating parallel to the road surface.

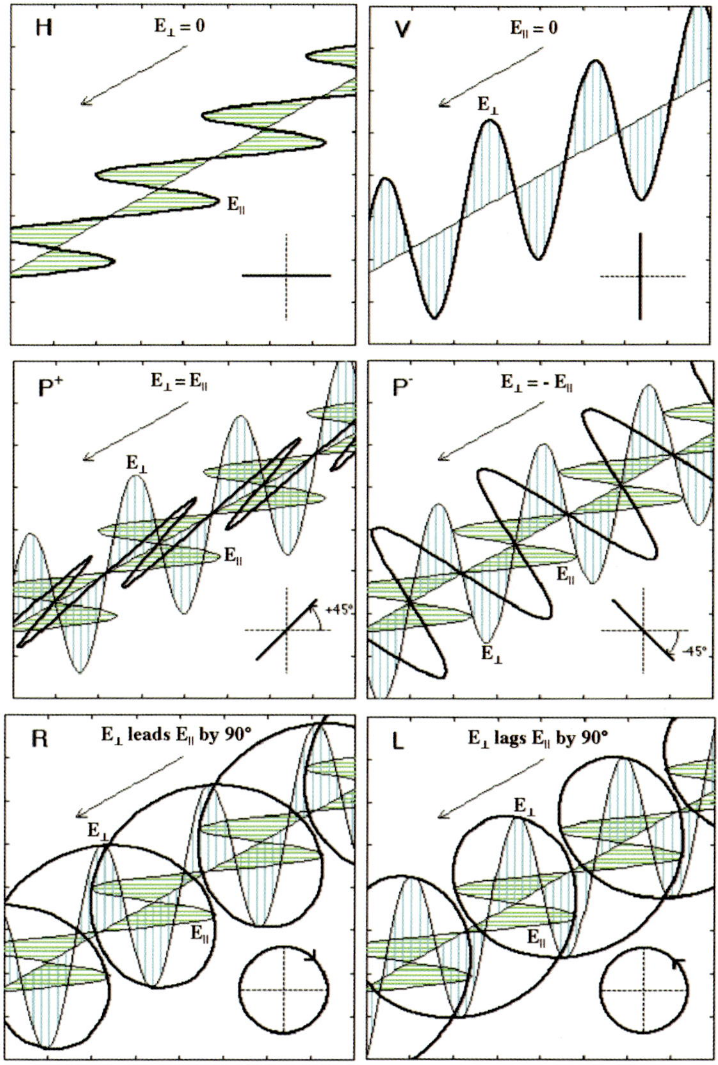

Figure 2. Electric fields for linear H, V, P^+ and P^- polarized light and for circular R and L polarized light. (a–c) show the $E_{\|}$ and $E_{\perp}$ components of the electric field as the wave propagates. The projection of $E_{\|}$ and $E_{\perp}$ on the x and y axes of Figure 1 are inserted in the lower right of each part (a–c) to illustrate how the total E field behaves as the wave approaches the observer, as viewed by the observer.

19.2.2 Scattering plane

When light is scattered by a particle, the frame of reference for subdividing E into $E_{\|}$ and $E_{\perp}$ is specified by the geometrical relation between the light source, the scattering particle, and the observer of scattered light. This light-source/scattering-particle/observer triangle is called the *scattering plane*. We follow the convention

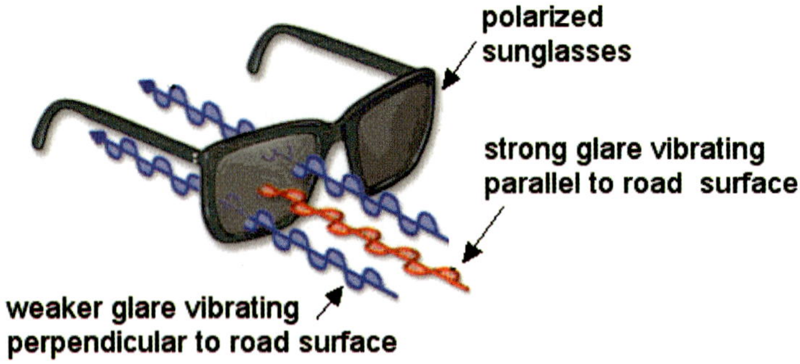

Figure 3. The strongest glare off a road surface is due to photons whose electric field is vibrating parallel to the road surface. Polarized sunglasses incorporate vertically oriented linear polarization filters that accept only photons whose electric field is vibrating perpendicular to the road surface and block photons vibrating parallel to the road surface. Hence, polarized sunglasses reduce glare off the road surface.

of van de Hulst [6] and Bohren and Huffman [7] who define $E_{\|}$ as parallel to the scattering plane and $E_{\perp}$ as perpendicular to the scattering plane.

For example, in light scattering experiments we commonly define the scattering plane as being parallel to the surface of our experimental table (Figure 4(A)). The $E_{\|}$ is aligned with the x-axis parallel to the table surface and $E_{\perp}$ is aligned with the y-axis perpendicular to the table surface. In this chapter, we use the term "horizontal" to refer to a plane horizontal to our experimental table, but more generally refer to a plane parallel to the scattering plane however it may be oriented.

Our second example is the polarized sunglasses to reduce road surface glare (Figure 4(B)). The scattering plane is defined as the sun/road/glasses triangle and $E_{\|}$ is parallel to this plane. The orientation of this scattering plane is perpendicular to the road surface, which is certainly not horizontal in the common sense. This example illustrates that the source/scatterer/observer triangle defines the orientation of $E_{\|}$ and $E_{\perp}$, not the observer's idea of "horizontal".

19.2.3 Mueller matrix

Experimentally it is easier to measure scattered intensities such as H, V, P^+, P^-, R and L, than to directly measure the phase and magnitude of scattered electric fields. The Stoke's vector is a description of the polarization state of light based on these measurable intensities. The Mueller matrix is a description of how propagation through a medium changes the Stoke's vector.

The polarization state of light can be characterized by the balances between the measurable intensities of these 6 types of light, H vs. V, P^+ vs. P^-, and R vs. L, as well as the total intensity I of the light. The convention for describing the polarization state is a Stoke's vector:

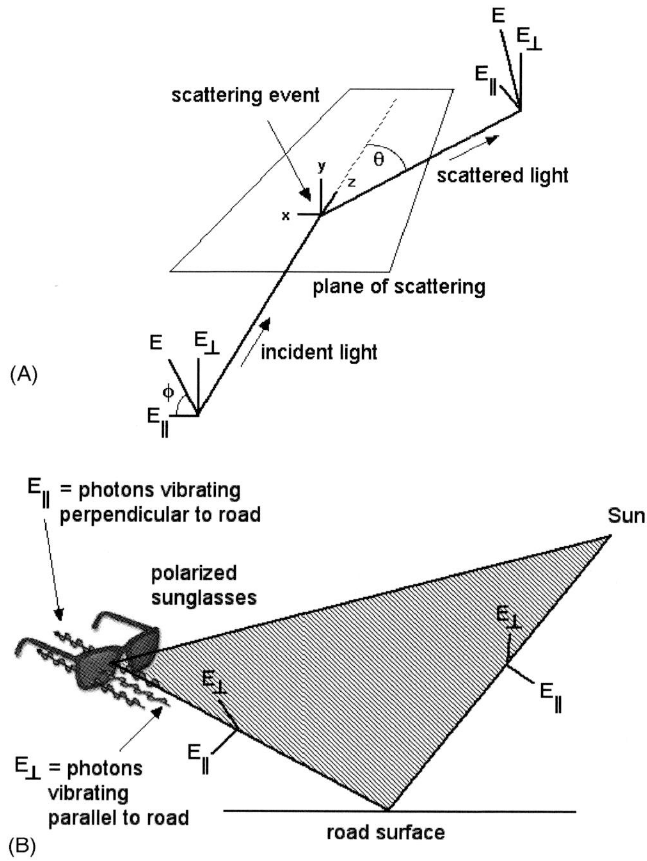

Figure 4. (A) Plane of scattering. The electric field E is divided into the two components $E_{\|}$ and $E_{\perp}$, such that $E_{\|}$ is parallel to the scattering plane. The deflection angle is θ. (B) Polarized sunglasses blocks photons whose E field is vibrating parallel to the road surface, i.e., perpendicular to the scattering plane defined by the sun/road/glasses triangle. Wearing polarized sunglasses, your eye is a V detector (see Fig. 5).

$$\begin{vmatrix} I \\ Q \\ U \\ V \end{vmatrix} = \begin{vmatrix} \text{Total intensity} \\ H - V \\ P^{+} - P^{-} \\ R - L \end{vmatrix} \tag{1}$$

If incident light is conditioned by suitable optics to yield an H, V, P^+, or R state of polarization, and the scattered light is conditioned by suitable optics to select the H, V, P^+, or R state of polarization before reaching a detector, then a 4×4 matrix of intensity transport, here called a Data matrix, is specified which relates the incident and scattered states of polarization (Figure 5). For example, an incident H polarization and detected V polarization is denoted HV.

incident \ detected	H	V	P	R
H	HH	HV	HP	HR
V	VH	VV	VP	VR
P	PH	PV	PP	PR
R	RH	RV	RP	RR

Figure 5. Data matrix for experimental measurements of polarized light using four incident states and four detected states. In this notation, *HV* is the intensity of detected V light in response to incident H light.

The Data matrix is converted into a second 4 × 4 matrix of transport, called a Mueller matrix, that relates the incident (input) and scattered (output) Stoke's vectors (Equation 2).

$$\text{Stokes Vector}_{\text{output}} = \text{Mueller Matrix} \times \text{Stokes Vector}_{\text{intput}}$$

$$\begin{vmatrix} I \\ Q \\ U \\ V \end{vmatrix}_{\text{output}} = \begin{vmatrix} M_{11} & M_{12} & M_{13} & M_{14} \\ M_{21} & M_{22} & M_{23} & M_{24} \\ M_{31} & M_{32} & M_{33} & M_{34} \\ M_{41} & M_{42} & M_{43} & M_{44} \end{vmatrix} \times \begin{vmatrix} I \\ Q \\ U \\ V \end{vmatrix}_{\text{input}} \qquad (2)$$

The conversion of the Data matrix into the Mueller matrix was specified by Yao and Wang [8] and is summarized in Equation (3).

$$\begin{aligned} M_{11} &= HH + HV + VH + VV \\ M_{12} &= HH + HV - VH - VV \\ M_{13} &= 2PH + 2PV - M_{11} \\ M_{14} &= 2RH + 2RV - M_{11} \end{aligned} \qquad (3)$$

$$\begin{aligned} M_{21} &= HH - HV + VH - VV \\ M_{22} &= HH - HV - VH + VV \\ M_{23} &= 2PH - 2PV - M_{21} \\ M_{24} &= 2RH - 2RV - M_{21} \end{aligned}$$

$$\begin{aligned} M_{31} &= 2HP + 2VP - M_{11} \\ M_{32} &= 2HP - 2VP - M_{12} \\ M_{33} &= 4PP - 2PH - 2PV - M_{31} \\ M_{34} &= 4RP - 2RH - 2RV - M_{31} \end{aligned}$$

$$
\begin{aligned}
M_{41} &= 2HR + 2VR - M_{11}\\
M_{42} &= 2HR - 2VR - M_{12}\\
M_{43} &= 4PR - 2PH - 2PV - M_{41}\\
M_{44} &= 4RR - 2RH - 2RV - M_{41}
\end{aligned}
$$

In summary, practical optical measurements of polarized light transmission or reflectance as intensities *H*, *V*, *P* and *R* in response to incident H, V, P and R light yields a Data matrix that characterizes the input/out transfer function. The Data matrix is converted into a Mueller matrix that characterizes the in/out transfer function of the tissue.

19.3 Experimental apparatus

19.3.1 Mueller matrix experiment

A Mueller matrix experiment involves a polarization assembly that conditions the source light incident on a tissue sample so as to generate either H, P^+, V, or R light. Each of these four types of light are sequentially delivered to the sample. Light that is transmitted or reflected by the sample is collected by a detector assembly. The detector assembly is a detector behind a second polarization assembly that conditions the light collected so as to detect only H, P^+, V, or R light. For each of the four types of light source, the four types of detected light are measured. Hence, a set of 16 measurements is acquired. The polarization assemblies are composed of optical components called a linear polarization filter (LP) and a quarter-wave retarder (QW). The QW is specific for a particular wavelength of light, so these experiments usually use either a laser or filtered light, yielding only a narrow band of wavelengths.

The three types of linearly polarized light, H, P^+, V, are obtained by passing unpolarized light through a LP that is oriented at an angle $\phi = 0°$, 45° or 90°, respectively, relative to the *x–z* horizontal plane of the experimental table (using the axes in Figure 1). This light is delivered to the sample. The detection of H, P^+ and V light scattered by the sample is achieved by collecting light through a LP oriented at $\phi = 0°$, 45° or 90°, respectively.

The right circularly polarized light, R, is obtained by passing unpolarized light through a LP oriented at an angle $\phi = 45°$ to yield equal magnitudes of $E_{\|}$ and $E_{\perp}$, which constitutes P^+ light. The light is then passed through a QW whose fast axis is oriented parallel to the table (horizontal) so that $E_{\|}$ transmits faster through the plate with less phase delay than $E_{\perp}$, such that after transmission the phase of $E_{\perp}$ leads $E_{\|}$, by 90°, which constitutes R light. This light is delivered to the sample. The detection of R light scattered by the sample is achieved by collecting through a QW retarder plate whose fast axis is oriented perpendicular to the table so as to allow $E_{\perp}$ to catch up with $E_{\|}$ and match phase, which constitutes P^+ light, then passing through a LP oriented at $\phi = 45°$ to select only P^+ light.

The result is a set of 16 measurements constituting a Data matrix (Fig. 5) which is converted into a Mueller matrix (using Equation 3). The Mueller matrix characterizes how the tissue transforms the polarization of any type of incident light. In other words, any incident light described by a Stoke's vector $S_{in} = [IQUV]_{in}$ is converted by the tissue described by a Mueller matrix into the collected light described by the Stoke's vector $S_{out} = [IQUV]_{out}$.

Such Mueller matrix experiments can be done with a narrow beam and a single detector, yielding a single Mueller matrix for one spot on a tissue. Alternatively, the experiment can be done with a broad illumination and a CCD camera as the detector, yielding a 2D array of pixels with each pixel being characterized by a 16-element Mueller matrix. A set of 16 images can then be displayed with each based on one of the 16 elements of the Mueller matrix. Combinations of these elements can also be used to generate images.

19.3.2 Simple polarization imaging

A simplified version of the Mueller matrix experiment is to use only the HH and HV elements of the Data matrix. A single H light source is used and two images are acquired, an HH image acquired through an H linear polarization filter and a HV image acquired through a V linear polarization filter. Figure 6 shows the experimental setup for such a simple polarization imaging system.

The HH image accepts light that is still polarized as H light. The glare (specular reflectance) off the tissue surface is also H light, so the illumination light is delivered at an angle and a plate of glass is placed onto the skin, coupled by a drop

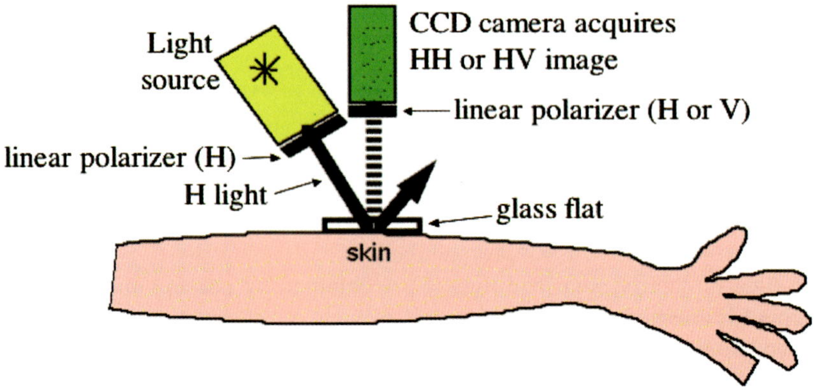

Figure 6. Experimental apparatus for simple POL imaging. A white light source passes through a linear polarizer oriented parallel to the source/skin/camera scattering plane (H light). A glass plate is optically coupled to the skin by a drop of water and the surface glare from the oblique illumination does not enter the camera. The light scattered from the superficial subsurface layers of the skin is collected by the camera after passing through a second linear polarizer oriented either parallel (H) or perpendicular (V) to the scattering plane.

of water or other index matching medium, to enforce a flat surface that deflects surface glare away from the camera. The camera views the tissue from above and collects the reflected H photons scattered from the superficial but subsurface tissue layers (here called S photons). The HH image also collects 1/2 of the multiply scattered light that is reflected from the deeper tissue surfaces (here called D photons). Therefore, HH = S + D/2.

The HV image rejects any H photons scattered from the superficial layers and accepts only 1/2 of the multiply scattered photons from the deeper tissues (i.e. 1/2 of the same D photons acquired by the HH image). Therefore, HV = D/2.

The difference image HH − HV equals (S + D/2) − D/2 = S.

The HH − HV image will cancel out the common multiply scattered light that contaminates both the HH and HV images, yielding only the superficially scattered photons S.

The polarization ratio POL is calculated:

$$POL = \frac{\mathrm{HH} - \mathrm{HV}}{\mathrm{HH} + \mathrm{HV}} \tag{4}$$

The POL ratio has additional advantages. The illumination light I_0 may not be uniform across the field of view. Also, there may be variable melanin pigmentation on the surface of the skin which acts as an attenuation filter T_{mel} describing the in/out transmission of light across the skin surface. The POL ratio causes these common factors I_0 and T_{mel} to cancel, yielding the ratio S/total where total denotes the total reflected light:

$$\begin{aligned} \mathrm{POL} &= \frac{\mathrm{HH} - \mathrm{HV}}{\mathrm{HH} + \mathrm{HV}} = \frac{I_o T_{\mathrm{mel}}\left(\mathrm{S} + \frac{1}{2}\mathrm{D}\right) - I_o T_{\mathrm{mel}}\left(\frac{1}{2}\mathrm{D}\right)}{I_o T_{\mathrm{mel}}\left(\mathrm{S} + \frac{1}{2}\mathrm{D}\right) + I_o T_{\mathrm{mel}}\left(\frac{1}{2}\mathrm{D}\right)} \\ &= \frac{\left(\mathrm{S} + \frac{1}{2}\mathrm{D}\right) - \left(\frac{1}{2}\mathrm{D}\right)}{\left(\mathrm{S} + \frac{1}{2}\mathrm{D}\right) + \left(\frac{1}{2}D\right)} = \frac{\mathrm{S}}{\mathrm{S} + \mathrm{D}} = \frac{\mathrm{S}}{\mathrm{total}} \end{aligned} \tag{5}$$

If a laser is used for illumination, there will be speckle due to interference of scattered photons. Such interference presents a random speckled appearance, and is usually stronger than the small S signal. Therefore, it is advisable to use an incoherent light source such as a filtered white light source or a light emitting diode.

Although this chapter discusses the use of HH and VV images for POL imaging, similiar imaging can also be accomplished by using RR and RL images based on right and left circularly polarized light (R and L, respectively). There are differences in how circularly polarized light propagates in tissues relative to linearly polarized light. The relative advantages and disadvantages of POL imaging based on HH and HV versus RR and RL are still under study.

In summary, two images, HH and HV, are acquired and used to calculate a new image POL which cancels variation in I_0 and T_{mel}. The POL image equals the superficially scattered light S normalized by the total reflected light S + D.

19.4 Current results

To illustrate simple polarization imaging, images of a freckle and a pigmented nevus were taken (Figure 7). The HH and HV images are nearly identical. The HV image is compared with the POL image for each case.

The HV image of the freckle clearly shows the melanin pigment of the freckle. The POL image of the freckle has canceled the T_{mel} factor thereby removing the freckle from the image.

The HV image of the nevus shows the strong melanin absorption of the nevus. The POL image shows a bright white nevus where white indicates a high POL value in the range of 0.1–0.3 relative to the 0.5–0.15 value of the surrounding skin

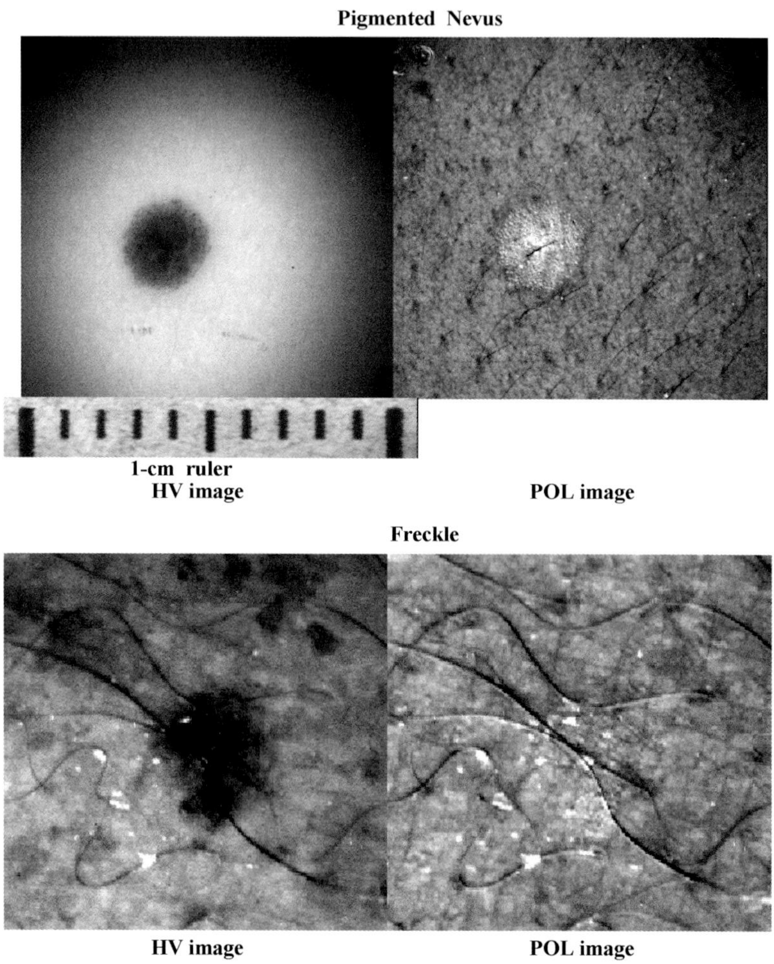

Figure 7. Normal light images (left) and POL images (right) for pigmented nevus. Note how the POL image makes the freckle disappear. (Reprinted from Ref. 3 with permission from *J. Biomed. Opts.*)

tissue. Notice how the region around each hair follicle in the POL image is darker, indicating where the epidermis has folded down into the hair follicle. The backscatter of H photons by this epidermal infolding is less than the backscatter of H photons by the dermis underlying the epidermis.

While the HV images suffer from the non-uniform illumination I_0, the POL images have corrected for variations in I_0 and are uniform in response except where there is insufficient light.

A second example (Figure 8) is a burn scar on the forearm of a young man, which was acquired by scalding with hot cooking oil as a youth. The skin site is pressed up against a glass plate with water coupling the skin to the glass. The edge of the water is visible in the images. The HH and HV images are similar, with the HH image showing the extra H photons scattered by the superficial tissue layers as well as the glare off the skin surface where the glass is not coupled to the skin by water. The burn scar presents a white appearance relative to the darker normal skin.

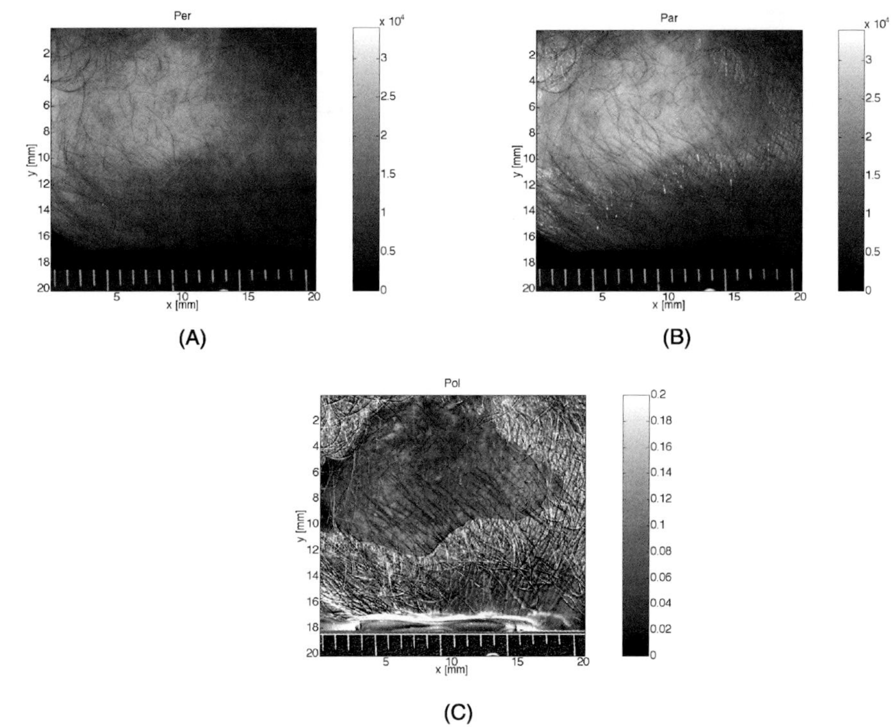

Figure 8. Images of a burn scar on the arm. (A) HV image called "PER" in units of CCD camera pixel values (counts). (B) HH image called "PAR" (counts). (C) POL image in units of polarization ratio (dimensionless). Note the water coupling of skin to the glass plate. When skin is not coupled to the glass by water, the specular reflectance from the air–skin surface overwhelms the tissue contribution to the POL image. The PAR and PER images show the scar as white relative to normal skin, while the POL image shows the scar as dark relative to normal skin. (Reprinted from Ref. 3 with permission from *J. Biomed. Opts.*).

In the POL image the burn scar presents a darker appearance relative to the normal skin. The burn scar is not reflecting H photons as efficiently as the normal skin. The POL image shows the variation in scattered H photons within the burn scar, probably indicating the variation in the depth of the original burn, giving rise to a variation in the size of collagen fiber bundles in the wound healing response.

The ability of linearly polarized light to penetrate the birefringent collagen fibers of the dermis is limited. Birefringent fibers present a long axis and a short axis, which allow different velocities of light. Hence, the horizontal and vertical components of linearly polarized light become dephased by the differences in velocities. This scrambles the linearly polarized light. We find that the POL images show roughly the upper 300 μm of skin, the upper 500 μm of muscle, and about 1200 μm of liver, and this depth of response is related to the strength of birefringence fibers in these tissues. The depth of POL imaging is not so dependent on the wavelength of light because the attenuation of imaging is dominated by the birefringence of the tissue.

19.5 Perspectives

The field of polarized light imaging is currently exploring the relationship between Mueller matrix images and the ultrastructure of a tissue. The relationship is likely to involve significant computational analysis of the 16 matrix elements for each pixel at each wavelength and angle used to acquire images. The goal is to characterize the size distribution and refractive index mismatch distribution of the tissue ultrastructure, in other words the size and density of nuclei, mitochondria, membranes, collagen fibers, and perhaps other structures. This work is still in its early stages of development in several laboratories around the world.

The simplified POL imaging seeks to create an image based on photons scattered from the superficial tissue layers, rejecting multiply scattered photons from deeper tissues. The image contrast is concentrated in the superficial tissue layers of interest. There is minimal computation involved in the POL image, leaving image analysis to the eye of the doctor. For example, the POL image of skin presents the complex fabric of skin structure, largely the structure of the papillary dermis and upper reticular dermis. Any disruption in this structure stands out like a flaw in a fabric easily detected by the eye of an observer. Hence, rapid real-time POL imaging is a practical form of imaging for surveying large tracts of tissue to find pathology. We are using POL imaging to find the margins of skin cancer in the Mohs surgery suite in order to guide surgical excision of the cancer.

19.6 Conclusion

Polarized light imaging can be implemented in a complete form using the Mueller matrix formalism to characterize each pixel of an image. Alternatively, the HH and HV elements of the Data matrix can be used to create a POL image that accepts only photons scattered from the superficial tissue layers.

Acknowledgements

We wish to thank Scott Prahl for many valuable discussions. This work was supported by an NIH grant (R01-CA80985).

References

1. R.R. Anderson (1991). Polarized light examination and photography of the skin. *Arch. Dermatol.* **127**, 1000–1005.
2. S.L. Jacques, J.C. Roman, K. Lee (2000). Imaging superficial tissues with polarized light. *Lasers Surg. Med.* **26**, 119–129
3. S.L. Jacques, J.C. Ramella-Roman, K. Lee (2002). Imaging skin pathology with polarized light. *J Biomed Optics* **7**, 329–340
4. V. Backman, R. Gurjar, K. Badizadegan, R. Dasari, I. Itzkan, L.T. Perelman, M.S. Feld (1999). Polarized light scattering spectroscopy for quantitative measurement of epithelial cellular structures *in situ, IEEE JSTQE* **5**, 1019–1027.
5. L.V. Wang, G.L. Coté, S.L. Jacques (Ed) (2002). Special Issue on Tissue Polarimetry, *J. Biomed. Opt.* **7**(3).
6. H.C. van de Hulst (1957). *Light Scattering by Small Particles*. Dover Publications Inc, New York.
7. C.F. Bohren, D.R. Huffman (1983). Absorption and Scattering of Light by Small Particles. John Wiley and Sons, New York.
8. G. Yao, L.V. Wang (1999). Two-dimensional depth-resolved Mueller matrix characterization of biological tissue by optical coherence tomography. *Opt. Lett.* **24**, 537–539.

Chapter 20

Ultrasensitive fluorescence detection at surfaces: instrument development, surface chemistry and applications in life science and medicine

Stefan Seeger

Table of contents

Abstract

Ultrasensitive fluorescence detection at surfaces has an impact on biology and medicine. To meet the demands of the most important applications new developments in optical design, surface preparation and chemical methods are necessary. Recent results and the potential of ultrasensitive fluorescence detection at surfaces in these fields are presented: fast single molecule counting to quantify molecular recognition events in immunology and molecular biology, a novel optical component based on paraboloid geometry for extreme reduction of the detection volume, application in single molecule sequencing, multiplex detection by time-resolved fluorescence spectroscopy for detection of cellular surface structures and single cell analysis during minimal invasive neurosurgery.

20.1 Introduction

Fluorescence spectroscopy is one of the most important detection techniques in biodiagnostics. The detection of biological molecules after tagging with fluorescent dyes on the protein and DNA level is now a standard procedure [1,2]. Even structural information is obtained by fluorescence experiments, e.g. DNA sequencing is performed by the "Sanger-technique" based on base-specific labelling with fluorescent dyes. However, new challenges are still open for research like higher sensitivity, faster analytical procedures, cheaper instruments, higher information content, etc.

Fluorescence detection very close to surfaces is of particular interest for application in life science, because analytical assays working in solution are restricted in the limit of detection, although they overcome all the inherent surface problems, like unspecific binding, light scattering, variation of surface properties, etc. Further, ongoing efforts to miniaturize chemistry, biochemistry and analytical chemistry, e.g. the lab-on-a-chip approach, makes surfaces more and more important: the surface-to-volume ratio increases when the system size is reduced since the surface decreases with the power of 2, the volume with the power of 3. Obviously the surface properties have to be taken into account.

This chapter gives an overview about our recent work on ultrasensitive detection of fluorescence at surfaces, which includes the development of new assay principles, instruments, tailor-made surfaces and applications in life science, mainly in protein and DNA analysis and medicine, in particular in vitro and in vivo tumour diagnostics in combination with minimal invasive neurosurgery. Literature is cited in the reference list and in the publications listed.

20.2 Single molecule detection of molecular recognition events at surfaces

The detection of single molecules clearly accomplishes the ultimate sensitivity of biological assays. About 20 years ago, researchers started to detect single molecule fluorescence in solids and liquids in the far-field and near-field modes with simple single-photon counting devices [3–27]. In this context, we demonstrated the ability to detect molecular recognition reactions between antibodies and antigens after capturing at surfaces [28,29]. Capture molecules are fixed at transparent glass slides after coating them with ultraflat cellulose-derived layers. Subsequent addition of a solution containing the analyte which is, or has to be, tagged with a fluorescent dye results in a specific binding at the surface. The surface is then scanned with a single-molecule detector, i.e. every molecule is counted.

To detect biomolecules on the single molecule level all background signal sources have carefully to be taken into account. Beside the electronic noise, which is no longer a limiting aspect, scattered light and autofluorescence are important origins for background signal. To show not only the single molecule detection event but also to demonstrate that very low concentration limits can be achieved, it is necessary to prevent unspecific adsorption of fluorescently labelled molecules at the surface, because unspecific adsorbed fluorescent molecules generate a false positive signal, i.e. the background level is increased.

The scattered light problem can be solved by reduction of the detection volume as it is done by confocal optics which includes usually a microscope objective with high numerical aperture to achieve small laser focusing and efficient fluorescence detection [29]. Here, the introduction of a pinhole in the detection part of the optics reduces the detection volume down to the femtolitre range (Figure 1).

The second noise source, autofluorescence, appears if biomolecules which show fluorescence in the spectral range of the emission of the fluorescent label are present. To overcome this background light, we used diode lasers in the red wavelength region to detect immunoreaction and further on the single molecule level. Biomolecule-generated autofluorescence appears mainly from the near-UV to the red wavelength region until 600 nm. The use of fluorescent dyes like the

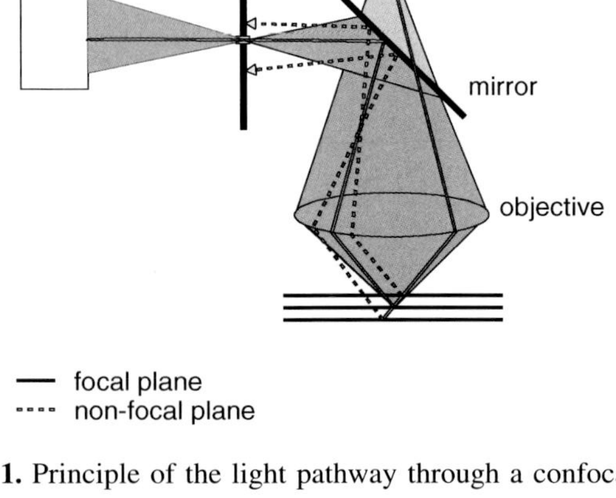

Figure 1. Principle of the light pathway through a confocal optics.

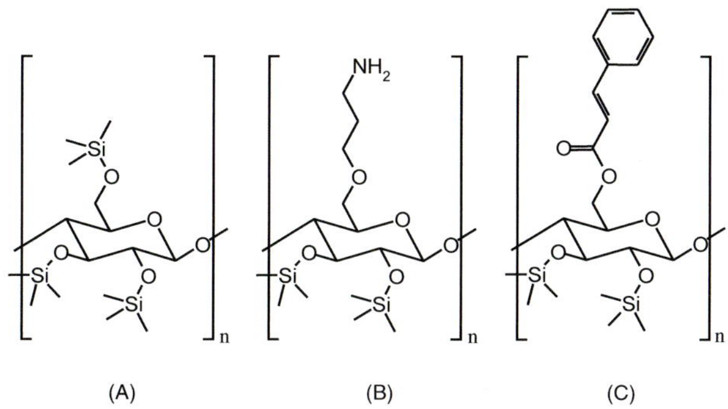

Figure 2. Examples of cellulose derivatives used for ultrathin coating.

cyanine Cy5 allows the excitation with diode lasers at 635 nm and the emission detection at about 650 nm. Here, autofluorescence can nearly not be observed.

The third major source of background signal results from non-specific binding of fluorescent-labelled molecules at the surface. Unfortunately, this adsorption process can be very efficient but strongly depends on the properties of the surface and the adsorbed molecule [30,31]. To prevent unspecific binding of fluorescent molecules which delivers false-positive signals we developed very well-defined surface coatings on a molecular scale which prevent unspecific binding nearly completely and at the same time allow covalent coupling of captured molecules, e.g. antibodies [33–39]. The principle of the technique is to coat a transparent surface with one or several monolayers of cellulose derivatives using the Langmuir–Blodgett technique [40]; some of the derivates are shown in Figure 2.

The molecules are adsorbed highly ordered at the surface (Figure 3).

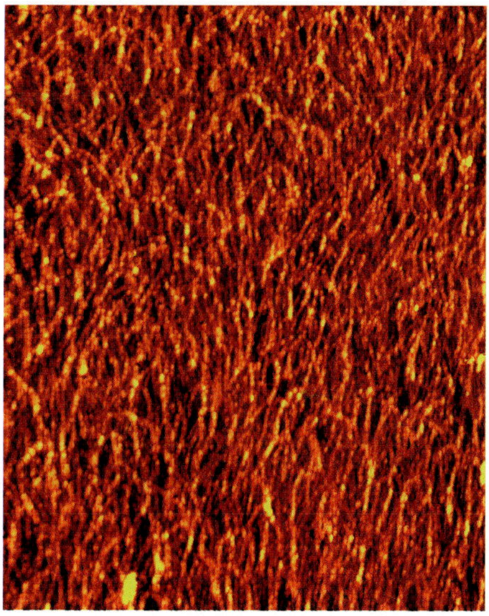

Figure 3. AFM image of a cellulose-derivative coated glass slide.

The functionalized cellulose coating can bind covalently biomolecules. Unspecific binding can almost be excluded.

Figure 4 shows data recorded from scanning the surface with a laser beam after antibody molecules have been attached and a Cy5-tagged molecule are specifically recognised and fixed by the antibody.

Figure 4(a) shows bursts after adding a solution of 10^{-9} mol l^{-1}. It can be seen that the binding is recognised and delivers a fluorescence signal (due to the scanning process the time scale can be translated into a line scale on the slide). The disappearance of the signal is due to photobleaching of the dye. The signals clearly disappear exponentially; this indicates that an ensemble of molecules was detected, indicating bleaching behaviour. Figure 4(b) was recorded after adding a

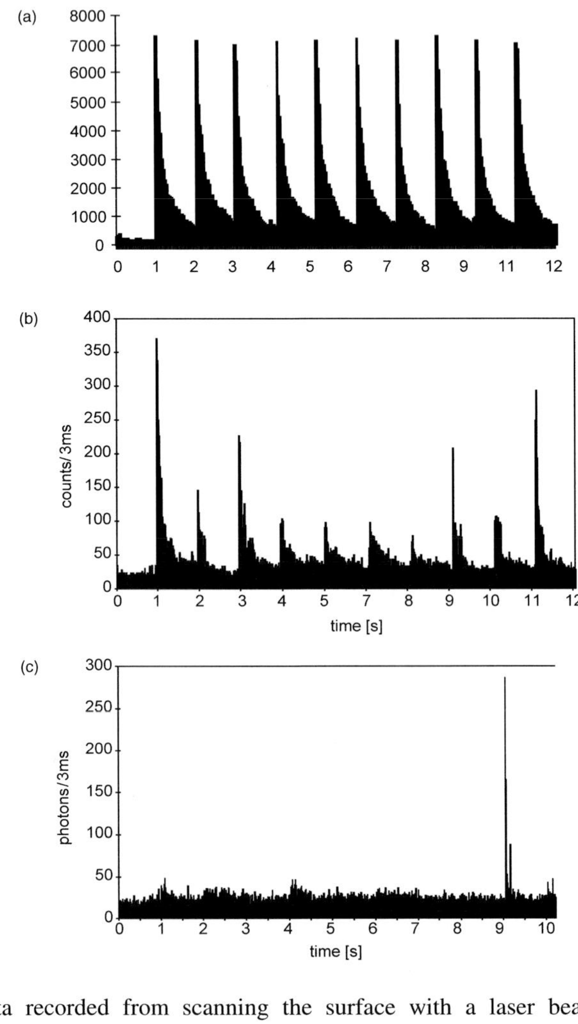

Figure 4. Data recorded from scanning the surface with a laser beam after antibody molecules have been attached. Bursts measured subsequent to addition of a solution of (a) 10^{-9}, (b) 10^{-12}, (c) 10^{-16} mol $^{-1}$.

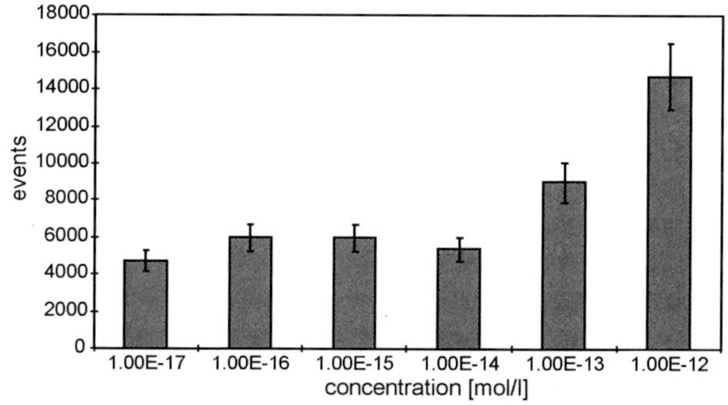

Figure 5. Behaviour of the burst size as a function of analyte concentration.

solution of 10^{-12} mol l^{-1}. The burst size decreases tremendously and the kinetic behaviour of the bleaching process indicates that a few molecules are present, i.e. it is not a clear exponential decay but also not a suddenly occurring darkness that must be expected from a single molecule. Hence, there are a few molecules in the focal spot. Finally, Figure 4(c) demonstrates the detection of a specific recognition event at a surface after adding a solution of 10^{-16} mol l^{-1}: a further decreased burst size and the sudden switching off the fluorescent light documents the detection of a single molecule. Further proof is the behaviour of the burst size as a function of analyte concentration (Figure 5).

Decreasing the concentration leads to a concentration range where the burst size does not further depend on the concentration, i.e. the worst signal-to-noise ratio is achieved: single molecules are observed.

Further developments in collaboration with Molecular Machines & Industries AG, Switzerland resulted in the first commercially available Single Molecule Detection System, which has been designed for surface generated fluorescence but is also applicable in solution (Figures 6 and 7).

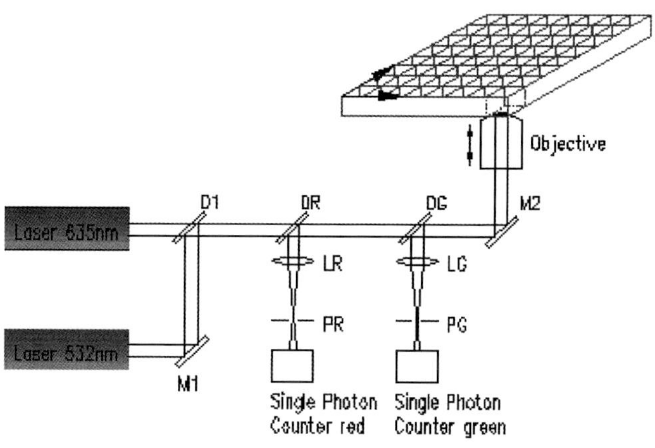

Figure 6. Optical set-up of the first commercially available surface-scanning single molecule detector.

Figure 7. Surface-scanning single molecule detection unit of MMI Switzerland.

This so-called LB8-Reader is a highly sensitive confocal two-color fluorescence detector for glass bottom microtiter plates. It can be employed to detect every kind of surface-based assay where fluorescence labels may be used. Under optimized conditions a limit of detection for specific identification in the attomolar range can be achieved. The system also can be used for fluorescence correlation spectroscopy (FCS) and kinetic measurements. The system supports 96, 384, and 1536 well-plates. It consists of two diode-lasers with different wavelength, a combination of filters, a microscope objective, a *x,y*-micropositioning table, a CCD-chip, an autofocus element, two single-photon avalanche detectors, an internal and external PC and application software.

Further, we developed coated microlitre plates in all three formats with a glass bottom of high precision and high optical quality. The coating of the glass bottom enables covalent binding of capture molecules and prevents unspecific binding (Figures 8 and 9).

Examples of the sensitivity are shown for protein assays (Figure 10) and a DNA-based assay (Figure 11).

20.3 Sizing of single DNA-fragments

Polymerase chain reaction (PCR) as well as DNA digestion by restriction enzymes produces DNA fragments of specific lengths. The determination of the size of the fragments is crucial for restriction mapping and DNA-fingerprinting. Electrophoresis is the most commonly used technique, but it is time consuming and difficult to automate. Therefore, numerous sizing techniques have been developed, including mass spectroscopy [41–43], flow cytometry [44–48], optical microscopy [49,50], capillary electrophoresis [51,52] and atomic force microscopy [53,54].

We developed an approach to size DNA-fragments by the detection of individual DNA molecules adsorbed on a coated glass slide [55]. The fluorescence signal of

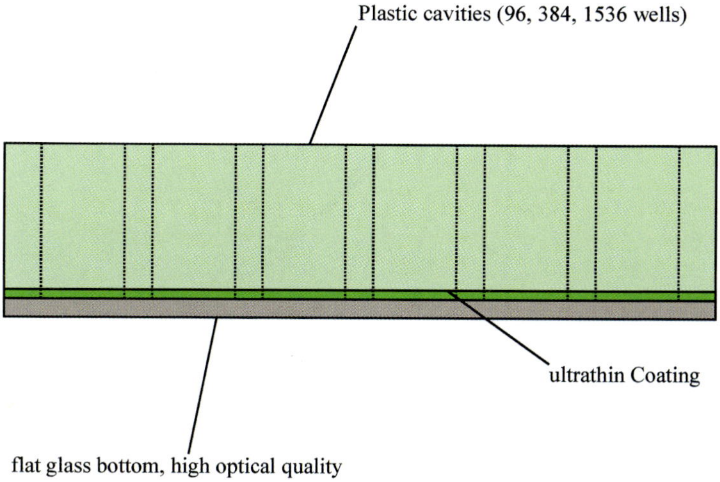

Figure 8. Coated microtiter plate with high precision ultraclear glass bottom.

surface-adsorbed TOTO-1 labelled fragments by a confocal scanning single molecule microscope has been detected and correlated with the size of DNA fragments. The amount of intercalated TOTO-1 dye molecules in single DNA fragments is proportional to their lengths [45]. This method consists of staining of DNA-fragments with TOTO-1, immobilization of the sample DNA to the surface and imaging of adsorbed DNA-fragments.

As an example, DNA-sizing of five different fragment lengths in a mixture has been performed. Further, by performing a restriction digest of m13mp18 RF1 DNA with the restriction enzymes EcoRI and PagI we demonstrate the higher speed of our technique in comparison to gel electrophoresis of the same sample. In addition to speed, DNA-sizing at the single molecule level naturally requires lower amounts of DNA compared with gel electrophoresis. Typically, electrophoresis requires 5 ng

Figure 9. Commercially available high-precision ultraclear glass bottom microliter plates (MMI, Switzerland).

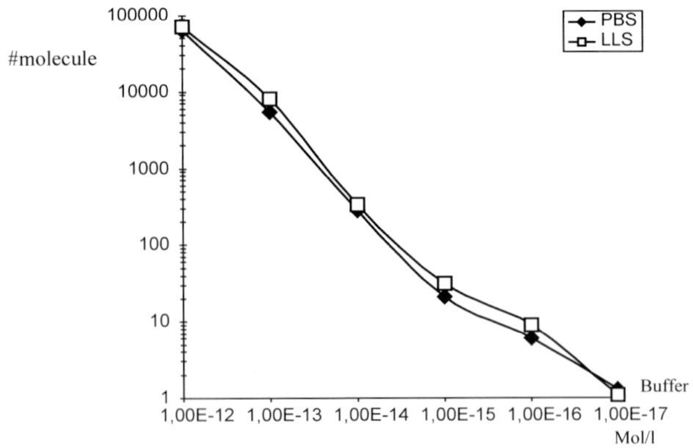

Figure 10. Example of the sensitivity of protein assays.

DNA per band, the DNA-sizing method described here has been carried out with just 1 pg per fragment length.

The adsorption of the DNA is performed by mainly electrostatic interaction between a positively charged surface coating (poly-L-lysine) and the negatively charged backbone of the DNA. The charge of amine-coated surfaces is affected by altering the pH and by the density of the surface groups. By shifting the pH of the DNA-solution below the pK_a, characterizing the basicity of surface, the non-specific adsorption of DNA molecules strongly increases. Figure 12(a) shows the fluorescence image of intercalated pBR322 DNA-fragments (4361 bp). The spots caused by single molecules emerge with very similar lateral extension and brightness. The single molecule images are fitted with 2D-Gaussians, the obtained maxima minus background are used for the histogram of Figure 12(b).

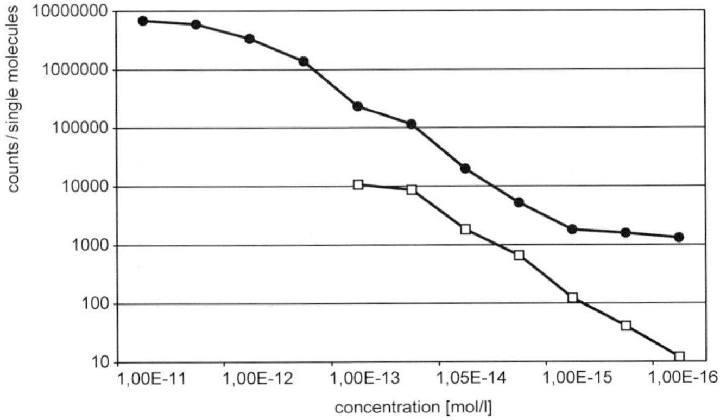

Figure 11. Example of the sensitivity of DNA-based assay.

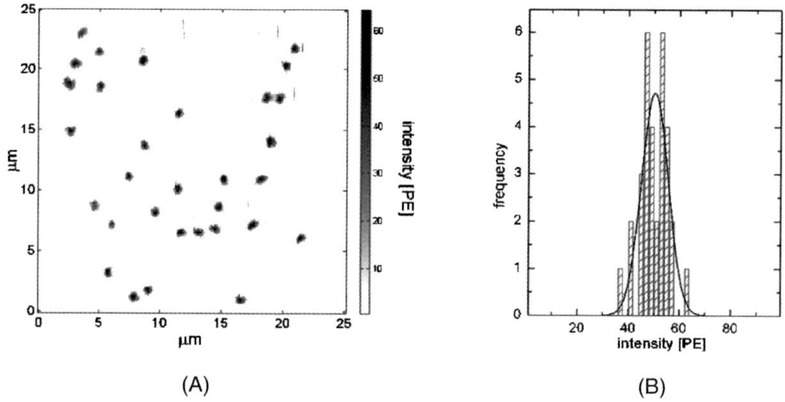

Figure 12. (A) Fluorescence image of intercalated pBR322 DNA fragments (4361 bp). (B) Histogram fluorescence intensity vs. frequency.

Further, DNA fragments of 1.9, 2.6, 4.1, 7.2 and 14.2 kbp were analyzed in the same way. In Figure 13 the average fluorescence intensity is plotted versus the DNA fragment length. For different sizes the CV varied from 6% to 14%.

An intensity histogram of five different DNA sizes is shown in Figure 14 out of a solution of 2×10^{-14} M of the fragment lengths 1.9, 2.6, 4.3, 7.2 and 14.2 kbp respectively. To obtain adequate statistics, the histogram contains the data of approximately 160 molecules, adsorbed at 5000 μm^2. Fragments of 2.6, 4.3, 7.2 and 14.2 kbp are clearly resolved, but the signals of 1.9 and 2.6 kbp fragments are superposed under these circumstances.

The method described is useful for rapid DNA fragment sizing. It demands a low amount of DNA needed, several orders of magnitude less than with gel electrophoresis. To ascertain the length of a particular fragment a quantity of approximately 10^{-17} mol was sufficient. The resolution achieved is sufficient even without further optimization for many applications and can still be enhanced by optimisation.

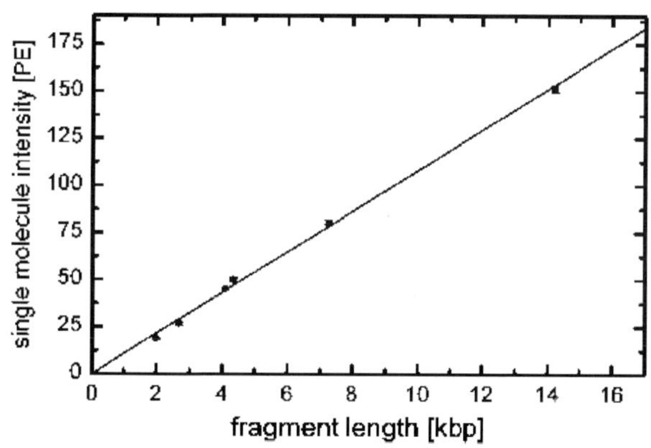

Figure 13. Average fluorescence intensity vs. DNA fragment length.

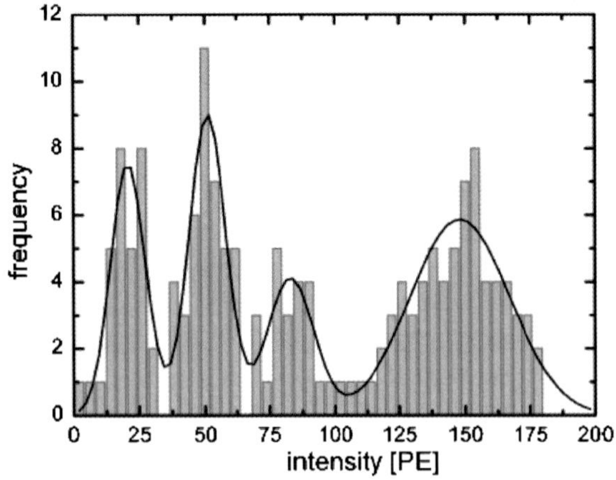

Figure 14. Intensity histogram of five different DNA sizes (1.9, 2.6, 4.3, 7.2 and 14.2 kbp) (see text for details).

20.4 New detection optics for ultrasensitive fluorescence detection close to surfaces

Many applications of ultrasensitive fluorescence detection shown here demand a selective fluorescence collection from surfaces. This can be achieved in part by confocal optics, but the discrimination between bulk and surface-generated fluorescence is insufficient. The spatial resolution of the detection volume is limited mainly in the z-direction, although in the x,y-direction it is close to the diffraction limit. In total, using confocal epifluorescence microscopy a detection volume of several femtolitre (10^{-15} l) can be realized, which is not sufficient for applications that demand higher concentrations of labelled species in the bulk.

The jump of the refractive index between aqueous solution and glass for detecting surface-generated fluorescence is used mainly in two ways, namely evanescent wave detection and supercritical angle fluorescence detection. Hirschfelds idea of illuminating glass–water interfaces using evanescent waves initiated several total internal reflection fluorescence (TIRF) sensors [56–59]. The emission behaviour of surface-generated fluorescence has been investigated thoroughly and can be described within the framework of electromagnetic dipole emission at a discontinuity of the refractive index [60,61]. The calculations show clearly that the spatial emission behaviour of dipoles close to a water–glass surface results in a strong maximum in direction of the critical angle $\alpha_c \approx 61°$ and above α_c into the glass half space (Figure 15).

In this chapter two related new optical set-ups for ultrasensitive detection of surface-generated fluorescence based on a parabolic lens are described, which collect, selectively, fluorescence emission at surfaces by a combination of confocal and total internal reflection techniques and supercritical angle fluorescence detection, respectively [61–64].

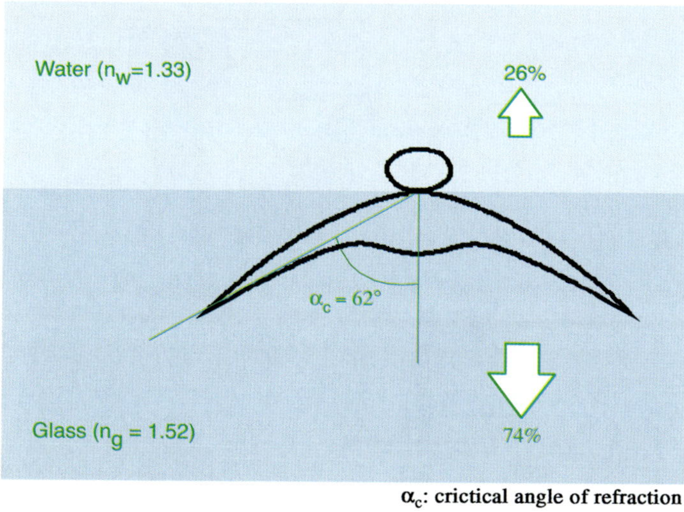

Figure 15. Spatial emission behaviour of dipole close to a water–glass surface. A strong maximum in direction of the critical angle $\alpha_c \approx 61°$ and above α_c into the glass half space is shown.

20.4.1 Confocal total internal reflection fluorescence (TIRF) based on a high aperture parabolic mirror lens

Total internal reflection fluorescence (TIRF) microscopy has been established to investigate surface-generated fluorescence without a background signal from molecules diffusing in the solution [66]. Although TIRF has been realized usually as a widefield configuration, a confocal TIRF, i.e. detection in sub-wavelength dimensions has not been developed due to the difficulty in achieving aberration free focusing at supercritical angles using a high aperture microscope objective.

Recently, we have shown how lenses with a parabolic geometry can be used for efficient collection of fluorescence photons emitted from single molecules [62]. A mirror with parabolic geometry has been shown to achieve an N.A. of almost 1.0 [67].

The set-up of the confocal TIRF system based on a parabolic lens is shown in Figure 16. The excitation light is guided through a cover slip. Hence, evanescent wave excitation is achieved which occurs only at illumination at high surface angles. The main part of the emitted light is directed in the medium of higher refractive index, i.e. in the glass surface (see Figure 15) [61,62]. The laser light is focused onto the fluorescent sample and focal emission is converted into parallel fluorescent light by the paraboloid lens. A further advantage of this concept is that all optical components are positioned below the surface, i.e. free access to the surface is guaranteed.

Supercritical diffraction limited focusing together with confocal imaging reduce the detection volume of the system to well below attolitres, more than three orders of magnitude smaller than achieved by conventional confocal optics. This means that, even at fluorophore concentrations of 10^{-6} mol l^{-1} in

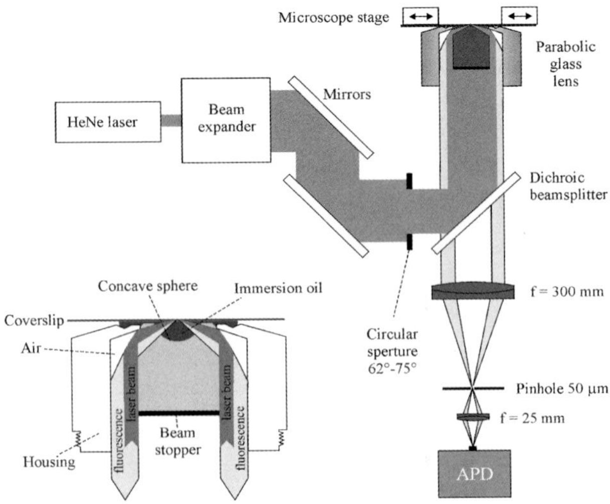

Figure 16. Set-up of the confocal TIRF system based on a parabolic lens.

the bulk solution, on average less than one molecule is situated within such a small volume. Hence, it becomes possible to perform single molecule detection at large dye concentrations, which is important e.g. to study enzymatic activity on the single molecule level.

Figure 17 shows the detection volume of the confocal TIRF microscope. The volume where molecules are observed above half-maximum intensity is $V_{obs} = 1.5 \times 10^{-3} \lambda_{vac}^3$, where λ_{vac} is the laser vacuum wavelength. For an excitation wavelength of 633 nm the detection volume is calculated to be 3.8×10^{-19} l.

Figure 18 presents an image taken by the confocal TIRF instrument of physisorbed Cy5-tagged immunoglobulin molecules and the enlargement of four of these molecules, which shows the spatial resolution of the instrument.

The confocal TIRF microscope combines the advantages of confocal optics into TIRF microscopy, such as excellent lateral resolution, a high signal-to-noise ratio for single molecule detection and strong suppression of bulk fluorescence. A detection volume well below the attolitre levels achieved and free access to the sample from above is guaranteed.

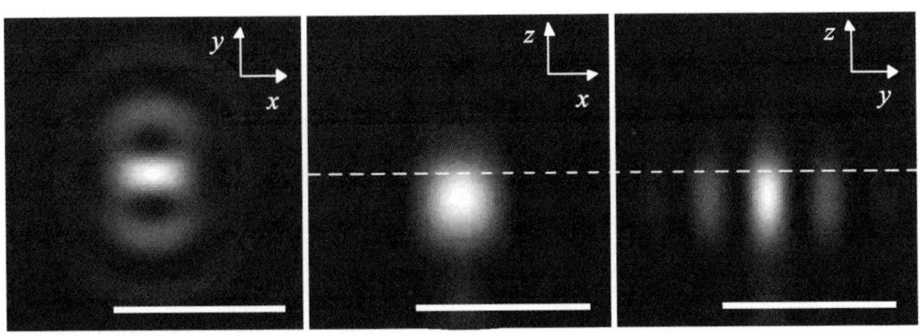

Figure 17. Detection volume of the confocal TIRF.

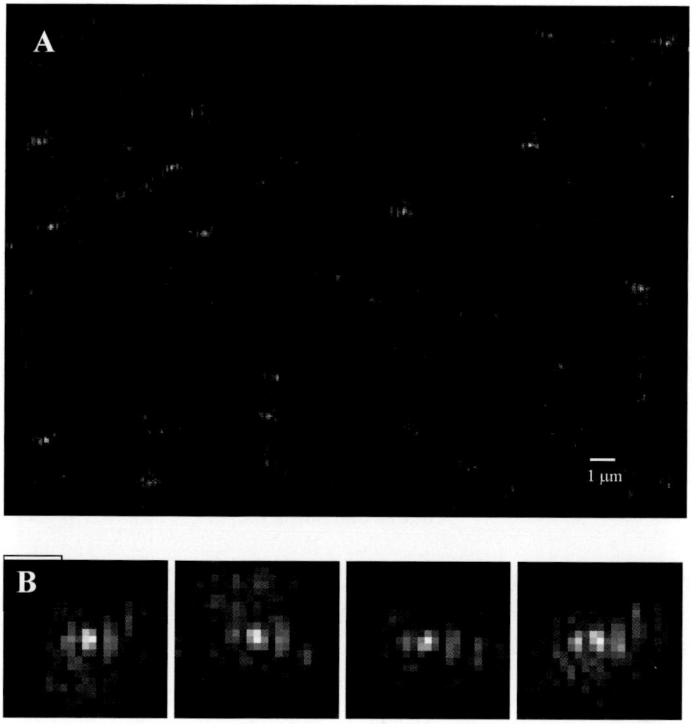

Figure 18. (A) Image taken by the confocal TIRF instrument of physisorbed Cy5-tagged immunoglobulin molecules. (B) Enlargement of four of these molecules shows the spatial resolution of the instrument.

20.4.2 Supercritical angle fluorescence detection for sensing of surface binding of biomolecules

Fluorescent molecules very close to a surface with a random dipole orientation emit 34% of all emitted fluorescent photons in the glass with an angle greater than the critical angle (supercritical angle fluorescence, SAF). SAF vanishes very rapidly when the molecule moves from the surface into the solution (Figure 19).

For example, if the molecules are 0.4λ from the surface, SAF decreases to only 10% of the value of molecules at the surface. Recently, we have introduced a parabolic glass element for the collection of SAF with very high efficiency [68].

Based on this novel optical component we have developed recently a biosensor for realtime measurement of molecular recognition events [67]. This instrument is quite easy to use, even an adjustment is not necessary (Figure 20).

The system is used mainly for straightforward kinetic experiments. In this set-up the detection volume has been expanded to about 60 μm, so that it is not really confocal, which decreases the signal-to-noise ratio, but with this set-up long time kinetics, of, e.g., antibodies or hybridization reactions can be studied, which is not possible by sharp focussing onto the sample due to the minimized time range

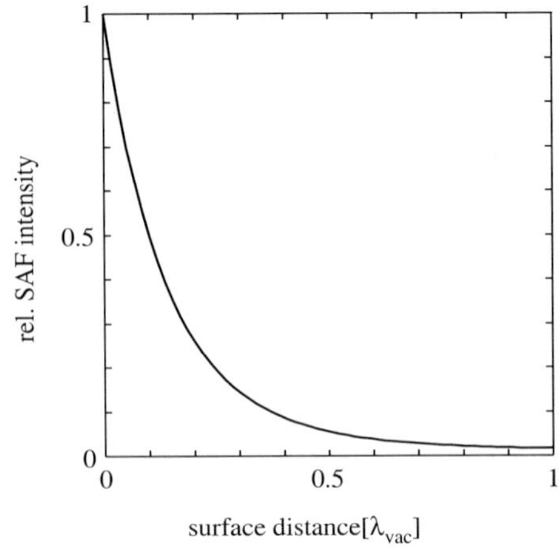

Figure 19. Supercritical angle fluorescence (SAF) as a function of the distance between emitter and surface.

and photobleaching. The kinetic data for an antibody–antigen reaction is shown in Figure 21. The dynamic range of this technique is shown in Figure 22. The sensitivity limit for the investigated immunoreaction is approximately 10^{-13} mol l^{-1}.

20.5 DNA-sequencing of single molecules

Recently, we described a new concept to sequence DNA on a single molecule level [68]. Sequencing at the single molecule level is of great interest, because

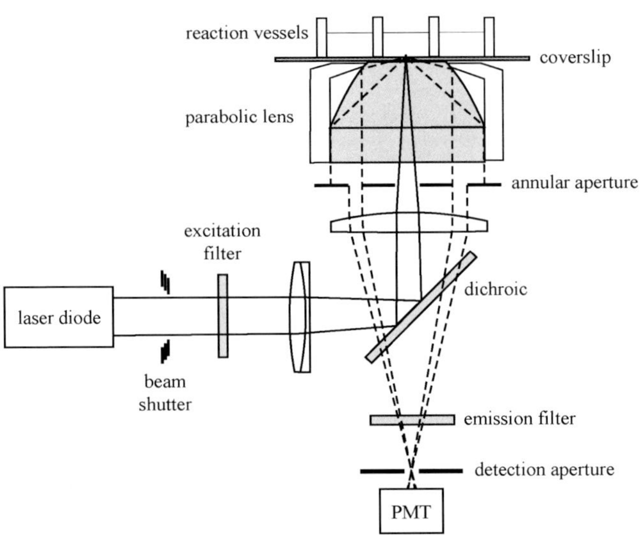

Figure 20. Optical set-up of the SAF-instrument.

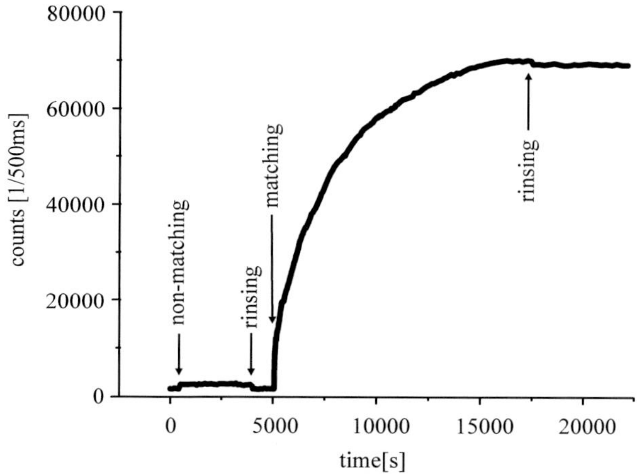

Figure 21. Kinetic data of an immunological reaction obtained by real-time SAF measurement.

time-consuming separation steps like electrophoresis can be prevented. An already proposed the concept is based on selective separation of fluorescently tagged nucleotides from a totally tagged DNA strand by an exonuclease suffers, e.g., from elaborate preparation steps before the sequencing step, difficulties in synthesizing a fully labelled DNA strand, etc. It has not been realized until now.

The concept we are working on is based on the real-time observation of the synthesis of a complementary strand with base-specific tagged nucleotides catalyzed by a polymerase enzyme. After incorporation of a tagged nucleotide, the fluorescent dye is bleached or photochemically cleaved (Figure 23).

It is quite important to discriminate between surface-generated fluorescence by the nucleotide during incorporation in the strand and fluorescence generated

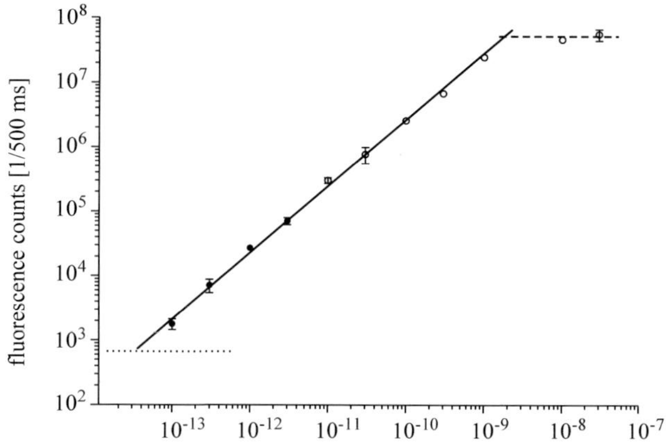

Figure 22. Fluorescence signal vs. antigen concentration.

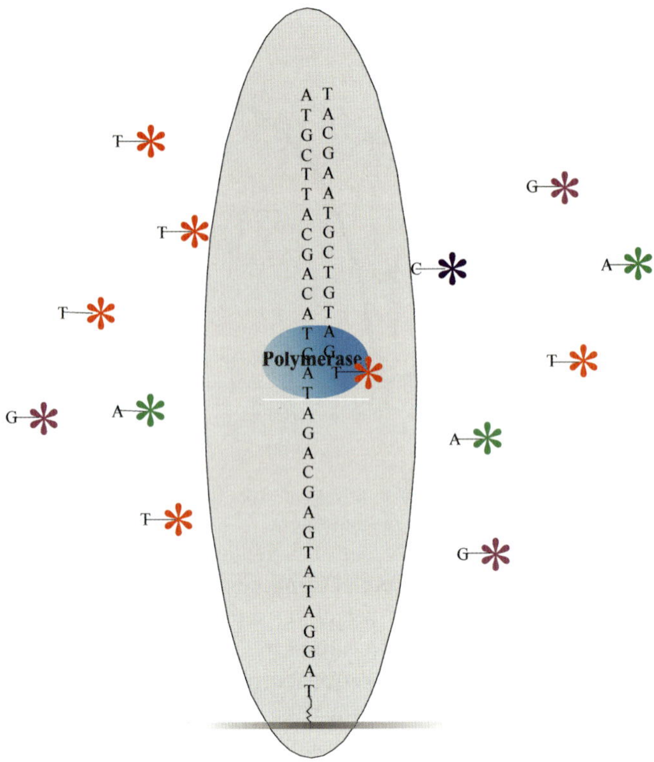

Figure 23. Principle of DNA sequencing based on the observation of single nucleotide incorporation during strand synthesis.

by tagged nucleotides in the bulk solution. Therefore, the described confocal TIRF system is ideal as an optical detection unit for this sequencing approach.

After real-time detection of nucleotide incorporation using the paraboloid detection system we are presently working on the single molecule detection of incorporation of tagged nucleotides to completely realize this sequencing concept [69].

20.6 Real-time detection of nucleotide-incorporation during complementary DNA strand synthesis

The synthesis of complementary DNA strands in vitro is a central step in all existing DNA sequencing procedures [70–72]. However, the base-specific real-time observation of nucleotide incorporation has not been described due to the high concentration of labelled nucleotides required in all fluorescence-based methods, which results in a high background signal. This does not allow specific detection of the incorporated nucleotides. In contrast, DNA synthesis measurements by surface plasmon resonance enables the detection of the complementary strand synthesis but lacks in the possibility to observe base-specific nucleotide incorporation [73–75].

Here, the real-time measurement of DNA strand synthesis of an ensemble of surface-fixed single stranded DNA (ssDNA) using Supercritical Angle Fluorescence (SAF) detection technique is described [69]. To prevent signal perturbation due to fluorescence dyes which diffuse through the laser focus, the detection volume has to be minimized. Therefore we used SAF, whereby the detection volume is restricted to a surface distance well below 100 nm [63]. We observed Cy5-labelled dCTP incorporation in the complementary DNA strand. Strands were chosen with one, three and zero guanine bases in the strand as reference (Seq1dCTP, Seq3dCTP, Seq0dCTP, respectively).

The dNTP mix was added to the surface-bound ssDNA with unlabelled dATP, dGTP, dTTP and Cy5-labelled dCTP. Afterwards, the enzyme was added to the reaction and one data point every 90 s was measured. The guanine bases in the immobilized DNA strand are separated by 9 thymine bases to reduce steric hindrance of the polymerase, due to the dye chromophore attached to the base-unit, which might prevent quenching of the dyes. In addition, to minimize effects due to the surface environment, primer elongation starts at the free unbound end of the strand.

An increase of fluorescence during complementary DNA strand synthesis of the Seq0dCTP-ssDNA could not be observed. However, we observed an increase of fluorescence for the incorporation for Seq1dCTP and Seq3dCTP (Figure 24).

The data clearly show the detection of the incorporation of dye-labelled dNTPs to the strand in real-time and the possibility to detect and compare the incorporation efficiencies of labelled dNTPs and different polymerases. This technique is an important step forward to direct sequencing via polymerase-catalyzed strand synthesis with fluorescently-tagged nucleotides as well as characterizing the activity of polymerases. Techniques for investigating a polymerases and dyes that allow complete strand synthesis are necessary to establish a single molecule DNA-sequencing method.

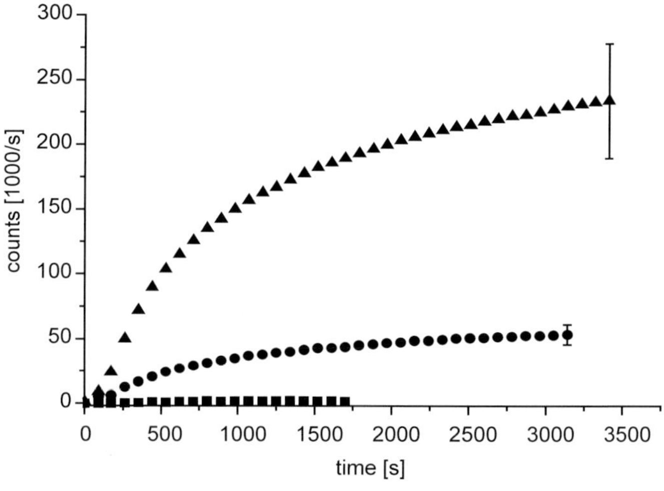

Figure 24. Fluorescence signal increase during Seq1dCTP (circles) and Seq3dCTP (triangles); control. Seq0dCTP (squares).

20.7 Multiplex principle in fluorescence based diagnostics

In 1993 we proposed a concept to enhance the number of analytes that can be detected and quantified simultaneously in one sample in one experiment [76]. Here, the fluorescence decay time of a certain fluorescence dye is used as recognition parameter, i.e. the dye is not only recognised by its colour but also by its characteristic fluorescence lifetime. Realising this concept, several dyes are detectable, can be quantified and can be used for multianalyte detection in one experiment at one wavelength. The advantage is, obviously, to combine the lifetime and the colour information, i.e. four colour groups of dyes, each with different fluorescence lifetimes provides a set of marker molecules that allows the detection of 16 different analytes in one experiment. This technique was first applied was in 1994 [77]. Here, two different immunreactions where detected simultaneously by a time-correlated single-photon counting (TCSPC) experiment, recognising two different dyes by their characteristic fluorescence lifetime.

In this experiment, mouse-anti-rabbit-IgG was labelled with 7-methoxy-coumarin-3-carboxylic assay, resulting in conjugate with a lifetime of 1.9 ns, whereas goat-anti-rat-IgG was tagged with 7-amino-4-methyl-coumarin-3-acetic acid (the measured lifetime was 4.6 ns). Antibodies directed against these conjugates were immobilised at a quartz surface. The lifetime was measured with a hydrogen-filled flash lamp based TCSPC system. Even with the poor full-width at half-maximum (FWHM) of the excitation pulse of 1.7 ns both immunoreactions could be identified simultaneously at the surface by fluorescence lifetime recognition. Subsequently, we also showed the combination of time-resolved detection of molecular recognition events with evanescent wave detection [78].

Although the first application of the multiplex principle was performed with a molecular assay, array techniques offer a much higher parallelization. However, array technologies can not be applied to the identification of molecular structures on cellular surfaces.

The identification of molecular structures on cellular surfaces is not only of great importance in drug discovery but also in clinical diagnostics. Multiparameter analysis by distinguishing different colours has already been introduced, see, e.g., [79]. However, the lifetime information was used recently by Lenenbach et al. to recognise different surface antigen structures at tumour cells [80].

Because of the heterogeneity of tumours not all the tumour cells of an individual patient express tumour specific antigens. For example, in case of mama carcinoma patients about 25–30% have a benefit from a monoclonal antibody therapy with Herceptin [81–84]. Therefore, there is a need for other possible specific tumour antigens that can function as a target for a therapeutic antibody.

We constructed a microscope for lifetime imaging, which is able to recognize different fluorescence lifetimes of different fluorescent dyes after binding of antibody conjugates to cell surfaces by data acquisition of multiexponential decay in every point [80]. The decay curve from every image point contains information about the partial intensities caused by fluorophores with characteristic lifetimes. Thus, the composition of a mixture of fluorescent dyes can be calculated for every measuring point by numerical analysis of a recorded data set of multiexponential decay curves. The calculated dye distribution within the sample provides the arrangement of the individual antigens.

To demonstrate the use of the described optical system for a time-resolved multiplexed assay we performed a time-correlated fluorescence measurement on a cultivated mama carcinoma cell line (SKBR3). This cell line is stained with two different antibody dye conjugates with distinct lifetimes. The fluorescent sample is excited with a pulsed laser diode delivering 100 ps short pulses at a repetition rate of 40 MHz.

Behind the Brewster telescope, which converts the elliptic beam profile into a circular one, a beamsplitter deflects the circular beam from below onto the rear aperture of the microscope objective. The sample is fixed on a microscopic glass slide and scanned by a microscopic scanning table. The fluorescent light from the stained cells is then collected by an objective and detected by a single-photon counting photomultiplier tube (Figure 25).

After recording a lifetime curve it was first stored in a separate memory of the SPC 300 card and the data in the memory transferred to the hard disc where the data for a completely scanned line was recorded.

The calculated intensity distributions for the labels used, DY-635 and EvoBlue30, supply the separate distributions of the two differently marked cell fractions. The image shows the total measured intensity at every measuring point to have a comparison to the calculated images. The red channel displays the total intensity image and shows separated cell structures, which partially overlap and which show a fine structure within one cell (Figure 26).

Figure 27 displays the calculated distribution on the surface of the EvoBlue 30 labeled cells. It shows part of the cells displayed in Figure 26. The cell pattern in Figure 27 matches that in Figure 26 very well.

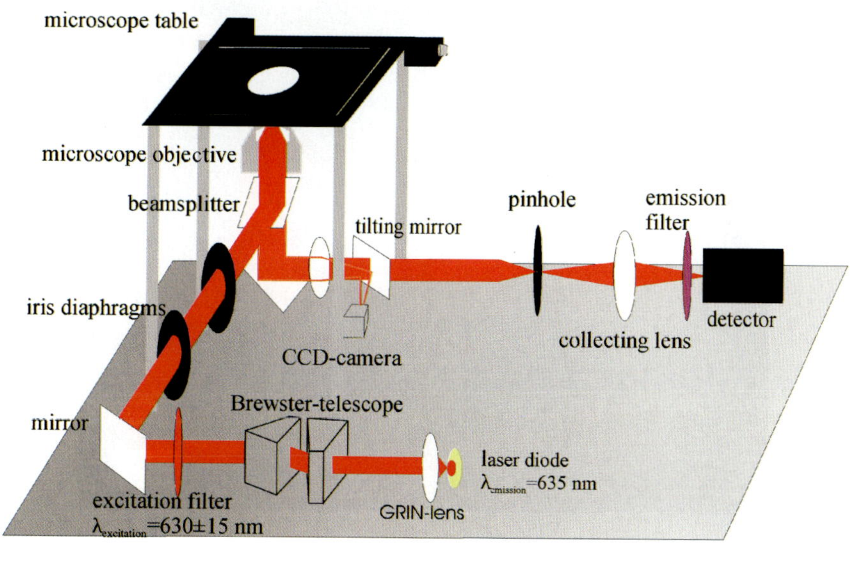

Figure 25. Optical set-up of the time-resolved image device.

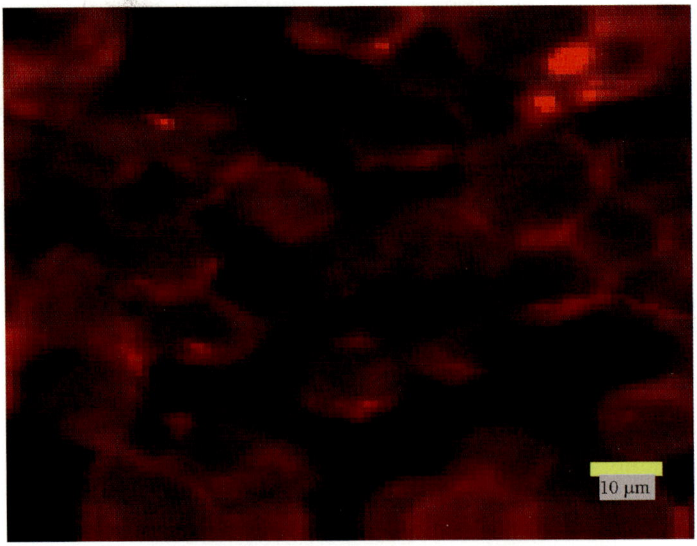

Figure 26. Total intensity image of cells.

The same is observed for the cells labeled with DY-635 (Figure 28). The corresponding intensity distribution is displayed by the green channel and delivers the complementary part of Figure 27.

An overlay of Figures 27 and 28 results in an image that is nearly identical to the total intensity image (Figure 29). Further, there are no artifacts in the calculated

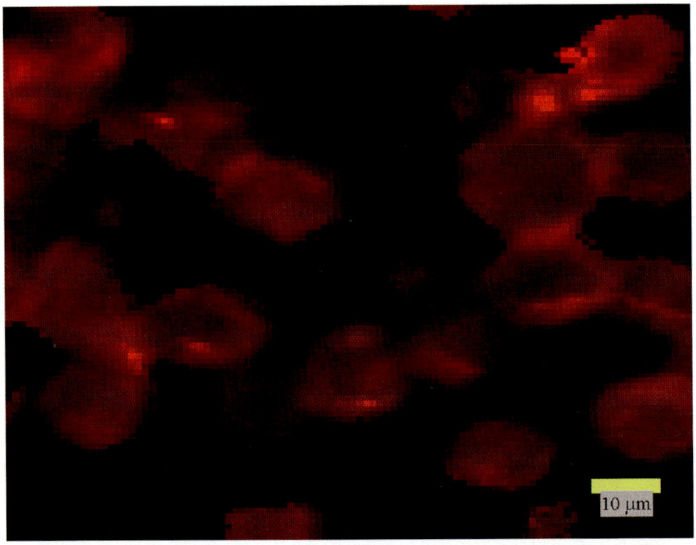

Figure 27. Distribution of fluorescence at the surface of cells obtained after tagging with EvoBlue-antibody conjugates. The image shows a part of the cells displayed in Figure 26.

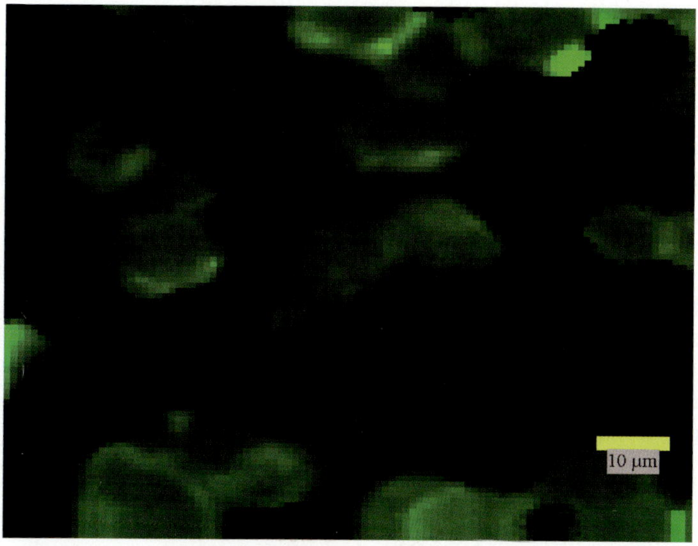

Figure 28. Distribution of fluorescence at the surface of cells obtained after tagging with DY-635-antibody conjugates. The image shows a part of the cells displayed in Figure 26.

images. No stripes occur in the calculated intensity distributions and the two differently labeled cell fractions are strictly separated. This means that none of the cells are displayed in both calculated images.

Lifetime images can be calculated very rapidly from a dataset of 10,000 decay curves. With the support of an automatic data evaluation, this method seems to

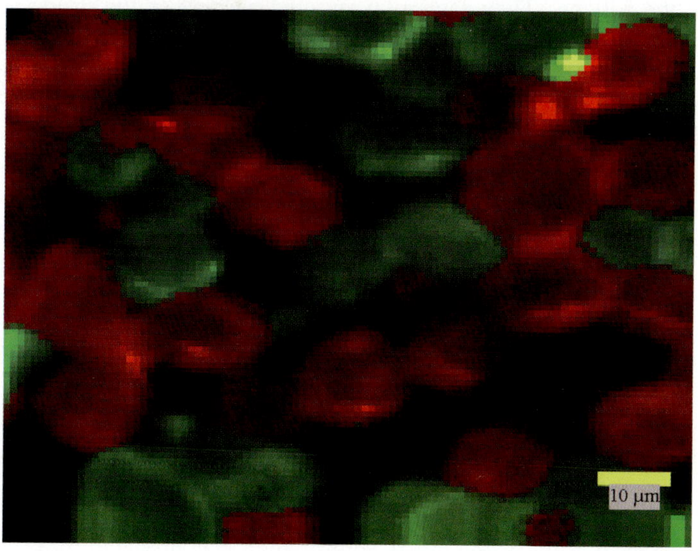

Figure 29. Image constructed by overlay of Figures 27 and 28.

be suited for clinical needs, for instance the findings in postoperative tumour diagnostics. With regard to the examination of surgically removed tumour, lifetime imaging could be used to detect different kinds of tumour-specific antigens simultaneously in a multiplex test.

For multiplexed assays of higher clinical relevance a distinction of more than two lifetimes is desirable. For the distinction of several lifetimes in a mixture these lifetimes should differ by a factor of 2. In our experiments all the used fluorophores exhibited lifetimes in the range of 1.8 to 2.2 ns when they were coupled to proteins. This was also the case when the lifetime of the uncoupled dye in solution was in the range of 1 or 4 ns. Hence, the availability of antibody dye complexes with lifetimes in the sub-nanosecond range or in the range above 4 ns is essential for a multiplexed test with a higher number of components. Because the decay times are measured in advance, a multiexponential fit can be performed with fixed times and variable amplitudes. When the labels with the necessary difference in their lifetime are available a multiplexed test with 3 to 4 different lifetimes could be performed, but only when the lifetimes are fixed during the fit.

However, a combination of spectral difference for the used fluorophores and different lifetimes could increase the number of different biological targets distinguishable by a multiplex assay.

20.8 In vivo detection of micrometastases during minimal invasive laser-neurosurgery by confocal laser-scanning fluorescence endoscope

Deep-seated brain tumours such as malignant glioma have a very bad prognosis for survival of the patients. They cannot be resected because of the strong lesions that would be caused by surgical intervention. Radiotherapy can be applied but shows poor survival time compared with cytoreductive surgery in combination with radiotherapy [85]. Therefore, a thin stereotactic probe (Figure 30) has been developed, which enables a very specific ablation of small tumour volumes with tolerable lesions and makes such unfavourably located brain tumours resectable [86–88]. The radicality of tumour removal can be increased if fluorescence spectroscopy is used as the diagnostic method [89]. Specific labelling of malignant cells by fluorescence spectroscopy enables the surgeon to excise the tumour borders specifically, without unwanted destruction of brain tissue.

A novel neurosurgical instrument has been developed to remove deep-seated brain tumours from the human brain, which are not resectable by conventional methods. It consists of an arrangement of three coaxial titanium tubes, with microlens optics integrated into the inner one and irrigation and a suction channel between the two outer ones. Laser surgery can be performed by a high intensity Nd:YAG laser that is focused by this optical system.

The tip of this laser-probe is inserted into the tumour and the laser focus can be directed onto each point of the target volume by a synchronized movement of the three tubes. The ablation process can be controlled by NMR tomography.

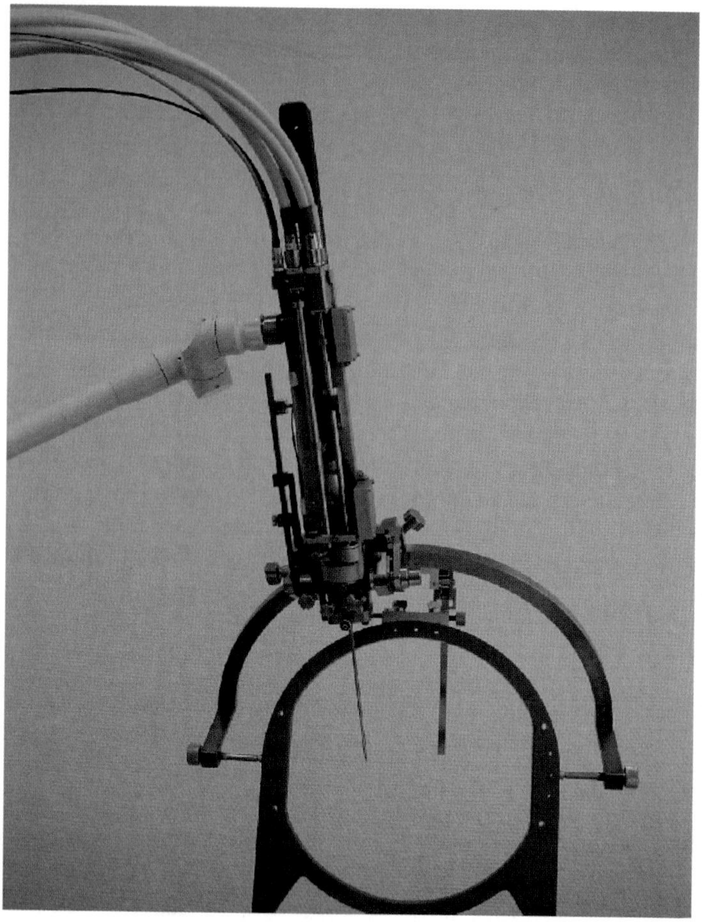

Figure 30. Stereotactic probe for very specific ablation of small tumour volumes.

The frame fixes the patient's head to avoid movement of the skull during neurosurgery. At the same time, the frame is connected to the operation table. Hence, fixed coordinates of the patient relative to the whole operating system are achieved. The two interspaces between the coaxial tubes make up two channels, one for irrigation and one for suction. This enables the ablated tissue fragments to be carried out in physiological saline and to be sucked out of the operation cavity.

The lens tube and the mirror tube can be moved along the symmetry axis of the probe. A synchronized movement of these two tubes with a constant distance between the deflecting mirror and the focusing lens guides the laser focus parallel to the axis of the tubes. Rotation of the mirror tube moves the laser focus on a circle centred to the z-axis. To adjust the ablation laser focus radially, i.e. perpendicular to the probe axis, the lens tube can be moved relative to the mirror. The whole ablation process is controlled by computer guidance. Because a tumour can have an arbitrary shape, a segmental ablation can also be performed, which acts in the same

way as a cylindrical ablation. The only difference is that the rotating mirror tube moves the laser over angular segments during an incomplete forward and backward rotation (Figure 31).

The geometrical restriction of a small lens diameter and a comparable long distance to the sample leads to a small numerical aperture of the system, which causes very low collection efficiency in comparison to conventional fluorescence microscopes. That is why we constructed a laboratory fluorescence microscope with this lens system of the laser probe to check its suitability for fluorescence detection of glioma cells.

This microscope was used to perform in vitro fluorescence experiments with a malignant glioma cell line, which was cultivated and incubated in 5-aminolevulinic acid.

The surgery instrument restricts the collection efficiency of the epifluorescence microscope to 0.25% of the emitted fluorescence into the half-space of the collecting optics. These circumstances made a careful optimization of all components of the detection system necessary to enable identification of the tumour cells.

For specific labelling of tumour cells 5-aminolevulinic acid (5-ALA) has been widely used in the last decade [90,91]. It has been intensively studied for its

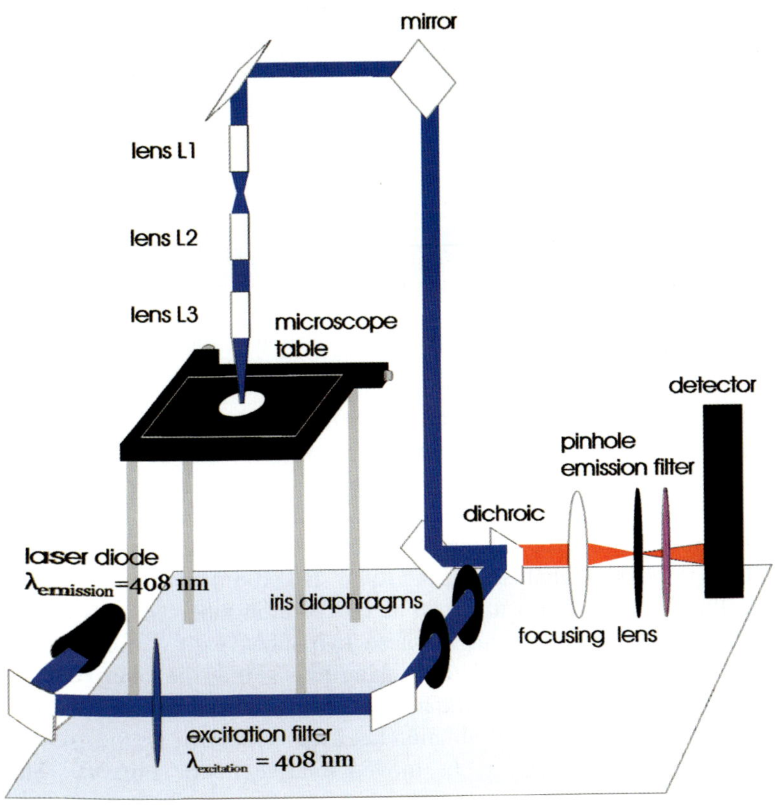

Figure 31. Laboratory set-up that meets all requirements of a neurosurgery instrument.

suitability in detection of bladder cancer [92–99], skin cancer, oral carcinoma, and for bronchial carcinoma [90,91], as well as colonic cancer [100] and brain tumours [101]. In all studies the local or systemic administration of 5-ALA improved the sensitivity for tumour detection. 5-ALA is absorbed by the carcinoma cells and transformed into fluorescent PPIX.

The laboratory set-up, which meets all requirements of the neurosurgery instrument, is shown in Figure 31.

A single cell could be estimated by observation under a light microscope to be between 20 and 30 μm. Thus, in the images of Figure 32 a cell is represented by 2×2 or 3×3 pixels.

There are many such structures of 2×2 or 3×3 pixels in the areas marked by a rectangle. They are all due to single cells, which have not yet formed bigger colonies. Every pixel of a single cell is represented by about 9000 cps and shows a strong contrast to the background that is 3 times lower and is displayed as dark blue.

Bigger cell colonies with a diameter of 200 to 400 μm show the same contrast at their border lines and a stronger contrast towards their centre. These circular areas correspond to 150 to 400 cells (marked by a circle in Figure 32). There, a signal of 15000 to 18000 cps, represented by yellow and light red, depicts a signal-to-noise ratio of 5 to 6 relative to the background signal. Both a single cell and a cell colony can be seen in strong contrast compared with areas without any cells and both are very small structures of a few microns, which could not be detected with any other diagnostic method.

The presented data show that the constructed epifluorescence-endoscope, in combination with the endogen staining method using 5-ALA, has a sensitivity that enables the detection of even single cells. The spatial resolution is 10 μm, which provides an image with a resolution of 4 to 9 pixels per cell. Presently, the system

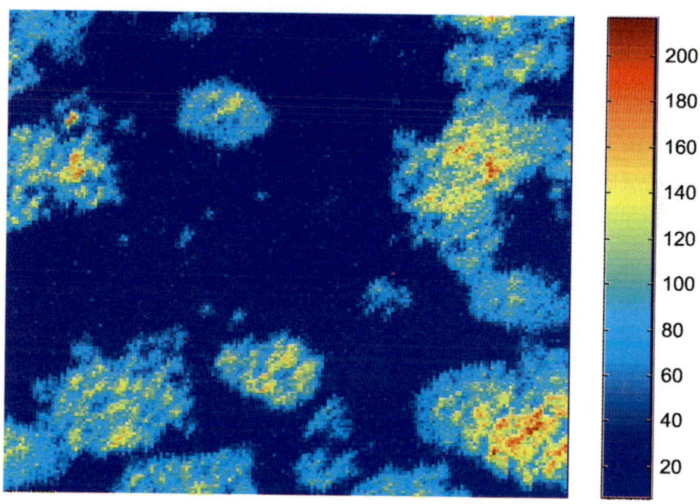

Figure 32. Image of a cell culture obtained by a low numerical aperture fluorescence detection system.

can be transferred directly to the in vivo situation because various studies show that 5-ALA can be systematically administered to patients without serious side effects [102–105]. There have also been clinical studies that tested the fluorescence contrast between healthy and malignant brain tissue during neurosurgery after application of 5-ALA. These studies show a significantly higher fluorescence for brain tumour tissue than for healthy neuronal tissue. Furthermore, there were no false positive results, meaning that no healthy brain cell had to be considered malignant. In addition, for in vivo diagnostics, we expect to find more than one layer of tumour cells in the focus, which would increase the fluorescence signal because several layers are observed simultaneously.

20.9 Conclusion

Ultrasensitive fluorescence detection at surfaces takes place in biology and medicine in many applications. Beside the optics, the structure and composition of the interface plays a crucial role for successful applications. Single molecule detection in vitro demands surfaces that prevent unspecific adsorption in order to reduce the background. A high sensitivity of the optical set-up is not sufficient for high assay sensitivity. The surface properties become even more important if the assay format is strongly reduced, e.g. in high-throughput screening (HTS).

If real-time detection is important the discrimination between surface bound and dissolved molecules is essential. Here, we have presented a new optical component, a paraboloid lens, which meets this requirement by drastically decreasing the detection volume to the attolitre level. Finally, even though diagnostics applications in vivo, like minimal invasive neurosurgery limit the possibilities of achieving high sensitivity, due to the required optical design, we have shown that single cell detection is still possible.

References

1. J.R. Lakowicz (1999). *Principles of Fluorescence Spectroscopy* (2nd edn). Kluwer Academic/Plenum Publisher, New York.
2. B. Valeur, J.-C. Brochon (2001). *New Trends in Fluorescence Spectroscopy*. Springer-Verlag Berlin-Heidelberg.
3. T. Hirschfeld (1976). *Appl. Opt.* **15,** 2965–2966.
4. D.C. Nguyen, R.A. Keller, J.H. Jett, J.C. Martin (1987). *Anal. Chem.* **59**, 2158–2161.
5. E.B. Shera, N.K. Seitzinger, L.M. Davis, R.A. Keller, S.A. Soper (1993). *Chem. Phys. Lett.* **174**, 553–557.
6. C.W. Wilkerson, P.M. Goodwin, W.P. Ambrose, J.C. Martin, R.A. Keller (1993). *Appl. Phys. Lett.* **62**, 2030–2032.
7. K. Peck, L. Stryer, A.N. Glazer, R.A. Mathies (1989). *Proc. Natl. Acad. Sci. U.S.A.* **86,** 4087–4091.
8. R.A. Mathies, K. Peck, L. Stryer (1990). *Anal. Chem.* **62**, 1786–1791.
9. R. Rigler, Ü. Mets (1992). *SPIE Laser Spectrosc. Biomol.* **1921**, 239–248.

10. Ü. Mets, R. Rigler (1994). *J. Fluoresc.* **4**, 259–264.
11. S. Nie, D.T. Chiu, R. N. Zare (1994). *Science* **266**, 1018–1021.
12. S. Nie, D.T. Chiu, R. N. Zare, (1995). *Anal. Chem.* **67,** 2849–2857
13. W.E. Moerner, L. Kador (1989). *Phys. Rev. Lett.* **62**, 2535–2538.
14. L. Kador, D.E. Horne, W.E. Moerner (1990). *J. Phys. Chem.* **94**, 1237–1248.
15. M. Orrit, J. Bernhard (1991). *Phys. Rev. Lett.* **65**, 2716–2719.
16. W.E. Moerner (1991). *New J. Chem.* **15**, 199–204.
17. W.E. Moerner, T. Basché (1993). *Angew. Chem. Int. Ed. Engl.* **32**, 457–462.
18. M. Orrit, J. Bernhard, R. Personov (1993). *J. Phys. Chem.* **97**, 10256–10268.
19. W.E. Moerner (1994). *Science* **265**, 46–52.
20. T. Plakhotnik, D. Walser, M. Pirotta, A. Renn, U. Wild (1996). *Science* **271**, 1703–1705.
21. E. Betzig, J.R. Chichester (1993). *Science* **262**, 1422–1425.
22. J.K. Trautman, J.J. Macklin, L.E. Brus, E. Betzig (1994). *Nature* **369**, 40–42.
23. T. Funatsu, Y. Harada, M. Tokunaga, K. Saito, T. Yanagida (1995). *Nature* **374**, 555–559.
24. J.J. Macklin, J.K. Trautmann, T.D. Harris, L.E. Brus (1996). *Science* **272**, 255–258.
25. M. Eigen, R. Rigler (1994). *Proc. Natl. Acad. Sci. U.S.A.* **91**, 5740–5747.
26. R.D. Vale, T. Funatsu, D.W. Pierce, L. Romberg, Y. Harada, T. Yanagida (1996). *Nature* **380**, 451–453.
27. J.H. Jett, R.A. Keller, J.C. Martin, B.L. Marrone, R.K. Moyzis, R.L. Ratliff, N.K. Seitzinger, E.B. Shera, C.C. Stewart (1989). *J. Biomol. Struct. Dyn.* **7**, 301–309.
28. F. Löscher, S. Böhme, J. Martin, S. Seeger (1997). Detection of single protein molecules at interfaces after antibody-antigen-binding. *Proc. SPIE* **3199**, 168–177.
29. F. Löscher, S. Böhme, J. Martin, S. Seeger (1998). The counting of single protein molecules at interfaces and the application in early-stage-diagnosis. *Anal. Chem.* **70**, 3202–3205.
30. J.D. Harding, R.A. Keller (1992). Single-molecule detection as an approach to rapid DNA sequencing. *Trends Biotechnol.* **10**, 55–57.
31. M. Rankl, S. Laib, S. Seeger (2003). Surface tension properties of surface-coatings for application in biodiagnostics determined by contact angle measurements. *Colloids Surf. B: Biointerfaces* **30**(3), 177–186.
32. M. Rankl, S. Seeger, in preparation.
33. A. Hartmann, S. Seeger (1994). Direct immobilization of antibodies on phthalocyaninato-polysiloxane photopolymers. *Thin Solid Films* **245**, 206–210.
34. A. Hartmann, D. Bock, M. Martin, S. Seeger (1994). Ultrathin antibody networks for evanescent detection of antigens. *Proc. SPIE* **2331**, 50–62.
35. A. Hartmann, D. Bock, S. Seeger (1995). One-step immobilization of immunoglobulin G and potential of the method for application in immunosensors. *Sens. Actuators* **28**(2), 143–149.
36. A. Hartmann, D. Bock, T. Jaworek, S. Kaul, M. Schulze, H. Tebbe, G. Wegner, S. Seeger (1995). Cellulose-antibody films for highly specific evanescent wave immunosensors. *Proc. SPIE* **2629**, 108–118.
37. F. Löscher, A. Hartmann, J. Ueberfeld, T. Ruckstuhl, D. Bock, T. Jaworek, G. Wegner, S. Seeger (1996). Cellulose-protein films for highly specific evanescent wave immunosensors. *Proc. SPIE* **2928**, 209.
38. F. Löscher, T. Ruckstuhl, T. Jaworek, G. Wegner, S. Seeger (1998). Immobilization of biomolecules on Langmuir-Blodgett-films of regenerative cellulose derivatives. *Langmuir.* **14**, 2786–2789.
39. F. Löscher, T. Ruckstuhl, S. Seeger (1998). Ultrathin cellulose-based layers for detection of single antigen molecules. *Adv. Mater.* **10**(13), 1005–1009.
40. I. Langmuir (1917). *J. Am. Chem. Soc.* **39**, 1848.

41. M.C. Fitzgerald, G.R. Parr, L.M. Smith (1993). Basic matrices for the matrix-assisted laser-desorption ionization mass-spectrometry of proteins and oligonucleotides. *Anal. Chem.* **65**, 3204–3211.
42. C. Henry (1997). Can MS really compete in the DNA world? *Anal. Chem.* **69**, A243–A246.
43. D.C. Muddiman, G.A. Anderson, S.A. Hofstadler, R.D. Smith (1997). Length and base composition of PCR-amplified nucleic acids using mass measurements from electrospray ionization mass spectrometry. *Anal Chem.* **69**, 1543–1549.
44. P.M. Goodwin, M.E. Johnson, J.C. Martin, W.P. Ambrose, B.L. Marrone, J.H. Jett, R.A. Keller (1993). Rapid sizing of individual fluorescently stained DNA fragments by flow-cytometry. *Nucleic Acids Res.* **21**, 803–806.
45. P.M. Goodwin, W.P. Ambrose, H. Cai, W.K. Grace, E.J. Larson, B.L. Marrone, J.H. Jett, J.H. Werner, R.A. Keller (2002). Single molecule nucleic acid analysis by fluorescence flow cytometry. In: J.J. Kasianowicz, M.S.Z. Kellermayer, D.W. Deamer (Eds), *Structure and Dynamics of Confined Polymers,* (pp. 351–370). Kluwer Academic Publishers, Dordecht.
46. A. Castro, F.R. Fairfield, E.B. Shera (1993). Fluorescence detection and size measurement of single DNA-molecules. *Anal. Chem.* **65**, 849–852.
47. H.P. Chou, C. Spence, A. Scherer, S. Quake (1999). A microfabricated device for sizing and sorting DNA molecules. *Proc. Natl. Acad. Sci. U.S.A.* **96**, 11–13.
48. A. Van Orden, R.A. Keller, W.P. Ambrose (2000). High-throughput flow cytometric DNA fragment sizing. *Anal. Chem.* **72**, 37–41.
49. D.C. Schwartz, X.J. Li, L.I. Hernandez, S.P. Ramnarain, E.J. Huff, Y.K. Wang (1993). Ordered restriction maps of Saccharomyces-cerevisiae chromosomes constructed by optical mapping. *Science* **262**, 110–114.
50. X. Meng, K. Benson, K. Chada, E.J. Huff, D.C. Schwartz (1995). Optical mapping of bacteriophage-lambda clones using restriction endonucleases. *Nat. Genet.* **9**, 432–438.
51. O. Mueller, K. Hahnenberger, M. Dittmann, H. Yee, R. Dubrow, R. Nagle, D.A. Ilsley (2000). Microfluidic system for high-speed reproducible DNA sizing and quantitation. *Electrophoresis* **21**, 128–134.
52. P.G. Righetti, C. Gelfi, M.R. D'Acunto (2002). Recent progress in DNA analysis by capillary electrophoresis. *Electrophoresis* **23**, 1361–1374.
53. Y. Fang, T.S. Spisz, T. Wiltshire, N.P. D'Costa, I.N. Bankman, R.H. Reeves, J.H. Hoh (1998). Solid-state DNA sizing by atomic force microscopy. *Anal. Chem.* **70**, 2123–2129.
54. T.S. Spisz, Y. Fang, R.H. Reeves, C.K. Seymour, I.N. Bankman, J.H. Hoh (1998). Automated sizing of DNA fragments in atomic force microscope images. *Med. Biol. Eng. Comput.* **36**, 667–672.
55. S. Laib, M. Rankl, T. Ruckstuhl, S. Seeger (2003). Sizing of single fluorescently stained DNA-fragments by scanning microscopy, submitted to *Nucleic Acids Res.*
56. R.M. Dickson, A.B. Cubitt, R.Y. Tsien, W.E. Moerner (1997). *Nature* **388**, 355.
57. S. Wennmalm, H. Blom, L. Wellermann, R. Rigler (2001). *Biol. Chem.* **382**, 393.
58. M. Lippitz, W. Erker, H. Decker, K.E. van Holde, T. Basché (2002). *Proc. Natl. Acad. Sci. U.S.A.* **99**, 2772.
59. H. E. Keller (1990). In: J. Pawley (Ed), *Handbook of Biological Confocal Microscopy.* Plenum Press, New York.
60. Q. Li, S. Seeger (2003). UV-laser-based fluorescence lifetime imaging microscopy of single molecules, submitted to *J. Phys. Chem.*
61. T. Hirschfeld (1965). Total reflection spectroscopy. *Can. Spectrosc.* **10**, 128.
62. D. Axelrod (1981). Cell-substrate contacts illuminated by total internal fluorescence. *J. Cell Biol.* **89**, 141–145.

63. D. Axelrod, T.P. Burghardt, N.L. Thompson (1984). Total internal reflection fluorescence. *Annu. Rev. Biophys. Bioeng.* **13**, 124–268.
64. P.B. Garland (1996). Optical evanescent wave methods for the study of biomolecular interactions. *Q. Rev. Biophys.* **29**, 91–117.
65. W. Lukosz, R.E. Kunz (1977). Light emission by magnetic and electric dipoles close to a plane interface. *JOSA* **67**(12), 1607–1619.
66. J. Enderlein, T. Ruckstuhl, S. Seeger (1999). Highly efficient optical detection of surface-generated fluorescence. *Appl. Opt.* **38**, 724–732.
67. T. Ruckstuhl, M. Rankl, S. Seeger (2003). Highly sensitive biosensing using a supercritical angle fluorescence (SAF) instrument. *Biosens. Bioelectron.* **18**(9), 1193–1199
68. T. Ruckstuhl, J. Enderlein, S. Jung, S. Seeger (2000). Forbidden light detection from single molecules. *Anal. Chem.* **72**, 2117.
69. T. Ruckstuhl, S. Seeger (2003). Confocal total internal reflection fluorescence (TIRF) microscopy with a high aperture parabolic mirror lens. *Appl. Opt.* **42**(16), 3277–3283.
70. T. Ruckstuhl, M. Rankl, S. Seeger (2003). Confocal TIRF microscopy of single molecules. *Proc. SPIE*, in press.
71. D. Axelrod (2001). Total internal reflection fluorescence microscopy. In: *Methods in Cellular* A. Periasamy (Ed), *Imaging* (pp. 362–380). Oxford University Press, New York.
72. S. Seeger (1998). Verfahren zur DNS- oder RNS-Sequenzierung. *Deutsches Pat. DE* 198 44 931.3.
73. A. Drechsler, M.A. Lieb, C. Debus, A.J. Meixner, A. Tarrach (2001). Confocal microscopy with a high numerical aperture parabolic mirror. *Opt. Express* **9**, 637–644.
74. A. Krieg, S. Laib, T. Ruckstuhl, S. Seeger (2003). Real-time detection of nucleotide-incorporation during complementary DNA strand synthesis. *ChemBioChem.* (4), 100–103.
75. F. Fanger, S. Nicklen, A.R. Coulson (1977). *Proc. Nat. Acad. Sci. U.S.A.* **74**, 5463–5467.
76. E.D. Hyman (1988). *Anal. Biochem.* **174**, 423–436.
77. L.T.C. Franca, E. Carrilho, T.B.L. Kist (2002). *Q. Rev. Biophys.* **35**, 169–200.
78. P. Nilsson, B. Persson, M. Uhlen, P.A. Nygren (1995). *Anal. Biochem.* **224**, 400–408.
79. M. Buckle, R.M. Williams, M. Negroni, H. Buc (1996). *Proc. Natl. Acad. Sci. U.S.A.* **93**, 889–894.
80. I.K. Pemberton, M. Buckle (1999). *J. Mol. Recognit* **12**, 322–327.
81. S. Seeger, G. Bachteler, K.H. Drexhage, G. Deltau, J. Arden-Jacob, K. Galla, K.T. Han, R. Müller, M. Köllner, A. Rumphorst, M. Sauer, A. Schulz, J. Wolfrum (1993). Biodiagnostics and polymer identification with multiplex dyes. *Ber. Bunsenges. Phys. Chem.* **97**, 1542.
82. K. Galla, J. Arden-Jacob, G. Deltau, K.H. Drexhage, M. Martin, M. Sauer, J. Wolfrum, S. Seeger (1994). Simultaneous antigen detection using multiplex dyes. *J. Fluoresc.* **4**(1), 111–115.
83. D. Bock, K. Galla, M. Martin, J. Wolfrum, S. Seeger (1995). Use of diode lasers in an evanescenct wave immunosensors for stationary and time-resolved detection of Antigens. *Sens. Actuators B* **29**(1–3), 293–299.
84. M.G. Ormerod (2000). *Flow Cytometry: A Practical Approach*, (3rd edn). American Chemical Society, Washington DC.
85. A. Lenenbach, S. Seeger (2003). Characterization of tumour cells in respect of their specific antigens using fluorescence lifetime imaging. In preparation.
86. R. Bell (2001). *Ann. Oncol.* **12**, 69–73.
87. N. Cornez, M. Piccart (2000). *Cancer* **87**, 847–858.

88. M. Piccart, C. Lohrisch, A. Di Leo, D. Larsimont (2001). *Oncology* **61**, 73–82.
89. I. Smith (2001). *Anti-Cancer Drug* **12**, 3–10.
90. F. Kreth, P. Warnke, R. Scheremet, C. Ostertag (1993). *J. Neurosurg.* **78**, 762.
91. S. Fischer (1986). *PhD-Thesis, Entwicklung eines konfokalen Laserscanning-Mikroskops zum Einsatz in der stereotaktischen Laser-Neurochirurgie*, Heidelberg.
92. M. Goetz, S. Fischer, A. Velten. J. Bille, V. Sturm (1999). *Phys. Med. Biol.* **44**(119).
93. J. Wahrburg, K. Schmidt, M. Goetz, K. Kappings, S. Goelz (1996). *Mechatronics* **6**, 479.
94. W. Stummer, A. Novotny, H. Stepp, C. Goetz, K. Bise, H. Reulen (2000). *J. Neurosurg.* **93**, 1003.
95. Q. Peng, T. Warloe, K. Berg, M. Kongshaug, K. Giercksky, J. Nesland (1997). *Cancer* **79**, 2282.
96. Q. Peng, K. Berg, J. Moan, M. Kongshaug, J. Nesland (1997). *Photochem. Photobiol.* **65**, 235.
97. M. Kriegmair, R. Baumgartner, R. Knuechel, H. Stepp, F. Hofstaedter, A. Hofstetter (1996). *J. Urol.* **155**, 105.
98. M. Kriegmair, D. Zaak, H. Stepp, R. Baumgartner, R. Knuechel, A. Hofstetter (1999). *Eur. Urol.* **36**, 386.
99. M. Kriegmair, D. Zaak, R. Knuechel, R. Baumgartner, A. Hofstetter (1999). *Urol. Int.* **63**, 27.
100. M. Kriegmair, D. Zaak, K. Rothenberger, J. Rassweiler, D. Jocham, F. Eisenberger, R. Tauber, A. Stenzl, A. Hofstetter (2002). *J. Urol.* **168**, 475.
101. R. Waidelich, H. Stepp, R. Baumgartner, E. Weniger, A. Hofstetter, M. Kriegmair (2001). *J. Urol.* **165**, 165.
102. D. Zaak, D. Frimberger, H. Stepp, S. Wagner, R. Baumgartner, P. Schneede, M. Siebels, R. Knuechel, M. Kriegmair, A. Hofstetter (2001). *J. Urol.* **166**, 1665.
103. D. Zaak, M. Kriegmair, H. Stepp, R. Baumgartner, R. Oberneder, P. Schneede, S. Corvin, D. Frimberger, R. Knuechel, A. Hofstetter (2001). *Urology* **57**, 690.
104. D. Zaak, E. Hungerhuber, P. Schneede, H. Stepp, D. Frimberger, S. Corvin, N. Schmeller, M. Kriegmair, A. Hofstetter, R. Knuechel (2002). *Cancer* **95**(6), 1234; *Cancer* **95**, 2580–2580.
105. K. Orth, K. König, F. Genze, A. Rück (1994). *J. Cancer Res. Clin. Oncol.* **120**, 657.
106. W. Stummer, S. Stocker, S. Wagner, H. Stepp, C. Fritsch, C. Goetz, A. Goetz, R. Kiefmann, H. Reulen (1998). *Neurosurgery* **42**, 518.
107. J. Dalton, C. Yates, D. Yin, A. Staughn, S. Marcus, A. Golub, M. Meyer (2002). *J. Pharamacol. Exp. Ther.* **301**, 507.
108. G. Popken, W. Schultze-Seemann, K. Seiler, M. Birkel, U. Wetterauer (2000). *Eur. J. Clin. Pharmacol.* **56** 241.
109. K. Rick, R. Sroka, H. Stepp, M. Kriegmair, R. Huber, K. Jacob, R. Baumgartner (1997). *J. Photochem. Photobiol. B Biol.* **40**, 313.
110. J. Webber, D. Kessel, D. Fromm (1997). *J. Surg. Res.* **68**, 31.

Subject Index